DIESEL'S ENGINE

DIESEL'S ENGINE

THE MAN, AND THE EVOLUTION OF THE WORLD'S MOST EFFICIENT INTERNAL COMBUSTION MOTOR

C. Lyle Cummins, Jr.

"We should not expect ever to utilize in practice all the motive power of combustibles. The attempts made to attain this result would be far more hurtful than useful if they caused other important considerations to be neglected. The economy of the combustible is only one of the conditions to be fulfilled in heat-engines. In many cases it is only secondary. It should often give precedence to safety, to strength, to the durability of the engine, to the small space which it must occupy, to small cost of installation, etc. To know how to appreciate in each case, at their true value, the considerations of convenience and economy which may present themselves; to know how to discern the more important of those which are only secondary; to balance them properly against each other, in order to attain the best results by the simplest means: such should be the leading characteristics of the man called to direct, to coordinate the labors of his fellow men, to make them cooperate towards a useful end, whatsoever it may be."

—Sadi Carnot on Heat Engines, 1824

Octane Press, Edition 2.0, April 2022
First edition published November 30, 1993, by Carnot Press as ISBN: 0-917308-03-4
Copyright © 2022 by Octane Press

On the front cover: A stained glass window at the MAN B&W Diesel AG administration building in Augsburg depicting Diesel's 1987 engine. *Courtesy of MAN*

ISBN: 978-1-64234-054-9
ePub ISBN: 978-1-64234-072-3

LCCN: 2021952427

Design by Tom Heffron
Copyedited by Maria Edwards
Proofread by Dana Henricks

octanepress.com

Octane Press is based in Austin, Texas
Printed in the United States

TO JEANNE

my wife

with grateful thanks and appreciation. For over two decades she has supported, encouraged, given a listening ear, and, as critic and editor, helped complete this shared adventure.

A WORD ABOUT UNITS

Most engineers and scientists are at home with both the English and metric systems, but custom often determines units of measurement used by the general populace. Fortunately, the modernized metric system, the International System of Units (SI), enables the industrial world to speak a common language.

In *Diesel's Engine*, the units used are those in which original data were taken or in the way that information was reported. It will present no problem because so few different kinds of units are involved. Clarification is given where confusion might otherwise result.

A few special equivalents: One German (metric) horsepower (Ps) = 0.986 English (and American) horsepower. No distinction is generally made in the text because of the small difference—and because it is probably within the experimental error of the measuring instruments. However, data originally taken in the metric system most likely will have used Ps for horsepower. Foreign writers may or may not have converted from Ps (Pferdestärke) to horsepower or the reverse when reporting in their own language. For pressure, 1 kg/cm^2 (14.224 psi) equals 1 "old" atm and likewise may have been used interchangeably (1 new atm = 760 mm Hg = 14.695 psi).

CONTENTS

PREFACE AND ACKNOWLEDGMENTS

Over two decades ago it was my intention to devote a few short years to writing a history of the diesel engine and then be off onto something else. Little did I realize that this detour from the life of a practicing engineer would turn into a new and totally fulfilling career.

Internal Fire, published in 1976, grew from an expected few chapters of introduction to the diesel story to its own history of the internal combustion engine up to 1900. It contained only a brief chapter on Rudolf's brainchild. *Diesel's Engine*, originally planned to carry his engine's progress to the mid-1950s, ends just after the exciting developments coming out of the First World War.

When feeling the heft of this book, the reader may rightfully ask how it can take so many pages to follow the trail such a short distance. At the beginning, this author's thoughts would have echoed the question. However, the deeper the search for answers to puzzling questions, the more I realized what a great wealth of unearthed material ought to be shared with others interested in the history of the engine and in the people and events who shaped its course of development.

Much of the material used in *Diesel's Engine* lay undiscovered in the archives of American and European companies and museums. To an English-only reader, there was a second roadblock to learning more of the story because so much historical information was not translated from the language of the countries where the diesel was born and spent its earliest years.

Most of the available lore on the engine had been handed down from publications of the time, which were absorbed by textbook writers of that and later generations, who, in turn, deleted the "flavor" and distilled the facts to suit their own academic purposes. The few accounts of those who actually fought in the development "trenches" have since become scarce. Except for the latter, writing from memory and personal records, almost nothing in print came from the primary sources of test journals, patents, newspapers, personal correspondence, and company documents.

Wherever possible I have used original sources and have quoted freely from them to capture the essence of their thinking and their passion. The reader will find extensive chapter notes containing not only fully cited references, but also vignettes too intriguing to omit. The notes, hopefully, will answer additional questions the text might have raised about where engines went and sometimes their fate. It has been difficult to forego including many insightful and delightful "gems" uncovered during my research. I'm afraid that at times I, too, fell into the trap so accurately described by the perceptive historian Barbara Tuchman: "Research is endlessly seductive; writing is hard work."

A most difficult task for me is to find a way to acknowledge with heartfelt thanks the many who have so willingly guided my efforts for the past twenty-one years. These colleagues and mentors in scores of companies, universities, museums, and libraries in the United States and throughout Europe have given of their time or opened their archives to my visits and continue to fulfill my requests. It is possible to only name a few who have sped me on this rewarding new career and have survived my insatiable search for details. Those whose help relates to material found in the succeeding volume will be in its acknowledgements.

Dr. W. James King directed me to Eugene S. Ferguson's clinching statement in his *Bibliography of the History of Technology*: "Although it is difficult to point to any other technical device that has so drastically changed the quality of life in the west, I know of no recent book in English on the general development of the internal-combustion engine." Thus ended my market survey! I have since thanked Gene Ferguson personally. Others who opened initial doors or were early mentors must include: Lewis E. Allsopp, Dorman, deceased; Dr. N. John Beck, BKM; D. J. H. (John) Day, British Patent Office, retired; Ernest Davies, Vickers, Barrow-in-Furness, deceased; Peter Davies, The Institution of Mechanical Engineers; Erik Eckermann, Deutsches Museum and author; John C. Ellis, Shell Oil Co., retired; Dr. Lamont Eltinge, Ethyl, Eaton Research, retired; Gregory Flynn Jr., General Motors Research Labs, retired; C. C. J. French, Ricardo International, retired; Lawrence A. Grander, ARCO, deceased; Søren Hansen, Burmeister & Wain, deceased; Hans-Ulrich Howe, Deutz, Goetze & AVL; Dr. John L. Hummer, University of Portland, retired; Dr. Walter A. Kilchenmann, Sulzer, deceased; Rodney Law, the Science Museum, retired; Heinrich Lippenberger, Daimler-Benz, retired; Professor Hans List, AVL; Professor J. Siegfried Meurer, MAN, retired; Hans-Jürgen Reuß, Deutz & MTU, author; C. G. A. Rosen, Caterpillar, deceased; Walter A. Schauer, Robert Bosch, retired; Friedrich Schildberger, Daimler-Benz, deceased: Robert E. Schultz, Diesel & Gas Turbine Publications; Dr. Warren E. Snyder, my former boss at American Bosch, retired; I. J. K. Summerauer, Saurer & Steyr-Daimler-Puch; Peter Thomas, The Institution of Mechanical Engineers; Arthur F. Underwood, General Motors Research Labs., deceased; Robert M. Vogel, The Smithsonian Institution, retired; Guenter Weisse, Robert Bosch; Bruce Wadman, Diesel & Gas Turbine Publications; A. R. (Will) Willems, SAE. As seen, many have since retired, a few are gone, and some continue coming to my aid when asked. They have all been special friends.

This book would not have been possible without the full cooperation of the MAN Historische Archiv in Augsburg. It was my great fortune in 1972 to meet Archivists Irmgard Denkinger and Eduard Klein, who opened my eyes to their treasure trove and who in turn gave me entrée to other museums in Europe. Irmgard, too, became a mentor and like a mother to me. We shared many an hour probing the mind of Rudolf Diesel, her "second love," through his correspondence and writings housed in Augsburg, long after she retired. I miss her. Klaus Luther, who succeeded her, likewise cheerfully and fully responded to my requests during his tenure.

Josef Wittmann, MAN archivist for the past seven years, has been absolutely invaluable to me with his great "finds" (for him, too), his untold hours assembling, making legible, and annotating literally hundreds of pages of drawings, documents, and bits of information that have made things clearer for me and that have answered long-unresolved questions. Josef, I salute you. *Vielen Dank!*

Similar support, freely and generously given, has come from Burmeister & Wain, now a part of the MAN Group. Through the urgings and interests of Søren Hansen, historical activities in Copenhagen are flourishing because of the efforts of Niels Egon Rasmussen and Anthony W. Woods, who have likewise inundated me with helpful material on their company's early people and engines.

Sulzer Brothers has been equally supportive, particularly through the "above and beyond" labors of David T. Brown, Sulzer's unofficial historian in Winterthur since the retirement of W. Bangerter. He has supplied me with copies of most valuable original materials from his time with Sulzer, which began in England. I am also indebted to Georg Aue, retired from Sulzer, who has lived with the diesel engine since his childhood in Russia where his father worked for that company before the Revolution.

Other friends, people who are working, retired, or now deceased, and museums, libraries, and companies who provided special aid in ways beyond the supplying of requested information include: Roy V. Anderson, Winslow Engineering, deceased; Per-Sune Berg, Volvo Truck, retired; Robert Bosch GmbH; Marie Rose Cochet, granddaughter of F. Dyckhoff; Robert Cox, Willans & Robinson historian; Michael E. Cross, Cross Manufacturing Co.; The Daimler-Benz Museum; Dr. H. J. K. Dessens, Nederlands Scheepvaartmuseum; Anton Dolenc, Steyr-Daimler-Puch; Danmarks Tekniske Museum; Detroit Public Library, National Automotive History Collection; The Edison Institute, Henry Ford Museum & Greenfield Village; Fiat Centro Storico; Jacques Gallois, SEMT Pielstick, retired; Mrs. Wendy S. Gulley, Nautilus Memorial Submarine Force Library and Museum; David H. Hamley, Westinghouse; Dr. Horst O. Hardenberg, Daimler-Benz, retired; Bruce Harling, The Institution of Mechanical Engineers; Dr. Rudolf

Heinrich, The Deutsches Museum; Alexandre Herlea, Conservatoire National des Arts et Métiers; Indiana Historical Society Library; Gert W. Kemper, MAN; Robert V. Kerley, Ethyl Corp., retired; Hans L. Knudsen, Cummins Engine Co., deceased; The Krupp Archives; Kung. Biblioteket, Stockholm; Dr. G. P. A. Mom, Autotechniek Institute; S. G. Morrison, The Institution of Mechanical Engineers, retired; Michael S. Moss, University of Glasgow; Multnomah County Library; Jan Norbye, author; Peter Nordby, Göteborgs Landarkiv; Sixten Nyqvist, Nohab, retired; Professor Edward F. Obert, deceased; Jacques Payen, Conservatoire National des Arts et Métiers; Dr. Franz Pischinger, FEV Motorentechnik; Rolf G. E. Qvarnström, Nohab, retired; San Diego Aerospace Museum; Peter Stephens, the Science Museum; Dr. H. J. M. Stevens-Hardeman, Werkspoor Museum; Dr. Bernhard E. J. Stüdeli, Sulzer Bros. Research Library; O. D. Treiber, deceased, and his daughter Mrs. Jeannette Kurtz; Jan Vegter, Brons material; The Vienna Technical Museum; Adriaan Willemsen, Saurer; Wisconsin State Historical Library; Frank O. Wright, Mirrlees, deceased.

For their help in translating hundreds of pages, I thank Per A. Bolang, Boeing, retired, Inga Rhodes, and Henry I. Willeke, deceased. Loren Bunnell, Graphic Arts Center, who suffered through *Internal Fire* with me and has been of such great assistance in the preparation of *Diesel's Engine*, also deserves recognition.

I am indebted to the companies, both those displayed at the end of the book and those who chose to remain anonymous, who so generously helped sponsor its publication.

For those whose names should have been listed but are inadvertently missing, I give my apologies.

To my many friends above and to others, particularly those members of SAE who have endured my sermons on the value of preserving the history of technology, mea culpa—but do it!

C. Lyle Cummins, Jr.,
Wilsonville, Oregon
October 1993

INTRODUCTION

Diesel was a man and is an engine. Everyday, millions drive past freeway signs reading *Diesel*. To most they signal availability of a special fuel. To some they conjure an engine burning that fuel. Few would imagine that, in fact, they are a deserved tribute to one whose creativity led to a new form of internal combustion engine and a new industry. No other power plant in such widespread service bears the name of its sire.

Diesel's Engine is in part the story of a struggle for survival and acceptance in its earlier years. It is a montage of Rudolf Diesel's life and ideas on which are superimposed the people and events shaping the destiny of his engine. These happenings are told through the letters and printed words of the players, major and minor, who recorded, interpreted, or spawned them. Thus, it is more than the technical history of a machine. *Diesel's Engine* is about a concept nurtured by passionate dedication and fostered by benevolent capitalism on an international scale. It tells of a desire to make an engine more fuel miserly than any predecessor.

Diesel's "Economical Heat Motor" was the first internal combustion engine to stem from thermodynamic theory rather than innovative technology. Its development required new materials, new manufacturing methods, and new applications of science to heat engines. A few of the ideas explored failed simply because the tools and techniques were not yet available. Consequently, the time to transform it from a vision into a useful machine lasted years longer than foreseen by optimistic inventor and financial backer. Two decades passed from the disclosure of an untested theory in Diesel's patent until, tempered by reality, that theory was transformed into a prime mover reliable and powerful enough to propel merchant ships across oceans.

By not swiftly living up to premature promises, the engine's acceptance came painfully and slowly. Its progress, although stimulated by the exigencies of war, was too often impeded by prejudice or by crises arising from personal and corporate overconfidence. The external factors of fuel quality, import duties, and price wars waged by the global oil giants of Rockefeller, Shell, and Nobel imposed yet another strong influence. But dedicated people persevered and saw it begin to fulfill its destiny. Novel designs would prove their merit and eventually allow the engine to capture markets traditionally held by entrenched descendants of James Watt's and Nicolaus Otto's brainchildren.

Rudolf Diesel's frenzied life was more than an obsession to build a highly efficient engine. This complex, sensitive, yet tenaciously determined man played a primary role in the formation of an international industry. His career endured conflicting interactions between dynamic personalities and marketplace failures. Few engineers have been so lauded or so damned by peers. Critics charged that Diesel either stole his ideas, that he never built the engine he patented and licensed, or that others made it a success after his own brief, initial effort. These accusations, as well as his own personal failures, exacted their inexorable toll.

In Diesel is seen the spiritual and financial rise and fall of a proud inventor and enthusiastic salesman who often allowed dreams to blur reality. Yet it was the forcefulness of his conviction, his brilliant stubbornness, that sold an unproven, highly advanced concept to a skeptical business community.

A century has passed since Diesel's engine made its debut. In trucks, farm equipment, locomotives, and ships, diesels are the acknowledged power of choice. One cannot predict the dominant form that internal (or external) combustion engines of the future will assume. Until then, the engine of Rudolf Diesel will have served well as an interim solution to the problem of conserving our finite crude oil supply.

CHAPTER 1

Go from Your Country and Your Father's House

"Go from your country and your kindred and your father's house to the land I will show you."

—Genesis 12:1[1]

A boy's diary for the year 1870 lies by his bedside. Random entries read:

- I saw more balloonists riding in their wicker baskets in the sky above Paris today. I still am not sure I know why hot air has the power to lift a weight.
- Father punished me today after I took apart a clock and could not get it back together again. He does not understand me.
- Another visit to the Conservatoire des Arts et Métiers. Lenoir's gas engine was operating. It is near Nicolas-Joseph Cugnot's monster steam-powered wagon. I am glad our home is so close to this wonderful museum.
- Father's leather goods business does not do well. I have few finished orders to deliver after school. Many of his customers are rich and live in beautiful mansions. Someday I am going to live in one, too.
- France and Germany are at war!
- We Germans are told to leave Paris at once. We want to return to Bavaria, but we cannot cross the battle lines.
- Father, Mother, my two sisters, and I take only what we can carry and leave for England. We go by train to the seacoast and are very sick all the way across the Channel.

- London is exciting, but so different from Paris. I am glad my mother speaks English well and that she taught me some, too. We have almost no money.
- I saw the Newcomen and Watt steam engines and the Stephenson locomotive in the Science Museum. I could spend all of my time there. It is much better than going to the English school.
- Father and Mother are sending me to live with a cousin in Bavaria. There is not enough money for all of us to stay together in London. I am to travel alone.
- It was very cold on the boat to Rotterdam and the trains to Augsburg. I changed trains often, sometimes riding with soldiers in freight cars. The trip took eight days, and I have a bad cold and earache when I get to Augsburg.
- My cousin and her husband make me feel welcome.
- I enter school where he teaches.

The diary is fictitious. The happenings are true. Twelve-year-old Rudolf Diesel had an unforgettable year.[2]

The Paris birth registry for March 18, 1858, records the arrival of Rodolphe Chrétien Charles Diesel, but fate changed the spelling of his given names and allegiance to the country of his birth. Twenty-two of Rudolf Diesel's fifty-five years of life were spent in Paris. His early boyhood was there, and it was an arena for the first decade of his professional career. But he never became a Frenchman; his ancestry, higher education, and remembered achievements—his worldwide and ongoing legacy—were all products of Bavaria.

The parents of this gifted boy appear an unlikely pairing. Gottlieb Theodor Hermann Diesel (1830–1901), a third-generation bookbinder, grew up in Augsburg in the kingdom of Bavaria. While he intended to continue on there in the family tradition, economic hard times following the revolution of 1848 compelled him, along with thousands of his countrymen, to seek more promising lands. Theodor and his brother Rudolf (d. 1885) decided to try their luck in Paris, where Theodor arrived in 1850. Leather was the principal material used for book covers at that time; his expertise with it allowed him to open a shop as a *maroquinier*, or maker of Moroccan leather goods and portfolios (fig. 1.1). Business was never prosperous (brother Rudolf gave up, returning to Augsburg), but Theodor was determined to remain and at times even employed several workers.

Elise Strobel (1827–1897) was born in Nürnberg, the eldest of eight children. She did not fit the usual mold for young German women of her day. Possibly for the same reasons that Theodor Diesel went to Paris, Elise tried London where she found a pleasant position as traveling companion for an older English lady. When her father died in 1850 leaving the seven younger Strobels without a parent, she immediately returned to Nürnberg and

Fig. 1.1. An early Paris business card of Theodor Diesel saying that his leather work included gun holsters. *MAN Archives*

stayed for several years to care for the children. Her mother had passed away five years earlier. While at home Elise earned money giving English lessons to emigrants destined for America. For unknown reasons she again left Nürnberg, this time choosing Paris, where because of her language fluency she supported herself by tutoring in both German and English.

Theodor Diesel and Elise Strobel met in their adopted city and after a brief courtship were married in 1855. However, the ceremony was performed in London and not Paris. Bavarian authorities refused to sanction a wedding in France but would if it were performed in England. (Twenty-two years later, after Theodor and Elise had returned to Munich, they were remarried in a police court.)

Three children were born to this couple in Paris: Louise (1856–1873), Emma (1860–1946), and Rudolf Christian Karl. (fig. 1.2)

Paris offered unbounded opportunities for the impressionable and precocious young Diesel to expand his mind. In addition to the vibrant atmosphere of life in a cosmopolitan city, nearby his home was Europe's most famous museum of technology and science, the Conservatoire des Arts et Métiers.

Rudolf's home life was in great contrast to the pleasures and excitement he derived from the city. Theodor, a strict disciplinarian steeped in the old tradition of child rearing, failed to understand or cope with his inquisitive and sensitive son. He did not suffer from lack of parental love, but the often harsh punishments seemed heavier than the deeds warranted. Because Rudolf and his sisters were alone at home much of the time, he could find ways to get into trouble. He once took apart a cuckoo clock and could not reassemble it before his parents returned from work. This resulted in his being tied to a piece of furniture while the family made a Sunday outing to Vincennes. Another vividly remembered chastisement followed his failure to tell the truth. He was sent to school with a sign around his neck reading *"Je suis un menteur"* ("I am a liar").[3]

Fig. 1.2. A 1943 photo of the Paris house where Rudolf Diesel was born.
MAN Archives

Elise Diesel recognized that her son was exceptional and did all she could to challenge his mind. In the Diesel family it was the mother who provided husband and children with an umbrella of security and stability to sustain them through often difficult, discouraging times. She was Rudolf's anchor (fig. 1.3).

In the mid-nineteenth century, two cities of the world stood in a class by themselves. Paris and London, each with over two million inhabitants, were more than double the size of their nearest rivals. Not even London could compete with the grandeur and spirit of the City of Light. The Emperor Napoleon III was devoting much of his reign to making Paris the most beautiful city of its time. Broad, straight boulevards, immense parks, majestic monuments and government buildings were matched by hidden engineering masterpieces beneath the ground. The water and gas mains and especially the gigantic network of sewers were envied marvels of construction.

The emperor was not to preside over the completion of all his grandiose plans. Goaded by a scheming Otto von Bismarck into thinking he could teach Prussia's chancellor a lesson, an overconfident Napoleon III precipitated an ill-conceived conflict that cost him his throne and humiliated his country.[4] The Franco-Prussian War of 1870–71 that began August 4 was short, yet it sowed seeds for two future catastrophes.

The war not only altered Napoleon III's plans, but also made German nationals unwelcome in Paris. Given other circumstances, Theodor and Elise,

Fig. 1.3. Theodor had expanded into making children's toys in the shop of the house where the Diesels lived when Rudolf was born. *MAN Archives*

as Bavarians, could probably have remained. However, their homeland was caught in the rising enthusiasm of Pan-German nationalism and sided with Prussia rather than remain neutral.[5] Consequently, only three days after the disaster at Sedan where Napoleon III was captured on September 1, the Diesel family closed up their small home and business, and sadly after all the years in Paris, fled the country.[6] They took only the barest of essentials, leaving everything else behind in storage.

For two reasons they went to London. The expensive journey to Theodor's hometown of Augsburg was highly uncertain, and in his mind the economic conditions there had not improved. No doubt their decision was colored by Elise's previous time in London and her fluency in the language.

Although Rudolf stayed in England only a short time, he experienced another side of the Industrial Revolution that differed from the technology he admired and studied in the Science Museum or on the London docks. He witnessed and never forgot the poverty and squalor smothering the lives of the average worker—adults and children—who were expendable toilers helping to forge Victorian triumphs in engineering and industrial enterprise.

Conditions for the displaced Diesel family were little better in London. Theodor became so discouraged that he soon wanted to return to Augsburg regardless of the consequences. However, his brother Rudolf wrote from there strongly advising against it. Instead, he proposed that his namesake nephew be sent to live in Augsburg with their cousin Betty (Brügger) Barnickel. Her husband Christoph (1833–1911) was a mathematics teacher in the Royal District Trade School (*Königliche Kreis-Gewerbs-Schule*) who could help the younger Rudolf continue his education.

Thus, his academic career would not be French, nor would he become a French citizen; because of his excellent lower school record, he had expected to enter that autumn in the École Primaire Supérieure, where attendance automatically conferred citizenship.

After only two months in London, Diesel said goodbye to his parents and began, alone, an adventuresome journey to his new home in Germany. He would not live with them again until they returned to Munich in 1878.

This talented twelve-year-old—bilingual in German and French, and thanks to his mother also competent in English—had already experienced more than most adults of his time (fig. 1.4). He had seen the best and the worst of the world's largest cities and had suffered the privations caused by war and poverty. Now he faced another exodus.

Augsburg offered new facets of technology to add to those that had awed Rudolf in Paris and London. While small by comparison, it was not, however, a typical Bavarian community. *Augusta Vindelicorum*, the name bestowed on it when founded by the Roman emperor Augustus, served in the high Middle Ages as an important trading center between the German and Italian states. Banking families like the Fuggers financed popes and emperors during the fifteenth and sixteenth centuries. Its gold- and silversmiths were famous throughout Europe. Later, a growing textile industry added to the city's regional importance and provided a source of craftsmen and factory workers. When the railroad opened in 1840 between Augsburg and Munich, a distance of about sixty-five kilometers, new companies were formed that could make heavier products from iron and steel and serve larger markets.[7]

One of these railroad-inspired companies was to play an important part in the life of Rudolf Diesel. Ludwig Sander (1790–1877), an Augsburg merchant, saw an opportunity to supply the textile industry with machinery and opened a machine shop the same year the trains first came to his city. Carl Buz (1803–1870), chief engineer of the railway, and Carl A. Reichenbach (1801–1883) leased the factory from Sander four years later and called it C. Reichenbach'sche Maschinenfabrik. Buz was the active manager from

Fig. 1.4. Rudolf Diesel at age twelve. *MAN Archives*

1845 until he retired in 1865, and his son, Heinrich von Buz (1833–1918), succeeded him.[8]

Reichenbach, a nephew of Friedrich König who invented the flatbed letterpress in 1811, was the means whereby the company began manufacturing his uncle's famous printing presses in 1845. As a result, Augsburg grew into a world center for printing technology—a reputation the city and the continuing company still enjoy.

Under the guidance of Carl Buz, steam engines and boilers, water turbines and power transmission machinery were also introduced. The company never made marine steam engines. Buz and Reichenbach bought the factory in 1855 and two years later changed the name to Aktiengesellschaft Maschinenfabrik Augsburg (Augsburg Machine Works, Inc.).

In 1870, a few years after Heinrich Buz had become head of the company, Professor Carl von Linde (1842–1934) published an important treatise on the theory and design of advanced refrigeration and ice-making equipment. Buz was impressed enough with Linde's ideas that he contracted to build the machines, which of course would be driven by his company's steam engines. This venture began in 1873. Augsburg industry also generated a need for strong technical schools that were ready to challenge the newly arrived young Parisian.

Since the German educational system developed so differently from that of the United States, it is worthwhile to describe the course Diesel pursued while receiving his schooling. Germany, along with other European countries, followed a "tracking" system whereby children, based on their achievements at quite early stages, were sent into paths that strictly determined future educational opportunities.

In the Bavaria of 1880, for example, tests decided whether a nine-year-old child was to attend a Latin school leading to a university-level education or continue on to a second checkpoint one year later. Some students then filtered into a *Realschule* where they continued until age fourteen to sixteen. Foreign or classical languages were not required, as this school prepared one for a vocational institute. Those unable to enter the *Realschule* left primary school at age thirteen and went into a work/study program for three to five years.[9]

Diesel's Paris education qualified him for either the humanistic *Gymnasium* or the more technically oriented *Realgymnasium*, which did not require Greek. Still, the latter school entailed taking courses in philosophy and literature. The *Gymnasium* led to a classical university, while the *Realgymnasium* prepared students for a *Technische Hochschule* (engineering university). Because Christophe Barnickel taught in the equivalent of a *Realgymnasium*, it was natural Rudolf would start there.

Rudolf Diesel owed much to the Barnickels, for not only the love and shelter they gave at a crucial point in his life, but also the intellectual and educational doors they opened for him that probably would have remained closed had he stayed with his parents. Betty (Brügger) Barnickel could sympathize with her cousin. Her own parents had died when she was young, and an uncle, Rudolf's grandfather, adopted her.

Emma Diesel, who married Christophe after Betty's death, years later shed light on details of Rudolf's arrival in Augsburg. According to her, Rudolf had expected to live with his uncle, and it "was most difficult for him" at first when he learned the arrangements were for him to stay instead with the Barnickels. Rudolf himself afterward recalled to her that his uncle's greeting was not very friendly. Emma went on to explain that the elder Rudolf and his wife were unable to take him into their home because of the size of their own family and their poor financial situation. Since the Barnickels had no children and often boarded students, Uncle Rudolf had agreed to pay the Barnickels the 1,000 marks annually, the same as was received from each of the other three boys living with them. It is probable that the younger Rudolf would later reimburse the Barnickels for what his uncle could not pay. Emma confirmed the very close relationship between young Rudolf and the Barnickels.[10]

He missed his own mother and father and sisters. His frequent letters to them reflect this.[11] Strong disagreements at times arose between Rudolf and his parents, but the familial bond always remained intact.

It was his parents' intention that Rudolf would live with the Barnickels only until his family returned to Paris after the war and he could again continue his schooling there. But this was not to be. He completed the rest of his education in Bavaria: five years of preparatory work in Augsburg and the technical university in Munich.

Testimony to Rudolf's brilliant mind and his stubborn determination were the scholarships he received and the money earned by tutoring in English, French, and other courses in which he excelled. However, he faced significant obstacles during the pursuit of these studies: he could count on no funds from his own family, his father did not want him to go beyond the three-year program at the Royal District Trade School, and, lastly, his French birth caused citizenship problems.

Rudolf was in Augsburg only two years when he made up his mind to become an engineer. In a letter to his father dated September 11, 1872, he informed them of this decision: "Now I shall be for one year in the Königlichen Kreisgwerbschule [sic]. Then I must go two years to the industrial school of Augsburg, and then I must go some years to the Polyteknikum of München."[12] The parents in turn were against it. According to his father, the sooner he quit school the sooner he could become a wage earner and repay the Barnickels for his stay with them; moreover, it was honorable to work with one's hands.

Regardless of parental opposition, the brilliant fifteen-year-old continued to demonstrate that he was more interested in developing his mind than in becoming an apprenticed craftsman. He completed his first studies in

Augsburg with very high marks and looked forward to entering the polytechnic industrial school. This two-year program, in Augsburg, usually led to Munich and an engineering degree.

Rudolf decided to visit his parents, who had returned to Paris shortly after the war, to convince them of his great desire for the additional schooling. The journey from Augsburg took forty hours by the cheapest train. Although he had been separated from his family for three long years, it was a bittersweet reunion.

Financial conditions for the Diesels were little better than when they were in London. Adding to the generally unhappy and tense visit, Rudolf's gifted seventeen-year-old sister, Louise, died suddenly a few weeks after his arrival. It was a deep and lasting loss to all the family because of her spirit and the joy she had given them. Even the money the musically talented girl had added to the Diesels' meager resources by giving piano lessons was denied the struggling family. This tragic death caused Theodor not only to grow more insistent on his son leaving school, but as time passed, to withdraw from reality into increasing belief in the mystical and supernatural. (Although the notions were different, Rudolf's grandfather also had lost some touch with the real world in his later years.)

Further complicating Rudolf's situation was Theodor's intention to become a French citizen, an act that according to Rudolf would have made a tremendous difference in his own career. Reading further from the letter to his father of September 11, 1872, he pens an impassioned plea:

> Please, please, and again please. I am a German and will be a German. When you have a French passport then I must be a soldier for the French army for nine years. Please do not become a French citizen. In Germany an ordinary soldier must be in the army three years, but if tests are passed then it is only one year. . . .
>
> And when I have finished my studies, I am twenty years old, and I must be a soldier. When you have a German passport, I must be a soldier for only one year and it is soon over, and I will not forget what I have learned. *Then I am twenty years old, and I can begin my mechanical way.* [Italics added.][13]

Rudolf finally convinced his father not to take this step, and later he himself became a German citizen on his nineteenth birthday, the earliest he could do so of his own accord.

During his travail with his parents in Paris, Rudolf never wavered in his determination to be an engineer, and again the Barnickels rescued him. They invited him to come back and stay with them until he could complete the final two school years in Augsburg. His parents were unhappy with the decision for differing reasons, yet ultimately accepted it. One must believe that the Barnickels refused to deprive Rudolf, with his demonstrated intelligence and maturity, of what they could contribute toward his education. They were the first of several mentors during Diesel's early years to throw him a lifeline in a crisis. Another helper at this time was the head of Rudolf's new school, who gave him enough money upon his return from Paris to support his forthcoming enrollment (fig. 1.5).

It was at the polytechnic school that Diesel discovered a new piece of laboratory equipment: a heavy glass-walled version of the ancient southeast Asian "firestick."[14] Akin to a tire pump, it demonstrated how, from the heat generated by the rapid compression of a single, inward pumping stroke, a piece of tinder placed at the fixed end of the cylinder could be ignited. He never forgot these demonstrations and once, years later, borrowed the *Feuerzeug* to explain to his own children how his engine operated (fig. 1.6).[15]

Fig. 1.5. Extracts from Rudolf's notebook made at about age fourteen. He filled it with family, school, and personal information, including these pages on scientific experiments and Thomas Jefferson's ten rules for living a useful life.
MAN Archives

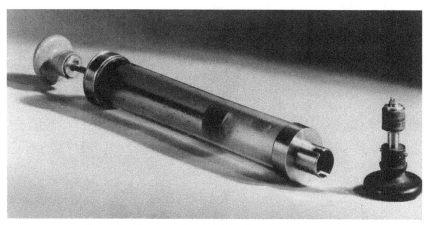

Fig. 1.6. Pneumatic tinder igniter (*Feuerzeug*) of
Diesel's Augsburg school days. *MAN Archives*

In addition to demanding studies, he continued to earn money by tutoring other students in French and English. Like his sister Louise, he had a love of music and learned to play the piano, a lifelong enjoyment for him. There was little time to spare for the usual boyhood amusements (fig. 1.7).

When Rudolf graduated from the Augsburg polytechnic school in 1875, the youngest of his class, it was with the highest marks ever given there.

Entrance was thus assured to the *Technische Hochschule München* (before 1868 the Munich *Polytechnische Hochschule*) if he could find the financial resources. The solution came by way of a professor at the Munich university who was also a government commissioner for education in Bavaria. After hearing about Rudolf's academic achievements and talking with him during an inspection visit to Augsburg, Professor Karl von Bauernfeind was so impressed with the boy that provisions were made for an initial scholarship of 500 gulden per year for the first two years. He afterward was granted additional funds to carry him through all four years of his studies.[16] (Diesel became a frequent and welcome guest in the Bauernfeind home where important German engineers met and marked him as a rising star.) Further help came from Theodor von Cramer-Klett, a Nürnberg industrialist who aided worthy students at the *Hochschule*.

The university years proved to be some of the happiest of Diesel's life. New friends with common interests and new areas of knowledge to explore gave him stimulation and satisfaction. Two of his student friends would play important but differing roles in Diesel's future: Lucian Vogel (1855–1915), a colleague and crucial supporter during the time of engine development, and Oskar von Miller (1855–1934), one of Germany's key electrical engineers but better known as the founder and guiding genius of the Deutsches Museum in Munich.[17] In later years Miller defended Diesel's name and reserved a place of honor for him in the museum.

Fig. 1.7. A pencil sketch by the artistic sixteen-year-old
Rudolf Diesel. *MAN Archives*

One professor above all others at the university greatly influenced Diesel, both when he was a student and during his professional career. This man was the same Dr. Carl von Linde whose ice-making machines were adding to the earnings of Maschinenfabrik Augsburg. Rudolf's earliest association with Linde came through his teacher's thermodynamics lectures.

Linde made important contributions during his long life. Born in northern Bavaria, he graduated from the *Technische Hochschule Zürich* and became a professor of machine technology at the *Technische Hochschule München* in 1868, the year the school was granted full technical university status. In 1879 Linde founded his own factory at Wiesbaden to make refrigeration equipment. He is probably most noted for his pioneering work in producing liquid nitrogen and oxygen from air in unlimited quantities.[18] His technology came to the United States around 1900 with the partial assignments of his patents to a Cleveland, Ohio, firm.[19] The Linde Air Products Company, also of Cleveland, was formed in 1907 to make oxygen and nitrogen by his processes. In 1917 the Linde Company was one of five combining to form the Union Carbide Corporation.[20] Also of interest is Linde's 1901 patent on a "method of utilizing liquid air in an explosion [internal combustion] motor."[21] He received numerous national awards as well as serving on the Deutsches Museum Board from the time of its founding in 1903.[22]

Under Linde the study of thermodynamics developed into almost an obsession with Diesel. This science dealing with the relationship of heat and work was a discipline ideally suited to his highly analytical and inquiring mind.

Captivating Diesel was a single work, *Réflexions sur la Puissance Motrice du Feu et sur les Machines Propres à Développer cette Puissance,* by Nicolaus Léonard Sadi Carnot (1796–1832). His brief but powerful 1824 treatise of "thoughts on the motive power of fire and on machines fitted to develop that power" described a theoretical cyclical process whereby the maximum thermal efficiency of a heat engine could be attained.[23] D. S. L. Cardwell, an authority on the history of thermodynamics, said of him:

> Perhaps one of the truest indicators of Carnot's greatness is the unerring skill with which he abstracted, from the highly complicated mechanical contrivance that was the steam engine (even as early as 1824), the essentials, and the essentials alone, of his argument. Nothing unnecessary is included and nothing essential is missed out. It is, in fact, very difficult to think of a more efficient piece of abstraction in the history of science since Galileo taught men the basis of the procedure.[24]

Since steam power was the only form of heat engine in Carnot's time (and essentially so during Diesel's student days), Carnot had written:

> The employment of very strong vessels to contain the gas at a very high temperature and under very heavy pressure [and] . . . the use of vessels of large dimensions . . . are, in a word, the principal obstacles which prevent the utilization in steam-engines of a great part of the motive power of heat. We are obliged to limit ourselves to the use of a slight fall of caloric [heat], while the combustion of the coal furnishes the means of procuring a very great one.[25]

When Diesel realized the large disparity in thermal efficiency between what Carnot proved theoretically possible and what was actually delivered in practice with the best steam engines, he wrote in his notebook under the heading "Mechanical Heat Theory": "Can one build steam engines which realize the perfect cycle process without their being very complicated?"[26] The entry date was July 11, 1878; the seed was planted.

Other comments in his notebook echo statements by Carnot. One written later that year refers to the amount of energy in a kilogram of coal, where he wonders if there is a way to do useful work without first having to generate steam. "But how is it practically feasible? That is precisely to discover!"[27]

Carnot's theories became a lodestar attracting Diesel when he embarked on his search for the most efficient heat engine. Tragically, the tantalizing dream postulated by Sadi Carnot proved for Diesel to be a nightmare haunting him for much of his life.

Rudolf expected to graduate in July 1879, but before the final examinations were given, he became seriously ill with typhoid fever, which was the epidemic ravaging Munich that summer. After recovering sufficiently, he was offered an interim job with Gebrüder Sulzer, a long-established machinery company in Winterthur, Switzerland, near Zürich. Professor Linde, whose refrigeration machines were in production there also, came to Diesel's aid by highly recommending him to the firm.

The short, but intensive, experience at Sulzer Brothers beginning in October gave Diesel a valuable opportunity to work in a factory environment. In addition to the Linde products, Sulzer was noted for its steam engines and boilers. At that time the factory employed 1,300 workers and was the largest business in a town of ten thousand. There Diesel saw machines in operation and served as an apprentice at a filing bench for four weeks. He experienced firsthand the toil of a shop worker; he came to understand the philosophy and concerns of his fellow laborers. His lodging was a boarding house shared with fifteen other men: the engineers and the office and factory workers at the Swiss firm.

At the beginning of January, Diesel returned to Munich in preparation for his rescheduled examination. The test was oral, consisting of questions from and discourses with his professors. On January 15, 1880, he passed with praise and congratulations from his teachers and received the highest grades awarded since the polytechnic university had opened its doors twelve years earlier (fig. 1.8).

Rudolf Diesel was ready to challenge his world. Against all odds he had achieved what most would have declared highly unlikely when the little refugee left Paris with his destitute family ten years earlier. However, someone

Fig. 1.8. Rudolf Diesel in 1880 at his graduation from the engineering school in Munich. *MAN Archives*

seemed to watch over the brilliant young dynamo. Rudolf had zeal and intelligence, vital ingredients for the success of any career or new idea, yet he also had the great fortune to receive support and resources at crucial times. This does not detract from his personal accomplishments, but reminds that no one is truly "self made." Few fulfill their dreams without the good fortune of an outstretched hand from another to sustain or guide in a helping way when their causes or aspirations appear doomed. Diesel was offered this hand.

The Bible verse Rudolf Diesel was asked to interpret during his Lutheran confirmation on March 24, 1872, came from Genesis 12:1:

Go from your country and your kindred and your father's house to the land I will show you.[28]

The following verse, not included, prophesies:

. . . and I will bless you and make your name great.

Notes

1. Eugen Diesel, *Diesel: der Mensch, das Werk, das Schicksal* (Hamburg: Hanseatische Verlagsanstalt, 1937), 26.
2. Ibid., 35–50. It is unfortunate that this definitive biography of Rudolf by his son has not been translated into English. Almost all works dealing with Diesel's personal life, including those in English, draw from this source with varying degrees of accuracy. The author also cites from *Diesel* and, hopefully, has avoided such mishaps.
3. Ibid., 33.
4. Erich Eyck, *Bismarck and the German Empire* (New York: Norton, 1964), 163–74.
5. Ibid., 179. In 1871 Bavaria became a part of the new German Empire, not because of Ludwig II's desire to be under Prussian domination but for the secret financial rescue by Bismarck of "Mad Ludwig's" debts incurred during the building of his famous Bavarian castles.
6. E. Diesel. *Diesel*, 46.
7. Ludwig Wegele, *Augsburg* (Augsburg: Perlach Verlag, 1970), 7–14.
8. MAN. Bulletin D.S. 315 E.I., in *A Brief History of our Works*, 2–6. Augsburg: 1961; MAN. Bulletin D.S. 342E, in *MAN Company History: Facts, Figures, Exhibits*, 3–11. Augsburg: 1977.
9. George S. Emmerson, *Engineering Education: A Social History* (Newton Abbot, UK: David & Charles, 1973), 87–89.
10. Emma Barnickel, "Answers to the 31 Questions of Eugen," in MAN Archives (1935).
11. Rudolf's letters preserved in the MAN Archives begin in 1870 with his arrival in Augsburg. Because of the young writer's maturity, they provide a degree of accuracy and insight rarely found in the reflections of one's early years.
12. Translation by Frau Irmgard Denkinger (1917–1991). Also preserved, besides the letters, are artwork from Rudolf's school days in Paris as well as watercolors and later sketches of his wife and children at home and in holiday scenes. MAN Archives.
13. E. Diesel. *Diesel*, 46.
14. Charles Singer et al., *A History of Technology*, vol. 1 (London: Cambridge University Press, 1957), 226–28.
15. Richard Lorentz, English patent No. 3,007 of Feb. 5, 1807, for a "Fire piston contained in a walking stick." It acted in the same manner as the Asian device.
16. Diesel. *Diesel*, 92. In the 1880s, 1 gulden = 2 old German marks, with the exchange about 4.2 marks per US dollar. Thus, 500 gulden = approx. $240.
17. Obituary, *New York Times*, April 10, 1934. And, Editorial, *New York Times*, April 11, 1934.
18. Obituary, *New York Times*, Nov. 17, 1934.

19. US patents 727,650 and 728,173, filed on July 7, 1895, and issued May 12, 1903. They were filed as a single application.

20. Linde—Union Carbide Corp., *Update,* March 22, 1982.

21. US patent 664,958, issued January 1, 1901.

22. Deutsches Museum, *Fünfzig Jahre Deutsches Museum München, 1903* (Munich, 1953), 6.

23. Sadi Carnot, *Réflexions sur la Puissance Motrice du Feu et sur les Machines Propres à Développer cette Puissance,* ed. E. Mendoza (New York: Dover, 1960), is a reissue of R. H. Thurston's translation of 1890. Prof. Mendoza has included works of Clapeyron and Clausius, along with a biography of Carnot and an explanatory introduction.

24. D. S. L. Cardwell, *From Watt to Clausius* (Ithaca, NY: Cornell University Press, 1971), 201. See also Lyle Cummins, *Internal Fire* (Austin, TX: Octane Press, 2021), 38–42.

25. Carnot, *Réflexions,* 49, of Mendoza edition.

26. E. Diesel, *Diesel,* 115.

27. Ibid., 116.

28. Ibid., 72.

CHAPTER 2

Ice, Ammonia, and Air

"How is an idea created? Maybe sometimes it strikes like lightning, but mostly it will develop slowly through intensive search under numerous mistakes . . ."

—Rudolf Diesel[1]

With great expectations, not as a visitor but as a resident, Rudolf Diesel made his way to Paris. The city was to become the site of his early career, the first home for his family, and the locale of his engine's genesis.[2]

Professor Linde provided the opportunity for Diesel's return to Paris. His former teacher needed someone to organize and then manage a newly licensed facility in Paris for selling and servicing the ice machines made by Sulzer Brothers. Linde believed that with Diesel's experience at Sulzer and product knowledge his former star pupil was the right choice. Because of an acceptance already enjoyed by Linde equipment in Germany and Switzerland, it was presumed that the unexploited French and Belgian markets should also prove profitable.

On March 20, 1880, Rudolf arrived in Paris challenged with his first professional assignment. He did not come with an expectation of immediate personal riches. Linde's offer stipulated Diesel would start as an apprentice or volunteer worker at a beginning salary of 100 francs per month. While not starvation wages, little remained after deducting modest living expenses to enjoy what Paris offered a young bachelor. However, Diesel's substantial job accomplishments were rewarded at the end of 1880 by being given an official title of "*Direktor*," or manager, of the Linde plant and by doubling his earnings. Eight months later they were again doubled to 4,800 francs per year. Using basic machine components shipped from Winterthur, Diesel oversaw design, fabrication, installation, and ice plant startup. Linde showed great faith in his twenty-two-year-old protégé to assign him such responsibility.

The Linde French license was initially owned by Baron Moritz von Hirsch, descendant of a Bavarian banking family. An international financier possessing great political influence, he helped the company and the young manager prosper. Yet, in time Hirsch lost interest in ice machines and sold his holdings to a large industrial firm that also handled "natural" ice, all that was available until Linde and others made ice on demand possible. This association with Hirsch and his brother Theodor, for whom Diesel later served as an engineering consultant, opened doors to the wealthy of Paris and provided valuable business experience.

Rudolf once again felt at home. The Paris of his childhood held treasured memories, and since he spoke the language like a native, he had little problem readjusting to his new environment. Freed at last from the worries of academic achievement and with few personal responsibilities, Diesel entered into the spirit of "Gay Paree" (fig. 2.1). While not slighting his job, as his performance showed, Rudolf did take advantage of his free time. Among his Bohemian friends were expatriates and struggling artists who existed on even less than his initial salary (fig. 2.2). A very special attraction was a young American girl whose claim to artistic talent lay in her somewhat unusual color shading techniques. She had been introduced to Rudolf by his sister in Munich, so the developing romance was not unknown to Emma. Nevertheless, he carefully avoided mention of this subject in letters to his parents. Toward the end of 1882, the girl decided to return to America, and her German friend almost went with her. Rudolf wrote to Emma that "America beckoned like a compass needle to the North Pole." However, Linde's ice machines won out.[3]

A difficult problem causing much anguish for Diesel was created by his parents from the time of his arrival in Paris. As in earlier days, they made

Fig. 2.1. A sketch from Rudolf Diesel's sketch book that
he began on his return to Paris. *MAN Archives*

Fig. 2.2. Diesel and friends in Paris, c. 1880. *MAN Archives*

demands he was unable to meet. The father could barely eke out a living in Munich and insisted that the son contribute more than Rudolf deemed feasible. If the money sent home had gone mostly for necessities, Rudolf would have understood and tried to comply with much less frustration. Instead, Theodor wanted the money to pursue "Spiritismus" and "Magnetismus," a form of escape he increasingly turned to after the death of daughter Louise.

Rudolf's mother irrationally demanded that he help his parents, citing religious claims to filial duty. The tragedies of her life had exacted their own tolls as heavy as those of her husband. Diesel's dilemma was shared in a letter to Emma, a confidante as well as a loving sister. He did not want to deprive his parents of their meager pleasures, but at this point little could be done. However, time and subsequent higher earnings alleviated parental torments, especially Theodor's constant inferences about Rudolf living like a king in Paris.

The perceptive young engineer noted Parisian factory conditions. He saw steam engines remaining vital for large factories in their ability to provide inexpensive power. However, smaller shops only had relatively inefficient steam engines available to them. He remarked a few years later that his goal was the creation of a compact engine for small industry to help in "innumerable end uses."[4] The need perceived by Diesel was ultimately fulfilled by the electric motor. Thoughts about motive power never left Diesel's mind.

Ice from the Linde machines under Diesel's care was not suitable for use in liquids humans consumed. Its primary purpose was to chill brewery tanks and to refrigerate meat and other perishable food products. This fact was not overlooked by Diesel. He soon realized there ought to be a sizeable market from the many Paris restaurants for table ice pure enough to use in drinks and food preparation. His first invention thus became what he called *carafes frappées transparentes* or "bottled clear ice." A French patent issued to Diesel on September 24, 1881, described a method of making potable ice in insulated bottles that could be delivered to customers. A second patent issued exactly a month later covered "clear ice" made in blocks.[5]

It was Diesel's hope to interest Linde in his ideas as his mentor had not yet pursued this market. Linde, and then Sulzer, both said no to Diesel, but Linde did suggest that Diesel contact Heinrich Buz at Maschinenfabrik Augsburg and offered to help in an introduction to him. Since Buz's company also made the Linde machines, it was an offer not to be refused. Rudolf visited Buz in Augsburg during February 1882 and was informed that Maschinenfabrik Augsburg might act as a development shop for the Diesel "clear ice" systems.[6] Since Diesel was already having parts made in Munich for his new inventions, he declined Buz's help.

Diesel built his potable ice system in the Paris Linde factory. However, he could not serve two masters. He learned that legally the inventions he patented were the property of Linde as they had resulted while Diesel was under contract with him. Linde did not want to stand in Diesel's way and offered the young man a solution to the problem: if Diesel agreed to become a Linde sales and service agent in the territory he already served, Linde would forego rights to Diesel's inventions. While this meant Diesel was no longer manager of the factory, he remained a board director with a base salary of 3,600 francs per year.[7] Diesel was not overly happy about this, yet he saw that it gave him the chance to go on his own to market his *Klareis* machines and have a small, assured income. (One must assume that, although the Linde company was under ostensibly French ownership, Carl Linde maintained considerable influence.)

The predicament left Diesel with a strong desire to maintain freedom from corporate involvement. He saw independence and personal wealth as mutually compatible, and based on an annual income during 1882 and 1883 of 33,000 francs, he must have considered himself well on the way to achieving his aims.

Widely held views about temptations awaiting young bachelors in Paris caused concerns in Rudolf's family back in Munich. When he had asked to borrow 200 francs from Emma to pay for patenting costs, it was feared the money might be spent on a mistress. Diesel's explanation to Emma eased at least her worries; also, their mutual American friend had departed for home.

It was Diesel's custom to visit with a German merchant and his wife each Thursday evening. On such an occasion in October of 1882, he was introduced to a very comely, blond and blue-eyed young lady also visiting his friends. Martha Flasche (1860–1944) came from a large family in

Remscheid, near Cologne. They were in poor economic circumstances because of the untimely death of her stepfather, a town notary. Since she had passed her teacher's examinations and spoke French and English well, it seemed that Paris might prove a good place to support herself. Martha and Rudolf were drawn immediately to each other, and a courtship soon began. In addition to their shared interests in the arts, music, and literature, Martha could not overlook the fact that her handsome suitor also seemed to have good financial prospects.

In the ensuing months, as Diesel traveled on business for Linde and *Klareis*, his letters to Martha became more adoring: ". . . no woman is ever more loved than you by me, no one, and no man has ever had more expectations for joy than I since you told me you wanted to give me your love, your heart and your life"[8] (fig. 2.3).

The wedding took place in Munich on November 24, 1883, in the presence of a small family circle. Diesel's parents welcomed Martha as a daughter-in-law, having come to love her as she did them during her visits before the wedding. Included with the family was Christophe Barnickel, the man who had done so much for the bridegroom. The former uncle was now a brother-in-law because of his own recent marriage the previous May to Emma after Betty's death in 1881. (Although a "September-May" marriage, with Christophe being twenty-seven years older than Emma, the two shared almost thirty years together.)

Fig. 2.3. Martha Flasche Diesel at the time of her marriage to Rudolf in 1883. *MAN Archives*

21

The honeymooners returned to Rudolf's modest bachelor apartment in Paris. His far from meager income all went for courtship and the pursuit of his ideas; fiscal prudence was not one of Diesel's virtues, and Martha exhibited similar tendencies.

Slightly disquieting to Diesel were the cooling relationships with his artist friends. They began to drift away because they either had hoped he would marry a French girl or believed he did not find enough time for them. (Diesel suspected the former.) The more important cares of establishing a home and of getting on with his career did not allow him to dwell on this. He soon had more to plan for with the coming birth of their first child, a son they named Rudolf Jr. (1884–1944). Two more children were born to the Diesels in Paris: a daughter, Hedwig (1885–1968), and a second son, Eugen (1889–1970).

During the first part of 1884, Diesel began in earnest his preliminary work on what he hoped would be an answer to the needs of smaller factories and shops. It was a motor using vaporized ammonia rather than steam as the driving force on a piston. He wrote in 1892:

> My original idea was to build a small engine that was always ready to start after only a brief heating period and would have ten to twelve hours of operation before "rewinding" again. Therefore, I selected liquid ammonia, which led me to the investigation of ammonia vapors, their absorption in different fluids, etc., and designed an ammonia engine.[9]

Since ammonia was also the refrigerant used in the Linde system, Diesel understood its thermodynamic characteristics. He believed such an engine to be more efficient than one driven by steam and that he could cope with any problems relating to the danger from toxic gases leaking past seals.

Commercial refrigeration and ice-making machinery by Mignon and Rouart had been sold, mainly in France, since the early 1860s. These were based on an ammonia absorption concept proposed by Carré in 1859.[10] No motive power was required, but the closed loop system entailed maintaining a fire under a pressure vessel to drive off ammonia vapor from the water-ammonia solution. Systems of this type are still in existence, but they are only feasible where, instead of directly heating the solution by an open fire, the heat is supplied from waste, low pressure steam. Otherwise, the energy requirements are too costly.

Linde's significant contribution to refrigeration technology was the use of an ammonia compressor,[11] in an era driven by small steam engines, which enabled both a more compact system and the use of pure ammonia. This latter point was important because the corrosive effects of a water-ammonia solution created serious problems. The "vapor-compression cycle," as it is known today, is the basis for most systems now in service, but it was an electric motor that replaced the steam engine.

The idea of using ammonia for the working fluid in an engine was not original with Diesel. An "ammoniacal" engine (by Fromont) exhibited at the Paris Exposition of 1868 ran on a heated gaseous mixture of 80 percent ammonia and 20 percent steam. The fuel consumed to heat the boiler was one-third that of a comparable size steam engine.[12]

Emile Lamm (1834–1873), who emigrated from France to the United States where he became a dentist, built such an engine in 1871 to power a streetcar.[13] The travel range of the tram depended on the volume of a liquified ammonia charge. The ammonia, sufficiently heated on board to gasify, passed only once through the system, including an engine similar to one run by steam, and back into a second tank containing liquid ammonia and water that was drained when the supply tank was recharged with pure ammonia. Lamm soon abandoned ammonia, switching to a portable, superheated water system that produced steam by surrounding a water tank with a heated calcium chloride solution. The transportable part of his system was simple, but the "filling station" was not. Since Lamm's work and others were reported in European technical journals, it can be supposed that Diesel was aware of at least some of these developments preceding his own.

Diesel spent not only every extra moment on his ammonia engine, but also more and more of his money. Unfortunately, his annual income began to drop at the same time due to the falling sales of his *Klareis* business and lessening prospects for the Linde machines. His income in 1884 and 1885 was about half that of the two previous years. Some of this he attributed to a very heavy supply of natural ice from a colder winter. Diesel remained the optimist, claiming it allowed more time for work on his ideas as long as he could not be selling. He wrote in 1887 that he "drew and calculated continuously on the projects . . . again discarding and beginning anew."[14]

An explosion one day in a laboratory area set aside for his engine experiments emphasized the destructive energy of the compressed ammonia as well as its very strong toxic effects. He visualized a possible battlefield weapon to incapacitate the enemy and went to the German military attaché in Paris with his idea. The idea did not reach Berlin.

Paris during the 1880s had an exciting quality of life, but often turbulent political conditions. When Diesel arrived in the city, the consequences resulting from the Franco-Prussian War were no longer visible to a casual observer. The heavy reparations payments demanded by Germany of five billion francs (the equivalent of one billion dollars) had been paid. Furthermore, numerous post-government crises had been staved off by compromises tolerable to enough of the electorate. The reparations debt, considered a stain on French national pride, acted as such a strong stimulant to the economy that the immense sum was delivered on schedule in 1875.[15] A spirit and momentum generated during these first years after the war continued and brought enviable prosperity.

There was talk as early as 1879 that Paris should host a world exposition ten years hence to acclaim the many industrial advances. Yet, as enthusiasm for such a display grew, the focus changed more toward a centennial celebration of the glorious years following the French Revolution. Gustave Eiffel (1832–1923), whose engineering achievements were renowned, was commissioned to create the world's tallest structure as a visual symbol of the exposition.

This reawakened French élan was to affect Diesel. For the first few years, his customers and associates appreciated his endeavors. However, by mid-decade he noted a rising nationalism that caused his German ties to become a burden. Although Linde's products were Swiss made, the name was associated with Germany.[16] Diesel himself might have passed for French, but his wife and family precluded this.

Despite his devotion to work, Diesel and his growing family spent summer holidays either at the seashore or in the mountains. It was a habit begun early in married life and would continue until his last days (fig. 2.4).

The Boulanger Affair of 1888 and 1889 was a political upheaval with potentially serious national consequences.[17] General Georges Boulanger (1837–1891), a former military hero, became the rallying figure for an overt nationalistic movement bordering on fascism. With memories of humiliating reparations, the loss of Alsace and Lorraine, fomenting actions in Germany by Bismarck,[18] and France's perceived strengths, only an empassioned leader

Fig. 2.4. Summer holiday scenes in Diesel's sketchbook, 1886 and 1887. *MAN Archives*

was required to kindle a fire. Fortunately for France, Boulanger's own faction eventually destroyed any chance it might have had to take over the country. This enunciated the unpleasantness Diesel had seen developing for several years and fueled a conviction that his deteriorating business was not due entirely to his own errors. Paris was no longer where he wanted to be.

Compounding Diesel's economic problems were recurring and severe headaches that had plagued him since he was a boy; as he grew older the pains seemed to intensify. A letter he wrote home in 1887 after suffering such an attack while on a train leaves little doubt of his anguish: "I have a headache—enough to drive one mad, really very frightening." In another letter he said: "Yesterday I had such a crisis from a headache that I left the Exposition, hired a carriage, and came straight to a doctor's office. . . . I am in this way so beset that I can barely do anything."[19] The affliction was diagnosed as "neurasthenia," a term now medically obsolete, and covered the symptoms of severe headaches, sensitivity to light and noise, and disturbances of digestion and circulation. It was said to be brought on by stress, fatigue, worry, and a sense of inadequacy, all describing Diesel's mental state for most of his life.

Diesel had reserved space at the 1889 Exposition Universelle (universal exposition) in Paris to demonstrate a working ammonia engine. He was excited that at last people were to see the results of his sacrificial efforts. But, unexpectedly, at the final moment he changed his mind and withdrew the engine from the exhibition. What caused this sudden reversal of plans is not positively known. It has been suggested[20] that Diesel recognized the significance of the latest advances in steam power. The new engines on exhibit utilized high degrees of superheat, which, in combination with triple-expansion cylinder designs, resulted in significantly better thermal efficiencies. Thus, Diesel may have decided the improvements he could offer were not enough to make this a judicious time for unveiling his first engine. His only recorded comment was: "To be delayed is not to give up. I go back to jump forward."[21]

Not long after the exposition, Diesel knew he must decide about remaining in Paris. He met with Linde in Munich during November 1889 where they discussed at length the deteriorating French business situation. Linde shared Diesel's concerns and offered him a job in Berlin to represent northern and eastern Germany for the Linde company. Linde asked what it would take for Diesel to come. In Diesel's words:

> I said 30,000 francs in guaranteed income—"Granted," [said Linde]—
> Now about the question of my invention! I said that I could not give
> up six years of work, the advantage of momentum and that this kind of
> work is my real aim in life.[22]

But Diesel did not get the answer he expected. Linde, with long experience as an inventor and businessman, explained that what Diesel expected to accomplish would not be easy. More significant, as well as a shock to Diesel,

Linde said the ammonia engine Diesel believed in so strongly was nothing out of the ordinary, and after years of struggle, Diesel had only a few ideas and studies to show for his efforts.

Diesel agonized over the decision. He realized what the Berlin job would make possible. It meant a profitable exit out of France, a chance to meet influential people in the German business and engineering world, and of course a return to what he called his homeland where he could rear and educate his children in a stable and comfortable atmosphere. Yet Diesel saw himself under the sword of Damocles—he would have to choose between a secure, middle-class life as a respected engineer or a creative career with all the associated risks and dangers.

Berlin won out. Diesel rationalized that it was not necessary to base one's entire future on a single decision, and he would see how forthcoming events might affect his cherished engine dreams. He would cross bridges as he came to them.

On February 21, 1890, Rudolf Diesel and his family arrived in Berlin. It was the first time he had ever been to the capital of Germany, a city so different from the one where he had spent over two-thirds of his life. Kaiser Wilhelm II (1859–1941), one year younger than Rudolf, had been German emperor and King of Prussia for three years. While a military influence had always been strong in Berlin, the new Kaiser, with his love of pomp and ceremony, added to an increasingly visible presence of the army and navy he loved. The stiff, often rude character of the average Berliner was in great contrast to the Parisians as Diesel remembered them.

Nevertheless, the initial months in Berlin were not unhappy ones for Rudolf and especially Martha. They moved into a house at the outer end of Kurfürstendamm Strasse, a location denoting one of the best addresses in Berlin. For the first time since their marriage, Martha felt herself accepted as part of the local society. Her outgoing personality soon helped her make many friends. Unfortunately, the high costs associated with the chosen address and Martha's involvement in Berlin life precluded saving for the future.

A change of cities had little effect on Diesel's habits. He was soon working at two jobs again—one for Linde and one involving his engine. His restless mind often worked through half the night trying to solve the riddles he created for himself about the engine that would change the world. These stressful times, in turn, brought on the dreaded headaches. He avoided almost all socializing, leaving that to Martha; every spare moment he devoted to The Engine. While at times believing that even his job was an intrusion, he did not knowingly slight what had been promised for Linde. His notebooks are filled with pages of comments and details involving refrigeration equipment for special installations in his sales territory.

It was not too long before a move to more modest quarters became a necessity as Diesel's extracurricular engine activities, along with family expenses, required additional financing. The importance of the engine in his life is evident from what he wrote during this period: "It cost what it wants, it goes as it wants."[23] In March 1892, the Diesels resettled in a *Gartenhaus* at 15 Brückenallee. Also of significance, the Berlin office for Linde was moved there, an indication that the ice business was not as prosperous as anticipated when Diesel came to Berlin two years earlier.

A further worry, the phenomenal success Nicolaus Otto (1832–1891) enjoyed with his four-stroke cycle engines, also forced Diesel to rethink the path he was taking. These relatively quiet and dependable engines built and licensed by Gasmotoren-Fabrik-Deutz of Cologne were sold throughout Europe and in the United States. Furthermore, the "Silent Otto" was applied in the same areas where Diesel hoped to install ammonia engines. Even before leaving Paris, he had seen an internal combustion engine, burning a liquid fuel, propel an automobile.

Diesel realized that his ammonia engine was only a variation of a steam engine; it still required an external heat source to produce superheated gaseous ammonia in a boiler. All such external combustion engines had inherent deficiencies and limitations that Diesel knew he could not overcome.

His reasoning went through a transformation in which he at last threw away the long-held belief that ammonia was the key to the most efficient engine. He was coming closer to what he so passionately hungered to invent.

Diesel recorded this evolution of thought in a sixty-four-page manuscript completed early in 1892:

> Both theory and practice had already led me to consider superheating of vapours, and my experience with the small engine I had built gave surprising proof of the advantage gained by superheating. I then developed a complete theory of a steam engine employing highly superheated ammonia vapor, and, by computation, found that it would afford a substantial advantage over existing steam engines. Moreover, the engine thus conceived was distinguished by extraordinary compactness in comparison with existing ones due to the fact that advantageous utilization of the process not only required the employment of high temperature, but also of very high pressure. In order not to go astray, I also made the calculations for engines which would operate on highly superheated *water vapor* and here again the necessity for high pressure became apparent because only a great pressure difference will permit the utilization of a steep temperature gradient [i.e., a large temperature difference] during expansion. It became apparent that all of the scientific material we have available on the behavior of vapor was inadequate for further treatment of the problem. I extended the range of Regnault's steam tables up to very high temperatures and found that under the circumstances the critical point was exceeded with the result that the liquid phase and the gas phase were no longer distinguishable. This gave me the idea to consider the vapor as a gas for

the sole purpose of obtaining a better theoretical approach. In doing so I discovered that there is practically no difference between vapor and gas, in other words, that I might also use gas or air while retaining the high pressure and temperature used in the previous investigation. With high temperature, however, advantageous utilization of a normal combustion process was not feasible. This suggested combustion in the highly compressed air itself. The pursuance of this idea led . . . to the concept of the engine.[24]

Diesel had built his bridge from superheated ammonia vapors to water and arrived at highly compressed air.

His mother received a letter dated November 15, 1891, in which he shared his almost jubilant state of mind:

I feel myself ready for the next step.

—I intend soon to come forward with the results of my twelve years of work and hope (hope!) to gain success, done in fact not with protection and support but through the worth of the cause alone.

—To err is human, I can deceive myself, but I have confidence in the cause . . . Twelve years have I with self-sacrifice tended a flower; now I will pick it and enjoy its fragrance.[25]

Notes

1. Rudolf Diesel, *Die Entstehung des Dieselmotors* (Berlin: Springer Verlag, 1913), 1.
2. Eugen Diesel, *Diesel: der Mensch, das Werk, das Schicksal* (Hamburg: Hanseatische Verlagsanstalt, 1937), 124 et seq.
3. Ibid., 143.
4. Ibid., 133.
5. Ibid., 135–36.
6. Ibid., 139.
7. Ibid., 152.
8. Ibid., 153.
9. Kurt Schnauffer, "Die Erfindung des Dieselmotors, 1890–1893," Part 1, in MAN Archives, trans. Henry I. Willeke (Augsburg, Germany: 1954), 2. Prof. Schnauffer made a great contribution toward the understanding of Diesel's work through his years of research and writing on the history of the engine. He deserves more recognition for these efforts.
10. Alexander C. Kirk, "Compressed-Air and Other Refrigerating Machinery," in *Heat in Its Mechanical Applications* (London: The Institution of Civil Engineers, 1885), 1886–88.
11. Aubrey F. Burstall, *A History of Mechanical Engineering* (Cambridge: MIT Press, 1965), 352.
12. T. Waln-Morgan Draper, "Anhydrous Ammonia Gas as a Motive Power," in *Cassier's Magazine*, vol. 4, no. 22 (1893), 308–10.
13. D. G. Tucker, "Emile Lamm's Self-Propelled Tramcars 1870–72 and the Evolution of the Fireless Locomotive," in *History of Technology 1980*, vol. 5 (London: Mansell, 1980), 115–16.
14. E. Diesel, *Diesel*, 133.
15. Joseph Harriss, *The Tallest Tower: Eiffel and the Belle Epoque* (Boston: Houghton Mifflin, 1975), 4–5.
16. E. Diesel, *Diesel*, 167.
17. Gordon Wright, *France in Modern Times* (New York: Norton, 1981), 251–55.
18. Erich Eyck, *Bismarck and the German Empire* (New York: Norton, 1964), 282–83. Germany

(Bismarck) had stirred up things in the Reichstag election of 1887 by saying France (Boulanger) wanted a revenge war against Germany. This put the French on guard and scared them. The whole Bismarck scare had to do with internal politics to change the government before Friedrich III came to power (p. 289).

19. E. Diesel, *Diesel*, 163, 170.
20. K. Schnauffer, "Die Erfindung," 2.
21. E. Diesel, *Diesel*, 170.
22. Ibid., 171.
23. Ibid., 187.
24. Kurt Schnauffer, *The Invention of the Diesel Engine: The Triumph of a Theory* (Augsburg, Germany: MAN Pub. No. SA364396E, 1958), 6. An address given in Augsburg, March 17, 1958, honoring the 100th birthday of Rudolf Diesel. Original in Schnauffer, "Die Erfindung," 5. where he quotes the concluding pages of Diesel's 1892 manuscript: *Theorie und Construktion eines rationellen Wärmemotors.*
25. E. Diesel, *Diesel*, 187.

CHAPTER 3

The Paper Engine—
Carnot's Legacy

"The genesis of the idea is the joyful time of creative work of the mind, as everything seems possible because it still has nothing to do with reality."
—Rudolf Diesel[1]

Rudolf Diesel created an engine on paper that proclaimed his invention of a highly efficient prime mover. It was the first heat engine, either of external or internal combustion, that from its conception had been based on the science of thermodynamics sired by Sadi Carnot.[2] His maxims became Diesel's inspiration, guide, and obsession. So radical were the ideas incorporated in the theory and design of Diesel's "Carnot Process" engine that only by acceptance of more realistic objectives could his proposal of 1892 be considered buildable.

Combustion State of the Art

Complementing Diesel's personal experience with steam and ammonia was his opportunity to glean from a growing storehouse of information about new gas- and liquid-fueled engine technology. In addition to more types and numbers of engines for examination, there were by now textbooks and technical papers on the subject. Dugald Clerk (1854–1932) in England, Aimé Witz (1846–1926) in France, and Otto Köhler (b. 1852) in Germany, to name but a few, were recognized experts who had written definitive works on internal combustion engines prior to 1890.

The predictions of these pioneers ran contrary to Diesel's thinking. Dugald Clerk, perhaps best known for his two-stroke cycle engines, in 1886 suggested:

By undue increase of compression, the negative work of the engine would be much increased, and the strains would become so great that heavier and more bulky engines would be required for any given power. Friction, due to this, increases more rapidly than efficiency; consequently, the gain in indicated efficiency [i.e., before deducting mechanical losses] would be more than compensated by loss of effective power. Improvement must be sought elsewhere.[3]

Clerk's analysis is not in error, but his chosen limits of then prevailing compression ratios of 3:1 to 4:1 would be proven far too low.

In a later edition, extensively updated in the mid-1890s to include liquid-fueled engines, he wrote:

The author accordingly considers a compression of 200 lbs. per sq. in. as considerably above the limit likely to be useful in a simple gas engine; to render such compressions possible he considers that compound engines will require to be designed. The gas engine, in the author's opinion, is now nearing the limit of advantageous increased compression [4:1], so that no further economy is to be expected there.[4]

Clerk's comments on oil-fueled engines, written at the time of Diesel's test program, reflect a consensus of then held beliefs:

The reader will have observed . . . that the oil engine is not so economical from a heat engine point of view as a gas engine; that is, the oil engine so far . . . does not convert so large a proportion of heat units into indicated work as a gas engine.

However, he adds a final note, leaving open the possibilities for future advances:

It is to be remembered, of course, that as yet engineers have had little experience in oil engines as compared with gas engines, and that probably with further development of detail the heat efficiency of the oil engine may yet be considerably increased.[5]

Another early authority spoke to the direction Diesel was heading. Otto Köhler offered his pertinent comments about the possibility of a Carnot process engine in an 1887 paper *The Theory of the Gas Engine*. Whether Diesel knew of the work and discarded its warnings or was unaware of it is not known. Köhler averred:

. . . that the complete process, disregarding the difficulty to actually execute it, is not suitable for practical application. Certainly, there is a possibility for high efficiency, but then the [cylinder] pressure increases so greatly while the mean effective pressure remains too small and the physical dimensions become so huge that the [efficiency] gain is absorbed in friction losses.[6]

Nothing in Köhler's paper suggested utilizing the high compression temperature as a source of ignition. To the contrary, he saw the elevated temperature causing a problem known to all engine designers. Köhler was correct in his comment about mean effective pressure (the average pressure exerted on a piston over the entire power stroke).

Those familiar with thermodynamics well knew that a higher compression (i.e., expansion) ratio increased thermal efficiency. However, there was no way to take advantage of this knowledge because the fuel charge preignited if compression ratios exceeded more than 3.5:1. Except for engines burning low heating value fuels from either blast furnace gases or inefficient coal-gas generators (these tolerated ratios as high as 6:1) the compression ratio (cr) barrier in commercial engines had been reached. Gasoline- and kerosene-fueled engines were limited to a maximum of 3.5:1 cr, a ceiling raised only modestly until tetraethyl lead was introduced in the 1920s.

Two significant developments of the early 1890s succeeded in extending compression ratio limits. However, these were directed toward preventing preignition of a premixed air/fuel charge that had entered the cylinder during a normal intake cycle.

Hugo Junkers tested a prototype of his later famous two-stroke cycle engines while working for Wilhelm von Oechelhäuser at Dessau in 1892 (chapter 14). This 100-hp, inwardly opposed-piston engine, burning low Btu gas (an important consideration), ran with a compression pressure of 19 atmospheres (19 kg/cm^2 = 270 psi or an 8.6:1 cr).[7] Test data revealed that he had reduced by 60 percent the gas consumption of good commercial four-stroke cycle engines using the same fuel, a remarkable achievement. Professor Junkers, who was teamed with a builder of state-of-the-art engines, incorporated the latest scientific discoveries in heat transfer and gas dynamics into his totally new engine configuration.[8]

A second interesting development came from Professor Donát Bánki (1859–1922) of the Budapest Technical University. Test data from an experimental gas engine built to his design in 1894 by Ganz & Co., also of Budapest, show a maximum compression pressure of 13 atm (185 psi—6.7:1 cr) and a brake thermal efficiency of 30 percent.[9] What made such a high pressure possible in an otherwise conventional engine was water injection into the incoming air/fuel charge to keep charge temperatures below the preignition point. First with gas and then gasoline ("benzine") fueled versions, water was sprayed into the air/fuel charge before it entered the cylinder. For the liquid-fueled engine, both water and fuel were sprayed into a vaporizing chamber adjacent to the intake valve via separate conduits. One- and two-cylinder, four-stroke cycle Bánki engines were sold throughout Austria and Hungary for several years in power ranges of up to 50 hp.

While engineers struggled to extract more power from compression-limited engines, Rudolf Diesel introduced a new dimension of thought light-years beyond the dreams of his peers. He proposed to build an engine with a compression pressure of 250 atm or over 3,550 psi—almost 100 times that of the early Otto engines!

The Carnot Process Engine

Diesel concluded that the only way a heat engine could extract the most work from the stored energy in a fuel was to closely approximate a theoretical operating cycle Carnot had proposed in 1824. The objective of Carnot's totally new concept was to explain the fundamental thermal processes at work in steam engines. The genius of his analysis was that it provided a standard whereby all real heat engine and refrigeration cycles may be compared with the theoretically perfect and with one another. It is necessary, therefore, to look briefly at what Carnot taught and how Diesel adapted those lessons to construct his own theories.

The "Carnot Cycle" is sometimes referred to as the "perfect thermodynamic cycle." Its four cyclic processes are "ideal" or reversible, a condition not attainable in nature. The derived property helping to demonstrate this impossibility is called entropy and forms the basis of the second law of thermodynamics. (In a "real" heat cycle there must always be a net increase of entropy.) The "second law" was needed to close a loophole in the "first law," which stated that energy can neither be created nor destroyed. Professor H.C. Van Ness clarifies this in simple terms:

> The First Law says that energy is conserved. That's all; you don't get something for nothing. The Second Law says that even within the framework of conservation, you can't have it just *any* way you might like it. If you think things are going to be perfect, forget it.[10]

(A more irreverent recitation of the two sacred statutes is: there is no such thing as a free lunch, and even if there were, you couldn't eat it all anyway!)

Carnot theorized that to attain the maximum thermal efficiency from a heat engine the following conditions must be met:

1. The temperature of the fluid [air] should be as high as possible, in order to obtain a great fall of caloric [heat], and consequently a large production of motive power.
2. For the same reason the cooling should be carried as far as possible.
3. It should be so arranged that the passage of the elastic fluid [compressed air] from the highest to the lowest temperature should be due to the increase in volume; that is, it should be so arranged

that the cooling of the gas should occur spontaneously as the effect of rarefaction [expansion].[11]

The single process just outlined corresponds to a portion of the power stroke of an engine having a potential ideal efficiency approaching, but never reaching, 100 percent. The other processes of Carnot's cycle as they relate to his theoretical engine follow. Also to be noted, Carnot retains the same gas in his cylinder, compressing and expanding it over and over; there is no intake of fresh air nor exhaust of "used" gas.

If the compression and expansion strokes of Carnot's cycle and Diesel's engine are each considered to be made up of four segments, then these segments can be represented on special graphs as a closed loop.

A circuit around the loop *a, b, c, d* of the two plots explains what occurs during each of the four segments (fig. 3.1).

a–b: Isothermal compression. Carnot required that the heat generated by compressing the enclosed gas during this part of the cycle must be rejected from the cylinder in order to maintain the gas at a constant, lowest-point temperature. Diesel proposed that during the first part of the compression stroke the air temperature could be kept from rising by injecting water directly into the cylinder.

b–c: Adiabatic compression. Carnot assumed a perfectly insulated cylinder, whereby the heat created due to further compression of the air, would be retained and not lost to the surroundings (i.e., adiabatic). A second assumption was that he could reverse this process and reconvert all the heat from the compression process back into useful work. (This meant no increase of entropy, and therefore, an isentropic process—a violation of the second law.) Diesel also desired adiabatic compression to complete his compression stroke, a reasonable assumption because in a real engine, compression occurs so quickly that there is little time for heat transfer during the compression process. Diesel did not consider Carnot's second assumption as he knew the restrictions imposed by the second law.

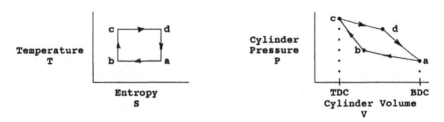

Fig. 3.1. Carnot's "perfect" thermodynamic cycle on Temperature-Entropy (T-S) and Pressure-Volume (P-V) diagrams.

c–d: Isothermal expansion. As expansion began, Carnot carefully added heat at a rate equaling the cooling rate due to the piston moving outward. The temperature of the cylinder gas thus remained constant and at the value of point *c*. Diesel similarly waited until the piston reached top dead center and then injected fuel to be burned at a rate and volume such that the temperature in the cylinder remained constant. The heat added as a result of combustion was intended to be as equal as possible to the cooling that would occur when the piston started on its power (expansion) stroke. *The maximum cylinder temperature was attained by the compression process and not through the combustion of fuel.*

d–a: Adiabatic expansion. The final segment of Carnot's complete cycle took place in a perfectly insulated cylinder and was also a reversible process. The temperature of the gas at the end of this expansion process dropped back to what it was at the beginning of the cycle. Diesel likewise assumed such a process, but in his theoretical engine, the expansion was carried out by using a very long stroke to allow the gas to cool almost to the incoming air temperature. For this to happen, Diesel also needed to close his intake valve before the end of the intake stroke so less than a full charge of air entered. He was convinced external cylinder cooling could be eliminated and even envisioned insulating the cylinder walls.

Diesel expected to create his maximum air temperature only by compression, a process calling for an extremely high-end pressure. He originally planned, therefore, to compress the air to 250 atm (3,556 psi) and by water injection control peak temperature at approximately 800°C (fig. 3.2). This would give his engine a theoretical 73 percent thermal efficiency, before mechanical losses, without violating other aspects of the Carnot cycle.

For such an engine to operate and avoid preignition, Diesel reasoned that he must keep the fuel away from the cylinder air until the end of the compression stroke. He also realized that the elevated air temperature resulting from rapid and high compression was far more than needed to quickly ignite the fuel charge when injection began at top dead center, the earliest possible time for his thermal combustion. *Compression ignition*, an idea often considered Diesel's contribution, was merely a bonus byproduct of the true invention.

Diesel suggested a form of engine based on the Carnot process in a manuscript entitled "Theory and Design of an Economical Heat Engine" that he completed in January 1892:

> It is easy to imagine a small quantity of gaseous or liquid, or coal dust, gradually introduced into a volume of compressed and highly heated air, and burning by spontaneous or by separate ignition. The piston is forced out at the same time in such a way that no increase of temperature takes place, because the heat developed by each particle

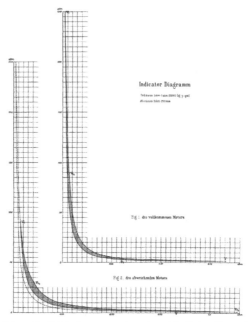

Fig. 3.2. Diesel's P-V diagram showing his "ideal" engine with a
compression pressure of 25 atm (3,556 psi) and a "compromise"
engine "deviating" to only 90 atm. Diesel, *Theorie, 1893*

of combustion is instantly absorbed by the cooling due to expansion.
Therefore, the whole of the heat developed will be transformed into
work.[12]

He believed there would be no need to deviate from his chosen process in a
real engine except for one possible modification: eliminate water injection
by reducing the peak compression pressure from 250 atm to 90 atm (a
reduction in compression ratio from approximately 52:1 to 25:1). At 250
atm and no water injection the temperature rises to almost 1,200°C; at 90
atm without water the rise is only to the desired 800°C. Diesel's calculations show that at 800°C the cylinder gas temperature drops to 120°C
at the end of the expansion (power) stroke, low enough to avoid external
cylinder cooling (fig. 3.3).

An important criterion for proper engine operation concerned the air-
to-fuel ratio. Diesel claimed that the heat units contained in coal dust were
far too great to be burned in his cylinder unless there was "an excess of air
for combustion . . ."[13] He accordingly based all calculations on an air-to-fuel
ratio of almost 100:1 ("99.324 kilos, air to 1 kg of coal"[14]). This extreme lean-
ness of the fuel charge became a Gordian knot for Diesel to publicly untie.

On February 11, 1892, Diesel sent a manuscript copy to his employer
Linde, as was customary. With great pride, he wanted to show his often-

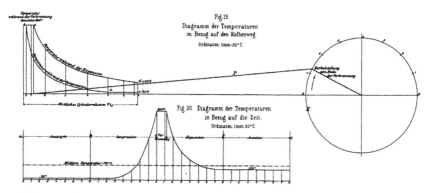

Fig. 3.3. Diesel's "compromise": 90 atm engine with a calculated peak cylinder temperature of 800°C using no cooling water. End temperature only 120°C at BDC of power stroke. Diesel, *Theorie, 1893*

skeptical mentor what had been accomplished since the ammonia engine, and even more important, to interest Linde in building a test engine:

> . . . and I have the pleasure to inform you that I have found an engine which theoretically uses only 10 percent of the coal which today's best steam engines would use. (I assume on the average 1 kg coal per hp-hr.) This result is not an assumption or hope. It can be proven mathematically exact in such a way that there is no doubt it can be achieved. . . . [15]

Linde's reply on March 20 was less than Diesel wanted to hear. Praise for the originality and thoroughness of the analysis was tempered with critical remarks about the practicality of the engine concept:

> . . . I want to confirm my verbal judgment that your direction is sharp and correct in your goal to use fuel for mechanical work in a manner which must be considered the best, taking into account the present physical knowledge and the state of machine construction.

> . . . I may not omit to add that in my opinion only in the best case perhaps one third of your theoretically calculated efficiency can actually be expected. . . . Nevertheless, the possibility exists to convert about 25 percent of the heat value of nearly all fuels into mechanical energy. That is somewhat more than is achieved today with special and relatively expensive fuels [in gas engines].

The thermal efficiency later attained on Diesel's first test engine was 25 percent! Linde also could see the value of attempting to burn a cheap fuel, such as powdered coal.

He continued with his praise and criticism:

> ... I recognize your course as the only correct one and that you found the relationships in the described processes. I also acknowledge that the design part of your work seems to be generally correct. However, I am far from acknowledging that the goal you aspire to is attainable.

The letter's conclusion also provided information about Diesel's future with the Linde company:

> That you are very well prepared for this work and that the goal is worth every effort I do not doubt, so I cannot dissuade you from devoting yourself to this cause.... When you decide to follow up your idea I see no other possibility than that you leave the company.[16]

Linde offered his personal help to Diesel, but he made it clear that Diesel could no longer serve two masters. A further disappointment was Linde's frank statement that he had no plans in his company for the development of the engine.

A second early appraisal came from another of Diesel's old professors at the Munich Technical University. Linde had passed on his copy of the manuscript to Professor Moritz Schröter (1851–1925), who sent a preliminary evaluation to Diesel dated March 29, 1892:

> I read your paper with much interest. I could expect from you that it would be on a sound theoretical basis. However, I must admit that I cannot share your sanguine expectations about how to overcome the practical difficulties. A functioning ignition at the high stresses and high temperatures (to mention only some) are difficult practical problems. They can only be solved by laborious and expensive experiments. At the moment I am so busy that I cannot study details. This would require a thorough study of your paper which is now impossible for me.[17]

The replies from Linde and Schröter were not positive enough for Diesel to cite them as solid endorsements when he approached potential manufacturers. However, Schröter became greatly interested in Diesel's project and later gave him valuable advice and support for both his work and in the promotion of the engine among technical people.

A letter from Schröter on May 2, 1892, gave the first hint to Diesel that his theory needed revising before transforming it into an operable engine. Based only on his visual study of the theoretical indicator diagram in Diesel's paper, Schröter questioned whether the engine could produce much power. He noted that the calculated compression and expansion segments in the pressure-volume diagram were already very close together, and if there were even minor adverse deviations, the area formed between the two curve segments could be even smaller (fig. 3.2). Schröter did not relate this small diagram area to the very lean mixture Diesel proposed. Neither did Linde specifically mention the overly lean ratio.

The pressure-volume, or P-V diagram, showing the relationship of cylinder pressure versus piston position has long been a useful tool to calculate the gross power released in an engine cylinder. John Southern, James Watt's assistant, invented a form of "indicator" mechanism in 1796 to trace on paper a permanent record of the events occurring in the cylinder of a steam engine. It soon became a universal way to report the power from a steam cylinder before deducting mechanical losses—hence the origin of the still-used term "indicated horsepower" (fig. 3.4).

The work produced during a cycle of an internal combustion engine is similarly represented on a P-V diagram. It is the area enclosed inside the lines denoting the compression and expansion (power) strokes, in other words, what remains after subtracting the work required to compress the air from the work produced during expansion. At an idle, the bounded area representing indicated horsepower is small, just equaling mechanical losses, so that the net output is zero. If this area becomes too small, an engine cannot even idle on its own, although the P-V diagram still shows a positive value for the indicated horsepower. The larger the area, the greater is the usable or "brake" horsepower. (Here is another term coming from the steam era when output

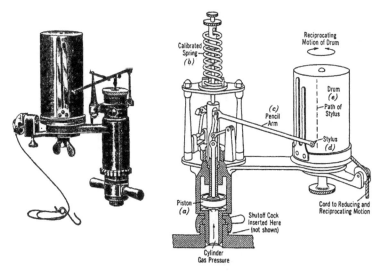

Fig. 3.4. Indicator for making a P-V diagram (indicator card). A pressure trace is made on paper wound on the drum as piston movement turns the drum. This provided a record of the cylinder pressure at each point in the piston stroke. *Hawkins' Indicator Catechism, 1903*

was measured by forcing a scale-weighted brake pad against the rim of an exposed engine flywheel.)

Schröter's mention of the overly thin curve area was the point when Diesel began to worry about how to "fatten" his diagram. But how to do it without sacrificing isothermal combustion, the cornerstone of his theory?

Meanwhile, in a reply to the professor's March 29 letter, Diesel informed Schröter of his decision to further reduce the maximum compression pressure to 44 atm (approximately 15:1 cr) from the original 250 atm, a move Schröter heartily concurred with, and which brought Diesel closer to the realm of the possible.

German Patent 67,207

Diesel immediately prepared a patent application after completing the manuscript. With his invention registered at the German patent office, he was free to safely publicize his new ideas. The heart of the application filed on February 27, 1892,[18] was taken from his manuscript. A granted patent,

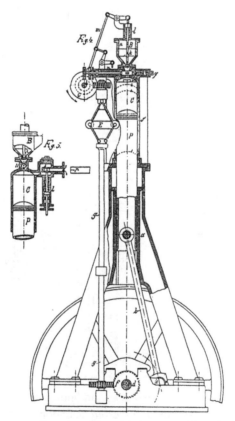

Fig. 3.5. Drawing of coal-burning engine design as shown in Diesel's 1892 German patent 67,207.

Process for Producing Motive Power from the Combustion of Fuels, was issued on December 23 of that same year and became famous. The claims in it proved to be a millstone around Diesel's neck.

The disclosure of the invention and a discussion of the prior art in the patent read much like a technical paper. He lectured on the theory of open- and closed-cycle hot-air engines and gas- and oil-fueled internal combustion engines, listing their potential efficiencies and deficiencies. Diesel stated his thesis on a Carnot process engine in which air is compressed to 250 atm and 800°C maximum temperature. Only once was the possibility of a lower compression pressure mentioned:

> If it be desired, for instance, that the later combustion shall take place at a temperature of 700°C, the pressure will be of 64 atmospheres; for 800°C, the pressure will be of 90 atmospheres and so on.[19]

He described a system for water injection but added that his "process" may be performed with or without it.

Although Diesel considered "liquid" and gaseous fuels, at this time he stressed finely powdered coal dust as his preferred fuel. It was introduced into the cylinder under the force of high-pressure air (fig. 3.5). For liquid fuel a "spray nozzle" and a "small pump" were used. Not mentioned in conjunction with liquid fuels was the use of air under high pressure to act as the injection force.

The design and operation of a compound engine was emphasized as Diesel expected this would be the ultimate form of his engine (fig. 3.6). It had three cylinders: two, high-pressure cylinders on each end where combustion occured and a center, low-pressure cylinder with its chamber above the piston

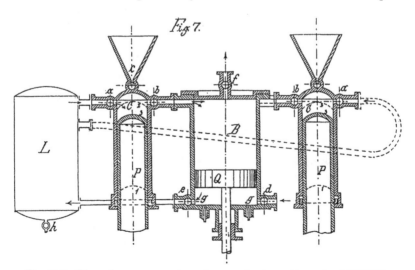

Fig. 3.6. Diesel's compound engine as disclosed in patent 67,207.

receiving the exhaust alternately from the high-pressure cylinders. The bottom side of the piston acted as a compressor to pump high pressure air into a reservoir that was the energy source for injecting coal dust into the first stage power cylinders. The low-pressure cylinder allowed for a maximum expansion of burned gases before they were exhausted into the atmosphere. The end cylinders operated as a conventional four-stroke cycle, and the center one delivered a power impulse with every revolution.

The Patent Claims

While an accurate description of an invention's structure and operation is necessary and important in a patent, it is the "claims" that define the protective boundaries within which an inventor is free to exploit his time-limited monopoly. Rudolf Diesel was entitled to build a strong wall of claims around his ideas, but he was blinded by a firmly held belief in the correctness of his theory. As a result, he failed to communicate to his patent attorneys the broad scope of his invention. Even the German Patent Office inadvertently limited Diesel to narrower protection than necessary. The claims in the initially submitted application,[20] and later in the issued patent, covered only isothermal combustion as applied to an engine operating on "almost exactly Carnot's process and this between high difference of temperatures."[21]

Diesel's application was prepared by well-known Berlin patent attorneys, F. C. and E. Glaser. As the contents were highly scientific, the attorneys must have relied on Diesel's expertise to explain the invention to them, and at this time he saw only the isothermal combustion. In his haste to file for patent protection, Diesel lost the advantage of learning from helpful friends weak areas in his analysis and incorporating this knowledge into the patent.

The evolution from the draft claims Diesel gave his attorneys to those ultimately granted was significant and can be traced. The first two of the original twelve draft claims are:

1. Avoiding the mixing of air and fuel—as this occurs not only in open and closed furnaces [steam and hot air] but also in explosion motors [Diesel's term for existing gas and oil engines].
2. Compression of the combustion air or the steam and gas mixture far above that used heretofore, so that the compression temperature rises far higher than the ignition temperature of the fuel. This compression can be purely adiabatic or first adiabatic or what is especially important, first isothermal and then adiabatic. The last process is more perfect.[22]

The draft claim 1 refers to drawing in and compressing only air, but Diesel failed to state it, and in claim 2 he should have inserted the word "inert" before "gas mixture" for greater clarity. Nevertheless, claim 2 defines the Diesel engine by its coverage of compression ignition in stating that the air temperature is *far higher* than the fuel's ignition point. He believed that such an elevated cylinder air temperature was new to internal combustion engines.

Draft claims 3, 4, and 6 are also of interest:

3. Selection of a mathematical relationship between the air volume and the heating value of the fuel so that the exactly predetermined maximal temperature was not exceeded.

4. Introduction of an exactly calculated small quantity of fuel per stroke into the compressed air volume. This shall be done in a mathematically determined way, so that the combustion process is as isothermic as possible. Without giving up this basic idea, it can also happen that when deviating from the strongly isothermal, the now warmer exhaust gases are used for heating or other purposes.

6. Completely avoid cooling the cylinder walls during both combustion and expansion—on the contrary, protect them against radiation.

The Patent Office returned the application on March 15, 1892, for rewriting because in the form submitted seventeen days earlier it outlined, according to the office, merely problems rather than offering specific solutions.[23] A revised application was sent on April 6 with a new title: *Process for Producing Motive Power from the Combustion of Fuels*. Although this new title was accepted, parts of the application were objected to several more times (a common occurrence). An interview at the Patent Office with the examiners on June 16 was followed by their sending claims acceptable to the patent office to Diesel on July 7. These were agreed to by him and incorporated word for word in his final resubmission of July 23. The patent was published September 3 for the public's right of challenge under German patent law. With no objections forthcoming, patent No. 67,207 was granted December 23, 1892 (fig. 3.7).

The problem encountered with the patent examiners was the same as had befallen his own attorneys. They did not understand what Diesel had invented. The Patent Office replied after the April 6 reworking that he ". . . is to take into consideration that the injection of fuel in compressed air is already known."[24]

Diesel could have no quarrel with this statement; he must have known of developments by others such as the work of Herbert Akroyd Stuart (1864–1927) in England.[25] Stuart's British patent No. 7,146 of 1890 included the following claim for his oil engine:

2. In an engine operated by the explosion of a mixture of combustible gas or vapour and air, forming said mixture by introducing the combustible gas or vapour into a charge of air under compression, substantially as described.

Fig. 3.7. Cover page of Diesel's German patent 67,207 of 1892.

This substantiates the statement of the patent examiner. However, Stuart's first claim describes a vital difference between what he did and what Diesel proposed:

1. In an engine operated by the explosion of hydrocarbon vapour and air, the employment of a vapourizer in direct communication with the working cylinder, which vapourizer is *maintained at the requisite degree of heat by the combustion of the combustible mixture therein, the said vapourizer also serving to ignite the combustible charge*, substantially as described. [Italics added.]

Neither in this or any patent of Stuart, nor any other patents covering engines where fuel was introduced into the combustion chamber, or an antechamber connected to it, is there reference to the heat from compression alone igniting the combustible charge.

It was Diesel's tragic mistake that he failed to object when the German Patent Office ruled that others anticipated his ideas. His invention was based solely on heating the air *far above* the ignition point of the fuel with the result that the engine did not depend on auxiliary heat for starting or running to initiate or sustain combustion.

His draft claims 1 and 2 do not mention isothermal burning, but this could be because he was trying to logically explain the cycle, beginning first with the compression stroke. Claims 3, 4, and 6, covering aspects of isothermal combustion, are in actuality dependent on fulfilling the objectives of his first two claims. The ideas set forth in his first claims were new and were entitled to acceptance on their own merit.

As a result of his own blindness and the apparent lack of comprehension of the total invention by his attorneys, Diesel acquiesced to the helpful suggestions of the patent examiners and accepted their claims wording. They in effect combined and reworked down to two the originally submitted claims so that Diesel's invention of highly heating the air to ignite the charge was forever linked with isothermal combustion:

1. The method of working combustion motors consisting in compressing in a cylinder by a working piston, pure air, or other neutral gas or vapour together with pure air, to such an extent, that the temperature hereby produced is far higher than the burning or igniting point of the fuel to be employed (Curve 1-2) [fig. 3.8]; whereupon fuel is supplied at the dead centre gradually, that on account of the outward motion of the piston and the consequent expansion of the compressed air or gas the combustion takes place without essential increase of temperature and pressure (Curve 2-3); whereupon, after the admission of fuel has been cut off, the further expansion of the body of gas contained in the working cylinder takes place (Curve 3-4) substantially as described.[26]

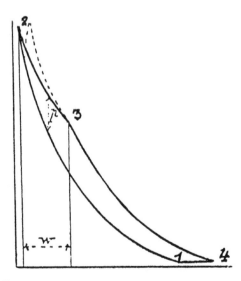

Fig. 3.8. Pressure-volume diagram of an isothermal combustion cycle as shown in Diesel's German patent 67,207 of 1892.

The granted claim 2 dealt only with the compound engine Diesel described in his patent.

Foreign patents based on this first German patent 67,207 were filed and later issued in Great Britain, the United States, Switzerland, and many other countries.[27]

Although Rudolf Diesel had a right to patent the compression ignition process, he held other views about its importance. Just before his death in 1913 he wrote on this subject:

> Frequently by laymen and also in scientific circles it will be mentioned that the chief characteristics of the Diesel process is the self-ignition of the fuel, the purpose of the high compression being that at top dead center the injected fuel ignites itself, and that the degree of compression required is for reliable self-ignition. Nothing is more incorrect than this superficial view that is directly contrary to the facts and especially the historic development.
>
> Motors with self-ignition of the fuel already existed. In my patents I have never denied the self-ignition nor in my writings mentioned it as a desirable goal. I sought a process with *the maximum heat usage,* and it happened by itself to be so that self-ignition was included. If by compression the air is heated beyond the ignition temperature of the fuel, then the ignition of the fuel in the air results automatically, but it is not the reason for this high compression. The self-ignition of all liquid and gaseous fuels occurs in an operating engine at pressures as low as from 5–10 atm, to a highest of 15 atm. Thus, it would be much simpler to build lighter and less expensive machines for these compression ratios, and the difficulty of the first ignition of a still cold engine could be overcome by a temporarily used artificial ignition. It was absurd merely for the ignition of a cold engine to build so heavy and complicated machines for 30–40 atm, as the motors, once running hot, would continue running as well with lower compression.
>
> The *purpose* of the process, that was sought for so many years and was realized with such difficulty, however, is quite different: the attainment of the highest possible fuel utilization; this purpose requires the highly compressed air. But as the latter ignites the *added* fuel too early, the self-ignition by compression, as was known for motors of that period, was an *obstacle* to the performance of the process. In this form it had to be avoided. *Air alone* by mechanical compression had to be so highly compressed that the desired heat utilization occurred.
>
> The degree of this compression is not only required by the ignition of the fuel, but also for obtaining the maximum economic fuel usage.[28]

These words of Diesel never registered in the minds of future generations. The terminology *diesel* and *compression ignition* became forever synonymous.

Patent No. 67,207, and its foreign cousins, were the first links in a lengthening chain of often unintended distortions of what Diesel believed and how his engines operated.

The Economical Heat Engine

Diesel embarked on two future tasks in 1892 after filing his patent application. The first was the search for a builder of the engine (chapter 4); the second involved editing his manuscript for publication. He submitted it to Springer Verlag in Berlin on October 2, 1892. They returned a contract on the eleventh in which they agreed to print one thousand copies.[29] By December 6, after many letters back and forth, the book was ready for the press. On the twenty-first, he received the first three copies, and then a week later Springer delivered to him eighty bound and twenty unbound ones for his own use (fig. 3.9). Bookstore sales began January 10. The meticulous drawings and detailed analyses could even suggest that Diesel's theories were indeed an accomplished fact (fig. 3.10). He also featured a design for liquid fuel injection (fig. 4.4).

The book analyzes and gives the advantages and disadvantages of known internal and external combustion engines cycles, but Diesel leaves no room for doubt in the introductory remarks as to his belief in the superiority of his own proposal:

> In the following work a new theory of combustion is described, and the conditions are deduced from it which ought to govern the process

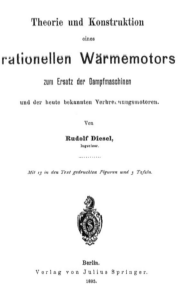

Theorie und Konstruktion

eines

rationellen Wärmemotors

zum Ersatz der Dampfmaschinen

und der heute bekannten Verbrennungsmotoren.

Von

Rudolf Diesel,
Ingenieur.

Mit 13 in den Text gedruckten Figuren und 3 Tafeln.

Berlin.
Verlag von Julius Springer.
1893.

Fig. 3.9. Title page of Diesel's 1893 book disclosing his isothermal combustion engine ideas. *His personal copy in MAN Archives*

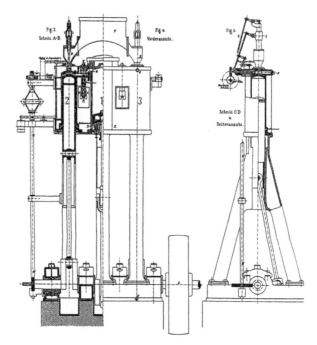

Fig. 3.10. Diesel compound engine in his 1893 book
Theorie und Konstruktion eines rationellen Wärmemotors.

in a motor cylinder, in order to obtain a maximum of work from the
total heat of combustion of the fuel. . . . This new motor bears a certain
resemblance to hot air or gas engines, because the process of combustion takes place in the cylinder. But the similarity is only apparent, and
the working principle, and especially the method of carrying out the
combustion, are entirely different. An examination of their theory will
show that gas and air engines are worked upon a defective principle,
and no improvement in them will produce better results, as long as this
principle is retained.[30]

In his rejection Diesel commented on contemporary engine products:

Motors in which air is compressed, and the combustible quickly
injected towards the end of compression, and simultaneously ignited
[Akroyd Stuart], show in the same way an increase of pressure, which
according to our theory, should be avoided. Here the maximum temperature is produced, not by compression but by combustion [constant
volume combustion].[31]

He next referred to George Brayton's early gas engine before Brayton
switched to liquid fuels:

There is another class of gas engines, in which the inflammable mixture is compressed into a separate receiver. From thence it is forced at constant pressure into the working cylinder, passing a flame on its way which ignites it, and the gases expand to atmospheric pressure. It seems as if the gradual combustion thus obtained presented some analogy to our second condition, [constant pressure combustion] but the similarity is only apparent.[32]

An important distinction is made here between Diesel's engine and the Brayton, which brought an already burning, pressurized charge into the working cylinder and closed the intake valve only after the piston was part way out on its power stroke.

Diesel demonstrated how his engine could also operate on constant volume and constant pressure combustion (the "second condition" referred to above), but his firm choice remained the isothermal combustion or Carnot's combustion at constant temperature.

As a result of Professor Schröter's concern about the unrealistically high compression pressure Diesel deemed necessary in the manuscript, the minimum permissible number was further reduced to 44 atm for the book, albeit as a passing comment:

It is interesting to note that the maximum pressures of this [constant temperature] cycle can be very much reduced if the highest temperature is diminished, as for example:

For max. temps. of	600°	700°	800°
… pressures about	44	64	90 atm
… thermal efficiency	60	64	68.4%

To diminish the pressures thus causes an additional sacrifice of thermal efficiency, but even if they be reduced to 44 atmospheres, the utilization of heat in the new type of motor greatly exceeds that in existing heat engines. This diminution in pressure will be found useful for certain types, but our aim must always be *to raise the temperatures and pressures to the highest working limit*. It would be easy to carry out the cycle with a much higher maximum than 800°C temperature, and 90 atmospheres pressure.[33]

In Diesel's ideal cycle the expansion process was carried out until the cylinder pressure fell almost to atmospheric. Realizing that this was not feasible in an operating engine, he showed both in the manuscript and the book a way to shorten the power stroke and not carry the expansion process so far. This was done by opening the exhaust valve with a cylinder pressure of 1.62

atm and a temperature of 187°C rather than the 130°C in the ideal process. He calculated that even at the higher exhaust valve opening temperature he still did not need to cool the cylinder. Further, air could be drawn into the cylinder throughout the entire intake stroke rather than just the first three-fourths of the stroke as was necessary with the longer expansion stroke of the ideal process. Only in this way could the air temperature be held to the desired 800°C maximum.

The critical problem of preignition with such a high compression temperature was uppermost in Diesel's mind:

> We have already seen, that if the air and the fuel are previously mixed, it is impossible to obtain very high compression, as ignition intervenes, and prevents the process from being carried out in accordance with theory. Another result of mixing air and fuel, is that combustion takes place suddenly, and is left to itself. Thus . . . the great rise in temperature (of combustion) necessitates considerable cooling of the cylinder. To mix the gas and air has the further disadvantage, that it is impossible to use the quantities of air necessary to take up the heat of the combustible, because ignition cannot take place with very diluted mixtures.[34]

He reiterates the above in an italicized statement showing the importance of not adding fuel till after peak compression pressure and temperature are reached:

> *There is here, we consider, a fundamental error, which cannot be avoided without abandoning one of the principles of the gas motor, namely that the air and the combustible are mixed.*[35]

Diesel concludes his small book with little room to defend himself if his theories could not be confirmed:

> All these considerations prove that none of our existing heat motors realize, even approximately, the conditions of rational combustion in their cycle. We may therefore reasonably hope that the new motor will give the economy of combustible claimed for it, as it is based on the principles of the perfect Carnot cycle.[36]

> *"History reports what has occurred and must be satisfied above all that it describe what someone has done and not, by way of opinion, what someone wanted to do. He who chooses to write a history based only on patents is like the author of a war history who decides to discuss just the battle plans."*

> —Conrad Matschoss[37]

Notes

1. Rudolph Diesel, *Die Entstehung des Dieselmotors* (Berlin: Springer Verlag, 1913), 151.
2. Sadi Carnot, *Réflexions sur la Puissance Motrice du Feu et sur les Machines Propres à Développer cette Puissance* (Paris: Bachelier, 1824), 118.
3. Dugald Clerk, *The Gas Engine*, 5th ed. (London: Longmans, Green & Co., 1894), 260–61.
4. Dugald Clerk, *The Gas and Oil Engine*, 6th ed. (New York: Wiley, 1896), 385.
5. Ibid., 462.
6. Friedrich Sass, *Geschichte des deutschen Verbrennungsmotorenbaues von 1860 bis 1918* (Berlin: Springer Verlag, 1962), 399.
7. Hugo Junkers, *Investigations and Experimental Researches for the Construction of my Large-Oil-Engine* (Berlin: Verlag fur Fachliteratur, 1912), 2–9. Further details of this engine may be found in Lyle Cummins, *Internal Fire* (Austin, TX: Octane Press, 2021), 244–46.
8. Junkers's US patent No. 508,833 (with Wilhelm Oechelhäuser) gives evidence of his pioneering opposed-piston engine work. Claim 1 reads: "A gas engine comprising a working cylinder having peripheral inlet and exhaust ports near its opposite ends, respectively, and reciprocally movable pistons adapted to control said ports, for the purpose set forth."
9. Both Bryan Donkin, *Gas, Oil and Air Engines*, 3rd ed. (London: Griffin, 1911), 512–16; and Hugo Güldner, *Das Entwerfen und Berechnen der Verbrennungskraft-Maschinen und Kraftgas-Anlagen*, 3rd ed. (Berlin: Springer Verlag, 1921), 682–84, describe the engine and give test results from several sources. Bánki's obituary (*Zeitschrift*, V.D.I. Band 66, Nr. 39, 30 Sept. 1922, p. 940) gives highlights of his internal combustion (I-C) engine work and his many contributions to several engineering fields. Note: "Old" atmospheres were used: 14.23 psi.
10. H. C. Van Ness, *Understanding Thermodynamics* (New York: McGraw-Hill, 1974), 1. This little book incorporates a series of lectures that are a delight to read.
11. Sadi Carnot, *Réflexions sur la Puissance Motrice du Feu et sur les Machines Propres à Développer cette Puissance,* trans. E. Mendoza (New York: Dover, 1960), 48. The equation expressing efficiency based on temperature differences is: $E = (T_1 - T_2)/T_1$ where T_1 and T_2 are expressed in absolute terms, in other words measured temperature + 273°C.
12. Rudolf Diesel, "Theorie und Construction eines rationellen Wärmemotor" (Berlin: 1892), 64. The original is in the Deutsches Museum in Munich.
13. Rudolf Diesel, *Theory and Construction of a Rational Heat Motor*, trans. Bryan Donkin (London: Spon Press, 1894), 30.
14. Ibid., 49.
15. As quoted in Kurt Schnauffer, "Die Erfindung des Dieselmotors, 1890–1893," Part 1, trans. Henry I. Willeke, in MAN Archives (1954), 21.
16. Schnauffer, *The Invention*, 22–23.
17. Ibid., 24.
18. The British patent No. 7,241 of 1892 (accepted October 8) was filed April 14. This provisional application, with no claims printed, is on the first six pages. It precedes the specifications and the ultimately granted claims received August 27 that are a close translation of the complete German 67,207. A word-for-word application of the provisional British one was filed in the US, via a New York City attorney, A. Faber du Faur Jr., on April 19. The US Consul General in Berlin attested to the invention being Diesel's on March 12. U.S. National Archives, Washington, DC, Record Group 241, Patent "file wrapper," Diesel No. 542,846.
19. British patent 7,241 issued 1892, 8.
20. If it is assumed that the Consul General, on March 12, saw an English translation of the application sent to London and Washington, DC, then the claims sent with the application would probably be identical to those first submitted to the German office on February 27. It is possible, however, that Diesel waited until the first reply from the German office on March 15 so some revisions could be made before sending the applications out of the country. This considers travel times of the period. (The "Blue Riband" steamship record from Southampton to New York City set in 1891 was five days, sixteen hours, and thirty-one minutes.) In any event, the claims filed in Washington on April 19 show isothermal combustion holding prime importance:

    1. A process of producing motive work from the combustion of fuels in a manner that, whilst the process is carried out, no heat is led away to the outside, and that the exhaust

gases do not lead away any notable quantity of heat, which process if characterized by the successive execution of the following single operations:

a) introduction into the cycle of a motor engine of a quantity of air determined according to the calorific power of the employed fuel and being so high a multiple of the air quantity theoretically necessary for combustion, that the temperature of combustion does not exceed a maximum permissible for the practical working of the motor without any notable leading away of heat [i.e., without any artificial cooling];

b) compression of this introduced air quantity to so high a pressure, that the temperature thereby produced in the compressed air is nearly the same as the previously fixed temperature of combustion;

There were two more parts to Claim 1 dealing with controlled admission of fuel (c) and cooling during expansion (d). Claims 2 and 3, dependent on Claim 1, are variations of the process. Claim 4, also dependent on Claim 1, is for a compound engine.

21. British patent 7,241, 4.
22. Schnauffer, *The Invention*, 7, 16, provides five of the draft claims as well as other valuable information not readily available on the first engine patent.
23. Ibid., 10.
24. Ibid., 16.
25. See Cummins, *Internal Fire*, 307–17 for more information on Stuart's life and engines.
26. A translation of these two claims is the wording used in Diesel's British patent 7,241, 13, 14. The final version of the complete specification of the British application was filed August 27, 1892, indicating that Diesel waited until his German application was approved. The latter was effectively done by his agreeing to the July 7 claims wording of the German examiners. The same version, also sent to the US Patent Office, had a filing date of August 26 and replaced the April 19 application, which was withdrawn. Hence the later filing date on the issued US patent.
27. US Patent 542,846, filed August 26, 1892, and issued as July 16, 1895. Switzerland granted Diesel a patent on April 2, 1892.
28. Rudolf Diesel, *Die Entstehung des Dieselmotors*, trans. Henry I. Willeke (Berlin: Springer Verlag, 1913), 3–4.
29. Schnauffer, *The Invention*, 22. Records show that as late as 1897 there were two hundred unsold copies. All were sold by April 1898 (p. 27).
30. Diesel, *Theory*, 1.
31. Ibid., 84.
32. Cummins, *Internal Fire*, 195–212, describes the Brayton engine.
33. Diesel, *Theory*, 68.
34. Ibid., 83–84.
35. Ibid., 34.
36. Ibid., 85.
37. Professor Matschoss (1871–1942) was noted for his history of steam engines. Eugen Diesel, *Erfindung und Priorität* (Invention and Priority), cites Matschoss in a lecture before the Hamburg district of the V. D. I., April 27, 1951, p. 18. MAN Archives. Trans. Henry I. Willeke.

CHAPTER 4

For Sale: A Dream

"Always only a part of the ambitious thoughts will be forced upon the material world. Always the completed invention looks quite different from the ideal originally perceived by the mind, otherwise it will never be accomplished."

—Rudolf Diesel[1]

After a ten-year search, Rudolf Diesel was convinced he had found *the* way to design an engine with the highest thermal efficiency. He believed his most difficult days were over and transforming ideas into reality should prove a simpler task: license a qualified manufacturer to develop and build the engine under his guidance and then await the forthcoming royalty checks. One company finally agreed to evaluate a test engine built to his design but gave him no financial support. Because of this limited commitment, he continued to promote his theories through the book based on his studies. Gift copies went to influential professors and companies deemed possible licensees. A few favorable academic endorsements resulted, but no new firms showed any interest. Meanwhile, when Diesel came to realize that his patented combustion process was unsuitable for a real engine, he quietly substituted another. The path of his endeavors still failed to follow his optimistic, short-range plan.

Maschinenfabrik Augsburg

Diesel wrote to Heinrich Buz, managing director of Maschinenfabrik Augsburg (MA), on March 7, 1892, one week after filing the patent application, to invite his company's participation in the development of the new engine (fig. 4.1). A copy of the January manuscript also went to the company. MA was one of two firms directly approached prior to publication of the book at

Fig. 4.1. Heinrich V. Buz (1833–1918). *MAN Archives*

the end of 1892. Diesel's previous contacts with Buz provided the entrée. Not hurting the relationship was the marriage of Lucian Vogel, his friend from university days, to Buz's daughter, Emma (fig. 4.2). Vogel went to MA after graduation and since 1888 had been chief engineer of the Linde refrigeration machine department.

Diesel explained in his letter that Buz and the MA engineers were spared a study of the theory outlined in the manuscript as that was already "acknowledged correct." They need only address those problems involving the engine's actual construction. Diesel thought it advisable, however, to point out a possibility of reducing the maximum compression pressure from his preferred 250 atm:

> . . . the test cylinder would suck in plain atmosphere air and compress without water injection. The desired temperature will be about 800°C, and with it the cylinder pressure rises to near 150 atm . . .

> . . . further suggestions during completion of the development will have to be made by you, as competent men and resources under your management alone form the reasonable approach.[2]

He also made it clear that he wanted someone else to turn the paper engine into a saleable product.

The manuscript and proposal were passed on to Josef Krumper (1847–1923), manager and chief engineer of the steam engine department, for comments. Krumper's report to Buz was negative. His belief in the engine continued to remain so for years, and he would become a "thorn in Diesel's side."[3]

Fig. 4.2. Lucian Vogel (1855–1915). *MAN Archives*

Buz's reply to Diesel on April 2 was both terse and disappointing:

We have carefully considered all aspects of the matter and come to the conclusion that the difficulties inherent in the realization of the project are so great that we cannot undertake the venture.[4]

It was an unexpected blow to Diesel, but he boldly wrote back to Buz on the seventh what could happen at M.A. if the engine fulfilled his prophesies:

Now I would like to stress here that [my proposal] dealt not about building a new system for gas and oil engines; in that case your refusal to take another course of action was perfectly understandable. It deals rather as a substitute for now-building steam engines, along with their boilers, with something much simpler and more perfect; from this standpoint, it seems to me a matter of salient interest, from business or other special consideration, for a distinguished firm of your worldwide reputation.[5]

Diesel knew Buz dared not overlook any new power form that might pose a threat to one of MA's major product lines, especially with steam engine sales so low at the time (fig. 4.3).

Nevertheless, the strong refusal from Buz gravely concerned Diesel about placing the engine at MA. A second choice was Gasmotoren-Fabrik Deutz, the leading builder of internal combustion engines. He had no direct contact there, but this was solved by asking Gustave Richard (1849–1912), a French authority on I-C engines, to introduce him to Deutz's managing director

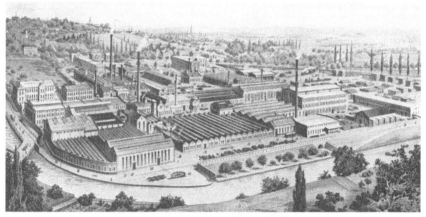

Fig. 4.3. Maschinenfabrik Augsburg c. 1900. *MAN Archives*

Eugen Langen (1833–1895). Diesel had corresponded earlier with Richard about a paper the latter published on German I-C engines that dealt at length with the successful Deutz product line. Richard complied as requested, and Diesel sent a letter to Langen on April 13 inquiring if Deutz might have an interest in his engine. Langen replied on the nineteenth with a "no thank you"; while he did not doubt the correctness of theory, he saw too many problems in making a practical engine based on the theory. Diesel wrote once more to Langen, trying to refute the arguments raised, but never received an answer. Thus, with Deutz turning him down, Diesel needed the support at Augsburg all the more.

Diesel continued a letter barrage to Buz without waiting to hear from him. This time he enclosed a copy of his reply to Professor Schröter in which he responded to the latter's concerns of March 29 (chapter 3) about the difficulties with such high compression pressures. In this reply, Diesel had suggested that even 44 atm should work. He hoped that the reference to a greatly lowered pressure would serve to change Buz's mind.

The answer Heinrich Buz sent to Brückenallee 15 on April 20 gave Rudolf Diesel his first, much needed encouragement.

Your 3 esteemed [letters] of 6, 9, and 13 instant to the undersigned could not be answered till today because he was away for a time on a trip.

Per your new statements, we are willing to undertake, under certain conditions, the completion of an experimental machine which must be of such construction to avoid all possible development complications. It should be only an initial step to determine *whether the system in general is practically feasible* without for now considering whether there is a considerable increase in efficiency in comparison with other motors. Only after startup problems are overcome, which in spite of everything will still show up, can we then gradually and step by step proceed.

We suggest for now that you submit construction drawings of such a machine before going ahead, and then we may invite you perhaps for further personal discussion.[6]

Although not an open-arms endorsement of Diesel's ideas, the Buz reply was logical. It was not a contract, but a foot-in-the-door proposal giving both parties an opportunity to discuss knowledgeably what M.A. was being asked to undertake.

Even so, Diesel could not afford to put all his eggs in the M.A. basket. The lack of a firm commitment from Buz and the disappointing negative response from Langen pushed Diesel to ask Professor Linde for help in promoting his ideas. Linde suggested a different strategy, that of asking several companies to form a consortium to study the engine. This also found no success, and his frustration was evident in a June 30, 1892, letter to Diesel.

I cannot recall a single time in which any new technological undertaking would have endured such an antipathy and such efforts to pry loose money from industrial concerns to invest in "safe" profits.[7]

The Design of the First Engine: 1892–1893

Diesel immediately started on the design of a test engine after Buz offered to evaluate his ideas. This kept him in contact with Lucian Vogel at Maschinenfabrik Augsburg during the ensuing months. His old friend, put in charge of the project for the company, became a staunch supporter of the new engine.[8] Vogel made many contributions to the first test engine. The preserved letters between him and Diesel during this period attest not only to the role he played, but also to his resignation that frequent revisions would become a way of life.[9]

Diesel had already sent a preliminary layout for a two-cylinder engine to M.A. on April 22. He claimed it would produce 50 hp at 300 rpm. It embodied separate intake and exhaust valves and long, close-fitting, cast steel "plunger" pistons. In lieu of piston rings, the combustion chamber was sealed by a leather packing gland and a pressurized oil film. An explosive cartridge started the engine.

Buz, however, told Diesel in an April 25 letter that M.A. wanted only a single-cylinder engine. All work thereafter reflected this firm decision.

Overall design responsibility soon went to Diesel. The tenor of questions raised by M.A. over Diesel's preliminary drawings forced him to defend every detail. Vogel recognized this and suggested Diesel be given the final word to avoid inevitable and interminable conflicts.[10] Vogel successfully argued that since it was Diesel's theory and not M.A.'s under evaluation he should be allowed to submit what he thought best.

Diesel in turn wisely made compromises and often accepted what Vogel and others at M.A. suggested because he recognized his own lack of practical engine experience. What he did not realize was that Maschinenfabrik Augsburg, a steam engine builder, and Firma Krupp, who joined the team later, had little knowledge of *internal* combustion practice. A detailed study of the engine assembled in July 1893 discloses the great depth of ignorance from which Diesel, M.A., and Krupp had to struggle upward.

On the other hand, the inventor's intuitive feel for injection and ignition events may be seen in the cylinder head of the April 22 design sent to M.A. It contained an antechamber surrounding a vertical injection nozzle (fig. 4.4), which, according to Diesel, would enhance spray atomization:

> At the moment the highest degree of compression is achieved, the needle lifts, and a sharp, thin jet of fuel enters into the hot compressed air with great velocity. This jet is hit from all sides with great force in the form of an air stream which comes out of the circular cavity; this air is nothing more than a part of the previously compressed air in the circular cavity axial with the center of and open to the cylinder. Because of the piston descent after compression and the reduction in

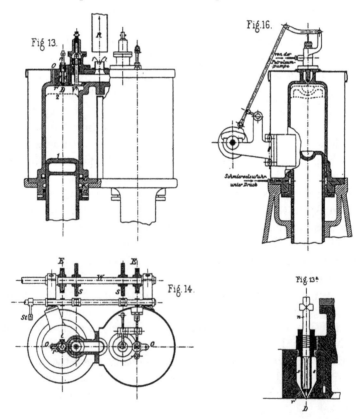

Fig. 4.4. Combustion antechamber and injection system. Details of Diesel's first engine design, 1893. Diesel, *Theorie, 1893*

cylinder pressure, the enclosed air in the circular cavity exits with great force and breaks up the jet of liquid fuel in the finest way. As a result the fuel is sprayed into the cylinder in a cone-like fashion.[11]

Diesel next modified the antechamber so that the connecting passages between the chamber and cylinder imparted turbulence to the air exiting and mixing with the injected fuel. He suggested that the air in the antechamber

... on the one hand effects a division and atomizing of the oil and on the other hand effects a swirling motion of the air mass and, therefore, distribution of the heat over the whole volume of air.[12]

The method of starting the engine was also changed. Compressed air, entering the cylinder through a separate, cam-actuated valve, turned the engine.

Vogel visited Diesel in Berlin during mid-June so they could plan their design responsibilities. Diesel would work primarily on the piston and fuel system. The rest of the engine, including crankshaft, connecting rod, and bearings, went to Vogel. This was logical as these parts mostly followed good steam engine practice and, except for the cylinder head, would later encounter only minor problems.

Diesel wrote Vogel the following exhortation shortly after his return to Augsburg:

From here on it is mainly your responsibility to support this project. I ask of you not only in my name, but also in the higher interest of Science and Industry, to concern yourself most kindly and bring this rapidly to its goal. *Swift* must be the solution, because you and your factory will be first in line to benefit from the invention . . . [and] if it is successful . . . it is an incomparable commercial operation.[13]

Vogel immediately immersed himself in work on the engine; he always did more than was asked of him.

Although Diesel said his process should work at 44 atm he still hoped to try for 90 atm. Knowing Diesel's intention, Vogel designed for 100 atm. The increasingly higher cylinder pressure in turn caused several bore and stroke downsizings from 200 × 540 mm to finally 150 × 400 mm.

Another compromise with the patented isothermal combustion process occurred during this time: Diesel asked Augsburg to change intake valve timing so that the air was drawn in over the entire intake stroke rather than the last three-fourths. This had been suggested in his book.

The injection nozzle was also modified. Instead of a central, axial orifice below the needle valve, Diesel now fed fuel from the nozzle through two, right-angle holes to place the jets directly into the air stream leaving the antechamber. In addition to improving fuel atomization, the air blast would have lessened carbon buildup in and around the nozzle holes.[14]

Vogel sent eighteen sheets of drawings and sketches to Berlin at the end of November. One progressive feature was the addition of balance weights to the crankshaft. Although not a new idea, MA had yet to use such weights on their own steam engines.[15] Diesel received these drawings while buried by the final editing of his book manuscript for the publisher, so a critique of Vogel's proposals had to wait until January. Then Diesel not only responded to what Augsburg had furnished, but also included all his own promised drawings in a package mailed from Berlin on February 6, 1893. Fifty more pages of calculations and construction details followed on the twenty-seventh.

First, Diesel insisted on a major revision to Vogel's design. He wanted a single-throw crankshaft conventionally supported between the two main bearings rather than Augsburg's proposed overhung shaft mounting. MA also had followed standard steam engine practice by making the connecting rod bearing length equal to the diameter. Diesel believed this was too conservative and asked that it be reduced to less than half the diameter of the journal, or 50 mm for a diameter of 120 mm; the final length was 60 mm. The shorter bearing also stiffened the crankshaft by placing the main bearings closer together.

An earlier drawing of Diesel's dated July 1892 showed a single, conventional poppet valve for both intake and exhaust. In the approved design it had two, conical-seat heads as found on some steam engines (fig. 4.5). Why he chose the dual function valve and then made it even more complicated is not known. In operation, the intake air was drawn into the cylinder through ports in the exhaust pipe flange where the flange bolted to the head. Exhaust gases were expected to simply blast past these ports and out the pipe. The double function valve was changed after testing began!

Diesel further agreed to eliminate the antechamber around the central injection nozzle. It may have been due to the reduced cylinder diameter leaving insufficient room for it in the head. A cam-lifted needle valve seating in the nozzle was also dropped. These deletions ended what little chance there was to adequately inject fuel by a mechanically generated hydraulic pressure. Sealing the coffin was the selection of a Körting nozzle, with no check valve, to spray directly into a deep cylindrical pocket set in the piston crown.

To be noted in the mailing of February 27 is that Diesel called for a hydraulic injection pressure of only 50 to 60 atm. This meant he must have accepted a maximum compression pressure of no more than the 44 atm.

Another revision difficult to understand was Diesel's drastic reduction in fuel metering pump volume. It is especially so in view of the pointed questions asked about the tiny fuel quantities he wanted to inject. He reduced the pump plunger diameter of 20 mm in an earlier sketch to only 4 mm. The stroke was likewise shortened from 30 mm to just 8 mm. This great reduction in chamber volume of about 94 percent returned Diesel almost to his original

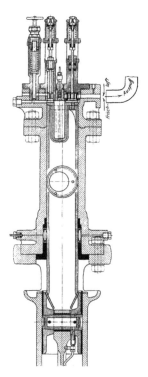

Fig. 4.5. Recessed piston and dual-purpose intake and exhaust valve.
Diesel, *Die Entstehung,1913*

theory of utilizing only one-ninth of the available air.[16] Was it to keep up the façade that he still followed the patent?

To directly refute this regressive step, Diesel wrote that such a minute pump chamber volume was probably not practical. He recommended using an air-over-fuel accumulator to pressurize the fuel: a hand pump raised the air pressure, and the fuel delivery rate was adjusted by means of a manually operated valve at the accumulator.

The camshaft was located about equidistant between the crankshaft and cylinder head (fig. 4.6). Large-diameter cam discs were keyed to this shaft. One disc actuated a lever that opened a spring-seated, conical needle valve to feed pressurized fuel to the spray nozzle in the head. The quantity of fuel to be injected was a function of needle lift duration and fuel line pressure.[17] All tubing fittings used an asbestos packing gland for sealing.

A second cam disc, slid axially during the starting cycle, opened the air starting valve. A third vertical valve in the head was an adjustable pop-off safety valve. The engine had no governor to control speed.

Integral with the cylinder head was the upper third of the uncooled cylinder. Both the extended head and the long, close-fitting and ringless plunger piston were of cast steel. Until then MA had little experience with

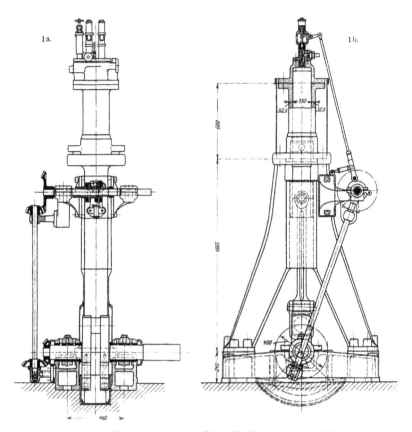

Fig. 4.6. Exterior views of Diesel's first engine, 1893.
Diesel, *Die Entstehung, 1913*

such material and called upon Krupp for advice about how to avoid porosity. Krupp eventually made these parts for the first engine.

The term "plunger" was apt for the piston as it had a length-to-diameter ratio of almost 8:1. To seal against compression loss, oil under pressure expanded a bronze diaphragm inset in the cylinder wall against the piston. A larger stepped diameter on the lower end of the piston held the piston pin. It, too, ran in a close-fitting cylinder and acted as a crosshead guide for the connecting rod.

M.A. accepted Diesel's calculations and countered with few changes to the drawings. On March 27 the factory began making parts.

Many letters passed between Augsburg and Berlin after machine work began and before Rudolf arrived to supervise the test program in July. Several addressed fuel type as well as Diesel's concern of possibly needing an auxiliary ignition aid due to the lower compression pressures.

Diesel had considered using only a liquid fuel from the time he began designing the test engine. Of special interest to him were crude oils:

. . . because as a refrigeration engineer I had kept busy for many years with the development of a procedure for the extraction of paraffins [C^nH^{2n+2}] from crude oil through cooling. With these tests I worked with crude oil from different lands and had an opportunity to thoroughly study this material.[18]

Experiments with a low grade of bituminous coal-based crude oil were discussed and then abandoned. Diesel referred to this viscous, tarry oil from the Pechelbronn, Germany, area as a "heavy, inflammable brown mass that would not flow through a pipe at ordinary temperatures."[19] Russian and American kerosene (*Lampenpetroleum*) and gasoline (*Benzin*) held the highest priority.

One additional fuel possibility was mentioned in his letter of June 28. A light Russian crude referred to in Germany at that time as *naphtha* came out of the well with about 5 to 6 percent natural gasoline.[20] Diesel believed that if his engine could burn this free-flowing crude he had a great advantage over other I-C engines even though thermal efficiency at the reduced compression pressures would be lower than touted in his earlier writings.

Diesel raised the idea in April of using a platinum wire, heated by compression, to aid in igniting the fuel spray. On June 16 he asked M.A. to set the cylinder head on a stand with legs before its assembly to the engine to see if such a glowing wire would ignite fuel injected in open air. This was successfully done.

Maschinenfabrik Augsburg wrote on July 12 that the engine was essentially completed, and they had met their contractual obligation. It was Diesel's turn to come from Berlin and demonstrate a potential product that would benefit the coffers of Krupp and M.A.

The Book and Its Consequences

Sandwiched in the middle of the design project was the final editing of Diesel's book manuscript before its publication (chapter 3). Not until it came off the press in late December 1892 could he again fully devote himself to the Augsburg project. However, the book was not simply an end in itself. Diesel saw it as a way to spread the gospel of his concept beyond the company walls of MA. At that point he dared not assume the engine had an assured home there.

Diesel dispatched copies of *The Economical Heat Engine* to selected professors and heads of other engine companies as soon as he received some from the publisher. Personalized letters accompanied those going to Langen at Deutz and Friedrich Alfred Krupp (1854–1902) in Essen. Richard in Paris and Sulzer Brothers in Winterthur were among the twenty-nine recipients of books personally sent during January 1893.

Professors Schröter and Gustav Zeuner (1828–1907), who had already provided helpful support to Diesel, also received copies. Schröter immediately wrote an article about "A New Heat Engine," which appeared in the February 4 issue of *The Bavarian Journal for Industry and Crafts*.[21] The article, prepared exclusively to support Diesel, contained highly favorable comments on the proposed engine, and there is little doubt that it helped in the negotiations then in progress with Maschinenfabrik Augsburg and Krupp. After describing the engine and supporting Diesel's theory that its operation followed the Carnot cycle, Schröter boldly stated:

> Of special importance is . . . the correct selection of air weight in relation to the heat value of the fuel. It is thus possible to control the . . . combustion temperature in such a way that the practical operation of the machine, the lubrication, etc., is possible without artificial cooling of the cylinder walls. . . . The surprise of the described solution . . . is the bold challenge of today's technique to work with compressed air of 250 atm pressure and 800°C temperature. . . .[22]

He concluded with "We have here the enjoyable example where the theory is ahead of practice and shows clearly the method how to reach the goal."[22] Schröter did not mention Diesel's 100:1 air-to-fuel ratio.

Zeuner, whose classic works were becoming the "bible" on thermodynamics in Germany, wrote Diesel from the Technical University in Dresden. Like Schröter, his response was very positive:

> Theoretically I am on your side and enjoy your approach. I have not read anything for a long time in our field that was so interesting. Both basic ideas of yours are absolutely new and correct: 1. Preheating the air by compression to combustion temperature . . . 2. Application of a large air quantity, which you calculated theoretically correct.[23]

Zeuner also failed to see (or point out?) the excessive air-to-fuel ratio. Diesel later cited this generous praise from the respected professor in the aid of his cause.

Richard's response apparently was unfavorable as can be deduced from Diesel's answer to him.[24] One of Richard's major concerns centered on the uncooled cylinder. Diesel's reply is of special interest as it closely described the "adiabatic" diesel engines under active development almost a century later:

> Concerning . . . your criticism, I want to insist on one essential point. It is that my cylinder is not cooled by waterflow. On the contrary, it is fully insulated. . . . Furthermore, the large expansion cools the gases nearly to room temperature. I therefore avoid two enormous losses of today's internal-combustion engines: the loss in the cooling water and the loss in the exhaust gases.[25]

Eugen Langen's reaction to the book remained negative, although less so than when he had previously reviewed the manuscript version. It is possible Langen or his people did not actually study the book because his letter of January 13 still referred to compression pressures over 200 atm.

Wilhelm von Oechelhäuser (1854–1923) of Continental-Gas-Gesellschaft Dessau, a maker of coal gas and large gas engines, also got a copy. Shortly thereafter, Diesel's engineer friend Otto Venator (b. 1854) went to Dessau. Oechelhäuser politely told Venator he would decline involvement as he doubted that Diesel's theory could be turned into a practical engine. Venator had hoped to talk with Hugo Junkers, head of the gas engine department, but nothing was recorded about this.[26]

Gebrüder Körting in Hannover responded to Diesel on January 14 with an offer to make tests but withdrew it two days later due to concerns over the high pressures:

> At the time being we restrict ourselves to relatively easy tests of a purely practical nature using gas engine models on hand. That means we can only go so far with compression without running into considerable difficulties in construction and operation.[27]

They apparently had assumed that one of their standard engines could be modified to test Diesel's theories but then quickly realized how highly stressed it would be.

Sulzer Brothers' chief engineer Wilhelm Züblin (1846–1931) visited Diesel in Berlin on February 2 to discuss his concerns raised by the book. An authority on thermodynamics as well as steam power plants, Züblin questioned whether an engine built to Diesel's theory could produce useful power. He argued that since work input during the compression stroke was critical relative to the power produced, it "may happen that the loss is as great as the output."[28] Diesel had asked Zeuner to meet with him and Züblin to allay the latter's doubts about the high air-to-fuel ratio of 100:1 limiting the power, but the two of them could not change Züblin's mind.

Diesel nevertheless heeded the point Züblin raised. In a new monograph called *Review of the Expected Effective Performance of the Diesel-type Motor,*[29] he used another approach to explain again the thermodynamic and construction advantages of his engine over steam. It went to Züblin on February 8 with copies to Schröter and Zeuner a few days later. Diesel's cover letter emphasized a new "calculation which is based on practical numbers about the anticipated results of my new motor" and that he had "striven to regard only actual conditions." He pointed to a larger fuel charge achieved by lengthening the injection time from the 7 to 8 percent of the piston stroke given in his book[30] to up to 40 percent. Putting in more fuel than this would have precluded isothermal combustion. While the longer injection period raised power by a theoretical 50 percent (the fuel increase was about 37 percent), the additional fuel needed to achieve the stated output of 25 hp at 300 rpm would have been closer to 800 percent! Now the engine at least might idle, but the revised air-to-fuel ratio of 72:1 was still a long way from being realistic. Diesel failed to alter Sulzer's decision to remain on the sidelines and await further developments.

The First Contracts

Although everything needed to build a test engine had been sent to Maschinenfabrik Augsburg, Diesel still needed to secure a satisfactory long-term commitment from his one interested prospect.

Rudolf's informal arrangement with M.A. merely called for a feasibility study, so while preparing the necessary drawings, he continued contract negotiations with Buz. After weeks of letters back and forth, a contract was finally signed with M.A. on February 21, 1893. Under its terms the company agreed to build a four-horsepower engine within six months upon approval of the drawings, which Diesel had already delivered on the sixth.[31] Further, for each engine later sold, Diesel was to receive a royalty of 25 percent. The contract, lasting the life of patent 67,207, assigned a territory covering the south of Germany (Bavaria, Baden, and Würtemburg).[32]

This was a significant first step, but Diesel neither received nor for a long time could expect to get any money. Since his employment with Linde ended when tests began at M.A., he was without income until royalties accrued from sales of production engines.

The solution to Rudolf's financial dilemma would come from Essen. Shortly after receiving his book, Firma Krupp invited him to a directors' meeting on January 31. He did not attend the conference with bowed head and hat in hand. In addition to building a test engine, his terms included:

- Payment of 30,000 marks per year during the test period.
- Payment of 500,000 marks when the engine was ready for production, less the sum of any annual payments already made.
- A 25 percent royalty on the sales price of delivered engines.[33]

Friedrich Krupp, sole owner of the company, received a report from his directors a few days later leaving no doubts that Diesel's demands were high. However, it stated that according to members Asthöwer and Schmitz:

. . . the theoretical basis of the project is correct and is assumed to be feasible. A motor of epoch-making importance will be created which surpasses the steam engine . . .[34]

Of concern to all at Krupp were not only the payments to Diesel but also the estimated 150,000 marks in development costs. It was a big gamble, yet they knew of Diesel's negotiations with M.A. and what would be lost if the engine proved to be a success without their involvement.

Influencing their decision was a visit two Krupp directors paid to Buz in Augsburg on February 14, one week before he signed M.A.'s contract with Diesel. They departed, convinced that Krupp should make an arrangement with Diesel, as they perceived Buz had already decided to proceed with the engine. They also thought there might be a way for both Krupp and M.A. to reduce each company's outlay through some kind of cooperative venture. Whether Buz intentionally planted the idea is not known.

Diesel added to this rapidly warming interest at Firma Krupp: he sent Schröter's latest favorable comments, said he had signed a contract with M.A. (his letter was dated three days before the actual signing on the twenty-first), and dangled before them more enticing words about the engine's potential:

> I must emphasize that neither the operation of my motor nor my patent depend on the use of coal dust, that nothing is in the way of a gas apparatus being added to the motor. It would be much simpler and safer to operate than today's gas motors and in doing so save enormous amounts of coal.[35]

Diesel had already made it clear that because of his engine's high cylinder pressures more of its parts would have to be of steel, Krupp's principal product. A possibility to run engines on the Diesel cycle with coal gas served as a further inducement because of a new subsidiary coming under Krupp's ownership.

He was soon to take over Grusonwerk, a highly respected company near Magdeburg that made steel castings, armor, and other massive products compatible with the Essen firm. Hermann Gruson (1821–1895) had signed a cooperative arrangement with Friedrich Krupp in 1892, leading to an outright purchase in May 1893.[36]

The hope of using gaseous fuels interested Krupp because Gruson was soon to begin production on a family of reliable gas engines bought from Buss, Sombart & Co. also of Magdeburg.[37] Thus, Grusonwerk would be a logical place to build Diesel's engine. Furthermore, by owning rights to it, he eliminated potential competition if gaseous fuels proved feasible. Diesel went to Magdeburg on February 25 to talk with Hermann Ebbs (1860–1932), Buss, Sombart's chief engineer who had assumed the same title when it became part of Gruson. Ebbs must have been favorably impressed by the engine as it was during this meeting that Gruson agreed to become involved with Diesel.

Firma Krupp and Diesel signed a contract in Essen on April 10, 1893, in which Diesel was to receive an annual payment of 30,000 marks until production began. He would then be paid a 37.5 percent royalty on engine sales until these reached a cumulative total of 500,000 marks, at which time the royalty rate dropped to 25 percent. In return, he granted Krupp the right to sell engines in areas of Germany not assigned to M.A. and in

Austria-Hungary.[38] Krupp also secured an escape clause: he promised to begin work when drawings were reviewed, but he had no deadline to meet and could cancel the contract without further obligation. Diesel now had an assured income from Krupp while development work progressed at Augsburg. Upon returning home from Essen in the late evening of that momentous day, he found Martha hosting a dinner party for her friends; it quickly became a time of champagne toasts to his success from the guests.[39]

Krupp's directors and Buz shortly thereafter made their own deal. Just two weeks later, on April 25, they agreed to form a consortium for the life of patent 67,207 whereby both companies were ultimately to share equally in all development expenses and in any future profits from German engine sales. There was also to be a free license exchange on all new patents issued to Diesel. The test work would be conducted at M.A., and Diesel's annual stipend would come from Krupp.

This was a coup for Diesel. Two of Germany's most respected industrial firms had committed themselves to bring his engine to fruition. Success seemed assured.

Gebrüder Sulzer still harbored doubts that the engine was all it purported to be; yet they, too, wanted to join Krupp and M.A. in case their fears proved unfounded. The Sulzers entered into an option agreement with Diesel on May 16, 1893, from which he received 10,000 marks. After the engine passed its acceptance tests, new obligations of 20,000 marks per year were incurred for five years. In exchange they obtained rights to Swiss patent 5,321, which corresponded to the German 67,207. Diesel was also to furnish Sulzer with all test data generated in Augsburg. They need not build engines and could cancel the contract up to the time production might begin.[40] The doubts raised by Richard and Züblin, their own engineer, cautioned Sulzer from going further. This conservatism proved a wise course.

Constant Pressure Combustion

After his marathon of activities with Krupp and M.A., Rudolf went on a needed vacation with his family to northern Italy. The relaxation during these several weeks while the test engine was being made in Augsburg freed his mind to tackle the one serious problem immediately at hand. He had not yet resolved to his own satisfaction, or to that of others, how he could effectively raise the engine's output without losing its claimed advantages.

Diesel returned to Berlin through Winterthur, where he signed the contract with Sulzer on May 16, and then visited at Essen for two days. The low power question was raised by both companies, since by then Krupp had also heard disturbing reports regarding the claims made in Diesel's book.[41] Correspondence reveals Krupp was particularly worried about the cylinder size required to produce useful power.

The critique precipitating Krupp's worry, and one Diesel dared not ignore, came in a lecture given by Otto Köhler, now a teacher at the Cologne school of machine design. Diesel's theories were thoroughly analyzed in the paper read before the Society of German Engineers (VDI) in Cologne on April 10, 1893.[42] Krupp's engineers most likely attended the lecture as Essen is less than eighty kilometers from Cologne. Köhler had been given a copy of Diesel's book by Venator in February during a visit Diesel requested. Diesel asked Venator to do this because Köhler and Venator had been friends at the University of Aachen. A blessing on the new engine by the respected engineer would be welcomed by Diesel.

Köhler had already replied to Venator on March 18, but with very negative comments. Beginning with praise of the "elegantly done" calculations, he went on to say that (1) neither the true Carnot cycle nor the modified Carnot cycle were feasible,[43] (2) the compound engine offered nothing new, and (3) self-ignition had already been used for gas engines! At the time, Diesel was not too worried about these damaging comments because his recent studies had eased his mind regarding the first point, and he was correctly convinced that Köhler was in error on the third point.[44]

Köhler based his comments about the Carnot cycle on his earlier study of the process. This work was printed in 1887 under the title *Theory of Gas Engines*, and his conclusions for the report to Venator and the VDI lecture were drawn from it:

> These examples lead to the conclusion that the perfect cycle process (besides the difficulties to follow it exactly) with air as the working medium is not suitable for practical use. A high thermal efficiency can be obtained, however, the initial pressures rise so much . . . that the cylinder dimensions and connecting rod assembly are out of proportion. Moreover, the thermal gain will be more than offset by the large friction losses.[45]

He further averred that Diesel's engine was identical to this process, and because of the high internal losses, "the indicated output is just enough to overcome the friction resistance. . . . So an output worth talking about can never be obtained."

One of Köhler's statements that "self-ignition, which requires high compression stresses, must be dropped" soon became a moot point. Another was a definite blow to Diesel:

> Therefore, one cannot agree with the conclusions of Diesel's paper that any other cycle than the perfect or nearly perfect one must be called wrong. . . . Although the result of this investigation is only approximate, it will be valuable as it may prevent long lasting and expensive tests which will be unsuccessful.

Köhler's complete lecture was not printed in a VDI publication until September 9, but Diesel must have been aware of its substance before starting his own new study.

That Krupp was more worried about power and cylinder size than M.A. and Sulzer may be deduced from a carefully worded letter Diesel sent to them on June 16.

In the meantime I have investigated more closely some theoretical points of my process that I could not do earlier due to lack of time. It results that by a *slight modification of the process the cylinders can be reduced considerably.* I hope to achieve about double the output as previously assumed.[46]

The "slight modification" mentioned in an almost offhanded manner was merely adding many times more fuel and cooling the cylinder with water. M.A. and Sulzer did not receive this news until several months later.

Diesel's letter to Krupp was based on an important private study begun after his visit in Essen the previous month. The study, titled *How the Engine Works Best,*[47] resulted in calculations leading to a further modified combustion theory.

The summary of the study opened by stating that "certain deviations from the perfect engine will have favorable effects" as Diesel "had repeatedly mentioned earlier." Among them were (1) not compressing the air charge on an isothermal-adiabatic line but only adiabatically, (2) not expanding to atmospheric pressure by reducing the piston stroke, *and* (3) enlarging the area of the indicator (P-V) diagram of a given cylinder. The latter point he said was the one most criticized by both "practical and scientific people" and the main one to be addressed.

Diesel had finally committed to paper a repudiation of his cherished isothermal combustion. He thus admitted that it could not produce the power needed for a practical engine:

Now the question comes up, in spite of earlier contradictory statements, if it is possible for larger diagram areas in the same cylinder with other combustion processes than isothermal.

Especially it is evident, a priori, that at the same compression, a combustion at constant pressure gives larger diagram areas than at constant temperature where the combustion curve immediately and steeply drops. Indeed at constant pressure the temperature is higher than that of compression. But we also saw that our cylinders can take higher temperatures. . . . We could achieve this higher temperature by higher compression. However, we would exceed the limit of 80–100 atm given to us. But if we work with constant pressure we do not exceed the pressure limit. We can even reduce it and in any case obtain far larger diagram areas.[48]

Fuel could now be added to utilize all the air charge rather than only the 10 percent under isothermal combustion.

Diesel further recognized a possibility of lowering the maximum compression to as little as 30 atmospheres, given this new opportunity to add so much more fuel. He surmised that proper combustion should still occur because the air temperature of 500°C at 30 atm was far enough above the ignition point of the fuels he considered using. He questioned whether "combustion can be carried out in the neighborhood of the ignition temperature."[49]

His conclusions explicitly stated that "combustion at constant temperature is *completely* excluded" and that "combustion at constant pressure is the only one to be selected." Diesel further commented about a third process, the one Otto (spark ignition) engines followed:

> The combustion at constant volume and lower compressions (30 atm) *can* have the same results as with constant pressure at higher compressions (64–90 atm). The cylinders will of course be something larger, the maximum pressures will be the same, but with a good design the results will be about the same. However, it must be assumed that the slightly increased mean temperature, and especially the very high maximum temperature, require artificial cooling.
>
> This will reduce the final efficiency compared with constant pressure. Besides, the precise operation at constant volume is difficult to perform, and in any event, falls under the conditions covered by my patent claims. Nevertheless, in this respect the competition must be watched. It may be needed to thwart them by supplementary patents.[50]

The statement that constant volume combustion (the Otto cycle) fell under the claims of patent 67,207 is difficult to understand, particularly when Diesel had repudiated this process in the patent.[51]

The solutions Diesel proposed in the above *How the Engine Works Best* were correct, but the quandary for him was how to publicly announce that the process he had patented and licensed was essentially worthless and that its replacement had no protection. Some of the clever arguments he used as camouflage only postponed attacks on these questionable comments. Although now was the time for him to be completely forthright, it is unfair to cast stones at him for the action taken. Pride and desperation often lead to irrational actions.

None of the results of these calculations, nor those made in the following weeks, could be incorporated in the test engine because by then it was ready for assembly.

Committing to paper that water cooling was essential required another soul wrenching admission. Diesel did this upon completion of yet a new series of calculations carrying a September 1893 date. (The first tests had ended in August.) The heat release was simply too great for the fuel now used. In a section entitled "My Opinion Concerning the Events in the Cylinder and the Effect of the Cylinder Walls," Diesel made these surprising statements, which he sent to Krupp on October 26:

> . . . that the cylinder walls and the cooling have no real effect on the process. . . . that for my engine artificial cooling of the cylinder walls will have no disadvantages, as cooling only removes heat, which must be removed anyway. My exhaust will just escape cooler.[52]

Krupp properly rejected Diesel's poor scientific reasoning and suggested any discussions on this subject should wait until after further Augsburg tests.

Neither Krupp nor M.A. became excited over the radical switch to constant pressure. They only wanted a workable machine. Nevertheless, another of Diesel's tenets was discarded, and his engine came down at last from Mount Olympus to join its mortal brethren.

In a final addendum to his original manuscript dated November 1893, one may read how far he had progressed in a year's time, and most important, how near his ideas were to future engines. Diesel drew a graph of specific output and volume vs. overall efficiency and commented:

> The points where output/volume and the economic efficiency are a maximum [lay] . . . between 30 and 40 atm compression, i.e., between 500 and 600°C, thus lower than we previously assumed. . . . The curves also show that higher compressions as mentioned are not advantageous because the gain in thermal efficiency is compensated by the loss in mechanical efficiency. . . . Therefore, it is recommended to run further tests in Augsburg at 30–35 atm compression and not to increase it before more detailed development work on the engine.[53]

The combustion process and finally the operating parameters were now feasible.

German Patent 82,168

Diesel prepared an application entitled *Innovations to Internal Combustion Engines* hoping that a patent covering the constant pressure process would give his licensees the legal safeguard needed to build a practical engine. That he waited until November 29, 1893, to file may be attributed to the time-consuming testing difficulties encountered at Augsburg (chapter 5). Another probable factor in the filing delay involved Diesel's determination to reconcile once and for all the arguments raised by Köhler that Diesel knew were valid.

Diesel was told by the German Patent Office that the constant pressure combustion process could not be directly patented because of his prior patent 67,207.[54] Its main claim (Chapter 3) had combined all the elements of the process, i.e., compression ignition, etc. with the phrase "combustion taking place without essential increase of temperature and pressure." The patent office thus insisted that the claims of the new application must tie to No. 67,207.

Diesel then restated his objective to gain part of what he needed, but the result fell far short of original hopes. The patent, which finally issued on July 12, 1895, showed this emphasis change even in the title: *Combustion engine with fuel injection that is variable in duration and pressure.*[55] Although the tight grip on the Carnot process was loosened, the patent really protected a specific method of fuel injection and left room for others to maneuver if they were so inclined.

Since the patent office requirement called for basing the new patent on No. 67,207, Diesel used the phrase ". . . without essential increase of temperature and pressure" to combine the combustion process of the first patent with what he wanted for the second. He claimed at idle the engine followed constant temperature and under load it switched to constant pressure.

To accomplish this, a fuel control was designed whereby the timing and duration of injection could be varied. Fig. 4.7 shows the expected changes to the indicator diagram as the engine came under increased loads.[56]

A third variable, increasing injection pressure while holding constant the fuel volume injected, reduced the injection time. Diesel mistakenly believed that a faster injection would of itself significantly increase power, so a means to effect this rise of injection pressure was also incorporated in the basic claim:

An internal combustion engine of the kind as specified in Patent No. 67,207, by which the modification of the output is done by modifying the shape of the combustion curve; a simple or compound fuel jet is injected into the compression space of the engine at variable overpressure and variable duration of the fuel injection.[57]

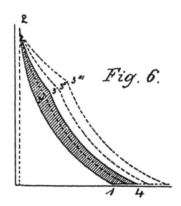

Fig. 4.7. P-V diagram shown in his German patent 82,168 that disclosed his idea of constant pressure combustion.

Neither of the patent's two claims refers to constant pressure combustion by name. (Claim 2 covered a means of using compressed air from the power cylinder as a source of air for injection purposes.) Diesel was better off than before but still did not free himself from the entangling web spun by the spider of patent 67,207.

Between the time of the manuscript's completion and the first engine tests at Augsburg, Diesel's initial concepts had been tempered by more than imposed facts of real-world requirements. Through the stubborn genius of his own analytical mind, he used logic and fundamental principles to transform his visionary and oftentimes impractical ideas into what was needed for his engine to become more than a dream.

Diesel's unique concepts, and his true invention, may be visualized by a simple graph of compression pressure plotted against year of invention (fig. 4.8). Before 1876, the date of Nicolaus Otto's first four-stroke cycle engine with a compression pressure of 2.5 atm, internal combustion engines had atmospheric (Lenoir), or less than atmospheric, pressure in the cylinder at the time of ignition (Otto & Langen). Then Otto created a discontinuity at that time—*invention*. From 1876 to the time of Diesel, the compression pressure curved only slightly upward to a fuel-limited maximum for production engines of about 4 atm. In the mid-1890s, Diesel created a second discontinuity in the compression pressure line. He moved the vertical line up to 34 atm, more than ten times the leap of Otto's engine. That was *INVENTION*. There have been no such vertical shifts in this line since Diesel's engine was born. After that the curve has taken only a gentle upward slope due to evolutionary technological advances.

The engine Diesel conceived is the only one not fuel-limited to a maximum compression ratio regardless of combustion process (constant temperature, pressure, or volume). It makes no difference that he began with a 12 or 13:1 CR—a vast compromise from what he first envisioned. He broke

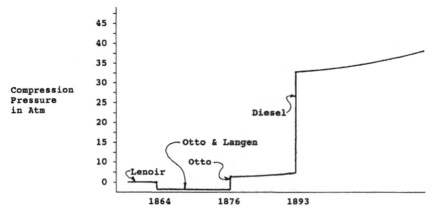

Fig. 4.8. Discontinuities caused by inventions in a curve of compression pressure advances made in the I-C engine.

through the "compression ratio barrier," something that no one else had done with a throttled-intake gas or gasoline engine or a vaporizing oil engine.

Notes

1. Rudolf Diesel, *Die Entstehung des Dieselmotors* (Berlin: Springer Verlag, 1913), 151.
2. Friedrich Sass, *Geschichte des deutschen Verbrennungsmotorbaues von 1860 bis 1918* (Berlin: Springer Verlag, 1962), 425.
3. "Lucian Vogel und seine Mitarbeit am Dieselmotor" ("Lucian Vogel and his collaboration on the Diesel engine"), MAN Archives (Augsburg: unpublished manuscript, 1940), 1.
4. Kurt Schnauffer, *The Invention of the Diesel Engine: The Triumph of a Theory*, MAN Pub. No. SA364396E, 1958, 10.
5. Sass, *Geschichte*, 426.
6. M.A.N., *Fünfzig Jahre Dieselmotor 1897–1947* (Augsburg: 1948), 15, copy of original letter.
7. Sass, *Geschichte*, 427.
8. Vogel's daughter, Frau Dora Meussdorfer, remembered that her father was enthusiastic about Diesel's idea and had promoted it to her grandfather both before Buz's April 20 letter to Diesel and after Vogel had studied a manuscript sent directly to him in July 1892. "Lucian Vogel," 1. Schnauffer, "Diesels erste Verträge," part 1, in MAN Archives (1954), 14. Schnauffer indicates Vogel had not gone over it. Vogel wrote Diesel on June 14, 1892, that he had been in Italy recuperating from an illness and was "till now not yet able to thoroughly study the treatise on the new motor."
9. Sass, *Geschichte*, 433 et seq.
10. Kurt Schnauffer, "Der erste Diesel'sche Versuchsmotor 1892–1893," in MAN Archives (1954), 5. Vogel's suggestion was indicated in a Buz letter to Diesel on July 2, 1892.
11. Schnauffer, "Versuchsmotor," 2.
12. Ibid., 4. The antechamber is also shown in figure 8 of Diesel's basic patent No. 67,207. Its description follows very closely the quoted passage, although "turbulence" is used rather than "swirl." This passage may have been incorporated in the patent application after Diesel sent the updated antechamber design to M.A. in July 1892.
13. Ibid., 6. See also Diesel, *Diesel*, 432.
14. In the 1920s, Cummins Engine Co. added what it called a "sneezer" to keep injector nozzle holes clean. Inserted in the center of the piston, this small air cell blasted a jet of air against the "cup" holes after every injection. Some diesel experts offered interesting theories about how this aided combustion, but its aim was simply to act as a "cup wiper." It was used until the mid-1940s.
15. According to Schnauffer, "Versuchsmotor," 12, this was the first use of balance weights in an I-C engine except for the 1886 Benz automobile engine. See Sass, *Geschichte*, 117, for a picture of the Benz crank.
16. With this great reduction in plunger diameter and stroke, the new charge volume was only 100 mm³. Thus, considering the low mechanical efficiency of the engine, the output would have been about that needed to idle the engine. This also assumes all of the pumped fuel was injected. (Based on 25 bhp at 300 rpm, 75 percent mechanical efficiency and a conservative 300 g/hp/hr.)
17. This was the forerunner of metering by "pressure-time" in a compression ignition engine.
18. R. Diesel, *Die Entstehung*, 156, note 6.
19. Ibid., 8.
20. A cross-check in fuels terminology for this period is in Hugo Güldner, *Das Entwerfen und Berechnen der Verbrennungsmotoren* (Berlin: Julius Springer, 1903). An English translation of his work can be found: Herman Diederichs, *The Design and Construction of Internal Combustion Engines* (New York: Van Nostrand, 1910). See chapter 8 for more on this.
21. Moritz Schröter, "Ein neuer Wärmemotor," in *Bayerisches Industrie Gewerbeblatt*, MAN Archives (1893), 3.
22. Ibid.
23. R. Diesel, *Die Erfindung*, 23–24.
24. Ibid., 38–9. The original letters from Richard are too damaged from water soaking to be legible.
25. Ibid., 39. Diesel engines, using ceramic pistons and insulated, uncooled cylinder liners, in conjunction with turbocompounding (an exhaust-gas driven turbine whose output is geared

into the engine crankshaft), are described in technical literature beginning about 1975. See: *The Adiabatic Diesel Engine,* SAE Pub. No. SP-543, 1983.

26. A January 10, 1893, letter from Venator to Diesel fully outlines Oechelhäuser's concerns, some of which involved his relationship with Langen at Deutz.

    Little is known about Venator who had befriended Diesel when they worked together in Linde's Berlin office. He became interested in Diesel's ideas and was helpful in keeping Diesel informed about new engine developments while Diesel was writing his manuscript in 1891. (Eugen Diesel, *Diesel,* 188). Venator's first initial was *A* in Eugen's biography of his father. However, in the above letter the signature initial has been verified to be an *O.* Schnauffer, "Die Erfindung," 50, said Venator (no initial) and Otto Köhler "evidently knew each other when they were classmates at Aachen, whose times there overlapped. The author thanks Dr. F. F. Pischinger, Aachen, for obtaining school records. Also a special thanks to Klaus Luther, who succeeded Frau Denkinger at the MAN Archives, for the Venator letter.

27. R. Diesel, *Die Entstehung,* 125.

28. Schnauffer, "Die Erfindung," 40. Züblin was awarded an honorary doctorate by the Swiss Federal Technical Institute (ETH), Zürich. Other accomplishments include his advanced work with steam turbines. The author is indebted to Marcel Züblin who kindly supplied biographical information.

29. The twelve-page, handwritten report and cover letter to Züblin are in the Sulzer Archives. Mr. David T. Brown of Sulzer most helpfully provided a copy. Diesel always gave the injection time as a percentage of piston stroke and not degrees of crank angle. He likened the delivery rate as "rather similar to the fall of sand in an hourglass." He was referring to the injection of powdered coal, but the analogy held true for liquid fuel since it also was pressure-time delivered.

30. R. Diesel, *The Theory and Construction of a Rational Heat Motor* (London: Spon Press, 1894), 57.

31. The 4 hp of the contract differs considerably from the probable 20–25 bhp at 250–300 rpm final target Diesel aimed for. Why this output was used is not reported.

32. "Vertrag Zwischen der Firma Fried. Krupp in Essen ein einerseits und Herrn Rudolf Diesel, Ingenieur in Berlin anderseits," in Sulzer Archives (Essen: 1893), 1–5. It was signed by Diesel and Krupp directors Asthöwer and Klüpfel.

33. Ibid. Diesel had agreed in his contract with M.A. to ask for 37.5 percent payments in future license contracts, but this was not done with Krupp.

34. Sass, *Geschichte,* 428.

35. Ibid.

36. Wilhelm Berdrow, *Alfred Krupp und sein Geschlecht* (Alfred Krupp and His Family) (Berlin: Paul Schmidt, 1943), 287. A history of the Gruson company is on pp. 283–95. A readable but somewhat flawed history of Krupp (all three references to Diesel and his engine are inaccurate) is William Manchester's, *The Arms of Krupp* (New York: Bantam, 1970).

37. Hugo Güldner, *Das Entwerfen,* 401. The engines Gruson built at this time followed designs by Hermann Ebbs when he was at Buss, Sombart in Magdeburg. Ebbs came to Gruson early in 1893 and immediately designed a series of new gas engines. See also Sass, *Geschichte,* 288–94, for more on Ebbs's work and life.

38. Krupp-Diesel contract. Krupp also agreed to pay Diesel's travel expenses to Essen and Magdeburg.

39. E. Diesel, *Diesel,* 203

40. Kurt Schnauffer, "Diesels erste Verträge," part I, in MAN Archives (1954), 18–26.

41. A number of comments were printed shortly after the engine tests began and after Diesel had realized the changes required in operating theory given in his book. One of note was by Professor J. Lüders at the University of Aachen whose critique was published in the August 15, 1893, issue of *Glaser's Annalen.* Diesel saw it before publication. While Lüders was incorrect in several conclusions and showed a surprisingly poor knowledge of engine design, he did appear to understand what Diesel had written. A comment from the article worth citing was that the fuel burned would be about the same as the lubricating oil probably also burned in the cylinder. Even then, he still did not directly question the high air-to-fuel ratio Diesel proposed. See Schnauffer, "Die Erfindung," 58.

42. Schnauffer, "Die Erfindung," 51. The lecture's title was "The Rational Heat Engine Compared to Other Engines." The Verein Deutscher Ingenieure, Germany's major engineering society, serves all disciplines.

43. The statement is not quite correct in that all heat engines are variations on the Carnot cycle, but this interpretation was not used until later.

44. Köhler was undoubtedly thinking of the Hornsby oil engine in England and the Capitaine-Grob in Germany. See Cummins's *Internal Fire* (Austin, TX: Octane Press, 2021), 318–23, for a description of the German engine. Dugald Clerk wrote on the ignition process of the Hornsby-Akroyd:

> In this type of engine the walls of the combustion space [i.e., the vaporizing chamber] are allowed to attain a temperature of nearly 800°C, sufficient to cause effective vaporisation and also to allow of ignition when compression is completed.... The air, therefore, is not heated up by passing through the vaporizing chamber; the exhaust gases of the previous explosion, however, are kept at a very high temperature in the ... [vaporizing chamber], which chamber is cut off to a certain extent from the main cylinder by the bottle neck.

Dugald Clerk, *The Gas and Oil Engine*, 6th ed. (London: Griffin, 1896), 468. The Hornsby was one of the earliest "hot bulb" engines.

45. Schnauffer, "Die Erfindung," 51–2. Otto Köhler, *Theorie der Gasmotoren* (Leipzig, Germany: Baumgärtner, 1887).
46. R. Diesel, *Die Entstehung*, 155.
47. Schnauffer, "Die Erfindung," 42.
48. Ibid., 43.
49. Ibid., 44.
50. Ibid., 46.
51. For those readers who either missed the "pleasure" of studying the constant pressure and volume combustion cycles, or did so too long ago to recall a picture in their minds, the following may be helpful. In understanding either Diesel's engine or diesel engines one should recognize a basic relationship between the theoretical thermal efficiencies of these cycles:

   - For any given heat input and given expansion ratio, the thermal efficiency is higher for the Otto (constant volume) cycle than the Diesel (constant pressure) cycle.
   - For equal heat input and any given peak pressure, the thermal efficiency is highest for the Diesel cycle.
   - For equal compression ratio and heat input, the thermal efficiency is highest for the Otto cycle.

   The diesel engine is more efficient than the spark ignition (Otto) engine only if it retains the advantage of a significantly higher compression ratio.
   The diesel has further advantages because (1) an engine usually operates at much less than full load, where for a spark ignition (SI) engine the effective compression ratio is less than at full load; (2) the SI "butterfly" (throttle) valve in the intake converts the engine into a stronger vacuum pump as the load decreases and the throttle closes—the worst condition (highest vacuum) for an SI engine, except when decelerating, is at an idle speed; and (3) as the specific output (kw/liter or hp/in^3) of diesel engines has greatly increased with time, the combustion process in turn more closely follows that of the constant volume cycle.

   See Edward F. Obert, *Internal Combustion Engines*, 3rd ed. (Scranton, PA: International Textbook Co., 1970), 172–74. The author, both as a student and teacher, is indebted to Professor Obert, who died in 1993, for his always lucid texts.

52. Schnauffer, "Die Erfindung," 61.
53. Ibid., 62a.
54. Ibid., 67. The application as filed is missing, and what it contained can only be surmised from correspondence between Diesel and the patent office, which dealt with objections and amendments.
55. This patent, No. 82,168, and a later German patent, No. 86,633, issued March 30, 1895, were combined to become British patent No. 4,243 of February 27, 1895, and US Patent No. 608,845 of August 9, 1898. The latter had a filing date of July 15, 1895. Diesel petitioned the US Patent Office on June 15, 1900, to withdraw the patent to add four new claims covering means for starting on compressed air as a two-stroke, then converting to four-stroke for power operation. It subsequently became "Reissued Letters" Patent No. 11,900 on April 2, 1901. See File Wrappers, Pat. 608,845 and Reissued Pat. 11,900, US Patent Office, Washington, DC.

56. Of interest are Diesel's ideas on how varying the beginning of injection would affect the combustion and p-v diagram. He disclosed this first in German patent No. 86,946, filed November 30, 1893. His US pat. No. 608,845, filed July 16, 1895, and issued August 9, 1898, covers the same idea (fig. 4.9). Claim 7 (US) protects opening the fuel valve before the end of the compression stroke.

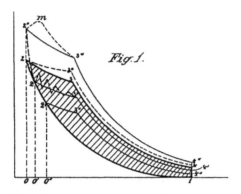

Fig. 4.9.

57. German Patent No. 82,168, Claim 1. It is equivalent to Claim 1 of British patent No. 4,243 of 1895.

CHAPTER 5

Trial and Tribulation: Augsburg, 1893–1894

"Tomorrow morning, Sunday, I set out for Augsburg to the most important and crucial moment of my life. Cross your fingers and pray for me!"
—Rudolf Diesel[1]

Rudolf Diesel penned these words to his beloved Martha on a summer's afternoon in Berlin's Tiergarten park. This peaceful respite while his family was on holiday freed his mind to contemplate what might lay ahead.

Maschinenfabrik Augsburg had written to Diesel on July 12, 1893, that the test engine was ready for assembly and that his presence would be expected when scheduled trials began on the seventeenth.

Almost a year had passed since work commenced on the design of a demonstration engine that would allow M.A. and Krupp to evaluate Diesel's concept. Some development problems were obviously anticipated, but it was naively assumed by all that these ought to prove of minor consequence. No one foresaw the seemingly unsurmountable and frustrating difficulties yet to come nor the time required to solve or circumvent them.

That Diesel and his coworkers finally developed a practical engine that surpassed all others in thermal efficiency is a tribute to their creative tenacity.

Diesel's engine demanded a technology not known, nor required, for contemporary heat machines. His challenges:

- Develop a high-pressure injection system to reliably meter, inject, and govern a tiny, timed fuel charge.
- Design a combustion chamber to deliver maximum fuel efficiency and power.

- Seal a combustion chamber under compression pressures fifteen times higher than usual practices.
- Minimize cylinder friction with such pressures.
- Make simple, leakproof connections for high-pressure fuel lines.

Diesel faced them all at once. The first one became his personal Pandora's box to be opened the day his engine breathed its first sign of life and narrowly missed killing its sire.

Common threads weaving throughout the early development years—and continuing to the present—were those of understanding the fuel injection and combustion processes and of converting that comprehension into practice. Misinterpretation of results led to time-consuming wrong turns and blind alleys. Diesel's dream of a simple fuel system became a complicated nightmare. Thirty years would pass before what he initially attempted, and then abandoned, was commercially produced. High pressure, hydraulic fuel injection systems had to wait until material and machining technology were perfected.

The Augsburg tests chronicle a partnership of the practical factory man with the textbook engineer. Diesel's engine came from the synergy of this combination. Eventual success was finally assured because there was also that vital "someone" willing to gamble corporate resources and reputation in what too often seemed almost a lost cause. The engine's evolution and the test program supporting it were remarkably well documented. Diesel kept meticulous records, drawings, sketches, and calculations. He himself would draw on them in 1913 when he gave an account of the entire development.[2] Original material and private papers that he did not donate to the Deutsches Museum in Munich were subsequently purchased from the Diesel family by M.A.N. Fortunately, almost all have survived time and the ravages of war to permit studies of those turbulent years.[3]

Diesel described that period between the engine's first firing in July 1893 and the acceptance trials in 1897 as comprising six distinct test series. The author has chosen to follow tradition and continue this logical grouping of experimental work. The first four test series occurred during 1893 and 1894. When retracing the step-by-plodding-step accounts found in Maschinenfabrik Augsburg test logs and Diesel's private journal, one must feel a strong empathy toward that small development team as it struggled through adversity or savored its brief triumphs.

The First Test Series—July to August 1893

Diesel suffered from a terrible cold when he registered at Augsburg's palatial Hotel Drei Mohren on July 17.[4] Nevertheless, he drug himself out to the factory. In his pocket was a letter from Martha that had awaited him at the hotel. She gave her support and shared her apprehension over the coming tests:

My heart trembles in my body when I think of your work and the arduous, difficult times of expectations. . . . There will probably be many things that won't work out at first, I mean, things not yet so ready to be. That I am eager and longing for news from you, beloved, you know, but I remain attentively patient till you have time and peace for your wife.[5]

He found the engine ready except for installing the compressor to supply starting air (fig. 5.1). A Linde refrigeration compressor was being adapted because of insufficient shop air capacity. This would not delay the start of testing since standard practice called for a lengthy breaking in of an engine before applying power. Tight-fitting bearings of that day often needed hours of such running to smooth off high spots. Diesel made compression tests during this time to check piston and valve sealing. A leather belt driven from factory line shafting and wrapped around the flywheel turned the engine.[6]

The engine was motored at 200 rpm on the seventeenth as scheduled. But even with the dual-purpose intake/exhaust valve held open, the piston quickly began to seize on the upper end where it rubbed against the cylindrical wall portion integrally cast with the head; steel does not like to run against steel. It was an ominous beginning.

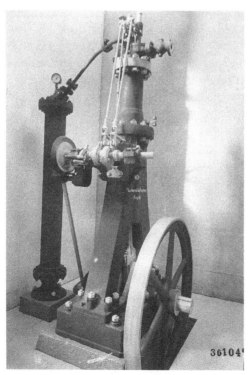

Fig. 5.1. Diesel's first engine at start of 1893 tests. *MAN Archives*

Diesel used the time waiting for repairs to allay his fears over achieving compression ignition at the new, lower pressure. First, he repeated tests M.A. had made with the hot platinum wire, using different fuels and mixtures. Both kerosene and gasoline were sprayed onto the glowing wire in open air. His test journal reports that ignition occurred if the wire was not directly in the path of the liquid jet because the spray "frequently extinguishes the glowing wire." Also, ". . . the glowing igniter piece should cause only ignition and not [merely] the steaming and heating up of the fuel."[7] A 90:10 mixture of kerosene and gasoline was "almost equal to gasoline in ignition capability."[8]

A second motoring test on July 21 also brought failure after an equally short time. Diesel realized that the piston must avoid a rubbing contact with a cylinder wall of the same material. Consequently, the piston diameter was reduced slightly where it entered the cylindrical head extension, and five non-tensioned rings in a much-shortened piston replaced the oil-fed, bronze diaphragm packing of the "plunger" design. Serving for the removed section of the piston was a rod connecting the new piston to a crosshead running in the larger diameter bore previously holding the stepped piston section. The cylindrical crosshead design remained unchanged through later engine modifications.

Krupp and Sulzer, who were consulted about the piston problem, preferred keeping the original plunger design and believed the scoring could be overcome. Diesel had already decided it must go.

The next trouble spot was air leaking past the double-headed intake/exhaust valve. It had caused headaches even on the static pressure tests, and after two weeks of trying to make it seal on a motored engine, both the valve and cage insert were further modified. Solving this problem—with the aid of a lot of grease—only pointed up another one.

Compression tests at 120 rpm gave a maximum pressure of just 18 atm due to a "mainly leaking piston." Under the urgency to produce some tangible result, Diesel simply lengthened the piston rod 6 mm with shims. The pressure only rose to 21–22 atm. In desperation he added 5 mm more to the rod and reduced the depth of the recessed insert in the piston. This lifted the compression pressure to 33 atm, but the calculated pressure should have been between 55 and 60 atm! It had taken twenty days to reach this point.

The moment of truth was at hand. With a bare minimum pressure at which he hoped a charge might ignite, Diesel opened the fuel valve for the first time on August 10, 1893.

He did not use lamp oil as the initial test fuel. The honor instead went to gasoline (*benzin*). This highly volatile liquid rewarded Diesel by igniting with such violent force that the pressure indicating mechanism blew off the cylinder and passed between himself and Vogel, narrowly missing them both. Injecting an excessive charge of low octane gasoline into a cylinder with a 13:1 compression ratio assured a robust birth announcement. The explosive cylinder pressure, which heralded proof of Rudolf's idea, was estimated to be over 80 atm, but fortunately, Vogel's "vertical cannon" could take such abuse. Diesel later wrote that after recovering from "the first fright, we then realized our great friend had proved automatic combustion as part of the method."[9]

In the next few days Diesel cautiously repeated the injection tests. These

> . . . again produced more or less heavy explosions without positive work and were intermittent with numerous misfires. . . . Black, sooty clouds came from the exhaust pipe with all of these tests. The [P-V] diagrams show the first demonstratable diagram areas. They were insignificant owing to still no positive work. Gradually [they] became larger and smoother, with [one] showing 2.15 ihp and a combustion line that was near isothermal. But this gave no running of the machine [by itself]. . . . After a brief time the entire engine became sooty, with it blowing from the valves and piston.[10]

Diesel soon realized that the violent, intermittent combustion was due in great part to air in the pipe between the fuel valve at the pump discharge and the open nozzle. The cam-operated needle valve to time injection was far from the nozzle. This allowed the long line between the needle valve and nozzle to partially fill with air from the power cylinder because there was no check valve at the nozzle preventing it. An open, Körting-type burner nozzle was used. After the needle valve closed, the air trapped in the line simply acted as a pneumatic ram to force out whatever fuel it held. There was no control over the fuel that might chance to enter the cylinder.

None of the engine power tests ever used fuel delivered by a metering pump. Already believing its capacity too small, Diesel had bypassed the pump plunger and added an air-over-fuel accumulator to keep fuel under pressure at the needle valve. The only reason that any fuel ever entered the combustion chamber was due to the large volume the accumulator could push through the line. The engine probably would not have fired at all with a timed charge from a positive displacement pump regardless of its size.

Diesel noted after the first explosion that the charge began burning when about a 20 atm compression pressure was reached. Since fuel was cam-timed to inject after top dead center, it must have ignited before top center because of an uncontrolled injection. The indicator diagram showed Diesel's first explosion may have followed Otto's constant volume combustion (fig. 5.2)! Subsequent diagrams gave evidence of a small indicated horsepower. This occurred when enough air had been purged from the fuel supply line during the continuous belting to give a semblance of a timed and metered fuel charge.

A critique of this situation in Diesel's journal leaves little doubt that he began to recognize some of the basic requirements of a fuel injection system for a compression ignition engine:

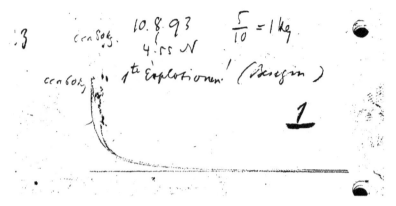

Fig. 5.2. Indicator diagram of first engine firing, August 10, 1893.
MAN Archives

The fuel needle must be directly situated at the entrance of the fuel into the cylinder since the open nozzle is an immense danger. It must have a *direct drive* injecting control, and there must be absolute safety in closing the needle following injection.

All air pockets in the fuel lines are to be avoided.[11]

Little more could be accomplished with the engine as it was except for a brief tryout of the starting valve mechanism. At almost 300 rpm the cam-operated starting valve would admit compressed air and could then be deactivated to allow the engine to run in a power mode. This first test series was concluded after thirty-eight days.

Most of the features necessary to demonstrate Diesel's concept proved woefully deficient in their design. In addition to its continual poor sealing, the common intake/exhaust valve also allowed a large volume of residual gases in the exhaust pipe to be sucked back into the cylinder. Separate valves in a new cylinder head were required. The open, check valve–less spray nozzle proved a total failure, and a completely new piston and ring construction was necessary to stop the compression leaks.

Finally, there was a matter of the uncooled cylinder and head. Although the little running with an occasional firing had caused no overheating, water cooling was seen as inevitable. Diesel had realized this when he changed his thinking to constant pressure combustion in June 1893. By then, however, it was too late for him to scrap the design that ostensibly followed his patented isothermal combustion, that is, the too-small fuel metering pump (that was never connected) and the uncooled cylinder. Nevertheless, as deficient as the engine proved to be, much was learned from it.

A report to Buz dated August 23, 1983, summed up the above, and Diesel, ever the optimist (and desperate promoter), prefaced it by claiming that "the feasibility of the process itself, even in this imperfect machine, can be observed as proven."[12] Although Diesel may have stretched a point in that the engine never ran, Buz was convinced enough to tell him to go ahead with the asked-for redesign. One may also assume that the enthusiastic Vogel also encouraged his father-in-law in this decision.

A chastened inventor survived his baptism of fire and returned home to Berlin. He wasted no time in beginning on the pressing redesign. As a first step he hired an assistant, Johannes Nadrowski, to help with the drawings. (Nadrowski later went to Augsburg and joined the engine design staff.) Rudolf's son, Eugen, recalled life in the Diesel home during this period:

> The middle-class spirit of work, study, and progress prevailed in our home. To cut expenses, there was also an atmosphere of organization as in an office. All the determination showed.
>
> In one room the engineer Nadrowski drew on a slanting board of fine wood strips. We children knew that the engine in Augsburg was associated with these numbers, pencil lines, drawing pens, and the gray tracing paper. The engine wreaked havoc like a demon in the house and exerted a forceful presence on us. Lucie von Motz had christened it father Diesel's "black mistress."[13]

Five months went by before the test engine turned another revolution. Rudolf spent the first two refining his original ideas and redesigning the piston, cylinder head, and fuel supply system. The rugged lower portion of the engine based on M.A.'s successful steam design had given few problems. Because the cylinder and head assembly were only bolted to an open crankcase A-frame, the parts needing the rework could be altered without affecting the entire engine (fig. 5.3).

A new, cast steel head housed an intake and an exhaust valve. Within the head of the exhaust valve itself was a small pilot valve opening first. This design resulted from a concern about excessive valve train loading. The engine still had no speed governor.

Although both Krupp and Sulzer insisted that a plunger piston would work, Diesel opted for a built-up piston having three grooves with two, two-piece (half-circle) rings backed up by a single stiff expander ring in each groove. He spoke of the expanders as tension springs. Several refinements to this were tried when the engine was belted over again, but none worked as expected. The backup springs generated too much rubbing friction. Furthermore, air cavities behind the rings, in combination with rough surfaces on the ring and groove end faces caused excessive compression leaks.

Cylinder lubrication was provided by oil thrown from grooves machined on an extension piece added to the bottom of the piston. The grooves refilled at bottom center when they dipped into a ring-shaped oil reservoir fed from a hole in the cylinder wall. All designs until 1897 used this method.

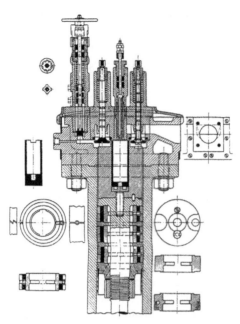

Fig. 5.3. Section view of 1894 engine, with piston details, after first rebuild.
Diesel, *Die Entstehung, 1913*

One of Diesel's tasks involved yet another look at fuel quantities. He
noted,

> Common oil engines need per hp per hr about 600 gm = 750 cc oil;
> therefore, for 10 hp 7,500 cc/hr. We would thus have to expect as a
> maximum with our motor the same quantity, and in 150 × 60 = 9,000
> injections/hr. The maximum per injection is 7,500/9,000 = 0.83 cc,
> and the minimum about half = 0.4 cc and below.[14]

He finally considered a realistic consumption. In order to pump such amounts,
the metering piston diameter was increased to 12 mm from the tiny 4 mm
of this unused pump plunger on the first engine. Volume per stroke went up
nine times and to where all the air might be utilized (fig. 5.4).

That same month Diesel outlined possible ways to inject a fuel charge.
Two had been on the engine during the first tests: the unused, too-small
positive displacement pump and the air-over-fuel accumulator. A third
method was to employ a blast of air both to atomize and to inject a measured
charge. This new proposal had great significance as it became Diesel's salva-
tion. Furthermore, the date of its disclosure in September 1893 lends further
credence to his being the inventor of so-called air injection.[15]

Diesel did not yet consider using an auxiliary compressor. Instead, he would
simply borrow a little combustion chamber air to do the job. Compression

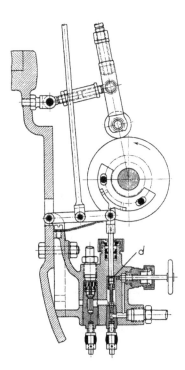

Fig. 5.4. Fuel metering pump and cam drive for nozzle valve, second design. Diesel, *Die Entstehung, 1913*

pressure was to open a valve in the cylinder head near top center to pressurize a small receiver in the head. When cylinder pressure dropped as the piston moved out, the air on the receiver would blast the fuel charge already waiting at the nozzle valve into the cylinder and atomize it as it was injected. Diesel's German Patent No. 82,168 of November 30, 1893, claimed the method as a way to achieve his constant pressure combustion process.[16]

A pre-1894 sketch had already disclosed a cam-lifted, fuel nozzle needle valve seating in and closing a single orifice opening into the combustion chamber. Diesel had begun to realize the absolute necessity to prevent combustion chamber gases from entering past the nozzle tip into the fuel passages. However, the fuel accumulator at the camshaft level remained too far from the nozzle.

Water cooling was added. Diesel had proved its necessity in his series of September calculations and so informed Krupp (chapter 4). This additional concession to cherished theory took the form of sheet metal jackets over the head and cylinder to form the water passages. Diesel specified them in his November 1 reply to M.A. when he approved the drawings they had sent him on October 17. One of his accompanying comments read that "Not only the head but also the cylinder jacket has to be cooled. For test purposes they must be cooled separately."[17] Drawings and sketches show the added covers, but for an unexplained reason Diesel failed to mention them in his test journal until well after adopting cast-in jackets. He then referred to the earlier sheet-metal ones causing "very long delays at every assembly and disassembly."[18] This pain was endured for a year.

Krupp's Gruson subsidiary in Magdeburg received another visit from Diesel in October. His growing relationship with Hermann Ebbs, able chief engineer for this gas and oil engine manufacturer, offered Diesel a chance to work with someone having a solid internal combustion engine background. Even though the two men occasionally differed in their solution to a problem, Diesel profited from his association with Ebbs.

Rudolf passed an especially enjoyable interlude with his family in Berlin while waiting for M.A. to make the new engine parts. Eugen later recalled that his father spent precious hours making doll furniture for Hedy and rabbit hutches for him and his brother, Rudolf Jr. The relaxing inventor even created a large Chinese "shadow theater" and the silhouettes to perform in it.[19] Eugen remembers this as one of the few times his father seemed able to forget the engine and act in a carefree manner.

The Second Test Series—January to March 1894

Diesel returned to Augsburg in time for testing on January 18. He did so with the belief that his engine's feasibility was a proven fact and that he must plan ways to license it in other countries.

The optimistic inventor had been invited by his sister Emma to live in the Barnickel home rather than a hotel. In addition to its warm family environment, his brother-in-law's residence built against the old city wall above the *unterer graben* (lower moat) was a convenient several-block walk from the M.A. factory (fig. 5.5).

Fig. 5.5. Barnickel home on Augsburg city wall where Diesel lived during early tests. *Author photo*

Rudolf's long separations from his family were not easy. His love for Martha, interwoven with his consuming obsession with an assemblage of inanimate ferrous parts, came to form the warp and weft of his life. He exposed this cerebral fabric in a letter written to her upon his arrival in Augsburg:

> I have much love, a truly fire-spitting mountain glows in me, but which only comes to the eruption when my motor succeeds and I can report to you: The moment has arrived.[20]

His moment was still to be denied him.

The test program began in its usual cautious fashion. All new castings were pressure tested with water before assembly, and their unexpectedly high porosity was sealed by a "water glass" (sodium silicate) treatment. Belting over of the engine began on January 25, but not for long; a heavy thumping noise proved to be the piston hitting the cylinder head.

While waiting for repairs to the piston and insert, Diesel made fuel atomization tests. With the cylinder and assembled head on an open pedestal, he could compare spray formations from the newest nozzle designs as they injected into room atmosphere. He used the larger capacity, positive displacement fuel pump as the pressure source. The first nozzle had an outward opening, conical seat check valve, which was lifted off its seat by fuel pressure. It showed good atomization but failed to properly seal.

A cam-lifted needle valve seating in the nozzle tip was tried next (fig. 5.6). The inwardly opening valve was a major step forward, and the principle is still commonly employed. While Diesel lifted his needle with a cam, the hydraulic injection systems introduced in the 1920s used pressurized fuel from an injection pump to raise the needle. Diesel wrote in his journal on January 30, 1894, that injection into air was "very precise" and had no "after dripping." One of his comments about the design was that the valve return spring must be strong enough to maintain needle seating under all conditions. Finally, he specified that needle lift had to be small—on the order of 1–2 mm.

Unfortunately, Diesel could not accurately meter a small fuel charge with a stroke-controlled plunger pump. He despairingly wrote,

> The design of a pump for so small an amount in such a short time by very high pressure bids almost insurmountable difficulties. It seems impossible with these small quantities to regulate a correct sucking and pressurizing pump.[21]

A return to the air-over-fuel accumulator solved this problem but created another and forced him further away from a simple injection system. His journal explains why:

Direct injection [with an accumulator] has the fault—really makes it impractical—that the injected quantity out the nozzle opening depends on the interval of time the nozzle is open so that for the same control setting more is injected at slow speed than at a rapid speed. The control would be correct for one rpm only.[22]

Rudolf then recalled placing a fuel charge in the nozzle and blowing it in with compressed air. During one of his reappraisal periods the previous year, he had suggested "borrowing" air from the cylinder at the end of the compression stroke to blast in the fuel.

If he used an auxiliary compressor, it would be possible to test blast air injection with the cylinder off the engine. The nozzle cavity above the needle valve seat remained in direct communication with a small air receiver charged by a Linde compressor. Fuel under accumulator pressure continuously trickled into the nozzle through a hand-regulated metering valve. The nozzle needle was cam-lifted to allow the air to blast in an atomized charge.

Air blast injection into the free-standing cylinder was tried on February 3. Diesel noted excellent atomization and that the nozzle jet was "like a steam cloud even with only 2 atm air over-pressure." He ran these bench tests using a maximum of 4 atm air and 10 atm fuel pressure. Maintaining a fuel pressure higher than that of the air was supposed to keep air bubbles out of the fuel line. However, as fuel entered the nozzle close to the blast air inlet port,

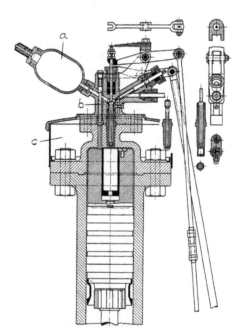

Fig. 5.6. Engine design with air-over-fuel accumulator for second test series, January–March 1894. Diesel, *Die Entstehung, 1913*

air probably found its way into the fuel line during some point in the cycle during actual operation.

Diesel did solve one plaguing nuisance problem at this time. The high-pressure fuel lines continually leaked past conventional packing materials and simple, threaded fittings. The latter could not be tightened enough, and the packings were soon softened by fuel until they also refused to seal. His solution was to forge-upset a conical taper on the pipe ends and to machine a similar conical seat in the mating pieces. A flat-ended clamping nut tight-ened against a parallel back surface of the upset pipe end to wedge-seat the mating conical surface. Rudolf's design is in common use today.

Compression tests ensued after the engine was reassembled and belted over on February 6. Only a 15 atm maximum pressure was attained because of air leaking past ring groove ends and joints in the multi-piece piston. The piston was again redesigned so that the ring carriers were fitted on a match-ing, tapered sleeve, which in turn slid over a rod integral with the piston head. The carriers then had their ring grooves ground after assembly onto the sleeve (fig. 5.7). The leaks were stopped at last.

Diesel made other tests while waiting for the new piston. The previous November he had sent M.A. a sketch of an "external oil carburetor." What they designed and made was a heavy vertical tank with the pressurized liquid fuel coming in at the bottom and a pipe to the injection nozzle exiting at the top. Connected to the cylindrical tank were the upper and lower ends of a vertical coil of copper pipe heated from below by a Bunsen burner (fig. 5.8). The purpose of the device was to inject pressurized fuel as a vapor rather than a liquid. However, applying heat to the coil failed to produce any vapor out

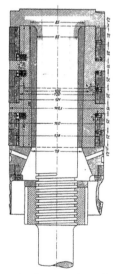

Fig. 5.7. Fourth piston design used on the second test series.
Diesel, *Die Entstehung, 1913*

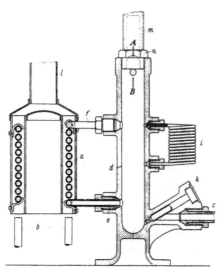

Fig. 5.8. External fuel oil vaporizer of early 1894: (a) pipe coils,
(b) heater, (c) fuel piping, (d) housing with liquid in lower part and
vapor in upper, (i) coiled pipe to determine liquid level by feel, (k)
external feed, (m), (n) safety valve and vent. Sass, *Geschichte, 1962*

the discharge line. Wrapping the apparatus in asbestos made no difference. This first unsuccessful vaporization experiment was temporarily set aside, but Diesel did not give up on the concept. He and others held the erroneous belief that in his type of engine, combustion did not begin until a high percentage of the fuel charge had first been vaporized.[23]

Direct injection of fuel from the accumulator began on February 15. When gasoline was injected at top center, there were the heavy combustion explosions as before, and the cylinder safety valve, set at 48 atm, repeatedly opened.

Diesel tried air blast injection for the first time two days later. The gasoline was under 40 atm pressure and the air receiver 34 atm. He found the combustion to be less violent using retarded injection timing. By beginning the injection at a little before 5 degrees after top center and ending at 16 degrees the engine ran quieter "without popping the safety valve." No indicator diagram was taken because "we ran out of paper."[24]

February 17 was a joyous day for Rudolf. He and Hans Linder (1858–1942), who had been assigned to the project by M.A., witnessed the first running of the engine. It idled on gasoline at 88 rpm under its own power for an entire minute before it had to be shut down because of a sticking exhaust valve.

Diesel's own words best describe the moment:

Since during the trials the engine was constantly turned over by belt, I myself did not notice this idling; but the mechanic Linder who helped at the metering valve from the wooden platform, noticed all at once that the belt was intermittently pulled by the motor instead of the motor being driven by the belt, and thereby recognized the first independent power manifestation of the machine. At this moment he silently raised his cap, and in that way I first came on the significance of the instant. In speechless joy I clasped his hand. There was no one else with us.

I then believed I was close to the goal and did not suspect that years of long, hard work still separated me from reaching it.[25]

Rudolf must have sensed that something momentous might soon happen because he had sent for Martha to come to Augsburg several days earlier. When he returned that afternoon to the Barnickel house on Springergäßchen he was "pale and trembling" and pulled Martha "at once into his room, took her in his arms and broke down in prolonged weeping."[26]

The following day Martha went with her husband to the test area. He had her move the lever clutching the upper drive belt pulley to the pulley that turned the belt over the flywheel. When the fuel valve was opened, the engine began to fire and run. She saw with her own eyes that Rudolf's dream was truly a reality.

Heinrich Buz, who was also in the test room, exclaimed with a shout, "The motor runs!"[27] Usually when he visited the room, the engine was either disassembled or not working. Buz's excitement on seeing the engine run by itself was understandable considering his growing investment!

Diesel worried about the explosive combustion following the injection of gasoline without the air blast. It was usually so violent that no indicator diagrams were taken. A misfire on the next cycle also followed because the high-pressure cylinder gases blew back the fuel into the supply line coming from the accumulator. Placing a check valve in the line made no difference. Diesel believed the knock phenomenon resulted from too large a fuel charge. However, the principal reason was detonation caused by uncontrolled ignition of gasoline in a high-compression engine. Injection with blast air quieted things down somewhat, probably because of a heat transfer effect from the air to the fuel droplets. Diesel's journal reads: "It appears then direct injection impossible . . ." The "dangerous tests" were stopped.

When Diesel duplicated the operating conditions of February 17, he could again idle the engine, one time for thirty-six minutes. The high indicated horsepower required just to idle proved that ring friction was far worse than expected. Diesel recorded 4.39 kg/cm^2 (4.3 bar or 62.4 psi) for the idling imep (indicated mean effective pressure).

Over one hundred diagrams were taken. Some showed relatively quiet combustion, but most had sharp, jagged spikes denoting rough and uncontrolled burning (fig. 5.9). Anyone who has heard the violent, sledgehammer-like pounding in a diesel engine when injecting an intermittent excessive fuel charge at low speeds can appreciate Diesel's concern!

33

Fig. 5.9. Indicator (P-V) diagrams showing violent and erratic combustion during the second test series. Diesel, *Die Entstehung, 1913*

The fuel metering valve leading into the nozzle was manually controlled during all these tests. It required someone standing on the wooden platform to reach under the several rocker levers to adjust the valve.

Until this point, the air for blast injection came from the Linde auxiliary compressor. Diesel next wanted to try his "self blast" idea as disclosed in Patent 82,168. The necessary parts, already incorporated in the cylinder head, consisted of a small, adjustable relief valve to let air from the cylinder enter the receiver and nozzle near the end of the compression stroke (fig. 5.6). The spring-loaded valve then closed, and as the piston started downward, a rapidly increasing pressure differential would be enough to blast in the fuel with air when the nozzle needle lifted. Diesel reported that the engine idled with a "nice blue smoke" out of the exhaust just as it did when using air from the Linde compressor. However, the maximum compression pressure dropped 2 to 3 atm, and combustion did not begin until after the cylinder pressure had dropped 10 to 12 atm from its peak. Adding more fuel than needed for idling only made the exhaust smoke blacken and the engine run slower.

Borrowing the compressed air from the cylinder caused another problem. The air temperature entering the receiver was hot enough to ignite the gasoline vapors in both the nozzle cavity and receiver. Diesel's assumption that the air would be sufficiently cooled through heat transfer to the water did not prove out. The resulting dangerous explosions interrupted the tests until the air receiver was wrapped in a wooden box as a safety precaution. Soot deposits were found even in the receiver.

More tests were made to compare the two blast air sources. The *kolossal* explosive cylinder combustion using the external compressor could not be softened. It was so bad that hot gases entered past the needle valve to raise havoc inside the nozzle.

Diesel hoped that better fuel atomization would improve combustion. One idea tried was an unsuccessful forerunner of the mechanical atomizers standard in most of the later air injection nozzles. It consisted of a small can screwed onto the nozzle and filled with a wad of fine iron wire. The end opposite the nozzle had drilled holes opening into the combustion chamber. The engine produced a little more power with Diesel's "basket" and had a cleaner exhaust, but the heavy knocking combustion remained. The wire in the atomizer soon melted because of the hot cylinder gases blowing back through the nozzle on a misfire that prevented injected fuel from acting as a coolant.

Oil, no doubt a kerosene, was burned in the engine for the first time on March 4. It started easily, using direct fuel injection from the accumulator and without the air blast assist. Diesel called attention to a noticeably quieter combustion but said that adding the air blast again caused the explosions in the nozzle and receiver.

Diesel's journal summed up that air injection by both self-blast and external pressure source were "useful" methods. It also outlined a frustrating state of affairs:

> A remedy must yet be found to dispense a constant fuel charge into the blast air stream; the test arrangement always gives too much [fuel] at the beginning and too little after the first explosion. The compressed, hot blast air must first be cooled in a coil until it comes in contact with the fuel in the nozzle.[28]

Unfortunately, this last comment was not remembered; repeated compressor explosions due to high air discharge temperatures would almost kill the entire program after production began.

A further journal comment hinted at thinking which was to lead Diesel on a wild goose chase. It offered an analysis of the events prior to ignition of the fuel:

> All methods of introducing fuel in liquid form have basic common drawbacks: Too much time needed for vaporization of the fuel, so that the diagram high point [peak pressure] cannot be attained, and the ignition first following vaporization follows much too late and with much too little pressure; the result is a much too low, lost combustion and insufficient development of the diagrams; all fuel methods yield much sooty combustion. Conclusion: All drawbacks will likely be avoided when one introduces the fuel in vapor form.

Years later he critiqued the above analysis with the almost painful statement:

> Under the dominion of these unhappy conclusions stand then the later following third test series, during which not much progress was made for a full 10 months. Because of starting from false assumptions, and because the correct basic idea of the process, the gradual direct introduction of finely dispersed fuels into the pressurized air, was abandoned.[29]

Another pertinent journal comment was temporarily forgotten in the coming months:

The ignition and combustion work without any auxiliary help and without interruption. Any type of assisting ignition device, even to start the engine totally cold, is superfluous. This is valid for gasoline as well as for kerosene.[30]

Diesel invited Dr. Gillhausen of Krupp to come on March 7 to witness the engine idle under both blast air methods. Afterward, Gillhausen pointed out that because the top ring was located far below the piston crown a significant air volume was trapped in the clearance space above the ring and was lost for combustion. He suggested placing additional rings spaced just below the crown. Gillhausen's point was proved when the compression pressure immediately rose from 33 to 44 atm. Even with a new, larger piston cavity the compression pressure could easily be held at 36 atm. All subsequent pistons had their rings positioned as high as possible.

This mostly disappointing tests series was halted on March 10 after Gillhausen's departure. Diesel then submitted a lengthy status report to Buz.

Not everyone viewed the recent events in the engine lab with concern. A March 17, 1894, article in Munich's *Münchener Allgemeine Zeitung* reported business news about Maschinenfabrik Augsburg. Its shares had recently jumped 30 percent due to a strong financial statement in spite of difficult times. The item ended with this prophetic comment: "Moreover, it is rumored of a valuable new invention in engine building for which all rights have been secured and which should promise great success."[31]

In the inventor's mind this success was just around the corner. With characteristic optimism Rudolf embarked on another foray to line up licensees, only this time he would seek them in France, where he retained influential friends, and in Belgium.

His first stop on April 15 was in Bar-le-Duc, a town about 200 km east of Paris. There he saw Frédéric Charles Dyckhoff (1855–1910), a French engineer with a factory and some engine experience. The timing was perfect as beautiful spring weather had also arrived. Rudolf and his lively host enjoyed several days walking in the countryside while making plans about the engine.[32] Dyckhoff agreed to build a 10-hp engine having a bore and stroke of 180 mm × 360 mm from plans that Diesel would furnish shortly. For a 25 percent royalty, his friend received sales rights to northeastern France.

Dyckhoff built an engine in 1894, which reportedly ran better than the one at Augsburg (fig. 5.10). Yet, in fairness to Diesel and M.A., the Dyckhoff engine incorporated the changes known to be needed after the second test series. If Diesel had not wasted most of that year chasing rainbows, he, too, would have been at least as far along as his French licensee.

Paris was the next point of call on Rudolf's itinerary where old friends made it a happy homecoming. They drank toasts to what had been a dream when he left them and now was to soon be a profitable reality. A few French companies were contacted, but nothing positive developed.

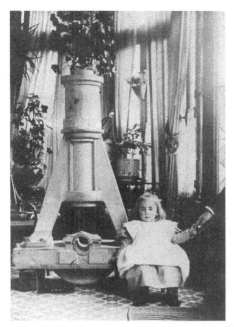

Fig. 5.10. A wooden model of Dyckhoff's engine (left) as exhibited in 1894 in the sunroom of his still-charming Bar-le-Duc home. Daughter Antoinette provides a measure of its size. The first Dyckhoff engine was built later that year in his nearby factory. *Marie-Rose Cochet Collection and MAN Archives.*

A nonexclusive contract was signed with the Carels Frères company in Ghent on April 30, 1894. If Diesel could supply them with drawings (the same as those going to Dyckhoff) by May 15, they would pay him 20,000 francs and build an engine by August 15. If Carels wanted rights to all of Belgium, they were to pay an additional 15 percent above the base 25 percent royalty. The minimum royalty per year would be 20,000 francs until a total of 180,000 francs had been paid. Diesel beat the drawing deadline, and Carels built an engine that never operated. Not until 1903 would Carels publicly demonstrate their own Diesel engine.

Diesel also visited the John Cockerill company near Liège. This respected builder of steam engines and heavy equipment was started by an Englishman in 1790 to make textile machines. Diesel signed no agreement with them.

A happy inventor returned to Augsburg with two foreign licenses in his pocket. He had enjoyed basking in the glow of a receptive atmosphere accorded him wherever he went.

The Third Test Series—June to October 1894

The next four and a half months were what Diesel remembered as "this most difficult of all test periods." The series mainly dealt with vaporizers and ignition aids, none of which markedly advanced the engine's development.

Diesel commenced with friction tests as a result of a June 22 conference in Essen with Krupp engineers. He found that 75 percent of the total friction came from the piston and rings, 18 percent from the camshaft and control gear, and the crankshaft bearings added 7 percent. The piston and ring friction was much higher than expected or should have been. These numbers were only relative as the data was obtained by throwing the drive belt off the flywheel and measuring the time for the engine to stop.

Although the external vaporizer so briefly tested in February had proved a failure, Diesel still could not give up the idea that gaseous, not atomized, fuel must be injected. He perhaps believed in the need for vaporization because of a late 1893 article by Richard on all known oil engines.[33] Diesel critiqued the article near the end of the second test series and compared his engine with those Richard described. The Brayton and Hornsby-Akroyd were carefully analyzed while the Capitaine received only casual mention (chapter 7).

The Hornsby hot-bulb vaporizer may have influenced him the most. His notes show just how far astray his current combustion difficulties were leading him:

> Akroyd is for me the most important, a very simple example for direct injection of the oil into a combustion chamber in the cylinder, with automatic heating of its own accord through the working process and self-ignition through the heat of the chamber walls.[34]

Diesel correctly cited the crucial difference between Akroyd Stuart and his own engine, yet added this startling note:

> My Augsburg engine is truly a combination of the Akroyd vaporization chamber and the Brayton injection device [an oil-fed burner that ignited the charge as it entered the combustion chamber under pressure]. I emphasize, however, that I did not learn of this design until November 1893, after the Augsburg engine was in operation for some time.

In the depths of his bewilderment, Diesel had forgotten, or chose to forget, that he described these two engines in his book well over a year earlier.[35] Fortunately, his experiments with this combination failed and thus ended a flirtation with what skirted near Brayton and Stuart.

The first vaporizer in the new test series was a small coil suspended from the cylinder head and protruding into the piston cavity. Diesel reasoned that the heat transfer from compressed hot air and combustion gases on such a compact vaporizer should be more effective than for the external design. Fuel under accumulator pressure entered the head, passed through the coil, and then a hand-regulated metering valve at the nozzle supply port (fig. 5.11).

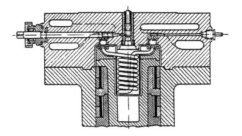

Fig. 5.11. Fuel oil vaporizer in combustion chamber used in the third test series. Diesel, *Die Entstehung, 1913*

Changes in the nozzle needle made the charge volume more independent of accumulator pressure and engine speed. It was an outward opening conical valve rather than one lifting into the nozzle. The supply port, near the nozzle's lower end, fed fuel into an annular groove on the needle. When the valve opened, the needle's full diameter moved to shut off the supply port so that the amount injected was a function of the volume of the groove below the needle shoulder, with the valve seated, and how far the needle moved before closing off the supply port. This length was adjustable from 4 to 8 mm. With a needle diameter of 15 mm, a stepped diameter of 10 mm and an effective stroke of 8 mm, the charge volume was about 785 mm^3.

Varying the accumulator pressure from 10 to 50 atm during bench tests yielded no significant difference in the liquid fuel volume injected per cycle.

Engine tests with the internal vaporizer, but without air blast injection, began on June 28. There was no relief from the hit and miss combustion of the earlier tests. Lamp oil produced the usual heavy cylinder detonation, and billowing white clouds of unburned fuel due to quenched combustion went out the exhaust. With gasoline, explosions occurred in the nozzle. Increasing peak compression to 38 atm did not reduce the misfiring. As might be expected, the heating coil had the opposite effect. It cooled the combustion air on the compression stroke to the point that the fire simply could not stay lit.

The mistaken conclusion that Diesel drew from these results led him up another blind alley. He aptly put it later that "as a consequence a fallacy had set in motion a series of further fallacies and tests in a vicious circle."[36] Diesel's journal suggested that, although there was complete vaporization of the fuel, it was not possible for ignition to occur. He said that either a spark or a flame ignition device would have to be added. So after only eight days of running he stopped the tests to look for an electric ignition system. A visit to Munich on July 5 resulted in his selecting one made by Zettler. It used a magneto with external breaker contacts to induce the spark to jump across electrodes in the cylinder.

This decision to add auxiliary ignition meant a different cylinder head as the existing one could not accommodate what Diesel had in mind. On July 6 work began on the new head of cast iron rather than steel. It would be more than two months before the engine ran again.

The enforced break gave Rudolf a chance to think about a family vacation during the last half of July. One place under consideration was the beautiful Harz mountains of north central Germany. Martha and the children would come from Berlin to join him there.

Rudolf felt a great need to escape his frustrations in Augsburg. This is seen in a June letter to Martha:

> Each day I think not just a little but very much of you and long to see you and to have some quiet weeks in a peaceful country stay alone with you and the children.[37]

Martha's comforting July 5 letter reveals her own feelings about how she interpreted Rudolf's state of mind:

> My most dearly beloved husband . . . But the worry of your health and the motor! When will come the first redeeming word! And how are you my dear, dear, beloved sweetheart? You sound so weary and exhausted in your last letters. Can you some day rest in my arms! I want to have and hold you and softly kiss your tired beloved eyes, and with my hand smooth your heavy-thoughted brow and say to you my dear, how dear you are to me. . . . You are so busy with the motor now that the yearning cannot also be ignited. But you make me ache when you long for and ask for me to be near you. Now I will exercise patience which is so difficult and not burden you with my own compelling complaints.

Diesel explained the thwarting situation in a reply to Martha on the seventh:

> The motor business at the moment is like this: I still cannot announce right now the redeeming word, but in the experts' view, the latest step is essential, for some alterations are now again necessary; I can positively say that I, by the next tests, will be triumphant. . . . In a week at the most I will be done (drawings to modify), then I hurry to you, and after the trip it will be finished here, as I believe in the final success.

This exchange of letters also provides a glimpse into a middle-class German marriage at the turn of the century. Otto von Bismarck was typical in his belief that the only legitimate interests of a *hausfrau* were covered by the derogatory alliteration *Kinder, Kirche, und Küche* (children, church, and kitchen). *Der Mann* was responsible for all else. Martha's deep concern over the summer plans involve "difficult" decisions about travel schedules, tickets, where to stay, and so forth. She is not sure what she should do, and her Rudolf somewhat irritatedly advises:

> I am sorry you have so much trouble arranging the stay in the country. Why not write the forest house where the game warden lives, the one we talked about . . .? If it does not work out then . . . come to me. . . . Also . . . for once be a man and make a decision.

Rudolf preferred that Martha handle the details—he had enough problems with the engine—yet he was prepared to help her. She knew this, but she also wanted to make the arrangements herself. Martha was caught up in the customs of the age where it was expected that the husband must always "do what he can for the wife."[38] She made the plans.

The days spent together again on this vacation in the Harz were what Diesel needed. A family photo is a captured remembrance of that special time (fig. 5.12).

Krupp contacted Diesel in early September to ask what he was doing as they had heard no news for over two months. Diesel immediately answered that the tests were stopped because of a wait for the new cast iron cylinder head. Krupp apparently was not satisfied because people came from Essen on the seventh to check on things for themselves.

Fig. 5.12. The Diesel family on holiday in the Harz Mountains, 1894.
MAN Archives

Diesel had begun to worry that his reputation at Krupp was being undermined by someone in Maschinenfabrik Augsburg. Not everyone at M.A., especially in the steam engine department, was enthused about the attention and money he received from Buz. He commented in his diary that "these are the difficulties and struggles which every prophet encounters . . . what a battle, still that is life."[39] Just before starting the tests he wrote, "Courage, only a short time yet, and I hope all will be right."

The ignition tests commenced on September 19. (They decided to proceed even though the new head was porous.) Diesel had conferred earlier with a Krupp engineer about spark ignition because he knew little about it. This may have been how he learned of Zettler.

But Rudolf then created a peculiar contrivance. To the Zettler electric igniter he added an asbestos wick to be fed by oil dripping past a valve. The wick end was positioned directly above the fixed electrodes protruding into the cylinder to keep it continuously burning (fig. 5.13). (One must assume the sparking across the electrodes was continuous since this was not an uncommon practice at the time.) The glowing wick was to be the primary ignition source for the vaporized fuel charge. Diesel recorded on September 25 that wick ignition did not work. "All that occurred with both kerosene and gasoline were thick, steamy clouds without ignition."[40] The wick would not stay lit, the sparking failed to ignite the fuel vapors, and the points quickly carboned up. Zettler's device was removed.

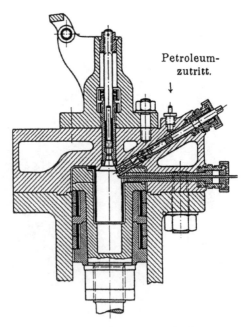

Fig. 5.13. Oil-fed wick dripping on electric igniter as an auxiliary ignition means, 1894. Diesel, *Die Entstehung, 1913*

Briefly entering the scene was another name to achieve fame. Robert Bosch (1861–1942) had opened the doors of "Werkstatte fur Feinmechanik und Electrotechnik Robert Bosch" in Stuttgart during 1886. His "workplace for precision tools and electrical engineering" began shortly thereafter making magneto ignition systems for the rapidly expanding Otto cycle engine business. Bosch's name arose out of correspondence between Diesel and Ebbs at Krupp-Gruson. Until this time, however, Bosch only had a low voltage "make-and-break" system, which generated the spark at contact points breaking open inside the cylinder. A system that might help Diesel was under development in Stuttgart.

An interim make-and-break method was devised because Bosch could not yet provide a low-tension magneto system. The stationary breaker point was located at the wick; the movable point was a spring steel bar attached to the bottom of the piston cavity. At top center the bar made contact with the fixed point and generated a spark as it pulled away when the piston started downward.[41]

Diesel tested this on September 29 while motoring the engine with the valves removed. He wrote that a fire would occasionally start using gasoline, but more importantly, the wick could be kept burning.

None of what Bosch ultimately sent to Augsburg made the engine run any better than it had with Zettler's apparatus. One problem with the Bosch system was compression air leaking past seals around the electrodes; they had not been designed for such pressures. Diesel gave up trying to borrow engine heat to vaporize a liquid fuel and ended these internal vaporizer tests October 3, 1894.

This led back to an external vaporizer that was based on what had been briefly tested in February. From mostly parts on hand, M.A. quickly built another, more complex version. Their drawing, modified from Diesel's sketch, resembled contemporary surface carburetors for gasoline-fueled, spark-ignition engines (fig. 5.14).

A change from the January design was a capability to use blast air, supplied either from the compressor or the engine cylinder, to enhance fuel vaporization. The hot air was filtered through a gravel-filled container and bubbled up into the kerosene in a small, heated, bomb-like tank.

Diesel approached these tests with the thoroughness found in all his work. Not having the needed data, he calculated the saturated vapor pressure curves for gasoline and kerosene and plotted them against the experimental results obtained from heating fuel in the vaporizer. The test results agreed closely with his calculations.[42]

This latest panacea added excitement: the first time the kerosene level in the heated "bomb" dropped too low, the ensuing explosion blew out a safety plug. A sight glass was added to avoid such a recurrence, but Diesel wrote of continual vaporizer explosions. The engine responded to this vaporizer as it did to the others with misfires, "restless" combustion, and P-V diagrams

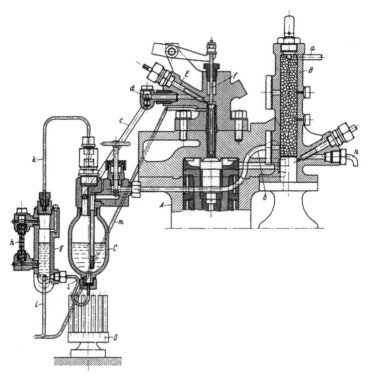

Fig. 5.14. A second external vaporizer to gasify fuel before injection.
October 1894: (*a*) line from fuel accumulator, (*b*) line to vaporizer *C*,
(*c*) line from vaporizer to hand-regulated throttle valve *E*, (*g*) leveling
bottle with sight glass *H*, (*i*) pressurized line from fuel pump,
(*m*) direct line from fuel pump to injector. Sass, *Geschichte, 1962*

showing a minuscule indicated power. Performance was the same either with
or without the electric ignition systems. The external vaporizer tests ended
October 12.

Hermann Ebbs and Krupp directors Gillhausen and Schmitz witnessed
the last of these tests. Following Diesel's lead, the consensus was to incorporate a more effective auxiliary ignition. Electric systems were not yet reliable,
but something similar to Daimler's externally heated "hot tube" was simple,
albeit hazardous, and might work. One made of porcelain was considered.
The visiting engineers, who were running out of suggestions, had no alternative other than to trust Diesel's supposedly greater understanding of the
combustion process in his engines.[43]

The Fourth Test Series—November 1894

Krupp's strong desire to use city lighting gas (coal gas) as a power source
effectively coerced the principals on October 12 to agree that the next series
of engine tests would burn this fuel. It was a logical extension of the recent

work on fuel vaporization. Diesel did not exactly like the decision, but it was one he had brought on himself as result of his earlier sales pitch to Krupp. He feared that such tests might result in other gas engine builders (Krupp-Gruson, for example) taking advantage of what was learned.

The changeover to a gaseous fuel was not difficult because of a commonality with the air injection parts. It mainly required the addition of a gas compressor that could be adapted from the two-cylinder blast air compressor. This single-stage, double-acting design was modified to pressurize in two stages: both ends of one piston and one end of the other increased the pressure to 3 atm and discharged into the cylinder of the remaining piston end, which raised the gas to a final injection pressure of 30 to 36 atm.

Diesel had first used the modified compressor to supply air to the external vaporizer of the third test series, which contained gasoline. He reasoned that while the original, single-stage compressor had barely raised the pressure high enough, the two-stage version might work better. In this one-day experiment with the vaporizer on November 6, the engine barely had the power to idle on its own, no different than eight months earlier.

The engine performed the same on coal gas as on the liquid fuels. Lowering the gas injection pressure from 34 to 30 atm quieted combustion but increased misfiring. One difference noted was that the exhaust smoke was clear when the charge ignited and blue to gray on a misfire.

For the gas tests the cylinder head had been converted back to the one with the separate air and fuel supply passages to the nozzle and the options for auxiliary ignition. Diesel hoped to test either gaseous or liquid fuel, or both, in combination with air injection. The injection air circuit to the nozzle became the gas feed. Gasoline, at accumulator pressure, entered the nozzle as before after passing through the hand-operated metering valve. The small wad of fine wire mesh remained sandwiched between the nozzle orifice and the perforated plate.

A failure to achieve consistent firing on coal gas prompted Diesel to study its ignition characteristics with the cylinder head off the engine. Using both spark and heated tube, he determined the following facts:

> . . . it would be justifiable to conclude about the above tests that a slowly entering gas stream ignites better than a vigorous stream, furthermore that warm gas ignites better than a cool one; the cooling effect of the gas stream is important because a glowing iron will become black by it. Further, with artificial ignition the ignition will take place in the gas stream itself and not at the side of it.[44]

Diesel observed that no ignition occurred when the gas jet missed the hot tube by just a few millimeters. Also, a jet impinging on the hot tube ignited only if the tube was "glowing white." There was no ignition at a dull red heat, and the incoming gas stream even further cooled the tube.

The point missed completely was that to attain burning, a gas-to-air mixture ratio within certain richness limits was needed. A low velocity gas jet came closer to providing an ignitable ratio because of more opportunity to mix with air.

Spark ignition did not initiate burning. Bosch came to Augsburg on November 12 to see for himself what the problem was. He found the magneto not working properly and the porcelain insulators shorting out if placed in the gas jet.

A compressor breakdown prevented a proposed test of injecting gas with assistance from blast air.

The versatility of test fuel combinations allowed Diesel to add gasoline as a small pilot charge. Backfiring stopped, and the engine consistently produced similar diagrams whether the nozzle needle was partly or fully opened. On November 14 the engine idled using "minute drops" of gasoline as a pilot fuel and coal gas for the main charge. This may have been the first dual-fuel engine test.[45] Both 1 mm and 2 mm nozzle orifices were tried, but the engine would idle only with the 2 mm diameter. A larger orifice than this produced a sootier exhaust and no increase in power.

A journal entry the next day reveals that Diesel now recognized one requirement for proper combustion in his type of engine. He referred to diagrams taken a month earlier with both injection of gasoline alone and gasoline blasted in by air. He wrote, "It seems that the direct injection does not produce a sufficient mixture of fuel and air."

Diesel tried to cure this deficiency by adding a mixer extension onto the end of the nozzle (fig. 5.15). It was a steel pipe screwed into the end of the nozzle and extending almost to the bottom of the piston cavity. Evenly spaced along its length were thirty, radially drilled 1 mm holes to uniformly distribute the charge. The nozzle orifice leading into the mixing pipe was increased to 3 mm diameter.

Diesel tested his "mouthpiece" (*Mundstück*) on the sixteenth by burning a gasoline/gas fuel mix. He reported that even with the large orifice the engine achieved the "principally correct diagrams" where the peak is "broad and stretches horizontally" (fig. 5.16). Erroneously noted was that the "jet itself pushed the air ahead instead of mixing with it" to cause delayed ignition and afterburning.

A second observation led to an end of the test series. After studying the pattern of combustion chamber deposits, Diesel understood that the fuel apparently was not mixing with enough of the compressed air charge. He calculated that at top center about half of the cylinder air was lost for burning due to it being trapped in either recesses or clearance spaces. Why no burning occurred there was wrongly attributed to "the flame pushing the air back into the lost spaces." He correctly concluded that the problem "unfortunately cannot be overcome without a redesign of the machine."[46] That same day he wrote Krupp about these latest findings and the need to rebuild the engine once again.

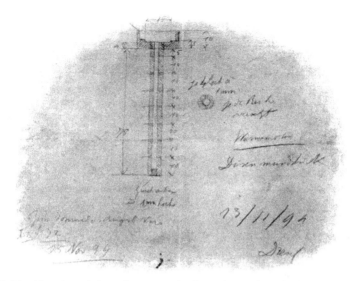

Fig. 5.15. *Mundstück* (nozzle spray pipe) on the end of the injector to inject fuel from horizontal holes into the piston cavity, November 1894.
MAN Archives

The number of holes in the mixer pipe was doubled to sixty. When injecting coal gas alone, the engine idled without misfiring, but failed to develop useful power.

A few more experiments, dealing mostly with auxiliary ignition, were made before ending the fourth test series on November 23, 1894. Krupp-Gruson's porcelain hot tube and a better-insulated Bosch "spark plug" added nothing and led to Rudolf almost dropping auxiliary ignition for good.

His 1913 account of this test series postulated "two of the important laws of Diesel engine construction:"

1. The law of self-insulation of the flame and the necessity of the blowing in of the fuel with air to ensure vaporization.
2. The law of the need to bring into play for combustion all the air in the compression chamber.[47]

One was invalid, as research later proved, and the other self-evident.[48]

Fig. 5.16. Improved indicator diagrams showing combustion with *Mundstück*, November 1894. Diesel, *Die Entstehung, 1913*

Progress—or the lack of it—toward the end of the fourth test series may be sensed from reminiscences by Lucian Vogel's daughter Dora. Although Rudolf lived with the Barnickels, many noon-time hours were spent in the Vogel home. She recalled these occasions:

> Herr Diesel often reported very satisfied, "Today the motor ran five minutes." "Three minutes," my father corrected him, and later the same thing came up with twenty and eighteen minutes. After lunch Herr Diesel did not rest but played the piano four-handed with my mother while my father did mathematics, which was his favorite recreation.[49]

Rudolf was ready for a Christmas respite from a year of no progress, so he welcomed the break until the next tests. Eugen recalled that his father "came to Berlin incredibly drained, some signs of fatigue and excitement, headaches and insomnia, undoubtedly sick."[50]

Of utmost urgency to Diesel in 1895 was the meeting of a deadline to send an operable engine to Vienna. Austrian patent law required that an invention's practicality be demonstrated to patent office examiners within a fixed time period from when the patent first issued. Because of last-minute improvements at M.A., Diesel could risk sending his engine rather than Dyckhoff's, which had been the backup. In early January 1895, the engine was dismantled and sent to Krupp-Berndorf, an Essen affiliate thirty-five kilometers south of Vienna. Metallwarenfabrik Berndorf had been established in 1843 by Hermann Krupp (1814–1879), Alfred's brother.

Rudolf arrived in Berndorf on January 7 to inspect air injection parts being machined there to new drawings furnished by M.A.; time had been too short to make them in Augsburg. Diesel went to the Austrian Patent Office the next day to arrange for the trials. However, a blowing snowstorm on the ninth shut down much of the city and stopped all trains. He spent several days of enforced stay in Vienna visiting museums and historic sites as well as enjoying needed rest. A sightseeing day trip to Budapest was squeezed in on the sixteenth.

The engine burned gasoline during the trials on January 17 and 18. It would only idle, and the flywheel drive belt remained connected to the factory power system. Operation was punctuated by explosive combustion and a cracking, smoky exhaust. Nevertheless, the examiners agreed that the engine did function and legally fulfilled the utility requirement enough for Diesel to retain his Austrian patent.

The modest Berndorf debut was an omen of better times ahead for the engine. Ostensibly, Rudolf had little to show for a year and a half of work except that his engine idled. Yet, as with all who create new technology, he was paying his entry fee for the privilege to fail in order that he might ultimately realize his goal.

"The execution is the time for establishment of all resources for realizing the idea. It is the still creative, still joyful time of overcoming the natural obstacles, from where one emerges stronger and elevated, even when one is defeated."

—Rudolf Diesel[51]

Notes

1. Eugen Diesel, *Diesel: der Mensch, das Werk, das Schicksal* (Hamburg: Hanseatische Verlag, 1937), 206. The German equivalent of "cross your fingers" is "press your thumbs."
2. Rudolf Diesel, *Die Entstehung des Dieselmotors* (Berlin: Springer Verlag, 1913), 158.
3. In the 1950s the German internal combustion engine industry commissioned the writings of their early history. This resulted in a classic work edited by Friedrich Sass, *Geschichte des deutschen Verbrennungsmotorenbaues von 1860 bis 1918* (Berlin: Springer Verlag, 1962). He based the major Diesel portion on an exhaustive study made by Kurt Schnauffer (1899–1981). Professor Schnauffer combed the Diesel papers and those of others involved with Diesel. Since the Sass book, and more importantly Schnauffer's manuscript, have not been translated into English, the story of the development years at Augsburg has only been told in that language through books and articles based mainly on Eugen Diesel's sensitive biography of his father. His book is not an in-depth technical history of the engine nor was it intended to be.
4. The Drei Mohren, with its prestigious address on Maximilianstraße, remains Augsburg's leading hotel. Because of severe bomb damage, it could not be restored to the grand style of Diesel's day. Next door is the Fürst Fugger Bank, whose founders financed Habsburg emperors in the fifteenth century.
5. E. Diesel, *Diesel*, 206.
6. Before the introduction of small electric motors, machine tools were driven by leather belts running off pulleys on wall or ceiling-mounted shafts. Power to turn the shafting came from a steam engine or water wheel. A large factory of that era resembled a forest of belts hanging from rows of long shafts.
7. Kurt Schnauffer, "Die erste Diesel'sche Versuchsmotor 1892–1893," in MAN Archives (1954), 20.
8. Ibid., 21. Schnauffer comments that Diesel anticipated the glow plug starting procedure now common on indirect injection diesel engines. Some anti-Diesel critics have averred that the use of an electric glow plug suggests such an engine more closely follows Akroyd Stuart's low compression oil engine, which depended on an external blow torch to heat a large vaporizing antechamber. This argument is invalid because the Stuart engine could not run at idle or light loads without the heating torch. When warm, the IDI diesel starts without a glow plug, and regardless of ambient temperature, idles without it.
9. Rudolf Diesel, *Die Entstehung*, 14. His comment is of interest because he says that compression ignition is only a part of the overall concept.
10. Ibid.
11. Ibid., 14.
12. Ibid., 15.
13. E. Diesel, *Diesel*, 213. Lucie von Motz was a close Berlin friend of Martha. Ibid., 178.
14. Schnauffer, "Versuchsmotor," 30. Abstract from Diesel's redesign calculations. Original in Deutsches Museum.
15. It has been conjectured that Diesel's assistant, Linder, came up with the air blast idea during the second test series. Diesel noted these injection methods on a drawing of the accumulator, and during the second series added a fourth method: "Through vaporization, that is (a) with internal carburetion or (b) with external carburetor housing." Ibid., 32.
16. Diesel's British patent No. 4,243 of April 27, 1895, which follows the German No. 82,168, specifically claims the use of compressed air from the combustion chamber.
17. Sass, *Geschichte*, 442.
18. Ibid.
19. E. Diesel, *Diesel*, 214.
20. Ibid., 217.
21. R. Diesel, *Die Entstehung*, 20.

22. Ibid., 21. Although Diesel gave up his "pressure-time" system, an advanced form used today accurately controls fuel pressure under all speed and load conditions. This is the Cummins "P-T" system introduced in 1954. Also of interest is an earlier Cummins system. Functionally similar to Diesel's pressure and time method, it was abandoned in the late 1920s when constant speed and variable load applications (one being a power shovel) caused excessive overfueling conditions.

23. Kurt Schnauffer, "Umbauten des Versuchsmotors und erster betriebsbrauchbarer Dieselmotor, 1894–1897" (Reconstruction of test motor and the first operational Diesel motor), (mss., 1958), 9. Deutsches Museum copy. Kurt Neumann was one of the earliest to explain how little charge vaporization occurred in a diesel engine. See: "Untersuchungen über die Selbstzündung flüssiger Brennstoffe" (Investigations of self-ignition of liquid fuels), *VDI-Zeitschrift* 70 (1926): 1071. A more recent paper on the process is: M. M. El Wakil, P. S. Myers and O. A. Uyehara, "Fuel Vaporization and Ignition Lag in Diesel Combustion," *S.A.E. Transactions* 64 (1956): 712-29. Researchers continue to explore phenomena relating to the onset of the diesel combustion process.

24. Ibid., 12. Diesel's test journal reported "a 1/5% [of piston stroke] phase lag after TDC. Snap away 2-1/3% after TDC. Needle stroke 2 mm."

25. R. Diesel, *Die Entstehung*, 22.

26. E. Diesel, *Diesel*, 218.

27. Ibid.

28. R. Diesel, *Die Entstehung*, 25.

29. Ibid.

30. Schnauffer, "Umbauten," 15.

31. E. Diesel, *Diesel*, 220.

32. Dyckhoff's stone chateau-like home is on a street by the Ornain River. The still-existing factory where he built the 1894 experimental diesel begins a hundred meters behind the house. The author had the privilege of spending two days in the spring with Dyckhoff's granddaughter Marie-Rose Cochet, who owns and lives in the family home. It *is* a lovely area, especially in the springtime!

33. Gustave Richard, "Les moteurs a pétrole depuis 1889," in *Bulletin da la Société d'Encouragement pour l'Industrie Nationale* (Paris: 1893). Cited in Schnauffer, "Umbauten," 54.

34. Schnauffer, "Umbauten," 57.

35. Rudolf Diesel, *Theory and Construction of a Rational Heat Engine* (London: Spon Press, 1894), 84–5. Schnauffer, "Umbauten," for an unexplained reason, agrees with Diesel's statement.

36. Diesel, *Die Entstehung*, 28.

37. Eugen Diesel, *Jahrhundertwende* (Turn of the Century) (Stuttgart, Germany: Reclam-Verlag, 1949), 97–98.

38. Ibid., 99. The expression "*auf den Händen trug*" idiomatically translated is "carried on one's hands" or "treat with great tenderness." From his own perspective of two tragic world wars, Eugen can express a sensitivity about this vignette in the life of his mother and father. He writes, "What inkling did mankind have at the time of the monster it was moving toward, of the necessity to have to decide in minutes to give up house, farm, home, to take this or that step, any of which could lead to destruction. But at the same time, we do not have the right to smile at that earlier era. Because as soon as only a breath of peace and order is restored, the thoughts of the same one who has been through the hell of death's door and destruction will again be concerned with the smallest worries of daily life, the envies and fears of the most insignificant things, and may even spend a sleepless night because he next morning must travel on the summer vacation. After such a sleepless night, Mother finally decided 'definitely for the Harz.'. . . 'I barely dare to look forward to a few weeks of a paradise-like with you.'"

39. E. Diesel, *Diesel*, 223.

40. R. Diesel, *Die Entstehung*, 29.

41. These early electric ignition systems are described in Cummins, *Internal Fire* (Austin, TX: Octane Press, 2021), 272–78.

42. R. Diesel, *Die Entstehung*, 157, gives helpful fuel data for the test fuels used. Oil: specific gravity = 0.793 at 18.7°C. Gasoline: sp gr. = 0.712 at 19.2°C.

43. Schnauffer, "Umbauten," 27, comments that "One is astonished . . . how the Krupp experts and MA even reinforced Diesel in his erroneous train of thoughts and opinions, instead of opposing them." His judgment was based on documentation in the MAN Archives.

44. Ibid.

45. Diesel was granted several patents on dual fuel engines after his more successful 1897/98 tests with coal gas. One, German No. 118,857 issued March 21, 1901, is very similar to US patent No. 673,160 issued April 30, 1901 (filed April 6, 1898). The German claim reads: "Combustion system for an internal combustion engine thereby is characterized, that in compressing and somehow igniting the working mixture a second fuel would be introduced in such a way that the process of combustion of the type and duration of the introduction of these fuels, i.e., of the control of the engine, will be determined."

46. Schnauffer, "Umbauten," 32–33.

47. R. Diesel, *Die Entstehung*, 35–36.

48. Schnauffer, "Umbauten," 34. Sass's comment on Diesel's observation is that "what is self-evident must also first be recognized." It is one well to remember when benefited by hindsight.

49. Letter from Mrs. Dora Meussdorfer as quoted in "Lucian Vogel und seine Mitarbeit am Dieselmotor," in MAN Archives (1940), 2.

50. E. Diesel, *Diesel*, 224.

51. R. Diesel, *Die Entstehung*, 151.

CHAPTER 6

Light at the End of the Tunnel: 1895–1897

"So inventing means to bring to a practical success, sifting out from a large series of mistakes through many failures and compromises, a correct basic idea."

—Rudolf Diesel, 1913[1]

The detours and dead-ends of the first four test series changed to a solid list of achievements over the next two years. Diesel knew at last that the growing light penetrating his long tunnel of despair was an affirmative ray of sunshine proclaiming fulfillment of a cherished dream.

Performance improvements, which resulted from another redesign, led to a new engine embodying all that had been so painfully learned. An improved air starting system severed Diesel's baby from its umbilical belt drive. The time-wasting diversions to explore fuel vaporization and auxiliary ignition were recognized as just that; using only the heat of compression to fire a better atomized fuel charge once again became the goal. Injector nozzles were greatly simplified, and the compressor supplying air for blast injection was made integral with the engine. Even the supercharging of combustion air was successfully tried. These promising events of the fifth and sixth test series preceded the engine's formal debut in 1897 to a host of skeptics.

The Second Redesign: The 220/400 Engine
The decision to modify again the much-abused test engine was made in November 1894. It only added to Rudolf's workload as he did not yet have someone reporting directly to him who could assume design and test responsibilities, something beyond the capabilities of the assigned mechanics who

had their hands full anyway. Vogel also fully supported the hiring of a backup engineer because he had his own department to run.

The man who would fill this role was Fritz Reichenbach (1868–1915). He had attended the Technical University (then the *Polytechnikum Maschinenbau*) in Stuttgart and was a competent mechanical engineer with research experience. He also happened to be a younger brother-in-law of Heinrich Buz and a son of the original founder of M.A. He joined M.A. on November 1, 1894, in time to witness the conclusion of the fourth test series.[2]

The strained political climate for Diesel's program within the company did not change when Reichenbach arrived. Krumper, as chief engineer in charge of steam engines, continued to belittle the engine's apparent lack of progress (according to Diesel). He appeared in Rudolf's small, closed-off area only a couple of times a year, and then often just to throw out a few sarcastic remarks. Buz still offered encouragement, and both he and Vogel were always "very pleasant." Diesel gratefully wrote of this positive support.

> Buz, especially, although he almost never said so, appreciated the difficulties, but also the significance of the business, and had himself proposed to design an entirely new motor in order to finally eliminate the remaining faults of the first, at which we so heavily labored.[3]

Diesel and Reichenbach were well along on the redesign before the Austrian patent tests. It involved everything above the A-frame support: cylinder, head, piston, valves, camshaft location, injection nozzle, and so on. The drawings and making of parts required over four months.

A major objective of the reconstruction was to have a minimum surface-to-volume ratio in the combustion chamber, that is, eliminate the nooks and crannies that trapped cylinder air and kept it from properly mixing with fuel. The new piston crown was flat and remained below the cylinder head when at top center. Recessed into the head was a cylindrical combustion cavity containing the pipe-like burner/nozzle (fig. 6.1). Diesel reported that the "lost spaces" had been reduced from 60 percent in the original 1893 design to 28 percent in the second and were now only 10 percent.[4]

The iron cylinder and head had cast-in cooling water jackets along with an increase in piston diameter. The 400 mm stroke was retained, but enlarging the bore from 150 to 220 mm almost doubled the displacement to 15.2 liters.

Three pairs of piston rings backed by tension rings sealed a further shortened, built-up piston. Cylinder wall and piston lubrication was still by oil thrown from the shallow grooves machined in the skirt end of the piston. The grooves refilled as before by dipping into an oil reservoir at bottom center.

Relocation of the camshaft to the top of the engine eliminated stretching or bending in the earlier design's long push rods. The rocker lever rollers acted

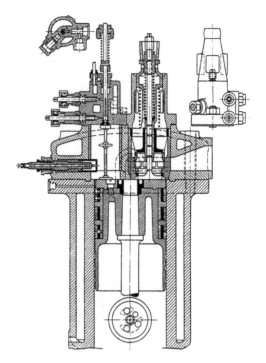

Fig. 6.1. Cross-section of the second redesign of the original
engine for the fifth test series, March 1895 to September 1896.
Diesel, *Die Entstehung, 1913*

directly on the cam discs. Diesel also paid close attention to the cam profiles
in order to prevent the rollers from lifting off the cams. He referred to this as
"quiet motion," something apparently not achieved in earlier running.

Diesel returned to a combined inlet air and exhaust poppet valve because
a larger nozzle body casting left insufficient room for two valve openings into
the cylinder. He recognized this as a backward step but hoped to minimize
mixing of the two gas streams from the converging inlet and exhaust passages
by adding a second, cam-actuated, sliding spool valve. It was concentric with
the poppet valve and alternatively moved to shut off one of the two passages
as required. A hollow stem on the spool valve acted as the guide for the solid
stem of the poppet valve. Each valve had its own return spring. When the
spool valve was in its lower position during the intake stroke, it closed off
exhaust ports in the wall of the cage assembly. Air was then drawn in through
slots in the valve spring cover (fig. 6.2) and past the opening created when
the upper end of the spool moved off a flat sealing seat. Conversely, during
exhaust the cage passages opened and the seated spool closed the air circuit.
An angled rocker lever, whose valve stem end nested around the end of the
poppet valve rocker, acted on the spool valve. The dual-purpose poppet valve
closed between the exhaust and inlet processes.

Fig. 6.2. Top of 220/400 engine cylinder head. *MAN Museum*

The new cylinder head was also machined for a spark plug; several types were tried while observing liquid and gas spray patterns from different nozzles in the freestanding head. The plugs failed to kindle a charge because an ignitable mixture was not passing the electrodes. The need for an auxiliary igniter was eliminated, however. Improved combustion chamber sealing ensured that even with a cold engine the heat of compression was enough to consistently burn an injected charge when starting.

As before, a hand-operated transfer pump raised fuel to the elevated accumulator. Compressed air from the tank went to the injection nozzle and the accumulator so that both fuel and blast air were kept close to the same pressure (fig. 6.3).

Fuel metering continued with a manually turned needle valve, which fed the nozzle valve cavity just above the conical seat of a cam-lifted needle. Parallel to and above the horizontal fuel valve was a second needle valve for adjusting injection blast air. The nozzle body held a pop-off safety valve to also supply starting air and a cast-in water jacket to preheat the fuel by hot water exiting the cylinder head.

A new sprayer Diesel termed a double star burner (*doppelten Stembrenner*) was screwed over the nozzle end (fig. 6.1). "Fuel fed through a common central pipe sprays from two discs having fine radial holes to provide a star-shaped burning pattern."[5] He positioned it "so that during the admission

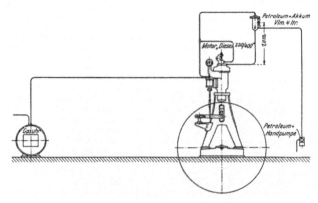

Fig. 6.3. Schematic of fuel and compressor air supply.
Diesel, *Die Entstehung, 1913*

period the entire air in the compression chamber must sweep over the burner." It proved to be a positive step. Diesel used the results with it as his base line when comparing engine performance and sooting characteristics against other nozzle designs.

A variety of atomizers were tried in conjunction with the double star burner. They merely screwed onto the threaded end of the nozzle projecting below the needle seat (fig. 6.4).

Air starting held a high priority for Diesel. His work led to a patent application entitled "Device to start four-stroke combustion engines through transforming the same into two-stroke air compressor machine." The idea entailed adding double-lobed exhaust and inlet cams, next to the basic ones,

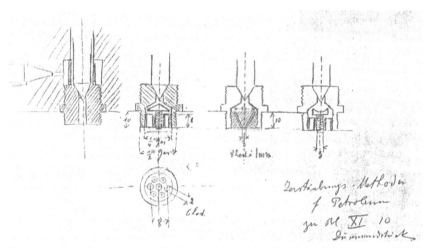

Fig. 6.4. Atomizers inserted below the nozzle and before the burners, 1895.
Diesel, *Die Entstehung, 1913*

so that compressed air could enter the cylinder on the down stroke of each revolution. When engine speed was high enough to inject fuel, a lever axially shifted the camshaft to disengage the starting cams and reengage the normal cams and the fuel cam. The engine then returned to the four-stroke running mode. Diesel was granted German Patent No. 86,633, which carried a filing date of March 30, 1895.[6] This system was much used even after the patent expired in 1910 as it was especially helpful for starting a single-cylinder, four-stroke engine.

The Fifth Test Series—March 1895 to September 1896

The long and productive fifth test series commenced with the previously mentioned spray and electric ignition tests on March 26, 1895. Initial water pressure tests showed that both the nozzle body and cylinder castings had excessive porosity. Impregnating the nozzle body with a water glass solution allowed it to be used, but the cylinder was too porous for this "fix" alone. As a temporary measure, the 300 mm thick cylinder wall was bored out to accept a 10 mm thick dry sleeve. Three cylinders were cast before getting one sound enough.[7]

Power tests with blast injection of gasoline began April 29 in which it was possible to maintain an engine speed of between 150 and 200 rpm. Diesel's journal entry on May 1 reads, "It is safe to assume that the correct diagram shape is now reached"[8] (fig. 6.5). Further noted was the invisible exhaust at an output of 14 ihp. The much higher power, due in part to a greater piston displacement, caused the crankpin journal bearing to overheat, and testing stopped from May 2 to the 17 for drilling a water passage in the crankshaft to cool the bearing. This expedient to forestall a major rework of the engine's lower end was successful.

Gillhausen of Krupp witnessed the May 18 tests. Then Diesel, who still hoped (as did Krupp) to use gaseous fuels, briefly tried burning with equally modest success coal gas assisted by the pilot injection of a little gasoline along with the gas. It entered the nozzle as before via the blast air circuit.

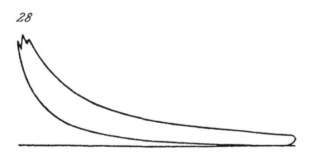

28

Fig. 6.5. First indicator diagrams, burning gasoline, when the second redesign of the engine was run, April 29, 1895.
Diesel, *Die Entstehung, 1913*

Fig. 6.6. First indicator diagram burning kerosene, taken May 30, 1895, with the second design of the engine. Sass, *Geschichte, 1962*

Kerosene (*Petroleum*) injection commenced on May 30, 1895. Although the first diagram (fig. 6.6) showed late ignition, it without question exhibited the desired flat, constant pressure combustion. The engine also ran noticeably quieter while retaining a clean exhaust.

With engine speed regulated to some degree, Diesel spotted a significant and unexpected result. Until this point, he had assumed output was a function only of injection duration, in other words, the time to add a given fuel volume at a constant rate. He understood at last that control over the injected fuel charge volume could be through a combination of both the injection duration *and* injection pressure. With duration held constant he could add more fuel to nicely "fatten up" the indicator diagram (increase power) simply by increasing the injection pressure, yet not unduly raise peak cylinder pressure. He took another important step toward a "modern" diesel engine.

Although indicated power kept increasing, 23 ihp at 200 rpm and 34 ihp at 300 rpm, the engine's mechanical efficiency was woefully poor. Internal friction severely ate into the gross power gains. The indicated mean effective pressure went from a no load 2.92 kg/cm² to a full load 6.85 kg/cm² at 200 rpm, which meant a mechanical efficiency of less than 58 percent. Neither did this figure include the power to run the auxiliary air compressor. Good contemporary oil engines of comparable power attained full load mechanical efficiencies above 80 percent.

A further three weeks were lost while Diesel experimented with an external atomizer. He wanted to find out if by mixing blast air and fuel outside of the nozzle he would improve atomization. A valve box with adjustable needles to control fuel, air, and the combined mixture (a throttle valve) was adapted to screw into the nozzle body. Tests showed power was lower with it than when fuel and air came together near the nozzle needle seat. This worthless atomizer again illustrates Diesel's inquiring mind and his thoroughness in investigating ways to improve performance. It is yet another example of why, with limited resources available for exploring so many unknowns, the engine required years to break out of the laboratory.

The new capability to run longer under higher load conditions pointed up a basic defect in the otherwise promising double star burner. (Diesel continued to call a spray nozzle a "burner.") In only a few hours the small, drilled passages coked up to a point where power drastically dropped off and the exhaust blackened. The burner simply became a victim of its total exposure to combustion chamber heat and no doubt to a backflow of hot gases into its many orifices without there being sufficient cooling externally from intake air and internally from injected fuel.

The orifice plugging problem launched Diesel into testing a series of burner/nozzle sprayers that today seem almost desperately fanciful (fig. 6.7). He describes them:

a) The star burner had suction rings and was similar to the double star burner [above], but at each star were vacuum rings which from the injector-like action of the fuel jets [pulled in and] carried along surrounding air.

b) Shows the so-called jet burner, whose injector-like suction working out of the entire compression chamber is clear without further explanation.

c) The fuel stream was thrown from under to over against the expanding air.

d) Shows a pipe burner with spiral winding which had the object to vaporize the fuel by strong preheating before entering the combustion chamber.

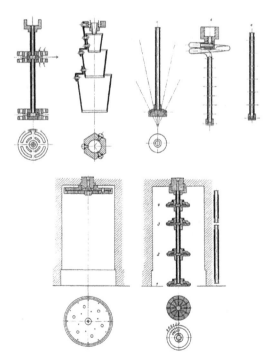

Fig. 6.7. Burner designs tested during the fifth test series.
Diesel, *Die Entstehung, 1913*

e) Finally shows a simple pipe burner similar to one already employed with the first gas tests.

f) Shows the so-called rain burner by which the fuel was injected in a rain form throughout the entire combustion chamber.

g) Shows a four-piece star burner by which the injection profile on each star was exactly set by vernier means.[9]

All were tested using both gas and kerosene, first in open air but encased in a sheet metal shroud to simulate the combustion chamber recess, and then on an operating engine. None did as well as the original double star burner in the engine.

June 26, 1895, was a milestone day for Diesel. After almost two years of testing, he finally measured horsepower output using a prony brake type of dynamometer.[10] The indicated thermal efficiency was a respectable 30.8 percent, but the brake efficiency came to only 16.6 percent—no better than good Otto cycle oil engines. The mechanical efficiency of 54 percent was even worse than the previous test when it was measured from an indicator diagram. This still did not include the Linde compressor. Fuel consumption was 382 g/bhp-hr (0.85 lbs/bhp-hr or 520 g/kW hr).[11]

Piston friction investigations started immediately. First found on static tests was that the piston sealed better without an oil film when the backup tension rings were removed. They apparently put an uneven side thrust on the piston to prevent it from centering as well. With no backup rings, the mechanical efficiency rose to 64 percent. Reducing ring width from 10 cm to 6.5 cm raised it another 3 percent. Fuel consumption dropped to 327 g/bhp-hr, and brake thermal efficiency climbed to 20.3 percent. Diesel's journal read, "Noteworthy is that with the three tests the kerosene consumption per *indicated* horsepower, in the amount of 206–225–211 g, is entirely stable and is less by far than one half that of all engines known till now."[12]

In conjunction with brake horsepower tests was a tryout of the air starting system based on his recently filed patent application. These proved so promising that the journal entry for July 3, 1895, states,

> Ignition and running immediately follows with the changeover of the control to the operating position. The starting question is now solved. *The motor is therefore operationally ready at every moment without any preparation.* [Diesel's emphasis][13]

Rudolf's deserved elation is found in his soul-baring letter to Martha written the same day:

> My motor still makes great progress. I am now so far beyond what all presently can accomplish that I can say I am first and foremost in this area of the technology of engine design, the first on our small

globe, the leader of the entire troops on this side and across the ocean. Does not your breast swell with these words? I would almost like to become haughty myself if I was so inclined. So again, I am happy, inwardly calm and serene, unassumingly satisfied in the knowledge that I have carried out a useful achievement, and fortunate that our future is secure, for this it now is.[14]

He had every right to be proud of how far he had come, but he failed to anticipate the distance yet to travel.

Diesel wrote to Krupp on July 8 telling of the air starting success and, more to the point, of the improved indicated specific fuel consumption. He particularly called attention to several recent tests on production German oil engines, which had brake efficiencies ranging from only 10.2 to 15.7 percent.[15] Diesel then almost casually proceeded to hint of a storm cloud looming on the horizon:

> Our tests still required the air injection of kerosene being run from an extra installed compressor. The air volume is so small—according to measurements—that the above results are not substantially affected. We are nevertheless now busy with plans for the engine to have its own air or gas pump.[16]

It was the last thing he wanted because of added cost in both hardware and mechanical efficiency.

The volume needed for blast air was also determined at this time. Diesel learned that the optimum ratio of injection air to cylinder air seemed to be about 5 percent. He also found that the compressor consumed almost 6 percent of the gross horsepower output.[17] Not all of this work was lost, however, because the separately compressed injection air became available for combustion after it entered the cylinder with the fuel. Since a little of the compressor work was regained, the net fuel consumption reduction due to an engine-driven compressor seldom exceeded much more than 5 percent at full load.[18] While this was an acceptable performance price for Diesel to pay, the cost to develop safe and reliable engine-mounted compressors was another story.

One day about the middle of July, when the engine was working hard under power, the piston rod crosshead seized in its guide. On disassembly, several other areas were seen needing attention: Small compression leaks led to the discovery that the cylinder had been bored out of round, a condition traced to a faulty machine tool spindle. There was also abnormal ring wear unrelated to the oval bore. These deficiencies taught afresh the lesson that the engine tolerated only the best in both materials and manufacturing methods.

The repairs, modifications, and new mechanism required most of two months. It also kept Diesel busy with other tasks, namely keeping Krupp happy, by applying for necessary foreign patents, and firming up the design of the 100-hp compound engine disclosed in his 1892 patent.

A single-stage, water-cooled compressor was attached to the engine at this time (fig. 6.8). The power takeoff for it was via a pair of oscillating drive links straddling the A-frame base. One end of each link pivoted on a lengthened engine connecting rod pin at the crosshead. The other end traveled in a slot that was guided by a stationary pin adjustably mounted to the engine. Positioning this pin closer to the crosshead reduced the compressor stroke. A two-piece compressor piston rod, joined by a turnbuckle to adjust its length, was journaled to a pin supported on each end by the links. Thus, both compressor discharge capacity and pressure were varied by altering stroke and clearance volume. The compressor had a volumetric efficiency of 71 percent and a power consumption of 1.3 indicated horsepower. Its operating losses were hereafter included in the overall mechanical losses of the engine.

Air for starting came from a tank charged by bleeding off a little compressed engine air through the valve at the cylinder. Further testing showed

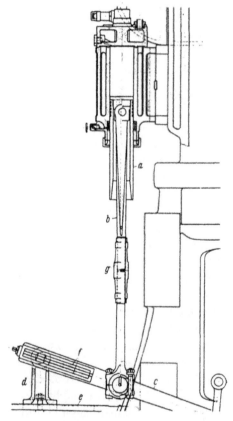

Fig. 6.8. Adjustable compressor drive mechanism added July/August 1895.
Sass, *Geschichte, 1962*

that combustion gases remaining in the cylinder air caused excessive carbon buildup in both the valve and tank. The bleed-off process also reduced compression pressure more than desired. Redesigning the valve allowed starting and injection air to come from the new compressor.

The engine continued to be without a speed governor. New instrumentation—thermometers and flow meters—were added to record water, air, and exhaust temperatures, compressor air consumption, and cooling water flow.

Johannes Nadrowski arrived from Berlin in mid-September to join Diesel's design team.[19] This additional help plus the building of an enclosed office in the engine test area gave an impression that things were beginning to happen at last. Yet, except for those directly associated with the intense, tireless inventor, most believed that his often smoky and noisy machine was a waste of Buz's money.

Testing began again on September 18, 1895, with the vernier star burner installed (fig. 6.7, lower right). Its disappointing indicated mep (mean effective pressure) of only 5.8 kg/cm² prompted a return to the original double star design. The mep immediately rose back to a respectable 6.8 kg/cm². Diesel made this notation about the fuel spray pattern's importance: A burner must

1. distribute the lamp oil evenly throughout the whole amount of air,
2. prevent spraying oil onto the cool walls of the chamber,
3. reduce the inner chamber space of the burner.[20]

By doing this he claimed there would be less afterburning and soot formation. These ideas resulted in German Patent No. 86,946, which covered "the modification of the combustion curves by retarding or advancing the beginning of injection."[21]

Although the engine reached a mep of 7.25 kg/cm² and gave a reasonably shaped diagram as well (fig. 6.9), Diesel found that the indicated thermal efficiency and fuel consumption had badly deteriorated. These results were recorded on October 11:[22]

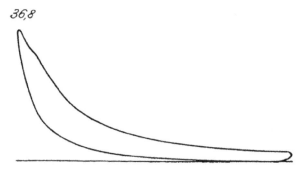

Fig. 6.9. Indicator diagram recorded on October 11,1895, after installation of air compressor. Diesel, *Die Entstehung, 1913*

Thermal efficiency	24.6%	(earlier 30.5 %)
Mechanical efficiency	67%	(same as earlier)
Brake efficiency	16.5%	(earlier 20.26%)
Fuel consumption/bhp-hr	356 g	(earlier 291)
Fuel consumption/ihp-hr =	238 g	(earlier 196)

Diesel sent a very positive, yet somewhat self-justifying letter to Krupp on October 14 inviting observers to Augsburg for a demonstration of a "self contained working machine." Rudolf spoke of the loss of efficiency, and that while he had proved his theory, there was still work to do to reach the expected goal of half the fuel consumption of other engines. The letter stressed that both "a further redesign of the engine and especially an analysis of the exhaust gas are essential." It concluded with the comment that "the now reached development stage is therefore to be considered only as a first beginning, but it nevertheless permits us to take up the struggle against the other systems having a total confidence in the success."[23]

Even though Diesel intended his last sentence to convey optimism, the transmitted results were less than what Krupp had expected after two years of testing. Rudolf did not receive the usual quick response to his letter, so he sent a telegram on the nineteenth: "Request information if you will look at heat engine." Krupp answered saying Schmitz and Gillhausen were coming to Augsburg on October 25. Instead of asking to see the engine perform on their arrival, they told Diesel that Krupp wanted to exercise its right to cancel the contract with the inventor. This completely stunned Diesel, especially when he believed he had made significant progress. Krupp's pulling out not only meant an almost total loss of income for Rudolf, but it also hit Buz hard. A clause in Maschinenfabrik Augsburg's own contract with Krupp read that all development costs between the two companies were not shared as they went along, but that Krupp would later reimburse M.A. only out of its profits from future engine sales. If Krupp quit at this point, M.A. received not a pfennig to cover the Augsburg-incurred expenses.

A compromise was reached after an intensive selling job by both Diesel and Buz on visiting Krupp directors. Diesel's contract with Krupp also called for testing his patented compound engine. Since it had yet to be designed, he received what amounted to a one-year breathing spell before Krupp could sever their contract with him. When the question briefly arose again in December 1896, Diesel had no difficulty convincing Hermann Ebbs of Krupp-Gruson, who was already on Diesel's side, that Krupp should remain a partner.

The Krupp reluctance to proceed with the engine was not entirely based on the October 14 letter. Some of the Essen directors were keenly interested in selling large engines fueled by blast furnace gas. Their disappointment with the results of Diesel's gaseous fuel tests became a frustration as they saw other builders offer low-compression production engines able to burn the waste gases.[24] Later penetration of such a specialized market by an unproven engine would be next to impossible.

Having satisfied Krupp, Diesel pressed ahead with modifications to the original engine and plans for a new one. He would use the old "work horse" for endurance testing and improvements to incorporate in the new engine. An engine-driven fuel pump was added in early November as a step toward unattended running. A plunger pump at the base of the engine operated through linkage off the compressor drive and fed the accumulator (fig. 6.10).

Even before the contract troubles, Diesel had stressed to Vogel his strong belief that a totally new engine was needed. Vogel was not at all in favor of expending this effort, and the two argued over his refusal of support. Diesel had Reichenbach and Nadrowski reporting to him, but the new engine was more than the three could handle by themselves, especially with the continuing work on the old engine. The outcome of this stormy time with his dear friend evoked an unhappy announcement by Rudolf during a visit home in Berlin:

Fig. 6.10. 220/400 engine in the MAN Museum. Configuration of 1896.
Note the fuel pump at the base. *Eduard Klein, MAN Archives*

If my weal and woe hangs in Augsburg, then I have thoroughly alien-
ated myself with Vogel; so once more must I keep quiet in looking after
my business, even to undertake to be at hand. . . . Thus, I must help
myself! To that end it is necessary to move right away to Munich.[25]

Diesel's temporary rift with Vogel might have forced this action, but a
residence change had been brewing in his mind. More than the demanding
technical and business details in Augsburg, it involved his psyche. He simply
could not continue being separated from his family any longer. Eugen Diesel
revealed that for all his father's drive and energy, he needed the support only
home could give him. "He seemed often like a sailor who looked in vain for
the harbor. For all his coolness and calmness he was an eternal refuge seeker."[26]

On October 1, Rudolf broke the news to his surprised family of his
decision to leave Berlin. Two months later they were with the Barnickels in
Augsburg and looking for a house to rent in Munich. The almost forty-mile
train commute would mean Diesel and his family could enjoy the lifestyle
of the large city. During mid-December 1895 they moved into Giselastraße
14, a short street ending at the beautiful *Englischer Garten* in the Schwabing
district.[27]

The move also placed Rudolf and his family in the same city as his
mother and father. This was not really a blessing for Rudolf, or for Martha,
who was to bear the brunt of the often-strained relationship between the
two generations. The younger Diesels had faced growing demands for
money from Theodor, who believed his son to be a rich man. Most of the
money given went to support Theodor's obsession with "magnetic healing."
Although Rudolf's financial situation had greatly improved, there had been a
long period when it meant a sacrifice by him and Martha to aid his parents.
Eugen Diesel spoke of the "middle class prejudices" that his grandfather held
against Rudolf. "During meetings between father and son the topic of money
always lay in the air, ready like a thunderstorm to break loose."[28]

Another momentous occasion by Diesel in October 1895 was the naming
of his engine. Both M.A. factory workers and those familiar with his book
already referred to it as the "Diesel Motor," and Rudolf saw the value of tying
in this existing usage with the engine when it became a product. However, he
still considered other names. In his contracts it was "Rational Heat Engine
Patent Diesel." "Beta" or "Deltamotor" held some attraction.[29] He even toyed
with the idea of "Excelsior," but Martha thought this rather trite. One day,
as the two sat in his Berlin office, Martha advised him to "call it simply
'Dieselmotor'" as it was already referred to. Her husband replied, "You are
exactly right," and that was the end of it.

The first endurance tests of the "Dieselmotor" started November 11 and
continued until near Christmas. In sixteen days of operation, the engine
accumulated a total of 111 hours. Franz Schmucker (1858–1924), who would
be of such help with the first production engines, tended it and kept the test

log. A check on November 18 showed a fuel consumption of 351 g/bhp-hr and a mean effective pressure of 7.6 kg/cm².

Carbon had to be cleaned out of the star burner's central pipe after about forty hours of running. First, a bronze and then a "soft" steel fuel pump piston both showed rapid wear, which made Diesel realize it must be hardened and then ground to size. The change from engine cylinder air to compressor air for starting (because of valve carboning) was successfully adopted.

Diesel wrote Krupp on December 21 to say that the engine was performing well, especially the piston ring sealing. An included indicator diagram showing the almost identical compression and expansion curves when motoring the engine (driving it by the belt) proved his point about the good seal. He said the only nuisance encountered was the necessity to clean the burner every four or five days. He called this operation "very simple (corresponding perhaps to the changing of the carbon rods in an arc lamp), so it is of hardly great importance."[30]

Engine testing recessed before Christmas 1895 to add a constant speed governor for unattended running. A drawing of November 14 shows the design[31] (fig. 6.11). A belt drive from a pulley on one end of the camshaft

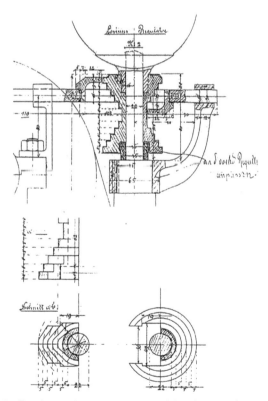

Fig. 6.11. Portions of governor control for the 220/400 engine. From a November 1895 drawing. *MAN Archives*

drove a vertical governor through right angle gearing off the pulley's driven shaft. Governor output twisted a control sleeve through a small arc with speed change. The governor drive shaft turned inside the sleeve.

The hourglass-shaped sleeve was round at its waist and had several cam rings of equal lift but with increasingly larger diameter base circles. These offered a choice of "softness" in speed regulation: For an equal lift, a large diameter ring had a shallower nose angle. This gave a softer regulation and hence less rapid "hunting" under sudden load changes. With a speed change, up or down, the sleeve turned one way or the other inside a slotted opening in a horizontal sliding bar. At the ends of the opening were adjustable "shoes." One shoe contacted the cylindrical "waist" when the other was set to ride on the nose of a pre-selected cam. A spring kept this shoe in contact with the cam. The cam nose determined the maximum fuel position. Through linkage controlled by the sliding bar, the fuel needle valve was moved either in or out. The governor linkage could be easily disconnected from the needle valve so it could be manually adjusted for starting or full-engine operation. It was recorded that this control method held speed within 2.5 percent of nominal. The only vestige of the system as the engine stands today in the MAN Museum is the belt drive pulley on the camshaft.

January 2, 1896, began with the arrival in Augsburg of a young man who would nurse Diesel's engine through those perilous years after production began. He was also to guide Maschinenfabrik Augsburg-Nürnberg (MAN) as it became a world-leading builder of the engine.

Imanuel Lauster (1873–1948), the son of a Swabian shoemaker, was encouraged by his father to seek a technical career. A first step was applying for an apprenticeship at G. Kuhn, a machine- and boiler-making company in Stuttgart. The story is told how fourteen-year-old Imanuel, accompanied by his father, was turned down by the elder Kuhn. The owner's son thought otherwise and called them back as the Lausters were leaving.[32] An intensive four years with Kuhn exposed Imanuel to the design office as well as the shop. During this time, Kuhn was licensed to build under the "Vulcan" name an oil engine patented by Langensiepen. Lauster was assigned to the project and had the job of keeping a Vulcan engine running when it was displayed at the Frankfurt Exhibition in 1891.[33] The engine proved less than a success, but it provided an invaluable experience.

At the end of his apprenticeship, Imanuel entered the *Königliche Baugewerkschule*, a state-run vocational school in Stuttgart. His five-semester record was outstanding enough for him to be recommended to a professor who sought an assistant at the *Technische Hochschule Karlsruhe*. Lauster's one year at the engineering school also afforded him an opportunity to attend class lectures.

It was in Karlsruhe that twenty-two-year-old Imanuel heard M.A. wanted to hire someone with technical expertise at the beginning of 1896. He applied for the job. Lucian Vogel, who as chief engineer handled the search, interviewed him first and quickly realized that the young applicant met all the requirements. He excitedly sent the candidate on to Diesel. Lauster later recalled that Diesel briefly greeted him and then said, "Please calculate the flywheel weight for the motor. I want to know it by evening." Lauster had not done this before but agreed to supply the figure. Diesel returned that evening and on hearing Lauster's answer replied simply, "That tallies within a kilo. I only wished to check if my calculation agrees."[34] The next day the two commenced on the new engine. Thirty-six years later, Imanuel Lauster became chairman of the management board (*Vorstand*) of M.A.N. (fig. 6.12).

Lauster soon learned why Diesel and Vogel had recruited outside the company. His memoirs explain,

> Little by little I came to realize that the entire group in the factory would offer slight confidence in the Diesel business. This had to be the reason why there was no interest from the available personnel at M.A. for the new position as Diesel's assistant; no one was willing to give up his secure position compared with the uncertain Diesel business.[35]

For some time, Diesel had wanted to find out how much, if any, unburned air remained in the exhaust. He did not know whether the inability to increase power resulted from inadequate air in the cylinder or from poor combustion due to the injection system, namely the nozzle and double star burner. He had looked in vain since October to find someone locally who could make exhaust gas analyses. Finally, Krupp loaned one of their qualified chemists by the name of Hartenstein to conduct the tests. He arrived on January 8

Fig. 6.12. Imanuel Lauster circa age fifty. *MAN Archives*

and two days later the first exhaust analysis pleasantly surprised everyone: 12.8 percent CO_2, 0.5 percent O_2, and 0.1 percent CO. At half load it was 6.4 percent CO_2, 11.4 percent O_2, and no measurable CO. Diesel was not entirely satisfied as he noted:

> We have clean combustion to CO_2 and thus almost total use of the available air. An improvement of the combustion process alone is accordingly unthinkable. If we want to achieve larger diagrams [more power], the only way it can happen is by the induction of more air, which could possibly be obtained by the separation of the intake and exhaust valve, for now as for the formerly given reasons, exhaust gases are surely sucked back into [the cylinder] . . . the consumption stays entirely constant.[36]

Diesel thought he had verified this exhaust dilution of intake air in a subsequent test. By applying full power at the instant the starting cams were disengaged, he attained a momentary indicated mep of 8.4 kg/cm². Power then immediately dropped off to the normal maximum of about 7.5 kg/cm². However, tests using better equipment from Essen still showed excess air under full-power operation: 9.6 percent CO_2 and 7.1 percent O_2. Diesel acknowledged the error in his journal and began seeking other ways to increase power while maintaining a clean exhaust.

Most unexpected was the engine's uncharacteristic part-load fuel consumption, something not known to have been checked until this point. Diesel and Ebbs, who came from Magdeburg on January 15 to witness the exhaust tests, observed that at half load the specific fuel consumption (fuel burned per horsepower-hour) was almost the same as for full load. Whether Ebbs explained to Diesel how this was an improvement over other I-C engines is unknown. The advantageous trait was quickly touted by Diesel:

> The tests show totally divergent conditions from all other known combustion methods. The fuel consumption with all other kerosene engines excessively increases 40–70% at half load in relation to full load. With the Diesel process it remains almost constant; down to ¾ load the consumption.[37]

He went on to write that the 28 to 29 percent brake thermal efficiency attained at half load

> . . . in this day and age is not even closely achieved with any system of heat engines.
>
> The explanation of the almost constant fuel consumption at half the load lies in the increase of the thermal efficiency whereby the decrease of the mechanical efficiency is largely cancelled out.[38]

These encouraging results were not kept from Krupp. Ebbs filed a lengthy, positive report praising the superiority of the engine and giving the impression that by taking care of a few troublesome details it was ready for production.[39] Buz also wrote to Essen on January 23 about these "exceptionally favorable" tests and stressed that because of a strong backlog of M.A. factory orders most of the new engines should be built at both Krupp's Grusonwerk in Magdeburg and Gasmotoren-Fabrik Deutz, who must again be urged to take out a license. Buz still thought that lower horsepower engines, a size the other two companies were better able to produce, would predominate.

Another milestone on February 6, 1896, was the taking of the 1,000th indicator diagram. On the fourteenth, the upper part of the engine was disassembled for inspection. Other than a small carbon deposit, everything appeared in order. The cylinder, piston, and rings showed no distress, something the hardworking team had not been treated to before.

The rapidly improving status of the engine, at least from Augsburg's viewpoint, led to a major strategy meeting in Essen on February 20. It was attended by Diesel, Krupp's management, including Ebbs and Vogel, who represented M.A.'s interests. Three areas were agreed on: First, and of greatest importance to Diesel, was reconfirmation of the tentative decision in October to proceed with the new single-cylinder, Series XV 250/400 (250 mm bore by 400 mm stroke) engine that he and Lauster had been designing. Next, drawings for the three-cylinder, Series XIV compound engine were authorized but not to be released to the shop until acceptance of the 250/400 engine's performance. Finally, the existing engine would be used only for proving out components in tests that Schmucker could handle. The plan pleased Diesel.

From March 14 to April 2 the old engine was run with the objective to "maintain exactly the working hours of the factory and that in no way were interruptions of the operation during working hours to take place."[40] Daily maintenance in this seventeen-day period included replacing the carboned-up star burner with a clean one at the end of every day. Diesel recognized the unacceptability of this necessary fifteen-minute operation and stopped the tests. He needed a solution to the burner sooting but believed the engine was otherwise "operationally safe enough."

A new "conical burner," tested on July 16, showed a radical, but crucial shift in Diesel's thinking (fig. 6.13). Its cone-shaped spray pattern from twenty-nine 0.75 mm diameter holes eliminated the bulky burners that extended far into the combustion chamber and rapidly plugged. However, after just a four-hour test with the new design, only five of the holes remained open and free of soot. The power produced when all the holes were open was similar to that from the double star burner. Of significance was a difference in the indicator diagram with this nozzle: It showed an almost complete elimination of late burning on the power stroke. Diesel perceptively analyzed this change in his journal:

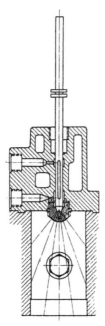

Fig. 6.13. First recessed "conical burner," July 1896; twenty-nine
0.75 mm diameter holes. Diesel, *Die Entstehung, 1913*

In any case the burner shows a significant improvement in contrast
to the previous designs; it especially shows that the holes which are
short and whose direction possibly coincide with the instreaming oil jet
[inside the nozzle] remain totally clean; therefore, one has to prevent
long holes and those which bend away from the direction of the oil jet.
Further, the burner has to be designed to be less unwieldy [*klotzig*], if
possible set totally in the head, and if possible well cooled because the
oil cannot then evaporate and thus no residue can form in the holes.[41]

Lighting gas tests using the twenty-nine-hole nozzle were made while
waiting for a new nozzle design to prove the above assumptions. There were
many initial misfires before attaining reasonably shaped diagrams (fig. 6.14).
Starting on gas without the pilot injection of a liquid fuel was also possible.
Fourteen years later Diesel wrote how this was "proof that the motor is just
as good for using lighting gas as for kerosene, that is, without any [engine]
alteration."[42] He never wavered in the belief that he had been denied an
opportunity to adequately prove the engine's full potential with gaseous fuels.
    Diesel and his colleagues came to see during the gaseous fuel tests that
they were up against another factor beyond their control. The chemical composition and thus heating value of the manufactured coal gas in Augsburg
varied substantially. Settings giving acceptable performance in one test might
not do so in another.

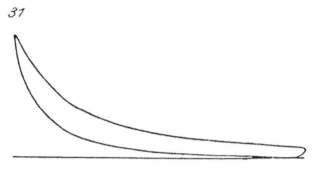

Fig. 6.14. Indicator diagram July 1896 with engine burning lighting gas only.
Diesel, *Die Entstehung, 1913*

An eight-hole conical burner that looked more like a modern nozzle was tried beginning July 24. The indicator diagrams with it held constant over fifty operating hours, and all the much shorter holes remained "completely clean and free" (fig. 6.15). This improvement coming in the obsoleted engine's final hours provided Diesel with vital knowledge.

A last, short test with gas on August 1 prompted the journal comment that "after twelve minutes a totally regulated working without misfire and delayed ignition" was attained in addition to good starting with a cold engine.

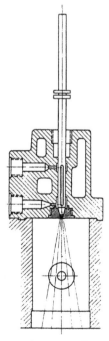

Fig. 6.15. Second "conical burner." Eight holes.
Diesel, *Die Entstehung, 1913*

Fig. 6.16. Diesel's first engine as photographed in 1896. *MAN Archives*

On September 7, 1896, the much-modified, much-abused machine was photographed and disassembled for storage (fig. 6.16). Today it holds the place of honor in the MAN Museum.

The lengthy fifth test series provided Diesel with valuable data and experience to apply to the new engine being designed. He later listed twenty-three "achievements" from this series.[43] Many were merely a brief recitation of a test procedure or objective. Others recognized the engine's inherent, yet important traits, such as the part-load fuel consumption advantage. Several, however, were positive and creative advances that brought nearer the efficient and reliable engine Buz and Krupp awaited.

The Series XV 250/400 Engine

The new engine design offered a rare opportunity as nothing was to be adapted from the existing hardware. Diesel and Lauster could start with a clean sheet of paper to incorporate what had been learned from years of work. In addition to the called-for mechanical improvements were several important lessons they would keep in mind:

- To cool and remove moisture from the air compressed for blast injection.

- To keep the surface-to-volume ratio of the combustion air at a minimum.
- To configure the nozzle location and spray pattern to reduce exhaust smoke by preventing hole carboning and wall wetting.

From a bore of 250 mm and stroke of 400 mm the expected output was to be 20 bhp at 170 rpm. The engine had a distinctive appearance due to a solid, machined bar serving as one side of the A-frame that supported the cylinder (fig. 6.17). No engines used the tripod-like leg configuration after 1897. The Series XV engine was the first to circulate water through the piston. Separate intake and exhaust valves, the high-mounted camshaft acting directly on the rocker levers, and a "flat" plate crosshead similar to steam engine practice were visible parts of the design. The crosshead permitted a short piston because of no side thrust and simplified water cooling of the piston.

Lauster began on the drawings at the end of January 1896 and by the middle of March had them almost completed. Then came a major revision that delayed their release to the shop until April 30. The change resulted in

Fig. 6.17. The Series 250/400 engine on test at M.A.N., February 1897. Its twin is on exhibit at the Deutsches Museum in Munich. Note the Prony brake dynamometer, the cooling water hoses leading to the piston rod for piston cooling, and the lube oil supply line at the bottom of the cylinder.
MAN Archives

the first supercharged internal combustion engine. Several interacting events led to this radical step.

After receiving a go-ahead at the Essen strategy meeting on February 20, Diesel, Vogel, and Ebbs went to the Gruson factory in Magdeburg for detailed planning of the M.A./Krupp-Gruson joint program. As engineers are wont to do, these discussions continued over a meal with ideas being sketched on whatever was at hand. In the *Restaurant Hohenzollern* Ebbs made a rough sketch showing how the piston underside could be used as an air scavenge pump for a two-stroke engine (fig. 6.18). Whether the conversation involved ways to increase available air for combustion is not known, but Diesel had been wanting to increase the air mass coming into the cylinder to raise thermal efficiency (power was not a consideration!). Ebbs's sketch triggered a way of doing this.

Shortly after returning to Augsburg, Rudolf filed a patent application on "The procedure for the increase of power in internal combustion engines." It was a means of supercharging. The patent drawings disclose a flat plate sealing the underside of the cylinder and holding cam-actuated suction and discharge valves. On each down stroke the piston pumped air into a small receiver. This in turn discharged into the combustion chamber past the intake during the normal intake stroke. The application filed March 6 was so broad that when it was published for "opposition" Deutz entered a protest saying the idea was disclosed in Daimler's German Patent No. 34,926 of April 3, 1885.[44]

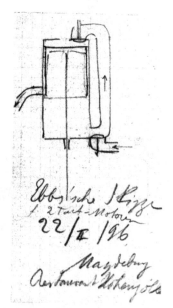

Fig. 6.18. Ebbs's sketch for using the piston underside to pump air in a two-stroke engine. *MAN Archives*

Consequently, the amended application was issued as Patent No. 95,680 on October 29, 1897, with only one limited claim tying the supercharging idea to Diesel's first patent:

> A method of accomplishing the characterized process of Patent No. 67,207 for the purpose of multi-stage compression to the combustion chamber of the one-cylinder engine, [the chamber] being connected to a pre-compression pump by an intermediate receiver, whereby the power can be regulated by changing the pressure in the intermediate receiver.

It is odd that Diesel would think of even partially controlling power by regulating the air rather than the fuel supply. He should have known by then that in his engine the air-to-fuel ratio need only fall within the limits of enough fuel for idling and enough air for a clean exhaust.

After a March 26 meeting with Ebbs in Augsburg, Diesel opted to add supercharging based on Figure 1 of the patent (fig. 6.19). The decision cost about six weeks, but the opportunity for more combustion air carried a higher priority. Ebbs received a set of the final drawings.

Cylinder lubrication initially was as on the old engine, that is, by grooved pieces on the piston skirt dipping into an oil reservoir. Supercharging quickly

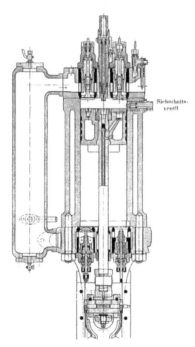

Fig. 6.19. Cross-section of 250/400 engine parallel to crankshaft.
Diesel, *Die Entstehung, 1913*

proved this to be unfeasible because of entrained oil spray being transported with the air via the receiver into the combustion chamber. A piston-type oil pump eased the problem. Operated by the air compressor drive linkage, it fed a metered quantity under pressure to the piston through four cylinder ports at ninety degrees to each other. The stroke of the pump was timed to supply oil only when the rings covered the ports. The rings sealed well enough and less oil did pass into the combustion chamber, but the volume entering from the air pumping chamber beneath the piston remained high enough to require the addition of a baffle-type oil separator in the receiver (fig. 6.20). Diesel reported that "only now are explosions excluded."

Close attention continued to avoid excessively high pressures, which might harm engine components and bystanders. "All dangerous locations" (mainly air tanks) held the "safety rupture valves" used earlier. These accurately machined cast iron discs were "shaped like a hat so the break could only occur by shearing on an exact circumference of a circle." Diesel claimed that their burst pressure "did not vary more than 1–2 atm." His comment regarding their performance bears repeating:

> By copiously applying such precautions in the five-year test period we were able to avoid even the smallest accident. How necessary this was can be seen from the high consumption of rupture valves. A part of the laboratory wall, at which the valves were intentionally directed, was full of holes.[45]

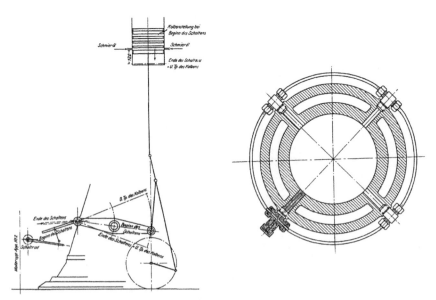

Fig. 6.20. Piston lubrication of 250/400 engine with supercharging.
Diesel, *Die Entstehung, 1913*

Metal machining chips were also strategically placed to prevent flash backs from the combustion chamber or air injection lines into the air tanks or compressor and cause explosions.

The starting valve was moved to the cast iron cylinder head rather than being combined with the safety valve in the large nozzle body of the 220/400 engine. The safety valve was placed horizontally at the top of the cylinder (fig. 6.21).

A much-smaller nozzle body contained only the needle itself. The conical portion inserted in the head sealed at the combustion chamber and at both the fuel and blast air supply ports by being ground to a matching tapered bore in the head. Later engines returned to the previous design of having air and fuel supply passages all contained in the nozzle body. High cost and the tender care necessary to prevent scratches and so on, which could easily cause ruinous leaks, prompted the dropping of a modern-looking design. (O-rings had yet to be invented.) The fuel metering valve bore and seat were machined in the head, and a "turning stuffing box" sealed the needle externally. Diesel said that several major European engine builders adopted and still used this sealing method at the time of his writing *Die Entstehung des Dieselmotors* in 1912.

Five cylinder heads were cast before a combination of improved material and casting methods produced one sound enough. Pressure tests on the first castings began in July and the delay due to the porosity problem postponed final engine assembly until early October. Diesel's journal entry on July 25 reported that the as-cast cylinders and compressor housings passed pressure tests of 80 atm yet showed serious leaks after machining. He urged all such finished parts be tested as a precaution.

The combustion chamber shape followed the dictum that Diesel rightly held to be so important. It was totally contained in the volume formed between a flat piston crown and head to eliminate all extra surfaces. One can recall that in the first design most of the chamber was within the piston and in the second it was recessed in the head. Direct injection of fuel into a compact combustion chamber was the first approach and remains the preferred configuration.[46]

Most of the effort devoted to a fuel injection system to this point focused on the nozzle. Diesel slowly and tediously learned how to create the pressure to inject the fuel charge into the cylinder through spray holes of the correct size, length, and direction. The time was at hand for a flexible and practical fuel metering control. So far it had been by manual operation of needle valves supplying fuel and blast air to the nozzle. The engine demanded constant attention in order to monitor engine speed under varying load conditions. The governor added near the end of the fifth test series was a stopgap measure to allow endurance testing. The air-over-fuel accumulator and needle valve were adequate for test purposes, but that was all.

As a first step Diesel mounted a fuel pump at the end of the camshaft where it was reciprocated by a stub crank journal on the shaft. This eliminated the hand pump supplying the accumulator placed above the engine. Fuel from the pump went to the accumulator and then to the nozzle (fig. 6.21). All early tests on the new engine refer to manual control of the fuel and not by a governor.

As before, fuel and blast air pressure were kept equal. The air volume needed for injection was regulated by a valve located at the starting valve. Tank air pressure fed both the starting valve and the valve spring chamber, which acted as part of the supply circuit for blast air. It then flowed past the needle seat, into an external line (not shown) and to a drilled passage in the head leading to the nozzle.

From the time the engine first ran under power in early December, Diesel knew the accumulator and hand-adjusted fuel metering approach were unacceptable. He said that "at a fixed position of the control valve the amount fed into the engine should be constant whether it runs fast or slow."[47] This was not the case.

On December 12 Diesel wrote that "if we can directly control the quantity of oil from the pump it can feed the nozzle exactly, and the accumulator

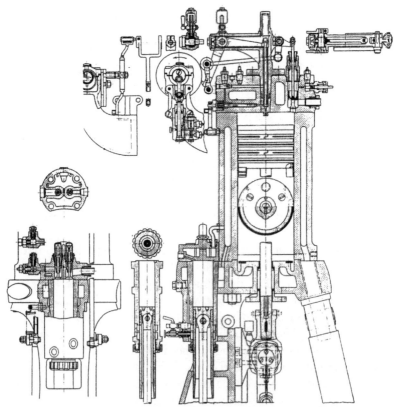

Fig. 6.21. Cross-section of Series XV 250/400 engine at right angle to the crankshaft. Diesel, *Die Entstehung, 1913*

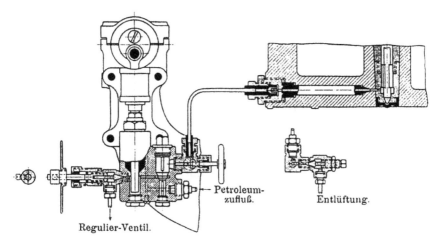

Fig. 6.22. Fuel pump with manual spill valve and check valve in fuel line on Series XV 250/400 engine. Used until the end of January 1897. Diesel, *Die Entstehung, 1913*

with its pipes, cocks, and complications falls away."[48] He suggested two ways to do this: either a variable stroke pump piston or a spill valve in the pumping chamber to bypass part of the charge back to the tank. He chose the latter for simplicity (fig. 6.22). A plug with a small axial hole replaced the fuel metering valve in the cylinder head, and a check valve in the fitting entering the head kept air out of the fuel line to the pump. Two valves were at the pump discharge: one bled air for priming and the other shut off fuel to the nozzle.

The hand-operated bypass valve with its 1 millimeter diameter spill orifice was added December 30 and used until the end of January 1897 when a flyball governor was added.[49]

Governor regulation finally provided "extremely precise" fuel control. A shallow wedge slid in or out by governor linkage acted on the end of a spring-loaded needle valve to vary how much fuel spilled past the overflow valve and back to the tank on each pump stroke (fig. 6.23). The linkage could be adjusted for a range of fixed speeds. A hand-operated lever could override the governor and place the wedge in any desired position. This method was used in the demonstration tests of 1897 and is on the engine exhibited in the Deutsches Museum. Diesel was granted a broad United States patent on the idea in 1900.[50]

A "spray burner" installed on December 12 had a much flatter spray pattern to fit the more compact chamber, although its sixteen holes were not all drilled at the same angle (fig. 6.24). Note Diesel's subtle change in nomenclature.

The new engine was served by a common air tank for both starting and blast injection. A single-stage compressor charged the tank. After the first demonstrations showed one would suffice, Diesel again reverted to separate tanks for generally improved reliability.

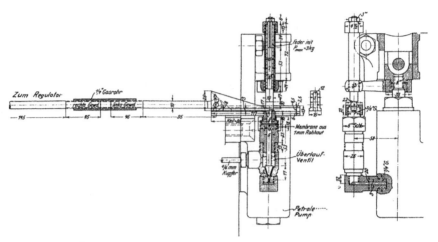

Fig. 6.23. Governor-controlled spill valve for Series XV 250/400 engine.
February 1897. Diesel, *Die Entstehung, 1913*

Ebbs meanwhile waited for a promised set of drawings as agreed on in February 1896 so Grusonwerk could begin to design and build Krupp's first engine. Diesel delayed sending them because the initial operating problems forced numerous revisions. The drawings finally left M.A. sometime in early 1897. Diesel had written Ebbs a long letter on December 21, 1896, outlining his various difficulties to date but closed on this optimistic note: "All in all, about the end of January 1897 we will have a completely mature, beautiful and economical motor which certainly will give us the victory."[51] His timing was not far wrong.

Two identical engines, designated the "A" and "B," were built to Lauster's drawings. Diesel used the "A" for Augsburg tests. The "B" went to Krupp-Gruson for a short time after its completion in mid-1897. It was to serve as a production prototype because Krupp planned to build the engines there rather than at Essen.[52]

The Sixth Test Series—October 1896 to February 1897

The new Series XV engine lived up to Diesel's expectations. It did so well that it created a euphoria that lulled him and others into making premature engineering and marketing judgments.

Between the first belting over of the engine on October 10 until almost the end of December, the engine endured a "shake-down cruise." Various fuel pump packing materials leaked; warped valve cages required new, stronger iron castings; cast iron main bearings with poured "white metal" had to be made. Diesel penned this emphatic journal comment: "Never use bronze bushings for main bearings."[53] Other modifications included quieter intake and exhaust systems. Separation of the cam followers from the cams led to new profiles and a reduction of nozzle needle lift from 8 to 4 mm. The evolving

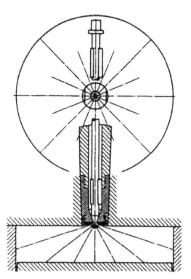

Fig. 6.24. Spray nozzle on Series XV 250/400 engine beginning
December 12, 1896. Diesel, *Die Entstehung, 1913*

fuel system, except for changes at the nozzle tip itself, has already been described. One piece of data coming out of this initial period was the 93 percent volumetric efficiency of the piston underside supercharging pump.

Performance with the "pre-compression" gave mixed signals to Diesel. The diagrams showed an indicated mean effective pressure of 10.6 kg/cm² (150.8 psi), the highest yet attained. However, he said they were "far too large" because the cylinder pressure was still 5.5 kg/cm² when the exhaust valve opened. He said that such incomplete expansion would make "the machine uneconomical." A reduction of air volume entering the under-piston pump (by modifying the suction valving) acceptably lowered the exhaust valve opening pressure and only modestly reduced the power. Adding thicker discs to the piston crown as the intake air pressure was reduced kept the peak cylinder compression pressure at about 34 kg/cm² (484 psi). Indicated mep reached 9.5 kg/cm² (135 psi) using the new spray burner (fig. 6.24) "without producing soot."[54]

The first brake horsepower tests on January 12, 1897, greatly disappointed Diesel. A full load indicated thermal efficiency of only 24 percent, along with a very poor mechanical efficiency of 65 percent reduced the brake thermal efficiency to a low 15.7 percent. As a consequence, the brake-specific fuel consumption (bsfc) rose to 396 g/hp-hr.

The long-lasting effect of these results can be seen in what Diesel wrote years later:

This test . . . also decided the question of the effect of the precompression. It is extremely damaging and will be abandoned from now on.

*So the standard four-stroke motor is established with direct aspiration from
the atmosphere as it is used still alone today.* [Diesel's emphasis][55]

He apparently forgot that one does not receive "something for nothing."
The precompression pump required energy that cost fuel. Its volumetric
efficiency loss reduced indicated mep, and the work to compress air lowered
mechanical efficiency.

The mechanism was removed, and naturally aspirated performance investigations began January 15. The following results are from a January 28 test:[56]

	Full load	Half load
Indicated thermal efficiency	31.9%	38.4 %
Mechanical efficiency	75	61.5
Brake thermal efficiency	24.2	23.6
Bsfc, g/bhp-hr	258	264

Diesel immediately notified Krupp of these impressive results. The pre-
compression pump was never reinstalled.

Air starting had improved to where the engine ran after only a few revolutions without initial misfiring. According to Diesel, it sometimes started with
a single turn.

Full-load exhaust still produced excessive "soot." The suspected problem
was insufficient atomization, so an internal atomizer in the nozzle was added
for the tests on the twenty-eighth. It comprised a wrapping of fine wire mesh
inserted between two washers having small holes drilled on a circular pattern. This assembly was slipped over the nozzle needle valve, just above the
conical seat portion, so that all fuel and blast air passed through it. (fig. 6.25).
Diesel reported these good results with the atomizer: "First the exhaust has
a completely different character. It is invisible at small and medium diagrams
and at large diagrams is whitish, steam-like, nearly without color."[57]

He later remarked that the "effect of the atomizer in the nozzle was a
great step forward." An invisible exhaust at 8 kg/cm^2 was something "we had
never accomplished before."

Word of the success (triumph!) at Maschinenfabrik Augsburg quickly
spread throughout the European engine industry. February 1897 saw an
almost continual procession of its representatives pass through the Diesel
laboratory. Dyckhoff, who led the parade on February 1, witnessed the engine
perform even better: 34.4 percent indicated and 26.6 percent brake efficiency
with a fuel consumption of 234 g/bhp-hp.

Gillhausen from Krupp and Hermann Schumm (1841–1901) from Gasmotorenfabrik-Deutz headed contingents arriving on the fourth. Diesel's daily
journal entry presents a picture of what must have been a nerve-racking time:

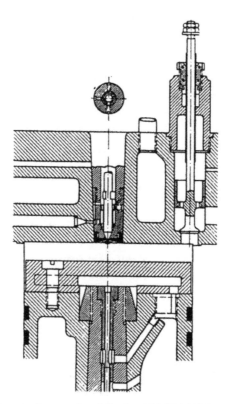

Fig. 6.25. Atomizer in the nozzle of Series XV 250/400 engine. Note the spacer plate on top of the piston. Diesel, *Die Entstehung, 1913*

What is not written in the report is that the gentlemen from Deutz placed the motor in various unfavorable situations in which other motors normally fail. However, the Dieselmotor passed all tests as a victor. In particular was a performance and brake horsepower test starting with an entirely cold machine, i.e., fully abnormal conditions. Also the load was suddenly decreased from full to zero and again to full without any variation in engine speed. Runs were made with the cooling water temperature down to 17°C. The fuel feed during the run was suddenly closed and opened again. Nothing could disturb the quiet, uniform operations of the motor.[58]

Principals of Sulzer Brothers visited Augsburg on the twelfth where they observed a similar performance.

The official acceptance trials were under the direction of Professor Schröter and his assistant, Dr. Munkert, also from the Technische Hochschule München. This detailed investigation lasting several days ended on February 17 and closed out the sixth and final test series.

Schröter obtained these full load results:[59]

Speed, rpm	154
Brake horsepower	17.8
Specific fuel consumption, g/bhp-hr	238
Indicated thermal efficiency %	34.7
Mechanical efficiency %	75.5
Brake thermal efficiency %	26.2

The conclusion of his report to Maschinenfabrik Augsburg and Krupp said,

> "So far I can summarize my judgement about [the engine] that it stands at the peak of all heat engines...."[60]

Schröter's closing statement about part-load performance is also worth noting:

> ...at half load with 9–10 hp the ... [thermal efficiency of] 22.5 percent was achieved. The mechanical efficiency is reduced from 75 percent to about 59 percent at half load, but—and herein lies a very characteristic and practical, significant feature of the Diesel engine— the thermal efficiency percentage of the *indicated* work *increases* from 34.2 to 38.4 at half load! Because of that the motor shows the peculiarity which cannot be shown with any other heat engine, that at half load the consumption per brake horsepower only increases ... 15 percent, a quality which for changing demands, the normal situation, is immensely valuable.

A Tribute to Buz

The February 1897 debut of Diesel's engine realized a goal coming only after many years of peer skepticism, personal sacrifice, and physical exhaustion. Yet in his moment of glory, he shared his triumph. In anticipation of Schröter's findings, Rudolf's February 10 letter to Heinrich Buz expresses a deep sense of gratitude to his long-suffering backer:

> Dear Sir,
> Although I am far from believing that with my engine I have already reached my aim, I may safely say that the results obtained so far are far in excess of what has been realized up to now, and that my principles signify a new era in engine construction.
> I am entirely aware of the fact that I have only been able to come thus far thanks to your kind, pertinent, and generous support. Surely you must have a great confidence in me and my idea, or you would not have assisted me to the extent you have. When no real progress was

being made, you have never even thought of losing your patience but have always tried to reach our aim by your wisdom and great experience.

To thank you for all this is the purpose of these lines. Please rest assured of the depth and sincerity of my feelings in this respect. All that is left to me to do is to express my wish that under your brilliant leadership the new enterprise may prove of great advantage to your firm.

Please pass on my most sincere thanks also to Mrs. Buz for the hospitality I enjoyed at your home and for the friendly words of encouragement and exhortation to be patient expressed on so many occasions and which have always touched my heart and helped me to overcome the difficult periods of experiments.

Assuring you of my highest esteem, I remain, dear Sir,
Respectfully yours,
Diesel[61]

An optimistic circular distributed by Maschinenfabrik Augsburg shortly after the Schröter tests proclaimed what Rudolf Diesel had accomplished:

Up to now we have executed only a single cylinder motor of 20 brake horsepower, which during many months of endurance tests with kerosene operation has in every respect been excellently certified and is completely marketable.[62]

"Therefore, every inventor must be an optimist. The power of the idea has impetus only in the soul of the inventor; only this holds the holy fire for realization."

—Rudolf Diesel[63]

Notes

1. Rudolf Diesel, *Die Entstehung des Dieselmotors* (Berlin: Springer Verlag, 1913), 152.
2. Ibid., 40, says Reichenbach did not arrive until the end of March when the fifth test series started, but MAN records verify the November date.
3. Eugen Diesel, *Diesel: der Mensch, das Werk, das Schicksal* (Hamburg: Hanseatische Verlag, 1937), 229.
4. R. Diesel, *Die Entstehung*, 40.
5. Ibid., 38.
6. Kurt Schnauffer, "Umbauten des Versuchsmotors und erster betriebsbrauchbarer Dieselmotor, 1894'1897" (1958), 39. A patent of addition, filed on January 18, 1896, disclosed a means to increase starting air pressure by adding fuel to the air during the starting cycle and then to ignite it by an externally applied igniter. This method was never used, mainly because of its complication. The German patent issued as number 90,544.
7. R. Diesel, *Die Entstehung*, 38. It is not known whether the engine on exhibit at MAN has the sleeved cylinder.
8. Ibid., 44.

9. Ibid., 41.

10. The prony brake is named after Gaspard Clair François Marie Riche de Prony (1755–1839), one of the founders in 1794 of the *École Polytechnique* and its first professor of mechanics. See older I-C engine textbooks, for example Obert, *Internal Combustion Engines* (Scranton, PA: International Textbook, 1950), 23–5, for its description. Practical only for slow-speed engines, it is much like a water-cooled friction band that holds wooden blocks against the outer rim of a flywheel. The band may be adjusted to exert a variable force or torque, acting through a lever arm, onto a weighing scale. (This writer had the pleasure of using a Prony brake to measure a steam engine's power at the Stanford University engineering lab in the 1950s.)

11. Fuel consumption is given in the original units unless otherwise noted. The first comparison figure, lb/bhp-hr, was used in the United States before adoption of the metric-based ISO standard, g/kW/hr.

12. R. Diesel, *Die Entstehung*, 46.

13. Ibid.

14. E. Diesel, *Diesel*, 235.

15. From tests by Professor Hartmann in trials for the German Agricultural Society (*Landwirtschaftsgesellschaft*) as cited by Schnauffer. Comparable English engines of that period, the Hornsby-Akroyd and Crossley, had respectively 14 percent and 16.7 percent brake thermal efficiencies. See also Cummins, *Internal Fire* (Austin, TX: Octane Press, 2021), chapter 13, for test data on these and other contemporary oil engines.

16. R. Diesel, *Die Entstehung*, 47.

17. Paul Burman and Frank DeLuca, *Fuel Injection and Controls for Internal Combustion Engines* (Springfield, MA: 1962), 10. Depending on compressor efficiency, even the three-stage, intercooled designs of later years had input horsepowers varying from 6 to 10 percent of brake engine output. Discharge pressure of these compressors varied from 600 psi at idle to more than 1,100 psi at full load. Diesel used air injection pressures of only about 5 atm or 71 psi above compression pressure in his early engines.

18. Harry Ricardo, *The Internal-Combustion Engine*, vol. 1, Slow-Speed Engines (London: Blackie & Son, 1922), 427, states that engines of this time returned about 60 percent of the indicated work (40 percent of the brake) as useful power.

19. Little is known of Nadrowski. There are brief references in Rudolf and Eugen Diesels' writings and correspondence in the MAN Archives. These 1895 letters between Rudolf in Augsburg and Nadrowski in Berlin were written when Diesel was trying to settle who would pay Nadrowski's salary if he came to work at the M.A. factory. Nadrowski was pressuring Diesel to make a decision because he had gone to work for AEG in Berlin and had to let them know when he was leaving. According to M.A.N. records, Nadrowski was never directly on their payroll.

20. Schnauffer, "Umbauten," 45. An October 5 journal entry by Diesel.

21. R. Diesel, *Die Entstehung*, 49.

22. Ibid. Dyckhoff, who witnessed a test on November 20, 1895, with Diesel, Reichenbach, and Vogel, published the taken data in a bulletin, *Notice sur l'essai au frein*, dated December 1895. The tract came to the MAN Archives via Augustin Normand many years later. It is the only known published performance data made before the 1897 acceptance tests. The results approximate the October 11 performance:

Speed, rpm	167.66
Indicated mean effective pressure, kg/cm²	6.7
Indicated horsepower, (metric)	19
Power to run compressor, hp	1.6
Friction, fhp	5.51
Brake horsepower, (metric)	11.89
Bsfc, g/bhp-hr (metric)	350
Mechanical efficiency, %	68.5
Thermal efficiency, %	24.7

The fuel was an "American lamp oil kerosene" bought in Augsburg with an LHV of 10,800 cal.

23. Ibid., and Schnauffer, "Umbauten," 46. It is worth comparing Schnauffer's selections with what Diesel in 1913 quoted from his October 14, 1895, letter. In turn, Sass excerpts another sentence to suggest Diesel painting an even rosier picture. See Friedrich Sass, *Geschichte des deutschen Verbrennungsmotorenbaues von 1860 bis 1918* (Berlin: Springer Verlag, 1962), 464.

24. By 1900, engines up to 2,000 ihp burning dirty, low btu blast furnace gas were sold. See Cummins, *Internal Fire*, 244, 273–91, et seq., for more on these engines.

25. E. Diesel, *Diesel*, 239.

26. Ibid., 238.

27. Giselastraße 14 today is a four-story residential building rebuilt after World War II.

28. R. Diesel, *Die Entstehung*, 225.

29. Ibid., 235. Diesel was thinking of De Laval's centrifuge separator (cream being one use), which the Swede had named the "Alphaseparator."

30. R. Diesel, *Die Entstehung*, 50.

31. The drawing calls for a "Cosinus Regulator," which turned at 250 rpm with an engine speed of 180 rpm. Fried. Krupp Grusonwerk lists the governor as one of their products on its 1893 stationery. Letter to Diesel in MAN Archives. The engine photo, taken before removal from its test stand (fig. 6.10), shows only the governor shaft support brackets. The governor and its control linkage were removed for the last gas test and not reinstalled for the photo or later when it went into the MAN Museum.

32. E. Diesel, *Diesel*, 246.

33. Bryan Donkin, *A Text-Book on Gas, Oil, and Air Engines*, 2nd ed. (London: Charles Griffin & Company, 1896), 364–65, describes the Vulcan engine as "working well" at 600 rpm when he saw it at the Frankfurt show. Sass, *Geschichte*, 468, says that when the wind blew from the wrong direction the engine's bad smelling exhaust "attracted very unpleasant attention." Lauster no doubt contributed the olfactory recollection.

34. E. Diesel, *Diesel*, 247.

35. Sass, *Geschichte*, 469.

36. Schnauffer, "Umbauten," 48.

37. R. Diesel, *Die Entstehung*, 51.

38. Schnauffer, "Umbauten," 49. This part-load advantage remains because mechanical friction losses in modern diesel engines are not all that different from SI engines. When speed is held constant and power is reduced in SI engines, the increasing flow restriction caused by the closing throttle valve in the air intake increases the air pumping losses during the intake stroke. At idle, the piston works like a vacuum pump. Diesels are "free breathing" engines. For an excellent treatise on pumping losses see R. V. Kerley and K. W. Thurston, trans. "The Indicated Performance of Otto-Cycle Engines." In *Society of Automotive Engineers*, vol. 70 (1962), 5–37.

39. Schnauffer, "Umbauten," 49.

40. Ibid., 51.

41. Ibid., 52. Diesel, *Die Entstehung*, 53, said seven holes were open at the end of seventeen hours.

42. R. Diesel, *Die Entstehung*, 53.

43. Ibid., 55–6.

44. The Daimler/Maybach valve in the piston crown allowed air from a pressurized crankcase into the cylinder only during the exhaust stroke to aid in scavenging. For details of this historic engine, see Cummins, *Internal Fire*, 254–57.

45. R. Diesel, *Die Entstehung*, 63.

46. All larger modern diesel engines use direct injection, that is, they do not inject first into a pre-chamber. Direct injection engines are inherently about 15 percent more efficient (less heat transfer and pumping losses), and as the emission problems of the smaller DI engine are being solved, its application is becoming almost universal.

47. R. Diesel, *Die Entstehung*, 72.

48. Ibid., 69.

49. The governor may have come from the old 220/400 engine. Diesel commented in 1912 that "The predecessor of this [new] motor and the numerous test objects which over the years had become a large collection have not been preserved." Ibid., 78.

50. US patent No. 654,140, filed September 10, 1898, and issued July 24, 1900. Claim 1 of the four claims broadly covers any governor means to return fuel; it does not mention an overflow valve or a suction valve. Wetmore & Jenner, Adolphus Busch's patent attorneys, stated this was a key patent because of the very broad Claim 1 covering any spill off to the tank. Wetmore & Jenner

to E. D. Meier, Busch-Sulzer papers in Wisconsin State Historical Library (March 15, 1907), 8.

51. R. Diesel, *Die Entstehung*, 70.

52. Ibid., 93, says the "B" engine was returned in March 1898, but letters and shipping papers in the MAN Archives indicate it came back on October 7, 1897. A historical footnote to the "B" is that recent research shows it to be the one exhibited in the Deutsches Museum and not the "A" as supposed. The Museum, in a January 13, 1904, letter co-signed by Oskar v. Miller and Carl v. Linde, asked Buz for the "first" engine. His response was positive. Buz said later, in an August 4, 1905, letter that in addition to an engine of the same construction as that tested by Schröter, he would also donate the 1893–1895 engine. He never spoke of the "A," which can only mean it no longer existed. Shipping papers show the "B" and the 1893–1895 engine (returned to MAN years later) left Augsburg on August 12. All documents pertaining to this transfer are in the Diesel papers at the Deutsches Museum. Since the "A" and "B" were identical, were built and performed similar functions at the same time, their historical significance should be viewed as basically equal. It then becomes understandable why, at the time, both engines were considered of equal merit. We thus must be glad that the museum sought to preserve one of the "twins." The writer thanks Dr. Ing. F. Heilbronner of the museum and Josef Wittmann of the MAN Archives for their cooperation in researching the author's question about the engines.

53. Ibid., 65.

54. When describing pressures, Diesel used "atm" (atmospheres) for proof test and cylinder pressures; for calculated pressures such as imep it was kg/cm². In his time these two were essentially equal: 1 atm = 1 kg/cm² = 14.224 psi. Today a standard atmosphere (at) is 14.695 psi, slightly more than the preferred ISO "bar": 1 at = 0.98066 bar. As the maximum difference between any of the above is less than 4 percent, the author chooses to use originally given units unless otherwise noted. Lab experimental errors probably "soften" the discrepancy.

55. R. Diesel, *Die Entstehung*, 71.

56. Ibid.

57. Ibid.

58. Ibid., 74.

59. Ibid., 84–85. Diesel's results are verified in Moritz Schröter, *Bericht über die Versuche ausgeführt im April 1897 an einem Versuchs-Motor mit Einem Cylinder von 20 Effectivpferden Normalkraft gebaut von "Maschinenfabrik Augsburg"* in MAN Archives (Augsburg: J. Walch, 1897), 13, 14, and 27.

60. Schröter, *Bericht*, 30.

61. MAN's English translation of Diesel's letter. MAN, *Fünfzig Jahre Dieselmotor, 1897–1947* (Augsburg, 1948), 22–23.

62. R. Diesel, *Die Entstehung*, 90.

63. Ibid., 152.

CHAPTER 7

Plaudits, Patents, and Purgatory

"The introduction is a time of the fight against stupidity and jealousy, laziness and malice, secret resistance and open struggle, the terrible time of conflicts with humans, *a martyrdom also if you are successful."*
—Rudolf Diesel, 1913[1]

The stormy years after the introduction of Diesel's engine may best be recounted by Charles Dickens's immortal prose: "It was the best of times, the worst of times, it was the age of wisdom, it was the age of foolishness. . . ."[2]

Enthusiasm ran high at Firma Fried. Krupp and Maschinenfabrik Augsburg after the engine yielded an efficiency almost double that of its nearest rival. Development expenses will be recouped, and profits will flow, they thought. Reality, however, failed to follow this script.

Rudolf himself anticipated the fulfillment of a cherished dream and soon basked in the recognition of his accomplishment. He seldom visited Augsburg anymore as his time was occupied by licensing activities and the promotion of the engine. From an office ensconced in his posh Munich apartment, he directed new projects and encouraged further development work at M.A.'s lab.

An unexpected legal attack led to a spirited defense of Rudolf's 1892 patent and triggered an intensifying struggle with his psyche. Intervals of seriously weakened mental health and a need for retreat from normal routine followed. Each new personal criticism or failure of his engine exacted a further toll.

For valid reasons, in his mind, Diesel formed a patent holding association to which he assigned all his rights in return for paper profits and freedom from day-to-day activities binding him to Augsburg. This decision would profoundly affect his own affairs and his licensees.

A disastrous involvement with a new engine company bearing Diesel's name cost him money, loss of reputation, and widened the growing breach in the once-close relationship between himself and Maschinenfabrik Augsburg.

Interwoven with Diesel's personal concerns were his efforts to create an international license network and to continue the ongoing work at Augsburg. These concurrent happenings to launch the engine are related in the ensuing two chapters.

The New Contract with Krupp and Maschinenfabrik Augsburg

The successful acceptance test by Schröter led the engine principals to reassess where they themselves stood. Diesel, Buz, and directors Schmitz and Klüpfel for Krupp clarified their mutual obligations with a revised contract signed in Essen on March 11, 1897. Its first clause stated that with the three parties "having built and tested a marketable motor of the Diesel system [they] shall as swiftly as possible begin the reasonable factory production of Diesel motors."[3]

Important for Diesel was a stipulation extending his annual stipend of 30,000 marks for at least five more years. This could increase to 50,000 marks as payments came in from other licensees. He in return agreed to support both firms in the development of the engine and to reduce their patent royalty obligations to him from 25 to 5 percent. They retained their assigned sales territories in Germany yet could jointly sublicense other German companies within the country.[4] Diesel continued to hold patent rights for all countries except the Austro-Hungarian empire, which he had earlier assigned to Krupp.

A follow-on agreement of April 22 further benefited Diesel: Krupp and M.A. were to reimburse him for his office expenses at 4,000 marks per year for two years, to assume payment of the heavy annual patent fees,[5] and to give him a 5 percent royalty on engines built for in-house use.

Chargeable development expenses for Augsburg and Essen to this date totalled 172,000 marks. M.A. incurred most of it, but since Diesel had received an additional 135,000 marks as a salary directly from Krupp, it meant each company had invested almost equally in the engine between 1893 and April 4, 1897.[6]

The Lectures

Diesel delivered several lectures in 1897 that served as public announcements of the engine's availability. From these peer encounters he also received praise for his achievement and a general acceptance of his theories.

The first lecture, in the M.A. Kantine on April 27, was given for Augsburg industry leaders. He repeated it the next day at a Munich meeting of the VDI (German Engineering Society).

A joint lecture by Diesel and Schröter in Kassel on June 16 was before a major congress of the VDI. They offered data proving the engine's superior thermal efficiency and spoke optimistically of its market potential. Diesel's segment, covering theory and construction, purposely avoided any mention of how the engine strayed from the patented concept. Schröter described his tests and ended with "As a representative of the technical sciences I heartily join in these congratulations and express the hope that this motor may be a point of departure to bring prosperity to the industry."[7] Rudolf was at the zenith of his career (fig. 7.1).

Another service was performed at Kassel. Diesel expressed his deep appreciation to Buz and M.A. for their support, and he singled out two co-workers: "I desire to thankfully acknowledge the assistance of Lucian Vogel and Fritz Reichenbach, engineers, with regard to the new motor. They devoted themselves to the experiments with never-failing perseverance, and thus most successfully advanced the work."[8]

These remarks were made as a favor to Vogel who had resigned from M.A. in April and had asked Diesel to mention "his work efforts in the development of the engine in an appropriate way." Vogel gave Diesel his reasons for leaving in an April 12 letter:[9] "The tension has existed a long time. Lately it has made me angry as I had expected some recognition when the final results were obtained with your engine. Instead I was blamed because the engine did not run immediately. . . . What you saw coming happened, though faster than I expected." Vogel cited a recent example of how he had been bypassed in a matter where he believed he should have been consulted. He also touched on his last confrontation with Buz. Years later Diesel recalled that "Vogel's departure was a heavy blow for me."[10]

Fig. 7.1. Diesel, Buz, and Schröter at the Kassel lecture. *MAN Archives*

Munich Life

Rudolf anticipated receiving a huge income from his licensees. With this pleasant thought in mind, he moved his family into a well-located Munich apartment in July 1897. Their building on Schackstrasse 2 was also home to important artists and musicians. Furnishings from the best Munich stores, the finest clothes, a manservant, and a French governess give evidence of the newly acquired trappings of wealth. Oil portraits of Rudolf and Martha, at 1,100 marks each, looked down from a living room wall. Noted in family account books are tickets for frequent visits to the opera, theater, and concerts, music lessons for the children, money for needy relatives and aspiring artists, as well as a dark room for Rudolf's new hobby of photography.[11] Eugen Diesel, an impressionable eight-year-old when the move was made, later recalled vases filled with flowers, festive dinners, and family outings.[12]

Diesel next rented an adjoining flat on the same floor to house an office for himself and a staff of engineers he would begin hiring. They were to serve as designers, run test programs at Augsburg, and when necessary, aid licensees in Europe, England, and the United States with their first engines. Most of these recruits to the *München Diesel Büro* went on to have distinguished careers in academia and industry. In addition to Diesel's old employee Nadrowski, there was Ludwig Noé (1872–1949), who joined in the summer of 1897. Also coming aboard that year were Rudolf Pawlikowski (1868–1942) and Anton Böttcher. Paul Meyer (b. 1870) and Karl Dieterichs (d. 1945), a college friend of Böttcher, arrived in January 1898. Paul Flasche, Rudolf's brother-in-law, acted as office manager for a time.

The outward trappings of success and security for Diesel do not portray the whole picture, however. Noé related an incident that occurred during his first interview with Diesel in Augsburg during the early summer of 1897: Holding his head, Diesel lamented, "I cannot go on. I cannot go on. I am finished, I can do no more. I need young people who can help me." Shortly afterward when the two were taking a walk, Diesel moaned and took off his hat, "Oh my head, my head." Then, "Can you tie yourself to a sick man for a few years? . . . I don't know what is going to happen. . . . Can you hang your life for some years on an idea?" Noé said he was deeply moved by Diesel and his plans and decided to work for the very distraught man.[13] This incident witnessed by Noé foreshadowed increasing anguish to come.

Emil Capitaine

A telling legal fray was unexpectedly thrust upon Rudolf in July 1897 by a tenacious new nemesis who had to be regarded with concern. Emil Capitaine (1861–1907), a Frankfurt engineer, had already made significant contributions to the early internal combustion engine. He had granted patent licenses to several German firms for his higher-speed, oil-burning engines. Thousands were built from his practical designs between the mid-1880s to after 1900.[14]

A family of these vertical cylinder models introduced in 1891 incorporated a much higher than normal compression pressure of 15.7 bar (228 psi). This served to raise the compression temperature to 350°C, just enough to marginally ignite gas oil (a mid-range distillate with a specific gravity of 0.84 to 0.87). A very long ignition delay, however, dictated that a heated tube be used to reliably initiate burning of the fuel within a practical time.[15] Diesel knew of these engines and considered them in his study of the oil engines described in Richard's 1893 treatise. He dismissed them as not anticipating his own invention (chapter 5).

Capitaine had followed Diesel's work closely and clearly saw that what was patented in 1892 differed substantially from the combustion method obliquely shown in the Kassel lecture. He knew he must act quickly in order to take any legal action, because under German patent law the five-year statute of limitations was about to expire. On July 31, 1897, he filed a "plea of nullity" against Diesel's Patent No. 67,207, which stated that it was invalid as the claims broadly covered what his own prior German Patents Nos. 60,801 and 60,977 disclosed.

Diesel chose not to dispute the case publicly because of pending license negotiations. Yet it was a blow to his pride and compounded his ever-increasing emotional stress.

"Phase 2" of Capitaine's campaign was his lecture, "Critique of the Diesel Motor," read before the Frankfurt district of the VDI on April 20, 1898.[16] In it he recited chapter and verse of how Diesel's real engine differed from that claimed in the patent. Capitaine brought into the open what Diesel had carefully tried to gloss over, namely the change from isothermal combustion and a consequent need to cool the cylinder:

The Process	The Machine
1. Establishment of the *highest temperature* through *compression not through combustion*.	1. Establishment of the *highest temperature* through *combustion not through compression*.
2. During the combustion *no* temperature increase.	2. During the combustion *temperature increase approximately doubles*.
3. Selection of the air charge *so that* the practical working of the motor without cooling water is possible.	3. Selection of the air charge *so that* the practical working greatly needs vigorous cooling of the cylinders.

Capitaine continued:

> Such an unusual contradiction forces upon every expert the very obvious question: Which elements cause Diesel to retract that evident non-agreement with his earlier work on his petroleum machine with those clear, practised claims of this theory?[17]

In addition to citing Köhler's analysis, he quoted Professor Meyer, who had written in the VDI journal that isothermal combustion *"must result in the most inferior conceivable efficiency."* Capitaine added, "And against this crushing judgement on the . . . working process, Diesel himself says nothing!" Further on: "But here comes Diesel recently [the Kassel lecture] with a definition of his claims that adjust his entire working process."

Capitaine offered the sage advice that "when an inventor frankly admits [that his goal] was reached in an entirely different manner he suffers no loss in his honor nor his actual earnings." He later metaphorically digs Diesel about changing his Carnot process into one requiring a cooled cylinder: "From Saul is suddenly becoming a Paul!"

A lengthy explanation was given as to how Diesel built on the work of others by merely raising compression pressure and temperature. The passage ended with more satire:

It seems strangely curious to the expert when he observes how a host of writers step forward . . . and emulate with their dithyrambics [enthusiastic statements]. Prepared eulogies have surely sung the great and important hygienic and societal significance of the Diesel invention, an invention revealing in the light of truth a simple reduction of the compression chamber, a feature long since recommended by others before Diesel became acquainted with petroleum machines.[18]

Professor Schröter was singled out as being one of those in the "host of writers."

Diesel delayed his reply to Capitaine until after he had received a most fortuitous legal decision from the German Patent Office. Its ruling on April 21, 1898, just one day after the Frankfurt lecture, stated that "The plaintiff's case partially rests on a mistaken interpretation of the significance of the challenged patent, and secondly partially on too broad an interpretation of the plaintiff's patents."[19] Rudolf was extremely lucky in the Patent Office decision because it had mistakenly assumed that his engine really did follow the Carnot process. He rightly achieved a victory but for the wrong reasons.[20]

Diesel reprinted the entire court document on June 25, 1898, and distributed it with a "To whom it may concern" cover letter entitled "Answer to Emil Capitaine's Critique of the Diesel Motor by Rudolf Diesel." In it he could not resist his own prick of a needle: "It is therefore enough that in the following announcement of this decision each individual assertion of Mr. Capitaine in a plainly scathing manner was rejected."[21]

The "skeletons" resulting from Diesel's earlier poor judgment were again returned to the closet, so Diesel hoped, but a few days later on July 4 Capitaine appealed the decision. He argued that the gradual injection of fuel into air compressed merely to a higher degree was not new.

At the same time, he issued a rebuttal in which he challenged Diesel's cover letter paragraph by paragraph:

No, Herr Diesel, the German Patent Office had not rejected each and every statement of Herr Capitaine in plainly crushing ways.

That is untrue![22]

After citing a part of the Patent Court's decision, Capitaine gave his own judgment:

The expert sees at once that the patent will be infringed only when combustion follows without a higher temperature, i.e., stays at a constant temperature.

The actual Diesel engine has more than twice as high a temperature, and thus has nothing to do with my challenging of Diesel's patent. Everyone can freely attain the Diesel engine without fearing a conflict with Diesel's basic patent No. 67207.

Capitaine, in the interim, covered his bets by offering to settle out of court with Diesel.[23] Although Krupp was against paying anything, the two men reached an agreement signed on July 12, 1898, in which Diesel paid 20,000 marks for Capitaine's dropping all litigation and all further comments on the patent or engine, in print or by word of mouth. He would pay a 20,000-mark penalty, per occurrence, for doing so.[24]

This private deal is a likely source for the derogatory comments made later by several British engineers who firmly believed that it was not Diesel but Akroyd Stuart, or Capitaine, who deserved the credit for compression ignition and the "heavy oil engine."[25] Arthur F. Evans wrote in 1930, without citing a source, that Capitaine "probably sold his invention to Diesel, and to remove any ambiguity he [Evans] will put this forward as an historical fact."[26] As late as 1944 the subject reappeared during his discussion of a technical paper: "I have it on excellent authority that Capitaine made the first Diesel engine and that the engine is stowed away in some museum in Germany, but . . . I have not been able to verify it."[27]

An armed truce remained between the inventors, and in 1907 Emil wrote several letters to Diesel and his lawyer. He argued that because others now said the diesel engines currently being built did not follow that 1892 Diesel patent, and operated as his own patents claimed, he should be free to break his imposed silence. Capitaine merely wanted "to be recognized as the inventor of the diesel process of today's diesel engines"[28] and did not claim title to Diesel's patented process. Rudolf ignored the request on his lawyer's advice, and that ended it.

Capitaine spent his last years designing marine gas engines with integral gas producers. These "suction gas" producers converted anthracite coal into a low heat value (Btu) gas, which allowed the use of a higher compression ratio. The overall result was an economical operation burning a solid fuel. His most notable engine/producer unit went into the sixty-foot yacht

Emil Capitaine built by Thornycroft of England in 1904.[29] William
Beardmore of Glasgow, Scotland, owned the British rights to Capitaine's
patents for the gas engine system. Emil developed a severe heart problem
while working with that Clyde area firm and was taken to Brussels where he
was cared for by a sister. He died there on December 14, 1907.[30]

Diesel Motoren-Fabrik (Aktien-Gesellschaft)

Augsburg, the birthplace of Diesel's invention, became the home for two
other factories building the engine, in addition to that of Maschinenfabrik
Augsburg. One, L. A. Riedinger, was a minor player and remained outside
the mainstream of events.[31] The other, while secondary in its capabilities, significantly affected the struggling infant industry and its principals during the
engine's crucial introductory period. It also bore the Diesel name (fig. 7.2).

The idea to form this new company came from someone other than
Rudolf, but he readily accepted it. The prospect of a handsome, quick return
on his initial investment, and a seat on the board of directors of the Augsburg
Diesel Engine Factory, Inc., were strong incentives to lend his presence.

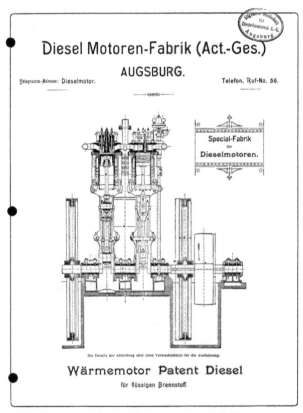

Fig. 7.2. Augsburg Dieselmotoren AG Brochure.
Diesel papers, Deutsches Museum

Neither could he overlook the prestige that would accrue. The offer excluded his active participation in engineering or management activities.

August Gerstle, owner of an Augsburg bank of that name, signed an option contract with M.A. and Krupp in November 1897 to secure German manufacturing rights for engines of 15 to 40 horsepower. As Gerstle did not disclose who were to be the other owners, the option contract stipulated that a subsequent license could not be assigned to anyone already in the business of making internal combustion engines. Concurrently, an unnamed third party bought an idle gear factory for 400,000 marks and turned it over to the organizing corporation.[32]

Diesel, Gerstle, and Bonnet, another Augsburg bank owner, formed a syndicate before the engine company came into being. Each would own one-third of the 1,200 total shares, to be valued at 1,200,000 marks, but needed to pay for only half of their 400 shares. The banks, acting as agents, were to sell the other half to interested buyers, and the proceeds would go to the principals. Only after these shares were sold could the three dispose of any of their remaining 200 shares. Christian Diesel (1857–1938), Rudolf's first cousin and rising Augsburg businessman,[33] and a Robert Jansen each bought twenty-five shares and were asked to serve on the board.

A payment to M.A. of 100,000 marks accompanied the signing of the license agreement on April 15, 1898. The agreed-upon standard royalty was 20 percent for engines 16 hp and under, and 30 percent for engines with outputs over that. The minimum annual payment was 20,000 marks.

Given the prevailing euphoria over the engine, one can see why Buz, who stood to gain immediate and future profits, was unconcerned about facing a local competitor. Reinforcing his contention of "strength through numbers" was a large Diesel pavilion about to open at a Munich exhibit; it would demonstrate engines made by four major German firms (chapter 8). The new diesel company's organizers peered through rose-colored glasses at their own profit potential. What none saw ahead was an impending disaster that at least Buz and Diesel were derelict in not anticipating.

At the core of the problem was a naive unconcern for the manufacturing skills required of a company wanting to produce a new and unseasoned, high-technology product. Even M.A. with their reservoir of capable people and facilities had to fight an uphill battle just to keep their first engines in the field (as will be seen in chapter 8). Diesel Motoren-Fabrik A.G., conversely, selected management, foremen, and shop workers with no previous engine experience. Their well-used machine tools were mostly inherited from earlier gear-making days.

Initial orders were not a problem; by October 1898 there were seven on the books. The first, a two-cylinder 40-hp engine, was shipped to the brewery of Gebrüder Glaubitz in Seelow (about forty miles east of Berlin) on March 23, 1899. Based on Diesel's advice, it was not built to M.A. drawings; he had heard of the troubles M.A. was encountering with their production design.[34] What these changes were are unknown.

Although the engine reportedly ran without difficulties in the shop, it immediately acted up after it went to the customer. From this point on the situation deteriorated. At the top of the problem list was the wire mesh fuel atomizer in the nozzle that was also giving M.A. fits. Many of the troubles paralleled what others were experiencing, but Augsburg Diesel Motoren-Fabrik (ADM) lacked the capability and the apparent understanding to react in a manner that gave the customer any confidence or compassion. A basic shortcoming was that, unlike M.A., ADM simply erected the engine, saw it run, and returned home. A major reason why M.A. salvaged their reputation and kept their engines from being ejected was a firm policy of requiring their service men to live with an engine, if need be, until it performed as guaranteed over a reasonable time. Adequate customer training of those who would ultimately maintain the engine was another M.A. credo.

The new company did manage to keep their first engine at Seelow, but others were eventually replaced with those built by M.A. In defense of ADM, however, two of their installed engines reportedly gave good service.

Nevertheless, the stories beginning to reach Diesel's office in Munich were alarming. Noé recalled years later that "the factory was in every respect poorly organized and managed. Never in his life did he ever again see such miserable workmanship."[35]

The outlook for ADM finally appeared so hopeless that on July 1, 1900, engine production ceased just two and a half years after the company was formed. It remained in limbo until 1906 when it went into liquidation; final dissolution came in 1911.[36]

What was the fallout of this catastrophic venture? It affected Diesel, M.A., and the ADM stockholders of course, and, more importantly, the diesel engine market.

The reason the upstart firm held such a disproportionate influence on the overall industry centered on the very one which enticed Diesel to join forces with it in the first place: his name could sell engines but in turn, under adversity, it could generate a severe backlash. Confusion, innocent or calculated, also arose as to who was who between the "Diesel Engine" factory and that "other" Augsburg firm that gave birth to the engine. Worried steam and gas engine competitors lost no opportunity to spread word about ADM's problems, and as is wont to take place, facts, and rumors based on distorted facts, compounded the plight of Krupp and M.A.

The latter should carry some blame for what happened to ADM as they, by conscious act or simple negligence, withheld information on the improvements that were easing their own engine problems. Such dissemination of developing knowledge was an emphasized part of Diesel's personally negotiated license contracts and were an expected requirement among all licensees. When ADM contacted Buz in September 1899 about a royalty reduction because of their problems, he replied that ". . . we never had these enormous difficulties and since recently we have experienced no difficulties at all."[37] Buz was not being honest in the first part of his statement regarding their first engine delivered to Kempten as will be seen in the following chapter. ADM did not find out until later that Buz's "since recently" referred to a retrofitting of all their engines in the field with an answer to the atomizer problem. Other improvements included narrower piston rings to reduce break-in time and raise mechanical efficiency, cam and linkage changes to reduce injector nozzle needle actuation noise, and fuel pump and compressor revisions.

The shareholders of ADM lost their money, but Rudolf suffered a greater torment than his monetary loss. He became the scapegoat for what happened, and there was little he could do about it. In a letter to Sulzer, he wrote that "the unheard of mismanagement by those in charge could not be openly pilloried because it would only appear that Diesel himself was making excuses"[38] and thus would make matters worse. The blow to his pride by the failure of "his" company was compounded with the criticism of his engine, which was also perceived to be a failure.

Yet neither can Diesel entirely shirk his share of responsibility for what happened. With his experience during the initial development years at M.A., it is difficult to understand how he could have been blind to what was occurring. In his defense, however, he was in virtual seclusion for months due to recurring bouts of illness during ADM's crucial times. When seemingly restored to health, he concentrated on his whirlwind schedule to expand the growing licensee family (chapter 9).

The question may also be asked if the organizational structure of German companies might not have made it difficult for him to change what he saw. The supervisory board (*Aufsichtsrat*) on which he sat presided over the management or operating board (*Vorstand*). As only one of several on the supervisory board who had numerous absences from its meetings, he may have lacked the influence over fellow members who were naively unconcerned about manufacturing and service deficiencies.

Buz used the troubling circumstances to lecture and chide Diesel over what the inventor did to the engine's market potential. He had a letter hand-delivered to Rudolf on December 14, 1900, which read in part:

> As long as such miserably built engines as in Breslau, Pensa and Arad are not removed from this world, they will be a steady brake on the whole Diesel program. . . . It is very deplorable that the competition particularly received this means to successfully fight the Diesel program. . . .[39]

A few verses from a scathing poem published in an Augsburg newspaper underscores another element resulting from the company's poor record (after a French engine received the Grand Prize at the 1900 Paris World Exhibition):

On learning of the great French award
achieved of late by Diesel's motor,
It's swiftly to my banker I call
With the joyous hope of what might befall.
But says he, giving his shoulders a shrug,
"The business goes into the hole it dug."
"Mere paper," tells he of my stock.
"The Paris news won't make it go up."
Now this becomes my only thought:
Taking with me the shares I bought,
Onto the Graf Zeppelin I hie
And make my ascent into the sky;
When I then to great heights attain,
I can end quickly all my pain,
Since upward again my shares will not go,
I plunge myself into the sea below.[40]

Satirical yes, but also a grim prophesy it was to become for Diesel, a major investor.

Allgemeine Gesellschaft für Dieselmotoren A.G.

The General Company for Diesel Engines, Inc., Diesel's brainchild, came into being on September 17, 1898. In exchange for shares in the "Allgemeine," a stock company to be owned principally by himself and a few others, he surrendered his patent rights, future royalties, and current ownership interests in any licensee firms (chapter 9). It was to cost him his fortune.

How much Rudolf's deteriorating health entered into his decision to create the Allgemeine cannot be known, but it must be considered a major factor. He had been emotionally drained by the patent fight with Capitaine, and he feared he could not continue under a simultaneous burden of business affairs in Munich and engine developments at Augsburg.

One worry Diesel did not have: lucrative license contracts had already made him a cash millionaire. The potential income from minimum annual royalty payments alone assured his further fiscal security.

On the other hand, the dismal performance reports of the first production engines must have given cause for concern. In addition to the loss of his reputation, it could easily translate into a delay or permanent forfeiture of royalties. Much credence has been placed on Diesel's acute awareness of this possibility and his consequent desire to quickly "bail out" at a profit before potential engine customers discovered just how dire its problems were. However, it is the writer's view that Diesel bore too much pride and blind faith in what he had achieved for him to truly doubt ultimate success. If indeed a hasty deliverance really was his motive, the expected engine-builder Allgemeine stockholders from Augsburg and Essen were privy to the same worrisome news he had. They stood to lose the value of their Allgemeine investment plus what they had already put into the engine.

Deeply seated, and often downplayed, was Diesel's long-held dream to independently chart his own course and free himself from the shackles of corporate obligations. This, along with discouragement over his mental state, had to be the prime catalysts for his escape by means of the Allgemeine. The tragic irony is that Rudolf was called upon to do nothing. He initiated the action to form a new company on his own and had to convince the participants of its validity.

What Diesel had in mind is explained in a letter he sent on July 8, 1898, to Berthold Bing (1847–1915), whom he had befriended during negotiations with the American licensee a year earlier. Bing had served Diesel well then and again later when a new German company was formed to license Russian interests.

The letter to Bing expresses strong feelings intermixed with barely camouflaged ulterior motives. It also recites the core of the contract Diesel ultimately signed:

> For some time financial people have urged me to convert my engine enterprises into a company; these opinions are frequently expressed recently by people I should take seriously, but I always have declined to do so because all financial affairs are unpleasant to me.
>
> Owing to the great growth in my affairs which have grown so much they can hardly be handled and threaten to drown me, and especially because of my last attack of nerves which caused, and still causes the worst fears, I am more and more thinking of putting my attractive and flourishing enterprise on a broader, more secure basis and to divorce it more from myself. I even regard it as my duty to all of my esteemed associates. As joining forces with financial circles does not agree with my personal views, I am considering the return of my entire enterprise to those who brought it into life together with me. . . .
>
> If the current owners of my patents and licenses form a company unto themselves, then all their expenses for royalties, etc. will flow back into their own pockets, and it can almost be said that the companies thus can build my engines without paying royalties, as my compensation for transferring my rights is very modest. It only takes into account the current status of business and does not anticipate the future.
>
> . . . I transfer into the enterprise all my shares [in other diesel companies] at par value—they are worth much more. Furthermore, I also transfer my existing but yet unexploited patents along with those contracts having guaranteed minimum royalties which are already far exceeded today. I will sell only this part of my business for cash or shares. All the rest, including royalties from all countries and other business interests, would represent nearly clear profit for the company,

as the minimum royalties and the value of the unused patents is already worth more than [if what I propose were to be liquidated].

I am of the opinion that my allied firms will be sympathetic to this proposition and they will see from it how thankful I am to all of them for their support of my affairs until now; especially will they acknowledge that I am turning over real, existing values with the result that, as indicated before, they can build my motors without royalties, and respectively collect fees from the other licensees according to their shares.

Of course, by this I waive my entire future profits, which I view as quite important, but I hope to be compensated by an adequate, yet large as possible share, to be determined by my partners.

The state of my health does not allow me to take part in these negotiations. You, my dear Herr Bing, already have successfully acted in some of my enterprises. . . . You know the situation as well as myself, therefore I have thought of you to handle this for me. . . . I know I could not put this matter into better hands.

. . . If my proposals will meet, as I hope, with the consent of the associated groups, I shall be liberated at last from commercial activities, which is not in my line, and which upsets my health and has already distracted me for more than a year from the central aim of my life—the technical and scientific promotion of my engine. It is my only wish to dedicate all my energy entirely to this aspect of enterprise.[41]

Poor health and a hope for more personal freedom were unquestionably honest motives, but Diesel was stretching the truth to say he had no interest in the business side of things. He was, of course, trying to sell Bing on acting as his agent in any negotiations, and this may account for some of the exaggerated comments.

Rudolf had already sent Heinrich Buz a memorandum on the Allgemeine concept before he wrote to Bing.[42] It stressed how Augsburg and Essen would benefit by their almost "paying themselves royalties." It was Diesel's strong desire from the outset that these two companies become major stockholders.

An unknown element in the evolution of the Allgemeine is the role played by Albert N. P. Johanning (1859–1911), Diesel's business manager after January 1898.[43] With his extensive financial and manufacturing background, Johanning would very likely have advised Diesel on the wording of the memorandum and the letter to Bing. He also could have been one of the "financial people" Diesel referred to.

The Allgemeine Gesellschaft für Dieselmotoren, A.G. began with a capitalization of 3,500,000 marks: 1,500 preferred shares at 1,000 marks per share and 800 common shares at 2,500 marks per share. The main preferred shareholders were Diesel with 250 shares, Buz with 200, Krupp with 200, M.A.N. with 100, and Bing with 150.[44] In return for turning over his patents, and so forth, Diesel received all 800 shares of common stock valued at 2,000,000 marks. With his 250 shares of preferred stock, his investment in the venture came to 2,250,000 marks.

Diesel was to receive 3,500,000 marks in cash by a separate agreement on the same date: 875,000 immediately and the rest in ten days. He used this to pay for his preferred and common stock and another 150,000 marks to cover all legal and other costs to form the company. His net cash gain upon the Allgemeine's creation was 1,100,000 marks.

When compared with Rudolf's previous engine-related income, this sum was not that significant. He had already received 1,820,000 marks through license contracts since his first one in 1893. All but 26,000 marks of it came in 1897 and 1898. Added to this were the 135,000 marks in annual stipends paid by Krupp and the 65,000 marks M.A.N. would give him in royalties on June 30, 1899.[45] Diesel's engine had thus earned him almost 2,000,000 marks prior to the forming of the Allgemeine. He had sold out for half of that amount.

Diesel was to assign all his future patents to the Allgemeine. He would, however, be reimbursed for patenting and "for experiments and travels and business expenses connected therewith . . ."[46]

The original supervisory board (*Aufsichtsrat*), chaired by Buz, included Diesel and Bing plus the major licensees: Adolphus Busch (America), Fr. Dyckhoff (France), L. Klüpfel (Krupp), Emanuel Nobel (Russia), Marcus Wallenberg (Sweden), and businessmen, Wilhelm Finck and Max Schwartz.[47] Johanning was the sole member of the management board (*Vorstand*) at its founding and served in that position until November 1903. That Diesel's fellow board members accepted "his man" to manage the Allgemeine is a sign of the cooperative spirit that prevailed at the start.[48]

Rudolf's expected business plan failed to materialize. First, his personal interest in the Augsburg Diesel Company and the French diesel company, valued at 700,000 marks in the capitalization, had to be written off in 1905 because of the Augsburg Diesel Company's demise and an inability to collect from France over a contract dispute. Between 1898 and 1905 there was a further loss of about 1,300,000 marks in unrealized income from struggling licensees. A few were refunded their money. This hemorrhage of slightly over 2,000,000 marks was blamed on Diesel.

To placate the other shareholders, he was "urged" to surrender his common stock whose par value almost equaled that sum. In return, Diesel was given 250 more shares of a preferred stock worth 250,000 marks. On March 31, 1905, the Allgemeine's capitalization was cut in half to 1,750,000 marks.[49] No one else contributed a pfennig.

Diesel signed a new contract on the same date that he thought freed him from any obligations to the Allgemeine. He also resigned from the supervisory board. Unfortunately, he interpreted the release differently than the Allgemeine, which led to charges, countercharges, lawsuits, and bickering between him, M.A.N., and several licensees. This fractious phase is more appropriately left to following chapters recounting the formation of Diesel's international license network.

The Allgemeine diminished in value over time because the expiration of Diesel's basic patents was fast approaching and because several major licensees refused to share their ideas and experience. This negated Rudolf's important tenet for the association of "one for all and all for one."

On February 27, 1911, the Allgemeine Gesellschaft für Dieselmotoren A.G. was liquidated. The original shareholders received 102.5 marks per share or 10.25 percent of their original investment.[50] Since they had been paid an annual interest of 5 percent, their net return over the life of the company was about 6 percent.

Diesel had shared in these annual payments, but some of his preferred stock had been "called" over the years.[51] He received 47,252 marks at the liquidation for his remaining shares. With annual interest, liquidation settlement, and payments for early share "calls" at 1,000 marks per share, Diesel received a total of 409,050 marks.[52] Even when this is added to the 1,100,000 marks netted at the founding, it still comes to less than half of what all thought to be "money in the bank" in 1898. It was solely Diesel's loss.

In addition to suffering his huge financial beating, Rudolf endured another bitter disappointment: no longer was he a major player on the engine's team; he had unintentionally, yet effectively, sidelined himself.

In hindsight one must ask "What if?" Would Diesel have done better without the Allgemeine? He could have fared worse. He was extremely fortunate that his basic patent survived 1898 intact. If he had been "one against the world," he might have faced further court actions. More importantly, the initial engine problems brought losses he would have had to write off anyway. Licensees could easily have used the troubles as an excuse to end their contracts (a few did) and leave him with less than he actually made. This second guessing assumes Diesel was a healthy man who would at least have fought a good fight.

A Dispirited Man

During the time of his engine's "childhood diseases," Rudolf went through a lengthy siege of his own personal hell. His high-strung temperament, years of intense, stressful work, and too-often combative interaction with colleagues and adversaries led to a prolonged treatment for his increasing periods of mental unease. Not to be forgotten is the probable inheritance of genes from a father and a grandfather who, in their own later years, lost touch with reality. A compelling argument exists that Diesel was a classic manic depressive.[53]

For some months Diesel's doctors had urged him to forego all business and engineering activities related to the motor, but a fear of how it might suffer and what its detractors might do if he withdrew from the scene kept him going. It was as if he had a premonition of what lay ahead when he wrote Bing about creating the Allgemeine.

In the fall of 1898 Rudolf committed himself to the Neuwittelsbach sanitarium on the outskirts of Munich. Except for a pleasant visit home during Christmas, he remained there until the end of January 1899. Eugen Diesel wrote of visits he and his mother made to his father's Spartan room.[54]

For the next stage of his "recovery," the doctors sent him to a secluded spa near Meran, a resort town 70 km south of Innsbruck in the Tirolean Alps, where he stayed until the end of March 1899.[55] For about six months Diesel was out of touch with the various Augsburg engine enterprises except for occasional letters. In addition to M.A.N. and the ADM, the Allgemeine had also moved to Augsburg from Munich the previous October.

Diesel and his family spent the following June and July near the northern Italian resort area of Madonna di Campiglio (fig. 7.3). His several letters to Augsburg from Italy show that he was returning to a normal state of mind.

Except for extended summer holidays, Rudolf never again went into periods of seclusion such as those at the Munich sanitarium and Meran. However, his hold on outward normality was tenuous.

Villa Diesel

Diesel began plans in the summer of 1898 for a new home commensurate with his envisioned wealth. His months in the sanitarium gave him time to pore over the drawings of a mansion to be built in one of Munich's most exclusive areas. He did this in secret as it was against doctors' orders for him to think about anything related to his work. A large design office for his engineers and business staff formed an integral part of the house.

Fig. 7.3. Diesel and children at Madonna di Campiglio, Summer 1899.
MAN Archives

Maria Theresia Strasse 32, a large corner lot overlooking the park along the Isar River, cost him 200,000 marks. The house itself added another 700,000 marks and took two years to complete. Nothing but the best went into its construction. Exotic materials from everywhere made up the decor (figs. 7.4 and 7.5). Diesel ultimately spent about 1,000,000 marks, one-fifth of his paper assets in the fall of 1898, on the new home by the time he and his family moved into it in the spring of 1901. *Villa Diesel* was conceived as an ostentatious monument to the products of the mind. Instead, it would become a mausoleum for the shattered dreams of an embittered man.

Fig. 7.4. *Villa Diesel*, Munich. *MAN Archives*

Fig. 7.5. *Villa Diesel*, interior. *MAN Archives*

Inscribed on a crystal medallion over the entry gate was the phrase, *Hic habitet felicitas, nil mali intret*—May good fortune live here and misfortune never enter.[56] The wrought iron garland that held the medallion remains, but the inscription is gone.

The young engineers who already worked for Diesel went through a transition. Karl Dieterichs and Ludwig Noé transferred to the Allgemeine in Augsburg shortly after it was formed.[57] Paul Meyer followed them there in January 1899.[58] Anton Böttcher, who made the first American, British, and Swiss engines perform, left Diesel to seek other employment in August 1899.[59] The circumstances of Pawlikowski's earlier departure in February 1898 are related in the next chapter.

The end of the century coincided with an ebbing of Rudolf Diesel's energy. Before, he drove himself almost without limit to work on his brainchild. After his illness and the release given him by the Allgemeine, his stamina and creativity were never the same. Diesel's major achievements were behind him, and too often he appeared by some to only bask in the afterglow of a sun already set. Yet the engine remained a vital part of his soul. He continued to defend and nurture it in the ways left open to him and with the internal fire he could still draw upon.

A Postscript: An Engineer's Utopia

The detachment from engine affairs resulting from his illnesses and his estrangement with the Allgemeine gave Diesel time for enforced reflection. An eccentric literary project grew out of this unwanted freedom. In 1903 he published *Solidarism, The Natural Economic Freeing of Mankind*, a book in which he solves the "social question" of labor.[60] The "Engineer in Munich" added after his name on the title page undoubtedly connoted his deeply held belief that a logical, engineer's mind could also solve the growing problems caused by the advance of an industrializing society. Rudolf probably began writing it soon after moving into the "Diesel Villa." His ideas can only be classed as utopian.

Much of the book's thrust draws from nineteenth century French social dreamers.[61] Rudolf's exposure to these ideas came during his years in Paris after his return. Among other ideas, he proposed that since workers were undisputed owners of their labor, they should enter into a lengthy contract spelling out their rights and responsibilities. It called for each worker to contribute a "penny a day" to an account managed by a safe savings institution. This could be withdrawn to finance individual work endeavors or be received in the form of periodic payments. Those failing to live up to the worker association's rules would be excluded from its benefits. Every aspect of his proposal was painstakingly detailed. He presented his thoughts in a manner much as he had in 1893 with his book *Theorie und Konstruktion eines rationellen Wärmemotors*.

Diesel annotated the title page of his personal copy with the words *Love, Brotherliness, Charity, Peaceableness, Veracity,* and *Justice* around a symbolic

Fig. 7.6. Title page of Diesel's copy of *Solidarismus. MAN Archives*

letter *S* (fig. 7.6). These relate to the "The commandments of Christianity that are also found in Solidarismus. The highest pursuit of the church is the realization of these laws by mankind."[62]

Contemporary reviews of the tedious book were mostly negative; more charitable ones referred to his thinking as naive. Ten thousand copies were printed, but only a few hundred were sold.[63]

Notes

1. Rudolf Diesel, *Die Entstehung des Dieselmotors* (Berlin: Springer Verlag, 1913), 152.
2. Charles Dickens, *A Tale of Two Cities*, the opening sentence of chapter 1.
3. R. Diesel, *Die Entstehung*, 89, 90.
4. A contract signed June 25, 1897, modified this as detailed in chapter 9. Kurt Schnauffer, "Lizenzverträge und Erstentwicklungen des Dieselmotors Im-In-und Ausland 1893–1909," in MAN Archives (1958), 5.
5. A German patent at that time cost 7,120 marks over its eighteen-year life: 30 marks per year the first four years and rising eventually in its last to 1,200 marks. Eugen Diesel, *Jahrhundertwende* (Stuttgart, Germany: Reclam, 1949), 82.
6. Schnauffer, "Lizenzverträge," 5, 28. His figures are taken from a "statement of development costs" dated June 16, 1900, in the MAN Archives. The exchange rate around 1900 was about 4.1 marks per US dollar.
7. "Diesels Rationeller Wärmemotor," reprint from *Zeitschrift des Vereines deutscher Ingenieure* (1897), 19. An English translation, *Diesel's Rational Heat Motor* (New York: Progressive Age Publishing Co.), was printed in 1897.
8. Diesel, *Diesel's Rational Heat Motor*, 14.
9. Diesel did this, but sadly, Vogel felt the praise was not strong enough. From here on Vogel became estranged from Diesel even though he still spoke well of the engine. Diesel's efforts did

help Vogel find a job in the engine design department at Nürnberg before Maschinenfabrik Augsburg united with it. However, in January 1899 he returned to Augsburg where he worked on special projects and reported directly to Buz. Vogel's daughter, when commenting on the estrangement between the two men said, "It is certainly not easy to work with my father." From "Lucian Vogel und seine Mitarbeit am Dieselmotor," in MAN Werk Archiv, (August 12, 1940), 7 pages. Parts of this story are also in Diesel, *Die Entstehung*, 89, and Diesel, *Diesel: der Mensch, das Werk, das Schicksal* (Hamburg: Hanseatische Verlag, 1937), 282.

10. R. Diesel, *Die Entstehung*, 89.

11. E. Diesel, *Diesel*, 274.

12. Ibid., 275–76.

13. Ibid.

14. Cummins, *Internal Fire*, 318–23. The Capitaine oil engines could only be started by first heating an external vaporizing chamber with a blow torch until it was hot enough to allow continuous ignition of the fuel charge sprayed into it. Enough residual heat was retained to sustain combustion when running under a load.

15. Hugo Güldner, *Das Entwerfen und Berechnen der Verbrennungsmotoren*, 2nd ed. (Berlin: Springer, 1905), 96. He uses the "old" atm (1 kg/cm²) or 14.22 psi.

16. *Kritik des Dieselmotors*, ten-page reprint of Capitaine's lecture. Diesel papers, Deutsches Museum.

17. Ibid., 2.

18. Ibid., 9.

19. "Urtheile des Kaiserlichen Patentamtes, Nichtigkeitsabtheilung," (Decision of the Imperial Patent Office, Invalidation Dept.), in Diesel papers, Deutsches Museum, 4.

20. Kurt Mauel, retired from the Deutsches Museum, nicely sums up the affair in "The Independence of Contemporaneous Inventions in the Field of the Combustion Engine (Otto engine, Diesel engine)," a paper read before the Society for the History of Technology (SHOT), (Washington, DC: December 1973), 33–4. He said that the facts were Diesel "*did* have genuine self-ignition because of the high pre-compression, which was not the case as far as Capitaine was concerned."

21. "Antwort auf Emil Capitaine's Kritik des Diesel-Motors," with the attached six-page reprint of the decision in MAN Archives (June 25, 1898).

22. "Antwort auf Rudolf Diesel's Kritik des Vortrages von Emil Capitaine," in MAN Archives (July 4, 1898), 2 pages.

23. Document signed by Capitaine and witnessed by Diesel's lawyer in Munich on July 12, 1898, 3 pages. MAN Archives.

24. Capitaine wrote to a lawyer in Munich on May 29, 1899, in which he avowed that it was he who went to Diesel and not the other way around. This letter was in response to one a few days earlier to Diesel from a Fritz Döpp, an engineer in Berlin, who flatly stated that Capitaine had told him Diesel wanted to pay hush-money (*Schweigegeld*) to settle things. Correspondence in MAN Archives.

25. Adding fuel to the fire was a story by T. H. Barton about a brief experiment at the Hornsby factory when he worked there in 1892. The vaporizing chamber on a Hornsby-Akroyd oil engine was removed, and instead, the water-cooled fuel nozzle, remounted in a plate forming a new cylinder head, sprayed oil directly into the main combustion chamber. These engines normally injected fuel into the vaporizing chamber during the intake stroke; it is not known if a fuel timing change was made for this test. An oak block was also inserted at the crank end of the connecting rod to raise the compression ratio. After starting difficulties, the engine reportedly ran under a load for six hours until the chief engineer stopped the test because of severe ignition knocking. No fuel measurements were taken. T. Hornbuckle, "The Hornsby-Akroyd Engine and Later Developments in the Heavy Oil Engine" (Herbert-Akroyd Stuart Lecture, University College, Nottingham, England, 1949), 15, 16.

26. Arthur F. Evans, *The History of the Oil Engine* (London: Samson Low, c. 1930), 50. Evans's quote first appeared in a paper delivered in London to the Institute of Marine Engineers, November 26, 1929.

27. Harry Shoosmith and Philip D. Priestman, *William Dent Priestman and the Development of the Oil Engine* in Diesel Engine Users Association (London, 1944), 15. Letter from E. R. Hillier included in the discussion of the paper.

28. Capitaine correspondence (from Dalmuir, Scotland) to Diesel and his Munich lawyer beginning March 22, 1907. The latter "stonewalled" it, and nothing further transpired. Capitaine file, MAN Archives.

29. The 75-hp engine and its gas producer are fully described in Cummins, "Suction Producer Gas Motor Yacht 'Emil Capitaine,'" in *The Practical Engineer* (1905), 431–33. See also Cummins, *Internal Fire*, 288–91, for more information on gas producers.
30. Found in MAN Archives. A January 1, 1908, letter to Diesel from Mrs. Bertha Riehl, Capitaine's sister, tells of Emil's last project and his death. She had asked Diesel to help her brother's family, who were left destitute.
31. L. A. Riedinger, Maschinen- und Broncewaaren-fabrik A.G., Augsburg, bought a license from MA on March 2, 1898. Vague records show they built only two 30-hp engines: one installed December 1899 in Augsburg's Hotel Drei Mohren after much effort, and the other in a Munich hotel in April 1901. The latter used a two-stage air compressor. Schnauffer, "Lizenzverträge," 18–19.
32. Ibid., 20. The purchased factory, Zahnräderfabrik Augsburg, was formerly under the name of Johann Renk.
33. Found in the MAN Archives, Christian Diesel's lengthy obituary in the *Neue Augsburger Zeitung*, February 5, 1938, makes no mention of the diesel company.
34. Schnauffer, "Lizenzverträge," 21.
35. E. Diesel, *Diesel*, 323.
36. Schnauffer, "Lizenzverträge," 23.
37. Ibid.
38. E. Diesel, *Diesel*, 323.
39. Schnauffer, "Lizenzverträge."
40. E. Diesel, *Diesel*, 325.
41. Diesel's letter to Bing, July 8, 1898. Stefan Loewengart, Berthold Bing's grandson, translated the entire letter in "From the History of My Family, The Bing Family of Nuremberg" (Israel: 1980), 17–18. The author has borrowed from this translation. The original letter and the interesting Bing history are in the MAN Archives.
42. Diesel's, "Stand der Diesel-Motor-Unternehmungen am 30 Juni 1898, Memoranda" (Position of the Diesel Motor Enterprise) in MAN Archives, eleven pages, was sent to Buz, who replied on August 27 that he was studying it.
43. Johanning's "Curriculum vitae" in MAN Archives.
44. Other shareholders were Hermann Pemsel, a Munich attorney with 70 shares, Martin Arendt, a Munich investor with 80, and three banks (one in Augsburg, one in Berlin, and one in Munich) with 700 shares. These men also joined the supervisory board. The Allgemeine "partnership contract" (Gesellschafts-Vertrag), MAN Archives, and Schnauffer, "Lizenzverträge," 115.
45. Summary of payments to Diesel ("Zusammenstellung der von Herrn Diesel . . . Zahlungen") in MAN Archives (February 9, 1937).
46. "Contract for Sale" (*Kaufvertrag*), September 17, 1898, between Diesel and the Allgemeine, Busch-Sulzer papers, Wisconsin State Historical Library, 5.
47. Allgemeine, "Bilanz . . . und Bericht des Vorstandes für das Geschäftjahr 1900." MAN Archives.
48. Johanning left to work full-time with the Nürnberg branch of the AEG. He had moved to Nürnberg in 1901 and managed the Allgemeine business from there as needed since it had become less than a full-time job. It was apparently not an entirely amicable parting. Johanning's story had a tragic ending. Diesel wrote to the Allgemeine on April 21, 1911, that Johanning "has recently taken his life and left his family in the most sad conditions." His last position had been director at the H. Büssing Automobilfabrik in Braunschweig. The Allgemeine's liquidation lawyers wrote to him on March 23 about two shares he was thought to own and wanted to pay him his due. Johanning responded by postcard two days later that he no longer had them. Documents and correspondence in the Allgemeine file, MAN Archives.
49. Allgemeine, "7 Ordentlichen Generalversammlung am 20 Februar 1906 für das 7. Geschäftjahr 1905." MAN Archives.
50. Allgemeine, "12 Ordentlichen Generalversammlung am 27 Februar 1911 für das 12. Geschäftjahr 1910." MAN Archives.
51. In his 1905 contract with the Allgemeine, Diesel agreed to a phased call of his shares at a rate varying from year to year. The 1898 "partnership contract" specified that this could be done by a lottery (*Auslosung*).
52. The total is also broken down by annual components in a letter of November 8, 1990, to the writer. Grateful thanks to Josef Wittmann, MAN Archivist, for his painstaking search through Allgemeine annual reports, liquidation settlements, correspondence, and contracts, which he

also kindly sent. He concisely summarized for the first time what Diesel had received after the Allgemeine's founding.

53. A good case for this is made in Donald E. Thomas Jr.'s, *Diesel: Technology and Society in Industrial Germany* (Tuscaloosa, AL: University of Alabama Press, 1987), 17, 223.

54. E. Diesel, *Diesel*, 318–19.

55. Ibid., 320.

56. Ibid., 353. In 1987 the author had the pleasure of touring the house. Externally it looks as it did when the Diesels lived there as it suffered almost no damage during wartime bombings. The ornate stairway and fireplace in the entry hall remain. Most bathroom fixtures, kitchen, and tiled pantry walls, and a few stained glass windows are still in evidence. The steam heating boiler was replaced, but the rest of the system is intact. The current lessee, a large recording company, appreciates the structure's past and has preserved what is known to be original. Unfortunately, a government agency leasing the house for years, had already altered much of the original interior.

57. September 28, 1899, letter of reference Diesel wrote for Dieterichs. MAN Archives. The original company headquarters for the Allgemeine was in Munich.

58. Paul Meyer, *Beiträge zur Geschichte des Dieselmotors* (Contributions to the history of the Diesel engine) (Berlin: Springer, 1913), 38. Meyer, by this time a professor at the Technical University in Delft, wrote of the difficulty in describing his former mentor: "What Schiller said of Wallenstein also fits him: 'Affection and hate are entangled . . .'"

59. Found in the MAN Archives, a letter entitled "Personal Matters" to Johanning from Diesel in August 1899 tells of Böttcher's difficult personality and how it upset both his own office staff and the people he worked with at the companies who were building their first engines.

60. Rudolf Diesel, *Solidarismus. Natürliche wirtschaftliche Erlösung des Menschen* (München, Germany: Oldenbourg, 1903), 124 pages. The writer wishes to thank Klaus Luther, head of the MAN Archives between Irmgard Denldnger and Josef Wittmann, for a photocopy of Diesel's own annotated copy in the MAN Archives.

61. Thomas, *Diesel: Technology*, 50–67, delves more deeply into the philosophical background and into Diesel's proposals. Diesel, *Jahrhundertwende*, 201–13, also explains cited passages.

62. R. Diesel, *Solidarismus*, 65.

63. E. Diesel, *Diesel*, 371.

CHAPTER 8

The Proof of the Pudding: 1897–1900

"Everything went quite well as long as a mechanic from Augsburg and an engineering school professor were permanently on hand."

An early engine owner[1]

The new engine Diesel and Maschinenfabrik Augsburg offered their first licensees and customers stopped far short of being a reliable product. Much of what befell M.A. after its earliest engines left the factory is expressed in the satire of the above quoted buyer.

Rudolf soon realized that the demonstration engine he had planned as a production prototype was too expensive. His modestly revised design, which M.A. broadened into a family of engines, endured their own trial by fire. Major and minor failures in the lab and in customer hands gave evidence of their unreadiness.

At the heart of the design and manufacturing flaws was an inherent deficiency in the injection nozzle. Several frustrating years passed before M.A. and others overcame this weakness, and the engine's future remained in jeopardy until then.

An extensive test program Augsburg started in mid-1897 was an unusual undertaking for a still-emerging I-C engine industry. It was vital because never before did an engine like Diesel's so push the state of the art. He encouraged many of the tests even though his personal involvement was minimal.

The old lab engines performed for visiting dignitaries and acted as guinea pigs for constant production revisions. More importantly, the engines proved their multifuel capabilities. Both gaseous fuels and a broad spectrum of liquid fuels were burned. Powdered coal was briefly tried.

Diesel's compound and Güldner's horizontal, two-stroke engines were tested at Augsburg but with disappointing results from both.

Diesel's Series XVI 260/410 Engine

The success of the independent tests in February 1897 convinced almost everyone that the Series XV 250/400 lab engine could go into production. One of the reasons Diesel sent its twin to Krupp-Gruson in Magdeburg was to have them use it as a manufacturing prototype.

After complaints that the design was too expensive, Ebbs from Magdeburg and Wilhelm Worsoe of Krupp-Essen met with Diesel in Augsburg the first of April to simplify it to reduce manufacturing costs. A family of engines ranging from 15 to 50 hp resulted from the meeting.[2]

The first member of this new family was the Series XVI engine, which Lauster designed during a six-week period beginning in mid-June. Functionally similar to the Series XV, its bore and stroke were increased to 260 × 410 mm. This 11 percent displacement change (to 21.77 liters) would raise output to a preferred 20 bhp at 160 rpm. The test engine had produced only 18.5 bhp at that speed. External modifications included a conventionally cast A-frame cylinder support and a smaller, faster turning (340 rpm at rated speed) governor mounted on the bedplate and driven *off* the crankshaft (fig. 8.1).

Diesel offered his early licensees drawings and a detailed design analysis.[3] Krupp's first Essen-built engine of November 1897 was one of this type.[4]

Because adequate injection air pressure at lower speeds had been marginal, the bore and stroke of the single-stage compressor went from 70 × 200 mm to 85 × 205 mm. An added "automatic" valve controlled the maximum air pressure.

Diesel insisted on an overload capability of 26 bhp at 200 rpm, so the intake and exhaust valve diameters were increased by 15 mm to 85 mm for improved breathing.

Water-cooled crankshaft main journals stayed the same size (130 mm diameter × 320 mm long), but the connecting rod journal was enlarged to 140 mm diameter by 150 mm long. Disk-shaped crank throws for the built-up crankshaft were retained so as "to give the shaft great stiffness." Diesel strongly advised "white metal" crankshaft bearings; bronze was *verboten* (forbidden). A large crosshead pin (90 mm diameter × 130 mm long) reduced an excessive bearing temperature, while the flat guide retained its bronze slides. Its loading was such that it never ran beyond "blood warm." Normal cylinder peak operating pressure was 35 atm, but pressures might go to 50 atm during starting or when "adjusting" the fuel and blast air valves. The safety valve was set to unload at 59 atm.

Flywheel mass depended on engine application; generators and "spinning frames" required less speed variation per cycle. Diesel's method of determining cyclic changes was to hold a tuning fork's vibrating tine against a strip of paper "pasted on the circumference of the flywheel." He said that such

Fig. 8.1. Krupp's 1898 Series XVI diesel at Munich Exhibition.
MAN Archives

"*practically* found" cyclic patterns tattooed on the paper "coincided exactly with those found by . . . formulas."[5]

Parts exposed to "compressed air or petroleum under pressure" were to be of nickel or nickel steel, otherwise they would "rust rapidly." This held especially true for the coiled wire mesh atomizer in the fuel nozzle.

Cylinder lubrication was as on the test engine. During the piston's down stroke, a common Mollerup pump metered oil under pressure to a ring of four cylinder ports (fig. 8.2). These were placed so that they fell between the top two rings at bottom center. A plunger in the lube pump housing pushed against an oil-filled cylinder. Oscillating linkage from the connecting rod slowly turned a ratchet wheel, which, through worm gearing, rotated a jack (lead) screw extension on the plunger shaft. On each crank rotation the plunger was ratcheted down a small distance to force lube oil against the piston. Diesel believed that piston sealing was "not because of a strong ring pressure but due to the trapped oil between the rings."[6]

Oils used for lubricating steam cylinders were to be avoided.[7] Diesel preferred a refined, straight mineral oil with a high flash point and free of the fatty and asphaltic constituents, which, under heat, caused deposits to

Fig. 8.2. Mollerup lube oil pump on Diesel's Series XV "B" engine in Deutsches Museum. *Author Photo*

form on the piston, rings, and cylinder. Steam practice tended toward heavy mineral oils compounded with up to 25 percent rapeseed "or other fixed oil." The rule was to use as little of the fixed (fatty) oils as possible and yet adequately lubricate.[8] These fatty oils, both vegetable and animal, released oleic acids (glycerides) that corroded the cylinder and formed "metallic soaps which choke the steam passages."[9]

The established starting sequence called for a nervous mechanic to stand on a platform close to what could become an awesome noise maker if the fuel system functioned improperly. After priming the fuel lines and placing the piston slightly past top center, the *monteur* moved a lever to axially shift a spring-loaded sleeve holding all the cams. The cam assembly was slidably keyed on the drive shaft (fig. 8.3). In this shifted mode the fuel and inlet valves remained closed, a double-lobed cam opened the exhaust valve on each up stroke, and another cam actuated the air starting valve at the beginning of each down stroke. After one or at most two revolutions, the cam assembly was returned to the running position to be held there by the spring. This deactivated the air starting valve, reengaged the fuel and inlet air valves, and converted the exhaust valve to four-stroke operation.

Fig. 8.3. Cam sleeve assembly. Diesel's Series XV
engine in Deutsches Museum. *Author Photo*

The Maschinenfabrik Augsburg Series XVII 300/460 Engine
Augsburg's first production engine was a two-cylinder version of the 30-hp
Series XVII 300/460 design. Its expected output was 60 hp at 180 rpm.[10] The
engine resembled the Series XVI, except for the governor at camshaft level.
Only the cylinder head of the starting/injection air compressor was water-
cooled; discharge air remained uncooled. Injector nozzles retained Diesel's
wire mesh wrap atomizer and slotted washers (*Lochscheiben*) of the Series XV
test engine (fig. 8.4. See also fig. 6.24). M.A. made the Series XVII available
to their own sublicensees.[11]

The Series XVI and XVII designs were improvements, but to their misfortune, retained viral flaws that erupted as near fatal diseases. Neither engine
underwent endurance or field testing prior to release. Even the old Series XV
engine had seen only a few hundred hours running time, all under the care
of perhaps "tenderhearted" operators—a too-common weakness of many test
engineers who tend to form paternalistic bonds with their inanimate "babies."
The consequence of such surprising naivéte by an experienced manufacturer
resulted in serious headaches with their first production diesel engine.

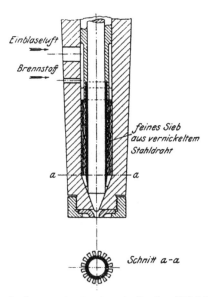

Fig. 8.4. Wrapped wire mesh atomizer in Series XV 250/400 engine.
Diesel, *Die Entstehung*, 1913

The Kempten Engine

Karl Buz, Heinrich's brother, headed Augsburg's United Match & Blacking Factories (*Vereinigte Zündholz und Wichsefabriken A.G.*). His company's Kempten branch became the guinea pig for the first Series XVII engine. Assembly began in Augsburg during September 1897, and it went into operation at Kempten on March 5, 1898. Only a good sibling relationship must have kept Karl from quickly returning it to his brother.

Augsburg heard less than two weeks later that lube oil from the air compressor was forming a combustible mixture with the hot discharge air in the delivery pipe and burning to cause the screen mesh atomizer in the injectors to quickly plug with carbon residue. The immediate fix was an intercooler consisting of an auxiliary tank through which the compressor discharge air passed. Even with this, the nozzle still sooted up after only two days of running.

Excessive blowby past loose-fitting pistons soon occurred. Escaping gases got so bad that women working in the room with the engine were sent home after becoming ill!

A long list of deficiencies reported on August 30 included broken air and cooling water lines at their fittings. Appended to the list was the threat that "in three weeks the engine will be quietly abandoned."[12]

M.A. did all they could, in Augsburg and in Kempten, to prevent this. Lauster led the troubleshooting groups at home while Schmucker almost lived at Kempten from March to October. By that time the worst of the problems had been remedied.

A man new to the diesel program also played a role in saving the Kempten debacle. Josef Vogt (1862–1904), who assumed Vogel's position after the latter left in 1897, came from Krumper's rival steam engine department. After his appointment as chief engineer of the fledgling diesel department, he bore the brunt of the difficulties M.A. encountered with their prematurely released engines. Vogt eventually became a strong diesel supporter in the few years left to him before an untimely death at forty-two.[13]

Imanuel Lauster carried much of Vogt's burden after he became ill in 1902 and succeeded him, at age thirty-one, two years later. Karl Dieterichs, who was in the lab during the Kempten struggle, said that Lauster never lost faith in the engine and worked to the point of exhaustion in carrying out his design, test, and manufacturing responsibilities as well as in dealing with unhappy customers and licensees. While not taking credit away from Rudolf, Dieterichs concurred fully with Buz's remark at the time that "without Lauster there would be no diesel motor."[14]

The Rugendas Engine

Number two on M.A.'s order list was the Rugendas Paper Cartridge Factory in Augsburg. By the time of its long overdue delivery to the *Papierhülsenfabrik Rugendas* in mid-October 1898, its builder had traveled much further down the learning curve. Although almost a twin to the Kempten engine, it was judiciously derated to 50 hp. The serious ailments it suffered, fortunately while still in the shop, included a seized piston, broken springs, overheated bearings, stuck valves, and glowing red injection air pipes. "Lauster . . . gave every imaginable effort. . . . In the quiet of each morning he . . . started the engine himself because the foreman did not have the courage for it due to the terrible explosions."[15] The engine was to provide over thirty years of service and is exhibited in the MAN Museum (fig. 8.5).

An improved fuel control, usually attributed to Lauster, was retrofitted to this engine a few months after it went into operation. In perfecting it, they also found the cause of the violent "cannon bangs" when starting the engine. These were so severe that mechanics, standing on the high platform, feared for their lives as they manipulated the cam sleeve lever during a startup. The heavy combustion knock was due to a too-stiff spring holding closed a suction valve in the fuel pump before the engine could come up to speed. When the engine was running, the valve's position was controlled by the constant speed governor (figs. 8.6 and 8.7). It then acted as a spill valve during the pump's delivery stroke. Depending on load, the excess fuel was returned through the valve into the supply line from the tank. If the valve was kept closed by an excessive spring force, the injector then blasted a fuel charge equal to the full delivery stroke of the pump into the cylinder with the consequences noted.

Diesel's reputation had suffered badly by the time the Rugendas engine entered service. More than a year had gone by since the Kassel lecture, and because of the publicity he created there, whatever news came out of

Fig. 8.5. 1898 Series XVII 300/400, two-cylinder,
50-hp engine for Rugendas. *MAN Archives*

Augsburg and Essen was closely watched—good or bad. He and his partners
needed to do something quickly to counter the bad press.

The 1898 Munich Exhibit

To allay the poor public relations, Diesel capitalized on a promotional event.
This would involve his major German licensees: Krupp, M.A., and two new
ones, Krupp's sublicensee Gasmotorenfabrik Deutz and M.A.'s soon-to-be
partner, Maschinenbau AG Nürnberg. He originally had wanted to use the
second Munich Power and Industrial Exhibition (*II. Kraft und Arbeitsmaschinen-Ausstellung München*) as a further way of parading his engines being
offered by major German firms. The exhibition, to open in June 1898, was
billed as the technical event of the year.[16] Rudolf had convinced the four
companies to finance the building of a large concession and to furnish operating engines for driving specialized machinery (figs. 8.8 and 8.9).

The costly Pavillion Diesel instead turned into a desperation move to
counter the adverse publicity being created by the Kempten engine and by
the onset of similar difficulties licensees were experiencing—well before
Augsburg Diesel Motoren-Fabrik AG engines reached the field.

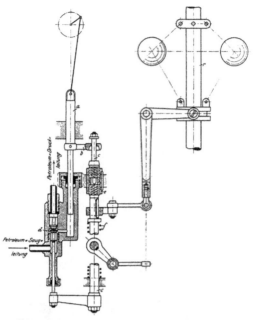

Fig. 8.6. Lauster's (Österlen's?) 1898 governor control of fuel pump suction valve. First used on the Rugendas engine. Diesel, *Die Entstehung, 1913*

Fig. 8.7. Lauster pump on Model DM engine in Deutsches Museum.
Author photo

Fig. 8.8. Diesel Pavilion at 1898 Munich Exhibition. *MAN Archives*

Fig. 8.9. Augsburg and Nürnberg operating engines at the 1898 Exhibition.
MAN Archives

Krupp displayed its second engine of 325/500 mm bore/stroke and an output of 35 hp at 170 rpm.[17] Its crosshead "burned out" on opening day so that no motor was actually running on that auspicious occasion.

Augsburg's was a 30-hp Series XVII. It, too, caused severe headaches. Paul Meyer, who was working for Diesel at the time, vividly described its frustrating proclivity:

> At about ten minutes from the time of starting, even before it had reached operating temperature, there would be such a frightful detonation that no one stayed around it. It went into operation in the earliest morning hours so as not to frighten the visitors. When the motor was stopped and started during the day, it could unhesitatingly start and run quietly, but could not be allowed to cool down in between. The reason for the violent combustion was not discovered during the exhibit. It was later found that only a small modification of the atomizer was needed.[18]

Nürnberg and Deutz each contributed a 20-hp Series XVI engine with their own sets of problems. (An expected two-cylinder Nürnberg never made it.)

Workers from the four companies plus Diesel's Meyer and Ludwig Noé valiantly kept the engines running during exhibit hours, often by working through the night (fig. 8.10).[19] Their efforts paid off, as an English reporter wrote that "The four engines at Munich ran very quietly, with no noise and

Fig. 8.10. Diesel, Noé, and support groups from Deutz, Krupp, and M.A.N. at Munich Exhibition. *MAN Archives*

scarcely any smell. The noiselessness of the exhaust was also commented on...."[20] Such an impressive showing by leaders of German industry diminished the rumors of the engine's demise.

The Fuel Atomization Dilemma

The diesel engine's Achilles' heel centered around the injector used in the air blast fuel system. For two years M.A. and the licensees had searched for a way to generate adequate fuel atomization during the injection yet retain atomizer durability and performance repeatability.

In modern hydraulic injection systems, atomization results mainly from an injection pressure up to thirty times, or more, greater than cylinder compression pressure. This pressure is created by a combination of nozzle hole area, duration of injection, and volume of the charge injected. In contrast, Diesel's air injection used only a two to three times maximum pressure differential between the blast air and fuel in the nozzle and that of the combustion chamber. As this pressure was not enough by itself to break up the fuel droplets passing through the nozzle holes, his solution was to pass the fuel through an auxiliary atomizer before it left the nozzle tip.

Diesel did this on the Series XV lab engine, as mentioned earlier, by wrapping a wire mesh around a tube slipped over the valve stem and positioning it axially between two slotted washers. The design's weakness was that the mesh either quickly carboned up, as on the Kempten engine, or was destroyed when fuel and hot air ignited in the nozzle itself. Further, when removed for cleaning or replacement, there was no assurance the mesh could be rewrapped and repositioned to give optimum atomization.

This was a major reason why fuel consumption varied so much from test to test.[21] A solution had to offer durability and an atomizing capability based on machined parts and not on the finesse of a skilled mechanic. Yet not for several years did a new design provide as low a fuel consumption as the wire sieve atomizer. Diesel had set a goal for others to reach.

The Plate Sprayer

Elimination of the screen wire was an absolute must, yet what would replace it? In January 1897 Diesel provided a clue with his labyrinth ring assembly that might have been the genesis of a plate sprayer (fig. 8.11). Unfortunately, it was only bench tested. In March 1898 Anton Böttcher reported that in desperation he had removed the wire screen entirely to make a Nürnberg engine perform at a New York Exhibition (chapter 9). On learning this, M.A. ran an engine minus the screen but with poor results. Burmeister & Wain, the Danish licensee, wrote on March 15, 1899, that they had a solution (fig. 8.12). However, by the time their drawing reached Augsburg, Franz Schmucker had placed M.A. on the right path a few days earlier (fig. 8.13).

The final design breakthrough is seen in an M.A. drawing of September 16, 1899 (fig. 8.14). For a long time Lauster was credited with it, but more

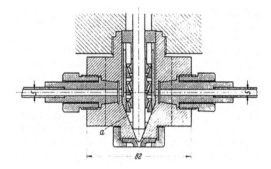

Fig. 8.11. Diesel's "labyrinth atomizer" with eight radial slots in a conical end piece (a). Only bench tested in July 1898. Sass, *Geschichte, 1913*

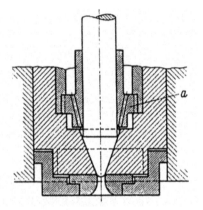

Fig. 8.12. Burmeister & Wain's April 1899 internal atomizer with small, drilled holes in the atomizer cone (a). Sass, *Geschichte, 1913*

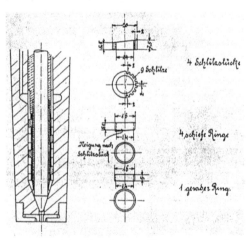

Fig. 8.13. M.A.N.'s May 1899 drawing of *Plattenzerstäubers* (plate sprayer). Sass, *Geschichte, 1913*

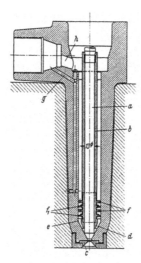

Fig. 8.14. M.A.N. September 1899 fuel valve with plate atomizer: (*a*) nozzle needle, (*b*) needle sleeve, (*c*) nozzle spray disk, (*d*) atomizer cone, (*e*) atomizer passage, (*f*) atomizer washers, (*g*) fuel supply, (*h*) blast air supply. Sass, *Geschichte, 1913*

recent proof shows that Schmucker was the true inventor.[22] By the end of 1899, the plate sprayer based on this was on all Augsburg engines. Without it, the diesel engine could easily have died a slow death. With it, the final obstacle to a commercially practical diesel engine had been overcome. Diesel himself contributed little to the solving of this problem.

Even with the new plate sprayer, several years passed before the engine totally overcame an unsavory name caused principally by a deficient fuel system.

M.A. received most of the flak fired at the new engine during 1898 and 1899 because it was the production leader. As a result, the resources they allocated for the diesel engine program went more to cure immediate problems than to developments with a delayed return. It was Diesel who urged Buz to invest in the longer term during the pre-1900 years.

Changing Times at the Augsburg Test Station
Maschinenfabrik Augsburg's *Versuchsstation* for Diesel's engine went through a transition during 1897 and 1898. It changed from an intense, close group to one even less organized and motivated. As Rudolf himself later said, the "creation time" ended and the "development time" was beginning.

Conflicting demands imposed on the lab's one available and worn-out Series XV "A" engine prevented sustained test programs. (The "B" twin was still at Krupp-Gruson.) Visits by potential licensees and other dignitaries meant returning the engine to a demonstrable condition. Frustrating mechanical breakdowns caused more lost weeks. Adding to these interruptions was hardware to be tested before going on the new engines concurrently being made. The only totally new engine tested in 1897 was Rudolf's compound type.

Because Diesel could spend little time in Augsburg, those assigned to the lab often fended for themselves as problems arose. His absence particularly affected the entries made in the test journal, something that had been his personal domain until this time. Where he might have entered several pages, Fritz Reichenbach, assigned this job by M.A., wrote but a line or two.[23] The fuel tests beginning in mid-1897 particularly had a reduced value due to sparse, sometimes unclear journal entries. Diesel lamented that almost a year was lost between the summer of 1897 and 1898 when a more organized test program finally began.

Diesel and Vogt, supported by Buz and Krumper, formalized an expanded fuel test program in mid-January 1898. A March 23 memorandum outlined procedures and lab organization: First the "A" and "B" engines were to be updated. The "A" was to be used for gaseous fuels and the "B" for liquid fuels. Except for validating new ideas or testing production modifications, this dictum was followed until the engines were retired in early 1900. The "B," returned to Augsburg in October 1897, did not run again until March 1898 at which time it closely equalled the "A's" standard performance.[24]

On paper, Vogt and Diesel jointly directed the engine program, but Vogt ran the program and wielded the authority, particularly in view of Diesel's lengthy absences. According to the memorandum, Diesel could alter or suggest tests.

Lauster ran the lab and oversaw drawings and engine manufacturing. His influence further increased after he assumed field service responsibility for engines sold. An engineer for each engine reported to Lauster: Karl Dieterichs from Diesel's staff ran the gas tests, and a Herr Grosser, for M.A., ran the fuel tests. The two did not begin their assignments until after the Munich Exhibition ended. An office aide and then a chemist were added. Recalling the earlier journal lapses, it was also ordered that a separate one be kept for each fuel and component test. Even a miscellaneous journal was decreed. The engineers faithfully complied.

The seeming equal responsibility of Vogt and Diesel caused a brief furor in early February between Vogt and Karl Dieterichs, who was unaware of Vogt's true position. Since Karl's pay came from Munich, he thought he answered only to his boss there. Vogt asked Dieterichs to restore his engine to standard for an upcoming demonstration, but the latter balked as he was about to test a new idea for an atomizer. Before he could take his case to Diesel, Vogt laid out the facts. Then everyone got into the act: Vogt and Lauster submitted reports to Buz, who curtly wrote Diesel telling him to put a stop to such things. Diesel by return mail said that he would.[25] The incident is also indicative of the tensions building between Buz and Diesel.

Another distressing and divisive episode had occurred a few days earlier. On February 4, 1898, Diesel learned of M.A.'s recent discovery that Fritz

Reichenbach, under a fictitious name, illegally secured at least two patents, one in 1896 and the other in 1897, based on notes and drawings taken from the test station. M.A. quietly fired him as of that date.[26] Diesel also was told that Pawlikowski, his own man, abetted Reichenbach. Pawlikowski was allowed to resign. The episode was not made public until Eugen Diesel wrote of it in 1949, and then he named only Pawlikowski as one of those involved.[27]

Reichenbach and Pawlikowski were good engineers who had performed well. Perhaps their indiscretions can be attributed to misplaced ambitions and a belief that their talents were not adequately appreciated. Diesel also unwittingly contributed to the discord. His drive and creativity infected others, yet often he did not, or could not, bring himself to recognize or properly reward them when they responded with their own inventive contributions. This was especially true of Pawlikowski who had many ideas of which Diesel considered only a few as having real merit.[28]

Diesel's Series XIV Compound Engine

Of great personal importance to Rudolf was a long-delayed test of the compound engine disclosed in his 1892 Patent No. 67,207 (chapter 3). He confidently expected it to provide the ultimate in thermal efficiency.

The final design generally followed the patent drawings except for the coal dust fuel system. This even included using the underside of the low-pressure piston as a supercharger compressor. However, the cooling water jets directed into this compressor were left out. The engine's two, four-stroke power cylinders of 220 × 400 mm bore/stroke flanked a two-stroke, low pressure expansion cylinder with a bore of 510 mm and the same 400 mm stroke (fig. 8.15). Each power cylinder exhaust valve was connected to an intake valve opening into the center cylinder via cast-in head passages. These four "hot" valves controlling the transfer of the gases were internally water cooled in a unique manner (fig. 8.16).

Diesel had begun work on the compound engine during 1894–1895 while still in Berlin. Nadrowski was the design engineer for it there and after coming to Augsburg. Because of the headaches with the single-cylinder engine, the compound version had been put on hold. When the drawings for it were completed in August 1896, the general configuration of its three cylinders bore a strong family resemblance to the Series XV. The power cylinders were direct copies.

Assembly of the crankshaft and main bearings began in mid-May 1897, but the final parts were not added until the end of July. Per standard steam engine practice, the crank and bearing assembly was first "run in" by an external belt drive before adding cylinder pedestals and upper engine. The first P-V diagram was made July 31, but meaningful tests did not begin until late October.

The engine was ill-fated from its beginning. Initial mechanical weaknesses had to be corrected: overheating power cylinders, warped valves, and a poorly fitting center piston, were among them.[29]

Fig. 8.15. Diesel's compound engine of 1897. *MAN Archives*

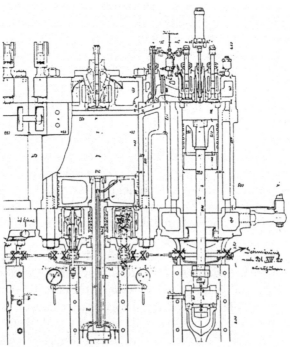

Fig. 8.16. Cross-section of Series XIV compound engine. Note the internally water-cooled exhaust valves and center piston. *MAN Archives*

At first the boost air tank placed between the compressor and power cylinders was heated to aid starting. This heating arrangement was not enough, so the tank was removed and the pumped air was sent directly to the intake valves. Even doing this, it was found necessary to add heat to the combustion air for starting. It seems odd that Diesel would continue with something so similar after the poor results with a boost design almost like that originally on the Series XV engine earlier the same year.

The two-plunger fuel pump was driven from a single eccentric, with both plungers moving in concert. Regulation was by the usual governor-operated wedge mechanism.[30] At first fuel distribution was poor because the two wedges were not rigidly held in alignment and quickly became cocked in differing amounts. After this was corrected, both power cylinders received equal fuel charges.

The greatest surprise, and what ultimately led to an abandonment of the engine, was the devastating heat loss occurring between the high- and low-pressure cylinders during the transfer of combustion gases. Tests showed that the loss to the coolant from the four water-cooled valves alone was 11.8 percent, or almost half of the heat one of the cylinders might convert to work. Added to this were the unmeasured heat losses from the connecting passages themselves and from the center, expansion cylinder piston.

Diesel was astounded because they were so much worse than his conservative heat transfer calculations indicated. He had once discussed this as a potential problem with Professor Zeuner. While not faulting his analysis, Zeuner warned Diesel that while he could not assign numbers to it, his intuition told him one should expect higher losses because of the turbulent and intermittent gas flow between the two cylinders.

Added to the heat losses were the pumping losses inherent in pushing the hot gases from one cylinder to the other. This resulted in a significant pressure drop between the cylinders.

The lowest fuel consumption ever attained was 499 g/bhp-hr—more than double that of the Series XV engine! Cylinder indicated mean effective pressures were as follows:[31]

Center cylinder underside boost	2.4 kg/cm^2
High-pressure power cylinder	19.4 kg/cm^2
Low-pressure center cylinder	2.74 kg/cm^2
Overall imep	4.48 kg/cm^2

Fuel consumption during the above data run was 524 g/bhp-hr. The highest recorded output was only 99 bhp, far lower than an anticipated 150 bhp.

Diesel had assigned Pawlikowski to supervise the assembly and tests; Böttcher and Reichenbach helped when they were available. Surprisingly, the journal shows Rudolf was in Augsburg only once in mid-August to witness a test.

The compound engine last ran on December 31, 1897, with Diesel's spirits badly crushed. He later wrote,

> Deeply afflicted, I thus had to unhappily bury my great expectations to substantially exceed the heat utilization of the one cylinder motor. Perhaps this short account of these tests and the scientific proof of the failure will prevent others from similar disappointments.[32]

The engine was quickly dismantled and scrapped because no one wanted it around as a reminder of a costly misadventure.

Gaseous Fuel Tests 1897–1900

From 1897 until 1900 the Augsburg lab tested many fuels other than the kerosenes. Alternative fuels were deemed necessary because the higher tariffs and transportation costs of these lighter lamp oils, mostly from the United States and Russia, were viewed as a sales deterrent. Krupp had always hoped Diesel's engine could burn coal-based gaseous fuels and, with apparent time at hand to do so, again urged him to test both illuminating and power gases. Rudolf likewise held great expectations for a gas fueled, compression ignition engine (chapter 5) and was happy to accommodate Krupp. He also wanted to try blast furnace gas, a steelmaking byproduct, as a possible fuel; this was never done.[33] Crudes and lighter oils similarly needed to be investigated as to their effect on combustion. Such varied tests would determine the engine's digestive tolerance for multifuels.

The original "A" engine served for both liquid and gas fuel experiments until after the lab reorganization and availability of the "B" beginning in March 1898. (These oil tests are related separately for chronological clarity.)

Diesel proposed several methods to introduce gaseous fuels into the cylinder:

1. Pressurize the gas in an external pump and blast it through a nozzle into air heated sufficiently by high compression to ignite the charge. Diesel termed it the "Diesel-gas process."
2. A corollary to 1. whereby the pressurized gas would come through the nozzle along with a little kerosene to assure reliable engine starting and charge ignition. With this the gas would also function as the atomizing blast for the oil. Neither 1. nor 2. required major engine changes.
3. Mix the gas and combustion air externally and suck it into the cylinder via the air intake valve where it would be heated by compression to just below the mixture's ignition point. A pilot charge of kerosene, which ignites at this temperature, would likewise be injected by blast

air at the proper time to begin combustion. Diesel called this the "Otto ignition jet method."

4. A more conventional gas engine method would use the Otto cycle adapted to a two-stroke, whereby a gas/air charge entered and would somehow be ignited, but *not* by compression ignition.

The "A" engine ran on Augsburg town (lighting) gas from September to the end of November 1897, with interruptions, using a conversion similar to that in the fourth test series (chapter 5). The gas was first pressurized in a two-stage Linde compressor to about 40 atm and stored in what had been the blast air tank. It followed all the circuitry normally taken by the blast air. The volume of injected gas was determined by the length of time the nozzle needle valve was held open and by the degree of lift. A special-profile cam follower could be rotationally adjusted by a manual control wheel (*a*, fig. 8.17) to alter lift and duration.

On gas alone (above Method 1) the engine often started and ran smoothly, but then intermittent afterburning and misfiring would set in. The engine might even come to a stop, especially if the nozzle holes were plugging. Discovered later, after a careful analysis of the Augsburg gas, was that over a period of time the gas composition would vary enough to cause these aberrations.

Local town lighting gas made from coal fluctuated not only in composition, and cleanliness, but also in heating value, commonly averaging about half that of natural gas (methane).[34] It usually contained about 50 percent

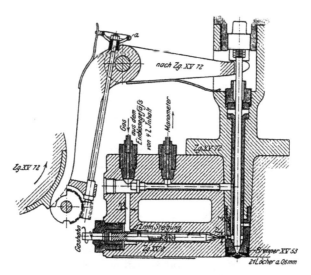

Fig. 8.17. Cylinder head and injector nozzle arrangement for operating with all-gas or gas plus ignition fuel. ("Nach Zg" = "After dwg") Diesel, *Die Entstehung, 1913*

hydrogen by volume. This was followed by methane at 35 to 40 percent and carbon monoxide at 6 to 8 percent. Because they burned with essentially a blue flame, the "illuminants" in town gas came mainly from the olefins ethylene (C_2H_4) and butylene (C_4H_8). The percentage by volume of these in town gas averaged only about 4 percent. The incandescent carbon particles burning in such "heavy hydrocarbons," as they were termed, provided the luminosity.

Nozzle hole clogging became a problem. Diesel reported that the holes plugged faster than with only kerosene due to the carbon in the "heavy hydrocarbons and especially the tar particles."[35] This proved to be the "final straw" as to why injection of gas alone was not feasible.

The next step added a little lamp oil at the start of injection (Method 2) as done in an earlier test series. Only then did stable combustion return, but the nozzle hole plugging was still present.

When tests began again in mid-March 1898 after the lab reorganization, they briefly tried gas alone and then followed with pilot ignition oil. At a compression pressure of 35 atm, this ignition oil could be reduced to 5 g/bhp-hr while maintaining good initiation of combustion. The power fuel consumption was a high 350 liters/ihp-hr (450 l/bhp-hr), and the output at this small pilot level was only an indicated 20 hp.[36] The lowest consumption achieved with pilot injection, 336 liters, had occurred the previous November. Town gas on that date was analyzed to have a lower heating value (LHV) of 5,000 cal/m³ at 760 mm Hg and 0°C.[37] The 37.1 percent indicated thermal efficiency with this rate of consumption equaled that of an engine running on lamp oil alone. (One hopefully assumes that the ignition oil was considered in arriving at the efficiency.)

Nozzle configurations played a large role in the test series. Hole number and size varied from 3 to 4 mm diameter (the final figure) to 70 of 0.5 mm (!). The latter gave the best performance, but obviously the holes quickly clogged. Another innovation tried, but having little influence on performance, was a conical spray deflector screwed into the crown of the piston.

The gas injection tests ended in June 1898 when the power cylinder became so badly worn that it had to be rebored. However, doing so uncovered serious porosity that required a dry sleeve to be inserted. The cylinder rework plus a casting problem with a new piston delayed further engine operation until that December. Doubts arose that the worn cylinder and piston adversely affected the last tests of Methods 1 and 2; they were briefly repeated in January with the same results. Diesel grudgingly admitted that injection of gas alone, or with an ignition fuel, was not practical as long as manufactured lighting gas remained the primary fuel.

The next series of tests (Method 3), lasting from April to June 1899, were based on an idea of Diesel's, which was filed in the German Patent Office on January 27, 1898.[38] It was issued on March 30, 1900, as No. 109,186 under the names of Krupp and MAN. (The latter's new initials). Although Diesel called it an Otto process with pilot injection, it sufficiently compressed the gas and air, externally mixed at ambient pressure, to reach a temperature high enough for the injected pilot charge of kerosene to ignite.

All modern "dual-fuel" engines that induct gas (sewer, natural, or refinery) and air and inject a pilot charge fall under the patent's single claim:

> Ignition and combustion method for internal combustion engines, characterized by the compression temperature of the working mixture not yet reaching its own ignition temperature, however the ignition temperature of a second, more easily ignited fuel or mixture, is reached or exceeded, so that an injection of the latter initiates the combustion of the mixture, whereby the method of combustion is determined through the way and duration of the injection of the ignition fuel, that is, control of the engine is determined.

Later generations were to make diesel engines run as this claim teaches, but its progenitor failed to do so primarily because he was shackled to lighting gas. He was never given an opportunity to try a gas that could have vindicated his beliefs and satisfied Krupp's hopes.

The culprit in lighting gas was its high hydrogen content. To avoid preignition of the hydrogen, whose ignition point was lower than the other constituents, the compression pressure had to be reduced. Dropping to as low as 21.5 atm to ensure a preignition-safe mixture richness would allow only a modest output. Moreover, ignition of the kerosene pilot charge at this pressure was very erratic, and a consequent misfire on one cycle led to preignition on the next due to leftover unburned gases causing an over-rich mixture.

On the other hand, if the compression pressure was raised to assure good pilot ignition, the mixture ratio had to be so lean that power output was very poor. At 27.5 atm, a safe minimum pressure for reliable ignition, only 20:1 gas-to-air ratio could be tolerated without the preignition. So much oil had to be injected to develop enough power that it became uneconomic to bother with the gas.

The next major test series involved burning coal gas made in an on-site generator. Gasifiers like this were widely used for I-C engines around the turn of the century because it cost less to make gas at the engine's site than it did to buy lighting gas delivered by pipe. Until the advent of such smaller, efficient producers, an economic power limit for gas engines was only 3 to 5 hp. Producer gas raised this to 100 hp and more. The gas's high carbon content, around 30 percent carbon monoxide, was about twice that of its free hydrogen. The heating value was one-third that of town gas, yet this proved not to be a disadvantage. When burning producer gas in an engine, the compression ratio could be raised at least two full numbers (5:1 to 7:1), which resulted in greater expansion and a higher thermal efficiency.

Diesel had earlier ordered a Lencauchez gas producer from France in October 1896. Although delivered the following February, it remained in storage until 1899 when engine tests using it finally seemed near.[39]

However, Buz was becoming alarmed over the cost of these seemingly fruitless gas tests. This was no doubt the reason he called a meeting on July 21 to find out what his money was buying. Max Enßlin, Dieterichs's able assistant, wrote a report, no doubt for the meeting, summarizing the successes and failures of the gas tests. Diesel could not attend as he was recuperating from another of his debilitating illnesses at a health spa in northern Italy.

Enßlin's initial report written in the end of June was harsh. (A supplement added just before the meeting mainly gave more technical details.) He itemized several strengths and weaknesses and then in all honesty had to say:

The engine lacks features which it must have to compete with the best existing explosion [Otto] engines: simplicity, cheapness, operating safety, and the ability to ingest large volumes of gas—generator or blast furnace gas.[40]

Buz suspended all further gas tests because of the meeting. His decision was final and no amount of impassioned pleading in a letter Diesel wrote on July 26, 1899, could change his mind. It was not entirely factual, but it provides a glimpse of Rudolf's overwrought condition.

Most Respected Herr Buz,
I can only say here again, what I have expressed recently to you, . . . that I am of the strongest conviction . . . we should make a new power-gas engine . . . which has justification. . . .

My plan was, after my final [health] restoration (in the present location) to carry out with all possible despatch the power gas tests . . . to settle the question for you and the licensors. It will at least work with the best and likely show a marked superiority. I have the belief that in the next five years all steam engine builders must take a hard look at large gas power engines. . . . In view of the development costs up to now, the completion of the [anticipated] programs will take few resources to complete, and I do not know how we can explain to ourselves and all licensors in the face of closing the gate on the best and most important tests, that we are no longer willing to continue. This is in spite of the installation being complete and preparations made. Honored Herr Buz, our work, that is so right and proper is also your work. To close down the power gas tests after making the investment is incorrect and risky, but testing alone will decide that.

Honored Herr Buz, I beg you, I beseech you, do not desert your great work now in a very *decisive* moment. Carry it through to the end, or at least to the end of the year. . . . Do not abandon this, otherwise it will lead to a lifelong uncertainty and uneasiness to have failed in establishing the bold, crowning work. . . . Please Herr Buz, have only a few months patience, allow me only a few weeks recuperation so that after my return I can myself direct the work and carry through.

Let not this decision take its final course.

Most respectfully yours, your grateful devoted
Diesel[41]
All to no avail!

Although the door had closed on actual testing, Rudolf continued with his hopes about a gas engine program at Augsburg. He filed a patent on October 3, 1899, under the alias "Heinrich Hornberger in Berlin," for a double-acting, two-stroke Otto engine not using compression ignition.[42] The patent issued June 13, 1901, as No. 121,009. His described engine had all valves instead of ports. Final scavenging came by a short blast of pressurized air from a valve into the cylinder. This occurred after the piston had shoved out through an exhaust valve as much of the spent gases as possible by the time it had reached top dead center. The exhaust valve then closed, a main intake valve opened to admit the pressurized gas/air mixture and it, too, closed shortly thereafter. Diesel stressed that all valves must open and close rapidly, something difficult to do unless the engine ran very slowly. He no doubt was thinking of the huge, low rpm gas engines being built in Germany and elsewhere, some of which burned blast furnace gas.

Diesel was convinced a good market existed for such an engine and urged Buz to consider it for M.A.N. Unfortunately, the firm's chief had his plate full with the liquid-fueled diesel engine. He turned it down, and gas tests became history at Augsburg.[43] Karl Dieterichs, sure that his future in the shrinking diesel development program would be short lived, also believed in the large gas engines. Accompanied by a strong letter of recommendation from Diesel when he left Augsburg in September 1899, Dieterichs spent the next four years working for an engineering company involved with the successful Oechelhäuser gas engines based on Hugo Junkers designs. He then became a professor at the University of Delft.[44]

Buz's decision to end the gaseous fuel tests was, in Diesel's distressed mind, another example of how M.A.N. was working against him. It proved to be a new wedge either purposely or inadvertently being driven between the two men. Diesel, unfortunately, could not see that Buz must stop chasing up seemingly blind alleys and concentrate on what could earn money for the company. Diesel also lumped Lauster in with the Augsburg "culprits" who had, in his eyes, sabotaged the gas tests.[45]

The Liquid Fuel Tests of 1897–1899

An investigation of liquid fuels ranging from alcohol to creosote began in July 1897 and lasted until the end of 1899. These were more than cursory tests on only the engine. Where possible, each fuel was checked as to heating value, physical characteristics, impurities, and how it burned when sprayed into free air. When used in the engine, the starting ease, combustion noise, and exhaust smoke and smell were noted in addition to power and fuel consumption. Some tests on a single fuel lasted weeks because compression ratio and nozzle configuration also entered the research.[46]

The importance of finding a substitute for the wire gauze atomizer became more and more critical as these liquid fuel tests progressed. This was frighteningly confirmed when the Kempten engine went into service in 1898. Much of the difficulty experienced, particularly with heavier oils, could be traced to the atomizer and to a still-evolving nozzle design (figs. 8.18 and 8.19). After the plate atomizer was adopted, the spectrum of useful oils grew considerably so that several of them became the fuels of choice within a few years.

Before 1900, and for years to come, I-C engines had to consume common, readily available fuels to survive, let alone be successful. Diesel's engine was no exception. By the turn of the century, however, knocking combustion in spark ignition engines was at least recognized as a serious problem, and a gasoline's molecular structure in relation to its effect on combustion began taking on significance. In engine textbooks before this time, it was merely reported *if* a certain gasoline could be tolerated using normal compression ratios rather than *why* it could.[47] Gasolines had once been a nuisance byproduct, but the exponential growth in consumption by automobiles diverted much research to this fuel class. Though years would pass before a practical additive (tetraethyl lead) was found to lessen the spark ignition engine's knock problems, the ongoing combustion studies did provide answers to some of the *whys*. Thus began a marriage of chemistry and engine combustion, and the spinoff from this work also benefited the diesel engine.[48]

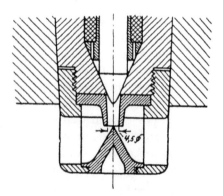

Fig. 8.18. Nozzle and spray diffuser, 1898–99. Hole sizes varied from 2.5 to 5 mm. The deflector was later dropped. Diesel, *Die Entstehung, 1913*

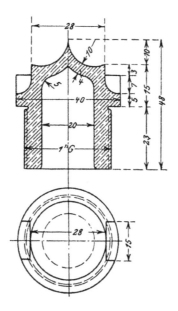

Fig. 8.19. Conical deflector screwed into piston crown.
Diesel, *Die Entstehung, 1913*

Organic chemists had begun to understand the molecular structure of petroleum and its distilled products in the late 1870s, and by 1890 the major crude sources were designated as being either more paraffinic (C_nH2_{n+2}) with "straight chains" as found in the Pennsylvania crudes, or more aromatic (C_nH_{2n-6}) with "closed chains." The Russian Baku area crudes fell into the latter category.[49]

Various "burning" (boiler and furnace) oils were also made from lignite, oil shale, brown coal, and the bituminous coals. Some of these were further processed into lighting gas.

Although such facts were known to Diesel and his colleagues, a big question mark remained as to what fuels might be compatible enough for use in the new compression ignition engine.

The Augsburg test fuels were referred to by either trade name or producing field, some of which no longer have a clear meaning. Thus, when looking at the liquid fuels of Diesel's time, the terminology and even spelling of both crude and processed products must be considered.[50]

Adding to this frequent fuzziness is that a fuel's trade or generic name in one country might be different in another. Lamp oil in Britain was a "paraffin oil" made from Scottish shale oil. In the United States the trade name for the light lamp oils was "kerosene," a derivative of the "kerosine" family; in Germany it was "petroleum oil." One exported US kerosene went by "American light" in Europe. *Naphtha*, or *Erdöl*, referred to crude oil in Germany. Here naphtha generally indicated a crude coming from Baku on the Caspian Sea, whereas in England and the United States, it usually meant a highly volatile oil lighter than "petrol" or "gasoline."[51]

In all Augsburg tests, kerosene was used as a control fuel to make frequent base comparisons of engine performance and, when required, for starting and warming up to operating temperatures. Then the engine was switched to the test oil through a valving arrangement that avoided introducing air into the lines or having to stop the engine.

Galician crude, "solar oil," "red" and "yellow" oils, and alcohol were first burned in the "A" engine while it was still being used for both gas and oil tests during 1897.[52] Only the Galician crude proved suitable at this time because the other oils caused rapid sooting in the atomizer and nozzle. Alcohol had other problems. The inadequate journal keeping negated much of the value of these first liquid fuel tests.[53]

More fuels became compatible in the engine, due mainly to improved atomizers and nozzles, as tests progressed over the next two years. With better atomization and optimized nozzle configuration, the heavier fuels could be broken up into smaller droplets and thereby ensure stable ignition and cleaner burning. The successful result was that the cheaper, heavier oils dominated in the first few years after 1900.

The findings of the fuel tests using alcohol are of special interest because of present work with methanol and ethanol. The lab learned that alcohol's ignition and combustion behavior was primarily a function of its moisture content. Operation was satisfactory only when the compression pressure was raised to 35 to 38 atm and the absorbed water stayed below 5 percent. (Samples received contained as high as 17 percent water.) Yet on one test with 10 percent water in the alcohol, an isfc of 288 g/ihp-hr was attained by a special tailoring of the nozzle valve lift and timing. The resultant indicated thermal efficiency of 39 percent equaled that of kerosene. This was based on the fuel's lower heating value (LHV) of 5,600 cal/kg. However, the engine could not start on an alcohol with this much water.[54]

The higher compression was used to produce a practical engine power. Because alcohol's LHV was only half that of the oils, a much larger liquid charge had to be blasted in by the air on each injection. With "normal" compression pressures the cylinder air was cooled by the incoming charge below the point of reliable ignition.

Alcohol's higher cost, plus the power limitation imposed by its air-cooling effect, even with higher compression pressures, ended its consideration as a desirable fuel.

The diesel engine's selling cost of twice that for a comparable power steam engine meant its payback had to be based on more than a miserly kerosene appetite. It must be able to burn cheaper fuels. Domestically, this almost ruled out the crude oil derivatives from producing fields in Germany as these were limited to relatively small ones in the areas near Hannover, Pechelbronn (north of Strasbourg) and the Tegernsee (Bavaria). France had almost no oil coming from the ground. The result was that the governments of these and other countries took advantage of the scarcity and imposed high duties on most imported crude and refined oils. The coal producers enjoyed this protection, which they no doubt promoted. Great Britain, Belgium, Denmark, and Sweden had no tariffs.

Diesel himself provides operating figures based on fuel costs in an 1899 technical paper.[55] He said that kerosene, with a duty of 7.5 marks, cost from 17 to 21 marks/100 kg. Based on an average fuel consumption, this translated into about 3.4 to 5.2 pfennigs/hp-hr. For cheap oils and residuals bearing no tariff, the fuel price dropped to 4 to 6 marks, giving an operating cost for fuel alone of only 1 to 2 pf/hp-hr—one-third that of kerosene. Rudolf was not the only one concerned about high oil tariffs. German engine builders had begun petitioning their government to lower duties for oils used as engine fuels in 1898, but not until 1904 did any reductions occur.[56]

Another factor in the crude oil equation were the chaotic price wars instigated by oil industry giants:

> Alliances were made, broken, re-made, faster than a competitor could learn that they were even contemplated. The Standard, the Rothschilds, the Nobels, and the Independents of both America and Russia, would be fighting in one country and leagued together in another. Indeed it was common practice for the contestants, by a price-fixing agreement on one market, to make such profits that they could indulge in a price-cutting war elsewhere. In Germany during the first part of 1899, the Nobels' Naphthaport was firmly allied with the Standard against the Pure Oil Company of the American Independents to the virtual exclusion of the Rothschilds. Germany, which had the highest consumption of oil in Europe, was the most important market.[57]

This double deterrent to engine sales of import duties and unstable prices elevated the German, and French, tar oils into potentially significant fuels[58] if the engine could digest such fuels coming from domestic lignite and bituminous and brown coals. Thus, the tests on these oils assumed major importance.

As already noted, the Augsburg lab at first found that the tar oil fuels, mostly processed from brown coal fields in Saxony, were not usable because of rapid sooting in the atomizer and nozzle. When these components improved, so did the engine's tolerance level for some tar oils, but not to the point where they were thought to be unqualifiedly acceptable. One time an oil gave passable engine performance and the next a supposedly equal fuel from the same supplier did not. Vendor to vendor variability also perplexed the lab. Impurities and "solid hydrocarbons" not bothersome in other applications clogged filters and pump passages.

Unknown at the time was how the design and inclination angle of the retort in which an oil was distilled affected its suitability as an engine fuel.[59]

The quantitative differences are evident in the following:[60]

	Lower heating value, Btu/lb	Free carbon %
Gas oil (Galician)	18,500	Nil
Paraffin oil	17,500	Nil
Tar oil (bituminous coal)/ Lignites (brown coals)	16,500	0.1
Tar oil (vertical retort)	16,000	2.0
Tar oil (inclined retort)	15,500	15.0
Tar oil (horizontal retort)	15,000	23.0

A growing demand for cheaper diesel fuels, however, eventually made it profitable for producers of refined tar oils to provide consistently clean and stable products. By 1914 the coal tar oils had become a major diesel fuel for Germany and France.[61] Unfortunately, they were highly aromatic ("closed chain") with a consequently higher ignition point. Therefore, to avoid using a higher compression pressure, the tar oils had to be mixed with 5 to 10 percent crude oil for good combustion. This blending was usually done in the injector prior to atomizing.

Several attempts to use benzol proved totally fruitless as this mostly benzene tar oil fuel was even more aromatic than the coal tar derivatives tested earlier. Benzol became a desirable gasoline with its higher natural octane, but its disadvantage was a high freezing point.

Conversely, the paraffinic tar oils made from lignite proved an excellent diesel fuel.[62] Their cost in Germany, the principal producer, was about the same as for imported heavier oils. During times of shortage, they became valuable substitutes for petroleum-based fuels.

Dyckhoff in Bar-le-Duc experimented with tar oils made from coke long after Augsburg had given up such fuels. His company's annual report of 1905 proclaimed a coke-based oil as a coming diesel engine fuel.[63] This meant always starting the engine on kerosene and then switching over. The oil Dyckhoff used was also so heavy that it required preheating by exhaust gases as well as mechanical stirring. Expert operators and frequent nozzle cleaning were a requirement.

Augsburg did not test any of the Scottish shale oils because it was already known that these were acceptable. The first British diesel ran on shale oil.

Licensees in crude oil producing countries had much to gain if the engine could burn their duty free, domestic oils. Nobel, the Russian licensee (chapter 9), was one of these. At that time, his company was by far the largest, fully integrated oil company in Russia, a country whose annual production in 1900 exceeded that of the United States.[64]

Besides several German crude-based oils, Augsburg tested those from Austria and Hungary (one empire then), Romania, Russia, and the US. Most performed well enough in the engine to pass as suitable for use. Good news for Diesel was that both Nobel's Baku crude and an American gas-oil sent by the US licensee proved to be excellent fuels. Gas-oil is a middle distillate coming after kerosene.[65]

Masut, a Baku residual much used as a boiler fuel in Russia, was so viscous it had to be mixed with intermediate oils if it was to flow through valves and the pump. Even a blend of Russian kerosene and creosote was burned, although the mixture ratio for operation was so diluted with the light oil (9:1) that it was uneconomic.

M.A.N. and all licensees were helped by the information gained from the liquid fuel tests (fig. 8.20). The engine's multifuel capability made larger and larger engines possible because they were no longer tied to the expensive

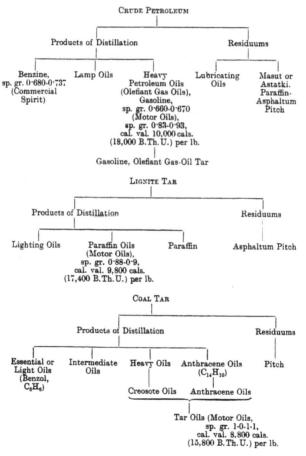

Fig. 8.20. Principal liquid fuels considered for all engines, including diesel in 1913. Wells and Wallis-Tayler, *The Diesel Engine, 1914*

lighting oils. Marine engines were to become a major beneficiary of this free-dom to bum a wide variety of fuels.

There are two postscripts to the liquid fuel tests. One concerns the 1900 Paris Exhibition when the French Otto Company, also a diesel licensee, ran a display engine on peanut oil. According to Rudolf, its performance was little different than when the engine burned a normal crude oil.[66]

The other involved Diesel's enthusiasm over the performance of the engine on Galician oils. He invested a significant sum in a producing company from Galicia because he saw it as a way to control a secure source of fuel. The venture lost heavily. He tried to interest Krupp and M.A.N. in the oil company, but they declined.

Diesel's deeply held belief in the future relationship of oil and the engine caused him to strongly suggest to the American licensee in 1901 about where they should locate:

It was also my thought you should set up your new factory in the midst of your crude oil regions, most likely in Pennsylvania in a site favorable to rail and traffic lines.

I should think that if you follow this broad idea of turning the Pennsylvania oil region into an industrial center, the special inter-ests, i.e., Rockefeller and his friends or other personalities, directors of railway companies in that district, etc., will give you the funds in quantity, and that you can obtain the interests of the great industrial and financial centers, possibly even government interests for such a business. . . .[67]

He wrote in vain.

Coal Dust Tests 1899

Coal dust was the last fuel investigation agreed on in early 1898. The tests lasted but a few weeks during December 1899. From the restrictive compro-mises made, it seems Diesel wanted to have on record he at least had tried using a solid fuel as specified in his basic patent. That M.A.N. had exceeded what they expected to spend on these end-of-century tests is borne out by Rudolf's later comment that he needed to use "all of his powers of persua-sion" to get M.A.N. even to try the fuel.[68] No doubt it is this brief test series, combined with the patent disclosure, that gave rise to the myth of his engine first running on a solid fuel. Coal was the *finale* and not the *debut*.

For cost reasons, the engine was set up to ingest the powdered fuels along with the incoming combustion air and not to inject it into the compressed air. Ignition was initiated by a pilot charge of kerosene injected along with the usual blast air. The engine always started on oil.

A commercially made Swiss coal dust feeder for boiler firing was adapted to the engine. It contained two vibrating and rotating sieves with meshes of 178/cm^2 and 494/cm^2. Their speed of rotation acted to control the volume of powder dropping from the hopper onto the sieves and then passing into the air inlet pipe. The coal feeder was driven off the end of the camshaft (fig. 8.21).

There was no fuel metering control since the main objective of the engine tests was more to learn how much power could be produced. About 0.25 grams of coal per revolution passed the second sieve when the engine ran at its governed speed. What was not drawn into the cylinder with the air fell into a sheet metal box.

The powders to be tested were from several brown coals and one bituminous coal. Diesel already knew that powdered coal could only be obtained from the few companies making it on-site for their own use in boiler firing. It was not uniformly ground nor fine enough. Open air burning revealed that while combustion was relatively clean, the ash residue on the test chamber walls consisted almost entirely of coke from the larger particles. The finer powders tended to bum sufficiently to avoid detection under a magnifying glass.

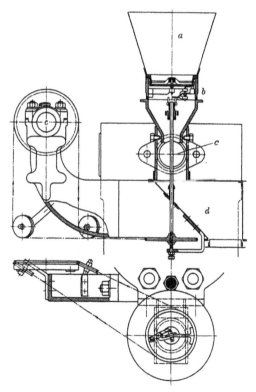

Fig. 8.21. Powdered coal feeding apparatus, 1899. (a) hopper, (b) vibrating screens, (c) air inlet pipe, (d) sheet metal box to carry away powder not entering inlet pipe, (e) drive from end of camshaft. Diesel, *Die Entstehung, 1913*

During the first test with the "A" engine on December 14, 1899, brown coal powder was fed by hand into the intake air. A small indicator diagram enlargement was noted. The next day bituminous coal was tried with similar results. When the powered feed was used for the first time on the sixteenth, the engine developed a mean indicated pressure of 6.4 kg/cm^2 of which 3.1 kg/cm^2 came from the pilot charge of kerosene. After only five minutes of running, the head and piston were removed for an inspection. The upper piston surfaces, particularly the area exposed to the copiously lubricated cylinder walls, were covered with adhered, unburned coal particles.

A run on December 20 lasting seven minutes gave a mean indicated pressure of 8.1 kg/cm^2. This was close to normal power. The piston crown and the space between the rings again were coated with a thick, sticky layer of coal dust.

Total running time before 1900 for a diesel engine burning coal dust was twelve minutes.

The closing journal report stated,

> The method cannot be accomplished in an engine of today's design. Remedies have to be found to keep away if possible the coal dust and the ash from the lubricated cylinder surfaces. In any case they should not get between cylinder and piston.[69]

To this day a diesel engine operating on powdered coal, regardless of particle fineness, sophisticated design, or dedicated effort, has yet to be commercially feasible. In Diesel's time the only one to take up where Augsburg left off was Pawlikowski, who began his work around 1912 using an early model M.A.N. engine. It would become his life's effort.[70]

Fuel Control Refinements

Fuel metering in the Series XV test engines was by governor control of a spill valve to return excess fuel from the fixed displacement plunger pump back to the tank (fig. 6.23). By this method the volume spilled from one stroke to the next could not be controlled accurately enough to ensure equal charge delivery to the nozzle under a constant load and speed.

Lauster had the idea in April 1898 of eliminating the overflow valve entirely and instead using the governor to hold open the pump suction valve to return the excess fuel back into the suction line[71] (fig. 8.6). He also replaced the unwieldy wedge by a threaded sleeve that the governor raised or lowered. This in turn positioned the suction valve so that it was held open through more or less of the pump plunger stroke. In the original wedge/delivery valve, as in this design, fuel spilled until the governor closed the valve so that the remainder of the pump stroke delivered fuel to the nozzle. Lauster received German Patent 136,050 on January 19, 1902, for the idea.[72]

The design was tested at length on one of the lab engines, where it performed so well that it was retrofitted on the Kempten engine. All future engines used the suction spill method. On later engines an eccentric replaced the threaded sleeve, which reduced the governor force required to effect a suction valve position change (fig. 8.7). Lauster attributed this idea to Vogt.

Injection Air from the Cylinder

One of Diesel's greatest disappointments was being forced to adopt the troublesome and costly compressor to supply air for injection of the fuel charge. He never abandoned finding a way to eliminate it. In 1898 Augsburg returned to an idea of the second test series (chapter 5) by which compressed air was taken from the cylinder at the end of the compression stroke to supply that required for starting and blast injection. Experiments carried out intermittently over a year's time again proved the idea was not practical.

Cylinder air discharged through a pop-off valve into a storage tank via a small, water-cooled heat exchanger attached to the engine. On its way from the tank to the injector, the air passed through a second cooler, placed after the tank, and a water separator.

Compression pressure had to be raised considerably to maintain enough heat for ignition. With a peak pressure of almost 50 atm, the cylinder pressure at the time of injection dropped to about 35 atm (fig. 8.22). The maximum output reached with this "self blowing in" system was only 13 bhp as compared to the normal 18.5 bhp. A specific fuel consumption of 305 g/bhp likewise was not good.[73] Because of the higher bearing and ring friction forces from the increased compression pressure, it was found that the overall mechanical efficiency was actually lower than with the auxiliary compressor.

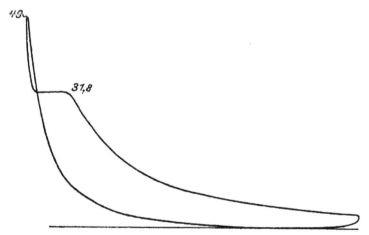

Fig. 8.22. Indicator diagram of engine using cylinder air to supply injection air. Air compressed to 49 atm before blast air valve opens. Diesel, *Die Entstehung, 1913*

Soot in the injection air caused rapid carboning of the nozzle and the piston wall above the rings. A "purifier" placed before the tank made little overall difference in cleaning up the air. There were also starting difficulties.

The use of the cylinder as a single-stage compressor to supply air for injection and starting was abandoned once and for all.

One worthwhile piece of information resulted from these otherwise unsatisfactory tests. The lab found that an optimum compression pressure was 32 atm, and that even at 28 atm, it was still possible to develop adequate indicator diagrams.

A two-stage hybrid version was developed and patented under M.A.N.'s name in Germany (1900) and under Diesel's name in the US (1902).[74] In this system the engine acted as the first stage of compression, and then a much smaller, second-stage compressor raised the pressure up to that needed for the injection air. Air from the cylinder passed through a timed valve toward the end of the compression stroke. The production version used a special cam to open the air starting valve after the camshaft had been shifted into the running mode (fig. 8.23). The compressor, with a bore and stroke of 40 × 60 mm, had only one-tenth the volume of the single stage it replaced.

Cost considerations seem to have been the overriding reason for M.A.N. to adopt this system. A letter of theirs said that ". . . power to idle the engine is comparatively reduced and production costs are considerably reduced."[75] A slight overall power loss (0.5 hp) did not affect fuel consumption enough to upset customers. Not stressed was the ability now to intercool the air between stages and thus greatly reduce the dangerous explosions caused by the single-stage compressor.

Experimental work on this concept was carried out in the latter part of 1899, and from 1900 to 1904 all M.A.N.-built engines incorporated it. They were not alone in its use. Diesel wrote in 1912 that "without careful maintenance [when using heavier oils] the cylinder discharge valve and compressor valves gave minor problems. . . ."[76]

The Swedish licensee's troubles were reported later in much stronger words:

[M.A.N.] should have understood that the contamination or risk of contamination was too high a price in comparison with the small cost saving.

The licensees were notified of the advantages of the high pressure pump but not of the soon to be discovered failure. This they had to find out for themselves—as well as finding solutions.

. . . The unfortunate test with the high pressure [second-stage] pump is one example of how a designer and engine builder should not proceed. If risks are to be taken the possible gain must be significant. That was not the case here.[77]

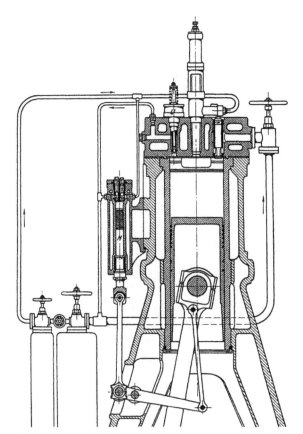

Fig. 8.23. M.A.N two-stage compression using cylinder air as the first stage of compression and a small, externally driven compressor for the second stage. Diesel, *Die Entstehung, 1913*

He further claimed that the horsepower loss when using the combustion air for the first stage cost more in monetary terms because of additional engine structure/weight than what a separate low-pressure pump would have. He said that this system was the last German contribution to the diesel engine in Sweden.

More conventional, two-stage compressors superseded the hybrid system. Other licensees, starting in 1901 with L. A. Riedinger Machine and Bronzeware Factory of Augsburg, slowly did this on their own when they realized that it would cure a major source of trouble.

The Russian licensee wrote in 1903 of only just learning about Riedinger's changeover.[78] License contracts called for a sharing of all ideas, but here was an instance of M.A.N. not acting with promptness to inform their "competitive partners" of what Riedinger had successfully accomplished. Such delays, or sometimes a total lack of sharing between the large licensees, were to happen with increasing frequency.

Güldner Two-Stroke Diesel Engine 1899–1901

A side excursion from the main efforts of the Augsburg development tests was a two-stroke diesel engine designed by Hugo Güldner (1866–1926), at that time an inventor and builder of gas engines. It cannot be confirmed if Güldner went to M.A.N. with the idea of having them build an engine to his design, but the suggestion certainly did not come from Diesel. On January 1, 1899, the Allgemeine Gesellschaft für Dieselmotoren AG (Chapter 7) hired Güldner as chief engineer for the two-stroke diesel development. The 20,000-mark budgeted cost was to be shared 50 percent by Diesel, 25 percent by the Allgemeine, and 12.5 percent each by Krupp and M.A.N.[79]

His horizontal engine had uniflow scavenging with intake ports in the cylinder and an exhaust valve in the head (fig. 8.24). The output was to be 12 bhp at 250 rpm from a bore and stroke of 175 × 210 mm.

Substituting for a crosshead was an equal piston diameter scavenge pump. The pump discharged into a tank in the base of the engine frame until the piston reached the last 10 percent of its stroke when it covered a port leading to the tank. The remaining stroke pumped air at 4 to 6 atm to a second-stage compressor for raising the injection air pressure to about 45 atm.[80]

Scavenge air, entering an air box past an automatic valve, flowed into the cylinder from the air box when the piston uncovered intake ports. The exhaust valve rocker lever was actuated by a face cam at the end of the control shaft. Rotating a hand lever caused the same cam and an extension on the rocker lever to open the air starting valve. Axially positioned on the control shaft were the governor, a crank throw for the second-stage compressor, an eccentric to operate the fuel pump, and a cam for the injection nozzle valve.

Güldner received three patents based on his Augsburg work, all issuing under the name of Heinrich Eckhardt, Berlin. Only one applied to the engine as built. Two air pumps with different discharge pressures used the first for scavenging and the other, in series with the first, for injection air.[81]

Because this was thought to be the first engine to inject at a right angle to the cylinder axis, a special "lateral" nozzle and slanted piston crown were previously tested on the "A" engine. (The concurrent American engine also had a horizontal injector.) Performance was similar to that with an axially located nozzle.[82]

Engine assembly was completed in mid-December 1899, and during the break in running, the power piston scuffed so badly it had to be replaced. The first test under power on February 17, 1900, resulted in a dismal 7 bhp at 251 rpm, a low mechanical efficiency of 47.6 percent and a bsfc of 380 g/bhp-hr. Not enough scavenge air was deemed the major problem. A Linde compressor, substituting for the engine scavenge pump, increased the air supply and thereby raised the indicated mep from 5.2 to 8.0 kg/cm². Increasing the injection air pressure to 60 atm still failed to "enlarge" the P-V diagrams because of undersized injection air passages.[83] An atomizer from the Swedish licensee, AB Diesels Motorer, was also used in the March tests. Although the engine ran well, its performance did not change.

The next step, to reverse the scavenging process by using the exhaust valve as the intake and the ports for the exhaust, gave no substantial improvement.

With a two-stage Linde compressor for the blast air, to make the entire discharge from the crosshead pump available for scavenging, the output went to 21.4 ihp. The brake power of 13.1 bhp showed the overall mechanical efficiency was only 57.2 percent. This gave a bsfc of 272 g/hp-hr. Raising

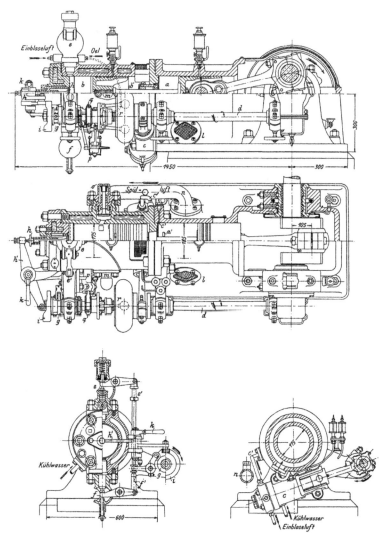

Fig. 8.24. Güldner 1899 two-stroke diesel engine: (a) scavenge air cylinder, (c) injection air compressor, (e) fuel nozzle, (f) starting valve, (h) exhaust valve. Güldner, *Verbrennungskraftmaschinen, 1905*

engine speed did not help, as beyond 260 rpm the valves gave even more trouble than usual.

A scavenge pump bore increase to 195 mm (a 24 percent volume increase) was the final attempt to eliminate the auxiliary scavenge compressor. Engine performance went up only to 9.47 bhp at 244 rpm. Brake thermal efficiency was only 21 percent, and the bsfc rose to 301 g/bhp-hr. This compared with the four-stroke test engines' best performance in 1897 of 211 g/bhp-hr and an overall efficiency of over 30 percent. The last Augsburg two-stroke tests, run in January 1901, clearly showed the engine was not competitive.

A second Güldner-type two-stroke was built in England under the auspices of the Diesel Engine Co. in London, an Allgemeine licensee. Scott and Hodgson near Manchester made it because the Diesel Engine Co. had no manufacturing facility. Improvements based on Augsburg's initial experience were incorporated.

Seventy percent larger than its German cousin, the engine had a bore and stroke of 7-7/8 × 10-3/4 inches and was rated 20 bhp at 212 rpm. The best bsfc at an output of 21.8 bhp was 228 g/bhp-hr. This gave a brake thermal efficiency of 27.8 percent when burning a Russian solar oil (10,000 cal/kg).[84] One reported problem was an undersized crank pin bearing that ran hot within minutes.[85] This was encouraging performance. However, its more expensive design and unresolved weaknesses, in comparison with the much improved four-stroke, forced the decision to end all the further two-stroke development. The engine made a final appearance at the 1901 Glasgow International Exhibition.

Augsburg was to end their own work on the Güldner project in March 1901 after learning of the English engine's better performance. Hugo Güldner left the "Allgemeine" the following October to pursue text writing and engine building (chapter 14).

Essen Goes to the Sidelines

Fried. Krupp anticipated their engine production would begin in 1897 when the Series XVI drawings were available. They gave all signs of becoming a major presence at the birth of the diesel industry. Yet none should forget that what had motivated them from the time of signing the contract with Diesel was the use of his engine as a way to promote sales of their steels and byproduct gas. Unfortunately, Essen's lack of experience with any kind of engine, unlike their partner in Augsburg, resulted in their always being rather tentative about the new power plant during its long development period.

When it became apparent that much of the engine could be made from cast iron and was limited to burning liquid fuels, the Essen directors lost even more devotion to the program. Then came the troubles of the first engines sold by Augsburg. To their credit Krupp initially opted to continue despite the difficulties and disappointing loss of opportunities.

After making one 20-hp Series XVI engine in Essen, Krupp followed the plan to produce all diesels at the Grusonwerk division. The first all-Gruson diesel, begun in June 1897, was the 50 hp, 350/550 mm bore/stroke engine sent to the Munich Exhibition, which opened in June 1898.[86] A tragic fire that same month destroyed the entire engine shop. No more engines, gas or diesel, were made at Gruson after that. Salvageable parts for two Series XVI engines went to Essen for assembly there. At the end of the year, Hermann Ebbs left Gruson to oversee design of large gas engines at Nürnberg.

Discouraged, Krupp ended Essen production and closed the engine lab in October 1899. Not until 1904, when its Germaniawerft division in Kiel began preliminary design work on a submarine diesel, did Krupp's engine program restart. Essen itself had completed no more than five engines: three 20-hp Series XVI and two 35-hp 325/500.[87] Yet there can be no doubt that without Krupp support Diesel—and M.A.N. —would not have seen his engine come alive in Augsburg.

The Reckoning

Krupp and M.A.N. ended their joint development contract in June 1900. Their total investment since the inception of the program in 1893 was approximately 600,000 marks, of which some 443,000 went primarily to run the test lab. Diesel had been paid about 135,000 marks from Krupp. Most of the 270,000 marks spent after the April 1897 accounting (chapter 7) was incurred by Augsburg to finance the fuel test programs. Within a few years, much of the entire development expense would be recovered through license fees and royalties.[88]

The paths taken by Rudolf Diesel's two partner companies during his engine's turbulent introduction widely diverged. By the turn of the century, the hazardous gauntlet was run and only M.A.N. continued. The right people in the right places had the faith, will, and staying power to make it happen. M.A.N. had a marketable product and a broadened understanding of its limitations and strengths.

Despite the problems M.A.N. faced during the *Kinderkranken* years of 1898–1902, the company still managed to sell engines. Production records in that calamitous period show that it was not in a holding pattern, but continued to seek orders:[89]

Year	Engines built
1897	1
1898	5
1899	4
1900	7
1901	17
1902	81

Notes

1. Eugen Diesel, *Diesel: der Mensch, das Werk, das Schicksal* (Hamburg: Hanseatische, 1937), 327.
2. Rudolf Diesel, *Die Entstehung des Dieselmotors* (Berlin: Springer Verlag, 1913), 82.
3. A forty-five-page handwritten tract with sketches, "Berechnungen zum Eincylinder Dieselmotor" (Augsburg: September 1897), was translated and printed by the Diesel Motor Co. of America: *Calculations for 20 H.P. Diesel Motor* (New York: March 16, 1898), 32 pages. Diesel papers, Deutsches Museum.
4. Wilhelm Worsoe, *Die Mitarbeit der Werke Fried. Krupp an der Entstehung des Dieselmotors in den Jahren 1893/97 und an der Anfangs-Entwicklung in den Jahren 1897/99*, 2nd ed. in MAN Archives (Kiel, Germany: 1940), 4. The first edition was published in 1933.
5. *Calculations*, 26.
6. R. Diesel, *Die Entstehung*, 150.
7. *Calculations*, 27.
8. Leonard Archbutt and Richard Mountford Deeley, *Lubrication and Lubricants*, 4th ed. (London: Charles Griffin & Company, 1920), 539. Many sources written between 1890 and 1920 specified cylinder oils for steam and I-C engines that were not always in agreement, possibly because lubricant choice was more often by personal custom than by scientific testing. Bearing oils usually contained higher percentages of the fatty oils and had lower flash points of around 175°C. See also George H. Hurst, *Lubricating Oils, Fats and Greases*, 4th ed. (London: Scott, Greenwood & Co., 1925), 284. First edition published in 1896.
9. Archbutt and Deeley, *Lubrication*, 111. See also Herbert Haas, *The Diesel Engine: Its Fuels and Its Uses*, Bulletin 156, Petroleum Technology No. 44, US Gov't. Print. Off., 1918, 95. Haas says "A suitable cylinder oil should have a viscosity of 9° to 10° Engler at 50°C., and should flow at -5°C. Its flash point should not be below 240°C. It should be a pure mineral oil free from asphaltum and pitch and resinous substances. . . ."
10. Kurt Schnauffer, "Die Motorenentwicklung in Werk Augsburg der M.A.N. 1898–1918" (1956), 5.
11. Between 1898 and 1902 (by now) M.A.N. built and offered its sublicensees the Series XVII through XXII, which spanned outputs per cylinder of 15 hp at 210 rpm to 35 hp at 170 rpm. Schnauffer, "Die Motorenentwicklung," 5. M.A.N. archivist Georg Strößner, during a July 13, 1959, visit to the Science Museum, London, informed that the Series XVI was not made by MAN. October 21, 1959, Memo. The Science Museum.
12. Schnauffer, 9. This recitation of engine problems differs considerably from an August 1898 sales testimonial: "The two cylinder, 60 hp Diesel Motor supplied by you has been at work at our branch at Kempten since 7 March this year and completely answers our requirements." Allgemeine Gesellschaft für Diesel Motoren AG, *Report No. 184, January 1900*. Such a comment provides strong evidence for using caution when evaluating research material. Nevertheless, personal experience of the present author having "sold in the a.m. and repaired in the p.m.," accepts the importance of both sales literature and service bulletins.
13. There is evidence that Vogt was not an advocate of the diesel when he first took over. Paul Meyer, who saw Vogt only in his first years with the new engine program, recalled that Vogt "rarely looked in" and that he "was considered as the silent adversary of the diesel motor." Meyer to the MAN Archivist, January 19, 1941. MAN Archives.
14. Letter from Karl Dieterichs to Georg Strößner, in MAN Archives (July 22, 1941), 2.
15. E. Diesel, *Diesel*, 327.
16. *Officielles Organ der II. Kraft und Arbeitsmaschinen-Ausstellung*, in MAN Archives (München: 1898), 103–7.
17. Worsoe, *Die Mitarbeit*, 55.
18. Paul Meyer, *Beiträge zur Geschichte des Dieselmotors* (Berlin: Springer, 1913), 37.
19. Ibid.
20. "Progress of the Diesel Oil Engine," in *The Engineer*, November 4, 1898.
21. There must have been a very good "wrap" during a trial for Burmeister and Wain on October 27, 1897, because fuel consumption was the lowest ever at 211 g/Ps hr (about 215 g/bhphr). It was the first time brake thermal efficiency passed 30 percent (30.2). A spiral winding was used. R. Diesel, *Die Entstehung*, 85.
22. Kurt Schnauffer, "Lizenzverträge und Erstentwicklungen des Dieselmotors im Inund Ausland 1893–1909" (Augsburg: 1958), 25–26. By the time of this paper, written two years after his study cited above, Schnauffer had found strong new evidence to convince him that Schmucker,

by then a test stand foreman, and not Lauster, had the idea for the drawing of May 12, 1899. Schnauffer interviewed two retired MAN chief engineers, Aechter and Ruf, who had worked on the engines in the 1899 period. Ruf first said it was not Lauster, and Aechter, who at the time in question ran the tests, confirmed it when he recalled that it was Schmucker who had told him to install the atomizer plate rather than the screen in the atomizing tests starting May 23. Schmucker had gone to work at M.A. in November 1893 and became salaried in 1909. No other information on him is available.

23. Kurt Schnauffer, "Arbeiten [Work] in der Augsburger Diesel-Versuchsstation 1897–1901," 6. Schnauffer cites examples of the journal sparseness in his references.

24. Diesel's handwritten report of this meeting is in the MAN Archives.

25. The written record begins with Vogt's "Report on the behavior of Engineer Dieterichs on the occasion of my given order, February 9, 1898." This and the ensuing repercussions are in reports in the MAN Archives.

26. An undated M.A. memo c. 1898 summarizes the episode. Fritz (Karl Friederich) Reichenbach went to work for a competitor of M.A. and later suffered a debilitating paralysis. One manifestation of his handicap was to become insanely jealous of his wife. This paranoia so overpowered him that he shot and killed her after a trifling argument; he then turned the gun on himself and died soon after telling a doctor what had occurred. A March 5, 1941, letter from the town office of Bensheim, along with a copy of the October 12, 1915, death report give the tragic details. MAN Archives.

27. E. Diesel, *Jahrundertwende*, 147–48.

28. Ibid., Rudolf's son relates his interview with Pawlikowski in 1934 and leaves no doubt as to Pawlikowski's deep bitterness over what had happened thirty-six years earlier. Eugen wrote that Pawlikowski "wanted to drive a thorn into my heart" by tying his father's suicide to Reichenbach's in saying "Your father had driven this man to the death!" Paul Meyer, who wrote one of Pawlikowski's obituaries, said that he had left Diesel in a dispute over whose name should be on patents Nos. 109,186 and 118,875. Pawlikowski thought that he was the inventor and not Diesel. *Zeit.*, VDI-Zeitschrift 87, nos. 15/16 (April 17, 1943).

29. Ibid.

30. R. Diesel, *Die Entstehung*, 135.

31. Schnauffer, "Arbeiten," 48.

32. R. Diesel, *Die Entstehung*, 140.

33. Schnauffer, "Arbeiten," 28.

34. Blast furnace gas, with a heating value of under 100 Btu/cu ft, was dirty and was available only near the industrial source. See Cummins, *Internal Fire* (Austin, TX: Octane Press, 2021), 291, for the application of this and producer gas to Otto cycle engines before 1900.

35. R. Diesel, *Die Entstehung*, 116–40.

36. Ibid., 116.

37. The heating value of gaseous fuels was usually expressed in either cal/m^3 or Btu/ft^3. For liquid fuels it was either cal/g or Btu/lb (1 cal/g = 1.8 Btu/lb). A fuel's heating value is specified as being "higher" or "lower" (HHV or LHV). HHV represents the full amount of heat that can be transferred from a reaction, and is what should be used to calculate efficiencies. However, this includes heat that can be only theoretically attained as it includes condensing the water formed as products of combustion. This is not possible in practice. Only the LHV is used in finding I-C engine thermal efficiencies, but when making efficiency comparisons this should be verified. The LHV is about 10 percent less than the HHV. The author's thanks for the above paraphrase to Obert, *Elements of Thermodynamics* (New York: McGraw-Hill, 1949), 208.

38. A follow-up patent to No. 109,186 was filed on February 8, 1898, and again issued in the names of Krupp and MAN on March 21, 1900, as No. 118,157 (US Pat. 673,160 of April 30, 1901). In this patent a second fuel could be injected along with the pilot oil to further control the combustion. One figure shows a double-seated poppet valve able to admit two fuels past the port into the cylinder. This idea was never tested.

39. See William Robinson, *Gas and Petroleum Engines*, vol. 2 (London: Spon and Chamberlain, 1902), 572–75 for a description of a Lencauchez gas producer.

40. Much of Enßlin's "Report about the activities of the test station in the month of July 1899," is quoted in Georg Strößner, "Frühe Versuche [early tests] mit dem Diesel-Gasverfahren," in *MTZ* (Stuttgart, Germany: December 1940), 385–94. MAN ten-page reprint from their Archives. The cited item, not given by Strößner, is in Schnauffer, "Arbeiten," 35, which also lists these deficiencies.

41. Strößner, "Frühe Versuche," 7–8 of the reprint. The entire letter is quoted. At the time Diesel was "on retreat" at Madonna di Campiglio, a resort in the Italian Alps.

42. Schnauffer, "Arbeiten," 56, and Sass, *Geschichte*, 504, attribute the "Hornberger" patent to Güldner, but the patent file wrapper in the M.A.N. Archives, which includes affidavits and numerous letters, convincingly shows that Diesel was the inventor.

43. The Nürnberg division of M.A.N., under Anton Rieppel's more venturesome leadership, became a major builder of large, horizontal, double-acting gas engines.

44. Dieterichs' letter to Georg Strößner, July 1941. In this same letter, he generally confirms what Strößer wrote about the gas tests. MAN Archives.

45. Yet the next July 27, Diesel wrote to Lauster thanking him for his successful efforts at the 1900 World Exhibition in Paris and enclosed a gift of 500 marks. Sass, *Geschichte*, 521. This is another example of the dichotomy of feelings in Diesel's tenuous relationship with Lauster after 1898.

46. R. Diesel, *Die Entstehung*, 107.

47. William Robinson, *Gas and Petroleum Engines* (London: Spon Press, 1890) is an early textbook in English giving the chemical composition of fuels from various crude sources, yet it tells little as to how they might affect combustion. For the nonchemist see, for example, Schlaifer and Heron, *Development of Aircraft Engines and Aviation Fuels* (Boston: Harvard Press, 1950). Sam Heron's "Technical Appendix B," 693–705, very understandably explains carbon-hydrogen molecular structures and how terminology for them came about.

48. Not until the late 1920s did the "octane" scale, based on a mixture of iso-octane (2,2,4-Trimethylpentane—C_8H_{18}) and normal heptane (C_7H_{16}) as reference fuels, come into use for spark ignition engines. Another ten years passed before the "cetane" rating of fuels was fully adopted for, at first, automotive diesels. The earliest to suggest the use of cetene (a straight chain paraffinic product, $C_{16}H_{32}$) were Royal Dutch/Shell chemists G. D. Boerlage and J. J. Broeze. See their paper "Ignition Quality of Diesel Fuels Expressed in Cetene Numbers," *SAE Transactions* (1932), 283–93. The reference fuel mixed with it was alpha methylnaphthalene ($C_{11}H_{10}$). The first reference suggesting cetane ($C_{16}H_{34}$) was T. B. Rendel, chairman, "Report of the Volunteer Group for Compression-Ignition Fuel Research," *SAE Transactions* (1936), 225–83. The reason for the change was a variance in cetane ignition quality from different suppliers. Alpha methylnaphthalene remained the blending reference fuel.

49. An excellent history of petroleum chemistry is in R. J. Forbes and D. R. O'Beirne, *The Technical Development of the Royal Dutch/Shell 1890–1940* (Leiden, Netherlands: Brill, 1957), 13–55. It cites Silliman's major 1855 report, which for the first time describes potentially useful products to be distilled from crude oil: light gasolines for dry cleaning, lamp oils, lubricants, paraffin waxes, and residuals like asphalt.

50. An especially helpful reference, in English, is by Boverton Redwood first issuing in 1896: *Petroleum* (London: Griffin). New editions were published as late as 1913.

51. Petrol was not a universal term for automotive fuels in England until after 1900.

52. Schnauffer, "Arbeiten," 18. The Galician crude gave too high a bsfc of 264 g/bhp-hr. Its specific gravity (sp gr) varied from 0.870 to 0.885. For comparison, the sp gr of American kerosene varied from 0.78 to 0.81; the Russian sp gr was 0.825. (The Galician oil field extended along the north slope of the Carpathian Mountains in what is now southern Poland.) Solar oil, also known as "pyronaphtha," was a common heating oil that fell between lighting and lubricating oils and had a 0.860 to 0.880 sp gr. It was usually made from brown coal and was defined as a nonvolatile tar oil. Both the "red oil" and "yellow oil" used in the Augsburg tests were also from brown coal sources in Saxony. See Redwood, *Petroleum*, vols. 1 and 2 for numerous references to these fuels.

53. The journal deficiency particularly occurred during the solar oil tests.

54. Schnauffer, "Arbeiten," 21, and Diesel, *Die Entstehung*, 111–12. Diesel's book and Schnauffer, who based his research on Diesel's book, journal entries, and engine drawings in the MAN Archives, are the basic sources for the liquid fuel tests.

55. Rudolf Diesel, "Mitteilungen über den Dieselschen Wärmemotor," in VDI-Zeitschrift (1899): 130.

56. "Petitionen wegen der Befreiung des Petroleums für den Motorenantrieb vom Einfuhrzoll," a report from the Reichsminister of Finance, Berlin, November 16, 1940, giving background of oil duties and later reductions. The information was requested by MAN. MAN Archives.

57. Robert Henriques, *Marcus Samuel: First Viscount Bearsted and Founder of the "Shell" Transport and Trading Company 1853–1927* (London: Barrie & Rockliff, 1960), 265.

58. A. P. Chalkey, *Diesel Engines for Land and Marine Work*, 4th ed. (New York: Van Nostrand,

1915), 50–1, wrote that the duty on some oils in Germany almost equaled the actual value of the oil itself. According to Giorgio Supino, *Land and Marine Diesel Engines* (London: Charles Griffin & Co., 1918), 25, the duty on "heavier" oils in France was 90 francs/ton (metric) vs. 36 marks/ton in Germany. Redwood lists oil tariffs for all countries. [Note: Ml = FF 0.8 = 1 English shilling (s) = 24 ¢ US, 20s = £1, 12 pence = 1/s.]

59. R. Diesel, *Die Entstehung*, 113.

60. Alfred Büchi, "Modern Continental Oil Engine Practice," in *Cassier's Magazine* (1913), 158.

61. Herbert Haas, *The Diesel Engine*, 89. Haas also affirms Diesel regarding the inclination angle of the retort being critical by saying that only vertical and not inclined or horizontal ones can be used (91).

62. Ibid., 77.

63. R. Diesel, *Die Entstehung*, 114.

64. H. F. Williamson et al., *The American Petroleum Industry: The Age of Energy 1899–1959* (Evanston, IL: Northwestern University Press, 1963), 254.

65. Gas-oil was "cracked" in red-hot retorts to form gases, which were then used to enrich town gas or industrial gases. Forbes et al., *Technical Development*, 52. In 1899 the US shipped 380,000 barrels of gas oil to Europe, 95 percent of the total gas oil exported. H. F. Williamson, *The American Petroleum*, 251. (Europe also received 13 out of the 17 million barrels of kerosene exported from the US that year, 246.)

66. R. Diesel, *Die Entstehung*, 115.

67. E. Diesel, *Diesel*, 334–35.

68. R. Diesel, *Die Entstehung*, 126.

69. Schnauffer, "Arbeiten," 39.

70. A photo of Pawlikowski and Adolf Hitler standing beside one of his coal-burning engines sometime in the mid-1930s was shown to the author when he visited the Deutsches Museum in 1972.

71. Diesel, (*Die Entstehung*, 101) says Fritz Oesterlen, a M.A.N. designer, invented the suction spill method in December 1898. Lauster, in his memoirs, said the idea came to him in April 1898 while on a train before Oesterlen worked for M.A.N. He added that because Diesel was gone so much during this period due to illness he could not have known for sure who did invent it. Sass, *Geschichte*, 510.

72. Lauster received US Patent No. 729,613, filed July 15, 1902, and issued June 3, 1903. Because of Diesel's prior US Patent 654,140 covering all ways to return fuel, Lauster was limited to the suction valve. It was, however, a strong patent and major deterrent to unlicensed builders. The concept would later be used by everyone after Diesel's patent expired.

73. Schnauffer, "Arbeiten," 41.

74. A German patent filed by M.A.N. on September 13, 1900, issued as No. 127,159. The US patent in Diesel's name and assigned to the Diesel Motor Co. of America was filed January 18, 1901, and issued as No. 708,029 on September 2, 1902. In a personal memo dated March 27, 1907, Diesel wrote that the lab journals show Reichenbach tested this idea of his before the first engine was sold. (A marginal note states it was May 1897.) He said M.A.N. made a big fuss about it being their idea, but that when the patent was filed in the US it was necessary for them to swear in the invention oath that the idea was in fact Diesel's. This is another vignette in the Diesel vs. M.A.N. saga. Diesel papers, Deutsches Museum.

75. M.A.N. to their licensee Maschinenfabrik Riedinger on the patent filing date. Schnauffer, "Die Motorenentwicklung," 17.

76. R. Diesel, *Die Entstehung*, 102.

77. J. K. E. Hesselman, *Teknik och Tanke* (Stockholm, Sweden: Sohlman, 1948), 113.

78. Schnauffer, "Die Motorenentwicklung," 18.

79. "Zweitakt-Diesel-Motor-Kosten im Geschaftsjahr 1 Juli 1900–30 Juni 1901." MAN Archives. Sass, *Geschichte*, 502, is incorrect as to the amount and expense shares.

80. Güldner, *Das Entwerfen*, 114–16.

81. German Patent No. 124,148 dated March 30, 1899. The other two patents were Nos. 109,562 and 111,302.

82. Diesel, *Die Entstehung*, 99–101, says the nozzle was a prototype for a horizontal engine, without saying it was for Güldner's two-stroke. There is no mention of the two-stroke in *Die Entstehung*, more proof that he did not initiate the Güldner program.

83. Schnauffer, "Arbeiten," 58.

84. A. Johanning, "Der Diesel-Motor, seine Entwicklung und volkswirtschaftliche Bedeutung,"

draft copy in Busch-Sulzer papers, in the Wisconsin State Historical Library (1901), 53. William Robinson, *Gas and Petroleum Engines*, 778, also has a description and performance data. Diesel himself, who later referred to the engine as a "complete fiasco" (*vollkommes Fiasco*) said Güldner gave false fuel consumption figures in his textbooks. Handwritten memo, c. 1910, in MAN Archives.

85. Ade Clark, "The Diesel Engine," in *Proceedings, Institution of Mechanical Engineering* (1903), 432–33. Discussion by George Wilkinson.

86. Worsoe, *Die Mitarbeit*, 42. The "Munich" engine was being used in a Krupp factory in 1924. It is on exhibit at the company museum in the *Villa Hügel*, the old Krupp family mansion on the outskirts of Essen and also repository of the Krupp company Archives.

87. Ibid., 55–60.

88. E. Diesel, *Diesel*, 250.

89. Production records. MAN Archives.

CHAPTER 9

Diesel's International Family: 1897–1900

"Therefore, every inventor must be an optimist. The power of the idea has its full impact only in the soul of the creator. Only this possesses the holy fire for the realization."

—Rudolf Diesel[1]

Diesel's main endeavor in 1897 and early 1898 was the bringing of licensees into the fold. His potential rewards were great as he owned the patent rights for most of the industrialized world. Only Germany and Austria-Hungary, which he had granted earlier to Krupp and Maschinenfabrik Augsburg, were excluded, but even here he would share in the royalties.

It proved to be a seller's market. In little more than a year, Diesel and his original sponsors had formed a network of over twenty licensees covering Europe and North America. The contractual terms were high considering that the engine had yet to leave its laboratory cocoon, but the tests witnessed by interested prospects confirmed the inventor's bold claims.

Diesel did not select only established manufacturers. His list also included bankers and entrepreneurs who were left free to contract with or sublicense builders within their assigned territories. Consequently, his large licensee family included the weak and marginal as well as those who became peers of the Augsburg company that had been midwife at his engine's birth.

One important contractual request by Diesel of all licensees was a sharing of patents and improvements. Such cooperation worked as intended at first, but human nature and latent nationalism gradually eroded an earlier team spirit.

Engineers from Maschinenfabrik Augsburg and from Diesel's office in Munich often had to "live" with licensees when they built and tested their first engines. The several plaguing design deficiencies detailed in a previous chapter added to normal startup difficulties.

Ingenuity and determination diligently applied by both licensor and licensee gradually improved the engine's dependability, but a few cautious ones either fell by the wayside or waited until its prospects looked more promising.

This chapter focuses on the corporate histories and some first engines of the risk takers who joined Diesel's licensee family prior to his surrendering of patent rights to the Allgemeine Gesellschaft für Dieselmotoren A.G. in September 1898.

Gasmotorenfabrik-Deutz

The world's largest internal combustion engine builder was the first major licensee in 1897. Gasmotorenfabrik-Deutz founded the industry and had been its undisputed leader for over thirty years.[2] Aided by its own licensees working under Nicolaus Otto's patents, Deutz dominated the gas and gasoline engine market in Europe and America. For this reason, Diesel and his corporate partners counted heavily on Deutz's participation. Deutz managing director Hermann Schumm (1841–1901) and his technical assistant Carl Stein had already put the test engine through a rigorous trial on February 4–5, 1897 (chapter 6). Immediately after that Schumm began talking with Krupp and M.A., who owned all German rights, and Diesel.

Schumm's worry over the validity of Diesel's Patent No. 67,207 was the first contract obstacle. He based this on Otto Köhler's 1893 critique of Diesel's book in which Köhler claimed that his own 1887 treatise on a Carnot cycle engine predated Diesel's idea for one. Deutz retained Köhler to evaluate the patent, and his conclusion remained unchanged. Schumm, who further viewed the Köhler paper as a bargaining ploy, sent a copy of it to Augsburg on February 19 along with his draft for a "plea of invalidity" against the patent. At Rudolf's urging, Lucian Vogel called on Schumm in the Cologne suburb of Deutz several times in an attempt to change his mind. All these events occurred within two weeks of the engine demonstration.

Neither Krupp nor M.A. bent under Schumm's direct attack. Buz telegraphed him on the twenty-second stating firmly and tersely that he "...deemed the Diesel patent indisputable. We cannot negotiate on the basis of contestability..."[3] Schumm knew the strength his potential opponents could exert in a court battle and dropped the patent issue in order to begin a more positive dialogue.

Köhler meanwhile used the opportunity later that year to threaten an action against the patent. However, he left open the door for an amicable settlement. Krupp, still fearing the patent's weakness, took no chances and offered Köhler a seven-year "consultancy" in the form of an annual retainer of 3,000 marks. He accepted.[4]

Deutz reluctantly signed a contract with Krupp and M.A. on July 15, 1897, which placed them in a noncompetitive position from the outset. Its

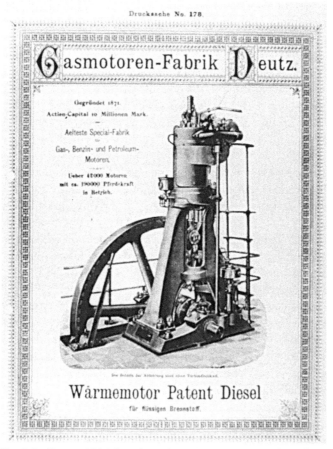

Fig, 9.1. Deutz 1898 brochure introducing its diesel engines.
Deutz Archives

high terms left them little chance to sell engines at a profit in Germany. In addition to an initial fee of 50,000 marks, Deutz would pay a 30 percent royalty on engines 16 horsepower and under. Above that power it dropped to 20 percent. In three years, a minimum annual royalty of 20,000 marks would begin.

They accepted this German market handicap with the hope of capturing Russia's potentially larger one. Rudolf himself thwarted Deutz here. Diesel originally was to let Augsburg or Krupp-Gruson in Buckau (Magdeburg) build all engines for Russia for a 5 percent royalty. Deutz, on the other hand, had to pay a 25 percent royalty on Russian-bound engines. In October 1897 Diesel gave them hope for a more favorable export license by assuring them that for a 100,000-mark payment he would reduce their 25 percent royalty to 10 percent. This still left a 5 percent penalty, but Deutz must have believed they could overcome it with their long experience in making I-C engines.

Diesel reneged on this implied promise to Deutz by granting Russian manufacturing rights to a new German-based holding company in September 1898.

After this rebuff Schumm decided to build only two of the planned twenty Type V, 20-hp (Series XVII) engines[5] (fig. 9.1). Parts already started for twelve were scrapped, but it is possible that some went to Austria for an engine being built there.[6] Schumm's chagrin may be read in what he told Hugh Meier, an engineer sent to Deutz as an observer by the American licensee:

> We intend to build the Diesel motor now, provided that Mr. Diesel will alterate [*sic*] and modify the conditions which are too stringent and which make our chances of a reasonable profit almost illusory.[7]

Schumm also complained about sharing information with licensees: "He [Diesel] cannot expect us to put our experience and future improvements at the disposal of allied firms."

The American licensee, who had yet to build an engine, bought Deutz's first one to display at a New York City electrical machinery exhibit in May 1898. Because even this engine would not be ready in time, a second ordered from the Nürnberg licensee was substituted. The Deutz diesel arrived in St. Louis on August 21.[8] The customer who later bought it endured various "mishaps," and after a threatened lawsuit, it was taken off his hands. Bad feelings were left all around.[9]

The reasons for the manufacturing delay, as explained in the Meier report, were from problems that seem surprising when considering Deutz's I-C engine background. However, they had never encountered such high pressures in combustion chambers, air compressors, or fuel systems. Deutz, like Augsburg, suffered a siege of porosity in the iron castings for the compressor and power cylinders.

Poor piston ring sealing added to the porosity problem. The use of more rings, stiffer rings, springs behind the rings, or small holes drilled down from the piston crown into the back of the upper ring groove might work for gas and gasoline engines but not the diesel. The trouble was mainly in the manufacturing technique. Not until a clever Deutz engineer could convince his own shop first to collapse the ring and then machine it to the cylinder diameter, did gas sealing improve. Deutz adopted this method for all their Otto-cycle engines.

Poor ring sealing created lubrication problems. The four small oil holes drilled through the cylinder wall to pressure lubricate the piston became clogged with carbon if there was blowby. Meier also reported that if the holes were too large all the oil came out the one nearest the supply line.[10]

Pistons went through several redesigns. One simple improvement was to machine an oil groove below the rings of a longer-than-standard piston and not cut it in the cylinder wall as Augsburg did. This cheaper method also eased piston removal from below because rings could no longer expand into the cylinder groove when the piston was pulled downward.

The wire mesh atomizer in the fuel nozzle was described but not condemned as Meier never saw an atomizer "destroyed by heat" while at Deutz. They used an electroplated steel wire cloth with one end soldered to the spool. The tightly wrapped assembly was "firmly twisted" into the nozzle bore so all fuel entering the cylinder passed through it before injection. Sieve length, 25 mm long in the 18-hp Augsburg test engine, was based on engine horsepower and fuel viscosity. This length performed well on kerosene, but "faint explosions" heard when burning "Solaroil" disappeared if the length was increased to 50 mm.[11]

The second Deutz engine benefited from what was learned on the one for America, although it still arrived a month late at the Diesel Pavilion in the Munich Exhibition. The engine went back to the factory after the show ended where it provided power in the foundry until being scrapped in 1917.

Despite the decision to drop diesel production, two experimental models were built in the 1898–99 period. A 20-hp "Model W" was based on the Series XVII design. Deutz added a long inlet air pipe to draw air from inside the base of the A-frame pedestal to reduce the suction noise.[12]

A "Model X" was the first European trunk piston (crossheadless) engine. Although difficulties were encountered with it at Munich's Technical University, it reportedly ran well enough for M.A.N. in the Augsburg lab (fig. 9.2). The 300 × 330 mm Model × produced 10 hp at 220 rpm. A low-mounted

Fig. 9.2. Deutz Model X (second from right) in M.A.N. lab. *MAN Archives*

camshaft placed perpendicular to the crankshaft actuated the valves through long pushrods. The engine is on exhibit in the Deutz museum as its earliest existing diesel model (figs. 9.3a and 3b).

Continued frustration over Diesel's actions and the license contract pushed Deutz into signing a new agreement with Krupp and M.A.N. on October 5, 1899. It lowered the royalty to 15 percent and postponed all minimum payments for two years. Nevertheless, Deutz canceled this one in 1902. Not until near the expiration of Diesel's patents, after beginning to build other types of kerosene burning engines, did Deutz again sell his. A token royalty of 1 percent was briefly paid when they successfully reentered the market in 1907.

Mirrlees, Watson & Yaryan Ltd.

A Scottish company quickly sought out Diesel as word spread about his engine. Sir Renny Watson (1838–1900), R. A. Robertson and John Platt, three directors of Mirrlees, Watson & Yaryan Ltd. in Glasgow arrived at

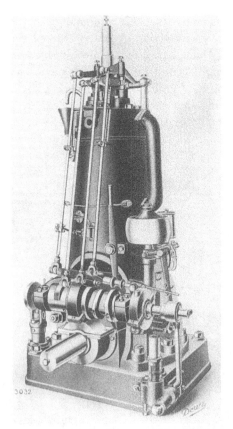

Fig. 9.3a. Deutz Model X trunk piston engine of 1898–1899. 200 × 330 mm, 10 hp at 220 rpm. Wet cylinder liner and camshaft at 90 degrees to camshaft. *Deutz Archives*

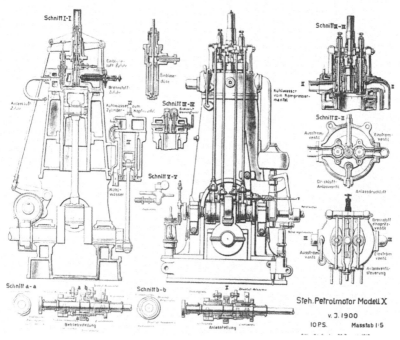

Fig. 9.3b. Deutz Model X trunk piston diesel 1898–1899.

M.A. the day after Schröter had finished his tests. They saw the engine perform from February 18 to 20, 1897. Mirrlees seemed a good prospect to Rudolf because of its long reputation for quality products.[13]

Robertson proposed that a license be negotiated in London where Diesel went after signing his updated contract with Krupp and M.A. in Essen (chapter 7). Several days of wrangling over specific language with both Robertson and his lawyers led nowhere. Rudolf wrote home on March 16 that just when he and Robertson

> are almost in agreement then come the lawyers to give the perfect wording. . . . The fatal result with it is that they fear about the worth of the patent and, as with Krupp, I must fight my way through the entire struggle once again. Evenings I fall in bed like I am dead and wonder how I will have enough strength to begin again on the next morning.[14]

To end the impasse, Robertson and Diesel agreed to visit Glasgow and ask Lord Kelvin (William Thomson, 1824–1907) to study the patent and the engine design. Rudolf was invited to stay in the Robertson home near the University, the site of James Watt's early work. Kelvin met with Diesel on March 17 to discuss thermodynamic and mechanical details. Two weeks later the famed scientist issued a report fully endorsing Diesel's ideas. His analysis of the patent and the engine offers what he believed both new and of value:

Diesel's process of heating the air, simply by compression, to a temperature far above the igniting point of the fuel, before the fuel is introduced to it, supersedes all use of flame or hot chamber for ignition, even when the engine, cold in every part, is started for its first stroke by energy stored in a compressed air vessel from its previous working. This capability for instant starting from cold, and for stopping and starting again with perfect readiness at any time, is, I believe, a very valuable item of superiority, for many practical purposes, over any gas or oil or other interior-combustion engine, previously made. Diesel's invention of introducing the fuel into air previously heated so far above the ignition point, merely by compression, and so causing the fuel to ignite, is, in my opinion, thoroughly original, and it has not been anticipated by any previous inventor.[15]

Lord Kelvin's statements led to an agreement on March 26, 1897. The death of Diesel's mother on the twenty-fifth may have hastened the contract's signing as he attended her funeral in Munich on the twenty-seventh.[16]

In return for existing and future patent rights in the British Isles, Mirrlees would pay Diesel £4,000 (80,000 marks) immediately, £1,000 in three years, and a 25 percent royalty. Mirrlees, Watson & Yaryan also agreed "to communicate to Mr. Diesel during the currency of the agreement and to give him the free use outside of the United Kingdom of Great Britain and Ireland of all improvements made by them whether patented or not in the design and construction of engines manufactured under this agreement."[17]

Another requirement was that Mirrlees build a Series XVI 20-hp engine for which M.A. sent drawings on August 9, 1897. Diesel was billed 110 marks for the set.[18] This engine, probably completed about the end of 1897, may predate the one M.A. delivered to Kempten. It was tested by William H. Watkinson (1860–1932), then a professor at the Glasgow & West Scotland Technical College, in April 1898. He later said the engine ran as well as any built in Germany—and suffered similar problems.[19]

Diesel paid his first visit to the Mirrlees factory in July 1898 to see the engine run. He expressed some negative opinions in a letter to Martha: "How it looks in its gloomy, dirty corner covered with dirt and neglected like a Cinderella." Nevertheless, he said it performed well and that Mirrlees were eager, "good people . . . but they are not motor-people."[20] He also feared that they would get the engine off to a slow start in Britain. Indeed, Mirrlees began to argue that it had been released too soon and halted the program when they perceived they were "nursing someone else's baby."[21] This first engine was rehabilitated in 1902 and ran for years on company premises. It is exhibited at the Science Museum in London (fig. 9.4).

That the Scottish company originally intended to exploit the engine is evidenced by two designs sent to Diesel for evaluation in the last months of 1897. The first was for a three-cylinder marine engine with a bore and stroke

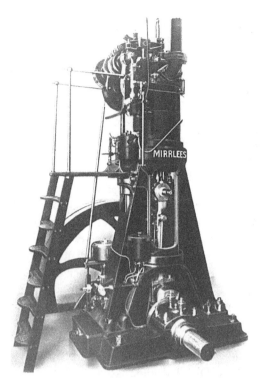

Fig. 9.4. Britain's first diesel made by Mirrlees, Watson & Yaryan Co. Ltd., Glasgow, 1897. Four-stroke, 260 × 410 mm bore stroke, 20 bhp at 160 rpm. The Science Museum photo.

of 255 × 305 mm. Speed and output are unknown. A second for marine use had a sliding cam to change engine direction, a method similar to a later reversing method.[22] Neither was built.

In January 1898 Mirrlees asked Diesel for a nonexclusive contract limited to engines sold in Great Britain. The request did not become a fact until November 7, 1899, when an agreement was signed with the Allgemeine, who had assumed Diesel's role.[23] Mirrlees would pay the Allgemeine, or a new English corporation that would assume the exclusive patent rights, £9,000 for this lesser privilege, but the royalty was eased to a 15 percent maximum. The negotiations were made through George Oppenheimer, a member of the banking family of Oppenheimer, Son & Co. Ltd. in London. He had hoped to gain control of the newly available British patents and form a company to exercise them.[24]

Diesel Engine Company Ltd.

Thwarting Oppenheimer was the Allgemeine's exorbitant price for the patent rights. Its first request was £150,000 in cash. Instead, an incorporation of the Diesel Engine Co. Ltd. on December 11, 1900, ended the impasse. The London company gained sales rights throughout the British Empire and in countries where no patents yet existed or were assigned to another licensee.[25] In return, the Allgemeine would be paid up to £75,000 in cash plus an equal sum in stock, based on the offering's success. The DEC was also required to build a new Güldner-type, two-stroke diesel engine. The Allgemeine never received more than £15,000 due to lack of investor interest.[26]

Company directors were Rudolf Diesel; William Oppenheimer, managing director of the Oppenheimer bank; Henry Hodgson of Scott & Hodgson, who built the two-stroke engine; and Norman Thompson, a prominent electrical engineer. The latter two resigned in 1902.[27]

The DEC was more a holding and sales company as it bought all its engines. These came first from M.A.N., then Sulzer and Carels, and finally almost all from Carels after it gained a majority interest. (Mirrlees could sell engines under its own name or to the DEC if requested by the latter.) Sulzer and Carels both received an initial order for twelve, 35 bhp engines in 1903. By the end of that year, Carels had delivered twenty. M.A.N. shipped none after this.[28]

A sales report to the Allgemeine on February 22, 1904, stated that the Diesel Engine Co. had up to then delivered a total of eighty-two engines with a cumulative output of 5,028 bhp. The ratio of domestic to export sales was about 4:1. Twenty-seven had come from Sulzer with about half being exported.[29] In 1907 the company reported cumulative sales of 11,550 bhp, and during that year's second quarter alone the monetary amount came to almost £75,000.[30] The Diesel Engine Co.'s contract with the Allgemeine ended March 30, 1909, and the company was officially liquidated in June 1914 after being absorbed as a going concern into a new Carels-owned venture.[31] DEC's later relationship with Willans & Robinson, another British licensee, is found in chapter 12.

Société Française des Moteurs R. Diesel

Frédéric Dyckhoff, Rudolf's old friend and earliest licensee, had learned enough during the long Augsburg development period to become one of the few selling diesel engines before 1900.

Diesel and Dyckhoff voided their old contract (chapter 5) so they could assign all French patent rights to a holding company, the Société Française des Moteurs R. Diesel, formed on April 15, 1897.[32] Rudolf, who attended the signing in Bar-le-Duc, received half of the authorized capital of 1,200,000 francs in stock, but no cash. Dyckhoff invested 25,000 francs and was given the right to build engines without paying a royalty. Ground was broken for a new engine factory in nearby Longeville the same year[33] (fig. 9.5). Dyckhoff ran that and also the Société (Sté.) Française.

Dyckhoff's philosophy differed from Augsburg's in his desire to make smaller engines. A February 1899 catalog lists sixteen engines of 4 to 20 bhp as having been built with nine of 8 bhp and under. This production was not accomplished without numerous difficulties requiring sustained help from Rudolf's own troubleshooter Karl Dieterichs.[34] For several years Dyckhoff

Fig. 9.5. Groundbreaking for Dyckhoff's Longeville factory. Frédéric is hatless and stands behind son Rudolphe in the front row, who is between daughters Antoinette (on left) and Charlotte. *MAN Archives*

marketed a 4 bhp at 280 rpm diesel with a 130 × 220 mm bore and stroke. It used an inserted "wet" cylinder liner (fig. 9.6).

Of special interest is his three-cylinder, reversible marine engine also built prior to 1900. It had a "square" bore and stroke of 132 mm and reportedly developed 15 hp at 600 rpm. A half-speed auxiliary shaft drove three individual fuel pumps and a single-stage air compressor. Instead of the conventional open A-frame supporting the cylinder, the crankcase was an enclosed box structure with large inspection doors. It is presumed to be the first working engine with a reversing mechanism as Mirrlees had made only drawings. One of its two camshafts could be selectively engaged for either forward or reverse (fig. 9.7). Dyckhoff's German Patent No. 107,395 of April 25, 1899, describes an "Apparatus for the reversal of . . . combustion engines."[35] Also constructed was a slightly smaller locomotive engine (130 × 130 mm), but neither it (fig. 9.8) nor the marine version ever left the factory.

An 80 bhp, two-cylinder Dyckhoff engine won the power division's Grand Prix at the 1900 Paris World Exposition (fig. 9.9). Unexpected repercussions of this award would cause a later controversy within the French Navy Department.

Sautter Harlé & Cie. bought a sublicense from the Sté. Française in March 1899. This company, which would supply special engines to Dyckhoff, including one for a canal boat, also designed and built marine and submarine diesels in the early 1900s (chapter 15).

Financial difficulties at Sté. Française beginning in 1901 brought Dyckhoff two years later to ask the Sté. de Moteurs à Gaz et d'Industrie Automobile

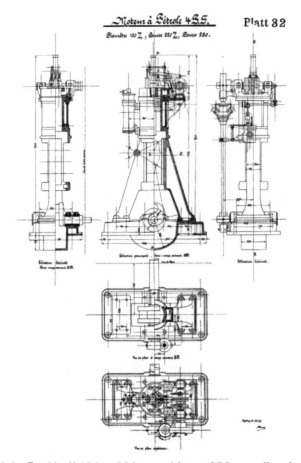

Moteur à Pétrole 4.3.5.   Platt 82
Diamètre 130 �%, Course 220 ⁷⁄₁, Vitesse 280.

Fig. 9.6. Dyckhoff 130 × 220 mm, 4 hp at 280 rpm diesel of 1899.
Note the wet sleeve liner. *MAN Archives*

(Marques Otto), a Deutz as well as an Allgemeine licensee, to take over the French Diesel patents. Nothing came of it.[36] How many engines the French Otto factory built is unknown, but the number must have been small.

In 1908 M.A.N. received from Sté. Française the right to sell in France for a 10 percent royalty. Dyckhoff is assumed to have sold the French patents soon after, because in June 1909 the Sté. de Chantiers et Ateliers Augustin-Normand of Havre was listed as owner. This company became a competitor to Sautter Harlé.[37]

Maschinenbau–Actiengesellschaft Nürnberg

Anton Rieppel (1852–1926), the enterprising *General-Direktor* of Maschinenbau-Actiengesellschaft Nürnberg[38] was another attendee at Diesel's lecture in Augsburg (fig. 9.10). Soon thereafter he witnessed the test engine perform and became a zealous believer in its potential. Unknown to Rudolf

Fig. 9.7. Dyckhoff 1899 reversible marine engine.
132 × 132 mm, 15 hp at 600 rpm. *MAN Archives*

at the time was that Rieppel and Buz were holding talks about uniting their two companies.[39]

Augsburg and Nürnberg combined to form Vereinigte [United] Maschinenfabrik Augsburg und Maschinenbaugesellschaft Nürnberg AG on November 24, 1898. This was welcomely shortened unofficially in 1904 to M.A.N. In 1908 the name became Maschinenfabrik Augsburg-Nürnberg AG with Augsburg and Nürnberg as divisions. The reader will note that the Augsburg company is always referred to as M.A. prior to the 1898 union and M.A.N. after this date.

Rieppel spent his entire professional career with the Nürnberg company. He was born in a northeast Bavarian village where his father sold forged iron goods. The elder Rieppel believed an elementary school education sufficed, but Anton had other ideas. His self-studies in math had prepared him for a trade school by age fifteen. After a brief attendance at the Munich *Polytechnikum*, the Nürnberg works lured him from school to design iron structures for its Gustavburg division. In 1876 he went there as its new manager. *Werk Gustavsburg* had been formed in 1860 to erect the first railroad bridge across the Rhine between that town and Mainz. It went on to build major rail bridges and canal structures throughout Germany. Rieppel so improved the operation that Nürnberg brought him back in 1888 to be technical director. The next year he was made a director and member of the management board. In 1892, at age forty, he headed the company.

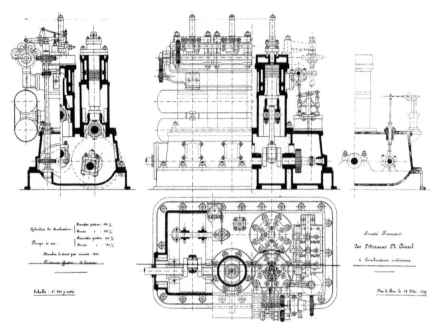

Figure 9.8. Dyckhoff reversible locomotive engine, 1899. *MAN Archives*

Fig. 9.9. The French-built engine that won a Grand Prize at the 1900 Paris World Exposition. *Diesel Collection, Deutsches Museum, Author photo*

Fig. 9.10. Anton von Rieppel (1852–1926). Head of M.A.N.
Nürnberg division. *MAN Archives*

On July 22, 1897, Nürnberg became a Diesel sublicensee. Because of the
pending Augsburg-Nürnberg merger, both M.A. and Krupp had agreed to a
more lenient contract. In return for all German rights, Nürnberg only had to
pay a 20 percent royalty and a 25,000 mark initial fee.

The first Nürnberg engine, a 20-hp Series XVII, was destined for the
Munich exhibit. Instead, it replaced the Deutz engine in Madison Square
Garden, where it was delivered on March 24, 1898. Anton Böttcher, Diesel's
field man, had barely preceded it, having just arrived from England after
helping with Mirrlees's first one. It took him several weeks to prepare the
engine, but the night before the May 7 opening the rings suddenly developed such bad blowby that a spare piston assembly had to be installed. The
engine drove a generator eight hours daily until the show closed on June 5[40]
(fig. 9.11). Voltage control was held to within 1.5 volts during sudden load
changes. The engine was reerected afterward at the US licensee's office on
24 West St., where Professor Joseph E. Denton of the Stevens Institute of
Technology ran a full performance test. Using an "ordinary colorless domestic
kerosene of 120° flash . . . and 0.784 gravity . . ." he obtained these results:[41]

Speed	187 rpm
Horsepower	20.8 bhp
Fuel consumption	0.586 lbs/bhp-hr
Indicated thermal efficiency	37.7%
Brake thermal efficiency	25.8%

Fig. 9.11. First Nürnberg 20-hp diesel engine at electrical
exhibit in Madison Square Garden, New York, May to June 1898.
Author photo from original in Diesel Collection, Deutsches Museum

This was similar to what Schröter obtained during his tests except for the higher indicated efficiency and slightly lower brake efficiency.

Morgan Construction Co. in Worcester, Massachusetts, bought the engine. Although in friendly hands, it was taken out of service after a time because they "found more or less trouble with it."[42]

Nürnberg's second diesel reached the Munich exhibit late due to unspecified problems; a planned-on two-cylinder, 40-hp engine never did arrive.

Rieppel had planned to make ten of the Series XVII engines because of the good New York performance. He rescinded that order after the Munich Exhibition troubles. Diesel managed to change Rieppel's mind, and by October 1898 six were readied for customers in Germany, England, and Sweden.[43]

All diesel production went to Augsburg from 1900 to 1904. Nürnberg engines already shipped, except those to America and Sweden, were returned and scrapped.

Before this centralization began, Nürnberg planned to develop automotive, streetcar, and locomotive engines. The innovative Rieppel blessed the effort, but one wonders how he could expect success in these most-demanding applications after the troubles encountered with stationary engines. Lucian Vogel was placed in charge of their design on his arrival in July 1897 and had this duty until he returned to Augsburg at the end of December 1898.

Vogel did not start with a clean sheet of paper. His configuration for the three mobile applications was based on a concept by Alfred Klose (1844–1923), a respected locomotive designer, who in 1896 had sold Rieppel on building gasoline engines for streetcars. There had been a concerted effort by urban governments during this period to electrify horse-drawn tram cars and to extend existing city lines into outlying villages.[44] Klose believed that I-C engines could compete with electrified trams and had contracted to design them.

The result was an inwardly opposed, two-cylinder, four-stroke engine with cylinders of 110/2 mm × 160 mm (i.e., each of the four pistons had a bore and stroke of 110 × 160 mm). An overhead yoke and a tie rod assembly above the vertical engine linked the upper pistons to the crankshaft. The axial motion of the upper tie rod was transformed into crankpin motion through a crosshead and connecting rod. An intake and an exhaust valve opened into an antechamber off the cylinder's midpoint. The gasoline engine was not built because of the diesel's arrival, and only a sketch exists of the gasoline engine. Vogel selected the following criteria for the diesels:[45]

Automobile:	Two cylinder, 5 hp at 800 rpm, 65/2 × 120 mm
Tram:	Two cylinder, 25 hp at 500 rpm, 135/2 × 200 mm
Locomotive:	90° V 4, 500 hp at 400-500 rpm, 320/2 × 400 mm

The automobile engine (*Kutschenmotor* or "carriage motor") was first. Injection air came from a vertical, single-stage compressor driven off a tie rod. A vertical camshaft lay on a plane between the cylinders and by bell-crank linkage actuated the vertical intake and exhaust valves, which opposed each other on a common centerline. A horizontal injection nozzle sprayed into the antechamber used by the valves. Figure. 9.13 is a similar design.

Rieppel kept Diesel informed on Nürnberg activities and sent him Vogel's drawings of a first design at the end of July 1897. Diesel's long response of August 13 to Vogel's work included suggested revisions, a rather strong criticism, and a valuable piece of advice.

Finally I come to the third reason to write you in a confidential way: Herr Vogel expressed to me that they should hold off on the railway motor until there has been the necessary practical experience learned from the auto engine. This plan is in direct opposition with what was

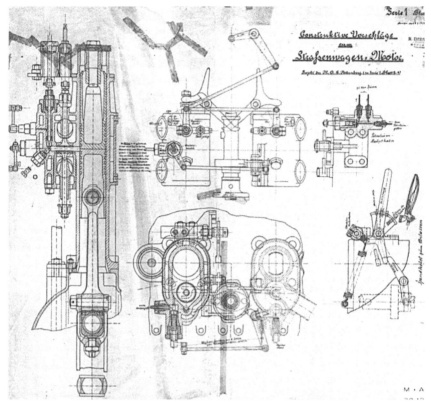

Fig. 9.12. Diesel's redesign of Vogel's first *Kutschenmotor*
at Nürnberg, January 28,1898. *MAN Archives*

discussed at the time in Nürnberg and which I certainly supported, namely production of railway motors first, and out of this carry over to the tram motor. Quite simply, the step from our Augsburg test motor to the almost equally robust railway motor is relatively easy and has a chance of success, while the leap to the tiny plaything of the auto engine holds great uncertainties. Moreover, the two exercises are different, and thus valuable time might be lost.[46]

Diesel received a revised design, which he again heavily scored, and with his response included a drawing dated January 28, 1898 (fig. 9.12). After at least five redesigns, the engine was built and run that December. Rieppel wrote in January 1899: "We have great difficulties with the engine, and I doubt that we will be successful without making a complete redesign . . ."[47] A May 1899 drawing shows its final form before the project was dropped (fig. 9-13).

The tram engine, a scaled-up version of that for the auto, was built and briefly tested. Only a cross-section sketch of it exists.

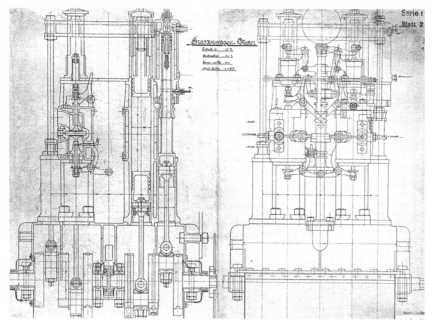

Fig. 9.13. Final design of Vogel's "carriage motor," May 13, 1899.
MAN Archives

Adolphus Busch, The American Entrant

Adolphus Busch (1839–1913), who is more often associated with beer and beneficence, was Diesel's last major licensee in 1897. He brought the engine to America and sustained it through a long siege of technical and financial reverses (fig. 9.14). At the time of his death, Busch was reputedly the wealthiest man in American manufacturing.[48]

Although a naturalized United States citizen, Busch never cut his emotional ties to his German homeland. Born in Mainz and educated through high school in Cologne and Brussels, he went to America in 1857 to join others of his family. His first job was clerking in a St. Louis steamboat office. When his father died in 1859, he and his brother spent their inheritance to buy a brewery supply company. The two men married daughters of Eberhard Anheuser, a good customer, and in 1879 the father-in-law's brewery became the Anheuser-Busch Brewing Association. When his fortune was assured, Adolphus made annual pilgrimages to Germany to relax in fashionable spas and on occasion to visit with hop merchant and good friend Berthold Bing.[49]

Bing sold German hops to Busch and relayed to the brewer news tidbits about his old country. One item imparted in the summer of 1897 was of a promising new engine from Augsburg that a Nürnberg company planned to build. Busch asked Bing to investigate, and he in turn called on Rieppel. A July 28 letter from Rieppel to Diesel telling of Busch's interest led to a

Fig. 9.14. Adolphus Busch (1839–1913).
Photo courtesy of Anheuser-Busch Companies, Inc.

meeting in Baden-Baden between Bing, Busch, and his son-in-law Hugo Reisinger with Diesel on September 6.[50] By the end of the next day, Busch had secured a free option on North American engine rights until November 15. If exercised, Diesel would receive 1,000,000 marks ($238,000) and a 5 percent royalty. (Bing's commission from Diesel was to be 50,000 marks.) Out of the momentous negotiating session grew a lifelong friendship and future business ties between Diesel and Bing that were evidenced during the formation of the Allgemeine.

The paths of Adolphus Busch and Edward Daniel Meier (1841–1914), St. Louis-born of German parents, would join and, through a long and mutually respected association, direct the course of early diesel engine development in America (fig. 9.15).

Meier's technical education had begun at Washington University, but he graduated from the *Polytechnikum* in Hannover, Germany. He served with distinction in the Union Army during the Civil War and afterward became a prominent railway engineer. This led to steel making and the design and construction of heavy machinery. In 1885 he acquired US rights to a new German boiler design and founded the Heine Safety Water-Tube Boiler Co. When and how he and Busch became acquainted is unclear, but both were active in St. Louis business circles. A fitting close to Meier's professional career was his election to the presidency of the American Society of Mechanical Engineers in 1911.[51]

After Busch heard from Bing about the engine, he cabled for Meier to come immediately to Germany and personally verify Diesel's claims.[52] Meier wasted little time because he arrived in Augsburg on September 27, just three weeks later. Interviews were arranged with Rieppel in Nürnberg and

Fig. 9.15. Edward D. Meier (1841–1914). *Photo courtesy of ASME*

with Diesel in Munich. During the interim Busch and Reisinger witnessed a demonstration of the engine on the fourteenth.

Meier wrote the following about Rieppel in his October 4 summary report to Busch:

> He is a very eminent bridge designer and a fine mathematician. . . . His criticism of the means and methods proposed by the men in immediate charge of the motor work are correct and incisive. He hears the pros and cons from all sides, and then renders the thoughtful and final decision. . . .
>
> Mr. Rieppel is entirely convinced of the correctness of Diesel's theories and calculations. He believes that the motor can be developed in special types for railway service, street car service, and for all kinds of road vehicles, down to tricycles. He sees no difficulty in regard to materials, temperatures or speeds which cannot be successfully met with the large experience of good steam engine work.[53]

Of even greater interest is the impression Diesel made on Meier during their two days together in Munich:

> Since the publication of his lecture of June 16, 1897, he has been fairly overwhelmed with praise and congratulations by the leading experts of Europe. Perhaps no better gauge of the strong common sense and real greatness of the man can be found than the fact that all this praise and adulation has not turned his head in the least. He remains the same simple, earnest, conscientious student that he was before his great invention dazzled the experts in his line of engineering.

I was particularly struck with the simplicity, and I might almost say humility with which he spoke of his great achievements. He considers his present success only as an incentive and a stepping stone to further work and progress. He is very accurate in all his statements, and thoroughly honest. . . . He patiently and good-humoredly answered all my questions, and was instantly ready to show me authorities for all his statements from the carefully chosen books of his extensive library. Unlike many an inventor, he has carefully considered the differences of the patent laws of various countries and the methods of their administration. Thoroughly conversant with German and French, he has a sufficient knowledge of English to carefully weigh every word in a specification and claim.

As an engineer he is of course more theoretical than practical, but has now at his command some of the best practical talent of Germany. Professionally well equipped and clear in his ideas, he is so simple and amiable of his manner that it is a pleasure to work with him, and these characteristics of the man promise to make the working together of the various interests in the many countries in which he has issued licenses satisfactory and beneficial to all, and will make a rapid progress in the practical development of the invention possible.[54]

The bond of respect that quickly grew between the two engineers eased them through difficulties in the years to come.

The engine performed well for Meier during the Augsburg tests on September 30. Fuel consumption was as low as 215 g/bhp-hr to give a brake thermal efficiency of 29.5 percent.

To allay doubt as to whether Diesel had "pulled the wool over" Meier's eyes, he wrote of his perception of Heinrich Buz and Buz's comments about the engine, which show the high hopes held for the engine by the Augsburg magnate.

Known as the Bismarck of German machine industry . . . [Buz] is a man of commanding presence, great firmness and decision. It is due to him, above all men, that Diesel was enabled to embody his theories in practice. . . . Mr. Buz, although not an enthusiast by nature, but a strictly practical man of affairs, thoroughly believes that the motor is the greatest advance ever made in dynamic engineering. He fully understands the difficulties that will arrive in going from type to type and from size to size, but looks forward to motors of the largest size, and for all purposes for which steam engines are now used. . . . Mr. Buz quietly but earnestly pointed to the probability of ultimately replacing the machinery of war ships even, by Diesel motors. He hopes . . . that he will have a motor of 800 or 1,000 H.P. running at the Paris Exposition in 1900.[55]

In concluding his report to Busch, Meier unequivocally stated:

I believe that the purchase of the Diesel Patents for America is as promising an investment as the purchase of any patent claim could be. Applying the same methods of giving licenses in the United States, which Mr. Diesel has so successfully used in Europe, we would soon create from among the largest and best engine works of the United States, an interest so strong that no one would dare to attempt infringement or evasion of our patent. Every practical builder will think too highly of the value of the long experience of the Augsburg Works in developing the first working Diesel motor, to think of trying to do without them at the risk of a patent suit.[56]

Note Meier's comment about licensing in the US. He may have planted the seed that Busch need not build engines, but it is more likely that the latter already had this intention.[57]

Based on this positive statement, Busch signed a contract with Diesel in Munich on October 9, 1897, obligating him to pay the inventor 1,000,000 marks within four months. He further agreed to a royalty increase to 6 percent from the original 5 percent, but he controlled all North America by gaining rights to Canadian Patent No. 44,611 of November 13, 1893. The rest of the contract followed those Diesel had with M.A. and Krupp.

Busch immediately ordered an engine from both Deutz and Nürnberg for shipment to the United States. To gain firsthand experience, E. D. Meier's nephew, Hugh, would observe the manufacture of the Deutz engine at their factory.

The Diesel Motor Company of America was incorporated in New York City on January 4, 1898. Busch, who had assigned his patent rights to this firm in July,[58] was president, with Meier chief engineer and Hugo Reisinger secretary/treasurer. The new company would only design and market engines and contract with others to do the manufacturing.

Concurrently, Busch had the St. Louis Iron and Machine Company build a two-cylinder, Series XVI engine that first ran on September 19, 1898. Böttcher moved out from New York to oversee the entire construction and testing of the engine. The following January it went in service at the Busch brewery (fig. 9.16). Böttcher took this opportunity to independently run unsuccessful experiments on the "brewery" engine using a single valve for both the inlet and exhaust functions.[59]

Meier was convinced that the offered German design was too costly to make and not suited to the American market. Böttcher strongly seconded him in this conclusion. In an August 1898 letter to Diesel, Böttcher spoke of the "amazingly high wages" paid to shop workers in the US and added that the first question people asked him was "How much is the engine?" Rudolf was surprised at the news.[60] As a result, Meier's group embarked on

Fig. 9.16. First US-built diesel engine by St. Louis Iron and Machine Co. off Diesel's drawings, 1898. Two-cylinder, 30 bhp. *MAN Archives*

the design of a unique engine configuration whose evolution and ailments are detailed in the next chapter.

The American company was kept well informed about the fuel tests being run in Augsburg. One letter of Diesel's to Meier regarding the gaseous fuel possibilities on March 22, 1898, stated that "On the strength of my experience, I feel safe in saying that the tests with the natural gas will be immediately crowned with the greatest success."[61] Such an optimistic statement greatly encouraged Busch and Meier, who had counted heavily on the use of gas. Kerosene worked against the engine in the United States because very unstable oil prices at that time discouraged buyers. Busch already knew of the problems with the first German and US engines, and on hearing that M.A.N. had abandoned the gas tests, he was sure he had bought a disaster from Diesel. His exasperation is revealed in a letter to Meier: "What did he do—nothing for two years, pocketed my money, lived well and pretended to be overworked."[62]

During this turmoil Meier himself confronted Diesel in a July 1899 letter detailing the problems caused by the engine's premature sale to licensees:

> It seemed to me from the beginning, and still does, that too much attention was paid, and too much time given, to the financeering when the most intense application in the development of the technicals was really required. There can be no doubt that great inventions require much of a strictly commercial nature to make them grandly successful, but it is after all an injustice to the commercial manager to start him on his work too soon. The mechanical and engineering work should always be completed first, at least to the extent of producing a machine which is ready for commercial exploitation. . . . The large sums of money paid for licenses can only be deemed to have been fairly earned when so much of them as necessary has been actually applied to the development of a motor which is a commercial success. The motor we bought was a mechanical success only.[63]

This chiding in muted tones of strong frustration was not untypical of the thinking of most licensees who built engines before 1900. Nevertheless, the Diesel Motor Company of America would continue its quest to develop reliable engines with stubborn independence and infusions of fresh capital.

Burmeister & Wain

The Scandinavian countries were next to join Diesel's international license family. Burmeister & Wain, a longtime Copenhagen ship and steam engine builder, bought the patent rights for Denmark in early 1898.[64] Although B&W waited until 1903 to begin production, they would become well known for merchant ship marine diesels by World War I.

Rudolf had sent David Halley, B&W's manager, a copy of his book *Theorie und Konstruktion eines rationellen Wärmemotors* in 1894 as part of his first sales campaign, but the transplanted Scotsman from Napier & Sons in Glasgow, replied that although interested he preferred to wait until he could see a running engine. Halley died the next year.

It was another Copenhagen firm who later contacted Diesel. Georg Garde, manager of A/S Titan, had read the Kassel lecture and asked Carl Winslow, a local consulting engineer, to visit Augsburg in October 1897. This investigation was done in cooperation with Burmeister & Wain who said they would consider a joint venture with Titan on the engine if it had merit.[65] A portion of Winslow's report tells of his visit at M.A.:

> When I entered the testshop . . . the engine was running absolutely regularly, and except for soft clicks from the valves, was completely silent. The exhaust was absolutely undiscernible, and in spite of a lack of silencer, the noise from the exhaust was . . . softer than a steam

engine. At the exhaust pipe outlet there was only an insignificant odor of kerosene, not enough to prevent keeping one's nose over the outlet for some time.[66]

Ivar Knudsen (1861–1920), who was made technical director of B&W in 1897, deserves much of the credit for his company's eventual success with the diesel engine (fig. 9.17). Knudsen came to B&W in 1895 from a municipal power plant after Halley's death and no doubt had known Winslow from his days at the plant where Titan supplied electric motors. He was a director and head of engineering at B&W until his untimely death from food poisoning while on a business trip in India.

Knudsen himself went to Augsburg in November 1897 to see the engine on test. Next, Martin Dessau (1865–1919), Burmeister & Wain's commercial director, visited Diesel in Munich on December 6 carrying a letter giving him authority to speak for both B&W and Titan.

Diesel signed license contracts on December 11 that gave A/S Titan and B&W joint Danish rights for engines of 20 hp and under.[67] B&W was granted exclusive rights for engines above that rating and inked its part of the contract covering the higher power range on January 28, 1898. The "20 & under" license cost 40,000 marks, and the "over 20" was 20,000 marks. There was a 10 percent royalty and a shared, minimum annual payment of 10,000 marks. Rudolf had received his 60,000 marks on December 20, over a month before B&W's signing.

Dessau had asked for an option on Diesel's Swedish patents, but this was preempted by a direct license with interests there.

B&W ended up as the sole Danish licensee when Titan withdrew after realizing B&W would not accede to its new request to build engines of up

Fig. 9.17. Ivar Knudsen (1861–1920), Technical Director of Burmeister & Wain. *B&W Archives*

to 30 hp As a result, B&W signed a second contract on March 4, 1898, taking over Titan's rights to the smaller sizes. Diesel added his name to the document on the seventh.

A set of Series XVI drawings came from Munich in January 1898, and the first 20-hp engine began running in the lab the following December. Although it suffered ailments similar to those built by the other licensees, B&W gradually cured most of them[68] (fig. 9.18). Especially helpful was the practical nozzle atomizer whose design was sent to Augsburg (fig. 8.13). Nevertheless, Knudsen followed a cautious development path. He and Diesel remained in contact during this period, and the two became friends despite B&W's delay in marketing engines.

I. P. Spangenburg, whom Knudsen chose to head the new diesel department in March 1898, was an engineer from the Danish Patent Office who had examined Diesel's patents. Spangenburg spent the summer of 1898 in Augsburg familiarizing himself with the engine and later wrote the 1899 test reports on the B&W lab engine. He had left the company in 1902 before production began.[69]

Fig 9.18. First Burmeister & Wain diesel, 1898. *B&W Archives*

B&W granted the Swedish diesel licensee in July 1902 a cumulative horsepower sublicense until the end of 1904.[70] It remained reticent about entering the market, not because of reliability concerns with the prototype engine, but of factors beyond its control.

Denmark levied no tariff on oil at the time.[71] However, the oils most wanted as diesel fuels were dominated by the East Asiatic Co. (*Østasiatiske Kompagni*), a Danish shipping firm, and the major oil importer. B&W feared the instability in oil prices resulting from frequent world market manipulations as well as high distribution costs. All fuel deliveries in Denmark would come from Copenhagen's free port where East Asiatic built a tank farm in 1902. These two factors alone could deter potential diesel customers from buying.[72]

Ivar Knudsen sought to bypass this worrisome predicament and investigated B&W's importing fuel directly from England. One avenue gave O. E. Jorgensen, Spangenburg's replacement, an occasion while in London to see George Oppenheimer, managing director of the Diesel Engine Co. He was briefed on DEC activities as well as the British fuels situation. Nothing came of the importing aspect, and not long after the England trip, East Asiatic seemed to have allayed B&W's qualms about fuel prices being a serious hindrance to diesel sales.[73]

Knudsen then proceeded in late 1903 with production plans for a line of stationary engines based on the promising prototype. The experience gained during the next decade laid a firm foundation for the coming leadership role B&W would play in marine diesel engines.

AB Diesels Motorer

The purchase of the engine's Swedish patents came through the Wallenbergs of the Stockholm Enskilda Bank and Oscar Lamm (1848–1930), an engineer who headed Nya AB Atlas. Knut Agathon Wallenberg (1853–1938), the bank's president, and his brother Marcus (1864–1943), its vice president and legal counsel, were also on the board of Lamm's New Atlas, Inc., a large Stockholm manufacturer of locomotives and railway equipment.[74] Marcus was the entrepreneur of the family and a friend of Emanuel Nobel in St. Petersburg. Nobel, who was being pressured at the time to secure a Russian license from Diesel, shared his knowledge with Marcus about the claims made for the engine. This led the latter to begin almost immediate negotiation for a Swedish license.

Diesel signed an option agreement with Marcus Wallenberg and Lamm on January 20, 1898, and later that day Marcus, Knut, and Lamm saw the engine perform in the Augsburg lab. Berthold Bing also participated in the talks.[75] This preliminary agreement called for the incorporation of Aktiebolaget Diesels Motorer, a Swedish stock company with a capitalization of 300,000 kronor. Diesel would receive shares worth 50,000 kronor and an equal amount in cash.[76] (A Swedish krona equaled 1.13 marks.) Future payments to Diesel included a 10 percent royalty for Finnish rights and a third of any sublicense fees.[77] The final contract was inked in Munich on January 25.

The company came into being on April 2, 1898, with production slated for a small factory under construction at Sickla, a railroad stop a few kilometers southeast of Stockholm.[78]

Work began in mid-1898 on an A20 engine (Diesel's 20-hp Series XVI) at Nya AB Atlas in Stockholm during this interim. Atlas did not understand the care needed to make the engine and met serious problems when it went on test. In mid-March 1899 the Allgemeine sent help through the services of Ludwig Noé, who fought with the engine until June before it performed to his satisfaction. Noé said a few years later that "his workshop experiences proved very useful to him, because he had to act not only as a lathe hand but also as a mechanic."[79] This trying time hastened the move from the ill-suited Atlas to the Sickla factory (fig. 9.19).

Of the twelve Augsburg-type engines, between 1899 and 1902 AB Diesel Motorer built six A20s and four, two-cylinder A50s of 50 bhp at 190 rpm.[80] The latter were made off drawings for the 275/420 mm, Series XVIII. Unlike the A20 by Atlas, those from Sickla reportedly satisfied their buyers. The first A50 went to AB Separator, De Laval's company at Tumba, Sweden, in November 1900 (fig. 9.20).

The embryonic firm had good, hands-on managers. John Schmidt (1866–1935), president until 1903, viewed his one-room factory and test area through a corridor window outside his office (fig. 9.21). Chief Engineer Henning Hallencreutz (1872–1935) began his Sickla employment with a stint of almost two years at Augsburg to gain experience. He returned in the spring of 1900 full of M.A.N. élan.[81]

Fig. 9.19. Exterior view of new AB Diesels Motorer factory at Sickla in 1900. Hesselman, *Teknik och Tanke, 1948*

Fig. 9.20. AB Diesels Motorer two-cylinder, 50 bhp, of Augsburg design in De Laval AB, Tumba, Sweden factory. *Diesels Motorer Catalog, 1906*

The fire of design genius at AB Diesels Motorer was to emanate from Knut Jonas Elias Hesselman (1877–1957). His chance coming to Sickla in January 1900 with little proven experience offered few clues to his future potential. Jonas's short professional career began with a two-year apprenticeship at a vocational school prior to his entering Stockholm's Royal Technical University (*Kungliga Tekniska Högskolan*). After graduating in 1899, he spent some months with a small company trying unsuccessfully to develop a kerosene-fueled engine. Six years later he became chief engineer when Hallencreutz left the company (fig. 9.22). He made numerous major contributions to fuel-injection and combustion systems in I-C engines at Sickla and in later years.[82]

Hesselman's already cited *Teknik och Tanke* (*Technic and Thoughts*), published in 1948, is more than a history of the early diesel engines in Sweden. *Technic and Thoughts* provides a wealth of material on contemporary diesel practice and perceptive, frank insights on people closely associated with the still-emerging engine. The book also relates his own activities at Diesel Motors, Inc., until departing in 1917.

Soon after coming to Sickla, Jonas experienced the diesel's prima donna character during a service call to an A20 engine installed in a candle factory. The sudden explosion of a fuel charge in the nozzle, destroying both atomizer and nozzle tip, "was like a cannon had been fired next to me and shook the entire foundation."[83] (A worm gear driving the cam had slipped on its shaft, which allowed the fuel valve to open before top dead center.) He wrote that "the 'bang in the candlestick factory' became a milestone between earlier tranquility and the future's hard reality."

Fig. 9.21. Interior view at Sickla showing four Augsburg-type A20 engines in test and assembly area. Hesselman, *Teknik och Tanke, 1948*

Economic reality led to another milestone event at Diesels Motorer after it was apparent that the Augsburg A20 and A50 designs could not be produced and sold at a profit in the Swedish market. Priced at 10,000 kronor and 25,000 kronor respectively, the engines were simply too expensive.[84] A new design grew out of this hard fact.

Lessening of dependence on Germany had also begun. Small successes gained here and there by the Swedish neophytes gave them confidence to undertake a major redesign that departed from the "gospel according to

Fig. 9.22. Jonas K. E. Hesselman (1877–1957). *Wärtsilä Archives*

Augsburg." Hesselman came to believe that tackling the new engine aroused professional and, to a degree, national rivalries between the companies. An example of his perception grew out of a fuel distribution problem on the multicylinder A50 engines. He had arrived at a simple solution adaptable to pumps for even three- and four-cylinder engines. Per the license contract, Sickla advised M.A.N. of it, and by so doing he felt it "probably laid a foundation for the opposition against me which prevailed for several decades into the future. It was not easily accepted that a greenhorn foreigner should stick his nose into their specialty field."[85] Of course this gives only his side of the story, which continues in a later chapter.

Maschinenfabrik Augsburg Moves into Russia

Rudolf Diesel and Maschinenfabrik Augsburg anticipated a great market for his engine in the empire of the Romanovs. Its unlimited supply of cheap and duty-free domestic petroleum would be the catalyst to sell many of the oil burning diesels.

Diesel's quandary was how he could reap royalty payments from M.A., who badly wanted access into that vast eastern land, yet be free later to deal with a Russian builder of his choice. It had not been his practice to let a German licensee sell engines into another country if a domestically licensed company already existed there. Austria-Hungary, for example, was licensed through Krupp. Nevertheless, on March 14, 1898, Diesel granted M.A. the right to send engines to Russia in return for a 100,000 mark initial payment and a 15 percent royalty on sales destined there.[86]

By a follow-up contract of the same date, Rudolf gained assurance that a future Russian licensee could study Augsburg's testing and manufacturing facilities. He firmly believed that he would face difficulties in attracting anyone from Russia who did not have access to M.A. technology in the beginning. Thus, under this new contract Diesel canceled M.A.'s 100,000-mark payment to him and reduced the royalty to 5 percent. In return, M.A. would allow a Russian engineer to spend up to five months in their factory over a three-year period. The visitation times were limited to three months during the first year and the remaining time as needs dictated in the last two years.

Maschinenfabrik Augsburg cooperated fully with these visitation rights and in return reaped handsome rewards from their Russian sales. Including the first two delivered in 1899 and the 89 delivered in 1908/1909, M.A. sold 398 engines into Russia during these eleven years. It was more than that of any single Russian builder and constituted almost half of the 809 total diesels made there up to 1910.[87]

Carlsund and the Nobels

The path to Russia began with the lectures given by Diesel and Schröter during the VDI Congress in Kassel. Anton Gustav Robert Carlsund, one of the chief engineers at Maschinenfabrik Ludwig Nobel, became so enthused that on returning home to St. Petersburg he urged his employer Emanuel Nobel (1859–1932) to buy the Russian patent rights. Carlsund (1865–1948) was developing a low compression kerosene engine and knew its limitations, particularly in light of the Kassel statements. Nobel reacted by sending him to Augsburg for more information. Carlsund later wrote he "had to work very long" before he could get Nobel to act on the engine.[88]

Carl Nobel (1862–1893), Emanuel's younger brother, ran the St. Petersburg factory until his early death from a diabetic coma. He and Carlsund had been classmates at the Royal Technical Institute, and it was Carl who had enticed Anton to come to Russia in 1891. Emanuel, however, ruled the huge Nobel ventures and dominated his siblings and their offspring.

Immanuel Nobel (1800–1872), the grandfather of Emanuel and Carl, had gone to Russia in 1837 where he started a munitions factory. It prospered for several years, but misfortunes and political unrest forced him into bankruptcy and a return to Sweden in 1858. The lure of Russia brought two sons back.

These two brothers, Ludwig Immanuel Nobel (1831–1888) and Robert Hjalmar Nobel (c. 1829–1896), were to found their own empire within Russia. Ludwig opened a St. Petersburg manufacturing company in 1862 to make armaments for the government and in time to supply machinery for his immense oil company. The latter's creation and growth was a product of Robert's foresight and the incessant prodding of Ludwig. Alfred Nobel (1833–1896) of dynamite and prize fame, and another of Ludwig's gifted brothers, was on the board of Nobel Brothers, but his own interests focused him away from Russia.[89]

Russische Diesel Motor Co., GmbH

Emanuel Nobel and Diesel met in Berlin's Hotel Bristol on February 14, 1898.[90] Nobel's friend Marcus Wallenberg who had just bought the Swedish rights and Berthold Bing, acting as Rudolf's agent, were also there. It was at Marcus's insistence that Emanuel had come.[91] A reluctant Nobel at first resisted Diesel's persuasive arguments, but the next day he acquiesced and agreed to form a holding company as suggested by Bing.[92]

The Russian Diesel Motor Co. opened its doors on February 16 to assume ownership of Russian Patent No. 261 and thereby act as licensor to Russia. Bing headed the company, which used his Nürnberg home as its office. Capitalization was 1,000,000 marks with stock valued at 1,000 marks per share. Nobel took 600 shares, Diesel 200, and 100 each went to Wallenberg and Bing. Diesel received 600,000 marks in cash from Nobel and 100,000 marks from each of the other two, but they were reimbursed that sum by Diesel for their help.[93] The income derived from a 5 percent royalty on engines built in Russia repaid Nobel's investment in ten years. Bing and Wallenberg likewise fared well, but Rudolf's shares went shortly to the "Allgemeine," which he later lost.

Maschinenfabrik Ludwig Nobel, St. Petersburg

The little-known Diesel engine activities at Maschinenfabrik Ludwig Nobel prior to 1917 were thankfully recorded by several of those who left Russia because of war and revolution.[94] The Nobels' diesel success was almost pre-ordained as all the needed ingredients for it were present: imagination, engineering competence, financial resources, and good reputation. As a bonus, they controlled energy sources and were themselves a potentially large engine user.

Carlsund had already gone to Augsburg when the Nobel factory received its manufacturing license on April 9, 1898. While there, he witnessed the fast-moving lab developments and Lauster's struggle with the Kempten engine. He later wrote this of his relationship with M.A.N:.[395]

> I personally got along very well with Lauster, and no secrets about production methods were kept from me. I was allowed to visit the workshops freely, and during my annual visits in Augsburg participate in their developments and new knowledge. These were applied to the benefit of the Nobel factory. We also received drawings from Augsburg which we studied, but I never directly built our engines from them as I redesigned them in our own drawing office.[95]

Upon returning to Russia, Carlsund ordered revisions to the Series XVI engine drawings to simplify the design and reduce manufacturing problem areas.[96] The timing of his visit to M.A.N. had been fortuitous because he could incorporate their newly released nozzle plate atomizer on the first Nobel engine. He retained, however, their two-stage injection air compression process, which used the power cylinder as the first stage.

His major departure was to cast the frame and cylinder jacket in one piece and to insert a separate, pressed-in liner (fig. 9.23). One may assume from the design chronology that the subsequent M.A.N. "DM" and the Swedish "K" models adopted this feature. Further revisions included:

- Elimination of the valve "cages" so the inlet and exhaust valves seated directly in the head. This enhanced heat transfer and significantly reduced costs.
- Making it possible for the operator to shift the injector and valve cams during a start without having to stand on the elevated service platform.
- Moving the platform itself to the opposite side so as to uncover the air compressor for easier servicing.
- Changing the compressor drive linkage so that the power and compressor pistons did not reach TDC at the same time and thus reducing the rod bearing load.
- Raising the compression pressure from 32 to 35 atm to ensure more reliable ignition when burning cheap Baku crude (Russian *Nafta*, German *Rohnaphta*) rather than kerosene.[97]

Fig. 9.23. First Nobel diesel, 1899. 20 bhp at 205 rpm. *MAN Archives*

The 260 × 410 mm engine went on test in mid-1899 and met its 20 bhp rating at 205 rpm. A specific weight of 375 kg/bhp compared well with the 425 kg/bhp of the Augsburg design. Carlsund reported it was so trouble-free that he released it to the factory in July 1899 where it provided power during the eleven hours a day work shift. No other licensee enjoyed such success with a first engine.

A second, and what became the production prototype, was begun in September and running by December. It was the first of the one- and two-cylinder A Series engines initially having two sizes of bore and stroke (fig. 9.24). The larger was based on M.A.N.'s 300 × 460 mm Series XVII and produced 30 bhp/cylinder at 180 rpm. A 1900 Nobel catalog listed sixteen A models from 15 to 100 bhp.[98] Nine were made in 1900, thirteen in 1901, and twenty-eight in 1902.[99]

Fig. 9.24. 1900 Nobel two-cylinder Model "A" engine. 40 horsepower at 200 rpm. *Diesel Collection, Deutsches Museum, Author photo*

Nobel had bought a 30 bhp M.A.N. diesel to learn from while their first one was being made, and Carlsund sent the results of an independent test using it to Augsburg on February 15, 1899.[100] Of special interest was its good performance when running on a 0.877 sp gr Baku light crude. For marketing reasons, it was very advantageous that the Russian engines be able to burn this cheap Nobel *Rohnaphta*. Carlsund was "especially pleased to confirm" that at full load the consumption of 250 g/bhp-hr when burning it translated into a fuel cost of only -.5 cent (0.7 kopecks) per bhp-hr.[101]

He also mentioned the seldom covered subject of fuel filtration. He warned MAN in his report that "deliberate care" must be taken to "filter well" in order to prevent the "fine, sandy powder" present in Russian crude oil from damaging the engine. The filter type was unspecified.

During this test a Nobel oil called "Viscosin" was used in the M.A.N. lubrication system. Hesselman later told of a sad experience in Sweden with it after longer usage. Nobel salesmen had sold it to them based on their own knowledge of it in their next-generation trunk piston engines. At room temperature the oil was a black paste and had to be heated before filtering and filling lubricators. To reduce this inconvenience, the Swedes had even cast a shelf on the exhaust outlet pipe on which to set an oil can. Some months after adopting Viscosin, an engine in the field was found to have a dry, shiny, and badly worn piston and cylinder bore. Detective work revealed that the oil in the pipes supplying the cylinder had solidified, and a residue of "tiny, hard crystals" in the bore clearance had acted like a fine grinding compound. The Viscosin was quickly replaced by a "more normal oil."[102]

Three "A" engines, a 20 hp and a pair of two-cylinder, 40-hp models, were driving power shafting in the Nobel factory by the end of 1900 (fig. 9.25). Their total output averaged 90 hp over the eleven-hour day with a fuel oil consumption of about 229 kg (14 *puds*). This was one-tenth that of an equivalent steam engine burning coal. Based on 300 operating days per year, the annual fuel saving by using the diesels was estimated to be 9,000 rubles or approximately $4,600.[103]

Fig. 9.25. Diesel power for Nobel's St. Petersburg factory.
Diesel Collection, Deutsches Museum. Author photo

Diesel's Last License Contracts

Preliminary contracts for Austria-Hungary, Belgium, and Italy were entered into by Diesel during his planning of the Allgemeine Gesellschaft für Dieselmotoren. It was as if he hoped to fill as many gaps in his licensee network as possible before starting in earnest with the Allgemeine negotiations. The contracts were a first stage leading to the founding of license holding companies patterned after the one in Nürnberg that sold Russian manufacturing rights. Diesel contracted with bankers to raise the startup capital, but the high payments due him at their incorporation turned off investors.[104] As a result, none of the companies in those three countries came into being before his patent assets went to the Allgemeine in September 1898.

Diesel had already assigned his basic Austro-Hungarian patent to Krupp in April 1893 at the time of his original agreement with them. It was done because their Berndorf, Austria, factory might become a potential manufacturer. During the interim, eight additional Habsburg empire patents had issued to him, but these ensuing ones held much less value without his having rights to the first one. On June 25, 1897, he signed a new contract with Krupp whereby both would share equally in earnings from a combined package of all nine patents. He now owned a worthwhile half interest in that basic patent, and for equity Krupp could look forward to a larger income than from it alone.[105]

A preliminary contract between Diesel and Krupp of May 28, 1898, created the framework for Diesel to form a holding company to sell sublicenses in Austria-Hungary. Both parties also entered into a second contract that day with Dr. Franz Klemperer, a Viennese with the reputed financial connections necessary to raise startup capital. In return for assigning the nine patents to him, they would, in addition to future royalties, split 1,250,000 marks to be paid by this licensing company at its incorporation.[106]

Before Klemperer could raise the money, however, the Allgemeine had assumed Diesel's interests and severed ties with the Austrian. It later did what Diesel had in mind but with more engine-oriented principals. The Aktiengesellschaft für Dieselmotoren was formed as a holding company in Budapest on February 21, 1899, which in turn licensed five manufacturing companies in Austria and Hungary:

1. Grazer Waggon- und Waggonfabrik in Vienna on February 25, 1899. (Graz Wagon and Railway Carriage Co.)
2. Johann Weitzer, Maschinen und Waggonfabrik in Arad on February 28, 1899. (Later bought by Grazer Waggon.)
3. Danubius-Schönichen-Hartmann Maschinenfabrik in Budapest on March 3, 1899.
4. Waffen- und Maschinenfabrik AG in Budapest on December 30, 1899. (Arms and Machine Factory, Inc.)
5. Leobersdorfer Maschinenfabrik von Ganz & Co., in Leobersdorf (Vienna) on April 2, 1901.[107]

Little was accomplished by these companies until the prognosis of the engine improved after 1900. At that time they began building to designs based on the M.A.N. second generation DM Series. This can be seen from the sales records submitted to the Allgemeine in number of cylinders shipped:[108]

	1899	1900	1901	Oct. 1902	Total end 1904	Total Mar. 1906
Grazer	—	—	—	—	45	88
Weitzer (Vienna)		—	1	3		
Weitzer (Arad)		—	1	—	2	6
Danubius	1	—	3	5	21	
Waffen ...	2	1	10	8	85	105
Leobersdorfer	—	—	—	1	39	

Discrepancies in the numbers exist, but it can be readily seen that until initial problems with the engine were solved, production in Austria-Hungary was almost nonexistent. Nevertheless, by the time the Allgemeine closed its records, the net return, after paybacks to licensees, was about 600,000 marks net from the companies in Austria-Hungary.[109]

Carels Fréres in Ghent still held a nonexclusive license dating from 1894, yet they balked at embarking on an engine program in 1897 after production appeared feasible.[110] This forced Diesel into doing the same thing in Belgium that he hoped for in Austria. By a May 21, 1898, contract he engaged Oppenheimer & Co., a German banking house in Wiesbaden to solicit capital for a stock company, which would control the Belgian patents. Rudolf was to receive 800,000 francs at the company's formation plus a 10 percent royalty on future engine sales. In addition to paying the bank an 80,000-franc commission, Diesel agreed to give it 20,000 francs in cash if it could coax Carels into building engines—at a 25 percent royalty.

Through another preliminary contract of July 1, 1898, Diesel asked the Oppenheimer bank to seek capital for a company in Italy like the one being promoted in Belgium. Here the payment due Rudolf after incorporation of the company was 700,000 francs. There was also to be a minimum annual payment of 25,000 marks the first year. This increased by 5,000 marks per year up to a maximum of 50,000 marks.[111]

Oppenheimer & Co. could not launch the holding companies, nor in the case of Belgium, change Carels' mind before the Allgemeine came into being and voided the original agreements.

Another minor contract, almost an aside, was one Diesel made with the Arthur Koppel company in Berlin on September 28, 1897. Koppel had asked

Diesel for rights to Egypt, but as there were no Egyptian patents yet, Diesel agreed to let him have patent drawings and specifications to apply for them in Koppel's name. Diesel was to receive a commission on future sales, and no money ever changed hands. The Allgemeine and Koppel mutually declared the contract voided by letters in June 1899.[112]

The shield of patent protection Diesel had created, and later expanded upon by the Allgemeine, presented a formidable defense against anyone attempting to sell the engine outside of a contractual relationship. By mid-1898 Rudolf had been granted 87 patents in 17 countries, and 141 diesel engine patents were in force in 37 countries by 1904.[113]

"Winners and Losers"

Whether it was through creative engineering talent, sufficient financial resources, market strengths, or a combination of these elements, a few of the licensees Diesel himself acquired in 1897 and 1898 produced saleable diesel engines at an increasing rate from their first ones built. Business travails curtailed the growth of some, and revolution sealed the fate of another. A few choosing the more cautious path also became highly successful. There was no one "right way" for the licensees to begin.

"Risk is a noble action."

Old Russian proverb.[114]

Notes

1. Rudolf Diesel, *Die Entstehung des Dieselmotors* (Berlin: Springer Verlag, 1913), 152.
2. The early history of Gasmotorenfabrik-Deutz may be found in Cummins, *Internal Fire*, 156–64.
3. Kurt Schnauffer, "Lizenzverträge und Erstentwicklungen des Dieselmotors im In und Ausland 1893–1909," in MAN Archives (1958), 9.
4. Eugen Diesel, *Diesel: der Mensch, das Werk, das Schicksal* (Hamburg: Hanseatische Verlag, 1937), 268. A copy of the letter with Krupp's offer is in the MAN Archives.
5. Schnauffer, "Lizenzverträge," 13. Deutz says three were built but gives no disposition for the third. Deutz letter of April 1, 1974, to the author.
6. Deutz believes a 20-hp Model W engine came from Langen & Wolf, their Vienna subsidiary. This may have been done to keep the Austrian patents in force as an engine had to be built there to do so. Gustav Goldbeck, "Aus der Frühzeit des Dieselmotors," in *MTZ* (March 3, 1958), 90. Worsoe confirms this and says the Krupp factory in Berndorf supplied the drawings to Langen & Wolf. Wilhelm Worsoe, *Die Mitarbeit der Werke Fried. Krupp an der Entstehung des Dieselmotors in den Jahren 1893/97 und an der Anfangs-Entwicklung in den Jahren 1897/99*, 2nd ed. (Kiel, Germany: 1940), 58. MAN Archives. The Vienna *Technisches Museum* where this engine is on exhibit, says it is by the "Berndorfer Metallwarenfabrik, Berndorf" although the cast name plate reads "Built by Langen & Wolf 1898–1899."
7. Hugh D. Meier, "Report on Construction of Diesel Motor at Deutz, Germany," (New York: 1898), 38. This gem provides an extremely valuable firsthand account of early engine design and manufacturing problems encountered by several builders. The US engine's s/n was 24901. Busch-Sulzer Papers, Wisconsin State Historical Library.
8. April 1, 1974, letter from Deutz to the author.
9. Letter from Edward D. Meier (Hugh D. Meier's uncle) to Adolphus Busch, March 29, 1909, referring to the "very unsatisfactory" first-imported engines. Busch-Sulzer papers, Wisconsin State Historical Library.

10. Ibid., 23. J. K. E. Hesselman, *Tecknik och Tanke* (Stockholm, Sweden: Sohlmans, 1948), 145–46, independently confirms what Meier found in his own experiences with cylinder/piston lubrication using a Mollerup or a plunger pump system. A closer look will be taken at his ideas regarding cylinder lubrication when covering the Swedish engine. Here it is enough to say that the problem centered around what occurred when lube oil, under a relatively low, intermittent supply pressure, opposed an unknown blowby gas pressure at any of the several equi-spaced lubrication ports around the cylinder.

11. Hugh Meier, "Report on Construction," 28–29. The Deutz nozzle had twenty holes of 0.75 mm diameter drilled at about a 45-degree angle, which remained carbon-free when using kerosene.

12. Schnauffer, "Lizenzverträge," 12, and an undated Deutz brochure.

13. The company was founded in 1840 as P. & W. McOnie to make sugar-refining machinery. James Buchanan Mirrlees (1822–1903) became a principal in 1848, and William Renny Watson (1838–1900) joined as a partner in 1868 what by then was Mirrlees & Tait. The firm was renamed Mirrlees, Watson & Co. in 1882. Homer T. Yaryan's name was added in 1889 when his steam evaporator company merged with Mirrlees, Watson. Derek A. Dow, *Redlands House* (Scottish Ambulance Service, Glasgow, 1985), a biography of J. B. Mirrlees and his association with Watson. Two unpublished biographies also on these men were kindly given to the writer by Mr. Dow, who is on the staff of the Archives of Glasgow University.

14. E. Diesel, *Diesel*, 260.

15. The Kelvin report was quoted in full by E. D. Meier in his lecture given May 18, 1898, at a meeting of the Franklin Institute. Meier, "Diesel's Rational Heat Motor," *Journal, Franklin Institute*, 146, no 4 (Oct. 1898), 251–53. A contributed article, "Lord Kelvin and the Diesel Engine," in *The Engineer* (1947): 186–87, and 210–11, tells of finding an original report copy and analyzes it in view of other issued patents and commercial practices of the time. Its writer believes Diesel did nothing new, an opinion held by a number of people in England. The article mentions a March 13, 1897, report on the Diesel patent by Wallace Fairweather of Cruikshanks and Fairweather. This was before Diesel's arrival in London. One may assume these were the lawyers Diesel referred to.

16. March 26, 1897, death notices of Elise Diesel in Augsburg and Munich newspapers. MAN Archives.

17. Verified copy of Mirrlees contract Article 4 attested to by Diesel on September 30, 1897. Busch-Sulzer papers, Wisconsin State Historical Library.

18. Drawing lists and sales book. MAN Archives. More recent Mirrlees company literature, "History and Development of Mirrlees, Bickerton & Day Ltd.," (Pub. No. 5062, Feb. 1955, 6 pages) mistakenly perpetuates the engine's size as 300 × 460 mm and that it produced 20 bhp at 200 rpm.

19. Discussion comment by Professor Watkinson in H. Ade Clark, "The Diesel Engine," in *Proceedings, Institution of Mechanical Engineer* (London: 1903), 445. A lengthy search (including archives of the Science Museum and Glasgow Universities) failed to find Watkinson's report. It has been traced to Charles Day and is assumed to have been destroyed after his death.

20. E. Diesel, *Diesel*, 293.

21. Clark, "The Diesel Engine," 444. Comments by Day.

22. Schnauffer, "Lizenzverträge," 56. These Mirrlees documents and drawings are in the MAN Archives.

23. Mirrlees-Allgemeine nonexclusive license agreement, November 7, 1899. MAN Archives. According to Day, Mirrlees had spent about £10,000 on the engine. Charles Day, "Address by the President," *Proceedings, Institution of Mechanical Engineers* (1934), 200.

24. Chas. Day letter to H. N. Bickerton, May 9, 1913, as printed in *Royal Commission on Fuels and Engines*, part 1 (British Admiralty, 1913), 106.

25. *Prospectus*, Diesel Engine Co. Ltd., December 10, 1900. On November 30 George Oppenheimer "resold" his rights "at a profit" to a trustee for the not yet incorporated company. Busch-Sulzer Papers, Wisconsin State Historical Library. The protracted chain of events leading to the final contract is found in Schnauffer, "Lizenzverträge," 58–60.

26. Schnauffer, Ibid., 60.

27. *Prospectus*, George Oppenheimer, the secretary and general manager, mentioned the stepping down in his *Report and Accounts, 31 March 1902*, Diesel Engine Co. Ltd. MAN Archives.

28. The Diesel Engine Co. began negotiating with Sulzer in 1902 and wanted to take much of its engine production even before the first one had been built. "Fifty Years of Sulzer

Diesel Engines 1897–1947" in *Sulzer Technical Review*, no. 2 (1947): 4. A search of M.A.N. records revealed that the DEC ordered no engines from Augsburg after 1903. A letter from Allgemeine manager Johanning to Sidney Whitman, London, August 7, 1903, complained of payment problems. He "fully conceded that the DEC can buy today engines just as good [as MAN's] from Carels Fréres or Gebrüder Sulzer, but it is unfair of DEC to treat the Allgemeine and M.A.N. in this way." MAN Archives.

29. DEC to the Allgemeine, February 22 and 24, 1904. MAN Archives.

30. "Report of the Diesel Engine Co.," *Times* (London), June 14, 1907, 14.

31. A December 1924 meeting held in the liquidator's office completed the "winding up" procedure. Document Ref. BT 31/16531/68054, Public Record Office, Kew.

32. Schnauffer, "Lizenzverträge," 45, and company prospectus dated April 1897. An earlier prospectus entitled *Société Anonyme des Moteurs R. Diesel* is dated October 1895. Stock for this company, capitalized at 1,000,000 francs, was never sold, however, the document was a model for the one in April 1897. Prospectuses in Diesel papers at the Deutsches Museum. *Sté. Anonyme* is equivalent to a joint stock company.

33. The factory at the east end of Longeville is next to the mainline railway from Mannheim to Paris. It has been used recently as a foundry, but the original office building and factory, seen by the author in 1992, is little changed.

34. Karl Dieterichs, working for Diesel at this time, is credited with helping Dyckhoff through his troubles in mid-1898. Schnauffer, "Lizenzverträge," 46, and references to Dieterichs in MAN Archives.

35. Dyckhoff's German Patent No. 107,395, was issued to him, Krupp, and M.A.N. Equivalent US No. 661,369 of November 6, 1900, was only in his name. Its main claim reads, "Apparatus for the controlling of explosion and internal combustion engines, thereby characterized so that for forward and reverse arrangements two different groups of cams or cam discs are slid on two or more control shafts, whereby the bearing of this control shaft can be rotated around its common operating shaft so that an optional cam group is able to move into the range of the rocker lever."

36. Two other early French diesel licensees were the Compagnie Française des Moteurs à Gaz et des Constructions Mécanique, Paris, January 15, 1898, and Paul & Augustin Farcot, Ingénieurs-Constructeurs à St. Quen, Seine, June 4, 1898. Schnauffer, "Lizenzverträge," 46.

37. In 1931 Augustin-Normand was granted a license from M.A.N. to build two- and four-stroke diesels. It was renewed after World War II but was terminated in 1963. MAN Archives.

38. A biography of Anton Rieppel and a history of the Nürnberg company are in Fritz Büchner, *Hundert Jahre Geschichte der Maschinenfabrik Augsburg-Nürnberg* (Augsburg: MAN, 1940), 59–112. Nürnberg's founder, Johann Friedrich Klett (1778–1847), opened his factory in 1841 to build some of Germany's first locomotives. Its product line grew to include large boilers, railroad cars, and steam and gas engines. Theodor Cramer (1817–1884) had been like a son to the founder and at Klett's death assumed control. He even added his benefactor's name to his own. Cramer-Klett continued with Klett's philanthropy, Diesel having been one of those receiving financial aid from him while a student.

39. Schnauffer, "Lizenzverträge," 14.

40. Böttcher recorded in his logbook that it ran "better and better and the noise is confined to the clicking of the fuel lever, the singing of the exhaust pipe, and a little shock in the crosshead which could not be eliminated for lack of time. . . . None of the other gas and oil engines exhibited can match the Diesel engine as far as quietness of operation is concerned. This has generally been conceded. . . ." Eugen Diesel and Georg Strößner, *Kampf um eine Maschine* (Berlin: Erich Schmidt, 1950), 54. A 1953 manuscript translation was kindly loaned to the author by Frau Irmgard Denkinger before her retirement as head of the MAN *Werkarchiv*.

41. Report to E. D. Meier, American Diesel Engine Co., from Professor Denton, Stevens Institute, Hoboken, New Jersey, Sept. 1, 1898, 12 pages. Diesel Collection, Deutsches Museum.

42. This engine, for which the American licensee paid Nürnberg $3,300, was sold to Morgan for $1,800. Diesel and Strößner, *Kampf*, 90. The quote is from E. D. Meier's March 29, 1909, letter to Adolphus Busch.

43. Kurt Schnauffer, "Die Dieselmotorenentwicklung in Werk Nürnberg der M.A.N. 1897–1918" (1958), 4. (Hereafter "Werk Nürnberg") Deutsches Museum Archives. The list is from an *Allgemeinen Gesellschaft für Dieselmotoren A.G.* publication.

44. Georg Siemens, *History of the House of Siemens. 1847–1914*, trans. A. F. Rodger, vol. 1 (Freiburg, Germany: Karl Alber, 1957), 333 pages. A business and technical history covering the rapid rise of communication, lighting, and power industries in Europe.

45. Schnauffer, "Werk Nürnberg," 7.
46. Letter in the MAN Archives. The unwarranted criticism Diesel heaped on the Vogel designs went beyond his letters to Rieppel. One must ask if the strained relationship between the two after the Kassel lecture prevented Diesel from making a dispassionate, unbiased judgment of Vogel's work. Frustration over these engine developments may have contributed to his return to Augsburg.
47. Part of Rieppel's January 10, 1899, letter to E. D. Meier, is in Georg Strößner, "Zur Früh-Entwicklung des Fahrzeug-Dieselmotors," *ATZ*, MAN reprint (March 1958), 6.
48. *Dictionary of American Biography*, XXI, Supplement I (1935), 141–43. A more autobiographical Busch entry in *The National Cyclopaedia of American Biography*, XII (1904), 23, has several inaccuracies.
49. Busch owned the "Villa Lilly," an estate near Wiesbaden (Bad Schwalbach) named after his wife. He died there on October 13, 1913, ten days after Rudolf Diesel's death. During World War II the home was used as a lying-in hospital for pregnant girls and then in the American occupation as the Naval Technical headquarters. Charles Lindbergh, *The Wartime Journals of Charles A. Lindbergh* (New York: Harcourt Brace Jovanovich, 1978), 963.
50. The complete Bing, Rieppel, and Diesel correspondence pertaining to the Busch license is in the Nürnberg branch of the MAN Archives.
51. Meier's lengthy vita is in *The National Cyclopaedia of American Biography*, vol. XXIII (New York: Jas. White, 1933), 103–4.
52. Meier first learned of Diesel's work through either Bing or Busch, probably in August. He then wrote "Nürnberg's" chief engineer Marx, and after a favorable reply about the engine, located a copy of Diesel's Kassel lecture in the US on August 9. It impressed him enough to urge Busch by letter on August 12 to pursue the engine. E. D. Meier, "Report on the Diesel Motor," (1897), 4. From a twenty-nine-page verbatim copy, typed in 1912 under Meier's "supervision" and so noted by him. Copy in MAN Archives. A look at the above dates and in the text show not only how fast communications were, but also how fast willing entrepreneurs could act.
53. Meier, "Report," 5.
54. Ibid., 16–17.
55. Ibid., 6, 10.
56. Ibid., 29.
57. If the Otto Gas Engine Works in Philadelphia, one of Deutz's major subsidiaries, had become a sublicensee or supplier, the early US diesel engine story might have been quite different. Why Busch and Meier did not more diligently pursue this avenue shows their lack of appreciation in the beginning for what was required to build the new engines, especially so after the reports of Hugh Meier on what the experienced Deutz had gone through.
58. Contract between Diesel and the Diesel Motor Co. of America dated July 1, 1898. MAN Archives.
59. Reference to this is in a November 9, 1898, Allgemeine letter to Meier dealing with "Combination of Suction and Exhaust Valves." Busch-Sulzer papers, Wisconsin State Historical Library. The "brewery" engine ran reasonably well in its original design state, but Böttcher had to continually clean or replace the wire mesh atomizers in the nozzles. Oil in the compressed air was the primary culprit. Böttcher did not ingratiate himself with the Americans, and Reisinger wrote Diesel that he was no longer welcome. Diesel later let Böttcher go. Letter from Rudolf to Johanning at the Allgemeine, August 1899. MAN Archives.
60. Letter quoted in Diesel and Strößner, *Kampf*, 113–19.
61. Letter in Busch-Sulzer papers. Wisconsin State Historical Library.
62. Richard H. Lytle, "The Introduction of Diesel Power in the United States, 1897–1912," in *Business Review History* 42 (1968), 125. Letter dated January 22, 1900.
63. Edward D. Meier to Rudolf Diesel, July 6, 1899. Letter in MAN Archives.
64. B&W roots go back to Hans Heinrich Baumgarten (1806–1875), who opened his workshop in 1843. Three years later he and Carl Christian Burmeister (1821–1898) entered into a partnership. After relocating to a site used for almost 150 years, Baumgarten & Burmeister began building steam engines and steamships. William Wain (1819–1882), a Manchester-born Englishman, came to Denmark in 1844 as chief engineer for the Royal Mail steamers. Upon receiving Danish citizenship, he was made second-in-command of the Naval Dockyards in 1858. Wain joined Burmeister as a partner in 1865 on Baumgarten's retirement, and it was

then that the firm's name changed to Burmeister & Wain. By becoming a stock company in 1871, B&W acquired capital to expand their engine works and shipyard. One of its pre-1900 vessels was an 1882 training ship, which later sailed as the *Joseph Conrad.* This famous square rigger is berthed at the Mystic, Connecticut, maritime museum. B&W also built warships for the Russian Navy and the Tsar's large royal yacht *Standart.* A broader company-commissioned history is in Johannes Lehmann, *Burmeister & Wain, Gennem Hundrede Aar* (Copenhagen, Denmark: B&W, 1943), 245 pages. His companion book, *Rudolf Diesel and Burmeister & Wain* (B&W, 1938), 150 pages, in English, covers mainly diesel activities. B&W's *Museum* (1959), 97 pages, further annotates company history.

65. C. Winslow, "Diesel's Motor," in *Ingeniøren*, no. 50 (1897), 345. Marstrand, Helwig & Co. changed its name to A/S Titan when it became a stock company to finance a possible Diesel license. Garde, too, wrote Diesel about his book in 1894. An October 1980 letter from H. Friis Petersen, B&W Engineering, to Klaus Luther, MAN archivist, summarizes that Diesel/Titan/B&W story. Diesel, *Die Entstehung*, 84–5, gives details of the test in which Winslow participated.

66. Ibid., 346. Trans. by Søren Hansen (1904–1993). The author is deeply indebted to this retired chief engineer of B&W whose engineering career there spanned from 1928 to 1977. He supplied numerous pieces of his company's technical and corporate past and translated them. His personal knowledge of B&W and other diesel companies, which he had shared with the author, since 1972 was of great help. He was a good friend.

67. The contract was based on Diesel's Danish Patent No. 38 of April 20, 1895, which had a fifteen-year life.

68. Letter, I. Knudsen (chairman of B&W) to Robertson at Mirrlees, Glasgow, April 25, 1899. The B&W business papers are in the Danish Trade & Industry Archives at Aarhus.

69. Biographical information in the Burmeister & Wain Museum.

70. H. P. Friis Petersen, "It Happened Like This," in *B&W Staff Magazine of 1964* (B&W), 32. English translation kindly furnished by Niels Egon Rasmussen, MAN B&W, retired. Petersen wrote numerous historical papers before retiring from B&W in 1975. Rasmussen continues with Petersen's work.

71. A. P. Chalkley, *Diesel Engines for Land and Marine Work* (New York: Van Nostrand, 1917), 50.

72. Friis Petersen, "The First Diesel Engined Ships," from a 1965 B&W house publication, page 18 of a 29-page English translation from N. E. Rasmussen.

73. Ibid.

74. Torsten Gårdlund et al., *Atlas Copco 1873–1973* (Stockholm, Sweden: Atlas Copco, 1974), 49–56. A copy was kindly sent to the author by Sixten Nyqvist, retired, Wärtsilä, Trollhattän.

75. Schnauffer, "Lizenzverträge," 72. His source, the test journal, refers only to the plural "Herren Wallenberg," which infers that Marcus and Knut had been with Diesel and later saw the engine run. No German-based accounts mention Knut in any of the negotiations.

76. Gårdlund, *Atlas*, 67. Of the 300 shares assigned, the Wallenbergs had 50 total, Lamm and Diesel 50 each, and Auguste Pellerin, a French business friend of Knut's, 40. Ten others owned the remaining 110 shares.

77. Schnauffer, "Lizenzverträge," 72. Norway, which was not independent from Sweden until 1904 was added in 1901 for a 5 percent royalty. The 33 percent sublicense fee dropped to 10 percent.

78. Sickla today is a suburb of Stockholm.

79. Noé's comments are quoted in a November 29, 1941, letter from Max Plochmann (1885–1969), a MAN chief engineer who had known Noé, to Dr. H. H. Blache of Burmeister & Wain. Copy in MAN Archives.

80. Jonas Hesselman, *Teknik och Tanke* (Stockholm, Sweden: Sohlmans, 1948), 77.

81. Ibid., 23–25.

82. Ibid., 12–15, and *Svensk Uppslagsbok* (Malmö, Sweden: Norden, 1949), 367–68, kindly translated by Per-Sune Berg, Volvo Truck Co., retired. Good minds ran in the Hesselman family. Jonas's older brothers Henrik, a botanist, and Bengt, a Nordic language expert, both merited citations in the above and other references.

83. Hesselman, *Teknik*, 18–19.

84. Ibid., 30, 31.

85. Ibid., 34.

86. Schnauffer, "Lizenzverträge," 81. In an April 1, 1898, letter to Diesel, MA asked him to let their German sublicensees also have the right to sell in Russia, but he did not agree to this.

87. R. Murauer, "Die Entwicklung und Verbreitung des Dieselmotors in Rußland und seine Verwendung als Schiffsmotor," in *Zeit, des VDI* (1909): 1,184.

88. Hans Flasche, in a December 5, 1947, letter to Eugen Diesel cites Carlsund for this statement. Copy in MAN Archives.
89. Nicholas Halasz, *Nobel* (New York: Orion, 1959), 284 pages, gives Alfred's biography and tells of nephew Ludwig's battle to create the Nobel Prize program. Robert W. Tolf, *The Russian Rockefellers* (Stanford: Hoover Institution Press, 1976), 269 pages, provides the Baku and St. Petersburg story.
90. Eugen Diesel, *Diesel*, 288–89. Nobel arrived a day late, but the delay allowed Diesel to meet with Friedrich Alfred Krupp during a special breakfast affair at the Hotel Bristol where Krupp was the honored guest. Krupp afterward invited Rudolf to his Essen Villa Hügel. That afternoon Wallenberg visited with Diesel, when he reportedly told Diesel he would use his influence to nominate Rudolf for a Nobel Prize.
91. Flasche letter to E. Diesel, 1947, quoting Carlsund. In a five-page memoir, in Swedish, n.d., Carlsund tells of Ludwig Nobel's visit in Stockholm with Marcus Wallenberg during January 1898 in which Marcus emphasized the diesel's Russian profit potential. This memoir, which amplifies details in his 1947 letters to Hans Flasche, was kindly shown to the author by Dr. Lorrie Holmin, who was living in Stockholm while researching the post-Ludwig and Robert Nobel families.
92. Carlsund also stated that the Wallenberg/Bing middleman involvement with the intermediate company "was unnecessary and tied our hands in various ways." He wished that he could have bought the patents himself. Flasche letter to E. Diesel.
93. "Gesellschaftsvertrag Russische Diesel Motor Co.," arts. 2 and 3. MAN Archives. Schnauffer, "Lizenzverträge," 82, tells of the reimbursement.
94. Of greatest value is Hans Flasche, "Der Dieselmaschinenbau in Russland von Anbeginn 1899 an bis etwa Jahresschluss 1918," (December 1947). Copy in MAN Archives.
95. Carlsund to Flasche, March 20, 1947. MAN Archives. He also wrote of the unhappy schism arising between Lauster and Diesel as quoted in Flasche to E. Diesel.

"We have within all of us a good share of vanity, and to correctly evaluate the individual and diverse work of others is a difficult matter; but when we do attempt to properly credit the efforts of others, it awakens dissatisfaction. However, it is often envy which is concealed under our struggle for fairness.

"Whether Lauster was driven by a wronged sense of justice or jealousy is hard to know, but I have a genuinely good understanding of him as a person and believe him to be honorable."

Carslund also spoke of Lauster's great abilities, faith in the engine, and "that without him progress during the critical years would have been very doubtful."

96. Hesselman, *Teknik*, 29. Marcus Wallenberg brought Emanuel Nobel and Carlsund to Sickla, and the latter became "a steady visitor." Carlsund always told what the other licensees were doing. Hesselman said that "Other foreign visitors very carefully absorbed what we had to show them but seldom gave anything in return."
97. February 1900 Carlsund report to Augsburg (Allgemeine No. 189). MAN Archives. Also, Schnauffer, "Lizenzverträge," 86–7. Carlsund gives weights for the individual parts. The flywheel alone weighed 1,450 kg.
98. Catalog in Diesel Collection, Deutsches Museum.
99. Murauer, "Die Entwicklung."
100. The test was performed by Professor G. v. Döpp and published in the *Protocolle des St. Petersburger Polytechnischen Vereins*, April 13, 1900, 112 et seq. Flasche papers, MAN Archives.
101. Carlsund letter to MAN, February 15, 1899. Copy in MAN Archives. He also made the intriguing observation that when the engine misfired at no load (200 rpm) he first thought the compression was not high enough. However, he found that by lowering the maximum compression pressure from 32 to 28 atm the misfiring disappeared! He theorized that the mixture was too lean at first and by reducing the available oxygen, even though the temperature for ignition was less, the flammability of the mixture would be higher with a richer charge. The governor maintained a constant minimum fuel delivery throughout. Carlsund reduced the compression pressure while the engine was running by simply inserting a hand-operated butterfly valve in the intake pipe! His theory was on the right track, but other factors had to have entered in. For example, the butterfly valve may have introduced some turbulence or other intake process changes to possibly induce a locally correct mixture

stratification and luckily cause consistent charge ignition. Since Baku crude was more aromatic, the higher charge air temperature should have been more beneficial. This may explain why their own first engine had its compression pressure raised to 35 atm.

102. Hesselman, *Teknik*, 142.

103. Döpp, *Protocolle*, 153. Carlsund reports the shop hours in the discussion of the lecture.

104. Schnauffer, "Lizenzverträge," 52, suggests that Diesel placed a high price tag on the deals to make his holdings look better as he negotiated for the Allgemeine. A financial spreadsheet of Diesel's, dated August 1898 (copy in MAN Archives), does show the as yet speculative money due upon establishment of the licensing companies in those countries included as future income. The amounts were before commissions. Yet M.A. and Krupp, who shared in the Austria venture, could not have been blind to the true conditions. Diesel would have appeared the fool to try to "pad the books." Big numbers were the order of the day in Augsburg and Essen as well as Munich!

105. The Diesel/Krupp relationship in Austria is detailed in the "Contract of Sale" between Diesel and the Allgemeine (Sept. 17, 1898), 1, 2. The German "Kaufvertrag" and an English translation of it are in the Busch-Sulzer papers, Wisconsin State Historical Library.

106. Schnauffer, "Lizenzverträge," 33. Klemperer's commission was to be 10 percent. In return he had to invest 50,000 Austrian guldens or about 80,000 marks.

107. Ibid., 37–38. See also Allgemeine 1899 Board of Directors Report for the involved financial details. MAN Archives.

108. Allgemeine records in MAN Archives.

109. Ibid., 39.

110. Ibid., 51–53. Schnauffer speculates that Carels was right to wait because of a very uncertain Belgian market.

111. Ibid., 99. The commission in Italy was 10 percent, and 30,000 francs would also be paid for "development expenses." Note: France, Belgium, Italy, and Switzerland, among others, had equal coinage values then. H. R. Kempe, "Foreign Moneys and their English Equivalents," in *The Engineer's Yearbook* (London: Crosby, 1901), 739.

112. Schnauffer, "Lizenzverträge," 63.

113. Ibid., 2.

114. As quoted in Tolf, *The Russian Rockefellers*, 172.

CHAPTER 10

Next-Generation Engines

"Marcus Wallenberg was without doubt deeply interested in the diesel factory, and I have heard him tell his friends about its orderly and steady growth. He likened it to a plant reaching upwards as a reward to the gardener. Those of us who worked with the plant sometimes felt that it would have developed better and faster if it had received more fertilizer in time."
—Jonas Hesselman[1]

Those of Diesel's first-generation licensees in Europe having the resources and resolution steadfastly continued with his engine. They soon realized, however, that while it could be made to perform, the design was too heavy, costly, and complex. This led to the development of their own second-generation engines with sometimes an occasional sharing, through the Allgemeine, of useful new features.

Weight reduction and structure simplification held top priority. Distinctive fuel systems and combustion chambers also evolved. Only stationary engines were considered at first, yet several designs had applicability for certain maritime uses. They remained almost unchanged, and the reversing requirements to maneuver were met by driveline devices. True marine diesels were yet to come.

The Augsburg DM Series Engine
M.A.N. was selling a fairly reliable power plant by the end of 1900, but Lauster knew its design was costly and rapidly becoming obsolete. That November Vogt had authorized him to start on what was to be the DM (Diesel Motors) series. Recalling their previous problems, M.A.N. chose an evolutionary path to pursue the new design. As evidence of its achievement,

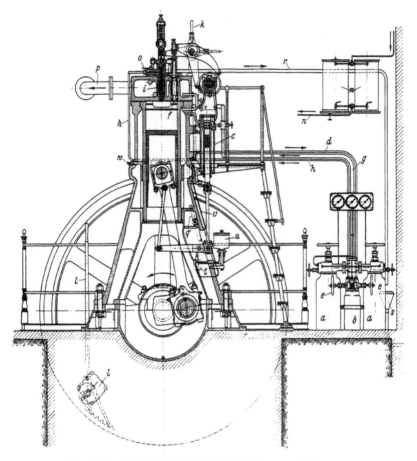

Fig. 10.1. 1901 M.A.N. "DM 70." 70 bhp at160 rpm,
400 × 600 mm bore and stroke. Sass, *Geschichte, 1962*

the resulting series would be made until 1912 and cover a product line from a single cylinder of 8 bhp to four cylinders producing 800 bhp.

The first engine, a DM 70, was delivered to a customer one year after work began on the drawings.[2] From a bore and stroke of 400 × 600 mm, the largest cylinder yet built at Augsburg, it produced 70 bhp at 160 rpm.[3]

Specific weights for the DM series were significantly reduced. A single cylinder, 30 bhp engine, for example, went from 310 to 265 kg/hp, and the two-cylinder, 50 bhp dropped from 270 to 220 kg/hp.[4] The DM 70 now weighed "only" a little over 14,000 kg without its external plumbing and tanks.[5] Selling prices likewise came down due to less iron and many manufacturing simplifications.

Eliminating the crosshead, a major change, lowered engine height significantly, but even with a new, longer piston skirt there remained a concern about piston failures due to excessive side loading. This fear came to a head in February 1901 some weeks before the DM 70 first ran. As a result, work began on a DM 8 (165 × 270 mm and 8 bhp at 270 rpm). It was believed that a much smaller piston would not encounter the side thrust problem, and neither did the DM 70 when it went into service.

The basic structure was completely redesigned: The DM 70's A-frame and external water jacket of the cylinder were cast as one piece and bolted to a separate bedplate (fig. 10.1). The DM 8 and the later DM 12 used a single casting incorporating the above and the bedplate (crankshaft support) (fig. 10.2). Both had a wet liner that was probably adopted from Nobel. The camshaft support bolted to the cylinder jacket instead of being cast integrally with it.

A second-stage air compressor on the DM 70 remained in its usual location at the side of the cylinder and was driven by linkage actuated from the top end of the connecting rod. The DM 8 had an offset pin on the end of the

Fig. 10.2. 1903 M.A.N. "DM 12." Note the lube oil pump and second-stage air compressor driven off the end of the crankshaft.
Photo, Collections of Greenfield Village and the Henry Ford Museum

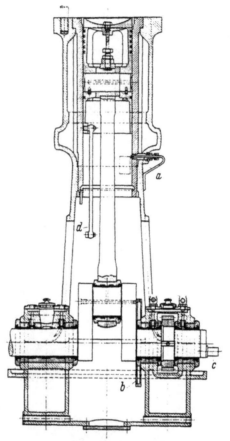

Fig. 10.3. M.A.N. "DM 8." Sass, *Geschichte, 1962*

crankshaft to operate a horizontal, second-stage compressor and lubricating oil pump (fig. 10.2). Until 1904 DM engines used two-stage compression in which the first-stage compression occurred in the combustion chamber prior to injection (chapter 8). A separate valve opened by its own cam timed the transfer of first stage air. There were five cams, rocker levers, and valves per cylinder: intake, exhaust, fuel injection, starting, and the compressed air transfer valve.

The DM 70 lube pump plunger was timed to discharge oil into a short, vertical groove on the piston between its lower ring and the wrist pin. An angled, downward hole from this groove aligned with pin passages for sending oil to the pin bearing. The cylinder supply port on the DM 8 simply opened into the wrist pin bore. Drip lubricators or grease cups supplied the other bearings. The method of oiling the connecting rod bearing carried over from the previous design: A hollow disc, centered on the crank journal and screwed to the crank cheek, was fed from the main bearing. Oil entering the rotating disc centrifuged out to an axial hole in the crank throw (fig. 10.3).

An improved fuel system, including placement of tank and filter above the pump, kept out air and lessened fluctuations in fuel delivery. As before, the pump plunger was driven by an offset pin on the end of the camshaft and drew in fuel past a suction valve, which lifted due to the reduced pressure above it. Different on the DM 70, and all but the smaller engines, was a float chamber integrally cast with the pump housing and opening into the old supply chamber under the suction valve and its closing spring assembly. This new chamber contained a vertically guided float and valve. The fuel supply entered the pump past the float valve, which maintained a constant head above the suction valve (fig. 10.4).[6]

A nuisance problem on the earlier design was the friction caused by the packing that sealed the governor-actuated rod passing vertically downward through the pump cover and acting on a finger to lift the suction valve off its seat. During the pumping stroke, this action returned fuel to the supply chamber under the valve that portion of the full load charge needed by the governor to hold the engine at a constant speed under varying loads. The more the suction valve was held open, the less the fuel volume going to the injector nozzle. By the addition of the float valve, it was possible to get rid of the packing gland on the control rod because any leakage was now returned to the float chamber. This greatly reduced friction in the governor linkage and allowed closer speed regulation.

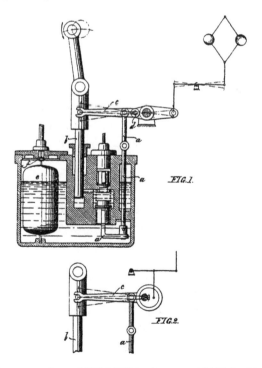

Fig. 10.4. Cross-section of "DM 70" fuel pump. *MAN Archives*

A second, hand-lifted rod, without a packing gland for the same reason, also raised the suction valve from underneath independently of governor action in order to shut off all fuel and stop the engine.

It became possible by 1909, as bore-to-plunger clearances could be held closer, to eliminate the packing gland sealing the pump plunger. Leakage returned to the float chamber.

Professor Meyer obtained the following performance data in June 1902 on a DM 70 and a DM 8:[7]

	DM	70	DM	8
Speed, rpm	158.8	160.5	270.3	276.3
Horsepower, bhp	69.6	34.9	8.62	4.68
Imep, kg/cm^2	6.78	4.06	6.64	4.33
Mechanical efficiency, %	79.1	66.2	77.0	63.2
Bsfc, g/bhp-hr	192	224	222	260
Brake thermal efficiency, %	32.6	—	28.0	—

The fuel burned was a Russian kerosene with an 0.806 sp gr at 18.6°C and a heating value of 10,300 cal/kg.

Total sales of the DM series during their eleven-year production life at Augsburg were a respectable 1,484 engines of which more than 400 were 20 bhp and under. Only six of the four-cylinder, 800 bhp engines were built.[8]

The acceptance of this second-generation series firmly established M.A.N.'s diesel reputation. New engine sales forecasts looked so promising to the Augsburg division for coming years that in 1908 they ended reciprocating steam engine production. Only ten years had passed after introducing the diesel engine until the division abandoned what once was a major product.[9] Success, however, brought complacency and a consequent unpreparedness in making ready more advanced designs.

Nürnberg Developments

All diesel activities were centered in Augsburg from 1900 to 1904, after which Rieppel insisted he be given an opportunity to develop his own engines. Lauster, however, was against this. He feared that since Nürnberg lacked knowledgeable diesel people there would be a crop of new problems put on the market. Augsburg had worked hard to erase the stigma created by the first troublesome engines, and he did not want another such harrowing experience.[10] The two men compromised and Nürnberg started with the line of proven "DM" engines Augsburg had been selling since 1901.

This trunk piston series was available by 1904 in single-cylinder outputs of 8 to 125 hp. Ultimately, two-, three-, and four-cylinder models were

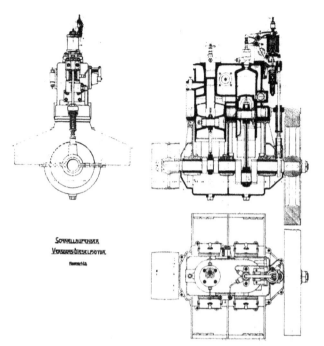

Fig. 10.5. 1905 Nürnberg S D 15 automotive-type, two-stroke
lab engine with 135 × 160 mm bore and stroke. *MAN Archives*

built with a maximum cylinder rating of 200 hp. During the first two years, Nürnberg built only the DM 25 (25 hp) and DM 35 and then added the 50 and 60. Two engines were shipped in 1904, eight in 1905, twenty-four in 1906, and forty-five in 1907.[11]

Rieppel had not forgotten about his smaller, automotive-type engines. He put his designers to work on the demanding little power plants soon after the smell of kerosene fuel once again permeated the diesel area in the factory.

The first test engine of this kind was a single-cylinder two-stroke with a bore/stroke of 135/160 mm. A two-stage, stepped-piston compressor operated off a second throw on the crankshaft. The lower stage supplied scavenge air for cross flow scavenging and fed the upper stage, which compressed blast air for injection. The two cylinders and heads were made in one casting (fig. 10.5). The engine ran, but no performance data was preserved.

Emil Vogel, Lucian's nephew, began experimenting on stationary engines at Nürnberg using "compressorless" air injection. It was Diesel's method (German Patent No. 62,198) whereby power cylinder air was compressed to a higher-than-normal pressure and bled off to blast fuel, at a retarded timing, back into the cylinder, which was at a consequently reduced pressure. He was no more successful than Diesel, and one must wonder why it was tried.

This "rob from the cylinder" idea was adapted to an automotive engine destined for a tractor in Russia. Nicknamed the "little Josef" after chief

Fig. 10.6. Two views of 1911 four-cylinder Nürnberg automotive-type diesel. Four cylinders, 160 × 160 bore/stroke, 50 hp. *MAN Archives*

designer Josef Wolfenstetter, it was built and tested between 1909 and 1911. The Model S V 5/4 (50 hp and four cylinders) was state of the art for its day and might have had a chance except for the source of blast air.

The engine used a new automobile practice of casting all cylinders in one block (fig. 10.6). There were two cylinders per head. The crank directly drove a camshaft on each side of the engine and operated, via push rods and rocker levers, the overhead inlet and exhaust valves and injectors. One camshaft opened the vertical valves and the other lifted the canted injector nozzle needles. The valves that opened downward into a combustion chamber space to one side of the cylinder required a redesign of the seating area for better water cooling. Seat distortion and loss of sealing occurred initially.

Of special interest were the four, individual-plunger fuel pumps driven by eccentrics and linkage off the camshaft ends (fig. 10.7). A hand throttle

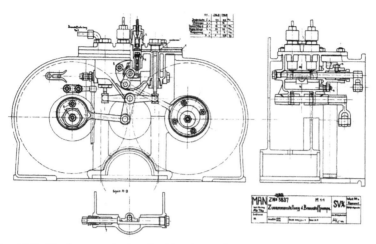

Fig. 10.7. Fuel pump design for Nürnberg S V 5/4 engine. *MAN Archives*

controlled the suction valve supplying the pumping chambers. Packing was not needed to seal the clearance between the plunger and its bore, which indicates the accuracy obtainable in bore and shaft surfaces.

The engine's downfall was reportedly its poor performance due to diversion of compression air to the injector and the subsequent delayed injection. The engine never left the lab and thus ended further thoughts of vehicular applications.

A menu listing the design creativity and diversity that Rieppel sponsored at Nürnberg between 1908 and 1914 contained these engine configurations: two- and four-stroke, horizontal and vertical, crosshead and trunk piston, and single- and double-acting cylinders.

Augsburg's Stopgap A V Series

When Diesel's main patents expired in 1908 others began building his engines. These included Benz & Co. in Mannheim, Güldner, and Körting in Hannover. Deutz also started producing again. M.A.N. quickly met this competitive threat, through what was a legal practice at the time, by getting the new entrants to agree on pricing their engines the same as Augsburg's. This was a good deal for M.A.N. because with its established name it could continue on almost as before. However, on June 30, 1910, it received startling news that the cartel arrangement had been scrapped by the others so that all were free to set their own price. Augsburg was apparently taken totally by surprise.[12]

Drawings dated the same day on which MAN learned of its competitors' action show that work began on modifications to the DM series engines. This counterstrategy mainly involved reducing bores and strokes and raising speeds to achieve equal power from smaller displacements. There was no time for a major redesign. The stroke of the new 35 and 100 bhp engines went from 460 mm to 400 mm and 680 mm to 630 mm respectively, while speeds rose from 190 to 205 rpm and 160 to 175 rpm. The smaller 8, 10, and 12 bhp models remained unchanged.[13]

These updated engines carried a new "A V" designation where instead of horsepower the number of cylinders and stroke were given.[14] The one-cylinder DM 70 became the A 1 V 57 where the old bore and stroke was reduced from 400 × 600 mm to 390 × 570 mm. Its speed went up 20 rpm to 180 rpm.

Lighter-weight engines resulted from smaller, shorter pistons and a reduction in flywheel mass. The reduced stroke also meant less metal in the crankshaft and the frame/cylinder casting. A dished piston crown, long used by the Swedes, was adopted throughout the model line.

The first "A V" left the factory in December 1910.[15] The series grew to include fifteen bore/stroke and thirty-two output combinations. MAN sold 513 engines between 1911 and 1914 with a few being delivered after the war as late as 1923.

The Modernized B V Engines

Not long after introducing the A V series, M.A.N. began to design more state-of-the-art engines. The resulting B V series with its "cleaner" look and worthwhile weight reduction entered production in 1912. Forty-seven B V engines had been delivered by 1914 of which three were the four-cylinder B 4 V 90 (900 mm bore) with an output of 1,000 bhp[16]

Most of the weight reduction came in the frame and cylinder castings; experience had shown where metal could be removed without sacrificing reliability. Narrowing the angle of the A-frame cut engine width, and a thinner water jacket reduced the diameter of the cylinder casting. A shorter connecting rod lowered overall height. For the first time the camshaft was encased throughout its length. A major change was adopting a two-stage compressor driven by the crankshaft on the front of the engine to serve all cylinders rather than have one compressor per cylinder as on the A V series. (M.A.N. was not the first to do this.) Bores and strokes remained similar. Specific weights were lowered 23 to 28 percent. Single-cylinder engines went from 234 to 155 kg/bhp. (fig. 10.8).

Except for the war years, M.A.N. built the B V series until 1930 by which time 314 had been sold. Their aggregate horsepower reached almost 118,000 hp. A six-cylinder, 640 × 900 mm (B 6 V 90) engine producing 1,500 bhp at 150 rpm became available during the 1920s.[17]

Dual-Fuel Injectors—1908

One small, but interesting development at M.A.N. begun in 1908 on an "A V" engine was a nozzle to inject first a pilot fuel and then a poorly igniting tar oil fuel.[18] Because of the high fuel import tariffs and occasional kerosene shortages, one of which was serious in 1907, Augsburg wanted an engine able

Fig. 10.8. 1913 M.A.N. 1,000 bhp B 4 V 90. Sass, *Geschichte, 1962*

to run on low cost and readily available tar oils distilled from German soft coals. This had not seemed possible when tests on such fuels were made in 1897–1899 (chapter 8).

An elaborate solution included a complicated injector nozzle using the standard blast air system but with the fuels supplied by two separate metering pumps. One pump was for a fast-igniting gas oil, and the other a tar oil the engine could not run on unaided. It proved difficult to design a nozzle accurate and reliable enough to inject sequentially the pilot fuel and then the tar oil for sustained, clean combustion. Fig. 10.9 shows the last design of 1912. Most of the tar oil passed through the atomizer portion of the nozzle, but there was also a timed needle valve lifting to ensure that some of it immediately followed the gas oil pilot charge into the cylinder.

Not long after this development, tar oils became available that engines burned without a need of first creating the small fire preceding combustion of main charge. Neither the percentage of pilot oil used nor the number of engines sold with this injector, if any, are known.

The M.A.N. "Horizontals"

In 1909 Nürnberg began adapting its horizontal gas engines to the diesel cycle. This established a successful line of huge engines; some with one meter bores and double-acting cylinders placed axially were able to bum even blast furnace gas. Hans Richter was their designer.[19] Nürnberg introduced the gas engines in 1904, and several companies, including Allis-Chalmers in the United States, had bought manufacturing licenses.

The design progression of the horizontal diesel versions at Nürnberg was first single-acting four-stroke, followed by a single-acting two-stroke, and lastly, in 1912 a double-acting two-stroke never put into production. Both

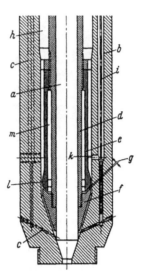

Fig. 10.9. Dual-fuel nozzle of 1912 that first injected a gas oil pilot fuel and then a tar oil with poor ignition characteristics. (*a*) Fuel valve, (*b*) bore for tar oil, (c) bore for pilot ignition fuel, (*d*) and (*e*) inner and outer bushings, (*f*) conical atomizer, (*g*) annular orifice of 0.18 mm, (*h*) blast air entrance, (*i*) needle valve to admit tar oil with pilot charge, (*l*) 0.15 mm annular clearance.
Sass, *Geschichte, 1962*

the four-stroke and two-stroke were made in multiple cylinders where the cylinders were paired next to each other and shared a common jack shaft to open the valves and regulate the fuel pumps. A four-cylinder engine was a double twin, with a flywheel, generator or output pulley placed between the two-cylinder pairs (fig. 10.10).

The first four-stroke, a one-cylinder model LD100, was delivered in May 1909. It had a bore and stroke of 450 × 680 mm and produced 100 bhp at 160 rpm. The smallest in the line was an LD 50, 50 bhp at 170 rpm, with a 350 × 530 mm bore and stroke. Largest of the four-stroke engines, was the 4LV70, 480 × 700 mm, 500 bhp at 167 rpm seen in figure 10.10.[20]

Between 1909 and 1914, Nürnberg delivered 160 of the LD and LV engines. The most popular of the seven offered bore sizes was the 160 bhp 2LV60 with seventeen being sold.

Two-stroke, single-acting engines had the same general cylinder/flywheel configuration as the four-stroke, and only came in two- and four-cylinder models. Uniflow scavenging was obtained by using cylinder intake ports and two exhaust valves in the head. A double-acting scavenge pump, driven by a crankshaft extension, lay parallel to the cylinders. The two-stage compressor was opposed to the scavenge pump, being driven by another, offset crank throw. Unlike the four-stroke with cylinder liners, the two-stroke cylinder was cast with the water jacket. Two-piece pistons were water-cooled.

Fig. 10.10. Nürnberg horizontal, four-cylinder, four-stroke, 500 bhp at 167 rpm, c. 1910. Two-stage compressor in foreground. Intake and exhaust valves eccentric operated; injector and starting air valve in head; governor above fuel pump with both driven off jack shaft. The opposite two-cylinders were a mirror image of the pair in the foreground. Sass, *Geschichte, 1962*

Both types of engines injected into a cone-like combustion chamber on top of the piston. The two exhaust valves, like the intake and exhaust of the four-stroke, were perpendicular to the axially placed injector and opposite each other. The fuel pump, injector, and governor mechanism remained unchanged.

This two-stroke LZ series was available in four cylinder sizes from 480 × 720 mm to 670 × 900 mm, giving outputs from 300 bhp at 167 rpm (two cylinders) to 2000 bhp at 150 rpm (four cylinders). No performance data has been found except for the bmep of 4.1 kg/cm^2. Eighteen of the LZ engines were delivered between 1910 and 1914.

An experimental, double-acting two-stroke diesel was similar in basic layout to that of the gas engine design. (Augsburg's four-stroke of figure 10.11 was comparable.) Air for uniflow scavenging entered through slots midway down the length of the cylinder as there was a combustion chamber on both ends. The exhaust valves were again vertical and opposite each other. Bore and stroke of this single-cylinder engine were 760 × 900 mm, and it was demonstrated in 1913 to produce 1,000 bhp. Engine speed is not known. Work on this engine was being done concurrently with that of vertical, double-acting marine engines at Nürnberg (chapters 16 and 17). Production of all horizontal diesels ended with the outbreak of the war and was not resumed.

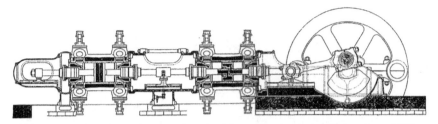

Fig. 10.11. Augsburg four-stroke, double-acting horizontal diesel, c. 1910.
Goldingham, *Diesel Engines, 1927*

Augsburg also began production of horizontal diesels based on the Nürn-berg gas engine design, but from the beginning opted for a double-acting, four-stroke configuration.[21] These came in two forms: a "tandem" where the flywheel was between a parallel pair of cylinders and a "double tandem." In the latter, each of the "tandems" had a common piston rod reciprocating two axial pistons in their respective cylinders (fig. 10.11). The very rigid rod deflected so little that it supported the pistons so that they did not ride on the cylinders—a feared source of wear with conventional horizontal engines. The rod ran on sliding bearing shoes through which water for cooling the pistons was supplied.

Oppositely placed intake and exhaust valves opened into small pockets in the upper ends of the cylinder walls. A fuel nozzle injected into each pocket in the same plane as the valve, but at an angle to it; the two injectors at each end of the cylinder were needed to provide good combustion (fig. 10.12). The castings containing the piston rod packing piloted into the cylinder on each

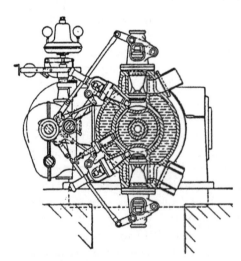

Fig. 10.12. Cross-section view of Augsburg double-acting horizontal diesel showing two injectors with each injecting into a pocket for the intake and exhaust valves. Ricardo, *The Internal-Combustion Engine, vol. 1, 1922*

end. An interesting design twist was their tapered, water-cooled projection further into the cylinder to somewhat shield the packing from the greatest heat created at the beginning of combustion.

The assembly of these engines consisted of mostly concentric parts, except for the base casting holding the main bearings and crosshead. This made for shipments of relatively smaller pieces, and erection was potentially easier than a vertical design.

The first of the Augsburg CL series was delivered in July 1910. This CL90, a tandem only, had a 650 × 900 mm bore and stroke with an output of 800 bhp at 136 rpm. Four cylinder sizes went from the CL80's 590 × 800 mm to the double-tandem CL140 with a 1,000 × 1,400 mm bore and stroke. Only one of these monsters was sold, being delivered in February 1914. Three, double-tandem CL90's with an output of 1,600 bhp at 136 rpm were made. One is shown as installed in a Halle power plant in 1910 (fig. 10.13).

Only twenty-eight of the CL series were built before the 1914–1918 war, and like the Nürnberg cousins, they were not produced again. The great strides made in vertical engines made this design obsolete.

The Swedish K Series

Jonas Hesselman at AB Diesels Motorer began calculations and drawings for a new engine in the fall of 1900, about when M.A.N. started on their DM series.[22] His similar goal to reduce weight was met in part by eliminating the

Fig. 10.13. Augsburg 1,600 bhp, horizontal four-stroke, double-acting diesel installed in Halle power plant, 1910. *MAN Archives*

heavy crosshead. This alone shed 584 kg from the Swedish two-cylinder A59, which was based on Augsburg's Series XVIII drawings. As with the M.A.N. "DM," physical size and costs were greatly reduced. President Schmidt and Chief Engineer Hallencreutz accepted young Hesselman's trunk piston design only after he allayed their fears over increased piston side thrust.

The Swedish and German trunk piston decision was made for the same reasons and at almost the same time in Russia by Nobel. The first American designed engine also had a trunk piston. Diesel himself specifically criticized the US choice and was disappointed by this growing trend. In his view the engine lost some of its mystique as it took on a more mundane look of the much cheaper Otto-type engines.

Discarding the accompanying crosshead pin also got rid of a common source of mechanical knock in early diesels caused by "bearing impact" during load reversals. Thus, a trunk piston was doubly advantageous for higher engine speeds and higher outputs for an equal piston displacement.[23]

Diesels Motorer's first K30 (*Kreuzkopflos*: without crosshead) had a 290 × 430 mm bore and stroke that was conservatively expected to produce 30 bhp at 250 rpm. It contrasted with the A50's 275 × 420 mm bore and stroke and 25 bhp/cyl at 190 rpm.

Hesselman had the good fortune of starting with a clean slate. His design included a cast, one-piece, tapered frame extending up and bolting to the cylinder head. It held a fully machined wet cylinder liner (fig. 10.14). As with Augsburg, the separate liner was likely borrowed from Nobel's A model. The Swedes, like everyone else, had encountered the same frustrating porosity and core problems resulting from a double-walled, one-piece cylinder casting.

The connecting rod was a five-piece assembly where the pin and journal bearing blocks were steel castings. This design permitted shimming between a bearing block and a rod end to easily adjust the compression ratio, an often-necessary task on early multicylinder engines. Manufacturing costs were also less in comparison to a one-piece forging. High-quality steel castings were not readily available, but Hesselman used them because "the Swedish foundries were ahead of their time."[24] Others copied the design.

A long-skirted piston followed conservative practice with a length 2.29 times the diameter. Five rings instead of the eight on the A50 reduced friction. Care was taken to have true cylindrical surfaces on the liner, piston, and rings to minimize blowby and scoring. Otherwise, Hesselman said, excessive blowby during the initial running-in of an engine, when it was being motored on compressed air, could cause enough ring movement so that they "sounded like a bunch of heavy metal washers falling down stone steps."[25] If cylinder irregularities were bad enough, this very disconcerting noise continued while under power.

Early builders recorded little about their engines' lubrication problems. Thus, we are fortunate that Hesselman provides a clear explanation of what turn-of-the-century diesel engineers thought went on and what they encountered.[26]

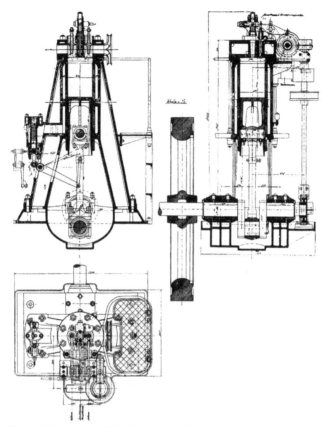

Fig. 10.14. Assembly drawing of first Swedish K30 engine dated June 1, 1901. The basic design would be used for more than thirty years. Hesselman, *Teknik och Tanke, 1947*

The K30 piston and rings were pressure-lubricated by a Mollerup pump of the type Augsburg used on its steam engines. The pump forced metered oil charges initially to six, 2 mm cylinder orifices, equi-spaced radially, and located axially on the liner so that oil discharged between the two topmost rings when the piston was at bottom center.

Most believed that substantially equal oil volumes entered via each port against the piston. However, the Swedes concluded that with the intermittently flowing pump, a small pressure differential between incoming oil and any combustion blowby precluded knowing when or how much oil actually passed through a given port. If a blowby path existed near a port, the oil pressure was believed too low to overcome the gases that could enter the port and force out more oil through other ports. If the blowby was bad enough, the external, horizontal pipe feeding the ports could be emptied. Upwardly angled ports expected to reduce gas backflow and even parts of the supply pipe were sometimes found plugged with carbon.[27]

A first revision lowered the ports to below the bottom piston ring. This sufficiently protected them from blowby exposure and kept combustion gases from contaminating the oil. The number of ports was reduced to four and then to only two, one of which supplied the piston pin. Doing this greatly reduced cylinder oil consumption. It was also found unnecessary to time the crank arm actuating the Mollerup to the piston's position. Finally, the supposedly required special cylinder oils could be changed to merely a cheaper, "good quality pure mineral oil" adequate for bearing lubrication.

Early K30 engines also used M.A.N.'s two-stage compression where first stage air came from the power cylinder. Porosity in second-stage compressor castings led to a wet sleeve for it as well. The compressor was cooled via a hole cast in the main frame against which it was mounted and which led into the water jacket of the engine cylinder. Sickla adopted a two-stage compressor at the same time as M.A.N. (fig. 10.15).

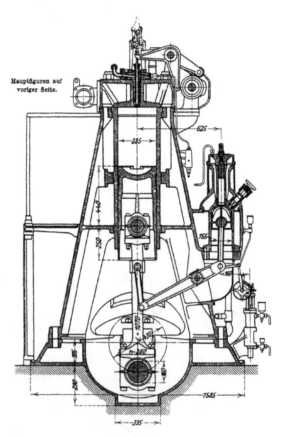

Fig. 10.15. AB Diesels Motorer two-cylinder, 90 bhp at 260 rpm engine c. 1908. Note the two-stage compressor. Güldner, *Verbrennungskraftmaschinen, 1914*

A high mount camshaft lay crosswise to the crankshaft, a placement soon changed when multicylinder engines were built. The starting air valve operated in a four-stroke rather than the usual two-stroke mode. Multicylinder K engines had a starting air valve on only one cylinder. Initially, rocker levers were of cast iron, and while slightly heavier needed no bushings as did steel forgings. Later competition from more aesthetically finished engines brought a return to the costlier forgings.

A new, much simplified governor control and an eccentric drive for the horizontal fuel pump were located on the vertical shaft between the crank and high-mounted camshaft. This shaft turned at crank speed that allowed the fuel pump and control assembly to be lighter and more compact (fig. 10.16). The pump housing was of cast iron rather than the usual bronze. The original suction valve seat machined in the iron was replaced by a hardened seat insert.

Governor output force raised or lowered a sleeve, free to slide on the vertical shaft, to control the distance the suction valve was held off its seat. As with the M.A.N. pump, the greater the valve lift during the delivery stroke of the plunger, the larger the percentage of the maximum fuel charge returned to the tank. Between the horizontal control rod and the vertical suction valve was a swinging link to transfer the rod motion to the valve (fig. 10.17). The link could also be manually rotated to override the control link and hold the valve

Fig. 10.16. Compact fuel pump and governor assembly, c. 1906, for the Swedish K30 series engine. It was functionally the same as that used on the 1902 model. Hesselman, *Teknik och Tanke, 1947*

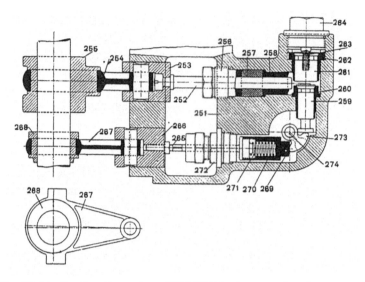

Fig. 10.17. Section through c. 1912 **K30** fuel pump showing control
rod (265), manually swingable link (274), and suction valve (259).
The design was functionally unchanged from the earliest pumps.
AB Diesels Motorer 1913 Instruction Manual

completely open to stop the engine. It was further positionable to provide more fuel during a start or to purge the pump of air after it had been drained.

By running at crank speed, a single plunger pump had the capacity to supply all engine cylinders with each stroke. Thus, a single pump casting and two plunger bore diameters served eight engine models having two cylinder sizes. Diesels Motorer was the only builder for several years to use a single pump plunger for multicylinder engines.

Although a single plunger provided the capacity, accurate fuel metering to more than one cylinder also required a distribution control. Hesselman's solution was a "manifold block" that held a disc with a calibrated orifice in the outlet line leading to each nozzle (fig. 10.18). This provided fixed resistances high enough to overcome the effect of unequal hydraulic resistances between the block and the nozzle fuel chambers. Soon added was a ball check valve seating against the outlet side of each "washer" orifice to prevent the inevitable backflow leakage out of a nozzle from affecting fuel metering to another cylinder or cylinders. A hand-adjusted needle valve at each outlet trimmed or shut off fuel to that cylinder if necessary.

The pump plunger was timed to the engine so that it did not deliver fuel to a nozzle whose needle was lifted for injection. In this way the single plunger could feed at least four cylinders with no problem, and while the idea was never tried, Hesselman claimed it might work on a six-cylinder engine.[28]

When tests began on the K30 in July 1901, no one foresaw any difficulty in its reaching the design rating of 30 bhp at 250 rpm. Bmep had been held to

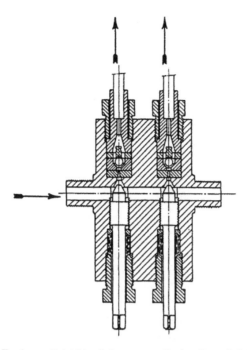

Fig. 10.18. Fuel manifold block for two-cylinder Swedish K engines showing calibrated resistance orifices. Hesselman, *Teknik och Tanke, 1947*

a conservative 3.81 kg/cm² at the higher speed versus 4.51 kg/cm² of the A50 at 190 rpm. Hesselman suffered great disappointment when the best attainable output with a relatively clean exhaust was only 25 bhp. It was then that he began to rethink the prevailing theories about how to initiate combustion.

Accepted lore originating in Augsburg, according to Hesselman, postulated that the angling jets of fuel and air exiting the nozzle and starting to burn would alternately ricochet off the flat piston crown and the flat cylinder head. After some unknown number of angling bounces, the unburned fuel eventually found all the oxygen in the cylinder to complete combustion. Having seen no evidence of this phenomenon on the piston crown, he began to question the theory. Out of it came the thought that "if the fuel will not merge with the air we will have to bring the air to the fuel and concentrate the air around the incoming jets."[29] The result was a concave piston crown as seen in a drawing dated September 13, 1901, (fig. 10.19).

Hesselman's intuitive first choice as to optimum bowl shape yielded a 100 percent increase in power to 50 bhp with a still smoke-free exhaust. He pushed it to 67 bhp, but the exhaust was quite black. As a result, the original 30 bhp design rating was forgotten, and the engine sold at 40 bhp. Jonas said he never again achieved such a performance gain for such nominal cost. One negative arising from his triumph was that ever after President Schmidt expected all major problems would be solved as cheaply![30]

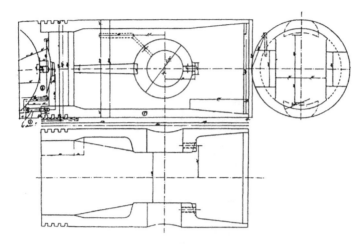

Fig. 10.19. First bowl-shaped piston crown on the Swedish K engine,
which increased output from 25 to 50 bhp, September 1901.
Hesselman, *Teknik och Tanke, 1947*

Hesselman wrote this about the "bowl-shaped" piston, which illustrates
an apparent rivalry between himself and some personalities at M.A.N.:

> It might seem that the arrangement was obvious, but it was not viewed
> that way at the start of the century. The new design was an immediate
> success in some circles; in others it was only slowly and reluctantly
> accepted. The technical director for diesel engines in Augsburg, Mr.
> Lauster, was for a long time opposed to this idea. However, when he
> was away on business trips, the employees experimented with the new
> combustion chamber only to be forced to set it aside again as soon as
> Mr. Lauster returned.[31]

Fuel consumption also improved when burning a lamp oil, that is, kerosene. On August 15, 1901, Diesels Motorer wired Rudolf, who was on holiday
in Berchtesgaden, to report that the "First test of crossheadless 30 hp with
high pressure air pump produced 193 grams [per bhp/hr] at 240 rpm." Diesel
quickly sent them his congratulations and added that they were well on their
way toward his ultimate goal of a 150 gm/bhp-hr specific fuel consumption.[32]

Surprisingly, fuel nozzles on the K series engines still used Diesel's wire
mesh wrap to internally atomize fuel prior to injection.[33] This was despite
a general acceptance of M.A.N.'s atomizer consisting of a multipiece stack
of slotted washers. During the late summer of 1902, Hesselman began
experiments on a one-piece atomizer, which resulted in his unique "Swedish
nozzle."

This type of atomizer (fig. 10.20), first tried in August 1902, had a thin
sleeve (*C*) separating the flow of air into the space (*E*) above the needle seat.

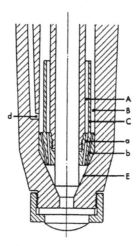

Fig. 10.20. First Hesselman one-piece atomizer, 1902.
Hesselman, *Teknik och Tanke, 1947*

Two concentric circles of holes (*a* and *b*) below the sleeve, where it screwed onto the end of the main needle bushing, connected the incoming fuel (*d*) and injection air with the needle seat. When the needle lifted, the outer, concentric air annulus above fuel inlet (*B*) partially atomized the fuel as both went through the circle of holes (*b*) into the space (*E*). The mixture was further atomized as it struck air coming through the inner holes (*a*). The design became a replacement for the wire mesh.

It had two weaknesses, however. Fuel did not have time to distribute itself equally around the outer sleeve before air forced the charge down the holes (*b*). Excessive fuel also found its way into the needle space (*E*) prior to injection, and unless the valve perfectly sealed, this fuel leaked into the combustion chamber to cause a smoky exhaust.[34]

The one-piece design gradually evolved into a type used for many years (fig. 10.21). Fuel entered an annulus at the lower end of the atomizer and rose into external, longitudinal slots and angled, upward slanting holes. These holes exited into the upper end of a tapered bore around the fuel valve. Injection air entered the bore and passed down a clearance into the tapered portion. Air also acted on top of the fuel above the outer, external slots. When the needle lifted, air on the outside forced fuel up the angled holes and into the high velocity air traveling down the needle clearance to thoroughly atomize the fuel charge prior to injection. Only fuel metered by the pump was injected, because air on the outside of the atomizer piece could push up the holes just when fuel rose above their outer, lower end. Some thought there was an "ejector" action whereby the high-velocity air passing down the needle also pulled fuel up the angled holes. Hesselman disputed this.

He also felt he stood alone in his worries about the entrained air in the fuel that came from the pump:

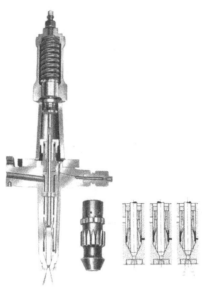

Fig. 10.21. Hesselman "Swedish" fuel atomizer, c. 1905. The left section view shows the fuel level after injection. The middle view is after a fuel charge has been deposited, and the right view is during injection. The strange nozzle hole shape was reportedly used when burning tar oil.
Hesselman, *Teknik och Tanke, 1947*

The air in the downward directed oil passage in the fuel nozzle was probably the most difficult of all the technical problems I encountered during my time at Sickla. Nobody has seen these air bubbles and nobody will ever see them, and besides they only appeared spasmodically.... To convince others about this was likewise not easy, and I am far from sure that I always was successful in this regard.... Many times in the future did I speak to deaf ears trying to explain that certain disturbances in small engines were caused by air bubbles which in one way or another had collected in fuel pumps.[35]

The multicylinder engines that followed in short order again showed the Swedes differing from accepted practice. They believed that greater weight savings and lower costs were best achieved by adding more but smaller cylinders rather than by building one- and two-cylinder models having increasingly larger cylinder displacements. Smaller cylinders permitted higher speeds and reduced cylinder specific weights. Lower manufacturing costs resulted from the less complex parts, which allowed simpler tooling and less setup time.

Along with the one fuel pump, a single, two-stage air compressor served up to four cylinders. The compressor was mounted on cylinder No. 1, the "base cylinder," and the additional cylinders were "bare."

Diesels Motorer shipped its first two-cylinder, 80 bhp model K 2 III in December 1902 and the first three-cylinder, 120 bhp K 3 III in January 1903.[36] An ever-evolving K series remained in production for over twenty years.

Graphic proof of Sickla's success in reducing weight is seen in these comparisons:[37]

A20 (20 bhp Series XVI)	8,500 kg	425 kg/hp
A50 (two-cylinder Series XVTII)	14,000	280
DM30 (one-cylinder MAN)	8,050	268
DM 2x25 (two-cylinder 50 bhp MAN)	11,000	220
DM 2x60 (two-cylinder 120 bhp MAN)	27,800	232
K 2 II (two-cylinder 60 bhp)	7,250	121
K 3 III (three-cylinder 120 bhp)	12,650	126

The last four are second-generation engines developed about the same time, but the Swedish weight per horsepower fell to about half that of Augsburg's. All engines had comparable equipment, and the weights included flywheels. The data is from a Swedish catalog of 1906 and a German catalog of 1909.

Production costs (parts and labor only) and selling price in 1902 Swedish kronor are equally interesting:

	Mfg. Cost	Price	
A20	8,560	—	(20 bhp)
A50	15,433	25,000	(50 bhp)
DM 2 × 60		29,000	(120 bhp)
K 2 II	7,500	—	(60 bhp)
K 3 II		25,000	(120 bhp)

Independent fuel consumption tests on two-cylinder, 80 bhp and three-cylinder, 120 bhp engines (fig. 10.22) were made by two Stockholm engineering professors in September and October 1902:[38]

	K 2 III	K 3 III
Speed, rpm	242	251
Power, bhp	81.5, 54.8	120.8, 61.7
Bsfc, g/bhp-hr	179.3, 193.6	173, 197

Fig. 10.22. K 3 III, 120 bhp Diesels Motorer engine
installed in the Östersund, Sweden electric plant.
Photo from a September 1906 AB Diesels Motorer catalog

These results show the improvement over the first K30 of a few months earlier in that part load consumption approximated the K30's bsfc at full load. Contrast these numbers with the 225 grams of the old A20.[39]

Acceptance of AB Diesels Motorer engines in Scandinavia came rather quickly even though the company had to travel down a learning curve of troubles as did the other builders. Helping greatly was a license monopoly for its area and Marcus Wallenberg's energetic salesmanship. Through his efforts Diesels Motorer acquired a short-term Danish sublicense from Burmeister & Wain in July 1902, giving it rights throughout Scandinavia for a few years.

The Vandal

European recognition was enhanced in 1902 by the Russian licensee who bought Sickla engines for a new, larger type of river tanker. This was done because a lighter-weight Nobel design was not yet ready.

Sickla shipped three, 120-hp K 3 III engines to St. Petersburg in January 1903 for installation in the triple screw *Vandal,* the first commercial marine application for diesel engines. (It was launched about the same time as Dyckhoff's much smaller *Petit Pierre.*) The *Vandal* was 244 ft., 6 in. long with a beam of 31 ft., 9 in. and a draft of only 6 ft. Cargo capacity was 820 metric tons of lamp oil[40] (fig. 10-23). The tanker entered service when the ice broke up in the spring of 1903 and operated until 1913 plying the 3,000 km route between the Caspian Sea and St. Petersburg on the Baltic.

Fig. 10.23. Nobel's Petroleum Company's triple screw, Caspian River tanker *Vandal*, entered service in 1903. Three mid-ship mounted, 120-hp AB Diesels Motorer engines turned propellers through an ASEA all-electric drive. *Wärtsilä Diesel AB Archives*

Each of the midship-placed engines drove a generator wired to a motor turning a propeller shaft. The electric motors were simply reversed for "astern." ASEA (*Allmänna Svenska Elektriska Aktiebolaget*, or General Swedish Electrical Limited Company) of Sweden designed and built the electrical gear for this first diesel-electric application by adapting controls from those used on electric street cars.[41]

Hesselman continued with his work despite a serious injury from an engine-caused accident in 1905 that left him physically handicapped.[42] Under his engineering leadership, AB Diesels Motorer began offering two-stroke and four-stroke reversible marine diesels in 1907 and 1908 respectively (chapters 15 and 16).

Several diesel-electric locomotives entered regular Swedish railway service in 1912.

Wallenberg never lost his enthusiasm for the diesel, and in 1913 he negotiated the purchase of a significant interest in McIntosh & Seymour of Auburn, New York. The American company, founded in 1886 by John E. McIntosh and James A. Seymour to build steam engines, had the facilities and capabilities to make diesels. For several years it would closely follow the Swedish designs.[43]

The first McIntosh & Seymour diesel was its model 4-A38, which had come from ABDM's model T3R. This four-cylinder, 18-7/8 × 28-3/8 inch bore and stroke engine produced 500 bhp at 164 rpm. It won a grand prize at the 1915 Panama Pacific Exhibition in San Francisco. About sixty of the A series diesels were built until its production ended in 1924.[44]

An engine based on the A-frame, Swedish K series became the M&S B series offered in two-, three-, and four-cylinder sizes beginning with a cylinder of 16 × 24 in. and an output of 90 bhp/cylinder at 200 rpm. A guaranteed fuel consumption not to exceed a full load bsfc of 0.42 lbs/bhp-hr covered engines from 500 to 1,000 bhp.[45]

McIntosh & Seymour continued to build stationary marine engines and then locomotive diesels during the 1920s. In 1930 the company was bought by the American Locomotive Co., also located in Auburn.

Progress in Russia

The second-generation Russian diesel introduced in 1903 was more evolutionary than those of the German and Swedish builders. Maschinenfabrik Ludwig Nobel in St. Petersburg had already solved a major casting problem with a wet cylinder liner. It had also gone further than the others in reducing cost on its original production model with the combined A-frame and cylinder jacket. The significant change in its B Series was a trunk piston. Shortly after the engine's debut, it became available in seven models with outputs of 10 to 90 bhp.[46]

The model B abandoned M.A.N.'s two-stage air compression method where first-stage air came from the engine cylinder. Nobel was the earliest of the major European builders to do so. Plaguing soot deposits from combustion gases in the second-stage compressor, and the lines and nozzle seem to have forced a faster decision toward a change in Russia than elsewhere.

"B" engines 25 hp and under had one stage of compression; those above that output used the extra stage. Single-cylinder engines over the 25 hp had a Nobel-designed two-stage, two-cylinder compressor. Each compressor piston was driven by linkage off the connecting rod and timed so that compression in the air cylinders was separated by 180 degrees. On two-cylinder engines above 25 hp, the first-stage compressor was mounted on one cylinder frame and the second stage on the other (fig. 10.24).

The sale of three, three-cylinder, 450 × 640 mm diesels to the St. Petersburg water works in 1903 opened new markets. Rated nominally at 300 bhp, the output was reduced to 180 bhp at 165 rpm because of the engines' vital service. The pumps were also made by Nobel. Even after one year in service, the engines exceeded the required water delivery rate by 10 percent.[47] Nobel-designed diesels were soon powering factories and generating electricity for privately owned utility companies in a widespread area of Russia. Cheap oil, burning at a miserly rate in a reliable power plant, made a potent sales combination, especially when both fuel and engine came from the same company (fig. 10.25)!

By 1908 ratings of the B Series had risen to 800 bhp at 150 rpm in a four-cylinder engine of 600 × 800 mm. These largest engines used water-cooled exhaust valves (fig. 10.26). The "Bs" remained a consistent seller until the outbreak of the war, with thirty-six delivered in 1908 and thirty-nine in 1913. However, the aggregate horsepower almost doubled from 3,532 for the earlier year to 6,280 for the last one. Accurate records of engines delivered after 1913 are not available (fig. 10.27).[48]

Fig. 10.24. Nobel Model "B" two-cylinder, 60-hp diesel. Photo taken March 1903. *Diesel collection, The Deutshes Museum*

Fig. 10.25. Nobel engine assembly area, c. 1910.
Diesel collection, The Deutsches Museum, Author photo

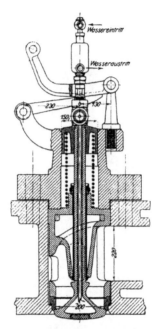

Fig. 10.26. Water-cooled exhaust valve, 1908 Nobel 600 × 800 mm, 200 hp per cylinder diesel. *Trans., VDI, 1909*

Fig. 10.27. Not all Nobel engines were trouble free! Note the cracked A-frame. *Diesel collection, The Deutsches Museum, Author photo*

Fig. 10.28. Hans Flasche (1875–1970), Nobel engineer
and Rudolf Diesel's brother-in-law. *MAN Archives*

An engineer recruited from Augsburg to help design the B Series was Hans Flasche (1875–1970), Martha Diesel's brother (fig. 10.28). After graduation from the Technische Hochschule Braunschweig he went to M.A.N. and was assigned to the new DM Series project. Carlsund brought him to St. Petersburg in 1901 where he stayed until the end of July 1914. After the war Flasche worked for a time at the Swedish Nobel company.[49] Did he leave Augsburg because of a growing animosity toward his brother-in-law?

Baku to Batum: Pipelines and Pumps

The Russian oil industry provided more opportunities to enhance the Nobel diesel name. One of these came in 1900 as a result of a government decision to reduce pumping costs on its new pipeline carrying kerosene from Baku on the Caspian to Batum on the Black Sea. This 560-mile-long pipe over desert and mountains lay alongside a railroad line that climbed to a summit of 2,485 feet. Its capacity was 48,000 gallons (probably Imperial) per hour.

The diesels replacing the steam engines powering the pumps were expected to reduce fuel usage by at least half. More importantly, the high cost and trouble to supply adequate, acceptably pure boiler water in that arid land would cease. Another factor favoring the diesel was that *masut*, the residual boiler fuel burned in the pumping stations to generate steam, would no longer have to be hauled to the pumping stations over the adjacent busy railroad. The new diesel engines could burn the kerosene already flowing in the eight-inch pipeline.[50]

Nobel delivered three, two-cylinder, 375 × 560 mm engines to the central pumping station at Baku in 1902. They were rated 100 hp at 165 rpm, and directly drove German-type Riedler pumps built under license. The operation was so trouble free that a decision was made the same year to immediately begin repowering all twelve pumping stations with four two-cylinder, 150-hp diesels per station. This order for forty-eight engines was more than the St. Petersburg factory could fill within the specified contract time.

The Kolomna Machine Works

Emanuel Nobel resolved his production dilemma by having the Russian Diesel Engine Co. in Nürnberg, which he controlled, to license a second company, the Kolomna Machine Works, located in the city of Kolomna 100 kilometers southeast of Moscow. Its existing product line included locomotives, pumps, and heavy machinery. Kolomna signed a contract on October 16, 1902, and using the Nobel design, successfully delivered the Baku pumping engines.[51] It would continue to be a major builder of diesel engines in Russia.

Nobel Two-Stroke Diesels

Anton Carlsund was the catalyst for a uniflow, crankcase scavenging two-stroke in 1902 based on a 10-hp A series engine (fig. 10.29). From a bore and

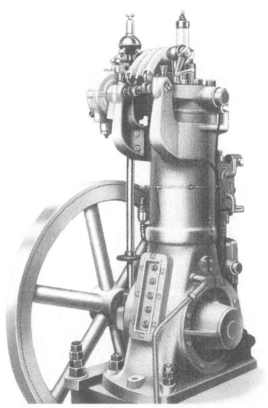

Fig. 10.29. First Nobel two-stroke diesel, 1902. 18 bhp at 300 rpm.
Wärtsilä, Trollhattän

stroke of 185 × 290 mm, it produced 18 bhp at 300 rpm. Scavenge air was drawn into a closed crankcase past ten "automatic" valves, with five set in a plate on each side of the A-frame. The piston uncovered cylinder intake ports after the exhaust valves in the head opened. (The intake from the earlier engine had become a second exhaust valve.) A slight supercharging occurred by diverting some first stage injection air into the crankcase. Scavenge air volume was about 1.2 times the cylinder displacement.[52]

After completing the initial tests, Emanuel Nobel asked that a Roots blower for scavenge air be installed, but it was soon removed for unspecified "technical operating reasons."

Performance tests on kerosene (*Petroleum*), *Solaroil*, and Baku crude (*Rohnaptha*) were made in May 1904. The following runs have comparable speeds and outputs:[53]

	Kerosene	Solaroil	Baku crude
Speed, rpm	266	266	266
Indicated mep, kg/cm^2	5.3	5.2	5.28
Indicated horsepower	24.4	24.0	24.4
Brake horsepower	18.9	19.6	19.7
Mechanical efficiency, %	77.5	81.6	80.8
Fuel consumption, g/bhp-hr	204	212	200
Air injection pressure, atm	53	54	55

The data show good multifuel capability and a very good fuel consumption for an early two-stroke. When first tested in 1902, it had a bsfc of 215 g/bhp, or less than the 240 g/bhp of its 10 bhp, four-stroke cousin. It went into the factory and during the 1905 war with Japan was said to have run eleven hours per day.

Carlsund built a three-cylinder, two-stroke diesel in 1905 with a bore and stroke of 260 × 400 mm. Its expected output was to be 200 bhp at 400 rpm, but due mainly to an inadequate air scavenging system, it could reach only 140 bhp. This deficiency, plus a disappointing fuel consumption, brought an end to further work on it as a two-stroke. However, its general design was the basis for the four-stroke D Series marine engine with a "sliding cam" reversing mechanism (chapter 15).[54]

The Ssarmat

Another man's genius added to Russia's diesel success. Karl Wilhelm Hagelin (1860–1953) offset the reticence of a more cautious Emanuel Nobel. Karl's father had come from Sweden to work for Immanuel Nobel, and chose to stay in Russia. The boy grew up as "a child of the Volga" where his father maintained river boats, and started with the Nobels as a pipefitter in the Baku oil fields at nineteen. Twenty years later, in 1899, Emanuel brought him to St. Petersburg to be technical director. As one of the five directors (vice presidents) he ran all Baku operations and its oil distribution system. Of the 12,000 Nobel employees, Hagelin was directly responsible for 3,800 in the oil fields and five refineries around Baku plus the 2,300 in the Nobel "fleet" and their associated 250 transport offices. He had great managerial ability, and was a creative, self-taught engineer. Above all, he was honest and someone in whom Emanuel could put complete trust. In 1906 the Swedish government appointed Karl general consul for the capital city of St. Petersburg.[55] He fervently believed in his native proverb that "Risk is a noble action."[56]

Hagelin saw the diesel engine as a way to overcome two related drawbacks to his oil transport system. Since the early 1870s, Russian steam locomotives and boats had burned the heavy *masut* residual oils.[57] However, their high fuel consumption cut into Nobel profits, particularly in the Caspian tankers and Volga side-paddle-wheeled barges and tugs. Secondly, oil reached the Baltic after transshipment into rail cars at the river's northern terminus. Hagelin wanted to move it all the way by water using an enlarged river, canal, and lake system open for several years, but no one had yet built practical, propeller-driven tankers of adequate size and power.

He silenced "can't go by water" skeptics the day one of his large Volga tugs steamed out of the Neva River and tied up at the factory dock in St. Petersburg. Emanuel then saw Karl's vision and authorized the design of two self-propelled tank barges, the first being the previously mentioned *Vandal* with its Swedish diesel-electric drive.

The *Ssarmat*, of identical size to the *Vandal*, began service in the summer of 1905 powered by two larger, Nobel E series diesels.[58] These four-cylinder, 320 × 420 mm trunk piston engines had been adapted from a stationary model (fig. 10.30). Output was 180 bhp at 260 rpm. They were placed near the stem and turned propellers via a semi-electric drive more efficient than the *Vandal* all-electric one. Also aboard was a single-cylinder, 10-hp auxiliary diesel. The *Ssarmat* ran with her original engines until 1923.

Although independently invented by Hagelin, the new drive was very similar to one patented about then by C. Del Proposto, an Italian engineer.[59] Several builders, including AB Diesels Motorer, later bought licenses and used the cumbersome drive until the advent of reversible diesels. The Nobel/Del Proposto system consisted of an engine-driven generator, a magnetic clutch, and a motor, all coaxial with the shaft. When going "ahead," the magnetic clutch was energized to couple the engine directly to the propeller. In "astern" the clutch was disengaged, and the generator drove the motor in the opposite direction. Nobel's first diesel tankers transporting oil on the Caspian from Baku to Astrakhan on the Volga also used the Del Proposto drive.

Fig. 10.30. Nobel 180 bhp diesel engines installed in Volga tanker *Ssarmat* in 1904. *Diesel collection, The Deutsches Museum, Author photo*

Korejvo Reversing Clutch

A Kolomna-powered side wheeler on the Volga River had a unique revers-ing method patented by Chief Engineer Korejvo in 1908.[60] The mechanical/pneumatic I-C engine control shows the complexity endured until the diesel itself could be made to run in either direction. This system was used in sev-eral river boats on the Volga and its tributaries.

The engine was installed parallel to a shaft supporting side-mounted paddle wheels (fig. 10.31). An air-operated clutch at each end of the engine could be alternately engaged to deliver power to either a Morse chain drive or to a helical gear set through a 6:1 speed reduction. The gears turned the paddle shaft for "ahead," and the chain drive rotated it in the opposite direc-tion for "astern."

An air valve controlled each clutch. It consisted of an inner piston moved by linkage from the governor and a hand-operated, ported sleeve serving as a cylinder for the piston but free to slide independently within a bore of the valve housing (fig. 10.32). To engage a clutch, the operator moved the valve sleeve to the "left." However, the governor weights had to be "down," that is, in idle fuel position, before a clutch could be engaged. At idle, the governor located the piston far enough to the "left" to uncover the sleeve ports and send air from the valve housing to the clutch. One would assume that the controls were arranged such that both clutches could not be engaged at the same time.

Friction elements in the water-cooled clutches consisted of the driving member's two cast iron surfaces, which ran against lignum vitae rings set in

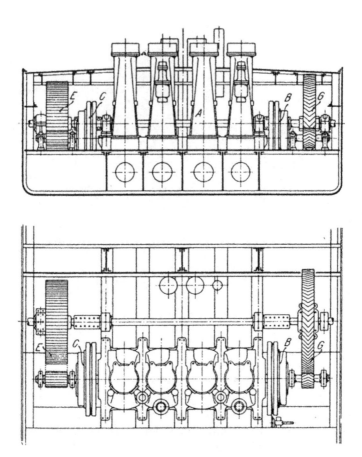

Fig. 10.31. Volga River side wheeler with 1908 Koreiwo reversing system.
Morse chain drive on left and helical gears on right end of engine.
Air-operated clutches alternately engaged the chain and gears.
Zeit., VDI, 1909

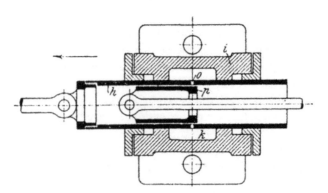

Fig. 10.32. Air valve for the Koreiwo reversing system.
Zeit., VDI, 1909

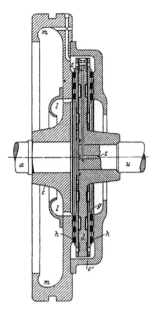

Fig. 10.33. Air-operated clutch for Koreiwo reversing system.
Zeit., VDI, 1909

both sides of the driven rotor (fig. 10.33). Fastened to the rotor was a hollow circle of 3 mm thick sheet copper to act as a diaphragm. A cast iron disc holding the wood rubbing rings was attached to a rubber backing which remained against the copper diaphragm. Air pressure forced the two discs outward against the clutch-enveloping driving member. A similarity of this clutch to a modern vehicle disc brake cannot be ignored, except that the lignum vitae rings expanded from the "rotor" rather than pads from stationary calipers.

Kolomna built stationary and marine diesels in both four- and two-stroke, and created a reputation for quality and dependability equaling that of Nobel. The two companies worked in close collaboration, and Kolomna engineers made their contributions to the success of the diesel engine in Russia. Only Kolomna remained a major diesel builder with its name unchanged after 1917.[61]

On Constant Pressure Combustion

One cannot admire enough Diesel's energy and perserverance during the years of his engine's creation. Only those who have taken part in a similar test and development program could understand his almost childish joy, when he on paper could catch a good looking and well developed indicator diagram.... It is obvious that his thoughts even a considerable time later circled around the shape of the constant pressure diagram as indicated by his son, Dr. Eugen Diesel . . . telling that this influenced the "D" in his personal signature. This is more of interest as an anecdote but may also serve as proof of how an inventor's work might influence everything in his life. His fate reminds of that King Midas, he who made everything he touched into gold.

—Jonas Hesselman[62]

Notes

1. Johann K. E. Hesselman, *Teknik och Tanke* (Stockholm, Sweden: Sohlman, 1948), 29.
2. Kurt Schnauffer, "Die Motorenentwicklung in Werk Augsburg der M.A.N. 1898–1918," (1956), 22. The DM 70 had a cylinder serial no. of 78.
3. The engine was designed for 60 bhp at 150 rpm, but ran so well at 190 rpm during April 1901 tests, that it was uprated to the 70 bhp at 160 rpm before leaving the factory. Allgemeine Gesellschaft für Dieselmotoren Report No. 217, end of April 1901, appendix, 3. MAN Archives.
4. Schnauffer, "Die Motorenentwicklung," 22.
5. Allgemeine Report 217, appendix, 1.
6. Allgemeine Report No. 214 (Feb, 18, 1901), appendix, 1, and Lauster US Pat. 729,613 of June 2, 1903. Hesselman, *Teknik*, 115, wrote that, in addition to problems caused by air leaking past the packing gland of the suction valve control rod, cyclic pressure-to-vacuum changes could slightly affect the volume of the metered charge.
7. Hugo Güldner, *Das Entwerfen und Berechnen der Verbrennungsmotoren*, 2nd ed. (Berlin: Springer, 1905), 474–75.
8. Schnauffer, "Die Motorenentwicklung," 29.
9. Ibid., 30. Nürnberg, however, continued making steam engines until 1956.
10. Schnauffer, "Die Dieselmotorenentwicklung," 13.
11. Ibid.
12. M.A.N. letter to its representatives on July 4, 1910, announcing cancellation of the pricing agreement. MAN Archives.
13. Engine data kindly furnished by Josef Wittmann, MAN Archivist and Museum Director.
14. The old "DM" model designations had become confusing. The first letter designating the edition, in this case "A," was followed by the number of cylinders (1, 2, 3, etc.) and "V" for *Vertikal* or "L" for *Liegend* (horizontal) cylinders. The last number was the stroke.
15. Schnauffer, "Die Motorenentwicklung," 33.
16. Ibid., 35.
17. MAN engine data, Schnauffer, "Die Motorenentwicklung,"
18. Ibid., 27.
19. Friedrich Sass, *Geschichte des deutschen Verbrennungsmotorenbaues von 1860 bis 1918* (Berlin: Springer, 1962), 330–35, has a short biography of Richter and describes his horizontal gas engines.
20. LD stood for *Liegender Diesel* (horizontal). When the two-stroke was introduced, the designations were changed to LV for *Liegender Viertakt* (horizontal four-stroke) and LZ for *Liegender Zweitakt* (horizontal two-stroke). Schnauffer, "Die Motorenentwicklung," 24–35, details the four-stroke and two-stroke lines at Nürnberg.
21. Schnauffer, "Die Motorenentwicklung," 64–6, describes the four-stroke line made there. The engines are not mentioned in Sass.
22. The design and operation of the K series are fully covered in Hesselman, *Teknik och Tanke*, 36 et seq.
23. Ibid., 41. Hesselman also debunks popularly held canons of the time on piston speed, which were promulgated particularly by Hugo Güldner's authoritative texts. The writer's father collided with Güldner's influence in this area as late as the early 1920s.
24. Ibid., 52.
25. Ibid., 47.
26. Ibid., 142–49.
27. Hesselman clearly explains that what some German diesel experts thought was happening did *not*: "It was just an illusion when they required that oil be forced in as the rings passed by the cylinder ports." He added with a little nationalistic barb that he found it hard to understand

why his experiences differed so greatly from the experts, but "It lies partly in the fact that the authority suffered from a much too strong belief in authority, which can be traced to Diesel's own designs (1896), and which are repeated almost unchanged by Güldner as late as 1914 and by (Friedrich) Sass (1929)." Hesselman saw why Diesel held such ideas but found it "more remarkable" that Güldner still did in 1914. Ibid., 148.

28. Ibid., 65.
29. Ibid., 123–25.
30. Ibid., 54–55.
31. Ibid., 57.
32. Copies of the telegram and Diesel's August 17, 1901, letter reply are in the MAN Archives.
33. Hesselman, *Teknik*, 120, spoke of Hallencreutz's preference for Diesel's wire mesh. He also referred to mesh atomizer troubles in late summer of 1902 while testing a new two-cylinder, 60 bhp engine. Ibid., 122.
34. Ibid., 127–28.
35. Hesselman interestingly described the air's action and consequences (131–32), including Prosper L'Orange's (chapter 13) explanation of it to him and how L'Orange came to understand the associated phenomena, 130–38. Anyone who has experienced the problems caused by entrained air in the fuel can empathize with Hesselman's dilemma and his difficulties making others understand the effects of the air.
36. The "K" cylinder diameters were classified by a Roman numeral designation. For example, a one-cylinder, 30 bhp K30 became a K III where the "III" referred to a 290 mm bore.
37. The data is from Schmidt's April 1903 report to the Allgemeine (MAN Archives) and Hesselman, *Teknik*, 81. M.A.N. brochure (No. 191) of December 1909 on the DM series gives a full load bsfc of 180 g/bhp-hr for the largest engine and 224 g for the smallest. This compares closely with the guaranteed bsfc of the 1910 Swedish engines. Hesselman, *Teknik*, 76.
38. The test results are from Allgemeine Reports Nos. 244 and 245, both of mid-December 1902. MAN Archives.
39. Hesselman, *Teknik*, 82.
40. Hans Flasche, "Der Dieselmaschinenbau in Russland von Anbeginn 1899 an bis etwa Jahresschluss 1918" (1947), 14. MAN Archives. He gives the dimensions in the original units, 244' 6" × 31' 9" × 6', and the capacity as 50,000 *puds* (1 pd = 36.1 lb). The dead weight tonnage was 800 "tons" (metric or English?). This well-documented, eighty-one-page manuscript is a gold mine of material on Nobel engines and their applications.
41. Ibid., 13.
42. Hesselman, *Teknik*, 76, and *Svensk Upplagsbok*. The author has heard that Hesselman was unable to walk after the accident.
43. John W. Anderson, "Diesel, Fifty Years of Progress," *Diesel Progress* (May 1948), 79.
44. Ibid., 80.
45. Lacey H. Morrison, *Oil Engines* (New York: McGraw-Hill, 1919), 251.
46. Flasche, "Der Dieselmaschinenbau," 9.
47. Ibid., 28.
48. Ibid., 6. Unfortunately, Flasche makes no distinction between two-stroke and four-stroke "B" series engines.
49. Ibid., p. A of foreword. From May 1919 to December 1925, Flasche was at the Swedish Nobel factory in Nynäshamn, and then until 1932 in Rotterdam at the Burgerhout factory, which built Swedish Nobel diesels under license. In 1932 he went to an engine company in Kiel, Germany, from where he retired in 1944. He died in Kiel at age ninety-three.
50. A valuable reference on the emerging Baku industry is found in Charles Marvin, *The Region of the Eternal Fire* (London: Allen, 1884), 413 pages. He gives historical and technical developments of the oil industry, and his geopolitical comments reflect the current problems facing that region. E. H. Foster, "A Russian Petroleum Pipe Line," in *Cassier's Magazine*, vol. XIX (November 1900), 3–16, details the "new" pipeline and refineries.
51. Kurt Schnauffer, "Lizenzverträge und Erstentwicklung des Dieselmotors im In- und Ausland 1893–1909" (1958), 85–86. Kolomna was to pay 50,000 rubles down and 5,000 per year for ten years. The royalty was 10 percent. E. Nobel profited through his Russian Diesel Engine Co.
52. Flasche, "Der Dieselmaschinenbau," 12.
53. Data sent by Carlsund to the Allgemeine on tests made from May 13 to July 17, 1904. Carlsund spells the crude oil as *Rohnaphta* while an Allgemeine-made copy uses *Rohnaphtha*. Original in MAN Archives.

54. Flasche, "Der Dieselmaschinenbau," 15, 57. French patent No. 346,453 of Nov. 28, 1904: "Improved system for a two-stroke I-C engine." Flasche said that besides a Russian patent there were other foreign ones, but he did not have accurate information as to the countries they were in.

55. Robert W. Tolf, *The Russian Rockefellers* (Stanford: Hoover Institution Press, 1976), 16–49. See also: *Svensk Uppslagbok,* vol. 12 (Malmö: Norden, 1949), 677; Ida Bäckmann, *Fran Filare Till Storindustriell* (Stockholm: Bonniers, 1935), 233–52.

56. Tolf, *The Russian Rockefellers,* 172.

57. Sydney H. North, *Oil Fuel: Its Supply, Composition and Application,* ed. Edward Butler, 2nd ed. (London: Charles Griffin & Co., 1911), 37–39.

58. Flasche, "Der Dieselmaschinenbau," 77. The Model "E" appears to have been made in only this one size and number of cylinders.

59. Del Proposto visited Hagelin later and questioned Nobel's illegal use of his invention. Hagelin proved to him that the drawings for the *Ssarmat* installation predated the issuance of the Italian's Russian patent. The two parted "as good friends." Hagelin memoir sent to Flasche in January 1947, 3. MAN Archives.

60. R. Murauer, "Die Entwicklung und Verbreitung des Dieselmotors in Russland und seine Verwendung als Schiffsmotor," *Zeitschrift,* Verein deutscher Ingenieure, (1909), 1187. The German spelling for the Polish engineer was *Koreiwo* as used in Murauer.

61. In addition to continuously building diesel engines, Kolomna supplied diesel-electric locomotives beginning in the early 1930s. Its product line in the 1950s included diesel and steam locomotives, diesel engines, space rockets, and potato-picking machines. J. N. Westwood, *Soviet Locomotive Technology during Industrialization 1928–1952* (London: MacMillan, 1982), 24.

62. Hesselman, *Teknik,* 118.

CHAPTER 11

The American Diesel Enterprise

*"He assures me there is never any complaint about the Augsburg engine.
. . . While I am ready to believe that great improvements have been made
in all the mechanical details of the engines now being built in Augsburg,
I cannot conceive that that alone would be sufficient to turn out such an
engine and have it run successfully without the most conscientious care and
management."*

Meier to Diesel, 1900[1]

Colonel Meier was right. Augsburg engines still needed tender care and
attention, something that American diesel engine users were not prepared
nor conditioned to dispense. The forgiving steam engine had spoiled them.
Thus, the American licensee was forced to march to a different drummer.
Simpified designs more adaptable to benign, or hostile, indifference would
depart from accepted practice.

The result of this philosophical divergence brought some successes, and
inevitable failures that could have been weathered, given the resources of the
licensee. However, a mistaken belief in how the American diesel industry
should be controlled would stunt the engine's growth. The time ultimately
wasted before these fundamental errors in judgment were recognized and
corrected enabled impatient manufacturers already in the gas and steam
engine business to enter as strong diesel competitors.

Adolphus Busch and Edward Meier

The American diesel saga, whose genesis was touched upon in chapter 9 is
a case study in the development of an American industry.[2] Adolphus Busch
and Edward Daniel Meier never wavered in their belief that they would make
a success of Diesel's engine in the United States, but to do so required Busch's

deep pockets and Meier's untiring efforts. Both at times had every reason to sell out or walk away from years of never-ending business frustrations and serious engine failures even though the causes for some of the setbacks must be laid at their own doorstep. Missed opportunities, unfortunate engineering decisions, and international business intrigue were all elements in their long struggle to survive.

Affecting the growth and direction of the diesel market in the United States above all else was Busch's strong belief in a business plan not suited to the product he wanted to promote. It was the patent owner's idea to create an American cartel of licensees, controlled by the Diesel Motor Company of America, who would build engines to furnished designs. His concept was not feasible for an unproven and temperamental power plant that was to compete against inexpensive, easy to maintain steam engines using readily available and relatively cheap fuels.

Meier, with the concurrence of Busch, opted for a new engine very different than those offered in Europe. His decision to build a unique diesel configuration was justified, but the resulting design bordered on the naive. Further, the new company and two direct successors never had their own factory. All engines came from contract builders, a situation providing the company with little control over manufacturing costs and quality.

Meier believed, and rightly so, that the available German design would not be accepted in the American market because of an inherently higher cost and because such engines required too much operator experience and dedication if they were to perform properly. It was for this reason that Meier and his small staff of engineers at 11 Broadway in New York City designed what came to be known as the "American" engine. Unfortunately, they lost almost a year during 1898 on studies where the engine would utilize natural gas, which prevented the start of a concentrated effort on the new engine.

The Frith Engine

Arthur J. Frith (d. 1914), chief designer of The Diesel Motor Company of America, and his assistant James D. Macpherson (1872–1911) drew heavily upon the successful Westinghouse gas engine. Their 11.25 × 20 inch, one-cylinder diesel was to produce 30 bhp at 170 rpm. It had an integrally cast head and cylinder, a trunk piston, and an enclosed crankcase. A housing mounted at the side of the cylinder head contained a horizontal fuel nozzle, an intake, an exhaust, and an air starting valve. A safety valve remained in the head (figs. 11.1 and 11.2).

Valve actuation was unique. Two eccentric shafts in the crankcase, acting through two-piece push rods, lifted the air start, exhaust, and fuel needle valves (fig. 11.3). Frith chose eccentrics to reduce the usual heavy clacking noise from cams. The intake valve, axially above the exhaust, was "automatic" in that the suction on the intake stroke pulled it open.

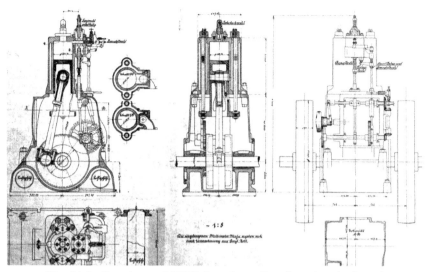

Fig. 11.1. Drawings of 1899 American diesel engine designed by Arthur J. Frith. Fuel pump not shown. *MAN Archives*

Fig. 11.2. 1899 Diesel Motor Co. of America 30 bhp engine. *MAN Archives*

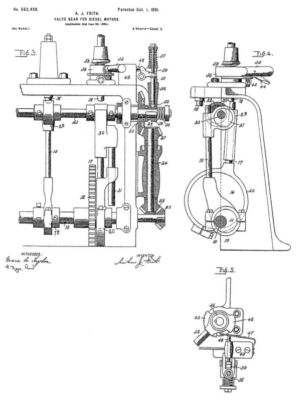

Fig. 11.3. Frith's US Patent **683,459** shows the valve
actuation of his 1899 American engine.

The air and exhaust valves opened in a two-stroke mode for starting;
the fuel and exhaust valves opened in four-stroke when running. Eccentric
shafts, one turning at crank and the other at half speed, acted on the guided,
split pushrods. A hand lever turned the upper part of the air/fuel valve rod
to either disengage a lifting finger under the starting air valve and engage
a finger under the fuel valve, or the converse, so that both could not open
together. The lever also declutched and stopped the upper, half-speed shaft
in a position for the two-stroke mode. When running, the eccentric action of
the two shafts caused the lower rod pieces to lift the upper parts every other
revolution for the four-stroke mode.[3]

Frith's method of valve operation raised an argument between Diesel and
Meier, with the latter receiving these comments from Munich in a letter of
August 31, 1899:

I cannot conceal that I do not consider the valve mechanism as a
simplification of our previous design because the use of six spur gears
and of four eccentrics is definitely more complicated than our four

spur gears and three cams. Besides, I think the lifting of valves by far distant followers a somewhat primitive method, which in particular will not permit high velocities without causing loud clattering of the whole engine. I should be very glad if my fears in both regards would not become true.[4]

Meier wrote this honest but slightly needling assessment to Rudolf over a year later:

As regards the Frith valve gear, you were correct in your criticism that it would not do away with the noise. It has in fact done away with all the noise but one, and that is a pretty bad one, viz. the loud knocking just as the needle valve is raised. But outside of this, the valve gear works much more satisfactorily in daily practice than the original Augsburg gear.[5]

A rationale for placement of the valves away from the cylinder bore was an American concern over a valve dropping onto the piston due to a failure in its retainer or in the valve itself. It is not known how great a problem this was.

The main and connecting rod bearings and piston pin, as well as the eccentric shafts were splash-lubricated from the rod end dipping into the crankcase oil sump. This simple method also came from the Westinghouse design (and was used in millions of Chevrolet automobile engines as late as 1952!).

Most unusual were two, single-stage compressor plungers bolted to a flange cast on the bottom of a split-skirted piston. These reciprocated in bores alongside the cylinder and were cooled by the water-jacketed cylinder. The air discharged into two tanks mounted under the crankcase. A major weakness in the design was that such a rigid assembly allowed no opportunity for the piston or its satellite compressor plungers to align themselves if the bores and moving assembly were not parallel or accurately located. More serious weaknesses having disastrous consequences showed up later.

Fuel metering was based on a Frith and Macpherson patent for a new method of regulating the length of time a pump suction valve remained open.[6] They used a governor to first slidably position a tapered cam on a rotating shaft that directly altered the stroke of the pump plunger (fig. 11.4). Two opposing springs acted against the suction valve located axially above the plunger. One, a weaker spring, held open the suction valve, and the other, stronger spring was between the valve and plunger end opposite the cam. The "stiff" spring when relaxed allowed the plunger and valve to move as a unit as the weak spring held the valve against it. The stiff spring began to compress when the valve seated at some point during the plunger's upward travel. The governor thus controlled the suction valve movement to meter the fuel via the cam, plunger, and stiff spring assembly. Note that the plunger

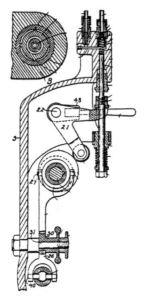

Fig. 11.4. Frith and Macpherson patent 672,477 drawing
showing the fuel pump on the 1899 American diesel.

delivered excess fuel even when at its shortest stroke; regulation was always more a function of the time interval the suction valve remained open to push fuel back into the supply line.

Although the nozzle's horizontal placement was the same as on Hugo Güldner's two-stroke engine at Augsburg, it is safe to assume that Frith and Güldner independently arrived at their ideas. Frith's US patent for horizontally injecting fuel into a cylinder had claims limited to a piston with a "channel" in the crown into which the fuel sprayed.[7] He may have been concerned about insufficient clearance for a spray to enter between the flat crown and cylinder head. The final crown design had only a slight cylindrical concavity.

Frith and Güldner gave their companies a chance to compete as to design and which engine would run first. Meier wrote Diesel that "I will be perfectly frank with you and tell you that I do not consider [Güldner's] an improvement over the original four stroke motor."[8] The Frith engine was first out of the starting block, but neither finished the race.

In February 1899 Meier sent drawings to twenty-four companies asking for quotes to build the engine in lots of ten. One of the few responding was Hewes and Phillips in Newark, New Jersey, who received the order. Diesel was sent a set of drawings in July and he, with Güldner, made an unfavorable, but not entirely inaccurate, critique of what they saw.[9]

An engine built to those drawings ran for the first time at Hewes & Phillips on October 11, almost two years to the day after Busch and Diesel signed their contract. Tests made at the factory in mid-December 1899 show the nominal 30 bhp engine reached a maximum of 41.6 bhp at 170 rpm. (No smoke level was recorded.) Injection air pressure at this output was 65 atm. Mechanical efficiency was 69 percent and remained about the same for a 35 bhp rating. At the lower power and 184 rpm, the bsfc was 249 g/bhp-hr, or very close to that of an Augsburg diesel.[10]

This first engine had been sold in August for $2,400 to Sieb Brothers, a Jersey City firm making wood products. It was with the stipulation that the two Siebs, who took delivery of it at the end of December, keep an accurate record of the engine's performance and deficiencies. V. B. Cowles & Co. Iron Works in Cleveland took the second engine. Number three went to provide standby power in a Sag Harbor, New York, electric plant located on the eastern end of Long Island.

A tale of woe is one way to describe the Sieb Brothers' daily log. During its first month of operation, the engine performed adequately on only one day. Over a nine-month period, there were several recorded failures in the valve train mechanism, but most of the problems centered in the fuel and injection air systems with air line, nozzle needle and spring failures, nozzle packing gland leaks, and so forth. Oil in the air lines and tanks sometimes caused minor explosions. The Cowles engine suffered a broken connecting rod.

Combustion of entrained oil in the air tanks under the crankcase also resulted in catastrophic explosions. At the Cleveland installation, both the original engine and its replacement were destroyed in such blasts. Hugh Meier, E. D.'s nephew, was tragically killed in one of them.[11]

Diesel received a copy of the Siebs' logbook in October 1900, and his analysis only confirmed the obvious. Too many features of the design were failures. Meier sadly concurred and broke the news to Busch that all fourteen engines built to that date should be written off as a loss.

An interim decision by Meier to keep the Sieb engine running was to eliminate the air compressor plungers and tanks from the engine and to remotely drive a two-stage compressor by an auxiliary electric motor. This gave the designers time to gain more experience with the engine itself.

It was now 1901 with no American engine on the market and the company in dire financial straits.

Frith's Reversible Marine Diesel Engine

Paralleling the development of Frith's 30 bhp engine was his reversible 5 bhp prototype marine diesel. Meier referred to it during a conference with US Navy officials in July 1899.[12] He wrote Diesel a week after his meeting that they were very unhappy with the USS *Plunger*, the navy's first submarine built in 1897. "The steam engines and boilers make it so terribly hot that nobody can live in the vessel when she goes under water."[13]

This smaller Frith engine bore characteristics of its untested sibling. Meier said it had operated in a range of 150 to 400 rpm.[14] Little is known of it except for a photo and Frith's patent (figs. 11.5 and 11.6). Claim 1 describes

Fig. 11.5. Frith-designed 5 bhp, reversible prototype
marine diesel that ran in mid-1899. *MAN Archives*

a manual means for admitting one or more charges of compressed air
while [the] fuel valve is out of operation to start the engine in a reverse
direction, and a cam mechanism . . . arranged to automatically shift the
time of operation of the fuel-valve and to interchange the functions of
the air-supply and exhaust valves after the engine has been started by
operation of the . . . compressed-air valve.[15]

The idea was not used again.

A second reversing patent evolving from this engine went to Walter W.
Scott, an engineer under Frith.[16] An axially shifting camshaft with spiral
splines engaged mating internal splines in a fixed sleeve so that a manual,
longitudinal movement of the shaft changed the relative timing of all the
cams to reverse direction.

Hindsight

Hugo Reisinger, secretary-treasurer of the engine company and Busch's son-
in-law, wrote Diesel on December 3, 1900, that "The fiasco of our design
must be attributed to a series of circumstances and mainly to the fact that
amongst our entire technical personnel we had not only no designer but
nobody who knew about practical engineering in general."[17] This damning
indictment by a nontechnical person contained a grain of truth.

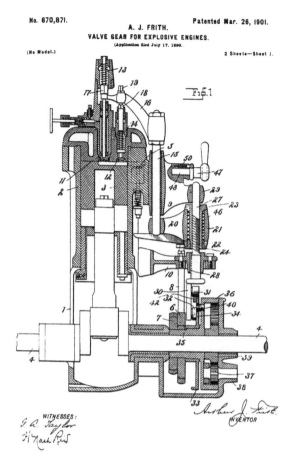

Fig. 11.6. Section drawing of 1899 experimental reversible marine diesel as seen in Frith's US Patent 670,871.

Frith received his "walking papers," effective January 1, 1901. It was foreordained that someone be a scapegoat for the engine debacle, and rightly or wrongly that lot fell to him. After his departure Frith joined the Armour Institute of Technology in Chicago.

In a final report to Meier before departing,[18] Frith told of his continued faith in the ultimate success of the engine and concluded by saying that "I trust that the efforts of those who have labored loyally for the success of this company and unselfishly to accomplish the best results that could be obtained will not be entirely forgotten."

Even Diesel came to Frith's defense in a January 8, 1901, response to news of his departure and final report:

Mr. Frith cannot be blamed for not having had practical experience in this matter because our engine was entirely new to him also.

...Therefore, I cannot refuse him any acknowledgement for his work, and I regret deeply that his industry and devotion were not rewarded with personal success.[19]

Rudolf reiterated in the same letter that the American company had moved too far too fast. They should have followed Augsburg's example and thoroughly tested all new developments before putting them in the field. He conveniently forgot that this was advice he himself had only recently failed to practice or encourage!

Busch called a special stockholders meeting in St. Louis on December 7, 1900, to apprise them of the serious technical and financial difficulties and to reaffirm his desire to forge ahead. He assumed some blame for the situation by admitting he had prevented the designers and builders from collaborating under one roof, where being better able to communicate might have shown up several problems. He also wrongly assumed that only a slight reworking of the European drawings was required.

One meeting action was the creation of a shareholders committee to study manufacturing alternatives. Its recommendation was that a search begin for a facility in which to make and assemble all engines. This could be financed by selling remaining shares in the company treasury. Parts should continue to be bought during the interim, but engine assembly had to be at a site controlled by the company.

Regardless of Busch's display of enthusiasm to the stockholders, both he and Reisinger were becoming more and more discouraged. Busch indicated that he wanted to wait until the European diesels achieved an unqualified success and then he would begin again; meanwhile, he intended to close the American company on May 1, 1901. Reisinger wanted to sell the US patents and abandon the diesel engine business.

Colonel Meier, however, believed they had learned much and that the Frith engine could be sufficiently modified into an acceptable product. In a February 5, 1901, letter to Diesel, he wrote of the decline in problems with the Sieb engine after removing the compressor and lowering output in order to reduce stresses on the fuel system mechanism.

Meier also referred to Diesel's January 8 letter in which Rudolf further criticized "our whole American type." It "created a panic in Mr. Reisinger" and discouraged Busch:

I never, for one moment, have lost faith in the invention. . . . I write you this fully in the hope that after due consideration of the matter, you will yourself be encouraged and recall to a certain extent your first too hasty criticism and condemnation of our type, and thus give us the encouragement which is sorely needed by all our friends, except
Yours very sincerely,
E. D. Meier[20]

Although Busch reluctantly kept open the company doors, a May 1, 1901, letter to all stockholders painted a dismal picture:

In the meantime our President, Mr. Adolphus Busch, has generously advanced the money necessary to pay all current expenses. These have been cut down by giving up our offices and reducing our force to one engineer-draftsman and one erecting engineer. Col. Meier, Engineer-in-Chief, has given his services without compensation since February 1st, while Mr. Reisinger has not drawn any salary for over a year.[21]

Attempts were made at this point to interest other firms in a joint venture with the struggling Busch enterprise. In the fall of 1901, an unsolicited proposal came from The International Power Company offering $250,000 in working capital and complete facilities for making engines. The IPC, incorporated in New Jersey by Joseph H. Hoadley, his brother, and Walter H. Knight, was a holding company either directly or indirectly controlling air compressor and steam engine firms, two of the latter being the Rhode Island Locomotive Works and the Corliss Steam Engine Co. in Providence.[22] Hoadley's entrance on the scene proved to be a very mixed blessing, but it did give Busch needed encouragement to reorganize and have his engineers begin a redesign.

The American Diesel Engine Company

Busch's new venture was the American Diesel Engine Company (ADE) incorporated in New York City on December 2, 1901. Its capitalization of $2,500,000 was underwritten by stockholders from the old company, with both Busch and IPC putting up cash. The same officers continued, but Hoadley and another IPC man sat on the board of directors.[23] A green light to design the new engine came after a manufacturing agreement was signed with IPC on January 8, 1902.

Not everything was resolved on the new design, as in December Diesel sent warnings to Reisinger about what would happen by not "copying to the smallest detail" the new M.A.N. "DM" engine. A December 20 epistle closed by saying that "Without the Augsburg engine example you will not succeed there."[24] In addition to letters, Rudolf gave the same lecture to Theodore H. Macdonald, who had been sent by the IPC factory to study engine manufacturing techniques at Augsburg, Budapest, and Bar-le-Duc.

Diesel's carping prompted a lengthy response from Meier on January 6, 1902, containing these excerpts:

My position was rendered more difficult by the fact that my friend, Mr. H. Reisinger, has no conception of the difficulties attending the development of a new invention. . . . Had you with equal frankness criticized our air pumps, we would not have fallen into that grave mistake. But I am now satisfied that our very troubles will in the end prove to be a great benefit. . . . It is that the injection air must be compressed in two stages, and for larger engines in three stages. . . . While I, as well as the very eminent engineers representing the International Power Co. . . . acknowledge the simplicity of the Augsburg plan we do not consider it sufficiently tested in outside practice to warrant our adopting it. They frankly prefer my plan of compressing the air for injection entirely outside of the cylinder of the main engine. . . . These gentlemen have a long experience of their own, and they will build compressors in the future. . . . Some months ago we sold Messrs. Sieb Bros. of Jersey City, a second engine, and Mr. Reisinger offered to import one from Augsburg. They emphatically refused to accept anything but an American type, just like the first one, i.e., with separate compressor, and this one will be put in within the next few days. From all these facts, my dear Mr. Diesel, you will see that a very good measure of success has attended my efforts, and I would be under obligations to you if you will kindly, on occasion, explain to Mr. Buz that he is in error in condemning, as he did in a recent letter to Mr. Busch, everything we have done here. We have made mistakes, it is true, but no more than were made by various other parties in Germany and France. . . . In regard to the direct adoption of any German engine or machine here, the experience of all who have tried it is against it. . . . Nothing would be more dangerous than to pursue the course you have frequently advocated to Mr. Reisinger, i.e., to place a number of Augsburg built engines in various places in the US . . . It is much better, my dear Mr. Diesel, that we should fully understand each other, and that your highly esteemed friend, Mr. Buz, should appreciate that it was only recognition of difference in conditions, not the vanity of a designer, that induced us to depart from his standard design.[25]

Philosophical and practical arguments continued between the American and German engineers, with often rather scathing comments shared among the American faction. Meier retorted at one point during a March 1902 discussion with IPC people:

While I have the highest opinion of Mr. Diesel's knowledge of thermodynamics, I do not consider him to have sufficient practical ability to build a wheelbarrow. His criticism of our first American engine was childish and absurd.[26]

A more serious circumstance arose soon after the new company went into business. Almost from the start Reisinger and Joseph Hoadley were at loggerheads with each other. Then Meier and Busch developed their own displeasure with Hoadley. The situation came to a head when Hoadley intrigued to wrest control of the diesel business from Busch. Precipitating the ownership crisis was Reisinger's startling action of selling his ADE stock to Hoadley in April 1902 and resigning from the company. Despair over the diesel business and his intense frustration with Hoadley was more than he wanted, and he saw a way out even if it meant Hoadley might succeed.[27] Just after Busch learned of this sale, IPC stock plummeted from $197 per share to $120 on April 30. A *New York Times* article the next day reported that the cause of the fall was not clear, and that Hoadley said his company "in the last few weeks has obtained control of the Diesel Engine Company, . . ."[28]

Busch was distressed by Reisinger's stock sale and incensed over Hoadley's untrue statement. During a May 12 ADE director's meeting, he let them know he was in command and told Hoadley to begin building the by-then redesigned engine. Ever-loyal Meier assumed Reisinger's duties in addition to his role as engineer-in-chief.

The relationship between Hoadley and Busch, Meier and others who followed at ADE became an armed truce for as long as this mismatched group had contractual ties. No one has a plausible answer as to why Busch let this gnat-sized frustration in his elephantine industrial empire continue to irritate him for six years. Meier clearly stated his opinion about how he felt in a September 1909 letter to Busch shortly after the last ties with the IPC group were severed:

> Our great drawback was our connection with the Hoadleys who have been successful speculators but have never created anything, and who lack the foremost qualities of an engineer: truthfulness and honesty.
>
> It has for years been a great task for me to continue in this Diesel work, because I felt that your good name and mine had to be pitted against those thoroughly unreliable scamps, and whatever success we have achieved for the Diesel engine has been in spite of them.[29]

The A Engine

Macpherson, who was given Frith's job, ably completed a redesign of the ill-fated original engine despite raging upper echelon storms. He had moved to Providence where most of the design work was done because the engines were to be built there in the Corliss works. His early relocation was requested to ease the beginning of production. Hoadley meanwhile had formed the American and British Manufacturing Co., an umbrella firm encompassing the Corliss works.

One must emphasize the word "redesign" because many of Frith's ideas went into the new family of "A" engines, which would be sold with only evolutionary changes until the end of 1913. The separate valve housing, enclosed crankcase, and detached compressor were carryover items (fig. 11.7). Early "A" engines also used the "automatic" opening intake valve. Thus, much lived on of the design for which Frith had received censure.

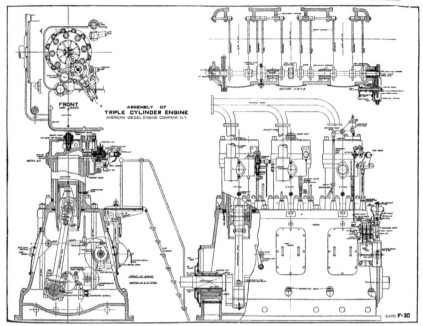

Fig. 11.7. American three-cylinder, 225 bhp "A" engine, c. 1907.
Diederichs, *Internal Combustion Engines, 1910*

Valve actuation switched from eccentrics to cams. Long cam followers pivoting from the opposite side of the crankcase lifted the pushrods. Bell crank linkage atop the fuel pushrod opened the nozzle valve.

Cams on a slidable sleeve, moved by a spring-loaded "starting handle," lifted either the nozzle valve or, when starting, the air inlet valve to crank the engine in a four-stroke mode. For a multicylinder engine only No. 1 cylinder was used. The operator was told to "Open starting air [shutoff] valve and hold the starting handle until you hear one ignition, then release it, and the engine is ready for work after coming to speed."[30]

Priming the fuel lines was necessary if the engine was to run when starting air cranked it over one or two revolutions. To do this a hand crank on the pump was pushed in and turned to rotate the pump shaft and send fuel to opened bypass valves at the nozzles. Pushing in the crank engaged a pinion gear with the governor gear and, by a "single tooth ratchet clutch mechanism," declutched the governor gear from the pump drive gear in the direction the engine turned. The pump gear turned freely on its shaft.[31] The pump was retimed automatically after priming by reengagement of the clutch dog.

The new fuel pump's governor acted directly on the suction valve to override a fixed motion imparted to it through linkage tied to the reciprocating pump plunger. The plungers were displaced by eccentrics on a sleeve keyed to the pump shaft. A circa 1907 pump (fig. 11.8) was little changed from that

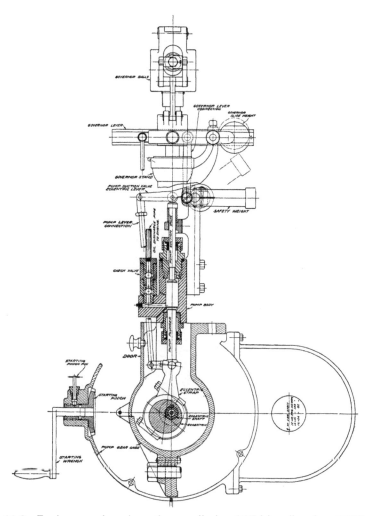

Fig. 11.8. Fuel pump, American three-cylinder, 225 bhp diesel, c. 1907.
Diederichs, *Internal Combustion Engines, 1910*

patented by Macpherson and used on 1902 "A" engines[32] (fig. 11.9). Speed changes caused the governor to shift the fulcrum of a lever acting on the suction valve and alter the interval the valve was open. Sluggish governor action sometimes made speed regulation for electric power generation difficult.[33]

A later "safety device" patent covered a weight acting on the suction valve lever to open the valve and prevent a runaway engine in case the governor failed.[34] One may infer that such a disquieting event had occurred.

The fuel nozzle of the first "A" engines had an atomizer based on the M.A.N. design. Inserted over the valve spindle was a stack of brass washers with drilled holes through which the fuel and blast air passed. Cooling water circulated in a jacket between a sleeve holding the washers and the cast

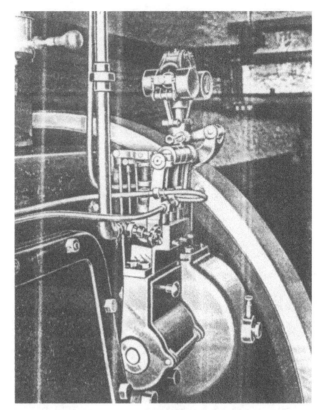

Fig. 11.9. Fuel pump and governor, 1902 three-cylinder American engine.
Power, 1903

housing. A later design used a one-piece inner sleeve whereby fuel entered the air stream just above the valve seat through a series of radial holes (figs. 11.10 and 11.11).

The "automatic" valve actuation was changed early on so that a rocker lever opened the inlet valve, but the dashpot on the top of the valve stem was retained to soften the valve's closing action (fig. 11.12).

Splash lubrication still served the main and connecting rod bearings as well as cams and rollers. The oil level in the crankcase was to be kept about two inches below the center of the crank pin at bottom center. A low-sulfur fuel allowed "several buckets" of water to be added in the "pit" under each cylinder; however, it was cautioned that if the water level got too high "it will work out the main bearing and doors and make a nasty mess on the engine room floor"[35] Consumption of a mineral-based oil averaged "about two pints in four hours."

Fuel specifications called for a minimum of 19° Baume gravity, a flash point of 140 to 240°F, and a maximum of 0.05 percent sulfur and water. Thirty-degree Baume was recommended.

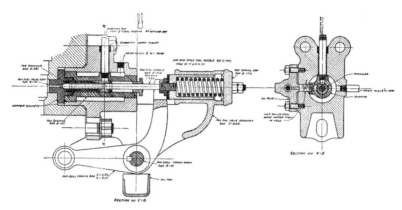

Fig. 11.10. Section through 1907 American fuel nozzle.
Diederichs, *Internal Combustion Engines, 1910*

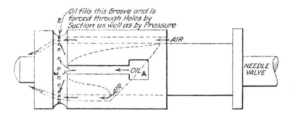

Fig. 11.11. Fuel atomizer, c. 1907. Morrison, *Oil Engines, 1919*

The two-stage compressor for starting and injection air was not an integral part of the engine, but the power to drive it by an electric motor or a belt off the engine flywheel was factored in to give a true brake horsepower and specific fuel consumption.

The model line originally was to include one-, two-, and three-cylinder engines, but no "pairs" were built. The one-cylinder engine, with a 16 × 24 inch bore and stroke, produced 75 bhp at 164 rpm. It weighed 43,000 lbs (the two flywheels weighed 11,800 lbs each), needed a floor space of 9 × 10 ft., and had an overall height of 12 ft. Busch gave IPC an order to build ten one-cylinder engines in May 1902 after successful tests.

A second order for ten 1072 × 15 inch, 75 bhp three-cylinder engines came in July. Twenty of this size were built.[36] Other three-cylinder engines eventually offered included these models:

12 × 18 in	120 bhp	at	220 rpm	33,000 lbs
14 × 21	170		200	60,000
16 × 24	225		164	80,000

More than half sold were the 225 bhp "triple" (fig. 11.13).

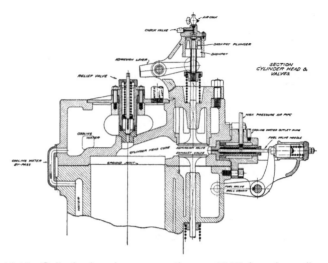

Fig. 11.12. Cylinder head cross-section, c. 1907 American diesel.
Diederichs, *Internal Combustion Engines, 1910*

A December 1906 acceptance test for a three-cylinder, 225 bhp engine gave the following typical results on a heavy oil with a 0.99 sp gr (7.42 lbs/gal) and an LHV of 19,500 Btu/lb. The compressor was belt-driven by the engine:[37]

Load	-1/3	-2/3	Full	12% Over
Test duration, hours	2	2	8	4
Brake hp	90.5	162.9	230.9	258.5
Speed, rpm	171.2	166.8	165.9	165.2
Bsfc, lbs/bhp-hr	0.566	0.481	0.477	0.476
Brake thermal efficiency, %	23.5	27.3	29.5	27.6

Engines burning distillate fuels had a slightly better bsfc averaging about 0.44 to 0.45 lbs/bhp-hr.

Of interest is the power requirement of the compressor:

Engine load	-1/3	-1/2	Full	Overload
Horsepower	13	15	22	25

Such significant numbers again illustrate the need to be sure that performance data on air injection diesels includes the compressor parasitic load. The compressor for many of the larger engines was eventually driven by auxiliary power, and it is not always clear if the stated net brake horsepower was for the engine less the compressor.

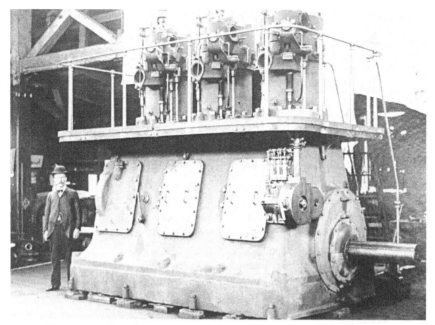

Fig. 11.13. American three-cylinder, 225 bhp A engine, c. 1907.
Diesel papers, Deutsches Museum, Author photo

Busch's Third Diesel Company

American Diesel Engine Co. sales remained disappointingly low. Between its introduction in 1902, when two were sold, and the end of 1907 only 151 "A" engines of all models went out the door. The best year was 1906 with 63; there were 27 in 1907 with a reported profit of $40,000. Only 8 were sold in 1908.[38] Engines often could not be shipped because of poor parts scheduling or defective parts, especially crankshafts.[39] ADE's manufacturing contract with Hoadley, however, kept him strongly in the picture to make most of the engines although others built them over the years. The Power and Mining Machinery Co. of Cudahy, Wisconsin, which was bought by the Worthington Pump and Machinery Company in 1907, had made ADE diesels as early as 1902.[40]

In referring to "the unholy alliance with Hoadley," Meier urged Busch to take legal action against him.[41] He wrote on September 8, 1908, that the bad shafts alone had cost $20,000, but this was

perhaps one-tenth of the indirect loss to us in prestige, in loss of trade, in concessions, etc. . . . Just consider how it would affect you if you had turned over the manufacture of Budweiser beer to some bad stuff that would have entailed a direct loss on you in replacement of $20,000. Would not your prestige for the Budweiser beer have suffered more than $200,000? That is what we have had and still have to contend against, and this fellow has the effrontery to place his claims as equal to ours.

Busch dispensed with Hoadley, albeit with an outlay of cash, by placing ADE in receivership on December 21, 1908. On the surface it was a consequence of the year's poor sales resulting from the recession of 1907. Such an action, however, provided him with a way to cancel his contract with Hoadley.

Busch had taken great pains to prevent him from learning that he was negotiating with Diesel and the Sulzers at the time with a hope of bringing in capital and needed new technology.

These talks first surfaced in a detailed, handwritten letter of August 28, 1908, to Meier from Busch at his German summer home.[42] He wrote of a meeting with Diesel about new engines and the formation of a company with himself, Diesel, and the Sulzers as principal stockholders. On September 8 Busch and Diesel signed an agreement involving the reorganization of the ADE by which "The American Diesel Engine Co. places its entire business, without restriction, into the new combination, while the syndicate (Consortium) Diesel-Sulzer places its entire assets in present and future patents and experiences regarding the internal combustion engine."[43]

Busch then bought the remaining assets of the old, still struggling ADE for $110,000, as soon as legally permissible, and became its sole proprietor. Accompanying its new identity of "Adolphus Busch, Purchaser of the American Diesel Engine Company" were management changes and a move to St. Louis. Failing health had persuaded Adolphus to turn the president's reins over to his son August Anheuser Busch (1865–1934). James R. Harris, general auditor of the Anheuser-Busch Brewing Association, was vice president and general manager. Meier stayed on in New York City as a deeply involved consultant.

August had not been enamored with the diesel business and in the beginning saw little future for it. His father at times had to generate enthusiasm in his son about its potential. Harris performed his duties diligently but with an accountant's view on how to run a business. He and Meier, who suffered the diminished influence of a semi-retired consultant, mainly disagreed over pricing policy; Meier's correspondence to Adolphus after the latter's move to Pasadena attests to this. The new group, including Meier and the Busches, nevertheless, all pulled together.

ADE versus the Allgemeine

An ongoing dispute between the American Diesel Engine and the Allgemeine Gesellschaft für Dieselmotoren in Augsburg began shortly after the formation of ADE in 1902 when the Allgemeine made requests for ADE to pay it a purportedly owed six percent royalty on sales. Meier, on the other hand, firmly believed that the unexpectedly high development costs and resultant heavy financial losses were caused by Diesel's premature release of the engine and reason enough to delay payments. Patent disagreements added further distance between the two parties in what was to become a tangled legal feud.[44]

In late 1904 Berthold Bing wrote Busch that the royalty matter could be settled if ADE paid 100,000 marks (about $24,000) in three annual installments.[45] Unbeknown to ADE were concurrent dealings between Diesel and the Allgemeine leading to his release from the contract he had with it (chapter 7). At this point waters became muddied over who was obligated to whom. Diesel on the one hand claimed a freedom to do as he wished with his patents; the Allgemeine said that Busch's Company had rights to future Diesel patents.[46] Busch and Meier were upset and refused to pay the Allgemeine the first 20,000 marks installment on the royalty settlement.[47]

After severing his ties with the Allgemeine, Rudolf began driving a deeper wedge between it and ADE. In a "strictly confidential" November 17, 1906, letter to Meier he suggests a way to place the Allgemeine in a position that prevents it from meeting contract terms.

> I would, for instance *urgently* advise you to demand from the "Allgemeine" the delivery of the shop drawings of the submarine boat engines, as they are at present built by the Société Française at Bar-le-Duc for the French Navy. As the "Allgemeine" is not in a position to furnish you these drawings, you will have a further reason to support your standpoint to be released also on your part of your obligations. The "Allgemeine" will no doubt offer to furnish you in lieu of the French drawings a drawing of a German marine engine under the pretext that it is the same engine. I can, however, only advise you, not to accept this substitute, but simply and strictly to demand a copy of the original French drawings. If you should accept a substitute, you would only weaken your standpoint.[48]

The ADE used this ploy as justification of its position to delay making the owed payment, and Diesel exacted a bit of retribution from the Allgemeine. More important was his desire to promote himself into the proposed US syndicate using his new ideas for locomotive, truck, and submarine diesels, in addition to the standard ADE stationary models already in production.[49]

While Busch, Diesel, and the Sulzers grew closer inch by agonizing inch to their own tripartite pact, the Allgemeine "affair" remained at a stalemate. Maybe it was a deeper awareness of his own mortality that caused Busch to soften his stand against the Allgemeine. His October 1909 message to Bing provides a window to observe this change of heart:

> Naturally I am entirely of your opinion to prefer a compromise to a prolonged suit. Besides Mr. Buz's friendship is dear to me, and I will willingly sacrifice thousands of marks, to maintain it.[50]

He went on to say that even though his attorneys told him he would win any suit to recover his original 40,000 marks he wanted to settle. This he did by paying another 25,000 marks on October 21, and the Allgemeine certified on November 11, 1909, that all US obligations to it were fulfilled and all ties severed.

Another year and a half passed before final agreement was reached between Diesel and Busch. Sulzer Bros. and Busch had their contract on December 7, 1910,[51] yet Diesel, no doubt wary after his Allgemeine experience, delayed things until July 12, 1911. He had signed a preliminary agreement on December 10, 1910, that Busch and the Sulzers thought would become final, but his concern over a few clauses frustratingly postponed his needed assent to form the new Busch and Sulzer enterprise.

Max Rotter (b. 1867), who had been chief engineer at Allis-Chalmers in Milwaukee, was to assume that same role at the new St. Louis company.[52] The immediate need of an experienced engineer resulted from J. D. Macpherson's debilitating illness, which struck him during a European trip in August 1910.[53]

Rotter went directly to Europe in the early spring of 1911 before starting in St. Louis. His objectives were to help Diesel quickly overcome the contract difficulties and to study the Sulzer models to be built in the United States. An anticipated brief stay lasted instead about three months because of Rudolf's seeming intransigence. Rotter used the time to visit other engine companies, including Nürnberg, Diesels Motorer in Stockholm, as well as an exposition in Turin where European diesels were on display. His frequent reports to Harris became more and more pointed about the need to begin work in St. Louis. At last, a July 15, 1911, cablegram summed up his relief after receiving Diesel's signed agreement:

DIESEL SETTLED[.] ENTIRE APPROVAL SULZERS[.]
SAILING NINETEENTH KRONPRINZ WILHELM
SINGING DOXOLOGY. ROTTER[54]

One may assume a chorus of praise echoed from St. Louis.

The Busch-Sulzer Bros. Diesel Engine Co. had already been incorporated in St. Louis the previous January 26 in anticipation of Rudolf's imminent joining, along with a commitment to build a factory. Busch contributed his entire ADE interests; the Sulzers assigned their US and Canadian patent rights and agreed to supply drawings for both two-stroke and four-stroke diesel engines. By means of the delayed July 12 contract Diesel canceled previous agreements with Busch and provided all his US patents. None of the three put up cash.[55]

With the stage set for a second act in the American diesel engine drama, what had taken place in the first? By the end of 1912 only 260 "A" engines of all models had been delivered. Thirty-six came in that last year of its production.[56] Fourteen irretrievable years passed without building engines in a facility controlled by the company marketing them. Country-wide sales efforts did yet not exist. The grand plan for a group of engine manufacturers under the control of, and building to the designs of, the American Diesel Engine Co. never materialized. Of more significance was that diesel technology in the United States lagged behind Europe by five years at the very least. The failure to achieve greater success cannot all be attached to Busch, and to Meier, but their business philosophy was wrong for the product they wished to market. Diesel erred by selling his rights for a major territory without first knowing how the licensee would exploit that area's potential.

Dr. Paul Rieppel, son of Anton Rieppel, would also be involved in the new Busch-Sulzer-Diesel enterprise. He had come to the US in 1907 to assist Allis-Chalmers with the building of the huge Nürnberg gas engines after A-C had bought a license from M.A.N. Disagreements arose between the two companies, and Rieppel left in 1908 but not before he became acquainted with Meier and the US diesel engine program.[57]

Plans were eventually made by Busch for the young engineer to work with Diesel for a while and then join ADE when Busch-Sulzer was formed. However, the long delay in creating the company made it impossible for Rieppel to come back to the US. Instead, he remained in Germany and worked for Blohm and Voss on large, Nürnberg-licensed marine engines. Rieppel later commented on his first meeting and experiences with Diesel:

> I learned how Diesel's theory was soon abandoned, and . . . how Diesel was forced down from his ideals into practical possibilities. . . . Still the impression grew in me that the final success was due to Diesel's strong creative will, which offset his lack of knowledge and his inexperience in design.
>
> . . . It was a very impressive moment for me when Diesel shook my hand in his friendly but slightly reserved manner. This slim man, his fine head and beautiful hands and his dignity and modesty gave no indication of his world-wide fame. Lively in his movements and concentrated in his talk, he exerted a suggestive force hard to resist. His balance of mind and body and his easy manner gave the impression of great strength.
>
> . . . I started with this work in Munich in the summer of 1908. Diesel owned one of the most beautiful houses in Munich, on the second floor of which he had installed a drafting room. Diesel's private office was on the ground floor next to the entrance. . . . Diesel came upstairs every day, when in town, and gave us his ideas and plans, discussed our work and often also general subjects. He did not share my opinion that simple sturdy engines were wanted for America. He proposed compound engines and small engines for automobiles, even airplanes. . . . He could not see the necessity for overall success with all its factors, reliability,

upkeep, life, he saw only the one factor, namely thermal efficiency. This led to much argument and disappointment on my part.

. . . In spite of the limitations in his deductions Diesel succeeded through his unusual smartness, his superior character and his tenacity. . . .[58]

A factory was finally built in St. Louis during 1912, to which Rudolf would come for the groundbreaking ceremony. Except for a very few more "A" engines, production began with Sulzer designs. The fast-growing need for submarine engines pushed Busch-Sulzer into supplying almost entirely these rather than stationary models. Nothing came of the "Diesel Syndicate" or the rail and road engines. The time had passed for such an idea, and land transport diesels were still years away.

With the expiration of Rudolf Diesel's first US patents as early as 1906, Busch-Sulzer Bros. would begin facing inexorable competition from other American companies.[59] It came slowly, but it was unstoppable.

Reflections

Busch held little in common with the first European licensees who signed with Diesel. Except for potential financial resources, his company was weak in critical areas where his German, Swedish, and Russian counterparts were for the most part strong:

- A depth of creative talent in product development.
- A solid commitment to become an engine maker.
- A willingness to venture into untapped markets.
- A workable relationship between licensor and licensee.

Busch's delay in becoming a producer was a serious error. His avoidance of a brick-and-mortar commitment cost far more in the end than if a manufacturing plan had been implemented at the outset. The conviction of the need for an "American" engine was correct, but the chance to capitalize on it was forfeited by not having design, test, and factory people working in close cooperation under one central authority. This decentralization severely restricted an ability to offer new technology in improved products. It was not so with the European brethren.

The American company's lost opportunities over many years cost it a leadership role and relegated it to being no more than another modest but respected builder among a growing industry.

A thought on technology transfer:

The idea of merely translating Sulzer's drawings into Christian and then building in accordance with same will be out of the question.
—Max Rotter to Harris, 1911[60]

Notes

1. October 5, 1900, letter, Edward D. Meier to Rudolf Diesel. MAN Archives.
2. This was done by Richard H. Lytle, "The Introduction of Diesel Power in the United States, 1897–1912," in *Business Review History* 42 (Summer 1968): 115–48. It is based on his master's thesis, "History of the Busch Diesel Companies, 1897–1922," (Washington University, June 1962). Both are valuable sources but do not delve into the engines' technical features.
3. Arthur J. Frith. 1901. Valve Gear for Diesel Motor. US Patent 683,459, filed June 22, 1899, and issued October 1, 1901.
4. Rudolf Diesel to Meier as quoted in Eugen Diesel & Georg Strößner, *Kampf um eine Maschine* (Berlin: Erich Schmidt, 1950), 97.
5. Meier to Diesel, October 5, 1900.
6. Arthur J. Frith and James D. Macpherson. 1901. Oil Pump for Explosive Engines. US Patent 672,477, filed September 21, 1899, and issued April 23, 1901. It built on Diesel's basic US fuel system Patent 654,140 controlling the suction valve duration of opening. There was no German patent on the idea. A detailed analysis of this and other US patents by Diesel, Frith, Macpherson, Lauster, et al., is in a March 15, 1907, report to Meier by Wetmore & Jenner, Busch's New York patent attorneys. Busch-Sulzer Co. papers, Wisconsin State Historical Library.
7. Arthur J. Frith. 1900. Internal Combustion Engine. US Patent 644,798, filed October 23, 1899, and issued March 6, 1900.
8. Meier to Diesel, October 5, 1900, Meier did not rule out the two-stroke diesel engine for marine use, but believed its time had not yet come.
9. Drawings for the new engine accompanied Meier's July 31, 1899, letter to Diesel. MAN Archives. A diary account of problems encountered by the first engines is in Diesel & Strößner, *Kampf*, 113–19.
10. "Allgemeine Report No. 183, end of Feb. 1900," 3, on the December 1899 test results reported by the US company. The Augsburg report also states that the US tests ended any fears about unsymmetrical thermal expansion and lubrication problems due to a long trunk piston. The comments confirm that the American engine was the first to run without a crosshead. MAN Archives.
11. Diesel and Strößner, *Kampf*, 119.
12. "Stenographer's Report of Examination of Col. E. D. Meier, before Committee of Naval Engineers, on the Substitution of Diesel Motors for steam power, in Holland Torpedo boats," July 25, 1899, Manhanset House, Shelter Island, New York, 27 pages, Wisconsin State Historical Library.
13. July 31, 1899, letter, Meier to Diesel. MAN Archives. The *Plunger*, with a surface displacement of 154 tons, was built by the J. P. Holland Torpedo Boat Co. at Elizabethport, New Jersey, in 1897 under an 1895 US Navy contract. William Hovgaard, *Modern History of Warships* (Annapolis, MD: US Naval Institute, 1971) 291. The sub was repowered by an Otto Gas Engine Co. gasoline engine.
14. "Stenographer's Report," 12.
15. Arthur J. Frith. 1901. US Patent 670,871 filed July 17, 1899, and issued March 26, 1901.
16. W. W. Scott. 1901. Valve-reversing mechanism for Explosive Engines. US Patent 682,757, filed November 20, 1899, and issued September 17, 1901. Meier wrote in his October 5, 1900, letter to Diesel that Scott's method was similar to one by Güldner in that both used spiral gearing, but Scott did it "internally" and Güldner used external gears. Güldner's idea read on an April 5, 1888, patent issued to E. D. Levitt and consequently was not patented in the US.
17. Reisinger to Diesel, December 10, 1900, as quoted in Diesel and Strößner, *Kampf*, 120.
18. Frith's December 10, 1900, partially quoted report. Ibid., 123.
19. Diesel to Meier, January 8, 1901.
20. Ibid., 124.
21. Quoted from a printed letter. Ibid., 133.
22. Lytle, "Intro. of Diesel Power," 126. Hoadley, more speculator than inventor, and Knight, who was an inventor, had teamed up in 1899 to promote their patented compressed air system.
23. Reisinger sent a printed letter to all stockholders of the Diesel Motor Co. of America, dated October 1, 1901, giving full details of the proposed new company. Wisconsin State Historical Library.
24. Diesel letters to Reisinger of December 7 and 20, 1901. Walter Knight at IPC was also perturbed by Diesel's letters and wrote Reisinger on December 30, 1901, that "the general

inference from Mr. Diesel's [Dec. 7] letter is that his engine does not follow the ordinary laws of mechanics, . . . and must have special dispensation from Augsburg before it can be made to feel the spirit of successful life." He closed by saying that if they did not know otherwise they would "consider Mr. Diesel's letter a fatal stab at his own child." MAN Archives.

25. January 6, 1902, letter, Meier to Diesel. MAN Archives.
26. Meier's "Notes of an interview between J. H. Hoadley, Walter Knight, and E. D. Meier about February or March 1902," 3. Wisconsin State Historical Library.
27. The outcome was too close for Busch's comfort as the final tally gave the Busch interests 6,076 shares, the Hoadleys 5,289 shares, and the company treasury 710 shares. "Statement of Stock & Subscription Accounts A.D.E. Co. June 1st 1903." Wisconsin State Historical Library.
28. *New York Times* (May 1, 1902), 5. *Times* articles on May 2, 3, 5, and 25 reported that the loss resulted from insider manipulation, which forced departures at IPC. The Hoadleys were not involved in the scandal.
29. Meier to Busch, September 27, 1909, letter. Wisconsin State Historical Library.
30. American Diesel Engine Co., *Oil Engines*, Instruction manual (c. 1907), 21. Wisconsin State Historical Library.
31. James D. Macpherson. 1908. US Patent 898,124, originally filed April 7, 1905, divided August 25, 1906, and issued September 8, 1908.
32. James D. Macpherson. 1908. US Patent 890,673, filed April 7, 1905, and issued June 16, 1908.
33. Lacey H. Morrison, *Oil Engines* (New York: McGraw-Hill, 1919), 160.
34. James D. Macpherson. 1908. US Patent 890,674, filed April 7, 1906, and issued June 16, 1908.
35. ADE Instruction manual, 23.
36. A "Memorandum of Engines ordered, delivered, and in stock or in course of construction," (March 21, 1905). Wisconsin State Historical Library. A three-cylinder, 10.5 × 15 inch engine was featured in *Power* (April 1903), 160–65, although no performance data was given.
37. Hugo Güldner, *The Design and Construction of Internal-Combustion Engines*, trans. H. Diedrichs (New York: Van Nostrand, 1910), 437.
38. Lytle, "Intro. of Diesel Power," 141. The 260 engines had a cumulative total of 47,000 hp at a selling price of about $55/hp. Estimated annual profits and losses from 1898 to 1912 are given on page 142. Only for 1906 and 1912 are the figures from official statements.
39. Meier to Busch (September 8, 1908), 2, 3. Wisconsin State Historical Library.
40. *The Worthington Double-Acting Two-Cycle Diesel Engine*, Bulletin S-173 (July 1927), 3. Meier to Busch, March 15, 1907, 5: Power & Mining Machinery could not make more than the two per month specified in the contract because of other manufacturing commitments. ADE production record, "Memorandum," does not specify where pre-1906 engines were built. Also, N.M.S.(?) at ADE to Max Rotter, December 14, 1910: ". . . larger engines [225 bhp] are being built by the Power & Mining Machinery Co. at Cudahy and the other engines by the American & British Mfg. Co. at Providence [Hoadley]." Wisconsin State Historical Library.
41. Meier to Busch (September 8, 1908), 2, 3. Wisconsin State Historical Library.
42. Busch to Meier (August 28, 1908), letter from "Villa Lilly," 2–4. Wisconsin State Historical Library. The new company would have others make and sell engines and pay a royalty at "so much per horsepower." It was deemed necessary to promote locomotive diesels to the Standard Oil Co. and show "them the enormous increase in consumption of coal oil and especially of all kinds of refuse crude oil . . . which they can not easily dispose of otherwise." Baldwin Locomotive Works and the American Locomotive Works will, according to Diesel, "cheerfully" join in. Submarine engines would also be an important product, and because "our good friend, Mr. H. H. Taft, will be elected President . . . we shall have a good chance, with the new Secretary of the Navy, to give our new submarine engine due care and consideration."
43. Agreement for the "Reorganization of the American Diesel Engine Company in New York," Clause 1, translation. Wisconsin State Historical Library.
44. The Allgemeine to ADE, September 15, 1903, extract. Wisconsin State Historical Library. ADE was told it only held nonexclusive rights to all US engine-related patents

taken out by other licensees, but still had exclusive use of Diesel and Allgemeine patents. A precedent for this had already been set when ADE agreed to a nonexclusive use of Lauster's US Patent 729,613 owned by M.A.N.

45. Minutes of "Meeting of Executive Committee at Mr. Adolphus Busch's Office" (St. Louis: November 12, 1904). Wisconsin State Historical Library.

46. Guggenheimer to Busch, translated copy, (October 28, 1905). Wisconsin State Historical Library.

47. A chronology and legal summary of all documents relating to the disputes, and the later status between Diesel and the American company, are in a June 1, 1910, report to Busch from William A. Jenner, the ADE attorney. A second legal opinion covering the same ground was made by Dr. Johannes Junck, a retired judge and Leipzig cousin of Meier's, dated July 10, 1910, trans. by Meier. Wisconsin State Historical Library. Jenner and Junck agreed that Diesel was obliged to give Busch's company his patents. However, by this time other events made the opinions a moot point.

48. Diesel to Meier (November 17, 1906), 3. Wisconsin State Historical Library.

49. Busch to Meier (August 28, 1908). Wisconsin State Historical Library.

50. Busch to Bing (October 11, 1909). Trans. in Wisconsin State Historical Library.

51. Dr. Hans Sulzer reported to Dr. E. Sulzer-Ziegler in an August 4, 1910, memo about his discussion with J. R. Harris in which all were in agreement that Diesel was only lending his name, and no one should expect practical and tested designs. Sulzer Archives.

52. Rotter was born in England and educated at the City of London School. He came to the US in 1891 and went to work for Fraser & Chalmers, a predecessor company of Allis-Chalmers. Rotter retired from Busch-Sulzer as vice president of Engineering sometime in the 1930s. *Trans.*, ASME, OGP-53-9 (1930), 117.

53. Macpherson was stricken after a reported "bottle of Moselle and copious champagne" from which he never again recovered his mental faculties. Trans. of "Medical Certificate" from the Psychiatric Klinik, Imperial University, Munich, August 7, 1910. He was reexamined in Paris and certified to have "commenced a general paralysis of the insane." James R. Harris, who had been with Macpherson, stayed with him in Paris for several weeks until a relative arrived who shortly thereafter brought him to New York. Macpherson died less than a year later. The file on this tragic episode is in the Wisconsin State Historical Library.

54. Rotter to Harris, cablegram, July 15, 1911. Wisconsin State Historical Library.

55. "Minutes of Special Directors' Meeting of the Busch-Sulzer Bros.-Diesel Engine Company, September 25, 1911." Wisconsin State Historical Library. Among the listed directors were Adolphus and August Busch, E. D. Meier, Rudolf Diesel, and Robert Sulzer who, with the exception of August, were not in attendance. The minutes show that the Sulzers received stock worth $250,000 and Diesel stock worth $50,000 for their contributions. Busch interest was valued at $1,000,000. Total authorized capital was $2,100,000, of which $850,000 was in preferred stock.

56. Lytle, "Intro, of Diesel Power," 141.

57. The unpleasantness was mainly over royalty matters. Meier to Diesel, November 13, 1908. Wisconsin State Historical Library.

58. Rieppel's lengthy, quoted interview is from a story by John Anderson in "Diesel, Fifty Years of Progress," *Diesel Progress* (May 1948): 73–74.

59. US Patent 542,846 of July 16, 1895, expired almost six years early on April 14, 1906, "coincidentally with the expiration of the British patent on the same invention and by reason of the limitation of the term of the American patent by the term of the British patent." Reissued US Patent 11,900 (original date Aug. 9, 1898, and reissued April 2, 1901) had expired on February 27, 1909, "to correspond with the British patent dated February 27, 1895. The term of these patents was fifteen years." Letter, Wetmore & Jenner to Meier, March, 15, 1907. Wisconsin State Historical Library.

60. Max Rotter to J. R. Harris (May 11, 1911), letter, 3. Wisconsin State Historical Library.

CHAPTER 12

The Prudent Ones

"I learned of the work done by [Mirrlees] on the Diesel engine, but was warned against having anything to do with it, as the engine had caused heavy losses and was considered dangerous."

—Charles Day[1]

The diesel's improved reliability after 1902 revived earlier enthusiasm among those first licensees who had set aside their plans for production after the engine's disappointing debut. A few others also came aboard. Several of these hesitant, or perhaps in hindsight, prudent firms that Diesel himself had mostly enlisted would become strong competitors of the stalwarts who started early and stayed the course.

Later, when patent protection ended for the sole American licensee, some of these established European builders saw a new opportunity and sold manufacturing rights to US businesses.

The diesel's future was secure.

Enter Gebrüder Sulzer

Sulzer Brothers, who had owned an option for a license since 1893, did not begin production in 1898. The Swiss were unsure about a market and the engine's state of development. This latter concern was borne out by its ensuing travails. Rudolf was greatly disappointed over the postponement because he held high hopes for this company that he had felt close to for so many years. Even an offer to give it his Russian patents went for naught.

The Sulzer family's long tradition of proven, quality products favored a more cautious business approach. Company roots went back to 1775 when a brass foundry was opened by Salomon Sulzer (1751–1807). It became an iron foundry in 1834, and during the 1840s his grandsons Johann-Jakob

Fig. 12.1. Jakob Sulzer-Imhoof (1855–1922). *Sulzer Archives*

Sulzer-Hirzel (1806–1883) and Salomon Sulzer-Sulzer (1809–1869) began making textile machinery.[2] Steam engines were added in the next decade, and centrifugal pumps and fans came along in the 1860s. By the 1880s Sulzer poppet valve steam engines set performance standards for the world. Linde's refrigeration machinery that had launched Diesel's professional career was produced at Winterthur after 1877. A second Sulzer factory, which opened at Ludwigshafen, Germany, in 1881, would also become a major supplier of Sulzer diesels.[3] By 1883 the Winterthur branch alone had 1,500 workers to make it one of the largest factories in Switzerland.

Jakob Sulzer-Imhoof (1855–1922), Salomon Sulzer-Sulzer's son, had been closely involved with the engine since his friend Rudolf introduced the concept to him. It was Sulzer-Imhoof (fig. 12.1) who held back on the manufacturing even though his company had to continue paying Diesel 20,000 marks per year if it failed to build engines after the successful acceptance test.

Two contract modifications were agreed to by the Allgemeine in December 1898, however. Sulzer's original 40 percent royalty payment was lowered to 15 percent, and until Winterthur made engines, M.A.N. could sell in Switzerland for a 15 percent royalty, of which 5 went to Sulzer and 10 to the Allgemeine.[4]

Sulzer-Imhoof firmly believed the engine's true future lay in the higher horsepower range. This contrasted with Diesel's early expectation that it would carve a power niche usurped instead by electric motors. Jakob wrote Rudolf in September 1897 he had told Maschinenfabrik Augsburg some time previously "that we are thinking of the large engines which are not yet being built."[5]

The Sulzers began, however, with a 20 bhp Series XVI 260/410 diesel in Winterthur, which went on test in June 1898. Although its poor performance was given as the chief reason to delay production, the development work continued. These efforts were to pay off.

The Diesel Engine Company in London had been buying its engines from M.A.N. for resale, and when Sulzer gained confidence in their own engine, they decided to compete for this business. First, on April 25, 1903, the Allgemeine granted Sulzer exclusive Swiss rights for a 6 percent royalty. This royalty also applied to countries where no patents existed. Of greater potential was a granting of export rights into countries where licensees already existed—if the licensees agreed. The Sulzers paid 60,000 marks up front for this, to be followed by a 3 percent royalty on engines imported into countries where there were patents. Great Britain was an important exception. Engines bought by the DEC were royalty free to Sulzer because of the Allgemeine's majority ownership of DEC.[6]

Gebrüder Sulzer had received an order for twelve four-stroke engines of 35 bhp each from the DEC on January 6, 1903, and such a large, confirmed sale no doubt hastened the signing of the April contract with the Allgemeine.

The first production engine, a single-cylinder model 1D40, was delivered to England in March 1904.[7] Its 310 × 460 mm bore and stroke had an output of 40 bhp at 190 rpm and a bmep of 5.46 kg/cm². Descendants of the original D Series were sold as late as the 1920s[8] (fig. 12.2).

Much of the engine's conservative design was borrowed from the M.A.N. DM Series, including the first-stage injection air coming from the power cylinder and the compressor for the second stage operated by linkage off the connecting rod. Its governor and fuel pump, however, were based on Hesselman's K Series design where a governor-positioned eccentric on the vertical drive shaft for the cams controlled the pump suction valve. The combustion chamber also resembled the Swedish configuration with its dished piston crown except for the added conical spray diffuser.

Thirty one-, two-, and three-cylinder Sulzer diesels of 40 bhp per cylinder, with a cumulative output of about 2,000 bhp, had gone to Britain by the end of 1904. The first of the few staying in Switzerland was a three-cylinder of 120 bhp delivered in September 1904. That same year a two-cylinder, 40 bhp (2D20) diesel with the Del Proposto reversing gear went into a Lake Geneva freight boat. Through December 1905 Sulzer had delivered 226 cylinders worth of engines, totaling 10,323 bhp. A year later this grew to 507 cylinders and 29,419 bhp. By the end of 1908, the numbers reached 837 cylinders and 51,710 bhp.[9] There were twelve cylinder sizes with outputs from 20 to 200 bhp, the latter being a three-cylinder engine rated at 600 bhp.

Max Rotter reported on a 3D200 he saw in May 1911. Since its installation in late 1908, the engine had operated an average of twenty hours per day to pump water from Lake Constance to St. Gallen some 12 kilometers away. It often ran continuously for several days a week with its belt driven pump working against a 340 m hydraulic head. Exhaust valves were cleaned every two weeks, and one cylinder liner had been

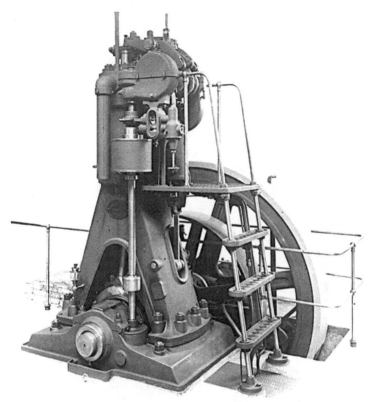

Fig. 12.2. First Sulzer 1D40 production prototype built in 1903.
Sulzer Archives

replaced due to scoring by loose rings. Rotter wrote that the "engine runs well and fairly quietly; but there is considerable oil leakage about same."[10]

A 3D100 Rotter saw in a wool spinning mill ran ten hours per day for two and a half years at an average load of 250 bhp and a peak of 330 bhp. Bearing lube oil consumption came to 1 liter per day while the cylinders used 3.5 liters per day. The lube oil filter was cleaned weekly, the valves monthly, the injection pump every six months, and the pistons once a year. Cooling water and exhaust passed through heat exchangers in series to heat water for washing the wool. Up to 4.5 m³/hr of water was warmed to 75°C.[11]

Fuel consumption for a three-cylinder, 400 bhp "D" engine was:[12]

Power	Full	-3/4	-1/2
Bsfc, tar oil, g/bhp-hr	197	203	213
Bsfc, gas oil, g/bhp-hr	175	188	189

The gas-oil fuel had an LHV of 10,000 cal/kg.

The Next Generation K Series

Sulzer introduced the updated K Series for stationary applications in April 1910. These departed from the A-frame of the D Series by having individual cylinder assemblies bolted to a box frame that enclosed the crankcase (fig. 12.3). Soon-to-follow marine engines also adopted this basic structure. Sulzer offered a "K" of 1,000 bhp at 187 rpm in January of 1913.

Both the bore and stroke of the "K" were considerably smaller, and it turned much faster than the "D" of comparable outputs per cylinder:

	K		D
50 bhp	310 x 350	mm, 333 rpm	340 x 510 mm, 190 rpm
100 bhp		—	450 x 660 mm, 187 rpm
125 bhp	450 x 510	mm, 250 rpm	—
200 bhp		—	600 x 840 mm, 150 rpm
250 bhp	640 x 760	mm, 187 rpm	—

All of the above were available in at least four-cylinder versions.

Fig. 12.3. 1910 Sulzer 4K50, 200 bhp stationary diesel.

Also borrowed from the D Series was a three-stage compressor driven off the crankshaft at the front of the engine. The crank and con rod bearings were pressure lubricated by an engine-driven oil pump. Drilled passages in the crank fed the rod bearings, but cylinder oil still came from a separate source. Contemporary "D" engines incorporated these same features.

Control of injection air pressure was automatic as loads varied. It had always been necessary to modulate this pressure, mostly by hand on older engines, to avoid a misfire or delayed ignition as the load decreased. If the blast air pressure required for good atomization at higher loads remained the same at idle and light loads, there was a strong tendency for the nozzle tip to be emptied of all fuel. This kept a normally retained residue from acting as the first fuel to enter on the next cycle and ensuring a sustained and predictable injection rate. Excess blast air would also cool down the cylinder air, which further increased ignition difficulties.

Sulzer controlled the injection air pressure by making the governor throttle the compressor's suction valve as well as the suction valve of the fuel pump. Maximum air pressure averaged around 60 atm.

An English diesel authority reported that the average bsfc of the K Series ran 5 or more percent higher than that of the slower running "D." Similarly, lube oil consumption was 0.015 to 0.02 lbs/bhp-hr versus the "D" with 0.01 to 0.015 lbs/bhp-hr.[13]

Sulzer Two-Stroke Stationary Diesels

As four-stroke sales grew, the Winterthur engineers also saw great promise in two-stroke diesels for both stationary and marine applications. Their initial effort in this direction led to a 90 bhp, reversible marine engine built in 1905 (chapter 15).

The first stationary two-stroke diesel, with a structure based on their trunk piston D Series, was developed during 1906 and 1907 and delivered to the Aarau, Switzerland, electric works in 1908. This 3Z133 (Z for *Zweitakt* or two-stroke) with three cylinders of 500 × 720 mm produced 750 bhp at 144 rpm.[14]

Air entered through intake valves in the cylinder head, and the piston uncovered exhaust ports located circumferentially around the cylinder to provide uniflow scavenging. In the head were four intake valves in cage assemblies, an injector nozzle, and an air starting valve, all cam actuated (fig. 12.4). The intake was timed to allow the pressurized scavenge air to slightly supercharge the cylinder. A double-acting scavenge air pump and a two-stage blast air compressor were both driven off an added crank throw at the front of the engine.

Piston cooling was essential as the 3Z133 had almost twice the power of its comparable displacement four-stroke. Telescoping tubes without seals carried water to and from the underside of the piston crown. Stationary tubes encasing those fastened to the piston caught the leakage and thus obviated a need for packing glands (fig. 12.5). This design was used for years.

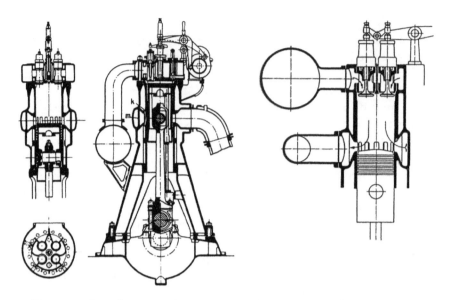

Fig. 12.4. First Sulzer stationery two-stroke, 1907. 750 bhp at 150 rpm.
Sulzer Bros. photo

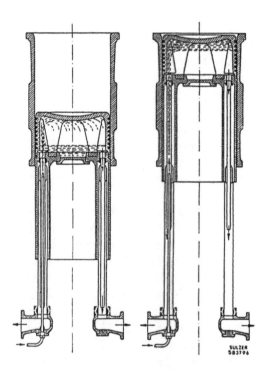

Fig. 12.5. Telescoping piston cooling tubes, first Sulzer two-stroke, 1907.
Sulzer Archives

Fuel consumption of the uniflow two-stroke Z Series was comparable to that of the four-stroke. Guaranteed performance for several models is found in a 1910 Sulzer price catalog:[15]

	4Z100	4Z133	4Z165	4Z250
Bore, mm	450	500	550	660
Stroke, mm.	660	720	780	900
Speed, rpm	165	160	155	145
Rated power, bhp	750	1,000	1,275	1,950
Bsfc, g/bhp-hr				
100% power	210	210	205	200
75%	210	210	205	200
50%	230	230	225	220

Weights and costs without flywheel were:

Weight, kg	77,000	99,000	120,000	175,000
Cost, Swiss Fr	128,000	158,000	187,000	264,000

Uniflow scavenging was phased out beginning in 1909 because of the complex cylinder head. The many valve openings and passages made it prone to thermal cracking. Marine engines (chapter 15) made the change first, and stationary models soon followed.

Sulzer adopted what they termed crossport scavenging, a form of loop scavenging, which had a unique aftercharging capability. Two rows of intake ports, one above the other, went partly around the cylinder and were opposite to a similarly partial single row of exhaust ports. A cam-actuated valve controlled the scavenge air supply to the upper intake ports (fig. 12.6). The system offered power outputs close to double that of equal displacement four-stroke diesel engines. Only Sulzer achieved such reliable two-stroke performance so early.

In operation, the descending piston passed the valve-closed upper intake ports and then the exhaust ports. The lower intake ports remained covered until cylinder pressure dropped low enough for scavenge air to purge most of the exhaust gases. Shortly before bottom center the cam-actuated scavenge valve opened and air then flowed through the upper as well as the lower rows of intake ports. Scavenge air continued to enter past the upper ports after the exhaust ports had closed in order to raise the pressure to near scavenge pressure and thus lightly supercharge the cylinder. The valve closed immediately after the upper ports were covered. Sulzer's original "double beat" scavenge valve was superseded in 1919 by a rotary valve.

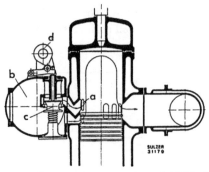

Fig. 12.6. Sulzer crossflow port scavenging of 1909. *Sulzer Archives*

The larger Z Series reverted to a crosshead design upon adoption of cross flow scavenging. The first, a four-cylinder 4Z200, had a 600 × 840 mm bore/ stroke and produced 1,500 bhp at 150 rpm. Two of these were delivered to a Calais, France, power station in 1912. They were the earliest Sulzer diesels to run on coal tar oil in actual service.

A 6Z300 also built in 1912 with cylinders of 760 × 1020 mm capped pre-1914 commercial developments. It once attained 4,500 bhp on test, but after entering service at Harland & Wolff's Belfast shipyard in 1916, it carried a rating of 3,750 bhp at 132 rpm.[16] These large, crosshead "Z" engines had individually cast frames bolted to the bedplate and supported the cylinder assemblies (fig. 12.7). Cover plates fastened to the frames enclosed the

Fig. 12.7. Sulzer six-cylinder 6Z300 of 3,750 bhp under assembly in 1912. Access plates removed. *Sulzer Archives*

crankcase. The crosshead was supported on both sides by a guide mounted on the frame bulkhead to allow service access from either side of the engine. A single guide and bearing was the normal practice. Four through-bolts extending from the top of the cylinder head to the bottom of the bedplate tied the structure together so that gas pressure loading against the head could be transferred directly to the bedplate to keep the cast iron cylinder in compression (fig. 12.8).

The Sulzer Diesel Department

A few of the engineers behind the success of early Sulzer diesels deserve mention. Some were at the height of their professional careers and others, while only at the beginning, also made significant contributions during this pre-1914 period.

Jakob Sulzer-Imhoof earned his diploma engineer degree from the Federal Technical Institute (*Eidgenössischen Technischen Hochschule*, or ETH) in Zürich and later was appointed an honorary Doctor of Technical Science by this prestigious engineering school for his work on two-stroke marine diesel engines. He was a partner in the Sulzer family company before starting his presidency of Sulzer Enterprises, Inc. (*Sulzer-Unternehmungen AG*) in 1914.[17]

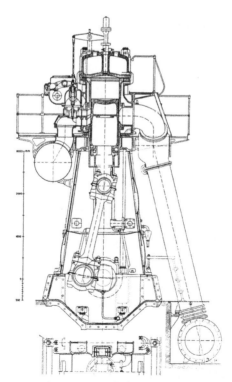

Fig. 12.8. Cross-section of 1912 Sulzer 6Z300. *Sulzer Archives*

Walter Schenker (1876–1965) came to Sulzer as an ETH Karlsruhe graduate. In 1903 he played a key role in the development of the scavenging processes for both the uniflow and the cross scavenged engines. He was technical director of the diesel department at Sulzer from 1910 to 1930.

Samuel Kilchenmann (1879–1969) was concerned with stationary engine design and later headed the stationary engine section.[18] Werner Tobler (1882–1959) began in the diesel department in 1909 and became its technical director in 1936. Wilhelm Hefti (1878–1936), who had started at Sulzers in 1907, also made important contributions to the diesel program from the time of his arrival.

Alfred J. Büchi (1879–1959) was never a stranger at Sulzer. His father, Johannes Büchi, had been works manager at the time Diesel visited there in 1898. Upon graduation from the ETH in 1903, he worked for three years at Carels in Ghent. (He later commented that "[Sulzer's] chief engineer came to Carels Bros. in Ghent and through an arrangement obtained the designs of the Carels Diesel engine."[19] Büchi's long career focused on exhaust gas turbines for supercharging. His theoretical research to harness exhaust gas energy, begun when he was officially employed by the Sulzers in 1907, would put the company in the forefront of these developments. His first turbocharger patent, German No. 204,630, was issued on November 16, 1905. Büchi was with Sulzer, except from 1918 to 1920, until he went to Brown, Boveri & Cie. in 1926 as a consultant. He opened his own engineering office in 1935.

Max Rotter provides windows into Sulzer as a company and its people through his copious correspondence to the US during his enforced stay in Europe. Though they are one man's opinion, his observations are nonetheless insightful. As time neared for the St. Louis factory to begin building European designed engines, Rotter shared his concerns about technology transfer as seen in three letters written to James Harris in late spring of 1911:

April 29:
As regards [Sulzer] shop methods: there are a few important items of information I can obtain. Taken as a whole, however, Sulzer's methods are quite antiquated. In spite of this they turn out about the best quality of work known in their lines of manufacture; but this is due to the skill of their men and the amount of hand-work they put on their machinery. We certainly could not obtain a sufficient number of workmen of this calibre, nor could we afford to pay for so much hand-fitting. . . .

May 11:
As regards Sulzers and Dr. Diesel. So far as I have been able to go into matters, I am inclined to feel, that Diesel's licensees are liable to run

into considerable trouble if they build engines strictly in accordance with his designs. He certainly does not figure as high a factor of safety as Sulzers considers necessary, on the basis of their past experience, . . . Nevertheless we ought certainly to obtain a number of useful ideas, as well as valuable patents, through our contact with Diesel.

In going over detail drawings with Mr. Büchi I have been very much surprised to note a lack of consistency, which Mr. Büchi explains as being due to the fact that the various types of engines are designed under different heads. . . . It is quite evident that we will have a lot of work to do in working out the best features from the various types and sizes, and combining them in a standard machine.

I was not aware that types K, Z, H [auxiliary] and S [marine] were so entirely new to Sulzers. In fact they are feeling their way with these machines, so that the design of these machines can scarcely be considered as having reached a final stage. The idea of merely translating Sulzer's drawings into Christian and then building them in accordance with same will be out of the question.

As I stated before, most of Sulzer's shop-methods are antiquated; however, we can learn a great deal from their foundry practice.

And finally, June 23:

It takes Sulzer so long to move, that I made up my mind the matter was not in such shape as would warrant my hanging around until things got going. . . .

Reverting to Sulzers: Lord protect me from a family business! Everybody has a finger in everything—everybody gives orders to everybody else: result—delays and broken promises. It is hard for Mr. Schenker to keep a promise regarding completion of drawings, when Mr. Sulzer-Imhoof bypasses him, and his next in command, and occasionally takes up the matter directly with the draftsman. Joyous arrangement, isn't it? The most level-headed men in the place are Büchi and Schindler.

I am interested in the fact that Sulzers are climbing down a bit in the speeds and ratings of their Z and S types; they seem to have bitten in a little too deeply.[20]

The Sulzer facet of the many-sided Allgemeine story has been touched upon, but not a direct confrontation between the two in 1909. The Swiss reasoned that since Sautter Harlé did not (or could not) supply drawings of a French submarine diesel engine then they, too, need not also share their designs. This led to a lawsuit won by Sulzer in an Augsburg court through a countersuit. On appeal, however, the decision went in favor of the Allgemeine. The two adversaries settled out of court per an agreement of January 8, 1907. The

Allgemeine voided all existing contracts and royalties in return for which Sulzer renounced a 5,000-mark credit and paid Sautter Harlé in Paris and an Italian firm 200,000 francs for various royalties owed. It is estimated that from 1893 to 1909 Sulzer spent about 350,000 marks for rights to the Diesel patents, a not inconsequential figure.[21]

In addition to their marine and submarine engines, Sulzers were involved also in the development of diesel locomotives. The first, a unique engine whose genesis in 1905 was from a design proposed by Diesel, did not see limited service until 1912 (chapter 18). Its 1,000 bhp unit transferred power to the wheels through a direct mechanical drive. Sulzers' own diesel-electric drive locomotives, which entered service in 1914, are also described in that chapter.

By 1913 Sulzer Brothers ranked as the second-largest diesel engine manufacturer after Augsburg.

Sulzer Technology Goes to St. Louis

The Sulzer K Series became the new B Series at Busch-Sulzer Brothers Diesel Engine Co. in 1914 (fig. 12.9). In addition to various minor simplifications to reduce cost, Chief Engineer Rotter made one significant design revision. He retained the same piston speeds but lowered rpm and increased the stroke/bore ratio from 1.13:1 to 1.28:1 because of his concern that the

Fig. 12.9. Busch-Sulzer Bros. Diesel Engine Co. Mods. 4B125 (520 bhp) introduced 1914–15. Haas, *The Diesel Engine, 1918*

Swiss engines would not have the desired service life. For example, the US 4B125 with its 19 × 24.5 inch (483 × 622 mm) bore and stroke developed 520 bhp at 200 rpm. The comparable 460 × 520 mm Sulzer 4K125 had an output of 500 bhp at 240 rpm. The piston speed of 815 ft/min (4.15 m/sec) was the same for both.[22]

BSDE built four other K-based Sulzer models in addition to the 4B125:[23]

4B30	10.5 × 13 in	120 bhp at 300 rpm
4B40	11.75 × 15 in	165 bhp at 277 rpm
4B60	13.75 × 17-1/2 in	250 bhp at 257 rpm
4B90	16.25 × 21 in	365 bhp at 225 rpm

Busch-Sulzer had intended to introduce a US version of the two-stroke Z Series, but the First World War changed that plan and, more significantly, the direction taken by the company. Submarine engines, based on the Sulzer U Series, became the most important product during the war (chapter 17). Not until its end did work begin again on stationary engines.

Carels Frères

Carels Brothers was no stranger to Diesel's engine, having become a non-exclusive licensee for Belgium in 1894. However, when others began diesel production in 1898, Carels justifiably believed that the prevailing high cost of fuel in Belgium meant there would be little market for the engine in that country. Rudolf himself could not alter this decision.

In 1838 Charles Louis Carels (1812–1875) founded the Werkplaats voor Mechanische Constructies with a foundry and machine shop in Ghent to build steam engines and hydraulic pumps for draining the marshy lands of the Low Countries.[24] Railway rolling stock was added in 1851 and steam locomotives in 1865. The company name changed to Société Anonyme des Ateliers Carels Frères after the founder's sons, Alfons (1838–1914) and Gustaaf (1842–1911), were handed the reins. They introduced the first Belgian compound steam locomotive in 1880 and superheat steam engines in 1898, and were *the Messieurs Carels Frères* who signed the 1894 contract with Diesel. Gustaaf's sons Georges August (b. 1873) and Gaston Louis (b. 1879) succeeded their father and uncle as the next generation of Carels Brothers. Georges became Diesel's close friend and would stand by him in times of need.

Work on the diesel did not entirely cease after 1898. Between then and 1902, Carels built a single-cylinder engine based on the M.A.N. Series XVII design.[25] Manufacturing began on both a one- and a two-cylinder engine "of 70 to 80 bhp" per cylinder using a standard Augsburg DM Series 400 × 600 mm design. They were destined for the Diesel Engine Co. Ltd. in London.[26]

Carels invited Professor H. Ade Clark, of Yorkshire College, Leeds, to test these engines in Ghent, which he did during February and March of 1903.[27] According to Clark the single-cylinder engine was still on the erecting stand and could not be put through as rigorous a pace when he tested it on February 7. The 160 bhp model was fully completed a month later and given a full performance evaluation. Under a maximum load of 165 bhp at 155 rpm, the bigger engine had a bsfc of 0.408 lbs/bhp-hr and a brake thermal efficiency of 32.3 percent. The fuel was a 0.922 sp gr Texas crude with an LHV of 19,300 Btu/lb. These results compared favorably to a similar M.A.N. engine. Clark had the opportunity to inspect the single-cylinder engine during its installation at the Temple Press in London a few months later. A tight bearing had to be fixed before it could attain the full 80 bhp promised.

In early 1903 the DEC gave Carels a multi-engine order followed by the one to Sulzers. Acceptance of the order resulted in a contract with the Allgemeine on September 14 of that year, which voided the earlier one signed with Diesel. Like the Sulzers, Carels would pay a 6 percent royalty for engines sold in Belgium and in countries where no patents existed. An additional 3 percent was owed if engines went to countries with patents. Similarly, Carels paid nothing on engine sales to the DEC.[28]

Proof that the Belgians could honor the DEC order of twenty-two engines is evidenced by shipments beginning in March 1903. A report to the Allgemeine on January 26, 1904, stated that all but seven had been delivered.[29] The mix of twelve one-cylinder, eight two-cylinder, and two three-cylinder engines were all of the DM 80 design. These had a 460 × 660 mm bore/stroke with an output of 80 bhp per cylinder at 160 rpm. At the time these were on the larger-size end of the DM Series. Not until 1907 did Carels add to their DM line by including engines of 50 and 100 bhp per cylinder.

Ensuing orders by the DEC brought the total cylinders destined for Britain up to 103 by June 30, 1906.[30] Up to that date, Carels had sent just thirty more cylinders to all other countries combined. Sixteen went to Romania. Only thirteen cylinders, seven engines worth, had stayed in Belgium, and one of these was at the Carels factory. The few sales attainable in their own country vindicated Carels's decision in 1898 to delay production. Royalty obligations to the Allgemeine ended on April 12, 1906, when Carels paid 20,000 francs to cancel the contract.[31]

An eyewitness account of Diesel activities at Carels comes from a Swiss engineer whose professional career ultimately brought him to the United States.[32] Jean Santschi was on his way to England in 1903, but "after sightseeing in Paris and Brussels" he "arrived at the English Channel 'with insufficient funds' to proceed further." He found a job in a new department at the Carels factory in Ghent, which was developing the diesel. Santschi drolly explained that his primary qualification was an ability to speak German, Flemish, and French, and someone was needed to translate German instructions for Belgian engineers and workers.

Standard practice for a newly assembled engine at that time was to belt it over with an electric motor, often for days, to seat the rings "so that a compression of approximately 450 pounds could be obtained." The fuel was turned on only after reaching this pressure, and when firing began "the belt would be thrown off."

Diesel was a frequent visitor at Carels. Santschi recalled the inventor having him file the cams lifting the fuel nozzle valves until the engine produced the combustion and expansion curves on the indicator cards that Rudolf wanted. (Büchi also spoke of Diesel's visits.[33])

Santschi was sent to India in 1906 to supervise the installation of eight two-cylinder, 160 bhp engines the DEC had sold there to run centrifugal pumps for an irrigation project. Diesel himself gave Santschi instructions before his departure. He lived in India for four years maintaining those and other engines sold through London.[34] After returning in 1910, Santschi went directly to the DEC, which placed him in charge of service for the engines being sold "all over the world."

Carels Charts Its Own Course

At the Liège Exhibition in 1905, Carels proved that it had the confidence to pursue its own diesel destiny by demonstrating the most powerful engine yet available.[35] This three-cylinder, four-stroke diesel, with a 560 × 750 mm bore and stroke, carried a 500 bhp rating at 150 rpm. A casual glance at its trunk piston design would reveal M.A.N. ancestry, but novel features are also evident (fig. 12.10). The engine was sold to SA d'Ougrée-Marihaye.[36]

Fig. 12.10. Carels Bros. 1905 stationary, three-cylinder, four-stroke diesel. 500 hp at 150 rpm. Postcard photo. *MAN Archives*

An uncooled piston had a dished crown with a rounded, conical node cast in the center. This modified "Mexican hat" profile acted to deflect the spray rather than to affect turbulence. A single exhaust valve was cooled by water flowing down the inner of two concentric pipes within the stem and into a hollow valve head. Heated water exited up the outer pipe. The fuel valve rocker lever was in two pieces to make for easy removal of the nozzle valve assembly. Other manufacturers were starting to adopt split intake and exhaust levers to aid in removal of the valve cages.

Cylinder lube oil conventionally entered under pressure from a Mollerup pump via a port located between the upper two of the eight rings when the piston was at bottom center. Metal shrouds enclosed the space between the A-frames to contain the oil mist created by splash lubrication for the bearings.

Like the Swedish and Swiss engines, the governor and horizontal fuel pump were driven directly off the intermediate vertical shaft between the crank and camshaft. If the engine exceeded 160 rpm, the governor tripped a lever that lifted the pump suction valve into a full open position and caused a shut down.

An electric motor driving two three-stage air compressors got their power from a 450 Kw, 550 v DC generator run by the engine. Thus, the stated 500 bhp rating as measured from generator output did not include the compressors' parasitic load.

A test made in mid-February before the engine left the factory gives the true brake horsepower:[37]

Load factor	Full	Half	Quarter
Speed, rpm	152.8	150.3	150.2
Max. peak pressure, psi	525	480	480
Blast air pressure, psi	66.3	50.9	35
Ambient air temp., °F	48	48	48
Exhaust temp., °F	806	496	275
Avg. imep, psi	97	56	25
Bhp w/o compressor	502	245	55
Compressor load, bhp	43	31	23
Net engine power, bhp	459	214	32
Net mechanical efficiency, %	72.3	58.8	19.9
Net bsfc, lbs/bhp-hr	0.451	0.481	1.415

The reported full-load smoke was probably due to very unequal fuel distribution. Peak pressure, as seen above, varied from 525 psi in the overfueled cylinder to 480 psi in the others. Full-load individual imeps were respectively 80.7, 93.9, and 115.6 psi.

Exhaust temperatures taken from an outlet close to the engine were read from a thermometer containing compressed nitrogen above the mercury "to prevent the latter boiling."

Three experts, including Boverton Redwood as referee, tested the heat content of the Galician fuel oil used, and finally settled on a value of 19,600 Btu/lb.[38]

In 1907 Carels built a single-cylinder, two-stroke test engine with a reversing mechanism.[39] After several redesigns it looked promising enough for the French Navy to have Carels develop a 1,000 bhp marine diesel based on it. For political reasons it was ordered through Schneider et Cie. of Le Creusot who would later build submarine and merchant ship diesels as a Carels licensee.

Carels scored another "first" in September 1909 by running the most powerful, two-stroke reversible marine diesel to date. The four-cylinder, uniflow scavenged two-stroke, with a bore and stroke of 450 × 560 mm, had an output of 1,000 bhp at 250 rpm.[40] A-frame and cylinder were a single casting (fig. 12.11). Problems with its trunk piston design kept the engine at the factory, and all future models of this size and above had crossheads. It is not known if the pistons were water-cooled.

Scavenge air entered through four poppet valves in the head. In lieu of a valve crosshead (bridge), each of the two cam-actuated rocker arms per cylinder directly opened the "near" valve. The rocker arm extended past the near valve and pushed on a "finger" lever to open the "far" valve (fig. 12.12).

Fig. 12.11. Carels 1,000 bhp prototype marine diesel engine, 1909. Chalkley, *Diesel Engines, 1917*

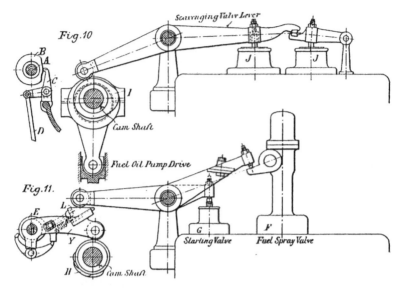

Fig. 12.12. Valve actuating mechanism for Carels's four-valve, uniflow scavenging, two-stroke diesels, 1909–1913. *Engineering, 1912*

The fuel valve was still lifted as before, but its rocker lever now rode on top of the cam to push down on a second, short bell crank whose pivot pin was in the nozzle assembly. The opposite end of this second lever then lifted the fuel valve.

Carels continued with the four scavenge valves and the above double-lever design for all their pre-1914 engines.

A single-plunger fuel pump and its governor-controlled suction valve followed the earlier practice of being operated by eccentrics on the vertical intermediate shaft. Adjustable orifices in pipes branching off the pump discharge line distributed equally metered charges going to the injectors. Larger engines after this one used a camshaft-operated plunger pump for each cylinder.[41]

A double-acting scavenge air pump and a three-stage Reavell compressor for starting and injection air were at the front of the engine. The scavenge pump was driven by a crank throw, and an offset pin at the end of the crank turned the compressor.

Reavell "Quadruplex" compressors, made in Ipswich, England, were of a compact, four-cylinder radial design with the cylinders at right angles to each other (fig. 12.13). The low-pressure cylinders were opposite each other in one plane, and the intermediate- and high-pressure ones were opposite in the other. The entire compressor was water jacketed, and the heated air traveling between the stages passed through coiled pipes in the water to achieve maximum intercooling. Reavell compressors were installed mainly on engines going to Britain.

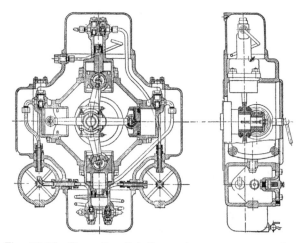

Fig. 12.13. Reavell radial, three-stage compressor for injection and starting air. Chalkley, *Diesel Engines, 1914*

Carels Frères concentrated development efforts on two-stroke stationary and reversible marine engines having increasingly larger horsepowers. The first of these, built at the end of 1911, went in the Canadian grain ship *Fordonian* and is described in chapter 16.

Vickers at Barrow-in-Furness, England (who started developing a submarine diesel in 1906) tested a Carels-based single-cylinder, 1,000-hp, two-stroke engine during 1912. During a 150-hour trial, it produced an average of 1,045 bhp for fifteen hours. The bsfc at this load was 0.587 lbs/bhp-hr.[42]

Carels supplied diesel engines for two major stationary installations from 1912 to 1913. An electric power plant in Bucharest, Romania, received the most powerful Belgian engines built before the war with their six cylinders of 700 × 1,050 mm and an output of 2,500 bhp. The total installed power was 7,000 bhp.[43]

The second, a pumping station for the Liverpool docks, was the largest diesel power plant in England. It had five four-cylinder, 1,000 bhp engines of 510 × 660 mm bore and stroke. Each drove a centrifugal pump at 180 rpm.[44] Engine details include telescoping tubes supplying water for piston cooling, a scavenge air pump driven off the crankshaft between cylinder number one and the Reavell compressor, and two fuel pumps, each with two pumping plungers. The engines had intake air silencers and water-cooled exhausts.

In 1912 Carels reorganized as a limited company under the name of Usines Carels Frères (Carels Brothers Works). They also bought a license for a "semi-diesel" sold by the ABC (Anglo Belgian Co.), a company that Georges Carels had established that same year. Gaston Carels became head of the Carels Brothers in 1914.

It was reported that by mid-1913 Carels had completed 342 engines, stationary and marine, having a cumulative output of 117,500 bhp.[45]

Carels suffered greatly as a result of the war. Not only was diesel work halted, but because almost all the machinery was taken elsewhere during the conflict's last days the factory was forced to begin anew.

After the war Gaston effected a merger with Société Française pour l'Exploitation des Procédés Thompson-Houston (Paris) and International General Electric (Schenectady) to form Société d'Electricité et de Méchanique, or SEM for short. The Ghent division, known first as SEM-Carels and then ACEC-Carels, continues to make diesel engines there but under the old ABC name.[46]

The Ipswich Connection

Carels formed a vital link with Great Britain through the Consolidated Diesel Engine Manufacturers Ltd. incorporated on March 27, 1912. Six hundred thousand shares were issued, at £1 per share, from an initial capitalization of £750,000. CDEM obtained half of the Belgian company's shares, plus one share, to give it control of Carels. In return, Carels Fréres received £150,000 in cash plus £100,000 in stock. The remainder of the proceeds from the sale of stock went "for the cost of building and equipping the new works at Ipswich, and providing additional working capital to meet the demands of the extended business."[47]

No legal evidence can be found, but contemporary accounts refer to the Diesel Engine Co. of London as having merged into Consolidated Diesel around 1912. It is probable that the DEC had already been under the control of Carels for some years.[48] John E. Thornycroft (1872–1960), of the company bearing his name, reported in December 1912 that "We had arrangements with the Diesel Engine Co. prior to their joining Carels."[49] Santschi told of his being transferred to the Ipswich factory when it opened and remaining there for more than a year.[50]

Rudolf Diesel served on the board of directors, ostensibly to lend more stature to the venture. His position and the block of stock he received came about through his friendship with fellow director Georges Carels.

An unfortunate issue arose because of Diesel's seat on the board, which had repercussions in the United States. Carels wanted to develop an American presence upon the expiration of the US basic patents, and the man chosen to be the "American representative," unbeknown to Busch-Sulzer, was W. R. Haynie, its own eastern sales agent. Busch-Sulzer learned of the duplicity on June 8, 1912, when this and other evidence was found in Haynie's New York office. It had received a letter from Diesel on June 13 advising that "he should have mentioned Haynie's name as Carels Frères licensee" when he had written James Harris on the fifth and further added that Carels "would issue a license in America."[51] Needless to say, Harris was furious and said "there is no question that Mr. Diesel should not be on the board of directors of the Consolidated Engine Manufacturers, Ltd." because it violated terms of his contract with Busch and the Sulzers by virtue of his also being on the board of Busch-Sulzer. The Swiss already felt this way. Harris believed the matter "should be decided as soon as possible." Nothing ever came of the Haynie affair except to let Busch-Sulzer know that Carels, and others, were approaching their shores. Haynie remained Carels's US representative for a few years.

Carels and Nordberg Find Each Other

The first Carels engines shipped to the United States were installed in a mill at the Burro Mountain copper mine in Tyrone, New Mexico, during 1914.[52] Bruno Victor Nordberg (1857–1924) also signed a manufacturing license with the Belgians in June of that same year. The two events are not unrelated because The Nordberg Manufacturing Co. in Milwaukee was the leading supplier of large hoists, steam engines, and other equipment to the American mining industry and was already building diesel engines.

Nordberg, a Finnish engineer immigrating to the United States in 1879, had worked for the E. P. Allis Co. before starting out on his own in 1886 to build steam engine governors. His product line grew to also include compressors, pumps, and blowing engines based on his seventy patents.[53]

As early as 1912, when it was legal to do so, Nordberg began making horizontal one-cylinder, loop-scavenged, two-stroke diesels with crossheads. The earliest ones with a bore and stroke of 11.5 × 15 inches were rated 50 bhp at 300 rpm (fig. 12.14). Next came a cylinder size of 14.5 × 18 inches and an output of 100 bhp at 260 rpm. A 1914 brochure shows a two-cylinder model producing 200 bhp at 260 rpm.[54] Fuel was injected without the aid of compressed air, a daring step to achieve simplicity. Delivery from a fixed-stroke

Fig. 12.14. Nordberg-designed 50 bhp diesel engine with solid-fuel injection, 1914. *Nordberg brochure*

plunger pump was regulated by governor action on a bypass valve. Fifty-eight of these models were built.[55]

It is not known if Haynie approached Nordberg or the latter initiated contact with Carels, but after signing the license agreement, either Nordberg or his son Bruno Victor Edward (1884–1946) went to Europe to study installations with Carels engines. Travels during that tumultuous late summer of 1914, which included Romania, were not the easiest.[56]

Carels placed Jean Santschi in charge of the Burro Mountain job and on its completion he transferred to Nordberg.[57] The two New Mexico Model EG engines under his care had a sea level rating of 1,250 bhp at 180 rpm. Bore and stroke were 525 × 660 mm. Because of the almost 6,000-foot altitude, the engines' scavenge air compressors were increased in size to 1,050 mm bore and 600 mm stroke in order to deliver air at a 2.5 psi gage pressure; this allowed a full load of 95 percent that at sea level. The reduction was due mainly to the power absorbed by the larger compressors. The engines each drove an 815 kva, 60-cycle, three-phase 6,600-volt generator, which provided for better than a 100-hp "cushion" under the maximum output of 1,200 bhp.

Nordberg began their production of Carels engines with the five-cylinder EG model (fig. 12.15). However, because World War I had started by this time, the company was denied anticipated assistance from Belgium. Santschi fortunately contributed his vast knowledge to complement their own experience gained from making the smaller, much less sophisticated diesels.

Nordberg soon built even larger models, and almost until their engine production ceased in 1973, the most powerful diesels built in the United States came from Milwaukee. Nordberg also bought the Busch-Sulzer Co. in

Fig. 12.15. Nordberg Model EG five-cylinder diesel of 1,250 bhp based on the Carels design, 1916. *Haas,* The Diesel Engine, *1918*

1946 but ended sales under that name after a few years and then made only service repair parts for B-S engines.

Mirrlees Begins Anew

Mirrlees, Watson Co. Ltd. reviewed their moribund diesel program in 1902. A more-aggressive general manager with engine experience had joined the company that year. Charles Day (1867–1949) was born in Newcastle-under-Lyme, northwest of Birmingham, and received his technical education in a Manchester area school. Although his apprenticeship included work on the "Stockport" gas engine made by J. & H. Andrew, much of his early career had focused on the steam engine. Before coming to Glasgow, he ran the engine department at Ferranti Ltd., a London company manufacturing complete electrical power systems.[58]

Day's earlier association with both gas and steam engines soon aroused his interest in the abandoned 1897 diesel, which sat in its dusty corner at the Mirrlees factory. Reports of other successes with the diesel engine led him and George Windeler, his future diesel chief engineer, to visit Augsburg during October 1902 and personally confirm M.A.N.'s progress.[59] His confidence in the engine's potential after returning to Glasgow was strong enough to convince his board they should reopen a diesel department.[60] Day's first action was to put the old engine in good working order by incorporating the latest compressor and fuel nozzle improvements.

By the latter part of 1903, the first new engine, which closely followed MAN's DM design, was running in the Mirrlees shop. The one-cylinder engine with a 12 × 18.25 inch bore and stroke produced 35 bhp at 200 rpm. A second, 25 bhp engine was started up around March 1904.[61]

A major complaint made to Augsburg about the operation of these first two engines was the necessary frequent cleaning of the second-stage air compressor. Mirrlees, like everyone else, was unhappy with the use of the power cylinder as the first stage of compression for the injection blast air. Day said in an April 1904 letter to the Allgemeine that they were making "a new air pump with a view to discontinue" the old system. Other changes included doing away with the fuel pump float chamber and enclosing the vertical shaft for "dirt and safety reasons."[62] The new, intercooled, two-stage compressor was vertical and driven through an offset pin on the front end of the crankshaft. Mirrlees was among the earliest to adopt the crank end placement in lieu of the more usual mounting at the side of the cylinder where it ran off linkage driven by the connecting rod. Later engine models had three-stage compressors.

The two Mirrlees diesels were wisely kept in the shop for testing and further development work. Production did not begin until mid-1904.[63] The first ones delivered were a one-cylinder 40 bhp and an 80 bhp two-cylinder. These were followed by a series of three- and four-cylinder models, all of which could be ordered with ratings of up to 50 bhp per cylinder at 250 rpm (fig. 12.16).

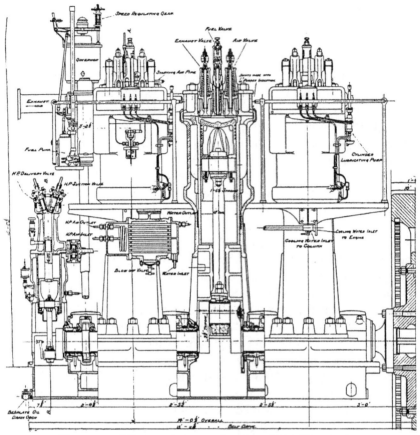

Fig. 12.16. Mirrlees A-frame diesel, c. 1912. The design
originated in 1903. Up to 50 bhp per cylinder at 250 rpm.
Wells & Wallis-Tayler, *The Diesel Engine, 1914*

Six from the first production lot were scattered around the world with
installations in Chile, India (two), Japan, the Malay States, and Singapore.[64]
Such widely spread deliveries of a brand new product must have been accompanied by great faith and a vagabond service team!

The Mirrlees dished piston resembled those of the Sulzer and Carels. It
also had the small, rounded cone in the center of the piston crown to deflect
the incoming fuel/air charge from a single, vertical orifice, which was aimed
along the cylinder axis. The "flame plate," with its rounded exit orifice, was
retained on the nozzle body by a threaded collar.

One internal atomizer design sufficed for each size nozzle. It was found
that a certain combination of M.A.N.-type stacked washers, which worked
well for viscous fuels, would also handle "less sticky oils" (fig. 12.17).

Each fuel nozzle of a multicylinder engine was served by its own pumping
plunger with all being contained in a single pump housing. The plungers did

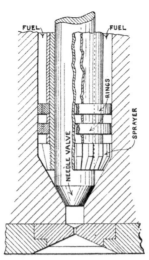

Fig. 12.17. Mirrlees atomizer in the fuel nozzle, c. 1912.
Wells & Wallis-Tayler, *The Diesel Engine, 1914*

not reciprocate in a sequence but were displaced by a single eccentric drive on the camshaft. The governor atop the vertical shaft superimposed its action to control the time interval the suction valves remained open (fig. 12.18).

A 1908 patent issuing to Mirrlees (Day and "others") points to a problem of excessive cylinder pressures occurring when too much fuel might be admitted on an engine startup.[65] The normal starting procedure called for moving a hand lever through a 90-degree arc to rotate the pivot shaft on which the fuel and starting air rocker arms were eccentrically mounted. This effectively lifted the fuel valve cam follower off its cam and engaged the starting valve cam follower to admit pressurized air into the cylinder. Day's patent, which was put to use, also covered a bypass valve inserted in the fuel lines from the pump. This valve, which was opened by a mechanism linked to the nozzle valve, remained open during the starting cycle to prevent a buildup of excessive fuel pressure at the temporarily closed fuel valve. It prevented an oversize charge of fuel from entering the nozzle when the engine reached runable speed.

Recommended injection air pressures (in psi) varied according to load and fuel used:[66]

	Crude oils	"Light" oils
Full load	940	850
-3/4 load	830	750
-1/2 load	720	650
-1/4 load	620	580
No load	520	520

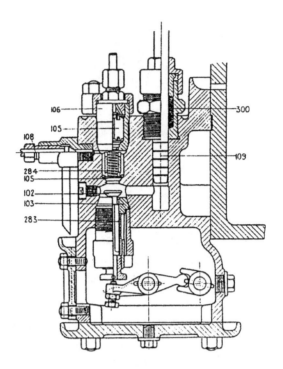

Fig. 12.18. Fuel pump, c. 1912 Mirrlees diesel.
Instructions for Erecting and Working Mirrlees-Diesel Oil Engines

Mirrlees warned that if the blast air pressure was too low for a given load the exhaust would be "dirty or blackish" and the engine would overheat (due to late burning) if allowed to run in this condition. The absolute minimum allowable pressure was 520 psi. A compression pressure of 450 to 500 psi was "sufficient to meet the needs of the worst oils, of slow running and of cold circulating water."[67]

Brake specific fuel consumption for the 250 rpm series of Mirrlees engines averaged about 0.45 lbs/bhp-hr at full load, 0.47 at -3/4 load, 0.53 at -1/2 load, and 0.7 at -1/4 load.

Maintenance schedules suggested that pistons be removed and cleaned every twelve months. The exhaust valves should be cleaned and, if necessary, reground every 180 hours when burning crude oil, or 400 hours if on "light oils." Once a week the fuel valves needed cleaning and their seats touched up by twirling the valves on them with a little lube oil to aid in the reseating.

Mirrlees Joins the Navy

When the British Admiralty designed the battleship *Dreadnought* in 1905, it asked for Mirrlees diesels rather than steam engines to power generator sets. Mirrlees met the tight delivery schedule for this famous ship, which was launched in 1906 and commissioned late that year. The two engines supplied bore little resemblance to the earlier M.A.N.-based configuration (fig. 12.19). These four-cylinder diesels had an 11.5 × 12 inch bore and stroke and produced 160 bhp at 400 rpm. Each cylinder assembly rested on its own individual, box-like upper crankcase housing, which in turn was bolted to the bedplate. An enclosed crankcase was formed by tying the housings together. Overall engine length was reduced by relocating the compressor to the right front side of the engine and driving it with a chain off the crankshaft. This had to be changed to a gear drive because of repeated chain failures.[68] The engine was directly connected to the generator.

Special tanks on the ship carried 120 tons of "patent fuel" for the diesels in addition to the coal and heavy oil for firing boilers.[69]

Because of several modernizations over the *Dreadnought*'s service life, it is not known if the original engines remained on board until the ship went into the reserve fleet in July 1918. (It was put up for sale in 1920 and scrapped in 1923.) This early application of diesel power on a capital ship speaks well of Mirrlees engineering and manufacturing capabilities, but its acceptance was not unanimous. A "Chief Engineer" wrote to *Engineering* in 1911 that "the Admiralty are still ordering steam turbines for this duty. There have been [diesel] troubles which have led to a parody running 'Up and down the voltage goes, Pop goes the Diesel'"[70]

Following the *Dreadnought* was a navy order for a few lightweight, four-cylinder derivatives to go in pinnaces, or ships' boats. These 9.75 × 12 inch, 120 bhp at 400 rpm engines drove the propeller through a clutch and reverse gear. They had aluminum bedplates and individual manganese bronze crankcase housings cast integrally with the cylinder jackets. One manufacturing difficulty encountered was a repeated cracking of the aluminum castings

Fig. 12.19. Mirrlees 160 bhp at 400 rpm engine-generator set in the *Dreadnought* launched in 1906. Evans, *History of the Oil Engine, c. 1931*

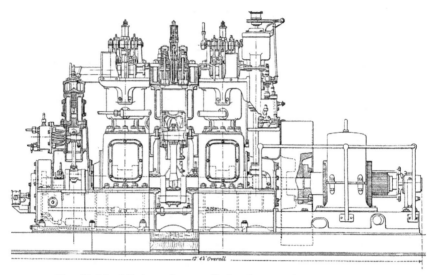

Fig. 12.20. Mirrlees three-cylinder, four-stroke generator set
on the Cunard liner *Aquatania*. *Engineering, May 29, 1914*

while cooling in the mold. The answer was to use less rigid cores, which could collapse slightly as the metal contracted during the cooling process.

Although the pinnace engines were designed in Glasgow, it is likely that they were delivered from a new factory opened near Stockport (by Manchester) in 1908.[71] Due to the diesel business's rapid growth, Mirrlees formed a separate company with a factory in Hazel Grove under the name of Mirrlees, Bickerton & Day Ltd. with Charles Day as managing director. (He later was chairman of both companies.) Henry N. Bickerton, founder and head of the National Gas Engine Co., joined the board of the new company to lend his name and experience.

Evidence of the British Navy's general satisfaction with Mirrlees ship-board gen-sets is seen in an October 1908 report stating that "every battleship and cruiser launched during the last few years carried at least one example" of the engines.[72] Later diesels had a one-piece crankcase housing to which the cylinders were attached, and their compressors also went back to the original location at the front of the engine.

An example of smaller Mirrlees pre-war marine generator sets is the 8 × 9 inch, three-cylinder diesel to drive a 450 rpm, 30 kw emergency unit placed aboard Cunard's *Acquitania*, which was launched in 1914[73] (fig. 12.20). This design eliminated wet liners so that the pistons ran directly in the cylinder castings. The fuel pump and governor went to the flywheel end of the engine (perhaps for torsional reasons).

A test of a 17.75 × 27 inch, three-cylinder Mirrlees A-frame stationary engine was made in October 1910 as installed in a Doncaster railway shop. Some results at varying load were[74]

	1/2	3/4	Full	10%+
Speed, rpm	190	190	190	190
Brake horsepower	179	262	346	380
Mechanical efficiency, %	59.3	67.9	71.3	70.3
Bmep, psi	42.5	54.8	72.0	79.3
Bsfc, lbs/bhp-hr	0.495	0.454	0.443	0.468
Air injection pressure, psi	746	782	863	917

In 1912 Mirrlees built their first six-cylinder model of this engine, which had a rating of 750 bhp at 200 rpm. By placing the vertical drive to the camshaft between the middle two cylinders they avoided the torsional problems often occurring in camshaft drive gears on longer engines when the gears were on the front end. However, an eight-cylinder engine made in 1919 passed through "unpleasant" critical speeds on its way up to operating speed and when stopping. A Lanchester damper solved the problem.[75]

The MS *Tynemount*, a 250-foot cargo vessel, launched in 1913, and it used two 300 bhp, six-cylinder Mirrlees diesels to turn a single screw through an electric drive. Although the power transmission system with its two generators and single motor proved disappointing, the engines performed well.[76] These smaller, nonreversible diesels followed the general six-cylinder design described above. They had a 12-inch bore and a shortened 13.5 inch stroke so that speed could be increased from 250 rpm with the standard 18.25 stroke to 400 rpm in order to drive their generators.[77]

Pistons went through a number of redesigns to improve cracking resistance and heat transfer.[78] The addition of inner ribs helped, but by 1915 engines with larger bores (above twelve inches) had a two-piece piston in which the underside support of the crown resembled a conical structure whose base circle rested on the sleeve piece containing the pin. The rounded top of the cone form protruding above the center of the dished crown became the spray deflector (fig. 12.21). Ricardo commented that it served to "spread out horizontally" the impinging fuel and that it would become such a hot spot the fuel could not "readily adhere to it" as it would on cooler cylinder walls.[79] Some engines retained the simple dished crown.

Mirrlees remained the most significant British manufacturer of stationary diesel engines prior to 1914 even though others had come on the scene by then. Cumulative production and orders up to April 21, 1913, included 388 engines with a total rated output of 58,670 bhp. Twenty of these were the 160 bhp engines driving generators on battleships.[80]

The company is the last of the very early British builders to retain name recognition following the many closings and consolidations of the British

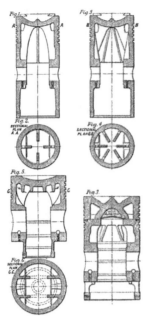

Fig. 12.21. Piston designs of Mirrlees diesels c. 1915.
Engineering, July 16, 1915

diesel engine industry over the last decades. Mirrlees diesels continue to be made at the original Hazel Grove site as part of Mirrlees Blackstone Ltd.

Burmeister & Wain

More than four years passed from the time when Burmeister & Wain signed a license contract with Diesel until they built production engines. A combination of concerns and circumstances pushed the Danes into this unanticipated delay. Once committed to make the engine, however, they backed their faith in it with resources and determination. Besides creating a major product line within the company, B&W would do much to enhance the diesel's budding reputation as a marine power plant.

B&W completed a Series XVI engine in late 1898, but it underwent no endurance testing at first. The engine only ran from time to time under load and then with a prony brake to measure the output. In January 1900 it was installed in the factory power plant to belt drive a generator. The diesel's intended function was to meet occasional light electrical loads during the night shift rather than bother with a steam-powered generator.

Such service still failed to add many test hours, although the operator became very adept in anticipating sudden shop overloads, which could, and sometimes did, stall the engine.[81] When a full night shift was added during 1901, the engine ran from six in the evening until six the next morning. This service lasted into 1902.

A dispute with the Allgemeine, beginning in 1900 and not settled until 1903, further delayed B&W's decision to start production. It stemmed from a standard clause in Diesel's original contracts that carried over into the Allgemeine's that called for licensees to reimburse his design expenses for any engines that they intended to develop. The payments were to be equally shared by those who pursued such a development.[82]

The problem arose when B&W took an interest in Güldner's ill-fated two-stroke engine. Because of drawings and other information sent north, the Allgemeine interpreted this action by B&W as invoking the repayment clause—and a way to recoup some Augsburg development losses. Unfortunately for B&W, no one else showed an interest in the engine, and the Allgemeine asked them to bear fully what were intended to be apportioned costs. This did not set well with the Danes.

The ensuing confrontation resulted in B&W giving the Allgemeine an option to buy back the Danish patent rights and resell them to some other party. No one, including the Swedes, stepped forward, and the option finally died a natural death in 1903. During the interim, however, the disagreement placed B&W's diesel plans on hold and curtailed the building of a second-generation test engine.[83]

A third, and perhaps more compelling argument against making diesels stemmed from prevailing fuel costs in Denmark. B&W's test engine burned only kerosene, and it was thought that production engines might also be required to do so. Because the price of kerosene was so high, Ivar Knudsen knew that diesels could not compete against steam engines when fuel savings were the main selling point.

The fuel situation changed dramatically in 1902 after East Asiatic Co. (*Østasiatike Kompagni*), a Danish shipping firm, built storage tanks at Copenhagen's Free Port. This happened because of what at first were unrelated events in London and Texas. Marcus Samuel, the creative head of Shell Transport and Trading Co., was forced unexpectedly to idle his fleet of new tankers. Soon thereafter he learned of an oil discovery in the Spindletop field and saw an opportunity. As fast as possible, and ahead of everyone else, he contracted with the oil's owners to buy most of the field's output. Within months cheap oil began arriving in Copenhagen by Shell tankship.[84] East Asiatic then urged B&W to test the heavier Texas crude as a prelude to its burning in what was hoped would be a growing population of Danish diesels.

B&W finally tried the new fuel in their test engine, but the higher viscosity oil could not be properly atomized under the pressure available from its old single-stage compressor. That ended the test as B&W had no other engine, and there would not be another until they settled with the Allgemeine. Nevertheless, a way presented itself because Swedish diesels at the nearby Free Port used the heavier oil.

This came about when B&W gave AB Diesels Motorer a temporary right in July 1902 to sell engines in Denmark until domestic production began.

Two three-cylinder, Model K III, 250 bhp engines had been generating electric power for the Port since the beginning of 1903 and had no trouble burning the Texas oil.[85]

The Spindletop oil had a special characteristic advantageous to its use in future Danish diesels. The oil import tax at the time was based on color and opacity rather than viscosity, so that the Texas oil's inherent darkness for the same viscosity yielded a lower tax than would ordinarily have been the case.

The fuel's high sulfur content presented a problem, but it was supposedly decreased by tapping the higher regions of a tank where the sulfur percentage was less. Knudsen later reported that B&W engines could live with up to 1.25 percent sulfur in the fuel.[86]

B&W's directors now felt justifed in beginning production. The Allgemeine's buy-back option, which expired in January 1903, no longer left a cloud hanging over the patent rights; the success of other licensees' engines could not be questioned, and the diesel had proven its capability to run on East Asiatic's oil supply. All B&W needed was an engine.

This came about under the direction of a new head of the diesel design department who succeeded Spangenberg in 1902. Olav Eskild Jorgensen (1875–1949) was only twenty-six when Ivar Knudsen gave him the job. Jorgensen had started at B&W as a design-draftsman in February 1899 after receiving a master's degree in mechanical engineering. He took time off in February 1900 for a year's "study tour" in the United States, and B&W put him in charge of the design office not long after his return. He would serve as chief engineer of the diesel program at Copenhagen from 1908 to 1912.[87]

Knudsen and Jorgensen realized that the crosshead design of the old Series XVI engine was undesirable because of its bulk and costly construction. The only two designs worth considering were Hesselman's K series and Augsburg's DM. While the K was superior in size, weight, and simplicity, the DM won out due to its more conservative design that Knudsen believed was necessary to ensure the reliability and durability vital for introducing a new product.

As it turned out, the ensuing Danish prototype design built in 1903 borrowed from both German and Swedish practice yet had its own distinguishing feature (fig. 12.22). The A-frame and cylinder resembled the DM Series, and the fuel pump and governor were pure Hesselman.

New, however, was a two-stage compressor driven off the front end of the crankshaft.[88] In B&W's view, the compressor's new location eliminated the attendant drive linkage and casting complications associated with the mounting alongside the cylinder. One should note that B&W was the first to use this placement and that Mirrlees, for example, legally borrowed the idea because B&W had dutifully informed its sister licensees of it under the "cooperation clause" of the license contract.[89]

An integrated, two-stage compressor provided enough clean and high-pressure air to atomize the heavier fuels. (M.A.N. probably managed to use

Fig. 12.22. Prototype Burmeister & Wain diesel built in 1903.
B&W photo from the Nohab-Wärtsilä Archives

the combustion chamber for first-stage air because their early engines mostly burned the lighter kerosenes.)

Concurrent with the prototype development was a test of Knudsen's airless injection idea. These experiments used the Series XVI engine with no modifications to the fuel pump and nozzle. Knudsen merely placed a spring-loaded accumulator in the fuel line between the pump and nozzle to maintain a constant hydraulic pressure whereby the volume of fuel injected per charge depended on pressure and opening duration of the nozzle needle. C. V. Kayser, the test engineer, recalled there was heavy detonation and a lot of smoke, but the engine was made to run under a light load. Knudsen filed a Danish patent application entitled an "Arrangement for engines with liquid fuel" on December 5, 1903. The Patent Office turned it down the next March because of German prior art.[90]

That same year saw B&W promoting the diesel through its lower fuel costs. In a summer 1903 issue of *Ingeniøren*, Knudsen gave figures for fuel savings with the diesel when compared with a comparable power, Schmidt superheat steam engine: 0.84 øre/bhp-hr with a diesel-burning Texas oil versus 1.27 øre/bhp-hr with steam. These were hotly disputed in later issues of the same journal by the Schmidt distributor who said the numbers were more like 1.15 versus 1.05 respectively. The operator of the Swedish diesels at the Free Port supported Knudsen statements,[91] but the discussion emphasized again how hard the struggle would be to gain entrance into the steam-dominated camp.

Prodding by an impatient customer led B&W to buy an 8 bhp DM 8 from M.A.N. while they made ready for production. This single-cylinder engine left Augsburg on September 10, 1903, and went into a Copenhagen workshop immediately on its arrival.[92]

That same year Burmeister & Wain sold ten engines in five cylinder sizes based on their own design. Five were delivered in 1904 even before the new prototype was fully tested. The order included engines ranging from 8 to 160 bhp with the largest having two cylinders of 80 bhp each. Production began in the fall of 1903 on a group of four engines, two of which were two-cylinder models. They had a bore and stroke of 320 × 430 mm and an output of 40 bhp/cyliner at 180 rpm[93] (fig. 12.23). All were reported running in 1938, and engine No. 1 from that original lot operated until it was donated to the B&W museum in 1943[94] (fig. 12.24).

These engines reveal the talents of O. E. Jorgensen and Ivar Knudsen, who stayed in close touch with diesel developments. With only minor refinements, the horsepower gradually increased to 180 bhp in 1906, 240 bhp

Figure 12.23. The second Burmeister & Wain designed diesel sold in 1904 to Klem & Krüger, Copenhagen. 80 bhp at 180 rpm. *B&W Archives*

Fig. 12.24. First Burmeister & Wain production engine delivered in 1904 to Larsens Vognfabrik. 40 bhp at 180 rpm. The engine is on display in B&W Museum, Copenhagen. *B&W Museum photo*

in 1907, and 360 bhp in 1908. The largest of the original engine family, a three-cylinder of 450 bhp and a four-cylinder of 600 bhp, were delivered to an Aalborg power plant during 1909. Their serial numbers of 179 and 180 provide a measure of B&W production over the first six years.[95]

A new design released that year reflected latest European practices yet bore its own characteristics (fig. 12.25). The upper crankcase housing on which sat the individual cylinders was a one-piece casting fully enclosing the crankcase. Long bolts whose heads clamped against flanges at the base of the cylinders tied the bedplate to the crankcase/cylinder structure. This directly transferred the tensile forces created by combustion chamber pressures into the bedplate.

A three-stage compressor with increased intercooling extended the time interval before having to clean carbon buildup off the compressor valves. The problem of excessive oil entrained in the compressed air still existed, and keeping the air as cool as possible was one way to live with it.

Pressure lubrication to the main and rod bearings and the piston pin ended the nuisance routine of filling oil cups and monitoring drip rates of the individual lubricators. Higher than expected leakage past the pressure-fed bearings allowed excessive lube oil to splash onto the cylinder walls and be

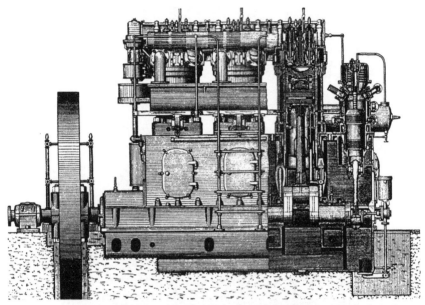

Fig. 12.25. 1909 Burmeister & Wain 240 bhp diesel with enclosed crankcase, pressure lubrication, and three-stage compressor.
Lehmann, *Rudolf Diesel and Burmeister & Wain, 1938*

carried by the rings into the combustion chamber. This consequent increase in consumption was stopped by simply adding a metal shield below the pistons to retain the oil in the crankcase. Pistons in higher output models were oil-cooled via drilled rod passages to the pins.

B&W made few changes to the fuel pump and governor but designed a new fuel nozzle, which remained unique to them for years. Heretofore, the rocker lever lifted the fuel valve upward off its nozzle seat, and the fuel/air charge entered through a single, vertical hole. B&W took the bold step to open a poppet-shaped fuel valve downward into the combustion chamber whereby the valve seat caused the fuel to spray as a conical sheet into the cylinder. It reportedly improved combustion over the superceded vertical spray without tending to carbon up as fast as did multi-hole nozzles.[96]

The fuel atomizer in the nozzle remained unchanged. It still combined the M.A.N. stack of perforated washers in conjunction with B&W's own design bushing with angled holes pointing toward the valve opening (fig. 12.26). The bushing was based on the idea B&W sent to Augsburg during the 1899 atomizer crisis.

Burmeister & Wain had wanted to offer marine engines ever since 1904 after seeing the success in Russia with the *Vandal* and *Ssarmat*. However, not until 1910 did B&W begin working in earnest on the large reversible engines, which began their dominance at sea over the next decade. This story is related in chapter 16.

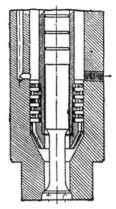

Fig. 12.26. Burmeister & Wain fuel atomizer.
Kennedy, *The Book of Modern Engines, c. 1912*

Werkspoor

The Nederlandsche Fabriek van Werktuigen en Spoorweg-Materieel in Amsterdam joined the diesel family in 1902. Werkspoor, as it was later known, made steam engines, locomotives, railway rolling stock, and sugar refining equipment. It had been building steam-powered machinery on the same site since 1840.[97]

Werkspoor entered the diesel business without ownership of patent rights. The Dutch Patent Office took a long holiday from 1869 to 1910, and no patents were issued during this interim either to Dutch citizens or to foreigners. M.A.N. took advantage of the legal hiatus in November 1902 by selling a manufacturing license to Werkspoor, which allowed them to build engines of 4 to 100 bhp for sale in Holland. There is some doubt as to whether Werkspoor pursued Augsburg for such a license.[98]

The contract called for M.A.N. to provide drawings, "know how," and, during assembly and test of the first engine, an engineer and an assembler. In return, Werkspoor agreed to pay 15,000 marks per year for three years plus a 10 percent royalty on all stationary engines. Opening up new areas was a clause referring to marine and locomotive diesels whereby M.A.N. would receive only a 5 percent royalty on such future engine sales. Royalties were to continue until 200,000 marks had been paid. M.A.N. further stipulated that nameplates read "Diesel Engine" and not "Patent Diesel."[99]

J. Muijsken, a director of Nederlandsche Fabriek from 1901 to 1928, assigned a promising young engineer to manage the new diesel program. The efforts of Dirk Christiaan Endert Jr. (1879–1951) over the next ten years greatly contributed to the success of the Dutch company's diesel engines.

Endert made the first of numerous visits to Augsburg in December 1902 to learn all he could about manufacturing techniques. During that three months stay, he met Buz and Diesel upon whom he seems to have made a favorable impression.[100]

Werkspoor began with one-cylinder DM Series engines of 12 and 15 bhp as well as a two-cylinder DM of 100 bhp. Progress was slow at first due to foundry problems and difficulties caused by using the power cylinder as the first stage of compression for the injection air. Only eight engines were able to be delivered or sold through December 1904.[101] In September of that year, Lauster had told Endert in Augsburg that he was afraid engine efficiency would drop if this design was abandoned and strongly advised against it.[102]

Nevertheless, in December 1904 Werkspoor switched to a two-stage compressor. The ones on lower-horsepower engines were still driven by linkage off the connecting rod. Larger models at some point during this period went to an offset pin drive on the front end of the crankshaft like the Burmeister & Wain.[103] These latter engines, incorporating Hesselman's horizontal fuel pump and a governor that operated off the vertical shaft, more resembled a Danish than a German engine (fig. 12.27).

Within three years the future of Werkspoor four-stroke diesels looked so promising that Muijsken told Endert he was recommending to his board that a special factory be erected to build them. It never happened.[104]

A totally new design originating probably from Endert in 1907 laid the foundation for marine diesels. (Werkspoor already supplied steam engines for vessels built in their adjacent shipyard.) Details including bore and stroke are lacking, but a circa 1908 stationary model shows the general configurations (fig. 12.28). The trunk piston, enclosed engine produced 250 bhp at 250 rpm from its three cylinders.

An important feature was the long, through-bolts extending down from the cylinder heads to transfer the combustion pressures directly into the bedplate. Werkspoor may have preceded Burmeister & Wain with the through-bolts, but in any event the design is contemporaneous.

A four-cylinder, 200 bhp engine of this type was installed in the schooner *San Antonio* as auxiliary power and drove a variable-pitch propeller because of not being reversible. Her maiden voyage from Amsterdam to Rotterdam was made November 5, 1909. The engine performed so well during a round trip to Stockholm from Amsterdam a few weeks later that the sails were left furled. Endert was in charge on these first trips. The *San Antonio* still plied the waters with its original engine thirty years later.[105]

The *Comelis*, a second schooner powered by a Werkspoor auxiliary, went to her owners in November 1910, but she had the misfortune to run aground on the second voyage through no fault of her engine.

At the Brussels Exhibition in 1910, Werkspoor exhibited a crosshead, four-cylinder diesel of 600 bhp at 215 rpm with a bore and stroke of 500 × 650 mm. Its total weight, including air tanks came to about 120 tons.[106] The crosshead configuration was chosen even though other makers continued using trunk pistons for four-stroke engines of similar power per cylinder. Full load bsfc was 0.37 lbs/bhp-hr, burning a fuel with a heating value of 18,000 Btu/lb.

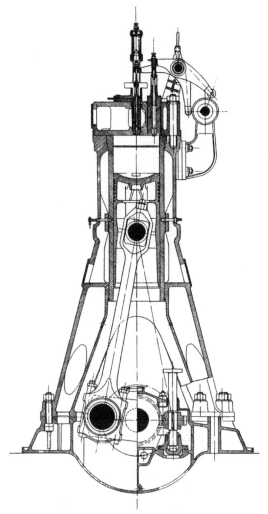

Fig. 12.27. Werkspoor c. 1904, 80 bhp trunk piston
diesel with front-mounted, two-stage air compressor.
Chalkley, *Diesel Engines for Land and Marine Work, 1917*

The frame consisted of three major castings running the length of the engine. A stiff lower one formed the bedplate. An upper one, to which individual cylinder heads were bolted, held the liners and served as the water jacket. Between these two was a deep intermediate casting that enclosed the crankcase. Tying the three elements together were ten long, vertical rods longitudinally spaced in pairs along the bedplate at each of the five water-cooled main bearings. The rods of each pair were placed as close together as possible. Side thrust loads from the connecting rods went through crosshead guides attached to the wall of the crankcase enclosure. Endert reported that several

Fig. 12.28. Circa 1908 Werkspoor three-cylinder,
250 bhp stationary diesel. *MAN Archives*

years earlier they had changed from the troublesome bronze to a "white metal" on crosshead guide bearings."[107] Large hinged doors on the enclosure offered easy access to the crankcase interior.

Pistons were water-cooled. Vertical tubes fastened to them slid in fixed supply and drain tubes attached to the crankcase enclosure. Stuffing boxes on the ends of the stationary tubes sealed against water leaks. This design gave trouble in service because excessive misalignment caused tubes to break off at their base. The problem was later solved by going to three tubes having greater clearances where the reciprocating tube now slid between fixed inner and outer tubes with the outer tubes leading to a drain line (fig. 12.29).

The piston rod passed through its own stuffing box, which was part of a plate forming a lid to seal off round holes in the top side of the crankcase casting. This kept out both cylinder heat and carboned oil drippings from entering the oil sump.

A pump driven off the crank pin for the compressor fed lube oil under pressure to the main and rod bearings and, via drilled rod passages, to the crosshead pins. Around 1912 this modern feature was dropped for a return to bearing lubrication by gravity feed from oil cups. G. J. Lugt (1885–1948), a gifted designer who became Werkspoor's technical director in 1916, commented on the reversion: We "wanted to keep out of the design everything that might scare off the engineer accustomed to steam engine practice." He expected to "feel such moving parts as were accessible."[108] A greater leakage past the pressure lubricated bearings hindered this dangerous caress of machinery. Being splashed with the oil flying around inside the enclosed

THE WERKSPOOR DIESEL ENGINE

Fig. 12.29. Werkspoor piston cooling with three-tube method, 1912–14.
Trans. Institution of Engineers & Shipbuilders in Scotland

crankcase was not to his liking. By 1918, however, Werkspoor gradually returned to pressure lubrication.

The fuel pump and governor of this stationary engine were still eccentric operated off the vertical shaft as on earlier Werkspoor engines, but there was a plunger pump for each cylinder with all discharging into a common pumping chamber. The four plungers were reciprocated by the same eccentric on the vertical shaft, and the throttle, with an overriding governor, controlled a single pump suction valve. After exiting the pumping chamber, the fuel entered a distributor box containing an adjustable equalizing valve for each cylinder and then into a separate line leading to the nozzle. One may recall that Hesselman had found it was possible to do this satisfactorily in four-cylinder engines.

Werkspoor employed eccentrics instead of intake and exhaust cams on the high mount camshaft to eliminate the noisy valve opening clatter associated with most diesels of the period (fig. 12.30). Higher speeds had only intensified this problem. As a valve eccentric rotated downward, it pulled on a short rod linked by a pin to a first intermediate rocker lever pivoting on the same shaft, which held the fuel and starting air rocker levers. The downward eccentric movement raised the first lever's opposite end against the underside end of a second lever that pivoted on its own shaft. This second lever's other end then pushed down to open the valve. Conventional cam designs returned with the advent of reversible marine engines.

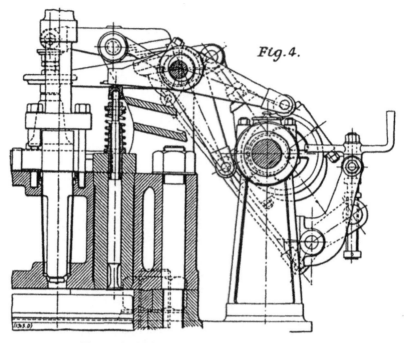

Fig. 12.30. Valve operating mechanism on
c. 1910 Werkspoor 600 bhp diesel. *Engineering*

An engine introduced in 1910 incorporated a new camshaft drive method. One reason given for the change was that the rotating vertical shaft caused assembly and operational troubles related to holding gear center tolerances for the paired gears at both shaft ends. The solution was to transfer motion through long connecting rods reciprocated by eccentrics. The crankshaft turned a short, parallel auxiliary shaft with the eccentrics that turned at camshaft speed. The rods drove a small crankshaft above which was an extension of the camshaft. At first there were only two rods driven by eccentrics spaced at 180 degrees that could impart a small compression loading. This was changed to three rods at 120 degrees in 1912 and later to four at 90 degrees, depending on the effort needed to turn the camshaft. Marine engines particularly required the extra rods due to the added turning effort of their reversing mechanism. Since the rods operated only in tension, they could be slenderer and without need of an intermediate support. On larger six-cylinder engines, the four rods were placed between the middle cylinders.

Open to See

Endert's boss and head of both the steam and diesel engine departments was Cornelis Kloos (1856–1931), a respected engineer with a degree from the Technische Hochschule (ETH) in Zürich and technical director of the Nederlandsche Fabriek. According to Endert, Kloos took little interest in the diesel during its first years at Werkspoor, and in time this would lead to both product problems and conflicts between him and Endert.[109] Kloos's design aim was to simplify major servicing procedures as much as possible by following a long-standing steam engine practice of offering ready accessibility to internal working parts. This applied particularly to the inspection and removal of pistons. Kloos was also swayed by a refusal of steam-bred marine engineers to accept anything foreign to their experience.

A new, more open engine structure reflecting his inclinations made its debut in 1912. It now allowed the pistons to be inspected and removed without a need to first take off the cylinder heads. Such ease of piston servicing became a much-touted Werkspoor feature and would be offered on all their new diesels whether crosshead or trunk piston.

An example of this design is seen in a stationary and marine three-cylinder engine with a 400 × 500 mm bore and stroke and an output of 200 bhp at 200 rpm[110] (fig. 12.31). Three integral cylinder head and upper liner halves were inserted into a casting running the length of the engine. This upper casting became a true beam supported only by shoulders on the long vertical rods tying the top beam to the under side of the baseplate for the direct transfer of combustion forces.

Between the rod-supported beam and the crankcase enclosure casting was an open space deep enough for a piston to be taken sideways from the engine. A separate and non-waterjacketed lower liner half was piloted in the

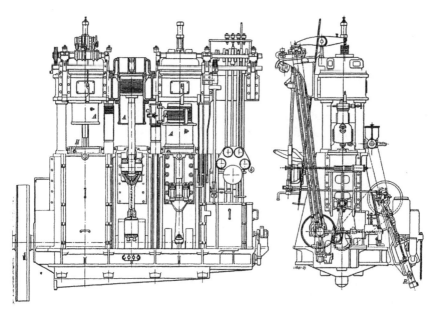

Fig. 12.31. Werkspoor three-cylinder, 200 bhp diesel. Note the two-piece liner whose upper half is cast integral with the cylinder head, 1912.
Engineering

upper half of the integral head/liner. The two liner halves were finish bored as an assembly to ensure concentricity and were made to be bolted together in only one position.

A built-in jack facilitated piston removal. Lead screws were machined on the "exposed" part of several tie rods. (Two sufficed for three- and four-cylinder engines.) A threaded sleeve on the lead screw that moved up and down by a hand crank formed a jack from which attached fingers could be inserted under the piston. With the crank throw positioned at bottom center, the piston was unbolted from its rod, and the lower liner half was detached from the upper. The piston and lower liner half, supported by the jack, were swung away from the engine. In order to merely inspect a piston without removing it, the lower liner half was separated from the upper and pulled down to expose the piston and rings. The upper cylinder wall was also visible for inspection at this time.

The trunk piston Werkspoor engines lacking the open area above the enclosed crankcase were also made to remove the pistons from below. In this design there was a similarly attached lower liner half, but it was also split vertically and bolted together along the lengthwise axis of the engine. After removing the finer segment adjacent to an open crankcase door, the piston was rotated out in a vertical arc along with its connecting rod where the rod, still fastened to its crank journal, acted as a pivot arm. The piston again had to be positioned at bottom center.

Kloos went one step further in his desire for a "visible engine." The next series of larger Werkspoor diesels had a frame completely exposed from the bedplate to the upper cylinder beam (fig. 12.32). In addition to the forged, vertical rods, diagonal tie rods ran from one side of the baseplate up to bosses on the opposite side of the cylinder beam to give the structure a transverse stiffness. Crisscrossing diagonal rods were placed on the ends of the engine and on the abutting ends of the upper beam sections when the upper beam was made in two pieces and bolted together as for a six-cylinder engine. This 1912 design was first used for a series of 560 × 1,000 mm, 1,100 bhp at 125 rpm stationary and marine diesels. It lasted into the 1920s.[111]

Taking the side thrust of the crosshead guide was a vertical support casting extending from the bedplate to the cylinder beam. The lower end was rigidly bolted to the bedplate, but the upper end was non-rigidly attached to the outside of the upper beam to allow a differential vertical movement between the support casting and the tie rods. Such a connection eliminated structural stresses resulting from their differing thermal and tensile expansions.

Not all went as planned with the Kloos design, according to Endert who, although a participant in it, had advised his boss against adopting such a radical new approach. Until this time Kloos had left the diesel program mostly in Endert's hands, but now overrode him by insisting on adherence to the open frame concept so dear to steam engineers. Endert wrote that "Regardless of the good ideas Kloos did carry out, this variation led to total

Fig. 12.32. Open frame design of 1912 design
1,100 bhp Werkspoor diesel. *Engineering*

failure and cost Werkspoor a vast amount of money."[112] Unfortunately, no record of the engine's actual weaknesses can be found, although it may be reasonable to assume that a general stiffening of the structure was necessary. There is one reference to "tremors" in the diagonals during a sea trial of the tanker *Juno*, although the problem was apparently solved before more engines were shipped.[113]

Endert was disheartened over the outcome of these engines. No compromise was possible in the diesel department because Kloos would remain his superior, but his employment contract ensured him continuity with Werkspoor. Muijsken placed him in charge of their general machine factory after a vacancy opened in that position. Endert said that he continued to offer his services to the diesel department during their endeavors to solve the initial problems with the first open-frame engines. Endert left Werkspoor in 1914 and became manager of the engine design division of the Rotterdam Drydock Society (*De Rotterdamsche Droogdok Maatschappij NV*). He was made a director in 1918 of this very important organization and over the years became its most influential management member almost until his death in 1951.[114]

Floating Fuel Tanks

The patented fuel metering system introduced on the open-frame engines differed from all others[115] (fig. 12.33). A single-plunger pump was still driven by an eccentric off the auxiliary shaft that also reciprocated the rods of the

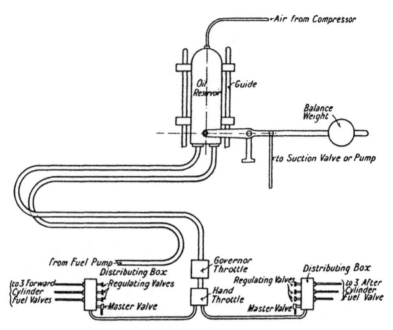

Fig. 12.33. Schematic of Werkspoor floating tank for fuel metering.
The Engineer, 1912

camshaft drive. New in the pump were ball suction and delivery valves rather than the conventional guided poppet types. The pump delivered fuel to a tee connection at the underside of a free-floating tank slidable in vertical guides. One tee branch went to the tank and another into a line leading to a governor-controlled valve in series with a hand throttle valve. Fuel exiting the throttle valve split into two lines going to two distributor "blocks" with each serving three cylinders of the six-cylinder engine. Three needle valves in each block individually calibrated the fuel flow to an injector in the usual fashion. The top end of the free-floating tank was connected to a line carrying injection air to the nozzles.

The system's novelty centered on the function of the free-floating tank. A rod extending downward from the tank acted on one end of a pivoting balance lever whose other end held a slidable counterweight. The weight was placed on the lever so that the tank "floated" in an equilibrium position when it contained enough fuel for about fifteen minutes of running at full load. If the load decreased by a closing of the hand throttle valve or the governor valve, the fuel level rose in the tank and lifted the weight. Attached to the balance lever between the weight and the pivot point was another rod extending down to a needle valve in the suction line leading to the pump. As the rod lifted it began closing off the suction line and reducing the fuel delivered to the tank. Vertical movement of the tank in actual service was very slight.

Advantages accrued from the system.[116] First was a need for only one fuel delivery pump. Of greater value was that with a wide-open throttle other cylinders could not be overfueled even if a distributor needle valve had to be completely closed to service a nozzle with the engine running. The fuel pressure in the lines going to the other injectors remained unchanged because the air pressure acting on the fuel in the floating tank did not change.

Werkspoor added to the system's reliability by placing a spare pump next to the one in operation. If that one failed or required maintenance, it could be valved off and the spare connected without stopping the engine because of the approximately fifteen minutes' worth of fuel in the floating tank to supply the engine until the switch could be made. The unsophisticated and relatively trouble-free system was used until the mid-1920s.

Production figures released by the Nederlandsche Fabriek in June 1912 show that from 1903 until that date the company had delivered 219 stationary diesel engines. These include land and non-propulsion shipboard auxiliary units of 8 to 1,200 bhp. There were twenty-two marine engines ranging from 200 to 1,100 bhp.[117]

Werkspoor licensed the North Eastern Marine Engineering Co. to build diesels in their Wallsend, England, factory, but the events of 1914–1918 delayed production until 1921. The Bofers Co. in Sweden, The Hawthorn Leslie Co. in England, and the Pacific Diesel Engine Co. in Oakland, California, were also licensed prior to 1920. All built numerous Werkspoor engines, which performed well according to most reports.[118]

A Werkspoor merger in 1954 with Koninklijke Machinefabriek Gebr. Stork en Co., at Hengelo, changed the company name, in shortened form, to VMF/Stork-Werkspoor. Stork-Werkspoor diesels were sold until Wärtsilä bought the company in 1990 when the new division became Stork-Wärtsilä Diesel BV. Although the Werkspoor name is gone, their tradition of innovation is not forgotten.

Fried. Krupp

When Krupp's dormant diesel program came to life again at their Germaniawerft division in 1904, the focus at first was on submarine engines. A later desire to reenter the stationary diesel market led to comparative tests between the two-stroke or four-stroke cycles to determine which held the most promise. The four-stroke won out, no doubt because of M.A.N.'s success with the DM series. Access to its drawings and manufacture through Krupp's active partnership in the Allgemeine was further reason for them to follow this design. Production of slow speed, stationary diesels began quickly in 1908 with eleven engine models having cylinder outputs from 30 to 125 bhp.[119] All were adapted to run on coal tar oils.

Krupp made several revisions to the "DM," one of which was soon abandoned. A pressed-in, "dry" cylinder liner made contact with the A-frame/

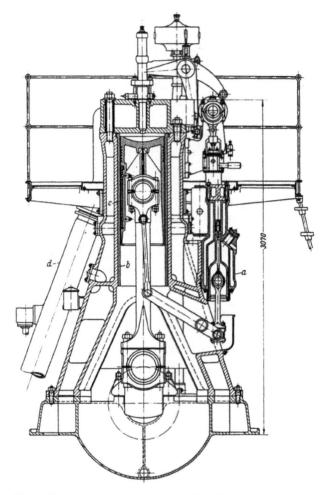

Fig. 12.34. Krupp Germaniawerft stationary diesel, 1908. Note the "dry" liner and "slide-shoe" *c* on piston. Sass, *Geschichte, 1962*

jacket casting over the liner's full length (fig. 12.34). Replacement difficulties were not worth any expectation of increased liner life.

A sliding shoe fastened into the thrust wall side of the piston acted to decrease piston wear. The results were promising enough for Krupp to patent the idea and several other companies to adopt it.[120]

Further development work made possible a reduction in engine-specific weights from an original 200 down to 150 kg/bhp. The largest engine was a four-cylinder of 500 bhp at 175 rpm with a bore/stroke of 475 × 700 mm. In all, Kiel delivered about 250 engines.

A fast-growing emphasis on marine and submarine diesels at the Germaniawerft (chapter 15) prompted Krupp to sell the stationary engine manufacturing rights to Eberhardt & Sehmer in Saarbrücken. This company, too, made a good number of them until the outbreak of war in 1914 ended production. Larger stationary engines were again made at Kiel after the war.

Willans & Robinson Ltd.

Willans & Robinson Ltd. of Rugby, England, was well known for its triple-expansion steam engines. These high-speed machines, capable of 500 rpm, were directly coupled to smaller and faster turning electric generators and thus did away with belt drives as required with conventional steam engines. Other company products included steam turbine governors and high-lift centrifugal pumps. Peter William Willans (1851–1892), Willans & Robinson's (W&R) gifted founder and inventor of the unique and acclaimed engine, had died in an accident before the diesel's advent.[121]

The company began negotiating for a manufacturing license from the Diesel Engine Co. in September 1904 and finally consummated one on the following June 21. It was an interesting agreement in that W&R had to fulfill a "First Trial Order" of six engines and then a second one of 100 cylinders, all made to drawings furnished by the DEC. After this the DEC agreed to fill 25 percent of all orders it received from W&R.[122] The license was weighted toward the DEC, and it proved a warning of what W&R would be up against the next few years. Creating the difficulties was a slackening of the DEC's United Kingdom engine orders, which meant that Carels, who by then owned a majority interest in DEC, were shipping fewer engines across the Channel.[123] Because of this, it is quite probable that only engines for the "First Trial Order" were built in 1906 and that few if any of those for second one were ever put into the field under the contract's terms. The Sulzers, who were in the same position as W&R with the DEC, sometimes conferred with the Rugby people over the problem of Carels taking the lion's share of the DEC's manufacturing orders. W&R never enjoyed a satisfactory arrangement with the DEC and fought an uphill battle until the agreement ended in September 1910. After that time W&R was free to sell engines under its own name. No sales records are known to exist.[124]

Early Willans & Robinson four-stroke diesels borrowed from MAN and Carels. A front-mounted Reavell compressor and a single-plunger fuel pump driven by an eccentric off the vertical shaft distinguished them from Mirrlees's similar looking A-frame design.

One small but significant departure in later W&R cylinder heads had a distinct advantage over its contemporaries. "Two valve" heads (one intake and one exhaust valve) were notoriously prone to crack across the short distance between the exhaust valve seat and the axial injector opening, which normally lay on a direct line between the valve seats. Conventional heads often contained enough volume directly above the nozzle opening to place a cast-in sleeve, which later held the injector. The sleeve served as a core in the mold. This extra material added heat-caused stresses across that narrow bridge between the valve seat and nozzle during both the casting process and engine operation. W&R eliminated the cast sleeve and its encasing bulk of metal by inserting a machined one into the head after it was cast. The water-cooled sleeve was held tightly against its seat at the nozzle opening to seal against combustion pressures.[125]

W&R exhibited a two-stroke, 75 bhp at 450 rpm diesel at the Royal Agricultural Show in Norwich during June 1911[126] (fig. 12.35). The three-cylinder 165 × 280 mm bore and stroke engine had eccentric-operated sleeve valves most likely based on the Knight design. Scavenge air at about 3 psi came from a pumping chamber formed between the lower, larger-diameter portion of a stepped piston and the end wall above it. There is no record of this engine ever going into production.

Later that year Willans & Robinson introduced a four-stroke, four-cylinder diesel with a bore and stroke of 15 × 22 inches producing 335 bhp at 220 rpm. It followed an earlier practice of using a single fuel pump plunger and a distributor "box."

The lubrication system was new. On the back of each A-frame column were three small plunger pumps moved by linkage from the connecting rod.

Fig. 12.35. Willans & Robinson two-stroke, 75 bhp diesel, 1911.
Engineering, June 30, 1911

One pump sent oil to a centrifugal ring on the crank pin, the second fed two opposed cylinder ports for piston oiling, and the third supplied oil to a lower cylinder port, which, via a piston groove, led to drilled holes opening onto the wrist pin. An adjustable, sight feed drip lubricator metered oil to the suction side of each individual pump. The claimed advantage was that such control over the oil prevented wastage and allowed the working parts of the engine to be left exposed.[127]

Most of the noise during engine operation reportedly came from the air being sucked into the cylinders. This was a common comment as well as complaint about diesels at the time.

The first three of these 335 bhp engines went to a nitrate mine in Argentina where it drove a Siemens generator. Because of the altitude at the site the maximum overload power rating was reduced to 230 bhp. Official tests of the next two engines included a twenty-four-hour run at 280 bhp. Full-load fuel consumption was 0.44 lbs/bhp-hr.[128]

A dispute between the Diesel Engine Co. and a customer in 1912 peripherally involved Willans & Robinson. When DEC sued for nonpayment, the defense countered that the delivered W&R 225 bhp/195 rpm engine was "inferior" to the Carels, which the customer had expected to get. A newspaper account said that DEC made no engines and bought them from other manufacturers, "including Willans & Robinson of Rugby. The engine in dispute was made by that firm under the Diesel patents. . . . [It] was similar to those supplied to several other firms which were working perfectly satisfactorily and evidence would be called to prove it."[129] The case was settled out of court.

By 1913 the Rugby factory offered thirteen engines in three cylinder sizes ranging from 50 bhp at 250 rpm to 960 bhp at 187 rpm. The largest had six cylinders with a bore and stroke of 510 × 720 mm. All were trunk piston, four-stroke with a one-piece casting for the A-frame and cylinder jacket. Indicated mep was about 103 psi at a sustainable full load, and the bsfc over the operating range compared favorably with Mirrlees engines. In the first five months of that year, W&R had received orders for seventeen engines. Over twenty-five were under construction in May.[130]

One clever design feature greatly simplified an often too-frequent removal of the fuel valve. As has been seen, others used either a two-piece lever or, like Carels, a double lever system. W&R still directly lifted the valve with a one-piece lever in the standard fashion by placing the cam follower on the "underside" of the cam. However, by simply casting a recess in the cylinder head, the cam end of the lever could be rotated into the recess and thereby lift the other end up and away from its contact on top of the fuel valve. Such little "tricks" were typical of the attention to detail in the design and manufacturing W&R gave its engines.

Willans & Robinson licensed the Dow Pump Co. of Alameda, California, on March 27, 1912, after its president George A. Dow had visited England and Germany in search of an engine to build. The next year a three-cylinder

W&R-type engine was on test at Dow. Between 1913 and 1923, Dow Pump & Diesel Engine Co. made about twenty diesels, two of which went into the 240-foot motorship *Libby Maine* in 1918. These six-cylinder, 12 × 18 inch, 320 bhp at 250 rpm engines were reversible and drove through a 2.5:1 reduction gear.[131]

In 1916 the Dick Kerr Co. acquired Willans & Robinson, and it in turn became a part of the new English Electric Co. in 1919. W&R was then known as the Willans Works, Rugby, and later simply the Rugby works of English Electric. It served as the main diesel engine factory until English Electric was absorbed into the British General Electric Co. in 1968.[132]

Conclusion

Of the accounted-for first-generation licensees postponing production, it was Sulzer Bros. (now Sulzer Diesel, a subsidiary of Wärtsillä) and Burmeister & Wain (now MAN SE) who would be among the market leaders. Nevertheless, M.A.N. enjoyed dominance prior to 1918. For example, they accounted for 45 percent of world diesel engine production at the time of Diesel's death in 1913. Werkspoor's D. C. Endert commented after a 1910 visit to Augsburg that M.A.N. had "already built some thousands of cylinders; Amsterdam was hardly to the 200th."[133]

"The Diesel engine made its first appearance while I was a boy, and I feel that I have grown up with it for all my life. . . . For some reason this perfectly matter-of-fact heat engine contrived for many years to weave around itself an atmosphere of mysticism. Distinguished engineers were wont to speak of it with reverence and even awe, and one was expected to stand bareheaded in its presence, while any criticism, however kindly meant, was regarded almost as blasphemy."

—Sir Harry R. Ricardo[134]

Notes

1. Charles Day, "Address by the President," in *Proceedings*, Institution of Mechanical Engineers (1934), 200.
2. *Sulzer Brothers Ltd.: Historical Retrospect, Technical Developments,* Sulzer Pub. 6410e (c. 1951), 5–15. Sulzer men traditionally added the wife's maiden name to their own. Salomon Sulzer-Sulzer married a second cousin with the Sulzer name.
3. On June 24, 1913, the family joint stock company was split into three entities: Sulzer Unternehmungen AG, a holding company owning Gebrüder Sulzer AG, Winterthur, and Gebrüder Sulzer AG, Ludwigshafen-am-Rhein. In 1939 the latter was sold to a German company. In 1941 the holding company, with the same management as Gebrüder Sulzer AG at Winterthur, was absorbed into that company.
4. Kurt Schnauffer, "Lizenzverträge und Erstentwicklung des Dieselmotors im Inund Ausland 1893–1909" (1958), 41–2.
5. "Fifty Years of Sulzer Diesel Engines," *Sulzer Technical Review*, No. 2 (1947): 4.
6. Ibid., and Schnauffer, "Lizenzverträge," 42.
7. Kurt Engeler, "1903–1952: 50 Jahre Sulzer Dieselmotoren-Produktion. Technische Statistik"

(June 7 1953), 3. He gives engine types, nomenclature, drawing numbers, and so on for Sulzer engines built to 1952. As of February 24, 1904, the DEC had ordered twenty-seven engines from Sulzer: thirteen with a combined output of 450 bhp stayed in Britain, and 14 totaling 860 bhp were for export. February 24, 1904, letter from the Diesel Engine Co. Ltd. to the Allgemeine. MAN Archives.

8. W. Bangerter, "Register, Dokumenten-Ordner," 1. He added to and amended Engeler's work with his own important "Documentation" dated January 26, 1967. David T. Brown kindly updated this in 1991. As one of Sulzer's unofficial historians, he has been most helpful to the author in providing archival information and in arranging productive visits in Winterthur.

9. Schnauffer, "Lizenzverträge," The MAN Archives have accurate engine records for Sulzer's early years because of their royalty payments to the Allgemeine.

10. Report enclosed with Rotter to Harris letter of May 27, 1911. He was shown Swiss installations having both four-stroke and two-stroke engines. Wisconsin State Historical Library.

11. Ibid., report on mill in Bürglen, Thurgau Canton.

12. Alfred Büchi, *Der Dieselmotor und seine neuzeitliche Entwicklung,* a lecture before Ruhr district VDI (April 15, 1914), 2.

13. A. P. Chalkley, *Diesel Engines for Land and Marine Work,* 4th ed. (New York: Van Nostrand, 1917), 98.

14. "Fifty Years of Sulzer," 8, and David T. Brown, *A Low-Speed Tradition in Stationary Plant* (1985), 8–9.

15. Sulzer, *Preisliste über langsamgehende Zweitakt Dieselmotoren (Modell Z),* in Sulzer Archives (June 1910), 3.

16. Brown, *A Low-Speed,* 13. Harland & Wolff bought three Sulzer "D" engines (sold through London's Diesel Engine Co.) in 1903 to drive generators. Michael Moss & John R. Hume, *Shipbuilders to the World, 125 Years of Harland & Wolff, Belfast 1861–1986* (Belfast, Ireland: Blackstaff, 1986), 113. The 6Z300 reportedly ran 16,000 hours before its first major overhaul. "Fifty Years of Sulzer," 9.

17. Extract from Sulzer biographical information on their management over the years, n.d., 7. Sulzer Archives. The Sulzer partners at the time of the change to a holding company in 1914 included Jakob Sulzer-Imhoof, Eduard Sulzer-Ziegler, Carl Sulzer-Schmid, Alfred Sulzer-Seifert, Robert Sulzer, and Dr. Hans Sulzer. From an October 15 contract draft between Busch-Sulzer Bros. Diesel Engine Co. and the newly formed Sulzer entities. This was part of a legal change needed in the US as well as in Switzerland after the death of Adolphus Busch. Busch-Sulzer papers, Wisconsin State Historical Library.

18. Sulzer Management, ibid., and Georg Aue, "Aus der Anfangszeit des Dieselmotors," *Industriearchäologie* (1987), 5. Walter Kilchenmann, Samuel's son, headed the Sulzer diesel department from 1948 until 1974. His interview with the author in 1972 gave valuable insight into the company's history and operation.

19. Alfred J. Büchi, "Personal Reflections on Forty Years of Diesel-Engine Development," in *Mechanical Engineering* (March 1939), 213–16. From Büchi papers, courtesy of Georg Aue.

20. Letters in the Wisconsin State Historical Library.

21. Schnauffer, "Lizenzverträge," 43–44.

22. Max Rotter, "The Diesel Engine in America," *Trans., Int'l. Engineering Congress, 1915,* San Francisco, Mechanical Engineering. Vol. 204. Brown, *A Low-Speed,* gives the 4K125 as 450 × 510 mm and 250 rpm. Sulzer could possibly have gone to this later.

23. Richard H. Lytle, "History of the Busch Diesel Companies, 1897–1922," MA Thesis (St. Louis: Washington University, 1962), 84. See also *Diesel Progress* (1948), 34.

24. For a Carels, ABC, ACEC history see: Gilbert Boerjan, "Zeventig Jaar Anglo-Belgian Company-Dieselmotoren te Gent," *Geschiedenis van Techniek en Industríele Cultuur,* No. 17 (1987): 40–55. Carels archival material is in the Museum voor Industríele Archologie en Textiel, Ghent, where Boerjan is a curator. (Ghent in Flemish: Gent; in French speaking Belgium: Gand) *Engineering* (1911), 112, has an obituary of "The late Mr. Gustave Carels," which adds to the Boerjan story.

25. A photo in Boerjan, "Zeventig," 40, shows the engine with a crosshead of the pre-DM series.

26. Carels, April 15, 1902, letter to the Allgemeine. MAN Archives.

27. H. Ade Clark, "The Diesel Engine," in *Proceedings,* Institution of Mechanical Engineers, parts 3–4 (1904), 395–455. This valuable paper has data from numerous tests of 1902–1903 diesel engines in England.

28. Schnauffer, "Lizenzverträge," 53.
29. Carels, January 26, 1904, letter to the Allgemeine. An August 7, 1903, letter from Johanning at the Allgemeine to an English acquaintance said that "he fully conceded that the [DEC] can buy today engines just as good [as M.A.N.] from Carels Frères or Gebrüder Sulzer . . ." MAN Archives.
30. Schnauffer, "Lizenzverträge," 54. The reported number of diesel engines in England by December 1904 was sixty with a combined output of 5,000 bhp. "500 Horse-Power Diesel Engine at the Liège Exhibition," *Engineering,* 79 (1905): 738.
31. Schnauffer, "Lizenzverträge."
32. Jean Santschi, "Memories of Dr. Diesel," a paper read at a Conference for Teachers of Diesel Engineering in Chicago, June 23, 1947. Santschi retired from Nordberg as Superintendent of Service & Installation near that date.
33. Büchi, "Personal Reflections," 213.
34. Gaston Carels wrote in 1913 that "in India, Diesel engines, aggregating many thousands of horsepower, are driving cotton mills pumping stations and electrical power stations. . . ." *Cassier's Magazine,* vol. 43 (London: 1913), 150.
35. "Liège Exhibition," 736–38.
36. SEM (Société d'Electricité et de Méchanique), *Evolution of the SEM-Carels Diesel Engine* (c. 1960), 2.
37. An extensive test made by Michael Longridge is incorporated in Chalkley, *Diesel Engines,* 158–67. The cylinder serial nos. 54, 55, and 56 indicate the engine was probably assembled during 1904.
38. Ibid., 163
39. Testimony of Gaston L. Carels, *Royal Commission on Fuels and Engines,* pt. 2 (London: British Admiralty, 1913), 409.
40. Ibid. and Chalkley, *Diesel Engines,* 223–25, and Carels, *Cassier's,* 147.
41. Giorgio Supino, *Land and Marine Diesel Engines,* 3rd ed. (London: Charles Griffin & Co., 1918), 206, helps us by stating that "when the fuel injection pump is driven from the vertical shaft, one pump usually serves to supply all cylinders, whereas if driven from the horizontal camshaft separate pumps are usually provided for each cylinder."
42. C. J. Hawkes, *Royal Commission on Fuels & Engines,* 12. Hawkes also wrote the summary for the lengthy report.
43. *Engineering* (August 1, 1913), 162; and Carels, *Cassier's,* 150.
44. *Engineering* (September 12, 1913). The capacity of the Worthington pumps was 58,000 gpm working against a static head of 48 feet.
45. "Internal-Combustion Engine for Navy Tank Steamer; Ghent Exhibition," *Engineering* (August 1, 1913), 161–62. See also March 20, 1914, 376–79, for more on these engines.
46. *Evolution of the SEM-Carels Diesel Engine,* 4; *The Motor Ship* (April 1970), 53; and *1991 Diesel & Gas Turbine Catalog,* 319.
47. 172,924 shares from the public subscription applied toward the land purchase. "Certificate of Incorporation," Public Records Office, Kew (London), and the *East Anglian Daily Times* (April 1, 1912), 5, a partial copy of which is in the British Museum Newspaper Library at Colindale (London). Your author was told that the missing part of the partially restored issue was destroyed as a result of bomb damage in the 1940–1945 war. Unfortunately, on the missing pages was a copy of the CDEM Prospectus, which has not (yet!) been found.
48. In the Allgemeine annual report for 1906, the sale of its stock interest in the Diesel Engine Co. was noted. *Allgemeine 8. Ordentliche Generalversammlung am 22 Feb. 1901 für das 8. Geschäftsjahr 1906,* 2. It is possible that Carels gained control of the DEC as a result of this disposal, but details of this and ensuing stock sales are unavailable.
49. *Engineering* (August 1, 1913), 161, and (September 12, 1913), 49, for example. Thornycroft made the comment during his testimony before the *Royal Commission,* 459.
50. Jean Santschi, "Memories," 7, 8.
51. J. R. Harris to attorney Eugene H. Angert (New York City: August 5, 1912), 11. All quotations on the subject are from this letter. Busch-Sulzer papers, Wisconsin State Historical Library. No further mention of Haynie was found there. Haynie resurfaces in 1916 when he became head of the Marine Oil Engine Co., which had the US rights to the Swedish Skandia oil engine. *The Gas Engine,* June 1916, 291, said that Haynie, "a gentleman of long experience in the oil engine business" had been with Bolinders, another Swedish company making a successful line of vaporizing, hot bulb (semi-diesel) oil marine engines.

52. Charles LeGrand, "Power Plant of the Burro Mountain Copper Co.," *Trans., The American Institute. of Mining Engineers* 55 (September 1916), 208–17, describes the installation that went in service around April 1915. Carels had received the order in 1913 from Haynie.

53. *Dictionary of American Biography*, vol. 13, (New York: Scribner, 1934), 546–47.

54. Nordberg Manufacturing Co., *The Nordberg High Compression Two-Cycle Oil Engine*, Bulletin 27 (December 1914), 10 pages.

55. *Diesel Progress* (May 1948), 74.

56. Ibid. The well-qualified John W. Anderson, a contemporary of the period and writer of the historical articles for this special issue, says the son went to Europe. R. W. Bayerlein, a retired Nordberg vice president starting employment there in 1917 at age seventeen, told the author in 1992 that it was definitely the father and not the son. Bayerlein also said that B. V. E. Nordberg had worked for him.

57. A story on the back of a Nordberg photo of the Burro Mountain engines gives the history of these engines and Santschi's early US involvement. Nordberg collection, MHT Archives, Smithsonian Institution, Washington, DC.

58. "Memoirs," *Proceedings*, Institution of Mechanical Engineers, vol. 162 (1950), 263–64. Mirrlees dropped the Yaryan from the company name when it reorganized in 1899.

59. Letters from Day to the Allgemeine dated October 18 and 28, 1902, bracket the time of the visit. This correspondence and numerous other letters in the MAN Archives provide much more accurate information as to dates and events than have been found heretofore in later articles on Mirrlees history.

60. Day, "Address," 200.

61. A March 9, 1904, letter from Day to the Allgemeine said that the 35 bhp engine had been running in the shop "for about six months." The second engine was "just going into operation in the plant." An earlier letter of August 21, 1903, spoke of problems with the drawings and asked them to allow Windeler to visit MAN again. Changes to the fuel pump, and so forth were reported in an October 29 letter. MAN Archives.

62. April 16, 1904, Day letter to the Allgemeine. MAN Archives.

63. In answer to requests by the Allgemeine, Day acknowledged in a September 29, 1904, letter that they had no brochures or price lists available yet but would within a few weeks. Of interest in the existing papers of this period are references by Mirrlees as to their difficulties with the Diesel Engine Co. (DEC). The low prices that the DEC wanted from Mirrlees precluded them from selling engines to London. Comments made by Day during a May 1904 visit at Augsburg as recorded in a M.A.N. internal memo dated May 16, 1904. MAN Archives.

64. Day letter to the Science Museum, London, March 14, 1947. Mirrlees file. The Science Museum Archives.

65. Charles Day. 1908. British Patent No. 6,606, issued 1908. Two other Day patents, Nos. 25,180 and 26,089 of 1909, dealt with reversing methods for marine engines. These were never put into production.

66. Mirrlees, *Instructions for Erecting and Working "Mirrlees-Diesel" Oil Engines*, No. 60/13. Inscribed by the factory is the Book No. 963 and the date March 31, 1916.

67. *Report of Public Lecture on the Diesel Oil Engine by Mr. Charles Day, Wh. Sc.* (Manchester: March 8, 1909), 14.

68. Day, "Address," 203.

69. John Wingate, *Warships in Profile*, v. 1 (New York: Doubleday, 1972), 23.

70. "Internal-Combustion Engines versus Turbines," *Engineering* (June 16, 1911): 790.

71. A Mirrlees builder's photo of these engines in a May 1911 catalog carries the Hazel Grove factory name: "M.B.D. No. 10." Busch-Sulzer papers, Wisconsin State Historical Library.

72. "Manchester Electrical Exhibition," *Engineering* (October 9, 1908), 485.

73. The unit was mounted high on the promenade deck "A" and would only have been used if the main 1,600 kw steam-driven sets were out. *Engineering* (May 29, 1914), 737 and Plate 106.

74. Dugald Clerk and George A. Burls, *The Gas, Petrol and Oil Engine*, vol. 2 (London: Longmans, Green & Co., 1913), 737.

75. Day, "Address," 205.

76. Ibid., 203. The *Tynemount* engines were later removed and put in service ashore and were still running in 1934.

77. A complete description of the *Tynemount* engines, electrical drive, and the ship is given in *The Engineer*, May 16, 1913, 516–19, and October 10, 1913, 380–82.

78. The evolution of Mirrlees piston design is outlined in the abstract of a paper given by Geo. E. Windeler, chief engineer at Mirrlees: "Cracked and Seized Pistons in Diesel Engines," *Engineering* (July 16, 1915), 57.

79. Harry R. Ricardo, *The Internal Combustion Engine*, vol. 1, *Slow Speed Engines* (London: Blackie & Son, 1922), 437.

80. *Royal Commission*, 93. H. N. Bickerton gives over twenty pages of very interesting testimony on Mirrlees and other builders' engines.

81. A very informative account of B&W diesel's first years is provided by Hans Friis Petersen's, "It Happened Like This," which was reprinted, with additions, from articles in the *B&W Staff Magazine* (1964). A forty-six-page translation, with annotations, by Niels Egon Rasmussen was kindly sent by him to the author. Although Petersen did not cite all of his references, the account should be considered quite accurate as few discrepancies are found when his work is compared against retrievable original sources. Petersen wrote numerous historical papers before retiring from B&W in 1975.

82. Clause 14 of Diesel's contract with B&W states that in addition to this payment, travel and consultation expenses by Diesel or his Munich staff and drawings at 20 marks per square meter were also reimbursable. (Adolphus Busch paid M 25/sq m!)

83. B&W also undertook a major machine tool modernization program, which ended in 1902. Without this, the production of diesel engines would have suffered. Petersen, "It Happened," 31.

84. Bringing oil into Denmark was only a small part of Marcus Samuel's global marketing strategy with Texas oil. He was prevented by a deal with the Rothschilds from selling Russian Baku oil in Europe, and the Texas oil gave him that ability. His advanced-design tankers were free to transport the Spindletop oil. They lay idle in port because of lost price battles in the ongoing war with Standard Oil and others. All went well for a couple of years until the Texas field suddenly stopped producing. The episode is in Robert Henriques, *Marcus Samuel* (London: Barrie & Rockliff, 1960), 336–39, 409–62. A shorter version with a different emphasis is in Kendall Beaton, *Enterprise in Oil, a History of Shell in the United States* (New York; Appelton-Century-Crofts, 1957), 38–45. For Standard's story see Ida M. Tarbell, *The History of the Standard Oil Company*, vol. 2 (London: Heinemann, 1905), 272–74, for a contemporary account.

85. Hesselman, *Teknik*, 78.

86. Petersen, "It Happened," 36. In a November 4, 1903, letter from B&W to Mirrlees, analyses of three oils from England showed sulfur contents of 1.35 to 1.57. The comment based on these numbers was that "the sulfur does not have so great an influence as people earlier were inclined to think, but that the effect on the iron or the steel in the valves is only produced when it is owing to the thick fluidity, due to the great content of asphalt and residium, cannot be sufficiently dusty for complete combustion, and consequently in drops is coming into contact with the valves." B&W Archives.

87. Biographical information kindly sent by Anthony W. Woods. He recently assumed the extra position of part-time archivist at MAN B&W Diesel A/S. Niels Egon Rasmussen had been B&W's unofficial historian since Friis Petersen's retirement and now that Niels Egon has himself recently retired—and will pursue B&W history in cooperation with Tony Woods—it is Tony who carries the "in-house banner." Niels Egon still consults with Petersen. The author is grateful to them and to B&W for its corporate resurrection of interest in the heritage of the Danish company.

88. Where the "prototype" engine was made is in question. No records exist for the engine at B&W or in Augsburg. After a thorough search of the complete delivery records in the MAN Archives, no evidence can be found that it came from Augsburg. B&W records show that they ordered a 25 bhp engine in 1903 "for test purposes," which was later sold to the Limfjorden Cold Stores Co. (*Fryseriet Limfjorden*), in Glyngre, Denmark. Sulzer did start delivery of the 25 bhp model in 1904. However, the "prototype" does not look like a Sulzer standard production engine. It is the author's belief, therefore, that the pictured engine was probably built in Copenhagen.

89. B&W sent drawings and information on their new compressor to Mirrlees (November 4, 1903) and to the Diesel Engine Co. in London (November 5) showing it driven off the end of the crankshaft. B&W Archives.

90. Petersen, "It Happened," 42–3. Kayser, who started at B&W in October 1903, ended his career as director of Holeby Diesel Engine Works.

91. Ibid., 34–5.
92. MAN records show this DM 8 was engine s/n 618, and B&W says it was sold to a Mr. Anchersen of Vejle who successfully demonstrated it in his shop at the end of September 1903. A 1904 report on the engine by C. V. Kayser gives this data: The standard cam-operated fuel valve lift was 1.1 mm, and the diameter of the single nozzle hole was 1.2 mm. The engine was also retrofitted with a two-stage B&W compressor. (B&W "Prver med Diesel Motorer; bog nr. 1.") This book, from which the report was kindly copied by Tony Woods, is in the Danish Trade and Industry Archives, Aarhus, where all of the early B&W business papers are located.
93. Johannes Lehmann, *Rudolf Diesel and Burmeister & Wain* (Copenhagen: B&W, 1938), 33.
94. The engine on exhibit in the B&W Museum, with S/N 1, was the first one built in Copenhagen. The excellent museum was to open in 1943 with the old engine as a showpiece, but damage during a British air raid to a nearby B&W factory power plant a few days before the scheduled event caused a judicious postponement. The actual ceremonies did not occur until December 1946. B&W Museum (Copenhagen: B&W, 1959), 45, 74.
95. Lehmann, *Rudolf Diesel*, 28.
96. J. W. M. Sothern, *Notes and Sketches on Marine Diesel Oil Engines* (Glasgow: Munro, 1922), 15, said, however, that combustion chamber heat tended to reduce life of the poppet shaped end on the outward opening fuel valve. Other contemporary sources do not mention this deterioration.
97. The factory site was a shipyard for the Dutch East India Co. as early as 1658, where in 1697 Tsar Peter the Great worked for a year as a shipwright. Two separate companies were formed in 1901, with one for ships (Nederlandsche Scheepsbouw) and Werkspoor the steam engines, and so on. Both cooperated together in building complete ships. "An Historic Yard," *The Engineer* (January 20, 1911): 64. See also G. J. Lugt, "The Werkspoor Diesel Engine," *Transactions of the Institution of Engineers and Shipbuilders in Scotland* (March 1926), 427.
98. Kloos wrote many years later that he and a Mr. Fenega in the shipyard division asked for and received from M.A.N. Dutch rights for ten years and that they assigned these rights to Werkspoor. August 21, 1992, letter to the author from Dr. Heleen Stevens-Hardeman, Werkspoor Archivist. The MAN Archives have no record of this, nor does the Werkspoor-M.A.N. contract mention Kloos/Fenaga.
99. Contract between M.A.N. and Nederlandsche Fabriek, signed in Amsterdam on November 1 by J. Muijsken and another and in Augsburg on November 4, 1902, by Lauster and Pfeiffer. Handwritten, four-page copy in MAN Archives.
100. MAN received papers in 1939 assumed to have come from Endert. Included were extracts "From the journal of the forgotten Holland land and sea Diesel engine pioneers" ("Aus dem Tagebuch des Vergessenen, Holländischen Land- und See-Dieselmotorpioniers"). These informative six pages relate his visits to M.A.N., engine contributions and strife with Kloos. Endert tells "his side of the story" without other confirmation, and thus some comments should be cautiously viewed. However, a 1939 article spoke of the "prominent Dutch expert who had proved his worth with the Dutch engine factory Werkspoor." "Generaldirector D.C. Endert 25 Jahre bei der Rotterdamsche Droogdok Maatschappij" ("25 Years with the Rotterdam Drydock Society), *Werft, Reederei, Hafen* (Shipyards, Shipping Lines & Harbors), vol. 21, November 1, 1939. MAN Archives.
101. July 1906 sales records show that about one hundred cylinders had been sold to that date versus eleven recorded through 1904. MAN Archives.
102. "Tagebuch," 2. Endert signed off a test report of a DM 35 (s/n 10) on February 22, 1905. Werkspoor Archives.
103. Ibid., infers that all engines made the change to a front-driven compressor, but a small, runnable Werkspoor engine in the Amsterdam factory of Stork-Wärtsilä, s/n 63 built around 1907 according to MAN records, still has a con rod-driven two-stage compressor. Figure 12.27 carried no drawing date, but because of the engine design it has to be well before 1909.
104. Ibid.
105. Ibid., 3, 4.
106. "600 Horse-Power Diesel Engine," *Engineering* (November 25, 1910): 731 and Plate LXXIX.
107. Endert wrote that during a December 1907 visit to Augsburg he learned that M.A.N. had already switched over to a white metal bearing surface having a high tin content. "Tagebuch," 3. Babbit bearing material is of this type. See for example, L. Archbutt and R. M. Deeley, *Lubrication & Lubricants*, 2nd ed. (London: Charles Griffin & Co., 1920), 430–39, for a good treatise on such bearings.

108. Lugt, "The Werkspoor," 429. Lugt, a graduate of Zürich's ETH, had been at Werkspoor since the early diesel days. He took Kloos' job when the latter retired. Lugt had an untimely death.

109. "Tagebuch," 5.

110. Lugt, "The Werkspoor," 429. The marine version is featured in "The Ocean-Going Oil-Engined Ship 'Semblian,'" *Engineering* (March 15, 1912): 348–49.

111. C. J. Hawkes provides a list of engines seen under construction during a visit at Werkspoor in August 1912. There were four for generator sets and five for single- and twin-screw tankers. Report of Engr.-Lt. C. J. Hawkes, RN, *Royal Commission*, 89.

112. "Tagebuch," 5.

113. "The Motor Ship Juno," *The Engineer* (November 15, 1912): 515. The 258-foot-long *Juno* was almost double the capacity of the *Vulcanus*. She was of 4,300 tons displacement and 2,675 dwt.

114. A record of Endert's accomplishments is detailed in a commemorative book published by the Rotterdam Drydock Society: *1902–1952, Een Halve Eeuw "Droogdok* (Rotterdam, Netherlands: 1952). Scheepvaartmuseum Library, Amsterdam.

115. Kloos was granted British Patent No. 8,867 of 1911, for example.

116. Lugt, "The Werkspoor," 433-34.

117. "Lijst van geleverde en bewerking zijunde" (June 1912), 8 pages. By comparison, a June 1910 M.A.N. publication announced they had delivered or built 1,636 engines (2,631 cylinders). Pub No. 209. Both in MAN Archives.

118. A circa 1921 Pacific Diesel Engine Co. brochure gives extracts from the ship's log for a January 1921 voyage of the *Charlie Watson*, a 250-foot, Standard Oil of California tanker powered by two Werkspoor-designed 800 ihp diesels.

119. Sass, *Geschichte*, 565–66. For more on this period of Krupp diesel history see C. Regenbogen, "Der Dieselmotorenbau auf der Germaniawerft," *Jahrbuch der Schiffbautechnischen Gesellschaft 1913*, reprint. Diesel papers, Deutsches Museum.

120. German Patent 251,509.

121. Henry W. Dickinson, *A Short History of the Steam Engine* (Cambridge: Babcock & Wilcox, 1938), 149–51, and L. T. C. Rolt, *The Mechanicals* (London: Heinemann, 1967), 69. Willans also developed the "Willans line" once used to calculate steam and I-C engine fuel economy. Harry R. Ricardo was an assistant to the head of the engine test department at W&R in the early 1900s before the diesel arrived there. Sir Harry Ricardo, *Memories and Machines: The Pattern of My Life* (London: Constable & Co., 1968), 80.

122. *Agreement for the Manufacture of Diesel Engines* (June 21, 1905), 6 pages. The document includes prices to the DEC, F.O.B. Rugby. Prices varied from a 40 bhp, one-cylinder at £525 to a three-cylinder, 400 bhp model at £2,990. By the end of 1907 the smallest engine cost £711 with the rest rising proportionally. GEC Archives, Rugby.

123. Robert Cox, Dorset, who has access to W&R material in the GEC archives, also provided the author with much valuable material, including the above W&R/DEC license agreements and catalogs, and cleared up numerous matters regarding W&R's involvement with the diesel engine. He is currently writing a comprehensive history of Willans & Robinson. The author greatly appreciates the sharing of his knowledge.

124. The author is indebted to Dr. C. C. J. French of Ricardo International PLC and Bruce Harling, Institution of Mechanical Engineering, librarian, who kindly provided a chronology of events in the corporate life of Willans and Robinson and its ensuing owners. *The Engineer* (May 23, 1913): 543, adds to the story. The DEC papers were destroyed when the corporation was dissolved in 1924. (Letter from Public Record Office, Kew, Surrey, to the author.) No W&R sales records exist in the MAN Archives.

125. This same solution, when reinvented by Cummins Engine Co. in the 1920s, used a rolled-in copper sleeve to reduce casting and operating stresses caused by mold cores shifting in that area. The faster heat transfer of copper helped to reduce carbon build-up in the tip end of a solid fuel injector, particularly around the spray holes. Detroit Diesel later adopted the same design for their engines.

126. "Exhibits at the Norwich Show," *Engineering* (June 30, 1911): 853. *The Gas Engine* (US), August 1911, 443, adds further information.

127. "335 Brake Horse-Power Diesel Engine," *The Engineer* (October 27, 1911): 430.

128. Ibid.

129. "A Diesel Engine Action Against a Braintree Firm," *East Anglian Daily Times*, June 7, 1912, 4. A second article on the tenth, on page 5, announced a settlement. British Museum Newspaper Library.

130. *The Engineer* (May 23, 1913): 543–46, also gives much detail on the design and manufacturing of W&R engines.

131. C. G. A. ("Art") Rosen (1892–1975), in a taped interview with the author on August 26, 1972, told of his participation in the test of the 1913 Dow engine while a senior at the University of California at Berkeley. He worked four years at Dow where he designed the reversing system for its two marine engines. He left in 1923 when the company failed. He recalled learning how to put his hand inside an A-frame while the engine was operating and then run it up the con rod to the wrist pin to get an oil sample. "You smelled it and looked at it to determine how much more to put on the lubricator feeds." Rosen became head of research at Caterpillar and in 1955 was president of the Society of Automotive Engineers (SAE). See Geo. A. Dow, "Performance of the *Libby Maine*," *Mechanical Engineering* (1919): 378, for engine and ship details.

132. References to these "Rugby" names are found in *Engineering* (July 21, 1933), and Willams & Smith, *The Oil Engine Manual* (London: Temple Press, 1939), 163.

133. "Tagebuch," 4.

134. Harry R. Ricardo, *The Progress of the Internal Combustion Engine during the Last Twenty Years*, International Engineering Congress (Glasgow: 1938), 4. From a preprint paper.

CHAPTER 13

New Ways to Light the Fire

"Very few inventions are 'the sole means,' and if they are, most inventions do not stay long in that class. If the price of a license is high, a great effort will be made to find a cheaper or even a better means to produce the result. Rarely does that effort fail."

—Clarence D. Tuska[1]

The protection afforded by Diesel's basic patents was one of those rare exceptions. They prevailed as "the sole means" for licensees and lived out their legal lifetime with few infringement worries.

Imaginative engineers outside of Diesel's club, however, wanted to emulate his engine with innovations offering equal fuel efficiency through simplified injection systems. Their goal was to end the need for injection air and its attendant high-pressure compressor. Their methods led to artful injectors and often complex pistons and combustion chambers. The results of such endeavors to evade Diesel's protected "sole means" varied from outright failure to acceptance only in specific applications.

Unanticipated spinoffs from a few of these clever ideas would lead to new techniques for initiating combustion and, in turn, new areas for the engine to capitalize on successfully. These had been optimistically prophesied by Diesel but could not have come to pass when using the fuel system he and the early diesel industry had adopted out of necessity.

Haselwander

Friedrich August Haselwander (1859–1932), whose education and early career were in electrical engineering, was the first to show a way of eliminating Rudolf's injection air compressor.[2] He applied for a German patent on October 20, 1897, a few months after the Diesel-Schröter lecture in Kassel.

It issued as No. 101,453, the first of several granted to him disclosing ideas whereby the piston itself created the air pressure to aid atomization and combustion of the fuel charge.[3]

What we know of Haselwander's early engine work is from these patents. They allow us to trace the progression of his concepts as he continued to learn from engine test results. Ideas from other inventors examined in this chapter are similarly viewed through their patents as no other records of their work are known to exist.

Haselwander's first patent shows a two-stroke engine having a conical step on a piston whose upper cylindrical projection fits into an atomizing chamber under the cylinder head (fig. 13.1). The end clearance in the conical annulus chamber formed between the two piston diameters is greater than that in the atomizing chamber. Dual nozzles vertically inject fuel into this upper chamber. As the piston nears top dead center (TDC) of the compression stroke, its upper "displacer" portion has entered the close-fitting bore, and the pressure in the upper chamber becomes significantly higher than that in the annulus at the time fuel is injected. Haselwander does not suggest that ignition begins in this upper chamber. He only says that its heated walls aid in fuel vaporization and that the sudden rush of mixture exiting the upper chamber as the piston descends serves to further atomize the fuel and aid burning.

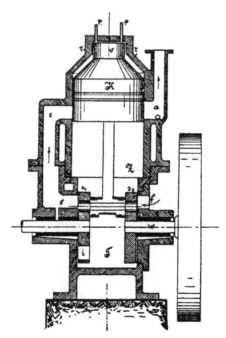

Fig. 13.1. The engine disclosed in Haselwander's first patent, German No. 101,453 of October 20, 1897. Hausfelder, *Die Kompressorlose Dieselmaschine*

His next patent, German No. 111,079 of May 6, 1898, does not mention such vaporization and offers a quite different approach in structure and function.[4] Here his engine is shown with a simple, two-diameter, stepped-piston whose upper projection again slips into a close-fitting bore to form what is now a primary combustion chamber (fig. 13.2). Different from before is that at top center the clearance in the lower, annulus-shaped chamber between the piston's larger base diameter and its end wall is much less than that between the "displacer" projection and its cylinder end wall. An air passage connects the annulus with the inside of the horizontal injection nozzle. This construction became a precursor of what is known as an "open nozzle."

In operation, the fuel charge is deposited in the nozzle's outer end during the intake stroke. As the piston moves inward on the compression stroke, the pressure acting on each end of the fuel charge remains in balance until near TDC when the displacer has slid into its mating bore to form the upper chamber. Because of the smaller end clearance in the annulus, the pressure difference quickly and significantly increases. This more highly compressed air acts behind the fuel in the nozzle to blast it out in atomized form through an orifice. The fuel ignites and burns first in the upper chamber and then in the main cylinder as the piston descends on its power stroke.

An independent test was made in 1900 on a Haselwander engine based on German patent 111,079.[5] The engine had a bore and stroke of 180 × 231 mm and produced a maximum of 5.6 bhp at 254 rpm. A major difference between this engine and Diesel's was that the peak compression pressure reached only 20 atm. Brake specific fuel consumption at the 5.6 bhp output on "American petroleum" (kerosene) was 343 g/bhp-hr and at 3.5 bhp was 384 g. The highest brake thermal efficiency attained was about 18 percent.

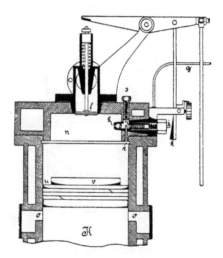

Fig. 13.2. Haselwander's second engine from German Patent No. 111,079 of May 6, 1898. Hausfelder, *Die Kompressorlose Dieselmaschine*

The Deutz-Haselwander Engine

Arnold Langen and other members of Gasmotorenfabrik Deutz management saw a Haselwander engine at a Mannheim agricultural equipment
exhibition in early 1902. This was about when they ended their Diesel
sublicense agreement with Krupp and M.A.N. (chapter 9). Because diesel
plans were on hold until the patent's expiration, Deutz wanted an alternative
engine to bypass the business barrier Rudolf had raised against them. As a
result, Haselwander's latest engine soon led Deutz to acquire a patent license
from him.

Unfortunately, after testing the engine in their own lab, they saw that its
very high fuel consumption had to be improved before it went on the market.
Indicator diagrams that showed a series of pressure spikes on the expansion
stroke suggested why there was such poor fuel consumption (fig. 13.3). A
too-slow buildup of air pressure behind the fuel caused a "creeping" injection
followed by delayed, uneven combustion. It was also believed that the fuel
ignited in batches due to the sudden pressure increases in the combustion
chamber. These set up back pressures at the nozzle exit higher than the discharge pressure, which deterred a steady injection. Fuel thus left the nozzle in
intermittent bursts rather than as a continuous stream. The test results sent
the engineers back to the drawing board.

Deutz devised a modest, but important modification that reduced fuel
consumption to, for example, an acceptable 255 to 259 g/bhp-hr on a 10 bhp
engine[6] (fig. 13.4). Instead of the passage from the annulus chamber connecting with the nozzle, it was moved upward to enter on the smaller-diameter
cylinder wall. Also, a turned groove on the upper piston portion remained
open to the annulus after the upper piston entered its mating bore. By this
simple means, the pressure behind the fuel stayed the same as that in the
combustion chamber until almost the end of the compression stroke. When
the back side of the fuel charge in the nozzle was finally connected to the
annulus, a much greater pressure differential existed. This delayed injection,
but greater pressure acted to blast out the fuel in one continuous spurt and
to eliminate Haselwander's slower, "oozing" pressure buildup in the nozzle.

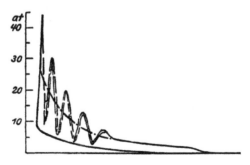

Fig. 13.3. P-V diagram showing Haselwander's "pulsed" injection
characteristic. Hausfelder, *Die Kompressorlose Dieselmaschine*

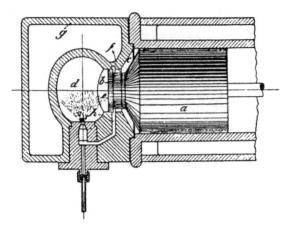

Fig. 13.4. Deutz's modification to the Haselwander design.
Hausfelder, *Die Kompressorlose Dieselmaschine*

Starting was always a problem because of the low compression pressure of about 20 atm. This necessitated the addition of a hot-tube igniter or an electric spark plug and then running on gasoline before switching to a heavier fuel oil. Another disadvantage was a frequent need to clean out the carbon in the blast air passage between the cylinder and the nozzle.[7]

A total of about 190 Haselwander engines in outputs ranging from 4 to 45 bhp were sold by Deutz from 1904 until the design was abandoned in 1906. Most went to Russia where they burned the country's light crude oils.[8]

Brons

N. V. Appingedammer Bronsmotorenfabriek traveled a path unlike others in the evolving diesel industry. This relatively small Dutch company pioneered a new form of compression ignition engine that became the first to successfully eliminate Diesel's nemesis of the high-pressure air compressor. However, the Brons engine was not classed as one of Rudolf's at the time, and some will still argue that it is not a diesel engine by present-day definition.

Brons and several of its licensees carved out a profitable niche market and enjoyed international success for more than two decades. Of greater significance than the sales achievement was a legacy of the Brons injection and combustion system that would later serve as a catalyst for innovation to a coming generation of diesel builders.

The genius behind this "upstart" of an engine was Jan Brons (1865–1954), a self-taught, mechanically inclined man with limited formal education[9] (fig. 13.5) He was born and raised in the north Holland village of Wagenborgen and lived his entire life within a few kilometers of there. Appingedam, the town where Brons would locate his engine factory, was on a river flowing through the nearby seaport of Delfzijl on the Ems estuary. Delfzijl is the major port for the province of Groningen.

Fig. 13.5. Jan Brons (1865–1954). *Volle Kracht Vooruit*

Tjako Brons was a carpenter and master builder in Wagenborgen who planned for Jan, the oldest of his four sons and two daughters, to follow in his footsteps. After finishing lower school, Jan studied the basics of architecture and at age seventeen drew plans for a new village church and tower, which were built in 1883.

The enterprising father also came forth with plans for mechanical products foreign to his regular business. Among the first of them to be made in 1887 by the new firm of T. Brons & Sons were portable, horse-operated threshing machines.[10]

I-C engines entered his thinking in 1890 when Tjako Brons placed an advertisement asking for a suitable motor to power the threshing machines in lieu of horses. A positive endorsement describing the Priestman oil engine gave Brons & Sons the urge to build a vaporizing oil engine of their own design.[11] Jan led in this new venture, which began in 1892. When their first engine was tested, however, it would run only on the highly volatile and unsafe gasolines. An improved vaporizer eventually allowed operation on the intended heavier lamp oils. The vaporizing chamber that contained a form of gas mantel was externally heated by a typical blow torch used with other oil engines. In 1895 Brons & Sons delivered a 10 and a 12 bhp "Safety Engine" from their new factory outside of Delfzijl.

The company had moved there in 1893 because opportunities for a general parts and repair business were much greater than in Wagenborgen nine kilometers away. By this time Jan carried most of the responsibility for T. Brons & Sons in Delfzijl, and his father continued with the earlier activities in the home village.

The horizontal, stationary oil engine, never a major source of revenue, was tried in other applications. Jan demonstrated his "Safety Engine" powering a large, automobile-like bus to the town *burgers* in 1899. That same year he also patented a rotary steam engine and later built a test model based on it.[12]

Nanno Timmer became Jan's partner in 1900 when Timmer supplied needed operating capital. The new company of Brons & Timmer continued with a limited engine production that now included a vertical model for auxiliary power in sail boats. Repair work and job shop machining still provided most of the income.

The diesel's growing promise did not go unnoticed by Brons, who envisioned a domestic market for small, relatively inexpensive engines based on it. He would have known there was no Dutch Diesel patent and thus could build an engine without infringement worries. Machining of parts began in late 1900 on an engine that still used air from a compressor to blow in the fuel. By the spring of 1902, his "double" (perhaps the compressor and power piston) engine was ready. The compressor failed shortly after the belt-started engine began firing, but to Jan's amazement, the engine continued running. Not until the head gasket failed and water entered the cylinder did the engine stop. This surprising discovery that he might not need a compressor led Jan to invent the Brons-type engine.

He next incorporated in the cylinder head a cam-actuated plunger piston that alternately formed and collapsed a small chamber ("A" in figure 13.6). Its intended operation was as follows: The small piston moved downward along with the power piston. The cylinder vacuum drew fuel into the newly formed chamber past a fuel valve timed to open during this vacuum period. At the end of the compression stroke, a spring quickly lifted the plunger piston when its cam follower dropped onto the cam's base circle. The rapid displacing action of the collapsing chamber forced the heated and partially vaporized air/fuel charge through orifices into the main chamber where ignition and burning occurred.

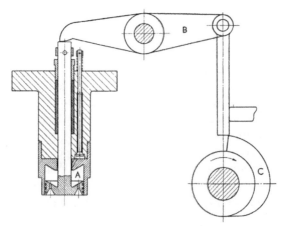

Fig. 13.6. Compressor/fuel delivery unit of Brons first engine, 1902. *Voile Kracht Vooruit*

398

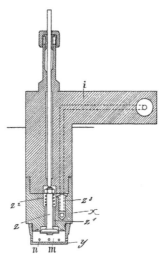

Fig. 13.7. Patent drawing showing the first Brons design where a cam-operated valve admitted fuel into a fixed-volume vaporizing cup. *British Patent No. 14,165 of 1904*

Practice did not follow theory. First, the plunger piston moved out farther than the cam lift would have imparted, which showed that at least some burning and expansion occurred in the chamber. Next, the plunger piston had welded itself to its guide wall in the outward position. When the engine continued to run, it proved that the pump chamber was superfluous.

Brons then attached a fixed-volume vaporizing "cup" onto the end of the fuel valve piece and drilled five horizontal spray holes through the cup wall (fig. 13.7). No supplementary air was admitted into the cup along with the fuel. This much simplified design performed as planned and became the prototype for engines that went into production in 1904 (fig. 13.8).

A quickly filed patent application in Germany on June 23, 1904, has its own little story.[13] When the German Patent examiners objected that such an engine as Brons disclosed could not work, he arranged a meeting with them in Emden to prove its feasibility. They quickly changed their minds after they saw that the supposedly non-operable motor had powered the canal boat that brought Brons from Delfzijl. Official word of acceptance came on February 10, 1905, and German Patent 167,149 was issued to Brons and Timmer on January 20, 1906—Jan's birthday.

The main claim of his British patent, which was similar to the German, explains what happened in the cup:

"In a working process for oil engines . . . [where] air is compressed, this compression thereupon ignites in the vicinity of the dead point [TDC] a small quantity of combustible [that was sucked into the cylinder] with a slight explosion, and this explosion drives the bulk of the combustible out of the chamber into the working cylinder, where during the descent of the piston and the expansion it is brought to complete combustion, essentially as described.[14]

One must assume that the fuel remaining in the cup had to be well heated, partly vaporized and "ready to go" by the time TDC was reached for Brons's explanation to be completely valid.

Brons came to understand that although his method certainly worked, it had an inherent deficiency. This was confirmed on the fourteen engines (182 total horsepower) built over the period of January 1904 to April 1905. His corrective British patent states the problem that was found with experience:

As the perforations in the sprayer [cup] require to be so small so as to render it possible to utilize the mixture in two separate periods, and as the walls of the sprayer are necessarily heated during the preceding explosion and combustion of that mixture so that the liquid combustible in the sprayer is rapidly gasified, it follows that the air contained in the working cylinder during the compression stroke of the piston is not able to enter the sprayer in a sufficient quantity and to mix with the

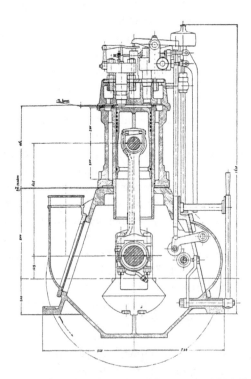

Fig. 13.8. Cross-section drawing of 12 bhp 1906–1910 Brons engine with vaporizing cup and cam-operated fuel valve. *Voile Kracht Vooruit*

gas developed from the combustible and nearly filling up the sprayer. The consequence is, that the mixture so formed is hardly explosible, so that the preliminary explosion taking place in the working cylinder in the proximity of the end of the compression stroke does not penetrate into the sprayer with sufficient power for rapidly expelling therefrom by shocks the gasified combustible through the perforations to the working cylinder.[15]

Brons solved his problem simply by adding an adjustable bleed valve to admit, through the fuel valve, a controlled volume of air into the cup, along with the throttle-regulated fuel charge. He reported in his patent that the supplementary air increased brake horsepower by 10 to 15 percent. It was the major additional feature claimed in the improvement patent.

Not long after filing the first patent in 1904, Brons had the great fortune to befriend Dirk Bonthuis Tonkes (c. 1878–1943), a younger Dutchman with a German engineering degree. Tonkes became a devil's advocate and advisor to Jan after Timmer left in April 1905 and Jan had taken over as sole owner.[16] On July 3, 1906, the two men organized a new company in which each was a co-managing director. Although Tonkes had to take a year's leave of absence in 1908 because of a serious illness, he still served in that capacity until 1927 when he moved to the supervisory board. He remained on this board until his death.

The name of the new N. V. Appingedammer Bronsmotorenfabriek heralded its move from the inadequate facilities at Delfzijl to a specially built factory on a site outside of nearby Appingedam.[17] Production began there on April 1, 1907.

One of the last engines to leave Delfzijl, a two-cylinder of 100 bhp, went on February 1 to a shipyard where the schooner *San Antonio* was being built. (The vessel's propulsion auxiliary was an earlier described Werkspoor diesel.) This Brons engine, one of the largest of its type ever sold, contained the supplementary air supply to the cup as did all engines from 1906 on.

From 1907 to the end of 1913, Brons made 496 of his type engine in Appingedam. The total grew to 1,002 through the end of 1919. The last engine sold with a Brons fuel system was one of 6 bhp in 1946.[18]

Deutz and Brons

A license granted to Deutz added another significant business milestone for Bronsmotorenfabriek in 1907. Deutz's interest in the engine had begun during 1905, and BMF correspondence files show that the road traveled to a final agreement was sometimes a rocky one. Arnold Langen, ever on the lookout for new engines to build (and no doubt disappointed with the Haselwander), came to Delfzijl in the spring of 1906. The ensuing negotiations broke off a few months later when Brons learned that Deutz had applied for a German patent on his idea to add supplementary air to the cup. Brons registered a protest yet failed to stop Deutz from being granted the patent in their name.[19]

Despite this apparent usurpation, Brons and Deutz still came to an agreement. Jan wrote his father on February 26, 1907, about the situation:

> There is someone here from the Deutz engine factory to button up anew the broken-off negotiations. They have made us an interesting bid for the German patent, but it is not yet good enough for us. If we can sell the other patents in proportion to what we are bid for the German, I'll be a man with a bonus.[20]

Shortly thereafter a deal was consummated with Deutz paying 35,000 Dutch florins, or about $14,000. Royalties were not mentioned. Excluded areas included Holland, countries where Brons patents existed, and the Dutch East Indies, which was developing into a major market for Brons. Ironically, Deutz reached an agreement with Brons at a time when they would soon be free to build conventional air injection diesels.

The contract clause excluding the East Indies was put to a legal test after Deutz started selling engines there. In this instance both German and Dutch courts, after two years of litigation, sided with Brons and ended the poaching.[21]

The most popular size "Dutch" Brons engine through 1910 was a one-cylinder, 12 bhp at 300 rpm model with a bore and stroke of 200 × 250 mm (fig. 13.9). A small air compressor provided the starting air that entered the cylinder through a cam-actuated valve. Brons production in 1910 was a little over 300 engines.[22]

A lengthy plague of porous castings visited Brons like all the other builders of high-cylinder-pressure engines. The Dutch company's solution to excessive scrapping of purchased castings was a decision to pour their own. Someone cautioned Jan that "if you want to live in anxiety start a foundry."[23] He ignored the valid warning, and from that day in May 1911 when BMF poured their first castings the venture proved to be worthwhile.

The Brons-type engine worked best in low horsepower power applications where speed and load were relatively constant. It was ideal for fishing boat and smaller coastal cargo boat use. A Deutz-Brons engine received first prize in two classes from the Berlin-based German Fisheries Association in 1911 after a year-long competition with other small marine engines in fishing boat service. One was a one-cylinder, 170 × 220 mm bore/stroke and 8 bhp at 350 rpm similar to figures 13.11 and 13.12. The second, a two-cylinder engine with a bore and stroke of 200 × 240 mm, produced 24 bhp at 340 rpm. A total weight of 2,215 kgs included a flywheel of 750 kg. Full-load fuel consumption on American kerosene averaged 238 g/bhp-hr. At half load it rose to 265 g/bhp-hr.[24]

Deutz built one- and two-cylinder models in four cylinder sizes of 6 to 16 bhp per cylinder. It also offered a three-cylinder of 50 bhp and a four-cylinder of 80 bhp before the war.[25] The smallest were started by first opening

Fig. 13.9. Brons 12 bhp, 300 rpm engine, 1906–1910. *Volle Kracht Vooruit*

a compression release valve and hand spinning the flywheel. The rest used the small compressor and a storage tank for air starting. Production of Brons engines at Deutz lasted until about 1926.

An earlier allusion that the Brons engine was not considered a true diesel needs explanation. If one holds strictly to the premise that Rudolf's engine had true constant pressure combustion, which it soon did not, and that fuel must be injected into the heated air, then the Brons was not a diesel. However, these diesel axioms do apply to the Brons if looked at broadly.

First, a fuel-quality-dependent compression ratio of 13:1 to 16:1 for the Brons equaled that of early diesel engines, and both started quickly when cold without supplying external heat. Further, P-V diagrams show that the pressure rise after ignition at full load is only about 20 percent above a peak compression pressure to make it similar to airless injection diesel engines appearing in the 1920s.[26]

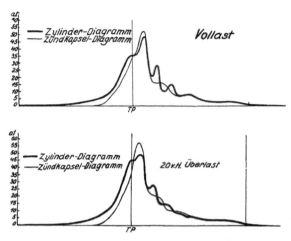

Fig. 13.10. Brons cup and cylinder full-load combustion curves on
a P-T diagram. Hausfelder, *Die Kompressorlose Dieselmaschine*

It is true that the fuel was deposited in the cup during the intake stroke, but most of the charge remained there being vaporized with the help of the supplementary air and made ready to ignite near TDC of the compression stroke. What vaporized fuel left the cup prior to the end of the compression stroke could in today's diesels be called a pilot ignition charge that had gone through a "fumigation" process.[27]

To some extent a similar event happened in diesels with blast air injection as the air came in contact with fuel in the warm nozzle cavities around and above the atomizer prior to injection.

The pressure relationship between the cylinder and the inside of the cup is shown for a typical Brons-type engine in the pressure-time (P-T) diagram of figure 13.10. One can see that cup pressure lags that in the cylinder until TDC and then exceeds the cylinder pressure during the first part of the power stroke. The P-T diagram also shows an uneven combustion later taking place.

A major weakness of all Brons engines was a sensitivity to fuel volatility. The fuels had to contain enough of the light, lower boiling point fractions for sufficient vaporization in the cup prior to ignition. The variation in compression ratios noted in the above paragraph show the spreads often dictated by fuel differences. Although the engine would run on heavier oils, the ignition problem often precluded their substitution on a given engine because of too low a compression ratio.

Brons attempted to solve an inherent difficulty of balancing outputs between the cylinders of a multicylinder engine with a patented "thermal regulator," which independently tailored the fuel rate of each cylinder. Overriding linkage from the throttle lever that lifted all fuel needles an equal amount, was a control for each cylinder based on the exhaust temperature of that cylinder. A pin inserted in the exhaust outlet at each cylinder changed length according to a temperature that was indicative of the fuel burned in that cylinder. As the vertical pin expanded or contracted, it determined the position of a lever whose pivot point was between it and the fuel valve. The common throttle linkage acted on the fuel valve through a pinned connection midway between the fulcrum and the valve for each cylinder. By this means a slight change in the pivoting of the lever independently altered, due to that cylinder's pin growing or shrinking, the lift the throttle control normally imparted equally to the valves of all cylinders. The governor also acted on the common throttle linkage to hold speed constant or to control overspeed. Despite this the multicylinder Brons engine was not known for its combustion smoothness at idle and light loads.

In addition to the original 1907 agreement with Deutz, Brons also gave them a Belgian license the next year. Other licensees included one to Lauren & Clement AG in Austria-Hungary (1909), AS Frichs in Denmark (1910),

Fig. 13.11. A typical Deutz-Brons marine engine, c. 1910, with a reversible pitch propeller. Note the fuel filter (*e*), governor housing (*i*), governor linkage (*l, m, n, o*), compression release lever (*t*), and crank (*u*) to control the propeller pitch. Sass, *Geschichte, 1962*

AG Gideon in Norway (1912), R. M. Hvid in the United States (1914), and probably The Iron, Steel and Engine Works Ltd. in Sormorvo, Russia, who built the engine.[28]

Bronsmotorenfabriek introduced the Type D engine in 1920, which they continued to build until 1934. The most popular model still had two cylinders of 230 × 350 mm and produced 50 bhp at 320 rpm. They began making a two-stroke, direct injection diesel in 1927 and then four-stroke diesels after 1935. In 1988 the Waukesha Engine Division of Dresser Industries bought BMF and continues to build engines in Appingedam under the American company's name.

Hvid—The American Brons Spinoff
Rasmus Martin Hvid (1883–1950), a naturalized US citizen from Denmark, received a patent in his adopted country for a Brons-type engine that must

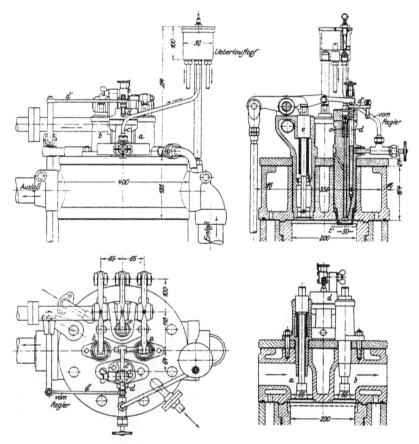

Fig. 13.12. Cross-section of 12 bhp Deutz-Brons cylinder and head showing the starting valve and the needle valve controlled by the governor. Güldner, *Das Entwerfen*

be minutely examined to see just how it differs from the Brons patent. Hvid ("Veed") began licensing engine builders in the United States as early as 1913, but as time passed, the structure defined in his original invention subtly changed to become so like the 1906 patent of Brons that on August 12, 1914, he bought a license from the Dutch company, most likely to avoid infringing its US patent.[29]

On February 2, 1912, Hvid filed for a patent, which did not issue until April 6, 1915. Its main drawing (fig. 13.13) shows two basic departures from the Brons construction: a tapering annulus chamber (*32a*) opening to the cylinder and surrounding the slightly sloping vertical walls of the cup, and a series of orifices (*33a*) at the top of the cup (*28d*) to connect the cup with the chamber (*32a*). His main claim recites all the Brons elements but contains phrases italicized by this author to make it patentable:

> The combination, with a cylinder and a piston of an oil burning engine, the cylinder having a combustion space, of an oil reservoir [32a] having an area exposed to the combustion space of said cylinder, and having restricted communication with said combustion space [cup holes 34a], *and a restricted chamber* [32a] *in open communication at one end therefrom with the said combustion space and in restricted communication* [holes 33a] *at its other end with the oil reservoir,* a portion of

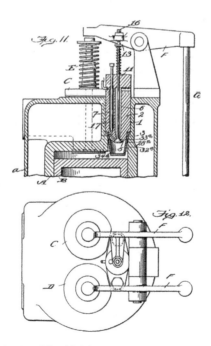

Fig. 13.13. The Hvid cup construction from his British Patent No. 4,370 of February 20, 1913.

the air admitted to the cylinder becoming heated during compression in the chamber to a degree greater than that of the air remaining in the cylinder, and sufficiently to act, *through the restricted communication between the chamber and reservoir,* upon the particles of oil in said reservoir which have become vaporized during compression and to thereby cause an explosion thereof.[30]

The addition of supplementary air supplied through the fuel valve is covered in his next claim.

Hvid's clever patent depended entirely on the upper cup holes connecting with the thin tapered annulus air space. When these holes and the air space filled with soot, as they undoubtedly did, or if the holes were never drilled in the first place, the engine became "Dutch." This inescapable closure must be what led Hvid to seek a license from Brons and thereby avoid legal problems. It is not known if anyone ever built engines having the upper row of cup holes, and no builder literature shows them.

The R. M. Hvid Co. was established in Battle Creek, Michigan, on July 25, 1912, with an initial capitalization of $50,000. In addition to Hvid, its directors included men formerly with the Advance Thresher Co. that had been taken over by the Rumley Co. when Rumley ended production of their oil-burning tractors. Both firms were in Battle Creek where Hvid had designed the engines for Rumley tractors. The Hvid Co. moved to Chicago in 1915.[31]

Hvid himself was not interested in making engines but only in selling licenses and collecting royalties. He never set up a test facility, and the absence of one meant his licensees had to do this necessary work.[32] The basic design supplied to licensees was typical of the myriad "hopper-cooled" engines sold for farm and light industrial use.

One of the earliest Hvid licensees was the St. Marys Machine Co. in St. Marys, Ohio, who announced their "HO" model in 1913.[33] The largest Hvid builder, Hercules Gas Engine Co. of Evansville, Indiana, contracted with Sears, Roebuck & Co. in 1917 to build the "Thermoil" farm engine line in several sizes of up to 7 bhp. By 1916 eleven companies had bought licenses, some of whom built only a few engines. Included in an overall list extending past this date was the Evinrude Motor Company in Milwaukee (fig. 13.14). A license sold in late 1918 to interests in Columbus, Indiana, put the Cummins Machine Works in the engine business a few months later.[34]

During their heyday, production of Hvid/Brons engines in the United States rose to thousands per year in sizes as small as 1 bhp and turning up to 1,000 rpm. A few were made as late as the latter part of the 1920s.[35]

The Trinkler Engine

Gustav Trinkler, a Hannover engineer, patented an idea after the turn of the century that advanced another way to atomize and inject the fuel charge in a compression ignition engine without an external high-pressure compressor.

Fig. 13.14. The Evinrude Oil Engine, c. 1917, from a company brochure.

His German Patent No. 148,106 of May 25, 1901, discloses a horizontal engine with a small, vertical piston located in the head[36] (fig. 13.15). This auxiliary piston, displaced by a timed drive, served as a second-stage pump to quickly boost compressed cylinder air high enough to force a fuel charge from the nozzle in adequately atomized form. Both end faces of the piston were exposed to power cylinder pressure during most of the cycle with the upper end always remaining under cylinder pressure. However, the resultant force acting on the upper end of the piston was less than that on the lower because of the piston area lost due to the actuating rod extension that passed through a stuffing box. The lower end of the bore guiding the "overpressure" piston led to a chamber in an open-type, horizontal nozzle located behind the fuel, which was deposited therein under low pressure during the intake stroke. The fuel stayed in the nozzle because of the equalized pressures acting on it.

Near TDC of the compression stroke the small piston closed off the port connecting its lower bore end to the main cylinder as it was mechanically forced downward. This rapidly raised the pressure behind the fuel in the nozzle high enough above that standing at the open nozzle orifice for the fuel to be atomized and injected into the combustion chamber where it ignited and burned.

Trinkler explained in his patent that if he injected fuel too soon he would have "explosive" combustion, and that by tailoring the start of the small

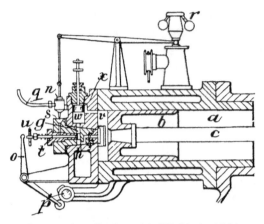

Fig. 13.15. The engine disclosed in Trinkler's 1901 patent with an open nozzle and an "overpressure" piston to boost the air pressure for injecting the fuel. Hausfelder, *Die Kompressorlose Dieselmaschine*

piston's movement he could have either constant volume combustion, a combination of constant volume and constant pressure combustion, or just constant pressure combustion.

An improvement patent by Trinkler, German No. 156,390 of October 25, 1902, reverses the way cylinder pressure acts on the overpressure piston so that at TDC the greater force at the start of piston movement acts in the direction of movement (fig. 13.6). He claims that the piston is not only displaced much more quickly through an improved, mechanical snap action, but also the new, additive pressure-derived force difference aids in its acceleration at the start of displacement.

No data exist for Trinkler sponsored engines, but it is most probable that he built a few examples to prove his theories and to interest potential licensees.

Gebrüder Körting

Gebrüder Körting AG of Hannover adopted the Trinkler design in an attempt to market engines that could compete with the diesel. In 1904 they signed a license agreement with Trinkler and began a development program.[37] It will be recalled that Diesel had sent a copy of *Theorie und Konstruktion* to them in 1893 because of their twelve years in the I-C engine business and that they had briefly shown an interest in building a test engine for him (chapter 4).

Ernst Körting (1842–1915) graduated from the Technische Hochschule in Hannover in 1864 and then spent several years working with a Vienna firm that made a successful line of locomotive feed water injectors sold around Europe and England. It was after gaining this experience that he decided to join with his brother Berthold in his native Hannover to found Gebrüder Körting in 1871 and start a similar business.[38] In 1903 it became Gebrüder Körting AG, a publicly owned company.

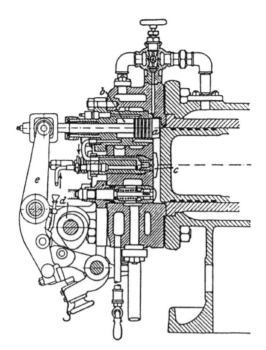

Fig. 13.16. Körting's compressorless Trinkler-based engine of 1904.
Sass, *Geschichte, 1962*

The Körtings began making vertical two-stroke gas engines in 1881 based on a design by Georg Lieckfeld.[39] Four-stroke engines of the same general configuration added during 1887 led to a line of increasingly larger horizontal engines. By 1900 they sold double-acting two- and four stroke models adapted to run on blast furnace gas. The largest, a two-cylinder, 2,000 bhp at 80 rpm monster introduced in 1904, had a bore and stroke of 800 × 1,400 mm.[40]

The Trinkler design, which Gebrüder Körting introduced in 1904, followed the general structure described in his second patent but incorporated numerous improvements the experienced builder deemed necessary (fig. 13.16). The engine would be sold in outputs of 12 bhp at 220 rpm up to 60 bhp at 180 rpm.

Its predictable Achilles' heel was an inability to cool the "overpressure" piston sufficiently to prevent scoring and seizing. The piston was too small to cool internally, and the addition of extra water passages around its sleeve proved inadequate. The potential rewards for trying to make the engine work were great enough, however, that Körting believed the effort worthwhile to do so. Its compression pressure of 28 to 30 atm yielded a low bsfc of 221 g/bhp-hr and a brake thermal efficiency of 29.2 percent, which made its operating cost highly competitive with the diesel engine at a much lower manufacturing cost.[41] Nevertheless, the Trinkler concept led up a dead-end street, and Körting delivered only fifty-four of these troublesome engines by the time their production ended in 1907.[42]

Körting Diesel Engines

The expiration of Rudolf's basic patent at last gave Körting a long-awaited opportunity to build a "true" diesel engine. It was ready for the event with adaptations of its very broad and well-accepted line of horizontal, single and multicylinder, single-acting gas engines.

A major difference between the Körting diesel and its competitors was a new open nozzle adapted from one patented in 1905 by Otto Lietzenmayer. Although the engine required a two-stage compressor, a costly, high-pressure fuel pump was not needed because the fuel went into the nozzle under low pressure past a simple check valve during the intake stroke[43] (fig. 13.17). A cam-actuated valve admitted the blast air to time the start of injection. The fuel's pressure balanced location in the horizontal nozzle ensured it would not be injected prematurely.

The first Körting model introduced in 1907 was soon expanded to ten different cylinder bore sizes of 190 to 650 mm bore and made available in one, two, and four cylinders (fig. 13.18). Output per cylinder ranged from 20 bhp at 250 rpm to 150 bhp at 170 rpm. Over two hundred of these horizontal engines were delivered by 1912.[44]

A four-cylinder Körting diesel was installed in a Palo Alto, California, municipal power plant in 1914. It had a 360 × 680 mm bore and stroke and produced 300 bhp at 180 rpm with a fuel oil consumption of 0.385 lbs/bhp-hr.[45]

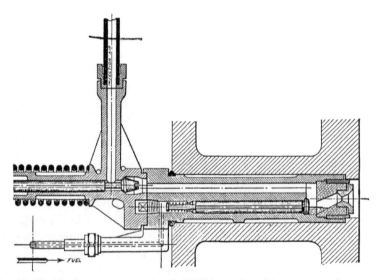

Fig. 13.17. Körting open nozzle of 1907 based on Lietzenmayer's patent.
Haas, *The Diesel Engine*

Fig. 13.18. Körting diesel based on their horizontal gas engine, 1907. It was soon made available in cylinder outputs of 20 to 150 bhp. Note the two-stage air compressor (*a*) and fuel pump (*m*). Sass, *Geschichte, 1962*

Gebrüder Körting did not build vertical diesels until almost 1914 and these went into submarines (chapter 17). Its reliable vaporizing oil engines, used in German subs since 1906, had become the standard power plant for the *Kriegsmarine's* U-boats until they were superseded by diesels in 1912.[46]

Early Deutz Prechamber Diesels

In order to trace the next developments in compressorless diesel engines, one returns to Gasmotorenfabrik Deutz. It had already begun work on such an engine as early as the time of its Haselwander license in 1902. Encouragements in this parallel program during the first years of its Brons efforts led to a line of commercially acceptable, lower- to mid-range size engines. These were not introduced, however, until a few years after Deutz had reentered the diesel market in 1907 with a design utilizing the blast air compressor.

The 1902 German patent issued to Deutz shows a form of lower compression oil engine more resembling one of the hot-bulb variety so prevalent at the time.[47] A long, horizontal and insulated tubular neck, axial with the power cylinder, connected the cylinder to a water-cooled box into which the intake and exhaust valves opened. Fuel was sprayed under modest pressure from a vertical nozzle into the neck at its midpoint. The engine required a heating torch when started. Test results were not promising enough to continue further development of this design concept.

The next compressorless step came in 1908 under the direction of Prosper L'Orange (1876–1939), who had come to Deutz in 1906 as head of its test lab.

L'Orange was born in Beirut (Syria at that time) where his physician father headed a hospital staff. The un-Germanic surname was derived from Huguenot ancestors who fled France in 1685 and had settled in East Prussia. Prosper was fourteen when he came to Europe with the dream of becoming a naval officer, but his poor eyesight forced him into other educational directions. He graduated with distinction from the Technische Hochschule in Charlottenburg and stayed on as an assistant in the heat technology lab. His post-academic career focused on I-C engines.[48]

One may assume that the catalyst behind a new idea for a compressorless engine came from the recently arrived L'Orange. It is seen in German Patent 196,514 of April 12, 1906, issued to Deutz, which disclosed an antechamber above and separated from a vertical power cylinder by a perforated partition.[49] A small cup sat directly under the holes (fig. 13.19). The idea was unsuccessful.

A further evolutionary stage of a practical prechamber is seen in Deutz's German Patent 238,832 of June 22, 1908, and attributed to L'Orange. Here the valves, again at right angles to the working cylinder, open into a narrowed extension of the working cylinder (fig. 13.20). The fuel nozzle sprays into a small passage leading from the valve area to an uncooled bulb or "burning chamber." No performance data or drawings of the engine have been found, and the figure is based on the patent drawing.

One serious problem L'Orange experienced was in producing an injection plunger and its mating bore. Only the artistry of a Deutz laboratory technician made it possible for the first time to create close enough operating

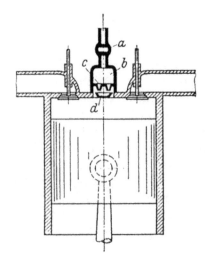

Fig. 13.19. Deutz's first antechamber design
after its April 12, 1906, German patent.

Fig. 1.

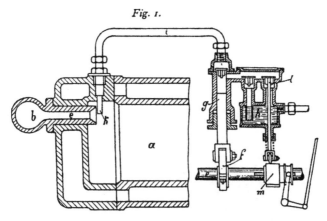

Fig. 13.20. A L'Orange prechamber based on the drawing
of a July 1908 Deutz patent. L'Orange, *Ein Beitrag*

clearances for controlling plunger leakage without need of an external packing gland.

L'Orange unexpectedly left Deutz at the end of 1908 and went to
Gasmotoren Fabrik Benz & Cie. in Mannheim. This ended work on
prechamber diesels at Deutz.

Deutz Air Injection Diesels

The chronology of compressorless events at Deutz requires a brief digression
to look at the advent of its air injection diesels. Ensuing compressorless variants were based on these engines.

Deutz prepared for the expiration in 1907 of Diesel's first patent knowing
that a compressorless system would not be ready in time. Its first air injection engine, therefore, was a typical one-cylinder, A-frame design resembling
the M.A.N. DM series that produced 30 bhp at 180 rpm. Because of the
engine's public demonstration in February 1907, almost to the day of the
patent's expiration, M.A.N. registered a protest with the argument that other
patents could legally be construed to block Deutz diesel sales. However,
Augsburg faced up to reality, and the two companies agreed on a pricing
practice whereby one would not undercut the other. Deutz finally shipped
its first post-patent diesel of 35 bhp in November. The line grew to include
two-, three-, and four-cylinder models of up to 200 bhp per cylinder with
reliability and fuel consumptions comparing favorably with its competitors.
A typical 50 bhp cylinder had a bore and stroke of 345 × 500 mm.

Concurrent work proceeded on a horizontal diesel derived from the configuration of its popular gas engines. This became the Model MKD introduced in 1909 (figs. 13.21 and 13.22). A two-stage compressor driven off the
end of the crankshaft supplied injection air, and the governor regulated a fuel
pump instead of a gas valve. A cam-actuated fuel valve, perpendicular to the

Fig. 13.21. Deutz 1910 Model MKD diesel engine with air compressor.
Deutz Archives

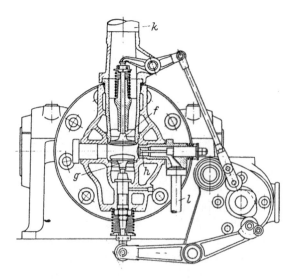

Fig. 13.22. Section through valve area of Deutz "MKD"
showing inlet, exhaust, and starting valves. *Deutz Archives*

cylinder axis, controlled the fuel flow to an open nozzle. L'Orange had used a development model of the "MKD" for his last prechamber work at Deutz.

Tests made in Nürnberg during 1912 on a 12 bhp "MKD" with a bore and stroke of 185 × 320 mm had this creditable performance:[50]

	Half load	Full load	Overload
Test length, minutes	37.5	71.3	19.6
Speed, rpm	281.7	280.8	277.6
Brake horsepower	6.2	12.1	14.2
Bsfc, g/bhp-hr	262	211	219
Brake thermal efficiency	24.1	29.9	28.8

Fuel was a Galician crude with an LHV of 10,162 Cal., a specific gravity of 0.858, and a flash point of 95°C.

The "Displacer Piston" Engine

Deutz traveled a different road than the prechamber to achieve a compressorless engine. The ancestry of its chosen displacer (*Verdränger*) engine may be traced to the pistons used in Haselwander's engines. It will be recalled that these had a cylindrical "knob" on the crown that penetrated a smaller diameter extension formed in the head of the cylinder (fig. 13.1). A rounded crown standard on the "MKD" piston possibly helped to direct thinking along the lines taken.

Drawings from two Deutz German patents—239,716 of October 1, 1910, and 250,216 of March 7, 1911 show a critical sequence of ideas (figs. 13.23 and 13.24). The "only" difference between the two is the location of the fuel nozzle. In the earlier patent, the nozzle was perpendicular to the horizontal cylinder axis. Thus, fuel was injected at almost a right angle into a segment of the air stream squishing past the displacer knob from the cylinder clearance volume and into the valve antechamber. (This chamber was also referred to as a combustion or burning chamber.) The intention was that the fuel charge would be sufficiently atomized for good combustion in two ways: first, during its initial entrainment, and then by the turbulence formed when a theoretical cone of displaced, squished air came together at its apex. The idea worked passably only when the engine burned the lighter kerosene-type fuels, but by this time it was imperative that any diesel sold in Europe had to digest the heavier, less volatile crudes and tar oils to be commercially successful.

The next, and logical, modification was to insert the fuel nozzle along the cylinder axis and inject directly into the turbulent, displaced air. (One may ask why this was not done first as this is where the "MKD" nozzle was placed. However, hindsight is perfect and the obvious is easily obscured in

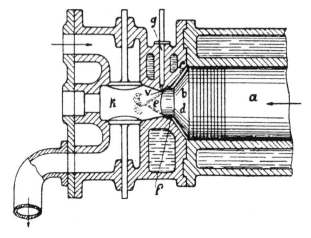

Fig. 13.23. Brandis patent drawing (Deutz) showing displacer piston with fuel sprayed at a right angle into air squish. *German Patent 239,716*

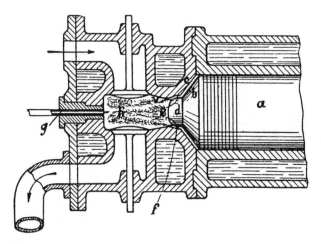

Fig. 13.24. Brandis patent drawing (Deutz) showing displacer piston with fuel axially sprayed into turbulent, displaced air. *German Patent 250,216*

the pursuit of proving a theory. The writer can painfully testify to this!) The main claim of the second patent, No. 250,216, broadly describes what was thought to happen:

> Injection combustion engine with a displacer on the piston, by which at the end of the compression stroke, a vigorous air swirl in the form of a hollow body is introduced due to the entrance of the displacer into a small, enclosed constriction of the cylinder, thereby characterized, that the fuel will be injected axially from the opposite end of the burning chamber in such a way toward the piston, that after its penetration

of the air swirl it strikes violently against the piston displacer, and in combination with the air swirl will be atomized.

The dependent Claim 2 refers to the air swirl that "receives the shape of a converging or truncated cone."

A little tale of deception lies behind the issuance of the above two patents because of the recent and sensitive memories associated with L'Orange's move to Benz. Deutz wanted to delay others from knowing where its developments were heading, so the patents were filed and issued under the name of Joachim Brandis, a research assistant working for Professor Hugo Junkers at the Technische Hochschule in Aachen. Deutz entrusted the discreet Brandis to pose as the surrogate inventor by taking advantage of an often-used loophole in German patent law. He was well known to them, having been an engineering employee and a friend of Hermann Schumm, their operations director who died in 1901.[51]

This simple relocation of the fuel nozzle gave Deutz a practical diesel without the need for a compressor. When introduced as the Model MKV Series in 1912, it became the first marketable compressorless diesel engine in Germany.[52] The MKV had little trouble burning Russian and other crudes, but the cheap German coal tar oils still required the assistance of a pilot charge of "ignition oil," at least on starting. A hand-operated valve at the nozzle adjusted the small volume of needed ignition oil once the engine was running (figs. 13.25 and 13.26).

These engines were built in sizes of 16 bhp to 125 bhp per cylinder until the middle of 1924. The total number produced of over 2,000 units would have been higher, but none were made during the 1914–1918 war. The largest was a four cylinder engine of 500 bhp. Crossley Bros. in Manchester, England, brought out engines very similar to the Deutz MKV in 1919.[53]

Professor Nägel tested an early 100 bhp MKV with results comparing favorably with similar size air injection diesels. Using an injection pressure of only 75 atm, he measured a bsfc of 199 g/bhp-hr.[54]

The Brandes Prechamber of 1907

A Swedish patent issued to one W. Brandes of Göteborg on February 20, 1907, was a precursor of the compressorless diesel. Brandes provided a transitional step between the traditional hot-bulb engines of Akroyd Stuart and others, and prechamber diesels that L'Orange would advance after his move to Benz.[55]

The patent discloses an externally heated, venturi-like throat having diverging, conical exits inserted between a water-cooled upper compression chamber and the power cylinder (fig. 13.27). Toward the end of the compression stroke, fuel is sprayed from an axial injector directly into the heated upper throat exit where it becomes partially vaporized on contact with the hot walls and is partially atomized by the high-velocity air stream passing through the throat.

Fig. 13.25. Deutz compressorless diesel, Model MKV 136, c. 1911, showing the displacer on the piston crown and a fuel nozzle adapted to burn coal tar oils with the help of an adjustable charge of a kerosene-type "ignition oil." *Deutz Archives*

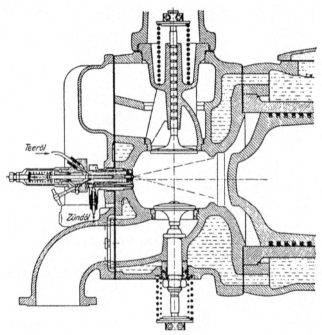

Fig. 13.26. Section through combustion chamber of a c. 1912 Deutz "MKV" compressorless diesel with dual-fuel nozzle and piston crown shaped to aid in atomizing the fuel. Hausfelder, *Die Kompressorlose Dieselmaschine*

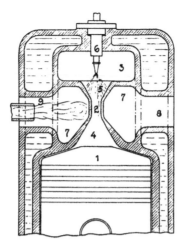

Fig. 13.27. Brandes 1907 Swedish patent drawing with the fuel spraying into a heated throat connecting an upper compression chamber with the cylinder. Hausfelder, *Die Kompressorlose Dieselmaschine*

Brandes wrote that the heated passage around the outside of the throat can be closed off by externally controlled valves while the engine is running so the throat may then be cooled by either air or water.

He referenced shaping the conical upper and lower divergent throat exits to accommodate either a "constant pressure or an explosion [constant volume]" combustion process. Although he is thus shown to have had the diesel in mind, it is more likely he intended his idea to be an improvement for lower compression, hot-bulb oil engines. None of these had the axial nozzle spraying on a heated intermediate throat prior to Brandes, yet within a few years several manufacturers adopted this idea.[56]

L'Orange at Benz

Only three months after Prosper L'Orange made his departure from Deutz to Benz & Cie. in Mannheim, his new employer applied for a patent embodying a novel prechamber, or *Vorkammer*. Future generations have credited this patent of L'Orange's as the basis for all prechamber diesel engines. While the assertion may be valid, his contemporaries, like Brandes, also searched for a practical prechamber-type diesel. The creative L'Orange enjoyed the good fortune to hold a responsible position in a company having resources to sustain a lengthy development program through oftentimes difficult stages.

Before proceeding further, the reader should understand that the term "prechamber," when applied to a diesel engine, is but one of several combustion chamber categories within a generic species more commonly known as indirect injection, or IDI. One distinction between the prechamber and its "antechamber" (swirl chamber, etc.) cousins is that the area of the passage between the auxiliary chamber and the cylinder is much more restricted. This

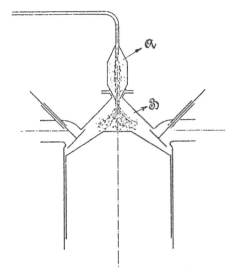

Fig. 13.28. L'Orange prechamber in his (Benz)
1909 German Patent 230,517.

passage can be either a single hole or multiple small holes. Also, the percent-age of the total air charge in a prechamber engine is less than required for combustion of all the fuel under heavier load conditions. High air turbulence induced during injection into the prechamber with its hot inner surface aids atomization and heat transfer. Surface vaporization and burning of individual fuel droplets is thus enhanced. IDI engines utilize an insulated insert form-ing part of the chamber, which heats up quickly and runs hotter to improve starting and combustion under load.[57]

The Benz/L'Orange German Patent 230,517 of March 14, 1909, refers not to a prechamber per se, but to a combustion process for a diesel engine. The drawing is a simple sketch of his proposal (fig. 13.28) on which a single claim reads:

> Combustion power engine for liquid fuels in which the fuel burns instantly on entrance in the engine, thereby characterized, that the liquid fuel will be injected across a hot chamber, whereby it in part burns completely, in part is atomized and in part is vaporized, and by this conversion the pressure in this chamber increases above the pres-sure in the working chamber of the cylinder, and whereby during the entire passage duration gases and vapors stream into the cylinder with the fuel and thereby atomize it.[58]

There is little doubt that L'Orange's claim falls within the current definition of a prechamber engine except for his not yet knowing that the charge does not pass through a vaporization phase. He does state that chamber volume and size of opening into the cylinder must be determined in order to obtain best performance.

One area where he had yet to make the complete transition to a modern prechamber was keeping the prechamber external to the engine and wanting to ensure ignition "possibly by heating the chamber with an external flame (especially on starting) as well as by the heat retained in the uncooled walls." However, this may be an unfair judgement because modern prechamber engines use electric glow plugs to perform the same function as the external torch for cold starts.

Deutz patent attorneys had surprisingly failed to notice an announcement of the patent when it was published for inspection prior to issuance. A reason given for this lapse is that the patent was indexed under a classification not normally scrutinized. By the time of discovery, it was too late for Deutz to raise the objection that L'Orange might have conceived the idea while still in their employ. A Deutz historian reported, nonetheless, that L'Orange "remained apprehensive about the strength of his patent."[59]

L'Orange built a test engine based on his prechamber patent in 1909 (figs. 13.29 and 13.30). It had a bore and stroke of 160 × 240 mm and a maximum speed of 400 rpm. One-fifth of the compression volume was in the prechamber. He reported the engine ran eight days without stopping and that the bsfc was 248 g/bhp-hr. The nozzle had filled with carbon by the end of the test, however.[60]

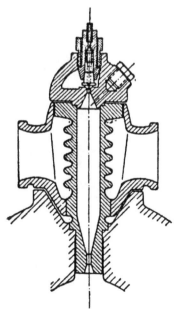

Fig. 13.29. Prechamber of 1909 L'Orange test engine.
L'Orange, *Ein Beitrag*

Ironically, Benz let the patent lapse for nonpayment of annual fees in 1915, a happening no doubt caused by a diversion of effort during the war. The German government later forgave those not paying such fees in that period, and Benz kept control of the patent until its original expiration date.[61]

An interruption caused by more pressing duties in Mannheim kept L'Orange from continuing with his prechamber endeavors until 1919. During this enforced hiatus, several German patents were granted to one H. Liessner. These new combustion chamber ideas that L'Orange most likely would have known about will be seen shortly.

L'Orange was first given the design responsibility for a small and rugged oil engine series Benz introduced at the end of 1909. These 2, 4, 6, and 8 bhp "P" models would achieve high-volume production for the times.[62] That same year he also became involved with revising the design of a marine diesel

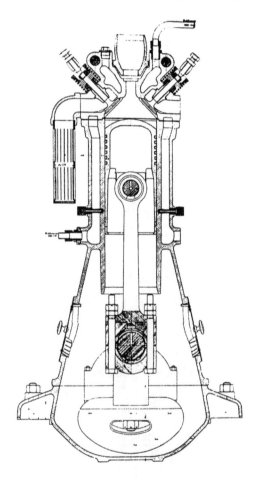

Fig. 13.30. Section through L'Orange/Benz 1909 test engine.
L'Orange, *Ein Beitrag*

engine that his company had licensed from Hesselman[63] (chapter 15). For the duration of the 1914–1918 war, L'Orange headed a department assigned to develop and produce Benz submarine diesels.

Benz had begun making conventional air injection diesel engines in 1908 after Diesel's second patent expired. It thus avoided any negative words from the Allgemeine. The A-frame design followed the standard practice of others except that the cylinders of the two-stage air compressor driven off the front of the crankshaft were horizontally opposed to each other. The first Benz single-cylinder diesel of 50 bhp at 120 rpm had a bore of 415 mm. By 1914 the line included 25, 40, 60, and 70 bhp models with the largest engine having four cylinders and an output of 280 bhp.

Evidence of L'Orange's reentry into prechamber research is seen in his German Patent 397,142 of March 18, 1919, assigned to Benz. It discloses another evolutionary step toward a modern design (fig. 13.31). The single claim of his so-called "funnel patent" deals mainly with an insert of that shape:

> Combustion power engine with higher compression, an ignition chamber situated in the cylinder head that is connected with the compression chamber by an inserted casing acting as a vaporizer and igniter enhancing means in association with one or more small passages, thereby characterized, that the casing is inserted in such a way that, with the exception of the well known ignition passage means in its lower end, the casing is sealed against leakage from combustion gases and is surrounded in its upper part by an isolated space.[64]

The prior art of Brandes, L'Orange's own 1909 patent, and the yet to be seen work of Steinbecker and Leissner were recited in the preamble (up to the "thereby characterized") of this claim.

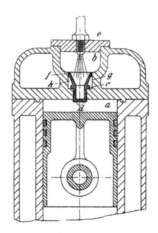

Fig. 13.31. L'Orange prechamber as shown
in his 1919 German Patent 397,142.

L'Orange states that one advantage of his new idea is the insert's longer life because previous designs were prone to a "complete destruction of the inserted piece, whereby their frequent and necessary replacement exerts special expenses and lost time, and it still cannot be prevented that the small, burned metal parts pass into the cylinder and destroy parts of the piston runway."

After describing the construction and disadvantages of hot-bulb engines, he said that "in contrast, because of the envisioned high-pressure engines with a compression between 35 and 40 atm, the ignition of the fuel will occur with certainty due to the compression temperature." L'Orange provided the insight that with his chosen placement of a mostly uncooled insert, "the engine, as has shown in practice, will perform over all outputs with reliable ignition and combustion." He attributed this capability at idle and light loads to less heat transfer away from the insert so that it remained hot enough to sufficiently vaporize and initiate burning of the fuel.

A later realized advantage gained by adding some form of prechamber, not known at that time, was a reduction of combustion noise caused by too high a rate of pressure rise in the cylinder after burning started. Air injection enjoyed an inherently low rate of pressure increase and hence quieter combustion. The opposite is true with airless fuel injection because it takes longer for the fuel droplets to find needed oxygen and then heat up enough to begin their complex burning process. This results in a greatly increased ignition delay with its associated high rate of pressure rise.

A production version of the engine based on his patent began in 1920, and in 1922 the Benz-Sendling two-cylinder tractor engine was introduced to become the first automotive-type diesel marketed. (Junkers offered a solid injection engine in 1921, but it was for non-mobile purposes.) To accomplish this, however, one must never forget the vital role played by the advent of a manufacturable hydraulic fuel injection system. These two parallel developments were inseparable.

As time passed L'Orange would devote more of his career to fuel systems. He rose to be head of engineering and research at Motoren-Werke Mannheim, a spinoff of the original Benz company resulting from economic changes after the war. (The automobile division would join with Daimler officially in 1929.) L'Orange left MWM in 1926 and the next year founded his own company. A few months before his death in 1939 he was awarded an honorary Doctor of Engineering, a tribute bestowed on only one person in Germany every year.

The Steinbecker "Back Fire" 1911

Karl Steinbecker of Charlottenburg, Germany, disclosed a prechamber idea in his 1911 patent, which at first glance looks very much like that of L'Orange's 1909 sketch: a small-diameter capsule with an even smaller throat opened axially into the main cylinder.[65] A major difference was that

Steinbecker's prechamber was water-jacketed and the fuel nozzle extended about halfway into it.

The fuel stream was injected directly into the throat just as the compression pressure in the cylinder reached a peak. His theory was that as fuel met the hot air stream entering the throat a little of it would find its way into the cylinder. It might be considered a pilot injection charge. The major charge portion that began burning in the prechamber on the trapped, back face of the spray raised the chamber pressure and ejected the remaining fuel/gases into the cylinder. The heat-prepared and greatly agitated mixture left the prechamber throat as a high velocity, intermixed secondary spray. It acted as a substitute for the atomizing air of blast injection. The idea to direct most of the fuel toward the prechamber's exit proved to be a rather important one because it later was found desirable to have only as much fuel in the main body of the prechamber as could find enough air to burn completely.[66]

This was the first of numerous Steinbecker patents issuing well into the 1920s.[67] A number of his injection ideas were adapted to Krupp's Germaniawerft marine diesels and, experimentally, on an automotive diesel.

Harry Leissner and the Ellwe Engine

Swedish engineer Harry Ferdinand Leissner (b. 1882) is little known outside of his native country, yet his contributions to early airless injection, prechamber engines deserve greater recognition. He was born near Stockholm and graduated from the Royal Technical University (*Kungl. Tekn. Högskolan*) in 1904. After receiving shop training in Germany and Sweden, he worked for AB Diesels Motorer from 1908–1911. He went to Ljusne-Woxne AB in Ljusne and began developing the prechamber, high-compression Ellwe (for L-W) engine, which went into production during 1916. Leissner, along with his engine, moved to Svenska Maskinverken at Södertälje (near Stockholm) in 1919.[68]

His three German patents may be seen as a bridge between the 1909 and 1919 inventions of L'Orange. While all the Leissner engines disclosed are two-stroke, their prechamber design applied to the four-stroke, a few of which were also built.[69]

Leissner's first patent, No. 282,739 of March 23, 1913, shows a simple, water-cooled antechamber into which fuel was sprayed (fig. 13.32). Unfortunately, an explanation of the ignition process in his main claim is flawed. He says that after "the entire fuel volume" has been injected at the end of the compression stroke when the "combustion air has already reached the ignition temperature of the fuel, . . . the fuel immediately ignites at the moment of injection in the working space of the cylinder" *before* it "has been able to condense on the walls of the pre-explosion chamber." It is difficult to see how he could prevent fuel from impinging on the cooled prechamber walls and causing severely quenched combustion.

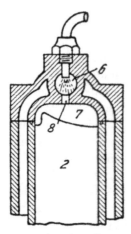

Fig. 13.32. Leissner prechamber German Patent 282,739
of 1913 with cooled chamber.

No doubt through further testing, Leissner takes a step forward in his second German patent 287,912 of February 28, 1914. Here an uncooled, replaceable intermediate chamber insert has small holes leading from it into the cylinder, which anticipates L'Orange's idea of 1919[70] (fig. 13.33). The cylindrical upper portion of the insert extends almost to the top of a cooled "pre-explosion" space surrounding it so that it is exposed to the heat of combustion gases on both sides of its thin walls.

Leissner goes from one extreme to the other as he says in his main claim that a portion of the fuel will ignite and burn in the pre-explosion primary chamber. When the pressure in it drops during the ensuing expulsion process, what passes through the inserted capsule will be partly "possible residue of fuel in the chamber" that will burn and part fuel that will "carry along." While he fails to mention any effect of the hot capsule walls on enhancing limited fuel preheating he gets closer to the solution.

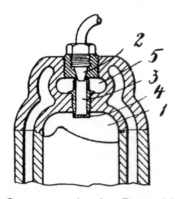

Fig. 13.33. Leissner German prechamber Patent 237,912 of 1914 with
replaceable insert containing small holes exiting into the cylinder.

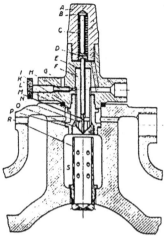

Fig. 13.34. Leissner prechamber in 1918 Ellwe engine
based on his German Patent 302,239 of 1917 with a perforated
wall to induce turbulence in the prechamber. *Motor, April 1918*

Finally, his third patent 302,239 of May 10, 1917, on which the production engine was based, is an improvement of the 1914 one wherein a series of angling holes are added in the exposed vertical wall of the insert (*R*) (fig. 13.34). The fuel/gas mixture passing through these holes into the primary "pre-explosion" chamber (*S*) imparts a strong and turbulent spraying action into the primary chamber which speeds up ignition and combustion there before the reverse flow into the cylinder begins.

In a 1918 technical paper, Leissner uses more engineering than patent terminology to say that:

> . . . the fuel which is injected into the nozzle [primary chamber] in atomized form, shall be ignited immediately and thus cause a high pressure in the cylindrical nozzle portion of the combustion chamber. This high pressure causes an injection of some of the oil into the cylinder and another smaller part into the chamber surrounding the nozzle, the pre-explosion chamber, partly through the perforated walls of the nozzle, partly through a slit between the upper edge of the nozzle and the upper wall of the pre-explosion chamber.[71]

Leissner also indicated that the quantity of fuel entering the outer chamber was automatically limited, and that the highest pressure reached was "when enough oil has entered to consume the oxygen present" and that "the gas mixture is flowing back through the nozzle and out into the cylinder." This "pulls with it the oil and gases remaining in the nozzle piece." The pressure in the pre-explosion chamber rose as high as 40 to 60 kg/cm^2 depending on load and type of fuel.

An admitted weakness in the Ellwe engine was a relatively rapid disintegration of the inserted nozzle piece. In normal operation the insert, made from a steel with 25 percent nickel to 100 percent nickel, lasted six months to a year although when the engine was run closer to full load it lived longer. Leissner attributed this to the fact that under lightly loaded conditions the nozzle was in a reducing atmosphere; it was also not cooled as much by the injected fuel.[72] Carbon deposits did not seem to build up in the insert under either loading cycle.

An eccentric on the crankshaft actuated a fuel pump plunger timed to lift the injector needle valve for proper combustion. A spring (C) in the injector returned the needle valve (M) to its seat "so that it can not drip in prematurely, and is excluding the air to prevent coke deposits."

Although constant pressure combustion was the goal, it could be only approximated in practice. Nevertheless, indicator diagrams showed that by tailoring the timing and prechamber configuration the preferred shape was reasonably attained under certain operating conditions.

Fuel consumption in g/bhp-hr compared favorably with contemporary M.A.N. diesels.[73]

	M.A.N.	Ellwe	
BHP	Guaranteed with 5% Error Margin	Normal Result	Best Result
7	235	235	210
10	225	215	200
15	220	200	185
20	210	200	185

While M.A.N. would have allowed itself a comfortable margin, the two-stroke Ellwe results of 1917 are still to be lauded.

According to Leissner, Ellwe engines were not fuel sensitive. Tar oils required injection timing advances, but the bsfc remained similar to the above results based on a lighter oil. His simple two-stroke engines, employing only crankcase scavenging, achieved a credible bmep of 4.8 kg/cm^2 (fig. 13.35). The smallest bore diameter was only 80 mm. A 20 bhp at 330 rpm model had a bore and stroke of 210 × 280 mm. It was this latter size engine that Benz bought for L'Orange in 1918 so he could thoroughly test it at Mannheim. A "rapid destruction" of the insert due to combustion heat in the prechamber along with fuel system deficiencies were L'Orange's reported major problems.[74]

One-cylinder engines in the 7 bhp to 20 bhp series used electric starters but required gasoline as the initial fuel. Leissner said that the engine burned

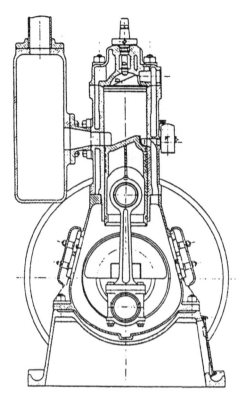

Fig. 13.35. Section view of 1918 Ellwe two-stroke engine with under-piston, crankcase scavenging. *Motor, April 1918*

about 10 g of gasoline per cold start. Engines available in 1918 offered air as a substitute for electric starting.[75]

During the 1920s Svenska Maskinverken produced Ellwe one-cylinder engines of 10 bhp at 625 rpm to 45 bhp at 425 rpm. Two-cylinder models went from 36 bhp at 550 rpm to 90 bhp at 360 rpm. These engines were reasonably popular as power for fishing boats, although they had been operated by the Danish Coast Guard as early as 1917.[76]

The Mianus Motor Co. in Mianus, Connecticut, bought a license for the Ellwe marine and stationary engine in the early 1920s to build one- and two-cylinder engines of 9 bhp to 30 bhp. Their fuel consumption was approximately 0.45 lbs/bhp-hr. It was reported that even though the insert in the American engine would burn out above where it screwed into the head, it ran "as well without as with the tube."[77] Leissner had mentioned this characteristic in his 1918 paper.

A L'Orange Reprise

Prosper L'Orange had adopted Leissner's small holes for the exits from his new prechamber. However, he importantly built onto this idea with his new "funnel" insert. This not only improved ignition but also eliminated gas flow restrictions within the prechamber itself. In addition, he ended heat problems with the insert by sealing his outer funnel wall from the combustion gases. Such an evolutionary progression of ideas is another example of patents protecting one inventor yet acting as a catalyst to another. This was so much so with L'Orange and Leissner that Benz and Ljusne-Woxna assigned each other free reciprocal license rights to their pertinent prechamber patents.[78] This gave L'Orange a green light with his development of a potentially more fruitful design without fear of legal entanglements.

Vickers 1910–1914

Vickers Ltd. at Barrow-in-Furness, England, pursued their own course toward an airless injection system uniquely differing from the displacer piston of Deutz and the prechamber of Benz. For the first time fuel, minus atomizing air, was successfully injected directly into a conventionally shaped combustion chamber. The inventions associated with the system have traditionally been attributed to James McKechnie (1852–1931), who was engineering director at Vickers during the time of its development.[79] Under today's laws others would probably have their names printed alongside his on the patents.

A major consequence of Vickers's work is that their efforts evolved into a hydraulic injection system first applied on pre-1914 submarine diesels. The theories on which the system was based swam against a strong current of prevailing thinking. McKechnie is to be credited for his determination and for ensuring the necessary corporate support during its lengthy development.

McKechnie was a Scotsman who rose through the draftsman ranks in the Clyde shipyards, and after years of experience in various marine engineering capacities was made manager of a Spanish shipyard. He came to Vickers around 1900 as engineering director. In 1910 he was elevated to the board of directors and four years later became managing director of the Naval Construction Works at Barrow.[80]

Vickers, whose company roots go back to 1828, had been an armaments supplier to the British military since 1888. Their Barrow shipyard, which built everything from battleships on down, often enjoyed a close relationship with the Admiralty Office. The navy had taken an early interest in submarines and encouraged Vickers to secure rights to the Holland sub patents owned by Electric Boat Co. in Connecticut. In this way Vickers, who did as asked in 1900, emerged as the principal sub builder for England.[81] This role further required Barrow to develop the diesels that had proven to be the only feasible power plants for subs. Chapters 15 and 17 look at Vickers's pre-1919 submarine engine program, except for the following on the evolution of its novel fuel system.

Other than patents granted to Vickers in McKechnie's name, little about the fuel system appeared in the literature until after the war. Military security had dictated that. The patents themselves are thus a prime source of information, because by war's end the developmental history of the system held little interest. Luckily, enough patents were filed for one to reasonably visualize the productive, and blind, paths taken by their designers and test engineers. Naval personnel reports add operational tidbits.

McKechnie's first British Patent 27,579 of November 26, 1910, was accepted the following November 27.[82] It formed a cornerstone on which the ensuing ones built by describing how to initiate a diesel combustion process through the direct injection of fuel under a "very high degree of compression" for atomization. Key to the invention was the broadly covered method of reaching, and sustaining, the hydraulic pressure to achieve adequate atomization. It was cumbersome and not altogether successful.

Of two methods described in the patent, the first is the one most often pictured. It was not likely used for very long, if ever, for several evident reasons (fig. 13.36). The system had three basic elements: a metering pump, a delivery mechanism, and an injector nozzle.

The pump, which functioned similarly to those on other contemporary diesels, had an eccentric-driven plunger piston (e^1) and a grooved sleeve (e^5), axially positioned by speed or load on the eccentric shaft (e^3), to control timing and open duration of the pump suction valve.

Movement of a delivery piston (C) was determined by a cam that transferred the motion through a cam roller (c^6). A heavy spring (D) held the piston-to-roller structure against the cam. This was not a true "elastic" accumulator as the delivery piston was displaced by mechanical cam action during its filling stroke. When the roller dropped off the cam lobe, the spring

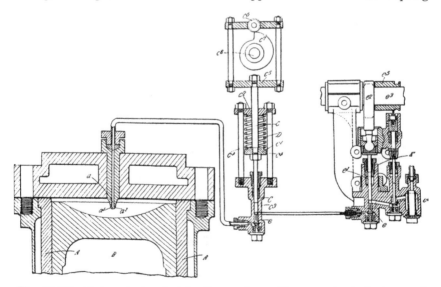

Fig. 13.36. Vickers hydraulic injection system with cam-actuated release of fuel from a delivery chamber. *McKechnie British Patent 27,579 of 1910*

quickly forced fuel under the piston (*C*) and past a check valve to the nozzle where it was injected. No check valve was shown at the nozzle tip.

The synchronized relationship between plunger eccentric, suction valve, and delivery piston established what became the injected fuel charge and what was forced back through the suction valve into the supply line prior to its closing.

While the patent states that fuel pressure could vary between 150 to 500 atm (2,200 to 7,350 psi), the maximum operating pressure was generally limited to about 4,000 psi.

The main claim of McKechnie's US Patent 1,079,422 of November 25, 1913, corresponding to the structure shown in figure 13.36, typically emphasizes this:

> In an internal combustion engine of the kind in which liquid fuel is injected into a charge of compressed air in the cylinder, a fuel pressure chamber having a resilient pressure applying member, means for supplying liquid fuel to the pressure chamber, and engine driven mechanism connected mechanically to the resilient member and adapted to gradually store energy in the resilient member and subsequently to quickly release it to suddenly inject the liquid fuel under extreme pressure from the pressure chamber into the engine cylinder.[83]

The term *resilient* means only that the piston/roller assembly is movable and not elastic.

A second design covered in McKechnie's basic British patent is an elastically expanded accumulator (fig. 13.37). The delivery piston (*C*) no longer moves in a timed relationship with the pump plunger and suction valve. Charge metering is now a function of the suction valve's closing time and the fuel pressure forces acting to push the piston (*C*) upward to compress the spring (*D*). This makes metering by the suction valve more controllable than in the first method.

Time of injection is determined by the lifting of a cam-controlled needle valve in the nozzle. When the needle lifts to inject the charge, all fuel suddenly displaced by the spring force acting on the delivery piston goes into the engine because the pump suction valve is closed. By separating into two distinct elements the functions of injection timing and injection force generation, the metering and timing over all load/speed conditions is more precise and more easily managed.

A simplified accumulator was Vickers's next step.[84] British Patent 26,227 of November 23, 1911, discloses an elliptical tube or tubes as the elastic element (fig. 13.38). Tightly inserted into the "spring-tempered steel" tube is a rigid, but grooved mandrel of the same shape. When fuel pressure is low, the tube wall rests firmly against the mandrel and then expands to a more cylindrical shape as pump pressure increases. The stored static pressure acting

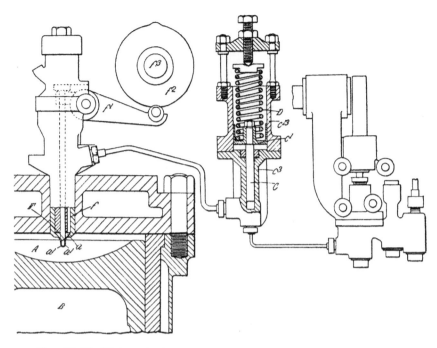

Fig. 13.37. Vickers fuel system with accumulator and cam-actuated injector nozzle. *McKechnie British Patent 27,579 of 1910*

on the distended tube is transformed into the injection pressure when the nozzle needle is lifted.

The so-called "Pulsator Tube" accumulator went on many sub engines, but it was later said to be "a doubtful success as after a few months in service the elliptical cross section had become truly circular" even without pressure acting on the walls.[85]

Two Vickers patents of 1912 and 1914 show modifications to the pump in which a spring-loaded pressure regulator was added to control peak pressures.[86] The intent was that with this simple regulator the pump pressure would be independent of pump speed to ensure better atomization at lower engine speeds. By this arrangement the suction valve no longer served as the throttle since it had a fixed, instead of variable constant opening and closing point. Added was a tapered, slidable cam to actuate a lever acting either to open a delivery valve at the pump or to lift an injector nozzle needle. The 1912 patent retained the tubular accumulator while the 1914 one eliminated it.

The "pressure regulator" pump apparently never saw production because reports on Vickers engines used in pre-1919 subs spoke of an adjustable cam to control metering by the suction valve.[87]

Vickers did extensive research on fuel sprays as disclosed in a January 1914 patent.[88] It emphasized the importance of a sharp corner, rather than rounded one, at the orifice exit and the great effect of orifice length—diameter

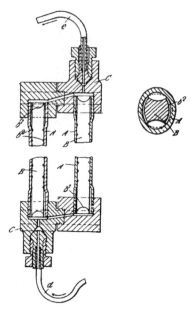

Fig. 13.38. Vickers tubular accumulator.
McKechnie British Patent 26,227 of 1911

ratios on spray patterns, and so on. While several novel nozzle tip designs were described, the production engines continued with conventional, inward opening needle valves on the standard 100 bhp per cylinder engines. The nozzle had five, 0.020-inch diameter holes pointing downward at 22 degrees from the horizontal.

This "nozzle" patent also shows a hydraulically actuated injector nozzle valve (fig. 13.39). A pressure-tight cap encased the needle's upper spindle end and valve spring. The force created by the differential pressure acting on the end of the spindle opened the valve. An air bleed valve for priming was in the cap.

Nonreversible submarine diesels from Vickers, including earlier eight-cylinder and derivative twelve-cylinder models, had a pump and accumulator tube for each cylinder. A later eight-cylinder, reversible engine used a row of four, suction valve-controlled pumps driven off the front of the engine. Each pump had a plunger piston with a diameter of 0.66 inch and a stroke of 1.25 inches. The pumps discharged into a single pipe running the length of the engine, which in turn fed all the cylinders. It was this "common rail" system that became a hallmark of Vickers's commercial and submarine diesels built after the war.

Important factors affecting combustion such as cylinder air turbulence and swirl were not yet considerations for the Vickers hydraulic fuel system. These to-be-forgiven omissions, along with "after-dribble," caused excessive exhaust smoke under higher loads, especially when fuel controls were not correctly synchronized for given operating conditions. Crews sometimes

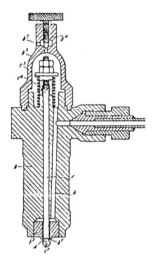

Fig. 13.39. Vickers mechanically lifted actuated nozzle needle valve.
McKechnie British Patent 1,059 of 1914

complained of the exhaust haze, and the enemy reported being able to track a sub's course from the smoke when it ran on the surface.[89]

Three independently adjusted handwheels determined injection timing, injection duration, and pump suction valve closing. They had an obvious interrelationship that required good crew training and discipline to achieve maximum performance with the least exhaust smoke. Later engines had controls combining injection valve timing and duration of opening (fig. 17.10).

Indicator cards showed that being able to independently alter the three variables allowed a changing of the P-V diagram from one resembling an air injection diesel's combustion with its soft, rounded crest at the end of the compression stroke to that having almost a constant volume combustion with a sharp, vertical pressure rise after compression. Optimum injection valve opening at full load was 16 degrees before TDC. Limited running with peak cylinder pressures of 700 psi were tolerated for emergencies, but the normal pressures were held to "a more reasonable figure."[90]

Vickers deserves great credit for being the first not only to build a practical airless fuel system, but also to apply it in what was a most demanding and critical service. The exhaust from the sub diesels was so dark that the enemy could spot them when surface running. A fuel consumption not exceeding 0.40 lbs/bhp-hr was comparable to or better than the sub engines using blast air injection. The removal of blast air had an important advantage of eliminating the refrigerating effect on the air in the charge as it entered the cylinder. This allowed a reduction in the compression pressure of at least 5 atm.

The Vickers's patents prophetically disclose most of the basic elements found in future pump-line-nozzle systems. Two elements were later combined when a helical groove was machined into a rotatable plunger so that this single element then performed both the pumping and metering functions. A hydraulically actuated nozzle valve needle, the genesis of a third to-come element, was anticipated in the above nozzle patent. Other creative people would make these evolutionary changes as Vickers was to continue with its common rail system.

Archaouloff's "Pneumatic" Injection 1915

Vadim P. Archaouloff, a professor at the St. Petersburg School of Engineering (Technische Hochschule), first tested his unique fuel system on a diesel in the Nobel factory during 1915.[91] The results were inconclusive at the time, and ensuing events in Russia prevented him from making further dynamometer work until some years later. He fled his native country after the 1917 revolution and settled in France where, during the 1920s and 1930s, he promoted his system with a degree of success to companies building large marine and stationary diesel engines. His early introductory work in Russia provides a fitting close to the story of airless injection developments during the first two decades of Diesel's engine.

What Archaouloff demonstrated as potentially feasible at the Nobel factory was how to multiply the peak compression pressure by means of a stepped-piston and, based on the differential area of the two piston diameters, provide enough hydraulic force to inject a liquid fuel charge into the cylinder.

A small air pipe from the power cylinder led to the larger side of an "intensifier" piston whose area was about twelve to thirteen times that of its opposite, smaller end acting on a delivery plunger. An attached metering pump deposited a fuel charge under the piston that ejected the charge into the cylinder when a differential valve in the injector lifted a needle.[92] The differential valve, a more mature form of the hydraulically actuated needle in the above Vickers nozzle patent, was being studied by a number of people at this time. Archaouloff was granted his first patents after 1918.

The few pioneers reported on in this chapter are but a more successful vanguard who led a small army of men dedicated to overcoming obstacles associated with airless injection. Attesting to this were the many patents issued in Europe and America during that time. Rudolf Diesel, in a patent applied for during January 1905, describes a unit injector employing airless injection (chapter 18). Nor are all injection problems solved yet as newly imposed circumstances demand ever-increasing performance and exhaust pipe cleanliness from today's diesel fuel systems. What intrigues is that, without realizing it, we are looking anew at lessons taught in Diesel's time that have current validity.

"Even as late as 1917, when I decided to try direct injection, a considerable distrust still prevailed among experienced diesel engineers towards the possibilities of this system in practical usage, I wrote to engineer P. L'Orange

and announced this decision. He expressed in his reply regrets that I, who had a certain success in diesel work, would enter an area where so many had already failed and which he in his flowery language described as 'the land of unlimited disappointments.'"

—Jonas Hesselman[93]

Notes

1. *Patent Notes for Engineers*, 7th ed. (New York: McGraw-Hill, 1956), 165.
2. For a biographical sketch citing sources, see Gustav Goldbeck, "Zum 100. Geburtstag von Friedrich August Haselwander," *MTZ* (February 1960), 58–9.
3. Ludwig Hausfelder, *Die Kompressorlose Dieselmaschine* (Berlin: Krayn, 1928), 47–54, describes the patents, cites claims, and gives operating deficiencies. Claim 1 of the first patent reads:

   A two-stroke engine for liquid fuels by which the working air valved into the crankcase becomes pressurized, characterized by the displacer (v) that sits on the piston in an appropriate form and towards the end of the compression stroke enters from the working cylinder into the displacer chamber (n), whereby from the said working chamber the pressurized air is displaced, and by means of the latter the liquid-injected fuel is atomized and is conducted into the combustion chamber of the working cylinder.

4. Ibid., 50. Haselwander's claims now read:

   1. A two-stroke engine for liquid fuel after the type of patent 101,453, thereby characterized, that the atomization of the fuel takes place due to compressed air from the ring chamber between the displacer and the cylinder wall in the central combustion chamber.
   2. A two-stroke engine for liquid fuel after Claim 1, characterized by a nozzle (d) in which the fuel is atomized by the compressed air entering from the displacer chamber and is swept away into the combustion chamber.

5. From Hugo Güldner, *Das Entwerfen und Berechnen der Verbrennungsmotoren*, 2nd ed. (Berlin: Springer, 1905), 126. Three fuels were tested, including Russian kerosene and Pechelbronn crude oil. No heat value was given for the American oil. The Pechelbronn, with a heat value (LHV?) of 10,018 cal/kg and a sp gr of 0.814, gave a higher efficiency than the Russian, with an LHV of 10,057 and a 0.815 sp gr.
6. Ibid., 127. It was patented under No. 182,767 of December 17, 1904. (The date given on European patents at that time was the filing date.)
7. Friedrich Sass, *Geschichte des deutschen Verbrennungsmotorenbaues von 1860 bis 1918* (Berlin: Springer Verlag, 1962), 587.
8. Ibid., and Gustav Goldbeck, *Kraft für die Welt* (Düsseldorf, Germany: Econ-Verlag, 1964), 105. Goldbeck was a historian at Deutz.
9. Brons biographical and company history is found in H. J. Stuvel, *Volle Kracht Vooruit: Bronsmotoren 1907–1957* (Appingedam, Netherlands: Brons, 1957), 150 pages. *Full Power Ahead* interweaves the story of the Brons family with that of the engine. Contemporary letters give it more the flavor of a biography rather than as many company histories overly influenced by the sales department. The story is also based on account books meticulously kept by Brons throughout his life. The author is indebted to the company for a copy of the book and to Fr. J. Hooyboer, C.S.C. for his translation help.
10. These machines were drawn out into the field by horses who were then put to work turning a shaft by walking around in circles as would be done for an old grinding mill. Instead, the rotary motion was transferred to a reciprocating, horizontal one that went to a separate unit doing the threshing. Larger machines used a pair of horses pulling on each end of the beam. See ibid., 16, for a picture of a four-horse model.
11. See Lyle Cummins, *Internal Fire*, 297–306, for a description of the Priestman engine.

12. Brons received US Patent 644,814 on March 6, 1900, for a "rotary engine." It was filed July 31, 1899.
13. Stuvel, *Volle Kracht*, 39.
14. British Patent 14,165 of 1904, filed on June 23, 1904, and issued February 23, 1905, has essentially the same main claim as the German. The claims of US Patent 868,839, filed July 1, 1904, and issued October 22, 1907, deal with construction and do not mention the combustion process.
15. British Patent 6,142 of July 25, 1907, was filed on March 14 of that year. US Patent 922,383 of May 18, 1909, covering the same idea, has a filing date of April 13, 1907.
16. Timmer set up his own engine company in Meppel, Holland, where he made a vaporizing oil engine with a "hit and miss" governor until 1928. Timmer also received a French patent for a two-stroke vaporizing oil engine. Jan Vegter of Siddeburen, Holland, who acquired the Brons company papers when it was bought by Waukesha, kindly informed the author about Timmer in a September 13, 1992, letter.
17. Stuvel, *Volle Kracht*, 60. N. V., *Naamloze Vennootschap*, is equivalent to Ltd. or Limited Liability.
18. Production records from Brons order book owned by Jan Vegter, Siddeburen, Holland, who kindly furnished them to the author.
19. Ibid., 62–3. German Patent 190,914 was filed by Deutz on April 3, 1906, and issued under that name. Only the fact that Deutz received the supplementary patent is mentioned in the Deutz history by Goldbeck, *Kraft*, 105, and Sass, *Geschichte*, 588. However, as Brons was granted patents in all other major countries of Europe and the US for the same idea, one can assume either that the contemporary Deutz experimenters were at best independent inventors or that Deutz, on realizing that Brons had not yet filed for a German patent, beat them to it.
20. Stuvel, *Volle Kracht*, 62. Sass, *Geschichte*, 588, mistakenly places the agreement date as the spring of 1906, which would fit the patent filing date of April 1906.
21. Stuvel, *Volle Kracht*, 63.
22. Based on 4,000 total horsepower produced and about 12 bhp per engine. Ibid., 102.
23. Ibid., 65.
24. Güldner, *Das Entwerfen*, 620.
25. Three- and four-cylinder engines are shown in Deutz brochure P. 2239 E. of August 1913, kindly furnished by Jan Vegter.
26. P-V diagrams and performance data are given in "The Brons Engine," *The Gas Engine*, October 1914, 597–601, and November 1914, 676–81.
27. See for example, P. H. Schweitzer, M. Alperstein, and W. Swim, "Fumigation Kills Smoke, Improves Diesel Performance," *SAE Transactions* 66 (1958): 574–95.
28. Details and performance of a one-cylinder, 40 bhp Russian Brons are given in *The Gas Engine*, November 1914, 676–79.
29. Hvid bought Brons's two US patents for $45,500. Brons could still sell engines in the US for a royalty of $1 per horsepower. From the 1914 contract between Brons and Hvid, a copy of which was kindly furnished by Jan Vegter.
30. Claim 1 of US Patent 1,134,857. A companion patent 858 was for a plunger in the cup that moved up and down with the fuel valve "for jogging the cup during the running of the engine." Its purpose was to keep the cup holes free of soot. It is doubtful that an engine with this cup ever went into production. Hvid's British Patent 4,370 of February 20, 1913, and German Patent 279,749 of February 25, 1913, combine the differing ideas of the two US patents in one. It is astonishing that Hvid's main British claim was allowed because it reads directly on the Brons structure of No. 6142 of 1907:

1. In an internal combustion engine of the character referred to, means for supplying air to the oil fuel vaporizing receptacle independently of the air supplied to the engine power cylinder, and means operating to cause ignition of the mixture of the air and oil fuel in the receptacle, whereby an explosion in the receptacle can occur preliminary to the main explosion in the engine cylinder. [!]

31. Hvid assigned all of his patents to the company. A November 17, 1913, letter to Brons from the First National Bank of Detroit provides this information. The company was formed "for the purpose of securing the rights on several firms of oil burning tractors and leasing these rights to the different manufacturers of this country." The bank told Brons that "the interested parties are all well regarded personally." Copy of letter from Jan Vegter. For further background consult the "Hvid Files," Local History Collection, Willard Library, Battle Creek, Michigan.

32. Comments made to the author by Hans L. Knudsen during a taped interview on August 24, 1972. A naturalized Dane, he was a design engineer for the St. Marys Machine Co. in 1914, and later became chief engineer for Hvid in Chicago. In January 1922 he moved to Cummins Engine Co. as chief engineer where he remained head of engineering until retirement in 1947.

33. "Another New Oil Engine," *The Gas Engine* 1913, 245.

34. For the trials and tribulations of building 1.5 and 3 bhp Hvid engines, see Clessie L. Cummins, *My Days with the Diesel* (Philadelphia: Chilton Books, 1967).

35. Later Hercules "Thermoil" literature states they had built 25,000 Hvid-type engines "since 1913." Glenn Karch, *A History of Hercules Engines* (Haubstadt, IN: Karch, 1989), 47.

36. Hausfelder, *Die Kompressorlose*, 43–46, provides scarce information on Trinkler. Neither Güldner, *Das Entwerfen*, nor Sass, *Geschichte*, show drawings from his patents. The main claim of Trinkler's first patent is quite broad:

> Device for the introduction of fuel into the pressurized air-filled combustion chamber by means of a controlled pressure-piston (x), thereby characterized that this controlled pressure-piston (x) in its associated housing (w), which itself is connected by two equalizing passages (v, g, h, i) with the combustion chamber (a), will move toward the end of the compression stroke, whereby one of the equalizing passages (v) closes, and fuel, through the other equalizing passage (g, h, i) will be atomized and dispatched into the combustion chamber.

37. As both Trinkler and the Körtings lived in Hannover, it may be assumed they would have been aware of each other's activities. The circumstances of their coming together are not known.

38. Sass, *Geschichte*, 654, references Johannes Körting, "Was das Gasfach unserer Familie gegeben hat," *Gas- und Wasserfach*, vol. 98, 833, for additional Körting history.

39. Körting sued to have the court declare Nicolaus Otto's 1877 basic four-stroke patent invalid in view of the Beau de Rochas patent. See Cummins, *Internal Fire*, 184–85, 299, 324, for the patent litigation story, and pp. 212–14 for details of Lieckfeld's small gas engines.

40. See Sass, *Geschichte*, 281–86, 311–18, for photos and cross-section drawings of these horizontal single- and double-acting engines.

41. Güldner, *Das Entwerfen*, 716. Professor Eugen Meyer who conducted the test on a 12 bhp engine at the end of 1904 used "crude naphtha."

42. Sass, *Geschichte*, 629.

43. German Patent 179,370 of October 28, 1905. Hausfelder, *Die Kompressorlose*, 157, spells the inventor's name as Lietzenmeier, and Supino, *Land and Marine Diesel Engines* (London: Charles Griffin & Co., 1918), 136, has Lietzenmeyer. The chosen spelling is from Güldner, *Das Entwerfen* 575–81, who describes and gives test data for horizontal engines (which resemble the Körting) built by Nicholson Maschinenfabrik AG in Budapest, which were licensed by the *Lietzenmayer'sche Gleichdruck-Motoren* GmbH (Lietzenmayer Constant Pressure Engine Co.) of Munich. He gives test data on a similar Lietzenmayer engine made by Gebr. Bromley, a Moscow licensee. For unknown reasons, Güldner fails to mention the Körting diesel.

44. Ibid., 630.

45. Pre-delivery test data for the engine by Belgian engineer M. R. E. Mathot show that it had a full load bsfc of 0.388 lb or 176.1 g/bhp-hr on a "California" fuel with an LHV of 17,740 Btu/lb. "A Diesel Engine Test," *The Gas Engine*, October 1914, 612–13. A. V. Youens, Palo Alto assistant city engineer, said they used the same fuel as that burned under their boilers. Because of the high asphaltic content in the fuel, the engine "was shut down every Saturday to remove the carbon deposit from one piston." They cleaned the valves every two weeks. The start of the 1914–18 war prevented good service information from reaching them, and they had to learn much on their own. From the Discussion in "The Diesel Engine in America," *Transactions, International Engineering Congress, 1915* (Mechanical Engineering) (San Francisco: 1915), 329. Original and operating costs are in *City of Palo Alto Reports, 1913–1914*, 32–33; *1914–1915*, 26–27, 30–33; and *1915–1916*, 32–33.

46. Sass, *Geschichte*, 631–35, describes these two-stroke, six- and eight-cylinder engines of 280 and 370 bhp. The submarine *U-1* of 1906 on display in Munich's Deutsches Museum has hull sections cut away in order to view its six-cylinder Körting engines.

47. German Patent 140,265 issued June 13, 1902.

48. Daimler-Benz compiled a biography of L'Orange dated February 24, 1939, 6 pages, at the time of his receiving the 1939 honorary engineering doctorate for Germany. Daimler-Benz Archives. He was to die only a few months later.

49. A good overview of I-C engine developments at Deutz is found in Gustav Goldbeck, "Entwicklung des Verbrennungsmotors in Deutz," *MTZ*, 25/10 (1964), 376–82.

50. Güldner, *Das Entwerfen*, 562. The test was conducted under the auspices of the Bavarian State Industrial Institute. The bsfc has the 1.04 correction from PS to bhp.

51. The most complete explanation about Brandis is in Goldbeck, *Kraft*, 114–15. He is quite adamant that Brandis was not the inventor. Hausfelder in 1928, *Die Kompressorlose*, 276, only says that the patents were "a direct forerunner of that design of compressorless engine."

52. Professor Nägel in "The Diesel Engine of Today," *Diesel Engines*, VDI (1922), 29, English translation gives full credit to Deutz for being the first in Germany with a compressorless engine.

53. Goldbeck, *Kraft*, 115.

54. Sass, *Geschichte*, 598.

55. Swedish Patent 24,041. The importance and implications of this patent, with rather differing interpretations, are given in Hausfelder, *Die Kompressorlose*, 159–63, and A. E. Thiemann, *Fahrzeug-Dieselmotoren* (Berlin: Schmidt, 1929), 127–28. The writer tends to side with Thiemann. See chapter 3 and Cummins, *Internal Fire*, 308–17, for comments on the Stuart patents and on the engine's design.

56. Neither Bryan Donkin, *A Text-Book on Gas, Oil and Air Engines*, 5th ed. (London: Charles Griffin & Co., 1911), nor Dugald Clerk and G. A. Burls, *The Gas, Petrol, and Oil Engine*, vol. 2 (London: Longmans, Green & Co., 1913), show oil engines prior to Brandes using his design. Thiemann, *Fahrzeug-Dieselmotoren*, 218, lists a number of engines who later adopted at least variations of the Brandes idea.

57. The reader is referred to such excellent texts as, for example, H. R. Ricardo and J. G. G. Hempson, *The High-Speed Internal Combustion Engine*, 5th ed. (London: Blackie & Son, 1972), for a more complete explanation and understanding of the types of IDI engines (pp. 30–44) and for the combustion process (pp. 4–11).

58. The patent, entitled "Combustion power engine for liquid fuels," was accepted after the usual practice of public inspection on February 1, 1911. The specifications and claim were on one page and the drawing another.

59. Goldbeck, *Kraft*, 113. Daimler-Benz historical publications obviously tend to downplay the idea's "early birth after the marriage."

60. Prosper L'Orange, *Ein Beitrag zur Entwicklung der kompressorlosen Dieselmotoren* (Berlin: Schmidt, 1934), 31.

61. Sass, *Geschichte*, 614.

62. L'Orange, *Ein Beitrag*, 32. Sass, *Geschichte*, 606–7, describes in detail the referred to "P-Motors."

63. Sass, *Geschichte*, 610, infers this was the engine used in Amundsen's ship *Fram* during his 1910–1914 Antarctic expedition; it was Swedish-made (chapter 15).

64. Benz German Patent 397,142, filed March 18, 1919, and issued on June 17, 1921.

65. Steinbecker received US Patent 1,154,816 on September 28, 1915. Of interest is that this patent, filed on September 11, 1911, was assigned to the General Electric Co. of New York.

66. Julius E. Wild, "Combustion-Chambers, Injection Pumps and Spray Valves of Solid-Injection Oil-Engines," *S.A.E. Journal* (May 1930), 3, as printed in the "Annual Meeting Paper."

67. Hausfelder, *Die Kompressorlose*, 204–22, looks at the many Steinbecker patents. Chas. E. Lucke, "The Rising Importance of Oil-Injection Type of Internal-Combustion Engine," *Trans., ASME* (1921), 1158–59, also describes his work.

68. Information on Leissner, the Ellwe engine, and the companies where it was made was found for the author by Per-Sune Berg. This came from the Royal Swedish Technical Museum in Stockholm and the Marinmotorhistoriska sallskapet in Umea.

69. Both Hausfelder, *Die Kompressorlose*, 119–28, and Thiemann, *Fahrzeug-Dieselmotoren*, 144–45, discuss at length the Leissner patents. Thiemann is unfairly harsh in his opinion of the inventor's interpretation of events: "Superficial . . . with insufficient knowledge and poor powers of observation." Further, "Only an entirely inadequate engineer, . . ." and so on. Hausfelder is more kind, which is perhaps why Thiemann's writing at a later date tends to contradict or upstage Hausfelder. Dr. Horst Hardenberg related in a November 25, 1992, letter to the author an unpublished story that the two "became enemies after having—equally unsuccessfully—competed for a professorship at a *Technische Hochschule*."

70. His patent corresponds roughly with US Patent 1,136,818 filed on March 18, 1914, and issued on April 20, 1915.

71. Harry F. Leissner, "Ellwemotorens Förbränningsprincip," *Teknisk Tidskrift* (1918), 46. From a paper read January 15. Translation by Per Bolang.

72. Ibid.

73. Ibid., 47.

74. L'Orange, *Ein Beitrag*, 32–4.

75. "Ellwemotoren fur ny startanording," *Motor* (1918).

76. H. F. Leissner, "Ellwemotorens," 48.

77. Lacey H. Morrison, *Diesel Engines* (New York: McGraw-Hill, 1923), 478–80.

78. L'Orange, *Ein Beitrag*, 34. The exclusive reciprocity was only for sales in Germany. Benz soon sold Deutz a license for the Benz/Leissner patents for engines up to 40 hp. Sulzer also bought rights to both.

79. Some degree of credit must also be given to W. Robertson, who headed the Diesel Section from about 1905 to 1920 and D. M. Shannon, a designer there during 1911–13. Robertson was brought to the author's attention during a 1972 meeting with Mr. Ernest Davies who had joined Vickers in 1925. Davies was Chief of Internal Combustion Engine Design at Barrow on his retirement in 1970. A reference about Shannon's contribution is found in "FIAT Marine Oil Engine, Mr. D. M. Shannon's Important Appointment," *Journal of Commerce & Shipping Telegraph* (London: March 1926).

80. "The Late Sir James McKechnie," *The Engineer* (October 1931): 397. His further activities can be found in J. D. Scott, *Vickers—A History* (London: Weidenfeld & Nicolson, 1963), which gives his strengths and weaknesses and remarks that he was mainly an able production-oriented man "who had ruled the shipyards and shops like an Eastern potentate" (p. 138). McKechnie was knighted for wartime services in 1918 and on his retirement in 1923 moved to the United States where he lived until shortly before his death. He was buried in Glasgow.

81. Scott, *Vickers*, 61–68. Another important aspect of the story is found in Gaddis Smith, *Britain's Clandestine Submarines, 1914–1915* (New Haven, CT: Yale University Press, 1964), 155, which covers US and Canadian ties to Barrow at the beginning of the war.

82. The patent leaves no doubts as to ownership. It begins, "I, James McKechnie, Director of Vickers Sons & Maxim, Limited, Naval Construction Works, Barrow-in-Furness . . ."

83. The US patent was filed on November 23, 1911, or about a year after its corresponding British one of 27,579. It would appear that the US Patent Office required the case to be divided so that the spring-loaded accumulator (fig. 13-36) was not included. Its structure appears in another patent.

84. The patent claims state that they are based on 27,579 of 1911. US Patent 1,066,936, issued July 8, 1913, also covers this design.

85. Comments to the author by Ernest Davies in prepared notes.

86. The first was British Patent 24,127, issued October 22, 1912. A comparable US patent was 1,147,806, filed on August 5, 1913, issued July 27, 1915. The second, No. 24,153, had the same date.

87. Two very informative articles: "British Submarine Boat Building during the War," *Engineering* (February 1919): 264–66, plus photos, and "Heavy Oil Engines for British Submarine Boats," *Engineering* (July 1919): 8–10, 16, plus drawings.

88. McKechnie's nozzle patent 1,059, issued January 14, 1914.

89. Hausfelder, *Die Kompressorlose*, 303.

90. *Engineering* (July 1919): 9.

91. Hans Flasche, "Der Dieselmaschinenbau in Russland von Anbeginn 1899 an bis etwa Jahreschluss 1918" (1947 ed.), 39. Copy in MAN Archives.

92. Ibid., 40. A more detailed explanation, in English, can be found in "Prof. W. Arschauloff's [*sic*] Patents," *Fuel Injector for Internal Combustion Engines* (Paris: c. 1922), 8 pages. The author was kindly given a copy by Sulzer Bros., Winterthur.

93. J. K. E. Hesselman, *Teknik och Tanke* (Stockholm, Sweden: Sohlmans, 1948), 223.

CHAPTER 14

Post-Patent Players

"I met an engineer just come up from the Argentine, who told me that there were no less than 20 agencies for European Diesel Engines, English, German and Swiss, in Buenos Aires alone."
—Edward D. Meier, August 1913[1]

The expiration of Diesel's first European patents in 1907 and 1908 was eagerly awaited by challengers ready to compete with the existing licensees. Benz, Deutz, and Körting are examples already seen of those wasting no time in doing so.

This chapter focuses on land-based engines of other players who entered the diesel field during that decade after the pioneer patents lapsed. A few, like Junkers, contributed to a rapidly expanding technology, but most followed the proven designs of established leaders. Several of these less adventuresome builders, like Tosi and Güldner with their evolutionary models, carved out profitable niches for themselves in Europe.

More US companies entered the diesel engine business using licensed European designs, while others dared to begin with creations of their own. These new builders, who would soon give Busch-Sulzer stiff competition, added the synergy that created an American diesel engine industry worthy of the name.

Hugo Güldner

Hugo Güldner (1866–1926) earned his reputation within the German internal combustion engine fraternity as an inventor, a text writer, and a manufacturer. Chapter 8 gave one example of his work with the development of the first two-stroke diesel in Augsburg.

Güldner was born in Herdecke south of Dortmund and received training at an engineering school in nearby Hagen. In the mid-1890s he interested several companies in a gas engine design that he had patented. None were built. With an investment of 50,000 marks from a Berlin businessman in 1897, he formed "Lüdeke & Güldner, Maschinenfabrik, Magdeburg." By the time this venture went into liquidation at the end of that year, Güldner had already filed for twelve patents and registered designs covering two-stroke engines. He spent 1898 with a German company licensed to make Hornsby-Akroyd oil engines, and from there moved to M.A.N.[2]

As almost a respite in his professional career, the thirty-five-year-old engineer took the time to write a classic text on I-C engines. *Das Entwerfen und Berechnen der Verbrennungsmotoren* (The Design and Calculation of Internal Combustion Engines) was published in 1903. This thorough tome became a "bible," strongly influencing a generation of German-trained engine designers.[3]

He reentered industry in 1903 as chief engineer of Maschinenbau-Gesellschaft München. One employment consideration was a requirement for them to build a 20 bhp, four-stroke gas engine based on a design he had just patented. The engine ran well. On February 15, 1904, his patent rights were transferred to a new company, Güldner-Motoren GmbH, which he headed. On his board were Carl von Linde and Dr. Georg Krauss (of Krauss-Maffei). Güldner occupied only a part of his previous employer's plant to build a line of gas engines and then leased the entire facility when the original company had to vacate the premises.

In 1906 the shareholders authorized a move to Aschaffenburg (east of Frankfurt) where Linde owned a branch factory that made his refrigeration machinery.[4] Before the end of 1908, all Güldner gas engines came from there. By 1912 power output of the largest of these engines was 850 bhp from four cylinders.

Güldner, who had waited impatiently for Diesel's patent to expire, built a four-stroke, 60 bhp test diesel engine that was completed by the end of 1907. Its design borrowed heavily from the successful features of others. A two-stage injection air compressor, with cooling after each stage, ran off the end of the crankshaft; a sturdy A-frame and cylinder jacket were in one casting, and a mid-height camshaft operated the valves and nozzle needle via pushrods. Each cylinder had its own fuel pump whose output was controlled by the governor acting on a valve in the discharge line.

His philosophy of conservatism resulted in heavy and slow—but reliable and long lived—engines with low fuel consumption. Such attributes gained the company a limited but loyal group of customers. The diesel engines had commonality with the strongly built gas engines to use the same tooling. In this way Güldner could offer an extensive line of diesels at minimum extra cost. A four-cylinder diesel introduced in 1914 produced 700 bhp at 160 rpm and weighed, with flywheel, 98,000 kg or 308 lb/hp.

Test results on the engines are found in Güldner's texts. The data gives typical state-of-the-art performance. A 1909 test of one-cylinder, 415 × 610 mm, 70 bhp at 170 rpm:[5]

Load	Full	-3/4	-1/2	Max
Speed, rpm	170.4	170.3	170.8	166.5
Power, bhp (PSe × .986)	71.3	53.8	35.2	88.7
Avg. blast air pressure, at.	59.1	59.4	59.0	63.3
Indicated mep, kg/cm²	6.98	5.90	4.67	8.00
Mechanical efficiency, %	70.7	64.3	55.3	78.0
Bsfc, g/bhp-hr	181.4	186.9	209.2	187.7
Brake thermal efficiency, %	35.4	34.3	30.7	34.2

Although the efficiency is based on a fuel with a heating value of 10,000 cal/ kg, that used was a Russian crude with an average of 9960 cal/kg. Heat values were unspecified as to being LHV or HHV.

A 1910 factory test on a two-cylinder, 535 × 780 mm, 300 bhp at 150 rpm engine yielded the following (this is the engine with four cylinders and 10 rpm more to give the 700 bhp mentioned above.):[6]

Load	Full	-3/4	-1/2	Max
Speed, rpm	148.5	150.4	152.4	147.7
Power, bhp (PSe × .986)	301.8	232.6	162.0	334.4
Indicated mep, kg/cm²	6.45	5.09	3.81	7.14
Mechanical efficiency %	81.8	79.1	72.6	82.5
Bsfc, g/bhp-hr	178.7	178.6	181.2	182.0

Here the fuel's heat value of 10,000 cal/kg is stated as the LHV.

A similar engine line of 50 to 1,000 bhp was advertised in 1923. Reported fuel consumption was as low as 166 g/bhp-hr.[7]

Hugo Güldner ran the Aschaffenburg works until his unexpected death after an operation in March 1926.

Franco Tosi

The Franco Tosi company of Legnano, Italy, had signed a one-year option contract with the Allgemeine in July 1903 for rights to Spain and Portugal as well as Italy at a cost of 12,000 marks. A stated reason for the hesitancy was a concern about the taxes to be imposed on imported fuels. The Italian Parliament failed to make a decision, and the lingering uncertainties over this caused Tosi to let the option expire. However, some months earlier it bought drawings for the 20 and 50 hp engines from M.A.N. to show its sincerity about wanting a license contract.[8]

Soon after Diesel's first patent expired in 1907, Tosi and others began making engines in Italy without communicating this fact to the Allgemeine. The "omission" caused displeasure in Augsburg because a companion Diesel patent would not expire in Italy until 1910. Nevertheless, the parties reached a settlement whereby for a payment of 10,000 marks a pending legal action was rescinded and Tosi received plans for a 150 hp M.A.N. engine.[9] Manufacturing legally began with designs based on the M.A.N. model "DM," but in time they were more influenced by what came from Sulzers.

Franco Tosi (1850–1899) received his engineering education at the Zürich Polytechnikum and worked in German and English machine shops before returning to his native Italy. In 1876 he joined a new factory in Legnano as technical director. He became a partner and, in 1894 as sole owner, changed the company's name to his own. By this time it was a leading Italian steam engine and electrical equipment manufacturer. Tosi founded homes for the sick, widowed, and orphaned. He was assassinated in 1899 by a former employee.[10]

International recognition of Tosi diesels came at the 1911 Turin Exhibition through the display of a four-cylinder, four-stroke, trunk piston engine.[11] It produced 600 bhp at 150 rpm from a bore and stroke of 535 × 770 mm, and during the show turned a 400-kw generator. A Reavell compressor for injection air was driven off the front end of the crankshaft (fig. 14.1).

Fig. 14.1. Tosi 600 bhp diesel at the 1911 Turin Exhibition.
Engineering, 1912

Individual A-frame and cylinder jacket castings rested on the bedplate and supported the cylinder heads. The governor, mounted atop the vertical shaft, controlled suction valves on individual fuel pumps that were driven by an eccentric on the vertical shaft. (It was a combination of MA.N. governor location and Hesselman-type pump actuation.) A lube pump also driven off the vertical shaft sent oil to the crankshaft and rod bearings and, through liner ports, to the pistons. The engine started on a light oil and during operation burned a heavier, cheaper crude oil.

Max Rotter visited the Turin Exhibition, and in a report to Busch-Sulzer on the displayed diesels said that Franco Tosi "is one of the largest and best machinery firms in Italy and has a world wide reputation of the highest character." About the engine and its intake air silencing method he said:

> The design of the engine is very similar to Sulzers' D Type, with a Reavell air pump. The chief differences from the Sulzer D are in the general design of the frames. While the workmanship is good (apparently as good as Sulzers'), the general design is not as sightly, the frames being somewhat slab-sided, the Hartung governor clumsy and the air pipes connecting each frame with each cylinder large and ugly. A further objection to the last mentioned arrangement is that it very materially increases the lubricating oil consumption, as all the oil vapor is drawn into the engine cylinders; as against this it practically deadens the sound of the suction.[12]

A criticism of early diesels was the air suction noise during intake. Tosi reduced the noise by drawing combustion air from inside the semi-enclosed A-frames. The air pipes mentioned above refer to this.

Franco Tosi began selling a line of stationary engines of this design with outputs up to 1,200 bhp. In several years this was raised to 2,400 bhp in six cylinders. Their two-stroke marine engines were introduced shortly after the Turin Exhibition (chapter 16). By the end of 1918, Tosi had built over four hundred diesels, including land and marine, with a total output of about 120,000 bhp.[13] The good reputation gained from these engines led to foreign licenses, two of which were the Fulton Iron Works in the United States and Scotland's William Beardmore & Co.

Franco Tosi Industriale SpA produced a line of well-accepted diesel engines until the end of the 1980s. In 1990 the company became a part of ABB Asea Brown Boveri SA.

Fulton Iron Works

The Fulton Iron Works of St. Louis opened its doors in 1854 to make steam engines for packet boats serving the Mississippi and Missouri Rivers. The company met the industrial growth in the Midwest by supplying steam engines for power plants. This was followed by the machinery and steam processing equipment used in sugar mills. Fulton became an important supplier of complete mills for North and South America, currently their sole product.

With Diesel's first US patents beginning to expire, Fulton saw a new business opportunity and inquired into an arrangement for a proven European engine to make in their own shops. They focused on the Franco Tosi diesel, no doubt as a result of the 1911 Turin Exhibition success and because the Carels, Diesels Motorer, M.A.N., and Sulzer designs were already spoken for.[14]

Fulton bought a manufacturing license from Tosi for four-stroke, trunk piston diesels in 1913, and the next year delivered their first engine to a power plant in Alamogordo, New Mexico. This three-cylinder, 13.5 × 19.5-inch model had an output of 150 bhp at 200 rpm and weighed 450 lb/hp. Brake specific fuel consumption at full load was reported to be 0.44 lbs/bhp-hr.[15]

The design generally followed Tosi's except for an axial, three-stage vertical air compressor driven off the front of the crankshaft. Instead of a pressurized system, piston and bearing lubrication came entirely from gravity drip feeds. Another change was necessitated because of potential legal complications.

The Tosi fuel pump, which metered by controlling the closing of the suction valve, was based on a patent by Lauster.[16] While his European patents had expired, the one in the US owned by Busch-Sulzer still had seven more years to run. Busch-Sulzer was advised by their patent attorney that Fulton would infringe this patent if it adopted the Tosi pump.[17] Fulton was also aware of the situation.[18]

The likely patent dilemma was resolved by Fulton on early production units by a pump that elastically varied the volume of its pumping chamber. During the first part of the plunger stroke, fuel pressure distended a diaphragm that came up against a stop positioned by governor linkage.[19] This controllable subtraction of displacement "absorbed" the excess, unwanted fuel pumped by the plunger's fixed stroke. At the beginning of the suction stroke, the diaphragms returned to their flat, relaxed position. In figure 14.2 the lever arms (L), lifted by the governor, rotated threaded shafts (R) in and out to position segmented stops (S). The outer ends of the segments pivoted on housing (C), and the inner ends engaged in grooves on the axially shifted shafts. The diaphragms (A) were well supported by the segments, and flexing was limited to no more than 0.030 of an inch. Diaphragm durability was not mentioned.

A distributor block with adjustable needle valves seating in calibrated orifices (as on the Swedish diesels) balanced flows and allowed the use of a single pump plunger. Indicator diagrams showed that performance differences between cylinders could be held to within 2 to 3 percent. This was deemed close enough to justify not going to the expense of a pump for each cylinder.[20]

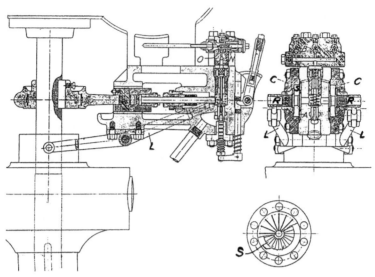

Fig. 14.2. Fulton diaphragm method of fuel metering.
ASME Journal, July 1914

When it was feasible to do so (and it appears to have happened before the Lauster patent expired), Fulton switched to suction valve metering. In the early 1920s a new design appeared with individual pumps incorporating a variable stroke plunger.[21]

Fulton installed two larger engines during 1914 in a Tucson, Arizona, power plant that were still in operation as late as 1948. These three-cylinder, 500 hp at 150 rpm diesels had a bore and stroke of 22 × 29.5 inches.[22]

The company continued to build a line of rugged and reliable diesels until a few years after the Second World War.

Langen & Wolf

Langen & Wolf of Milan was another Italian firm to enter the diesel market in 1907, and because of no prior contact with the Allgemeine, was selected to be the first of the new Italian builders sued for patent infringement. It settled in 1909 on the same terms as did Franco Tosi with the Allgemeine.[23]

Società Italiana Langen & Wolf, Fabbrica di Motori a Gas "Otto" was a long-established engine builder. Deutz's Austrian subsidiary organized it in 1888 as a branch factory for building Otto gas engines.[24] The Germans had brought it under their own management in 1890 but later sold a minority interest to Italian investors. By 1910 Langen & Wolf was independently owned. Max Rotter commented about the company in his review of the engine at the 1911 Turin Exhibition: It was "backed by some of the largest Italian banks. It builds a cheap class of machinery, which it is able to dispose of at good prices in cases where the bank influence is strong; but sells at very low prices where there is open competition."[25]

Rotter did not seem too impressed about the engine. He wrote that the four-cylinder, 500 bhp engine he saw was "not in operation, and it is reported that it does not run very well. It looks like a cheap job." The fact that it looked similar to the older Sulzer Type D with horizontal injection pumps at one end indicates it was not of the latest design because of what was written a year later.

Langen & Wolf had adopted a conventional four-stroke, A-frame stationary engine design, but a few of its features as described in 1912 are worth noting in view of the state of the art at that time.

The vertical shaft leading to a high-mount camshaft extended below the helical gears at crankshaft level to drive a rotary-type lube oil pump at crankcase level. A two-way valve at the pump discharge led to two filters so that one might be cleaned while the other remained in service. Surprisingly enough, this simple idea was not at all common and was remarked upon as much needed because "either no provision is made for filtering or there is only a single filter."[26]

A fuel metering pump was located at each cylinder instead of a single larger pump plunger with a distributor. Arguments for and against individual pumps on multicylinder engines were still going on, especially for the smaller power sizes. A single failure with separate pumps allowed an engine to keep running minus that cylinder. On the other hand, as the debate went, a larger, single pump meant a more rugged mechanism less prone to fail. The jury remained out on which was best.

The upper part of a two-piece piston held the rings and bolted onto the lower part containing the piston pin (fig. 14.3). Less heat transfer to the lower piston walls and pin, and hence better lubrication, were given as reasons for the design choice.

An exhaust valve face of cast iron was inserted over a steel stem (like a thick washer) and screwed against an integral flange on the end of the stem. To ensure concentricity the face piece piloted into the flange. Valve life was reported to be longer, but manufacturing difficulties also increased.

The standard 400 bhp at 250 rpm model had a bore and stroke of 440 × 480 mm. It was also available with a 560 mm stroke.[27] Engines 400 bhp and below used a two-stage horizontal air compressor, while higher output models had a vertical, three-stage compressor. Both were driven off the front of the crankshaft.

Langen and Wolf built only stationary engines with a number of them used on shipboard for auxiliary power. As of 1912 the largest engine made was a 600 bhp, four-cylinder engine being installed at the time in an Italian battleship. By this date the company had delivered a cumulative output of about 20,000 bhp. A reporter's visit to the factory found that "the firm appeared to be very busy" with "at the very least forty engines of various capacities from 12 horse-power [*sic*] and upwards in different stages of construction."[28]

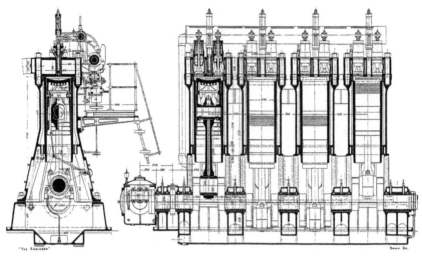

Fig. 14.3. Langen & Wolf (Milan) 400 bhp stationary engine of 1912.
The Engineer, July 1912

It is not known exactly when Langen & Wolf ended diesel production, but it may be fair to assume that the engines were a victim of the Depression of the early 1930s.

Hugo Junkers

The most imaginative diesel engines emerging after Rudolf's patents expired were the inventions of Hugo Junkers (1859–1935) (fig. 14.4). They resulted from his creative genius and his exhaustive research into the dynamic and thermodynamic elements of internal combustion engines. His pathfinding investigations of combustion and heat transfer examined what heretofore had been either neglected or not yet considered.

Junkers is more remembered for his pioneering work with all-metal aircraft and not for his contributions to marine, automotive, and aircraft diesels. Other products of his that are recognized in their own specialized fields include gas calorimeters and water dynamometers. Middle-class Europeans born in the first half of the previous century will also recall Junkers as one who provided them with safe and efficient gas heaters for bathing.

The Junkers family had lived for several generations near Rheydt, now a suburb of Mönchen-Gladbach in the Rhineland. In 1820 Hugo's grandfather opened a cotton weaving mill in Rheydt, which prospered and expanded under his father's management. The mill was a place where Hugo, the third of seven sons, could explore and study its maze of machinery.[29]

Hugo's father wanted him to have a university education and made sure he followed the traditional course of instruction leading to that goal. He attended engineering schools in Charlottenburg, Karlsruhe, and finally Aachen where he received his degree in 1883. His academic and lab competence while at the

Fig. 14.4. Hugo Junkers (1859–1935) when he was a professor at Aachen. Sass, *Geschichte, 1962*

Technische Hochschule Charlottenburg did not go unnoticed by Heinrich Slaby, one of Germany's distinguished thermodynamics professors. This recognition led to a fortuitous opening move in Hugo's professional career.

Slaby highly recommended the new graduate to his friend Wilhelm von Oechelhäuser, who was chief engineer of Continental-Gasgesellschaft in Dessau. The company designed and made gas-producing machinery for the electric generating plants and the nineteen gas works it owned in Germany and Poland.[30] In 1883 Junkers joined Oechelhäuser as a research assistant. It was in Dessau where Junkers first experimented with opposed-piston gas engines and where in 1892 he received a broad patent for that type of engine.[31] Junkers also undoubtedly read Diesel's book at this time (chapter 4). Although Junkers remained officially under contract with Oechelhäuser until 1894, he had already embarked on new ventures two years earlier. The two men parted ways as friends only because of differing philosophies regarding basic research versus production. Hugo much preferred the former while his boss, with a line of engines to sell as a result of Junkers's work, favored the latter.

Junkers blazed a logical trail of innovation stemming from the immediacy of his own work and had the intuitive gift of perceiving the applicability of basic principles to unrelated purposes. He further recognized that if one of his inventions met his own needs there should be a market for it as others would have similar needs. Examples of this technological/entrepreneurial interchange are deserving of attention.

While working on gas engines under Oechelhäuser, Junkers found that available calorimenters did not give precise enough heating values of the various gases burned in the research engines. The result was his invention of a more accurate gas calorimeter than any yet known. He founded Junkers & Co.

in Dessau in 1895 to make it from designs first patented in 1892 (German patent 71,731). Many gas companies and major gas users in and outside of Germany were customers.

The calorimeter, however, was not the first product to come from the new company. It followed the "gas geyser," a clever water heater spinoff having greater immediate sales potential. The calorimeter, with its cumbersome instrumentation, measured the heat transferred from burned gas passing through flue pipes into water circulated around the pipes. A water heater on the other hand only needed to make the easy heat transfer. Out of Hugo's recognition of this came a broad line of safe, simple, and self-contained water heaters that tapped into gas lines already serving homes and hotels. The original units merely hung above a bathtub where water, heated on demand, sprayed from a shower head. These miniature "Old Faithful" gas geysers gained wide acceptance, and the profits they generated for Junkers & Co. helped to defray costs Hugo later incurred during his private research on opposed-piston diesel engines.

The focus of Junkers's endeavors shifted from manufacturing when he gladly accepted an offered professorship at Aachen's technical university in 1897. The post gave him opportunities to teach and direct much-preferred laboratory activities. Before leaving Dessau, however, Hugo married a woman seventeen years his junior. Therese would become the mother of his twelve children.

Junkers taught thermodynamics and supervised laboratories at the engineering school until 1912. He greatly expanded its test facilities, which originally emphasized mining and metallurgy, to include more I-C engine support. As his personal interest in aviation grew, a wind tunnel was added. (He received German Patent 253,788 in 1910 for an all-metal flying wing, which housed both crew and engines.)

With money from water heater sales, Junkers created his own engine research facility adjacent to his home on the grounds of an estate he had bought in Aachen. He named it the "Workshop for Innovations."[32] A wind tunnel was also added to the "workshop" at a later time. Junkers's own aeronautical group designed a series of all-metal, non-aluminum airplanes, the first of which flew in 1915. Several hundred saw service during the 1914–1918 war.[33]

Research at the school had begun in earnest during 1906 on specific heats of gases at high temperatures, heat transfer at high temperatures and pressures, and the effect of cylinder turbulence on heat transfer—all work about which little was known.

Junkers's methodical heat transfer studies began with a "bomb" calorimeter (the igniting of a combustible charge in a strong, enclosed vessel), followed by a simulated engine combustion chamber that expanded and contracted by movement of a reciprocable "pendulum" piston.

A two-stroke, compound engine was built next to determine heat transfer characteristics of an actual cylinder under very high pressures when burning

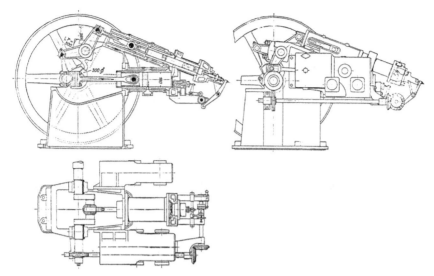

Fig. 14.5. Junkers 1906 compound engine for studying heat transfer of a cylinder under high pressure. Junkers, *Investigation, 1911*

gaseous or liquid petroleum fuels (fig. 14.5). A P-V diagram with an imep of 45 atm shows a peak pressure of 121 atm in the high-pressure cylinder while running on the diesel cycle and burning kerosene (fig. 14.6). A maximum recorded pressure was 250 atm (approximately 3560 psi).

Junkers found that heat transfer was highly dependent on gas pressure and cylinder air motion. Further, the "extraordinary intensity of heat-transfer" immediately after combustion showed him the importance of doing everything possible to reduce heat losses if he were to attain the highest thermal efficiency. The intense severity of heat transfer under pressure also pointed

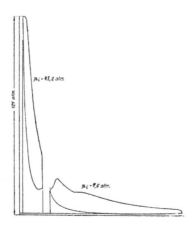

Fig. 14.6. P-V diagram of Junkers compound engine.
Junkers, *Investigations, 1911*

out why there were so many thermal failures in exposed combustion chamber surfaces, in other words, cracked cylinder heads.[34]

This painstaking research led Junkers to his classical solution of a two-stroke, opposed-piston (o-p) engine that reaffirmed his earlier work while with Oechelhäuser. Here the two pistons in a single cylinder, when nearing outer dead center, first uncovered exhaust ports at one end of the cylinder and then scavenge air ports at the opposite end. Such a configuration allowed for a potentially longer stroke to give greater gas expansion and reduced to a minimum the surface area exposed to peak cylinder pressures. Both factors contributed to higher thermal efficiencies.

The only thermal stress points came from three small holes at the cylinder's midpoint into which two injector nozzles and an air starting valve were inserted. Elimination of the cylinder head not only reduced heat transfer losses but of course the prevalent head failures. The simple cylinder design also permitted much-higher operating temperatures and pressures than otherwise possible.

An additional benefit of the opposed-piston design was that for an equal rate of compression and expansion the piston speed was only half that of a single piston acting against a fixed cylinder head.

The main bearings and engine frame absorbed no stresses from the combustion-created forces usually transmitted through the crankshaft and cylinder head. These forces now acted only on the three crank throws per cylinder which, by means of side rods and a yoke, connected the outer throws to the outer piston, which formed a force frame tying the two pistons together (fig. 14.7). The unbalanced dynamic forces were also much less than in a conventional engine configuration. All Junkers diesel engines would use this design except those for aircraft, which had two crankshafts joined by a gear train at one end of the engine.

The patent claims allowed Junkers for his o-p diesel engine were cleverly based on a combination of Diesel's constant pressure combustion and the

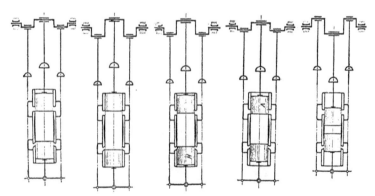

Fig. 14.7. Schematic of power transmission with the Junkers opposed-piston engine. Junkers, *Investigations, 1911*

o-p configuration. The German claim described a marriage of the two and mentioned heat loss reductions. His single US patent claim used a different approach with a minimum reference to structure:

> The method of operating two-cycle slow combustion engines, which consists in compressing the air from two opposite directions simultaneously until it reaches a temperature higher than the igniting point of the fuel employed, injecting it into said air when almost compressed to its smallest volume a quantity of fuel in a finely divided condition, whereby the fuel is immediately ignited by the heat of the air, continuing the injection of the fuel during a part of the working stroke, expanding the gases in opposite directions simultaneously and driving out the waste gases by a straight continuous air stream entering the working cylinder at one end and leaving it at the opposite end.[35]

The US patent would likely have been declared invalid in later courts because it merely claims known art in a combination that is not "new and novel."

Junkers built his first experimental "double piston" diesel based on the above design criteria in 1908, but trouble-free operation did not come quickly. He said that "the first cumbersome and exhaustive investigations had been conducted with miscarriages at first in regard to fuel injection and complete combustion and only after surmounting great difficulties were the tests eventually concluded with very satisfactory results"[36] (fig. 14.8).

Enough confidence was gained with the 1908 engine to modify and extend it in 1910 into what Junkers called a "tandem cylinder." This sophisticated but functional design consisted of a duplicate, coaxial cylinder with its two pistons similarily tied to yokes and side rods (fig. 14.9). As before, all the combustion-generated forces passed through the old and new linkage

Fig. 14.8. First Junkers opposed-piston diesel test engine of 1908.
Junkers, *Investigations, 1911*

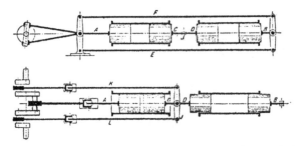

Fig. 14.9. Schematic of linkage on Junkers 1910 tandem cylinder diesel. The vertical piece tying side rods *A, E,* and *F,* to a central connecting rod acts as a sliding crosshead. Similar sliders tie rods *K* and *L* to their respective connecting rods. Junkers, *Investigations, 1911*

into the three crank throws. Drawings of a 1,000-hp experimental tandem marine engine built in 1912 show in two views how the pistons were linked to the crank (fig. 14.10). Scavenge pumps straddled the outer cylinder, and the four stages of compression for blast air injection were split into two, two-stage compressors straddling the inner cylinder. The scavenge pumps and compressors operated off the central piston yoke.

As the Junkers engine was capable of operating under very high cylinder pressures compared to those having cylinder heads, it held the promise of answering a frequent complaint raised by the new steam-turned-diesel operators. The criticism stemmed more from tradition rather than from an engine weakness mainly because of steam's notorious capability to run at high overload powers. Shipboard engineers, spoiled by the steam engine's forgiveness when called on to give its utmost, were particularly vocal about being unable to do the same with the new I-C engine. It might withstand a 10 percent overload for a reasonable time without damage, yet no diesel owner was going to buy a larger, derated engine and give his crew a chance to boost power by up to 40 or 50 percent.

Junkers's method of running at these high overloads, and still allowing the engine to survive, was to supercharge by raising both exhaust back pressure and scavenge air pressure as more fuel was added. He first did this by inserting a butterfly throttling valve downstream of the exhaust silencer. A later design (German Patent 299,198) used a spring-loaded valve that closed off the exhaust line and against which back pressure acted. This pressure increase caused a subsequent and equal rise in scavenge pressure and degree of supercharging. The indicated output increased by 50 percent under an attainable boost pressure of 1.5 atm. Scavenge pump work needed for this went from 3.9 percent of the positive work at a normal full load to 9.5 percent when supercharged.[37] The positive and negative work increases are shown in the P-V diagrams (fig. 14.11).

An example of this overload capability is seen in the 1,000-hp test engine built about 1912 and referred to above. As with the first tandem engine, its

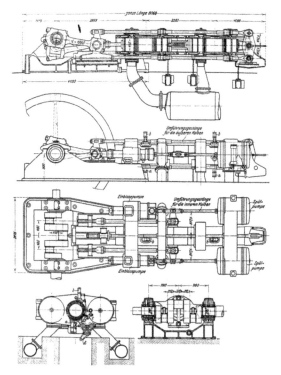

Fig. 14.10. Junkers 1000 hp at 120 rpm tandem, two-stroke,
o-p diesel, c. 1912. *Einblasepumpe* = injection air pump,
Spülpumpe = scavenge air pump, *Umführungsgestänge für die äusseren
Kolben* = guide rod for outer piston. Güldner, *Das Entwerfen*

normal full-load imep of 10 atm (142 psi) increased to 14 atm (203 psi) at
the maximum supercharge. The engine produced its 1,000 hp at 220 rpm
and could run smoothly at speeds from 25 to 220 rpm. Bore and stroke were
450 mm by 2 × 450 nun.[38] Its fuel controls and reversing mechanism are
described in the next chapter.

Absence of a cylinder head allowed compression pressures as high as 48
atm (almost 700 psi), or 25 percent above that of conventional diesels. The
higher air temperatures generated by such pressures made it possible to readily ignite tar oils without the pilot injection of more volatile kerosenes during
starting and warmup phases. After the war began in 1914, Germany would
depend even more on these tar oils derived from coal.

Junkers devoted much effort to improve the injection process. His perfected design utilized two fuel nozzles that were parallel and opposite but
offset to each other and at right angles to the cylinder. Cam-actuated linkage
lifted the fuel valves shortly before TDC, and injection occurred over 30 to
40 degrees of crank angle. Each nozzle sprayed a flat fan of fuel into the space
between the piston crowns.

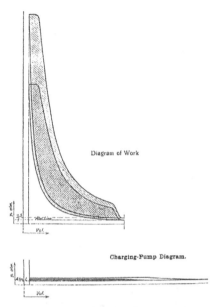

Fig. 14.11. P-V diagram showing increase in positive work
with a 1.5 atm supercharge. Junkers, *Investigations, 1911*

Although the engine needed more piston-to-crankshaft linkage than
with conventional practice, this was offset by fewer valves and their actuating
mechanism, and of course the head. For a given output, the length and overall
height of a single cylinder (non-tandem) engine compared surprisingly well
with other engines. The very light frame and an inherently good fuel efficiency were other advantages in its favor.

A test in 1912 on the first tandem engine yielded a brake-specific fuel
consumption of 216 g/bhp-hr at a power rating of 212 hp. The fuel was a
Mexican crude oil with a rather low LHV of 9700 cal/kg (17,460 Btu/lb).
The same engine on a good crude had a bsfc of 170-180 g/bhp-hr when
running with the injection air compressors reciprocating but unloaded. The
bsfc increased to 230 to 240 g when burning tar oil.[39] Similar results were
obtained from the 1,000-hp test engine. Its bsfc was 176 g/bhp-hr using an
oil of 10,000 cal/kg (0.387 lbs/bhp-hr and 18,000 Btu/lb). A fuel with an
LHV of 19,000 Btu/lb would equate to 0.366 lbs/bhp-hr.[40] The 1,000-hp
engine used scavenge air pumps and an air compressor 50 percent larger than
required to reach the desired 40 to 50 percent overloads.

It is worthwhile to compare the specific power of several sizes and types of
Junkers engines made in the 1910 to 1915 period against comparable output
production engines from M.A.N. and Sulzer.[41] The differences are striking
even if one takes into consideration the lower ratings placed on commercial
applications, particularly when recalling the high overload capacity possible
with the Junkers design:

	Date	No. of pistons	Displacement in liters	Specific power Bhp/1
Junkers "tandem" 200 hp	1910	4	50.3	3.98
M.A.N. A4V49 4-cyl 200 hp	1911	4	183.2	1.09
Sulzer D-60 3-cyl 180 hp	1918	3	166	1.08
Junkers "tandem" 700 hp	1914	4	201	3.98
M.A.N. B4V85 4-cyl 800 hp	1913	4	929	0.86
Sulzer D45 6-cyl 750 hp (4-stroke)	1918	6	706	1.06
Sulzer Z133 3-cyl 750 hp (Valve scavenged 2-stroke)	1906	3	424	1.77
Junkers "tandem" 1,000 hp	1912	4	286	3.49
M.A.N. BV490 4-cyl 1,000 hp	1913	4	1158	0.86
Sulzer Z200 3-cyl 1,125 hp (Port scavenged 2-stroke)	1912	3	712	1.58
Junkers "double" 250 hp (Vertical 3-cyl tugboat engine)	1915	6	66	3.79

Hugo visited St. Petersburg, Russia, in the spring of 1910 where both he and Rudolf Diesel lectured at technical universities. The main purpose of his trip was to form a business relationship with Emanuel Nobel. An historic photograph shows Junkers and Diesel with Nobel in a large gathering of notables at one of the St. Petersburg affairs (fig. 19.2). Junkers also had a 1,000-hp horizontal tandem diesel operating in a concurrent industrial exhibition. Not long after the meeting, Nobel acquired a Russian license and built a stationary 300-hp tandem engine.[42]

Numerous companies sought licenses from Junkers following a major technical paper he gave before a meeting in Berlin of the German Marine Engineering Society in November 1911.[43] He soon licensed several German and foreign firms, only one of which would persevere to build a truly successful marine diesel. This was Doxford of Sunderland, England, who began o-p engine development work in 1913. The war stopped these efforts, and they did not put an engine into production until 1920.

A great disappointment over the less than successful engines built by some of Junkers's first marine licensees (chapter 16) led him to organize his own diesel factory in 1913. Junkers Motorenbau GmbH. (Jumo) opened its doors in a leased building at Magdeburg where he started production and continued the ongoing making of test engines destined for Aachen. The outbreak of war ended all work here. He moved the engine program to Dessau 60 kilometers to the southeast where most of his growing manufacturing facilities were located.[44]

Constraints imposed by the war limited the building of double-piston models, both marine and stationary. The focus instead was on the 100 hp (M22) to 200 hp (M24) range. There was little point in considering engines for ocean-going merchant ships.

Preliminary investigations into solid injection led to prototype designs for aircraft diesels in 1916. The first of these, the FO1 and FO2 (F for *Flugzeug,* or airplane), had two crankshafts joined through a gear train. They were the genesis of Jumo aircraft diesels produced beginning in the late 1920s.

Work also started on the "K" or *Kleinmotor* (small motor) series by the end of the war. A two-cylinder, 60 hp 2 K130, which went on the market in 1919, was quickly followed by the HK models. These had a scavenge pump above each double-piston power cylinder. The HK series would evolve into a line of automotive engines.[45]

Hugo taught at Aachen until 1912 but devoted more time to his own engine and aircraft research. The diesel airless injection developments for aircraft engines was done here. He and his family as well as his research activities moved to Dessau in 1915, and the only thing remaining in Aachen was the manufacture of water brake dynamometers, which also later moved east.

Postscripts to the Junkers story include Hugo's "polite" ouster from his company in 1933 and seizure of his financial interest in it. His ethical and moral philosophy was not compatible with the new Nazi regime. He died two years later. Charles Lindbergh, as part of a Naval Technical Mission, inspected the Junkers engine and aircraft factories in Dessau only days after the Second World War's end in 1945. He specifically commented on the almost total destruction and looting of a very extensive company museum and archives.[46]

New European Stationary Diesels

High fuel taxes in France limited production by diesel engine licensees almost entirely to the marine field. Not until after 1918 did they or new French firms begin to sell meaningful numbers of stationary models.

Fuel costs in Great Britain, although lower than on the Continent, still inhibited sales. One English diesel user reported in 1913 that the price of his next load of fuel would increase from 45 shillings 3 pence per ton to

74 shillings 6 pence ($11.07 to $18.12 per ton). "This fact alone goes to the root of the whole trouble with regard to the extended application of the Diesel engine to-day [*sic*]."[47]

The established British diesel builders dominated the land engine market, but two additional firms entered it after 1912 with their own engines.

The Westinghouse Company, in both the United States and Manchester, England, had for years built dependable and advanced design gas engines. The British branch introduced in 1913 a line of three-, four-, and six-cylinder, four-stroke diesels with powers ranging from 180 bhp to 750 bhp.[48] The four-cylinder, 485 bhp at 200 rpm engine had a bore and stroke of 18 × 26 inches. Westinghouse had delivered only five engines by mid-May of 1913.

Referred to as the "Wiesel," it followed the general configuration of Westinghouse's gas engines where cylinder heads were bolted to flanged liners that in turn rested on a box frame (fig. 14.12). A flange on top of a thin water jacket was inserted between the head and liner. A rubber ring between the lower end of the jacket and the liner sealed the lower end. The box frame casting enclosed both the crankcase and the gear set driving the camshaft. Starting and injection air came from a Reavell three-stage compressor running off the front of the crankshaft.

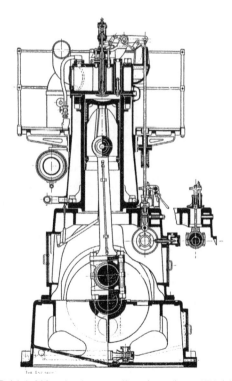

Fig. 14.12. 1913 British Westinghouse diesel engine. 450 bhp at 200 rpm.
The Engineer, May 1913

Forced lubrication fed main and rod bearings as well as pistons at a pump pressure of about 20 psi. The camshaft and other lubricated points had their own oiling sources. Neither the pistons nor the exhaust valves were internally cooled.

All valves were opened via pushrods and rocker levers. Sleeves cast on the cylinder heads guided the upper rod ends. Rollers on the rocker levers rested on the pushrod ends.

Each cylinder had its own eccentric-driven fuel pump with the governor acting on the pump suction valves. Fuel consumption over a 425-hour test, with the load varying from 25 to 100 percent, averaged 0.447 lbs/bhp-hr.

A noticeable offset of the cylinder axis with the crank centerline reduced piston side thrust on the start of the power stroke—another practice borrowed from the gas engines. This offset design obviously precluded the engine's use in marine applications where reverse running was required.

Belliss & Morcom of Birmingham, the second British company, was founded in 1852. It began making conventional A-frame engines in 1912 and during the 1914–1918 war supplied submarine engines based on a design furnished by the navy. As late as 1957, Belliss & Morcom sold four-stroke diesels of 120 bhp to 1,500 bhp using three to eight cylinders and three sizes of bore and stroke.[49]

Three additional German companies who started work on diesel engines soon after the basic patents expired also deserve a brief mention. All had prior experience building gas, gasoline, or various forms of oil engines. Their diesel production continued until recent times.[50]

Hamburger Motorenfabrik Carl Jastram, founded in 1873, started with small, four-stroke kerosene-fueled engines in 1887 shortly after Nicolaus Otto's patent had been declared invalid in Germany. After 1900 a line of Carl Jastram gasoline and kerosene marine engines were sold. A decision to enter the diesel market with a two-stroke, uniflow design led to many difficulties, and it was three years before the company could offer an engine for sale. The air injection engine had an intake valve in the head and exhaust ports in the cylinder. Jastram later changed to a four-stroke cycle.

Carl Kaelble GmbH in Backnang was established in 1884 and began making gasoline engines the next year. Hot-bulb oil engines followed. The company used these engines in their own stone crushers, powered saws, and mobile compressors. In 1908 Kaelble offered a gasoline-engined road roller. These, along with farm tractors that later were all diesel powered, filled a niche market in Germany between the wars. Kaelble's first diesel, also built in 1908, was a four-stroke, horizontal design with a compressor for blast air injection. Their numerous patents over the years covered several diesel injection and combustion innovations.[51]

MODAG (Motorenfabrik Darmstadt GmbH) manufactured conventional horizontal gas engines with hot-tube ignition a year after its founding in 1902. Gasoline-fueled engines ensued, and in 1908 the company bought a license to build the English Hornsby hot-bulb oil engines. A diesel program waited until 1912 when a horizontal model with air blast injection was introduced. MODAG built a line of two- to nine-cylinder engines of up to 1,200 hp under that name until the late 1950s when it became DEMAG AG, Duisburg. Production ended sometime in the 1960s.

The American Newcomers

E. D. Meier told the Electric Boat Co. (EBC) in 1910 that "the original Diesel method was open to the world," but nonetheless, Busch still owned "very strong patents covering the introduction and regulation of the fuel supply, in fact much stronger than the European patents."[52] Electric Boat received this warning because it was already building sub diesels for Vickers and had launched a program based on two-stroke M.A.N. marine and sub engines (chapter 17). A license agreement between EBC and MAN just discovered by the American Diesel Engine Co. greatly pained Busch, who threatened both companies with a legal action that never materialized.[53] These EBC engines are described in the marine diesel chapters.

US firms like Fulton, Nordberg, and McIntosh & Seymour, previously mentioned in connection with the European companies from whom they secured licenses, started their diesel programs only after patent worries became moot even though the supposedly vaunted fuel pump patents owned by Busch had some years yet to run.

Lyons—Atlas

The Atlas Engine Works of Indianapolis, like Electric Boat, also launched a diesel program in 1910, but it is doubtful if an engine was delivered before 1912. The guiding hand at Atlas was Norman McCarty who had been the sales engineer for Busch's American Diesel Engine Co. (ADE) in New York. McCarty was regarded as a very capable engineer who correctly believed there was a market for diesels larger than what ADE offered.[54] He arrived in Indianapolis during 1910 and by the next year had an engine of his design running. It was the most powerful diesel yet built in the US. McCarty's awareness of the fuel metering patents allowed him to design around them and avoid the wrath of Busch attorneys.[55]

Atlas appeared suited to build the new diesels. Founded in 1872 as the Indianapolis Car & Machine Works, it had grown by the early 1900s to be the largest factory in the city. Under the name of the Atlas Engine Works since 1878, it made portable and Corliss steam engines. A major innovation in the late 1880s was its adoption of standardized engines and boilers, which made them very cost competitive. During its peak shortly after 1900, the company employed 2,200 workers and through mid-1912 had built 35,000 steam engines and 45,000 boilers. By then it had also delivered 15,000 gasoline engines for automobiles.[56]

The steam business suffered a precipitous decline after the turn of the century because of the incursion into factories and mills by electric motors and gas engines. As a result of that and a serious general economic situation, Atlas went into receivership in 1907. A gasoline auto engine venture, however, offered hope for the future, and the boiler shop area of the twenty-seven acres under roof was turned over to I-C engines. When the James W. Lyons interests of Chicago bought control in 1912, Atlas was making both poppet valve and, under license, Knight sleeve valve engines. Two years later the company entered the automobile business with the introduction of Lyons-Atlas and Lyons-Knight cars, neither of which achieved renown. The chief engineer for the automotive-type engines was Charles E. Sargent (1862–1934) who had already made a name for himself as the inventor of the Sargent gas engine.[57]

McCarty, who had joined Atlas Engine Works while it was still under court appointed management, was retained by Lyons. A newspaper account of the purchase mentions a McCarty diesel being "in service at the Atlas plant."[58] The four-stroke, three-cylinder engine with its 21 × 30-inch bore and stroke produced 450 bhp at 180 rpm. The Lyons-Atlas diesel line was to include two, three, four, and six cylinders, all with a base 150 bhp per cylinder. They had a trunk piston in conjunction with a heavy, single casting for the A-frame and cylinder.

One carryover from the Busch engines McCarty once sold (and in common with steam engine practice) was the method of setting main bearing clearances. A movable wedge under the lower half of each bearing could be adjusted laterally in or out as required to maintain the desired clearance.

An electric motor ran a separately driven Ingersoll-Rand air compressor. One must assume that the brake horsepower ratings included a deduction for the power to operate the compressor.

All valves were actuated by eccentrics from the "camshaft" located slightly below cylinder head level. Each eccentric rod oscillated a pivoting, oval shaped wiper cam that acted on the flattened end of a rocker lever opposite the valve (fig. 14.13). An advantage of the arrangement was a quick, quiet opening of the valves. This contrasted with the American Diesel Engine, and most early diesels, which were noted for noisy valve operation.

Also new to American practice were the water-cooled exhaust valves and pistons. Pivoting, "grasshopper leg" tubes carried water to and from the pistons.

Individual fuel pumps for each cylinder were enclosed in a single housing placed directly above the camshaft. McCarthy avoided both the Busch and Hesselman fuel metering patents by a third means. The new method also used a fixed-stroke delivery plunger, but the governor now controlled the stroke of a second metering plunger instead of positioning a movable rod to hold open a suction or bypass valve. Both vertical pump plungers were eccentric driven from the camshaft positioned directly beneath them. The

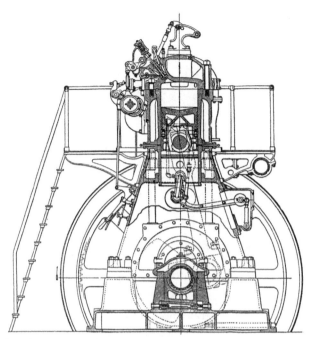

Fig. 14.13. 1912-era Lyons-Atlas diesel engine of 150 bhp per cylinder. It was the largest built in America at the time. Haas, *The Diesel Engine, 1918*

metering plunger delivered its charge to the pumping plunger chamber when that plunger was on its downward, suction stroke. Check and delivery valves controlled the flow of fuel from one plunger chamber to the other and to the fuel nozzle valve.[59] The fuel valve sat in the cylinder head at a 45-degree angle.

The first installation of a three-cylinder engine was in a Casper, Wyoming, electric plant[60] (fig. 14.15). Two engines went to Hawaii, both of which became at the time of their delivery the largest American diesels yet built. A four-cylinder, 600-hp engine pumping irrigation water for the Hawaiian Commercial & Sugar Co. was the first of the two. Acceptance conditions required that it remain in operation for a month with less than ten hours for down-time maintenance and that it burn a high-asphalt California fuel oil. The combustion chamber showed no deposit buildup after a factory test of 144 hours on this fuel, and the engine was shipped in the fall of 1914.[61] A six-cylinder, 1,000-hp model also reportedly went to Hawaii.[62] Lyons-Atlas was still listed as building diesel engines in 1921,[63] but only five engines of the A-frame design were built. Only one still ran in 1923.[64] The other, well-established diesel manufacturers offered engines of much greater horsepower by this time.

McCarty also designed, and Lyons-Atlas built, an experimental double-acting, four-stroke, reversible marine diesel in 1915, which eliminated the

Fig. 14.14. Lyons-Atlas 150–175 bhp per cylinder diesel, c. 1912–14.
Haas, *The Diesel Engine, 1918*

need for a piston rod to pass through a stuffing box to seal one combustion chamber. Instead, there were two vertical, coaxial cylinders whose open ends faced each other. They were separated by the length of the piston stroke plus the thickness of a yoke fastened to the midpoint of a rod tying the two pistons together. The yoke ends were pinned to vertical, reciprocating rods linked to a second yoke below the lower cylinder. A connecting rod joined the lower yoke to the crankshaft. All valves for a given direction of rotation were actuated by a single eccentric. A second eccentric-controlled valve opened when the engine turned in the opposite direction. Reversing was done by shifting a lever via Stephenson linkage to disconnect one eccentric and engage the other. The bore and stroke of each piston was 11 × 15 inches to give a designed output of 100 bhp at 150 rpm. The engine never left the factory.[65]

That Lyons-Atlas aggressively pushed diesel sales was borne out in a December 17, 1913, Busch-Sulzer internal memo reporting that Mr. Lyons himself had made "desperate attempts to secure the business" of a Florida company who was contracting for four 500-hp engines. Evidently there was customer concern "in regard to the financial ability of [Lyons-Atlas] to meet its contracts."[66] One of the customer's agents ranked the Atlas engine second in performance and quality to the Busch-Sulzer A Model.

Sales of McCarty's engine were discouraging, but the company, probably through his efforts and contacts, built other diesels under contract. Lyons-Atlas made development engines for the Southwark Foundry & Machine Co.

in 1913 to 1914.[67] During the war they subcontracted to supply Nelseco (Electric Boat) submarine engines for the US Navy.[68]

Upon the merger in 1918 with the Hill Pump Co. of Anderson, Indiana, Lyons-Atlas became the Midwest Engine Co.[69] For several years the company, under the old and new names, built Hvid engines under a license. These included the typical horizontal style as well as three- and four-cylinder vertical models. The three-cylinder engine had a bore and stroke of 9.5 × 10.5 inches and produced its rated 60 bhp at 315 rpm. The bsfc at that rating when burning Standard Oil Co. fuel oil was 0.48 lbs/bhp-hr.[70]

During the early 1920s the Indianapolis factory also made a small, walk-behind tractor known as the Midwest "Utilator."[71] In 1922 the company was reorganized again, and finally in the early 1930s, the Midwest Engine Co. was permanently laid to rest.

The American Horizontals

Horizontal cylinder gas and gasoline engines predominated in America, as in Europe, during the early 1900s. Many vaporizing, low-compression oil engines followed suit. Thus, engine companies already experienced in this configuration adopted it for their first diesels.

A totally new application for the diesel developed at this time. An exploding population of automobiles with their insatiable appetite for gasoline forced the oil industry to find another way to transport vastly increased amounts of oil from well to refinery sometimes hundreds of miles away. Their practical solution was to construct pipelines requiring many pumps, often remotely located, that were reliably powered first by steam and then by internal combustion. An improving economy of operation insisted upon by the oil companies was a further inducement for a few manufacturers to add diesels to their horizontal gas and oil engine lines. These already made the pumps.

Otto Gas Engine Works

The Otto Gas Engine Works in Philadelphia, a subsidiary of Deutz since 1894, built the first horizontal diesel in the United States, possibly as early as 1910.[72] This was a year after the four-stroke, air injection Deutz model MKD, on which it was based, went into production in Germany (figs. 13.21 and 13.22). The American version carrying a "DH" designation produced 50 bhp.

Snow—Worthington

The first American-designed horizontal diesel came from the Snow-Holly Works of the International Steam Pump Co. in Buffalo, New York.[73] The parent company was no stranger to the diesel engine. Another of its subsidiaries, the Power and Mining Machinery Co. in Cudahy, Wisconsin, built engines for the American Diesel Engine Co. between 1902 and 1912 (when the Busch-Sulzer plant opened).

The genesis of International Steam Pump Co. were steam feed water pumps, direct and then duplex steam pumps, all patented by Henry Rossiter Worthington (1817–1880). From these and a duplex water works pump, Worthington created a large and successful pump and steam engine business.[74] He was also a respected engineer. An organizer and founding member of the American Society of Mechanical Engineers, he served as its first vice president until his death in December 1880. Illness had prevented him from accepting the presidency.[75]

International Steam Pump Co. (ISP) came into being in 1899 after several pump builders, including Worthington's, combined to form a stronger enterprise. The ISP name lasted until 1916 when it became the Worthington Pump and Machinery Co. with corporate headquarters in New York City. ISP had gone into receivership in 1914 and emerged with the name of its most prominent ancestor.

The Snow Pump Works opened its doors in 1889 to build steam-powered pumps for Standard Oil Co. It was one of the original groups joining together to form ISP ten years later. In 1900 Snow added horizontal gas engine-compressor units. Holly Manufacturing Co. merged with Snow in 1902.[76]

Power and Mining Machinery Co. in Cudahy, started in 1901 by ex-Allis-Chalmers engineers, became a part of International Steam Pump in 1907. A. Niedemeyer, who was manager at Cudahy when it made the ADE diesels, saw great promise for the engine. In 1911 he talked ISP management into sending himself and S. B. Daugherty, head of gas engine design at Snow-Holly, to Europe to investigate its diesel industry. Niedermeyer went to Buffalo on his return with the assignment to run a new diesel engine program.

Development commenced in March 1912, and the first engine went on test August 1. The horizontal, one-cylinder, four-stroke followed a conservative design practice, which included air injection and a crosshead piston. It had a bore and stroke of 18 × 27 inches and had a rating of 110 bhp at 180 rpm. An additional throw on the end of the crankshaft drove a horizontal, two-stage air compresor. Larger engines used three stages. The open fuel nozzle was based on the Lietzenmayer system.

This first "Snow Oil Engine" left the factory in February 1913 destined for the Saltillo Electric & Power Co. in Saltillo, Mexico. It and those that followed operated on heavy, asphaltic crudes from California and Mexico as well as the lighter oils from the Eastern and Mid-Continent fields. A multi-fuel capability was typical of this vintage horizontal-type diesel engine.

The second engine built, with a 13 × 20-inch bore/stroke and 55 bhp at 225 rpm, went to Giliad, Texas, in May 1913. These two cylinder sizes, plus a third with a 19 × 33-inch bore/stroke, were sold in one-, two-, and three-cylinder combinations. Snow guaranteed a full-load bsfc of 0.50 lbs/bhp-hr, but actual performance was better. A 300 bhp engine had a bsfc at full and three-quarter loads of about 0.42 lbs/bhp-hr. Brake thermal efficiency with this consumption averaged 33 percent.[77]

ISP/Worthington instituted a parallel program for a horizontal two-stroke, loop-scavenged diesel. The first, a two-cylinder with a bore and stroke of 19 × 33 inches, produced 450 bhp at 150 rpm. It followed after the first four stroke to Saltillo exactly two months later. This was even before the second four stroke was shipped. Field testing never seemed to happen close to home!

Credit for the first American oil pipeline pumping unit going into service also belongs to the Snow-Holly Works. The Prairie Pipeline Co. put a twin-cylinder, four stroke on line December 15, 1915. The 19.5 × 33-inch bore/stroke produced 300 bhp at 165 rpm engine and ran on the pipeline fuel. It drove a 2,000 barrel/hr National Transit duplex, double-acting pump.

Worthington had sold 329 engines totaling 74,000 bhp when horizontal diesel production ceased in 1928. Of these, 299 (65,400 bhp) were four stroke. Two-stroke production ended in 1923.

Vertical diesel development started in 1918 with a six-cylinder, four-stroke marine engine that weighed 430 lbs/bhp. The 29 × 46-inch cylinders produced 1,760 bhp at 120 rpm. It was abandoned in favor of a two-stroke, double-acting design begun in 1921. Nevertheless, large four-stroke stationary engines would become an important part of the Worthington product line. The vertical engine story is reserved for the next volume.

Allis-Chalmers

When the company founded by Edward P. Allis (1824–1889) of Milwaukee, Fraser & Chalmers, Chicago and two other firms joined to form the Allis-Chalmers Co. in 1901, Edward's sons exercised management control. William W. (b.1849) became chairman of the board and Charles (1853–1918) its president.[78] The Edward P. Allis Co. was already renowned for its huge steam engines, pumps, and other heavy machinery when it built the famous West Allis factory in 1901. That same year Henry C. Holthoff, manager of mining machinery, and other Allis-Chalmers (A-C) executives resigned after being confronted with their less than "loyal to the company" affairs. The result of this departure was their formation of Power and Mining Machinery in nearby Cudahy.[79]

One new post-1900 A-C product line was the immense horizontal engines burning blast furnace gas. The addition was fortuitous because of a severe dropoff in large steam engine business. By 1904 the West Allis plant was making tandem cylinder gas engines of up to 6,000 hp (fig. 14.15). The frame casting alone for the largest of these engines weighed more than 90 tons.[80] A-C bought a license to build the large Nürnberg gas engine during these years but abandoned it after encountering too many troubles making the first one.[81]

Allis-Chalmer's four-stroke, horizontal diesel engine program begun in 1912 proved to be almost a footnote to its other product lines. Similarities to the Worthington were a two-stage, horizontal compressor and a

Fig. 14.15. Seventeen 2,500 kW, Allis-Chalmers "Twin Tandem,"
blast furnace gas units, c. 1915. Note the man standing in
the middle of second engine. A-C, *Gas Engines, 1915*

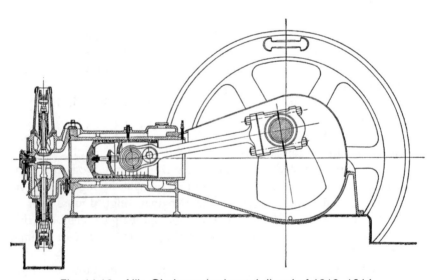

Fig. 14.16. Allis-Chalmers horizontal diesel of 1912–1914.
Haas, *The Diesel Engine, 1918*

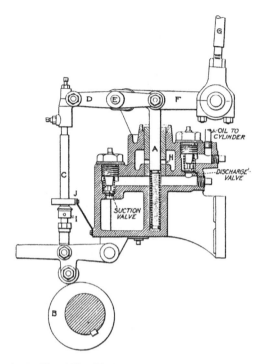

Fig. 14.17. The Allis-Chalmers fuel pump was typical of pumps
on most horizontal diesels in the period. Governor control rod (*G*)
changed fulcrum (*E*). Morrison, *Oil Engines, 1919*

Lietzenmayer-type open fuel nozzle (fig. 14.16). A variable stroke plunger in the fuel pump delivered fuel to the nozzle during the engine's intake stroke (fig. 14.17). Differences were in the trunk piston and the valve arrangement. "Over and under" (vertical) inlet and exhaust valves, actuated by eccentrics, opened into a recess in the cylinder head. This was a practice borrowed from horizontal gas engines[82] (fig. 14.18).

Engines came in one, two, three, and four cylinders. The single and twins had a one-piece frame, the three-cylinder used a single and a twin frame combination, and the four-cylinder had two twin frames. The flywheel was placed between the frame units on the three- and four-cylinder models.

Shell Oil Co. placed an order for perhaps as many as one hundred diesel pumping engines to push crude through a new ten-inch-diameter pipeline running the 428 miles from its Cushing, Oklahoma, field to its refinery in Wood River, Illinois. The line finally went in service during July 1918 after almost a year's delay caused by pipe laying and other problems.[83]

Allis-Chalmers left the diesel business in the early 1920s and did not make another such engine until it bought the Buda Co. in 1953. Before that time other engine companies supplied A-C with the diesels needed for its construction equipment.[84]

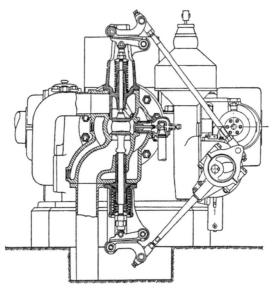

Fig. 14.18. End section of Allis-Chalmers horizontal diesel showing the eccentric drives for inlet and exhaust valves. Haas, *The Diesel Engine, 1918*

As a postscript, a brief mention should be made of Allis-Chalmer's apprentice program in Milwaukee for graduate engineers. Both Carl Breer and Fred Zeder of Chrysler Engineering fame went through this program in 1910 before launching their automotive careers.[85]

National Transit and Others

The National Transit Pump and Machine Co. in Oil City, Pennsylvania, also entered the diesel market with a horizontal model shortly after 1912. It was a logical step considering the company's long-standing importance in supplying steam engines and pumps for the oil pipeline business. National Transit Co. itself was a pipeline company established by John D. Rockefeller in April 1881 when he was fighting Tidewater Oil Co. in the battles to control the transportation of oil from the Pennsylvania fields to East Coast refineries.[86] The pump and machine company was National Transit's manufacturing adjunct.[87]

There was little innovation in the four-stroke, trunk piston engine. It had a two-stage horizontal compressor and a valve arrangement similar to the Allis-Chaimers. An open-type nozzle received fuel from a pump with a constant stroke plunger and a governor-controlled bypass valve. The engine was offered in one and two cylinders and with a horsepower range of 50 to 300 hp. The 15.5 × 24-inch cylinder engine turned at 180 rpm.[88]

National Transit built a limited number of engines with most going into pipeline pumping. While diesel production stopped in 1918, the company continued to build "semi-diesel" oil engines until into the 1930s.[89]

McEwen Brothers

Typical of several limited production, horizontal diesels in the US during the 1912–1918 era was that by the McEwen Brothers Diesel Engine Co. Little is known of it except for one source.[90] Spanning the 65 to 300 bhp range, the engine was a four-stroke, trunk piston design with "over and under" valves. The 65 bhp model had a bore and stroke of 14 × 22 inches. In an October 1916 test it gave a full-load bsfc of 0.42 lbs/bhp-hr and a brake thermal efficiency of about 33 percent.[91]

The McEwen fuel pump was simplicity itself. A fuel cam reciprocated a pump plunger via a rocker lever acting on the plunger. Suction and delivery valves controlled fuel flow in and out of the plunger chamber. Inserted at a right angle through a slot in the plunger was a bar with a recessed wedge that was positioned by the governor (fig. 14.19). The farther the wedge moved into the plunger the less was the plunger's stroke and the volume of the fuel charge. Injection air from a two-stage compressor blasted in the fuel charge through an open-type nozzle.[92]

Only "about a dozen" were made before the company reverted to lower compression semi-diesels.[93]

The Standard Diesel

A novel two-stroke, horizontal diesel of trunk piston design has been attributed to C. H. Blanchard. It was first built by the Standard Fuel Oil Engine

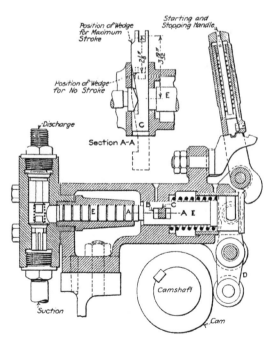

Fig. 14.19. McEwen fuel pump, c. 1916. Morrison, *Oil Engines, 1919*

Co. in 1913; other companies, including the Hatfield-Penford Co., offered it later. A final version came out after 1920 in multicylinder, vertical form.[94]

The Standard had a stepped-piston whose larger diameter doubled as a scavenge pump and as the first stage of three-stage compression. A two-stage compressor mounted horizontally on the side of the engine completed the process. Cylinder inlet and exhaust ports provided loop scavenging.

Piston and cylinder construction differed from usual practice. The "power" piston consisted of a crown piece containing the rings that was piloted on and bolted to an intermediate cylinder of the same diameter. That piece in turn was fastened to the larger-diameter scavenge piston portion of the assembly. Two axial water passages cast inside the intermediate piece connected telescoping tubes to the piston crown for cooling. The power cylinder was a separate casting piloted into the engine frame, which formed the cylinder bore for the scavenge pump piston. A 60 bhp engine, the smallest size, had a scavenge bore of 30 inches.

An eccentric on the camshaft operated a fuel pump plunger supplying metered charges to an open nozzle. The governor varied the eccentric's lift by rotationally changing the offset of the eccentric's axis to the centerline of the shaft. This determined the metering stroke of the pump plunger. Because the plunger pumped twice as often as in a four stroke, it was reported that the suction and delivery valves had to be reground more frequently.[95]

Injection air pressure to the nozzle was controlled by the constant speed governor in a unique way. The scavenge pump discharged air varying from 7 to 11 psi to the cylinder ports and to the first stage of the external two-stage compressor.[96] This intermediate stage discharged into the starting air tank and not directly to the final compressor stage. Inserted in the line to the tank was an air "bypass" valve whose discharge did lead to the final stage of the compressor. A reciprocating plunger in the valve had a governor-adjustable stroke like that for the pump plunger, except when load increased, the stroke shortened instead of lengthening. Thus, as load went up and higher injection air pressure was called for, the plunger stroke decreased to allow a discharge port in the plunger bore to remain uncovered longer. This in effect raised the initial pressure in the cylinder of the final stage before its piston began the compression stroke.

Injection air pressure requirements for typical open nozzle engines held relatively constant, depending on the engine, at about 650 psi from no load to near 40 percent load. They then linearly increased to approximately 1,000 psi at full load. This was 50 to 100 psi lower than closed fuel nozzles with needle valves wanted.[97]

Full-load fuel consumption for a 60 bhp Standard diesel was 0.48 lbs/bhp-hr to give a brake thermal efficiency of 28 percent.[98]

In the Wings

Numerous American companies building gasoline and/or low and medium compression engines did not "graduate" to diesels until a time beyond the scope of this book. These include Buckeye, Chicago Pneumatic, Cooper Bessemer, De La Vergne, Fairbanks-Morse, Ingersoll-Rand, Rathbun-Jones, Superior, and Waukesha. Some have dropped by the wayside while others were absorbed or merged into conglomerates. Several US diesels mainly for marine use noted in chapter 16 are Atlas Imperial, Craig, Fulton (Erie, PA), Southwark-Harris, and Winton.

One person meriting mention before closing the American stationary diesel story is William T. Price, a Canadian-born engineer. His initial work occurred prior to 1919, but its results did not come to fruition in true, higher compression diesels until the early 1920s. His designs helped De La Vergne, Ingersoll-Rand, and Rathbun-Jones enter the diesel industry. The Price combustion system with two, angling fuel sprays at right angles to each other (minus an ignition source placed between the sprays) would be used on diesel engines built by the above three companies for years.

Notes on Semi-Diesels

Numerous makes of four-stroke, horizontal engines of the so-called semi-diesel type were sold up to 1918 and beyond. These oil-burning, lower-than-diesel compression ratio engines with compression pressures in the 250 psi range gave good service and reasonable fuel economy. Considering their much lower first cost, they offered a practical stopgap power source between gas engines and a more developed diesel. The semi-diesel, as thought of in the US, ran on a cycle whose indicator diagram resembled one more like a constant volume engine in that the pressure rose significantly immediately after combustion at the end of the compression stroke and before the piston moved outward very far on the expansion stroke. After being aided by an auxiliary heat source when started, the cylinder then retained sufficient heat to augment the heat of compression so that the engine ran without the applied heat. Injection pressures were low enough to obviate a need for blast air and thus to allow a less costly, solid injection fuel system. The lack of a compressor also contributed to a higher mechanical efficiency that brought fuel consumption closer to that of a diesel.

This concept differentiates from a low-compression, hot-bulb vaporizing oil engine, such as the Hornsby-Akroyd that sprayed fuel under low pressure onto the hot surface inside a cylinder antechamber. These engines were often mistakenly upgraded into the "semi-diesel" category. They would generally not run at an idle nor under light loads without having to apply external heat to the vaporizing "bulb." The choice of the term *semi-diesel* was a most unfortunate one as it caused confusion among users and sometimes resentment among the builders of "true" diesels.

Peculiar to Britain, the terminology of *diesel, semi-diesel,* and *oil engine* was a hotly debated issue during this period and for years after the First World War.[99] Passionate views by some builders and ardent nationalists stoutly held that Akroyd Stuart was the true inventor of Diesel's engine type. It was definitely not Rudolf from the land of the Teutons. Stuart himself, who was living in western Australia, added his fuel to the fire by writing "Letters to the Editor" to various English publications and even by trying to embroil the British Admiralty in the controversy. These rather passionate epistles were dispatched almost until his death in 1927. The Institution of Mechanical Engineers formed an "Oil-Engine Nomenclature Committee" in 1920 as an attempt to resolve the dispute. It met for two years, and in a final report of December 16, 1921, stated that for high-compression engines the term *diesel* was (sadly for some) really a *fait accompli*. It especially considered "that the term *semi-diesel* as applied to Oil Engines is indefensible." Such a term implied that Diesel received credit for something rightfully belonging to Stuart. A suggestion was to use "surface ignition" for lower-compression engines and "compression-ignition" for the diesel. The report sagely concluded "it is improbable that any system of nomenclature could be devised to cover all divergent points of detail or to meet practical requirements."

> *"Right here, allow me to quote from a private letter giving a statement of Mr. [Samuel] Vauclain of the Baldwin Locomotive Works; the latter told my informant that when the depression in business struck their works, the first engine to be shut down was a Hornsby-Akroyd; then came the steam engines, leaving the Diesels to run the plant, and that when the Diesel is shutdown, the plant will close."*
>
> —Edward D. Meier, 1909[100]

Notes

1. Edward D. Meier letter to J. R. Harris, Busch-Sulzer Bros. Diesel Engine Co., August 19, 1913, 5. Busch-Sulzer papers, Wisconsin State Historical Library.
2. Friedrich Sass, *Geschichte des deutschen Verbrennungsmotorenbaues von 1860 bis 1918* (Berlin: Springer, 1962), 598–99.
3. Springer Verlag, Berlin, published all editions: 1903, 1905, and 1914. Reprints of the last edition came out as late as 1921. An English translation, aimed at the American engineer, was adapted from the second edition by Cornell professor H. Diederichs: *Design & Construction of Internal Combustion Engines by Hugo Güldner* (New York: Van Nostrand, 1910), 672 pages.
4. A 1953 catalog of German diesel engine builders lists Güldner, but as a division of the "Company for Linde's Ice Machines AG." Diesel engine production ceased during the 1960s.
5. H. Güldner, *Das Entwerfen und Berechnen der Verbrennungsmotoren*, 3rd ed. (Berlin: Springer, 1920), 568. The test was made by Güldner engineers.
6. Ibid., 569. Test conducted by Professor A. Staus in August 1910.
7. Advertisement in *Diesel-Engines*, VDI (Berlin: 1923), XIII.
8. Kurt Schnauffer, "Lizenzverträge und Erstentwicklungen des Dieselmotors im In- und Ausland" (1958), 100–101.
9. Ibid., 102–3. Diesel's second Italian patent was the one in dispute. Originally issued as No. XXIX/38290, it was renewed under LXXV/132 until March 31, 1910. Records in MAN Archives.
10. Obituary, *Zeitschrift*, VDI, part 1 (January 1899), 54.
11. "The Turin International Exhibition," *Engineering* (1911): 234–35. The large exhibition was intended not only to show the world the industrial progress made in Italy but also

to emphasize the unification of the country. A later article described the engine in detail: *Engineering* (May 1912): 696–97.

12. Letter, Rotter to Harris, May 31, 1911. Busch-Sulzer papers, Wisconsin State Historical Library.

13. "Tosi Diesel Engines," *Engineering* (January 1919): 25.

14. The above May 1912 *Engineering* article is cited in a June 16, 1913, letter to Busch-Sulzer from Wetmore & Jenner, patent attorneys. Busch-Sulzer papers, Wisconsin State Historical Library.

15. *Diesel Progress* (1948), 68.

16. US Patent 729,613 issued June 2, 1903. See chapter 8 for a description of Lauster's design.

17. Letter from Wetmore & Jenner.

18. Fulton's chief engineer, in a June 3, 1914, technical meeting said that "the present patent situation in the United States will undoubtedly greatly restrict the use of such [suction valve control] pumps for some time to come." H. R. Setz, "Recent Developments in the Manufacture of the Diesel Engine," *Trans., A.S.M.E.* (1914), 430.

19. Ibid., 431.

20. Lacey H. Morrison, *Diesel Engines* (New York: McGraw-Hill, 1923), 334–36, describes and illustrates the pump.

21. Setz, "Recent Developments," 431.

22. *Diesel Progress*, 68.

23. Sulzers's Italian subsidiary was a third company to be put on notice by the Allgemeine, but never settled with them. Schnauffer, "Lizenzverträge," 103.

24. Gustav Goldbeck, *Kraft für die Welt* (Düsseldorf, Germany: Econ-Verlag, 1964), 227, 228. Langen & Wolf in Vienna was established in 1872, the same year Deutz started companies in Paris and Philadelphia.

25. Letter, Rotter to Harris, May 31, 1911.

26. "The Langen & Wolf Diesel Engine," *The Engineer* (July 1912): 85. Wells & Wallis-Tayler, *The Diesel or Slow-Combustion Engine* (London: Crosby Lockwood, 1914), 207–9, also describe the engine, but their data is from *The Engineer*.

27. Ibid.

28. Ibid.

29. A thorough, but often flowery, biography was written by a man who had worked for Junkers in his last years until his death and who had an understandable reverence for his subject. See Richard Blunck, *Hugo Junkers, der Mensch und das Werk* (Berlin: Linpert, 1942), 300 pages. The Junkers name, according to Blunck, indicates an ancestry stemming from an aristocratic family displaced to the Rhineland during the Thirty Years' War.

30. Oechelhäuser's father (also a Wilhelm) had managed the company since shortly after its founding, and when he moved up to the supervisory board in 1890, Wilhelm Jr. filled the vacancy. For more biographical information see Gustav Goldbeck, "Wilhelm von Oechelhäuser und der Gegenkolbenmotor," *MTZ*, no. 5 (1973): 139.

31. Ibid., 140–41, and Cummins, *Internal Fire*, 244–47, for information on these engines. Junkers received British Patent 14,317 issued August 8, 1892, and US Patent 508,833, filed November 5, 1892, and issued November 14, 1893. US Claim 1 is extremely broad: "A gas engine comprising a working cylinder having peripheral inlet and exhaust ports near its opposite ends, respectively and reciprocally movable pistons adapted to control said ports, for the purpose set forth."

32. Vignettes of Junkers and his activities while at the Aachen Technische Hochschule, and later, are found in Theodore von Kármán's fascinating autobiography *The Wind and Beyond* (Boston: Little, Brown & Co., 1967), 376 pages. Von Kármán succeeded Junkers at Aachen as chair of the aeronautical institute in 1913. He was an aerodynamic design consultant for Junkers on the J1 airplane in 1915 and on numerous times after that. He said Junkers was "a great and versatile teacher-industrialist-scientist. . . ." (p. 75)

33. A recent book in English, Gunter Schmidt, *Hugo Junkers and His Aircraft* (Warrendale, PA: SAE, 1988), 224 pages, is an abbreviated biography and covers Junkers planes from 1915 to 1945. See also Turner and Nowarra, *Junkers, an Aircraft Album* (London: Ian Allen, 1971), 128 pages.

34. Hugo Junkers, "Investigations and Experimental Researches for the Construction of my Large-Oil-Engines," *Jahrbuch der Schiffbautechnischen Gesellschaft 1912*, translated reprint, 18–19. This landmark paper covers much of Junkers's engine research at Aachen and includes his proposals for large merchant and naval vessels.

35. German Patent 220,134, issued September 27, 1907. Quoted US Patent 1,117,498, filed August 28, 1908, and issued November 17, 1914. A companion US Patent 1,117,497 with the same dates covers the engine's lower scavenge air temperatures.

36. Junkers, "Investigations," 28. Grammar and word arrangement of the quotation have been slightly edited to improve the translation.

37. Ibid., 43.

38. Philip L. Scott, "Construction of Junkers Engine," *SAE Trans.*, part II (1917), 407. The author of this twenty-page paper worked in Aachen with Junkers during most of 1916. He provides insights and explanations missing in German papers and texts.

39. Güldner, *Das Entwerfen*, 715, 716.

40. Scott, *Construction*, 415.

41. Data compiled from Scott, ibid., and Güldner, *Das Entwerfen*, MAN and Sulzer Archives.

42. Hans Flasche, "Der Dieselmaschinenbau in Russland von Anbeginn 1899 an bis etwa Jahresschluss 1918" (1952), 56. Copy in MAN Archives.

43. Junkers, "Investigations."

44. Paul Kratz, "Aus der Geschichte der Junkers-Motorenfabrikation," *Der Propeller*, nos. 6/7 (1936): 85–87. A small, but valuable, collection of these Junkers company publications is in the San Diego Aerospace Museum Library. See also Blunk, *Hugo Junkers*, 73.

45. The automotive diesels were licensed to several European engine builders. The C.L.M. division of Peugeot built engines under a Junkers license from 1928 to 1934 and continued to make opposed-piston, automotive diesels until into the 1950s.

46. Charles Lindbergh, *The Wartime Journals of Charles A. Lindbergh* (New York: Harcourt Brace & Jovanovich, 1970), 981–89. His comments on this entire investigative mission are most insightful.

47. "Some Stationary British Diesel Engines," *The Engineer* (May 1913): 487.

48. See Dugald Clerk and G. A. Burls, *The Gas, Petrol, and Oil Engine*, vol. 2 (London: Longmans, Green & Co., 1913), 164–68, for a description of the gas engine model preceding the diesel. The diesels are described in E. Mortimer Rose, *Diesel Engine Design* (Manchester: Emmott, 1917), 199–200, and "Some Stationary British Diesel Engines," 484–87.

49. *British Diesel Engine Catalogue*, 2nd ed. (London: BICEMA, 1949), 16, and ibid., 22.

50. The early history of Jastram, Kaelble, and MODAG is given in Sass, *Geschichte*, 637–46 with details of other engines made in addition to diesels. MODAG is listed in *Deutsche Verbrennungsmotoren* (Frankfurt/Main, April 1953), as making diesels in that year, and from information furnished by Robert Bosch to the author in October 1972, Jastram was still buying Bosch injection systems for their diesel engines. By that time Kaelble was using Daimler-Benz diesels for heavy tank haulers for the military. They went out of the vehicle business in the 1980s. (Dr. H. O. Hardenberg to the author, January 13, 1993.)

51. One Kaelble patent, German No. 890,146, issued July 22, 1951, was for an engine retarder. It caused weeks of effort by the author and others to convince the German Patent Office that his father's idea for an engine brake did not read on this patent. During contractual difficulties for years the patent was again wrongly cited. The name Kaelble is embedded in his memory.

52. February 23, 1910, letter from Edward D. Meier to J. R. Harris reporting on his remarks to the Electric Boat Co. Another Meier letter of March 22, 1910, said three US patents were involved: Diesel 654,140, issued July 24, 1900; Frith & Macpherson 672,477, issued April 23, 1901; and Lauster 729,613, issued June 2, 1903. Busch-Sulzer papers, Wisconsin State Historical Library.

53. Adolphus Busch sent this cablegram to Rieppel at Nürnberg on March 17, 1910, to open an often-heated dialog between M.A.N. and the Busch interests: "Colonel Meier advises you sold American license for your Diesel marine engine[.] cannot believe report true on account of our personal relations as well as my strong patents[.] cable full information at my expense." Busch-Sulzer papers, Wisconsin State Historical Library.

54. Remark by Charles E. Beck, engineering head of Busch-Sulzer, who had known McCarty. Charles E. Beck, "Historical Development of the Large Modern Diesel Engine in America," *The First Fifty Years of the Diesel Engine in America*, ASME (1949), 9.

55. Meier to Busch-Sulzer, August 31, 1911, on receiving the brochure "Atlas Crude Oil Engines, 'Diesel Type.'" from McCarty: "This is probably an engine designed by Mr. McCarty in the attempt to avoid conflicting with our patents." Busch-Sulzer papers, Wisconsin State Historical Library.

56. "Whistles of Atlas Engine Works now calling Hundreds of Indianapolis Workingmen back to Benches of Big Shops," *The Indianapolis News*, October 6, 1912, 27. Copy in Indiana State Library.

57. *A Chronicle of the Automotive Industry in America 1893–1946* (Cleveland: Eaton Corporation, 1946), 35. Sargent was chief engineer at Lyons-Atlas/Midwest Engine Co. from December 1912 to 1921. He held seventy patents for many types of inventions. Obituary, *Indianapolis News*, September 23, 1934, and correspondence with granddaughter Gail Mawhinney, 1975–76.

58. "Whistles of Atlas," 27.

59. Herbert Haas, *The Diesel Engine; Its Fuels and Its Uses*, Bulletin 156 (Washington, DC: US GPO, 1918), 77. This is the only source found that too briefly describes the McCarty pump. A 1925 automotive diesel fuel pump by Franz Lang also had two plungers of similar function. See *Bosch und der Dieselmotor* (Stuttgart, Germany: Bosch, 1950), 26.

60. *Diesel Power* 25, no. 5 (1947): 82. The author of this extensive and generally accurate twenty-fifth anniversary historical issue is possibly L. H. Morrison, who was associated in earlier years with the magazine and who wrote his first diesel text in 1919. Enough nuisance errors exist, however, to require the use of caution and to seek verification.

61. "Largest Diesel Engine Built in America," *The Gas Engine* October 1914, 611.

62. *Diesel Power* and *Diesel Progress* (May 1948), 55. John W. Anderson authored the articles for this fiftieth anniversary issue of the industry.

63. *A Fifty Year History of the Diesel and Gas Engine Power Division of the American Society of Mechanical Engineers*, ASME (1971), 5.

64. Lacey H. Morrison, *Oil Engines: Details and Operations* (New York: McGraw-Hill, 1919), 44. Although intended as a "hands-on" textbook when written, Morrison's intimate knowledge of the diesel engines being introduced and revised in that brief time period make it as close to being an accurate "primary" reference source as one can hope to find. He goes into great depth as he covers the many small design differences between almost all of the US diesel engines made through 1918. He thoroughly describes and illustrates the pistons, connecting rods, valve cages, fuel pumps and nozzles, and so forth of each engine. His next edition, *Diesel Engines*, in 1923, not only shows the progress made in those four years, but it also gives clues as to who no longer is in business.

65. *Diesel Progress*, 78–79. Illustrations of the engine, unfortunately, are too weak to reproduce.

66. C. G. Cox, Busch-Sulzer sales manager, to J. R. Harris, December 17, 1913. Cox was reporting on the bids of several US diesel builders for engines to be installed in the Prairie Pebble Phosphate Co., Mulberry, Florida. The Snow engine built by the International Steam Pump Co. (later Worthington) received the bid on price (about $70,000 each for the four, 500 bhp engines) and, more importantly according to Cox, on somewhat incestuous boardroom connections. Busch-Sulzer papers, Wisconsin State Historical Library.

67. Beck, *The First Fifty Years*. Beck saw the engines on test during a visit with McCarty at the Indianapolis factory. He said they built them in 1913–14.

68. *Diesel Progress*, 77. Anderson, author of the article, was an engineer at Electric Boat in 1918. (See John W. Anderson, "Discussion of Certain Problems in Regard to Marine Oil Engines," *ASME Journal* (November 1918): 941–43.

69. *Indianapolis Star*, June 23, 1918.

70. Chas. E. Sargent, "A Type of Heavy-Oil Engine for Automotive Purposes," Trans., SAE., 222–34, 1918.

71. Charles Shelton, "Midwest Utilitor," *The Gas Engine Magazine*, July 1989, 19–23.

72. The Otto Gas Engine Co., *The Otto Cycle* (1913), 13, makes the claim of being the first. *Diesel Progress*, 68, gives the date. Although the author has no primary source for the 1910 date, the chronology fits. In any event, it is before 1912 when others built horizontal diesels in the US. For early history of the Philadelphia company, see Cummins, *Internal Fire*, 190, 278.

73. *Worthington and the First 50 Years of the American Diesel Engine*, Worthington Bulletin S-500-B44, May 1948, 5.

74. Obituary, *New York Times*, December 18, 1880. He died on the seventeenth, his sixty-fourth birthday.

75. Frederick R. Hudson, *A History of the American Society of Mechanical Engineers from 1880 to 1915* (New York: ASME, 1915), 12, 135, 139.

76. *Diesel Progress*, 63.

77. Morrison, *Oil Engines*, 251–52.

78. Walter F. Peterson, *An Industrial Heritage: Allis-Chalmers Corporation* (Milwaukee Co. Historical Society, 1978), 105. This 448-page book is a comprehensive corporate history.
79. Ibid., 137, 146. They were quietly selling stock in a new company to be called the Gray, Holthoff, Leuzarder company, which became P&MM. Business was to be diverted from A-C.
80. Ibid., 126. A few 10,000-horsepower engines built some years later had cylinders of 60 × 64 inches.
81. Edward D. Meier to Adolphus Bush, March 9, 1909, 5. Busch-Sulzer papers, Wisconsin State Historical Library, Madison. Max Rotter, who had been a chief engineer with Allis-Chalmers prior to his leaving for Busch-Sulzer in 1911, may have given Meier this fact.
82. Haas, *The Diesel Engine*, 70.
83. Kendall Beaton, *Enterprise in Oil: A History of Shell in the United States* (New York: Appelton-Century, 1957), 144–45. Beck, *The First Fifty Years*, 9, provides the approximate number of engines.
84. Ibid., 349.
85. Carl Breer, "The Three Engineers: Zeder-Skelton-Breer," n.d., 60–6.
86. Ida M. Tarbell, *The History of the Standard Oil Company*, vol. 2 (London: Heinemann, 1905), 12–13.
87. National Transit Co.'s Oil City facility was incorporated as National Transit Pump & Machine in April 1915. "National Transit Co., Its Rise and Fall," Oil City, PA. *The Derrick*, August 14, 1971, 5. Information kindly sent by the librarian, Oil City Library.
88. Technical references to the engine are in Haas, *The Diesel Engine*, 74-75; Morrison, *Oil Engines*, 20, 251, 273–75; Morrison, *Diesel Engines*, 40; and *Diesel Progress*, 169.
89. *Diesel Progress*, 169. Morrison, *Diesel Engines*, 40, provides the production number.
90. Morrison, *Oil Engines*, 20, 251, 273. This is the only reference the author has found that not only mentions but describes the McEwen Bros. diesel.
91. Ibid., 252. The LHV of the fuel is not given.
92. Ibid., 168.
93. Morrison, *Diesel Engines*, 40.
94. Both *Diesel Power*, 83, and *Diesel Progress*, 56, mention Blanchard. Morrison, *Oil Engines* and *Diesel Engines* thoroughly detail the engine but make no mention of Blanchard. *Diesel Power* gives the later builder as Hatfield-Penford and Morrison, 1923 says Hatfield-Penfield.
95. Morrison, *Oil Engines*, 179.
96. Ibid., 203.
97. Ibid., 207.
98. Ibid., 253.
99. Reports in Library, Institution of Mechanical Engineering, London. The argument continued on as is seen in the concluding comments by Sir Lynden Macassey and Herbert Cowpers, "Compression-Ignition Engines: Britain's Priority in Development." (London) *Times Review of Industry*, October 1948: "As [Dr. Diesel] was only one and not the first of many workers, the custom of naming after him the whole range of modern compression-ignition engines is unwarranted and unjust to the many other engineers whose contributions exceeded his, both in relative importance and in priority. It was typical German propaganda." Reprint copy kindly given to the author by Ray Hooley, Lincoln, England.
100. Meier to Adolphus Busch, March 9, 1908. Busch-Sulzer papers, Wisconsin State Historical Library.

CHAPTER 15

Prophesy Fulfilled:
Marine Diesels, 1903–1910

*"The man therefore who will take his courage in both hands, and run a
ship or ships fitted with Diesel or other internal combustion engines on a
regular service around the Horn will do more to hasten the coming of the
internal combustion engine afloat than anything else could possibly do. . . .
Difficulties, objections, or whatever they may be called, do exist, but some
will be cured by the mere effluxion of time, some are in the process of being
cured, and none are hopeless."*

—*The Engineer,* 1909[1]

Steam reigned supreme on the seas, and any invasion of its domain by an alien
upstart would not go unchallenged. Nor did it. Only the faith of determined
diesel builders in cooperation with courageous ship owners and desperate
admiralties put Diesel's engine in front of a propeller.

Rudolf foresaw marine applications in 1892 as a way to increase a ship's
cargo capacity, and Edward D. Meier discussed the possibility of submarine
diesels with the US Navy as early as 1899.[2] Frédéric Dyckhoff that same year
patented a reversing method for diesels whose basic principle would find
shipboard use. However, these ideas were but harbingers of events leading
to tolerated, if not totally satisfactory, engines that slowly converted a few
skeptics during the first decade of the present century.

An urgency fed by the jingoistic nationalism of an armaments race fostered the creation of new marine diesel technology at an ever-accelerating
rate. The submarine was a prime example of causing this to happen. What
once had been relegated to the role of a bothersome auxiliary of the mighty

dreadnought would in a few more years become feared as a powerful weapon in its own right.

Marine diesels faced new design challenges. A flexing ship's hull, working in heavy seas, was far different from a solid, stable foundation so taken for granted ashore. A rolling and pitching vessel encouraged enclosed crank-cases with the adoption of pressure lubrication. Salt water and harbor silt troubled water-cooled pistons. Access to make major repairs, often within a cramped engine room, added other exacting demands. Fuel cost was not a major consideration by navies, but fuel availability as well as digestibility in an engine were. Marine engineers and poorly trained crews long used to a steam engine's forbearance of mistreatment and poor maintenance were unprepared to accept the diesel's craving for care "by the book."

Pre-1911 engines evolved in distinct stages. First was the direct adaptation in 1903 of stationary diesels driving through cumbersome reversing systems on inland waters (chapter 10). Next came the nonreversible, marine four stroke designed specifically for surface and undersea operation. Two- and four-stroke reversibles followed in 1906–1908. Diesels powering ocean-spanning merchant ships emerged after 1910.

Dyckhoff and Bochet

Two French engineers, Frédéric Dyckhoff, Diesel's first licensee, and Adrien Bochet (1863–1922) teamed up to develop the first marine diesel engines. Bochet was chief engineer at Sautter Harlé & Cie., a Dyckhoff sublicensee since 1899, where the engines would be built. One may assume that Frédéric kept his friend Rudolf informed about its progress in the Paris factory.

Dyckhoff had anticipated a marine diesel with the reversible engine he built and patented in 1899 (chapter 9). By now using Sautter Harlé's resources, he expected to offer a small diesel for powering barges along the extensive canal network near him in northern France and in neighboring Belgium. These flat-bottomed, shallow-draft barges required an engine configuration different from the typical low speed, vertical types that transmitted unbalanced forces into the hull. Rhythmic pounding on shallow, flexible keel structures had led to hull failures when conventional kerosene and producer gas engines were first installed. Deutz had eliminated the problem by a horizontal, opposed-piston design.[3]

The Petit Pierre

Bochet was acquainted with Diesel's engine through the published technical papers before he saw Dyckhoff's prize-winning example at the Paris Exposition of 1900. He concluded that an existing Sautter Harlé gasoline engine could be adapted to run on the diesel cycle.[4] It also happened to be horizontally opposed. Whether Dyckhoff was a party to the decision for the diesel conversion is not known, but it certainly was an engine layout acceptable in canal barges. Also not known, yet plausible, is that the gasoline version had been used as a barge power plant.

Its uniqueness lay in a configuration whereby the axes of its single cylinder and crankshaft intersected at a right angle midway along the crank throws (fig. 15.1). The combustion chamber serving the two coaxial and inwardly opposed pistons was a connecting passage over a tunnel through which the crankshaft passed. Rocker levers on a camshaft parallel to the crank actuated the vertical valves and fuel nozzle opening into the bridging passage. Connecting rods linked yokes on the outer piston ends to the four crank throws. Spring-loaded supports under the bisected cylinder's outer ends acted as dashpots to damp vertical loads coming from the rotation of a trunnion-suspended engine structure. A photo (fig. 15.2) of what is probably the converted prototype diesel, circa 1901, shows a low and compact layout with the piston yoke, base-located air tanks, and an air compressor driven off a jackshaft. Size and performance are unknown.

An engine of this design was installed in the canal barge *Petit Pierre*, which began operation in September 1903 (fig. 15.3). Dyckhoff chose the boat because it plied the Marne-Rhine canal that passed through Bar-le-Duc and would enable him to easily monitor performance. The midships-

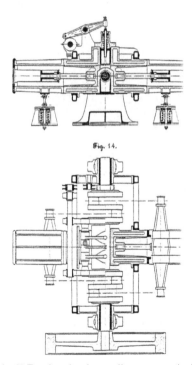

Fig. 15.1. Dyckhoff-Bochet horizontally opposed piston marine diesel engine of 1901–1903. The crankshaft passes between the two piston crowns. *Zeit., VDI, no. 38, 1903*

Fig. 15.2. December 1901 photo of probable prototype engine designed by Bochet in the Sautter Harlé factory. *MAN Archives*

Fig. 15.3. *Petit Pierre*, the first marine diesel-powered vessel, 1903. *Marie-Rose Cochet photo*

placed engine had a 210 × 300 mm bore and stroke (for each piston) and developed 25 bhp at 360 rpm (fig. 15.4). Reversible propeller blades provided maneuvering capability. On September 30, Dyckhoff sent Diesel a postcard with a picture of the *Petit Pierre* to report that the boat had traveled the eleven kilometers between Bar-le-Duc and Commercy in two and a half hours and they had "breakfast on board." All of his family signed the card.[5] A month later Diesel accompanied his friend on the boat for a day trip. The engine's service life is not known, and because no reports on it exist after the initial days of operation, it is assumed to have been not long.

French Submarines 1903–1910

A less than satisfied French Navy Department wanted a power plant to replace the discouraging steam and gasoline engines installed in their first

Fig. 15.4. Engine room of *Petit Pierre*, 1903. *MAN Archives*

submarines. The diesel, with its starting ease, low fuel consumption and ability to burn safer fuels appeared to offer a promising engine alternative for the *sousmarin*. These advantages had been demonstrated ashore with the 1900 Paris Exhibition engine. Rudolf himself then met with French naval officials in 1902 regarding the use of diesels and informed Buz of his meeting. Although a French delegation visited M.A.N. on Buz's invitation that May, a cooperative program with Augsburg never materialized mainly due to a Navy Department reluctance to buy foreign engines. Sautter Harlé, however, responded to its navy's need with what it thought would meet submarine requirements.

The navy contracted with Sautter Harlé to build an engine based on the *Petit Pierre* layout for their new submarine *Z*. Design work began in 1901, but the engine was not installed until 1904. It consisted of two, larger cylinder units, now vertical, similar to the *Petit Pierre* tied together and with the flywheel inserted between them (fig. 15.5). Projected output was 120 bhp at 400 rpm. The engine proved to be unacceptable during sea trials because of excessive torsional vibration in the shaft connecting the two separated cylinder unit assemblies. A next-generation Sautter Harlé diesel replaced it.[6] This engine and the ensuing nonreversible ones used the electric generator and motor drive for surface maneuvering.

Before the *Z* had ever put to sea, a French Navy Department decision prohibited Dyckhoff's Société Française from sharing Sautter Harlé drawings with the Allgemeine if the Paris company wanted to supply sub engines to the navy. This precluded the Allgemeine and Diesel from making a single

Fig. 15.5. Sautter Harlé 120 bhp, opposed-piston diesel engine installed in the French submarine *Z* during 1904. *MAN Archives*

franc out of what could have been a lucrative venture. The 1903 decision had further ramifications. As it prevented a fulfillment of the basic license contract, the ban became a further wedge widening the previously mentioned rift between the Allgemeine and Diesel, the American Diesel Engine Co. and Sulzer Bros.

The French were the earliest to commit almost entirely to diesel-powered submarines. (Experiments continued with steam turbines in a few large subs.) Their diesel fleet led the world for over a decade. By 1911 the French Navy had commissioned at least sixty diesel subs while the British came in second with only thirteen. Neither Germany nor the United States had one in service.[7]

While the Z engine experiment was a disappointment, a new contract was already awarded to Sautter Harlé to supply more four-cylinder, four-stroke vertical diesels of conventional configuration for additional submarines under construction. The first of these, the *Aigrette*, launched about the time of the Z in 1903, had a surface displacement of 175 metric tons, and an engine of about 200 hp.[8]

A minor scandal arose in France over the *Aigrette* during the next year because it was claimed that their Navy Department had "rendered Germany an incredible service by the communication of recent experiences with submarine engines" in this boat.[9] The unfounded charge stemmed from that informal visit by French naval engineers to MAN.

Differences of opinion among navies as to the best submarine hull shape prompted the *Aigrette* to be matched against the repowered Z in a series of tests during 1905. The former was a Laubeuf "submersible" design with a boat-

Fig. 15.6. 350 Bhp Sautter Harlé submarine diesel built in 1906.
Edgar C. Smith, *Short History of Naval and Marine Engineering, 1937*

shaped outer hull to give better surface performance, while the *Z*, a true "submarine," had a cigar-shaped hull. (Laubeuf was chief engineer of the French Navy design department.) Although the *Z* was slightly faster submerged, the *Aigrette* design won due to surface seaworthiness and a greater resistance to pressure hull damage in a collision.[10] All earlier boats, including the Holland designs in America, had been of the "submarine" type, and not until the advent of the nuclear age did undersea vessels return to this originally conceived profile.

Sautter Harlé next introduced a vertical, trunk piston, four-cylinder diesel of 395 bhp at 340 rpm for the 400 metric ton (surface displacement) subs *Opale, Emeraude, and Rubis* (fig. 15.6). During October 1908 the *Emeraude* made a surface run of 692 nautical miles in 80.5 hours at an average speed of 9 kts.[11] Her sister boats also made equally long open-sea voyages.

New sub engine designs in 1907 under Bochet's direction led to six-cylinder, four-stroke diesels of around 700 bhp at 400 rpm. Two of these at 25,000 kg went in the *Mariotte*. Engines for the *Pluviose* class subs were powered by two 350 bhp diesels.[12] That same year the French installed M.A.N. diesels (covered later in this chapter) in their subs *Circe* and *Calypso*.

Adrien Bochet left Sautter Harlé in 1908 when Dyckhoff sold his ailing diesel enterprise in Longeville to Augustin Normand. He moved there himself after reportedly aiding Dyckhoff in the sale.[13] One of Bochet's first projects in Le Havre was to install Nürnberg two-stroke engines in the *Quévilly* (described later in the chapter).[14] Augustin Normand was also beginning to make a vertical, four-cylinder, four-stroke diesel in Le Havre as illustrated in an A-N 1909 brochure.[15]

Société des Moteurs Sabathé

Société des Moteurs Sabathé, St. Etienne, was starting to build a line of small marine engines that would grow into much larger ones for submarines. All had a unique fuel system described in chapter 17.

Delaunay—Belleville

Société Anonyme des Etablissements Delaunay-Belleville, St. Denis sur Seine, received a license from Diesel on December 2, 1909.[16] The association with the inventor led in September 1910 to very preliminary layouts by Rudolf, acting as a consultant, for a pair of 2,400-hp submarine engines.[17] He had proposed that the structure of these single-acting two strokes be mostly of aluminum. Their ten cylinders would have a bore and stroke of 440 mm. Four scavenge pumps and compressors would be used as he believed that Nürnberg's stepped pistons were too heavy. He had ruled out a four stroke because of excessive length. Delaunay-Belleville soon ended this project.

Louis Delaunay-Belleville (1843–1912), who for a short time built the Belleville automobile, also looked at Diesel's truck engine design[18] (chapter 18).

Maschinenfabrik Augsburg-Nürnburg

While both the Augsburg and Nürnberg divisions of M.A.N. began making marine diesels before 1910, it was Augsburg who started well ahead of Nürnberg and who led by far in the number of engines produced. If not for prolonged delays by the German Navy Office, Augsburg diesels would have been installed in U-boats years before the actual date.

Included in Diesel's above mentioned 1902 report to Buz was a hint about patriotic duty inferring ". . . if one has come that far in France, so hopefully will the authorities of our Fatherland then gain the same opinion of the importance of our engine."[19] This suggestion of new market opportunities reinforced by the visit of French Navy engineers to Augsburg was not lost on M.A.N. However, the German Navy, who had yet to order its first submarine, saw no reason to consider an unproven engine for a nonexistent vessel in its fleet. So it was that M.A.N. engineers struck out on their own in December 1902 to design an experimental engine without *Reichsmarine* support.[20]

Augsburg's Submarine Engines 1903–1920

This first Augsburg nonreversible marine diesel carried the model designation SM 4 280/300 (four cylinders and a bore/stroke of 280 × 300 mm). It had a projected output of 140 bhp at 400 rpm and a target specific weight of 40 kg/hp. Such a light, yet more rigid engine meant a leap into the unknown because M.A.N. stationary engines of equal output weighed over 200 kg/hp.

The resulting trunk piston design included a baseplate to support the crankshaft and a rather shallow crankcase enclosure on which sat modified A-frame cylinders (fig. 15.7). Borrowed from the still new DM Series was an air compressor on the side of each power cylinder that was driven by linkage

Fig. 15.7. 1904 M.A.N. (Augsburg) experimental marine diesel
of 140 bhp at 400 rpm. Compressor side. *MAN Archives*

off the connecting rod. In the usual fashion, it acted as a second stage of compression after receiving first-stage air from the adjacent power cylinder. Each cylinder was also served by its own fuel pump whose housing formed a part of the camshaft support bracket (fig. 15.8). An eccentric on the camshaft operated the pump plunger, and the governor controlled the pump suction valves' duration of opening. In order to keep engine height to a minimum, the governor sat atop its own short vertical shaft driven off the crankshaft.

Fig. 15.8. 1904 M.A.N. (Augsburg) experimental marine diesel
of 140 bhp at 400 rpm. Camshaft side. Del Proposto,
Machines Irréversibles, 1906. B&W MAN Archives

The German Navy declared a willingness to buy the engine in June 1903, but it was not delivered to the Imperial Wharf in Kiel until mid-October 1904. No data exists on performance, including fuel consumption, and the final specific weight.[21]

A 1903 order from the Russian Navy to Krupp's shipyard for several diesel-powered subs would lead to Augsburg's involvement. Germaniawerft engineers drew up a preliminary design in January 1904 for two 200 bhp at 500 rpm engines and approached M.A.N. about building them. Augsburg declined, and the Russian order was not filled. Consequently, Krupp asked Körting to supply low compression, oil-burning engines. These went in the *U-1*, Germany's first sub, and the next seventeen *Unterseeboote* built at Kiel.[22]

Krupp pursued a cooperative sub diesel venture with M.A.N., and in March 1904 the two agreed to jointly design an engine. Wilhelm Worsoe, director of the Germaniawerft engine manufacturing section, then came to Augsburg where he spent six months working with Lauster and his engineering group.

The product of their efforts was an open crankcase, four-stroke, four-cylinder engine having a bore and stroke of 270 × 300 mm and designed output of 200 bhp at 500 rpm. An upper, cast frame on which the cylinder assemblies sat was in turn supported by diagonally braced pipe columns. The open design, done no doubt for weight reduction, followed a steam engine practice Werkspoor, as has been shown, later adopted. Navy concerns over insufficient power, not enough stiffness, and the open crankcase caused it to turn down the proposal and end the joint project. Nevertheless, its interest in diesel power was sharpened enough for it to suggest that a new, higher output design would be considered.

Augsburg accepted the sub diesel challenge, and work began in early 1905 on a 300 bhp at 400 rpm, four-cylinder engine with a bore and stroke of 330 × 360 mm. New features of this model SM 4 × 360 included: a cast crankcase enclosure on the bedplate and supporting the cylinders; cylinders with cast on water jackets (earlier "sub" types used sheet metal); inserted liners; and a single, two-stage compressor driven off the front of the crankshaft. The four valves per cylinder (intake, exhaust, fuel, and starting) were in the head and actuated by a high mount camshaft through rocker levers. Two, double plunger pump housings were attached on each side of the center camshaft support bracket (fig. 15.9).

Two design features came from the Krupp-M.A.N. effort: the governor sat atop its own short vertical shaft, and a hand-rotated auxiliary shaft ran the length of the engine to effect an offset of eccentrics and thereby cause the air starting valves to open in a timed sequence. This latter feature would be used on many future Augsburg marine diesels.

Jointed, oscillating pipes ("grasshopper legs") fed and drained cooling oil for the pistons via passages extending upward from the skirt to the crown. Water pipes circulating through the crankcase cooled the oil sump.

Fig. 15.9. Augsburg four-stroke sub engine for the French Navy, 1906–07. Model SM 4 × 360, 300 bhp at 400 rpm. *MAN Archive*

Specific weight of the engine came to only 33 kg/bhp, and the fuel consumption was a respectable 195 g/bhp-hr. The bmep at rated speed reached 5.5 kg/cm².

When the Torpedo Inspection Department, who still held sway over German Navy submarine matters, inquired in October 1905 about the disposition of these engines, it was informed the French Navy had already ordered two, and then four, to go in a pair of their own subs. Surprisingly, Kiel still did not place an official order and continued to hang back. The first pair of engines were completed in January 1906 without a reversing mechanism, which was retrofitted in September. The second pair was built with it. Delivery came during August and September 1907 followed by an immediate installation into the *Circe* and *Calypso*. The two subs, commissioned before the end of the year, would see active duty in World War I.[23]

The four SM 4 × 360 engines for the *Circe* and *Calypso* had a simple reversing mechanism very similar to the one intended for the 1903 sub engine. It allowed only one cam per valve and did not require rocker levers to be lifted off their cams. At the top of the vertical shaft transferring motion to the camshaft were two facing bevel gears slidable on the camshaft. Only one or the other could mesh at the same time with a mating gear on the vertical shaft to determine engine direction (fig. 15.10). With this design the camshaft always turned in the same direction regardless of the crankshaft's rotational direction. The method's reliability proved to be less than expected in service and was installed only on the four submarine engines sent to the French.

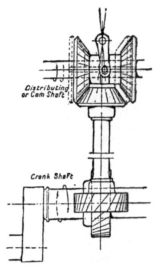

Fig. 15.10. Sliding bevel gear arrangement to reverse M.A.N. SM 4 × 360 sub diesel. Supino, *Land and Marine Diesel Engines, 1913*

One requirement in changing the direction of rotation is that the fuel pump not deliver fuel during the reversing process. Also needed is a control to engage the air valves to admit pressurized air into the cylinder. This can be to either start from a dead stop or, if running, to first brake the engine to a stop and then get it moving in the new direction.

The reversing sequences are best pictured when there is a single control, that is, a handwheel, to perform them. The wheel has a "stop" position at its midpoint of rotation. Next, in either direction from "stop" is a "start" position where air is admitted. At full rotation is "ahead" or "astern." Fuel is shut off between these last positions. As the engine turns forward or in reverse under the force of the starting air, the handwheel is then fully rotated to "ahead" or "astern" and fuel flows to the injectors.

Early engines did not have such a single, or convenient, reversing control, but used separate levers and wheels to shut off fuel and admit starting air. For this reason, achieving minimum reversal times during ship maneuvers required skilled engine room operators possessing deftness and finesse. Not observing specified control sequences or adequate pause times could lead to dire consequences. Later, more sophisticated mechanisms allowed an engineer to turn a control wheel from "ahead" to "astern" and back with fewer concerns.

The good test results of the diesels for the French subs in 1907 finally persuaded the German Navy to place an order in December 1908 for a single, much larger model. This new, six-cylinder engine, designated the SM 6 × 400, would produce 850 bhp at 450 rpm and have a cylinder size of 400 × 400 mm. In many ways it was a scaled-up version of the four just built.

Its specific weight, however, could not exceed 26 kg/hp, a stringent requirement Augsburg learned of in July 1908. Meeting this meant the difficulties of casting from steel a thinner crankcase enclosure and a three-piece bedplate. Two two-stage air compressors were mounted on the front of the engine (fig. 15.11).

Reversing took place by turning a handwheel to first lift the rocker levers off their cams and then axially slide the camshaft to bring into play another set of cams for running in the opposite direction. The shifting camshaft idea was not new, but M.A.N. had its own system of eccentric sleeves and rods to lift the levers. Positioning the handwheel as required also stopped fuel metering and activated the starting air valves in the sequences explained above. The air valves were engaged by rotating an eccentric shaft similar to the mechanism used on the "French" engines.

Tests began in April 1910 and by the next month the engine had produced 900 bhp. Overall weight was 20,100 kg. At the called-for 850 bhp rating, this yielded a specific weight of 23.6 kg/bhp, or 2.4 kg/bhp under the limit. The fuel consumption had to meet a maximum of 200 g/bhp-hr plus 10 percent, but the test results showed a bsfc of 192 g/bhp-hr.[24]

On August 4, 1911, the navy ordered seven more engines, and all eight left the factory between July 1912 and May 1913. They were installed in the subs *U-19* to *U-22*, which went in service over a three-month period beginning in July 1913. The boats had a length of 62.4 m and a surface displacement of 650 tons. Cruising range on the surface at 8 kts was 7,600 nautical miles. Maximum surface speed was 15.4 kts.[25]

These SM 6 × 400 engines began a series of four-stroke submarine diesels that grew rapidly in horsepower output over the next years. Their story is in chapter 17.

Fig. 15.11. Augsburg diesel destined for German sub *U-19*.
850 bhp at 450 rpm, 1909. *MAN Archives*

Nürnberg Sub and Marine Diesels 1907–1910

Nürnberg's bold undertakings into naval-oriented marine diesels after 1907 resulted from Anton Rieppel's enterprising and sustaining leadership. His vision, viewed possibly as foolhardy by a few contemporaries, led to failures as well as successes. Nevertheless, the new technology his engineers created by striking off into uncharted areas greatly extended the knowledge and potential of the diesel engine. If Augsburg's many achievements are seen as being evolutionary, its sister division in Nürnberg might then be considered as having taken revolutionary steps. Nürnberg's later involvement chronologically places it after others yet to come in this chapter, but its corporate tie to Augsburg suggests its progress be recorded at this point.

By 1907 Rieppel believed the time was ripe for Nürnberg to make its own mark in the diesel industry. It had only built the Augsburg-designed DM series engines, which, in Rieppel's mind, placed it at disadvantage. As long as Nürnberg made what was conceived in southern Bavaria, it could never be accepted as a full member of the new industry. Buz and his management did not exactly look forward to this happening, and when they learned what Rieppel had in mind, they tried to talk him out of it. But he was his own man and saw it as a duty to profitably expand his company.[26]

The first diesel engines Nürnberg developed on its own were intended for both ships and submarines. Its decision to go two stroke was based on Rieppel's prerequisite that any new line of engines must be distinguishable from the Augsburg diesel products. No doubt reinforcing the decision was Sulzer's recently publicized progress made with its land and early marine two-stroke diesels.

Little is known of Nürnberg's first experimental two-stroke diesel built in 1907. It was probably designed under the direction of Hermann Ebbs, who had succeeded Lucian Vogel at Nürnberg.[27] The engine produced 100 bhp, most likely in a single cylinder, and had a bsfc of 260 g/bhp-hr. It used a stepped piston for pumping scavenge air, a feature to be retained.

The test engine's performance was encouraging enough for the company in July 1908 to offer multicylinder engines of 600 to 850 bhp. The latter engine was obviously intended to directly compete with Augsburg's four-stroke model of the same output. A strong Nürnberg inducement was its guarantee to the German Navy of a specific weight of only 15 kg/bhp and a fuel consumption of 210 g/bhp-hr, plus 5 percent. Such a bold weight pledge caused alarm in Augsburg where the struggle was going on just to lower weights below 30 kg/bhp. Discussions between the two potential rivals in the same family led to Nürnberg's slight relaxation of fuel consumption to 220 g plus 5 percent for an 850 bhp diesel.[28]

First built was a six-cylinder, 150 blip at 550 rpm marine engine having a bore and stroke of 175 × 220 mm. This S D 15/6 (Ship, Diesel, horsepower/ cylinders) continued with the test engine's stepped piston as a scavenge pump. Other features included uniflow scavenging through a single valve in the head, cylinder exhaust ports, and a central, overhead camshaft whose lobes actuated small levers pivoting on pins anchored to inclined air valve inlet cages and fuel nozzles. The injection air pump was driven off the front of the crankshaft. Figure 15.12 is similar to this first engine.

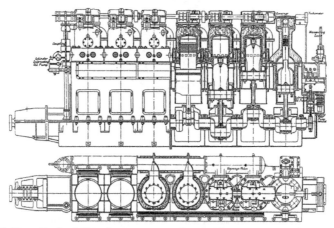

Fig. 15.12a. Typical Nürnberg two-stroke, uniflow scavenged, submarine diesel, c. 1909–1914. 450 bhp example built by Nelseco, 1912. Magdeburger, *"Diesel Engines in Submarines,"* 1925

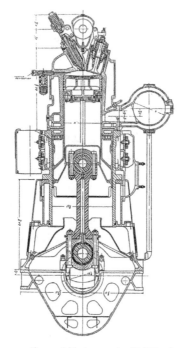

Fig. 15.12b. Cross-section of Nelseco-built Nürnberg sub diesel, c. 1909–14. Magdeburger, *"Diesel Engines in Submarines,"* 1925

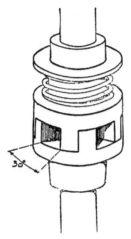

Fig. 15.13. Three-dog coupling to alter timing for engine reversing with Nürnberg two-stroke diesels, 1908. Supino, *Land and Marine Diesels, 1918*

The differential pump and power piston were of one piece except for the dished crown, which held the rings and was bolted to the top of the power piston skirt. Cooling oil, under pressure from the connecting rod and piston pin, was carried to the crown. A seal on the end of the piston pin in the scavenge piston skirt kept excessive oil off its cylinder wall. Oil escaping past the piston clearance went down a tube extending to the lower end of the scavenge piston skirt where it discharged into the crankcase.

Scavenge air pumped by each stepped piston fed into a common receiver from which individual pipes led to the inlet valves in the head. Scavenge air pressure averaged about 1.6 kg/cm^2.[29] The exhaust manifold was on the same side as the air receiver.

Air and fuel valve timing were changed when reversing engine direction by means of a spring-loaded, three-dog (tooth) coupling inserted in the vertical shaft to the cam. A 30-degree angular clearance separated the engaging dogs (fig. 15.13). As the engine began turning in a new direction by a separate air starting system, the lower coupling-half dogs turned through the clearance angle before contacting the dogs in the upper coupling half. The resultant timing change was enough for the engine to run well in either direction, although optimum timing favored running "ahead."

A rather sophisticated pneumatic system opened the starting air valves (fig. 15.14). A control lever with positions for "ahead," "stop," or "astern," determined the action of four servo valves, an air supply valve, two cam-reciprocated piston valves, and the air valve in the head. When the lever contacted a selected end stop, it positioned a pair of servo valves to simultaneously connect the starting air tank with the air valve in the cylinder head and push the piston valve against its cam lobe. This allowed cam action to determine the timing of the air valve in the head. As soon as the lever

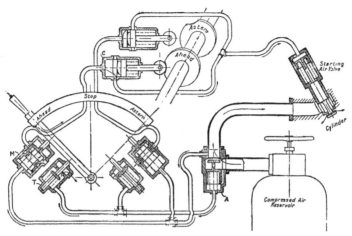

Fig. 15.14. Pneumatic control system for starting and reversing Nürnberg two-stroke diesels, 1908. Supino, *Land and Marine Diesels, 1918*

broke contact with the servo valves, control system air was vented to cut off air from the tank and retract the operative piston from its cam.

Two of these 150 bhp engines were delivered to the Portuguese Navy for a gunboat installation in May 1910. Because the scavenge pump worked with a very large air surplus—a 48 percent greater cylinder volume than that of the power cylinder—their bmep was only 3.76 kg/cm². (The next engines reduced the excess volume ratio to around 33 percent, which raised the bmep to about 4.3 kg/cm².) Fuel consumption is not known.[30]

An S D 85/8, concurrently developed with the 150 bhp engine, was intended for sub service. Its bore and stroke were 310 × 340 mm and produced 850 bhp at 450 rpm. There were two scavenge air valves in the head and two injection air compressors on the front end of the engine. Specific weight was only 18 kg/hp. So many problems were encountered with the engine that in the spring of 1910 it was decided not to try to sell it to the navy. Three, however, were eventually delivered in 1913 with one going in an Italian ship and two in the Dutch submarine *K-1*.[31] They apparently performed satisfactorily in service.

Smaller engines of 200 bhp (S D 20/4) and 300 bhp (S D 30/6) had cylinders of 240 × 260 mm and produced their rated power at 500 rpm. An S D 20/4 and a three S D 30/6 were installed in Dutch U-boats, and three of the latter also went in Austrian subs.

The Quévilly

Another pair of the S D 30/6 were those for the four-masted *Quévilly*, the first French merchant ship to be powered by diesel engines. This bark-rigged tankship built in 1897 was 94.3 m long and had a 6,200 metric ton displacement. Augustin Normand received a contract in 1910 to repower the ship with two Nürnberg reversible, 300 bhp, two-stroke diesels for auxiliary propulsion.[32] Sea trials took place in January 1911 with a maiden voyage to New York City starting in March.[33] Adrien Bochet, who was responsible for the installation, described this round trip to the US as relatively uneventful. However, another report stated three pistons had to be replaced at sea during the first two transatlantic crossings.[34]

Denmark bought six and Austria two of the S D 40/6 engines for their subs.[35] These 400 bhp at 500 rpm diesels had a 270 × 290 mm bore and stroke. There were two scavenge air valves per cylinder and two blast air compressors.

Of special significance is that the license for the Nürnberg engines was the first in the United States after that Diesel sold to Adolphus Busch. The Electric Boat Co. bought a license from Rieppel in 1910 and established the New London Ship & Engine Co. to build the German engine. The EBC/Nelseco story is in chapter 17.

Gebrüder Sulzer

Sulzer Brothers' two-stroke evolution began with the design of an experimental marine diesel in mid-1905. It was also during this period that two of their stationary four-stroke engines were installed in small inland vessels.

The *Venoge*, a Lake Geneva freight boat, was thought to be the first, and only, marine application of such a Sulzer diesel (fig. 15.15). However, a recent discovery has shown that in May 1905 a 30 bhp, two-cylinder Model 2D15 went into a Leeds-Liverpool canal boat. The engine was ordered through the Diesel Engine Co. in London during March 1904 and was shipped from Sulzer at the end of that year. The boat is assumed to have had a reversible propeller. Sometime in late 1906 the engine was removed and ended its days powering a sawmill.[36]

Fig. 15.15. The *Venoge*, a Lake Geneva freight boat with a Sulzer diesel installed in 1905. 35 × 6 × 1.9 m. *Sulzer Archives*

Fig. 15.16. Engine room of the *Venoge* with a Sulzer 2D20, 40 bhp diesel and Del Propos to electro-mechanical reverse gear. *Sulzer Archives*

A two-cylinder Model 2D20 in the *Venoge* had a longer service life (fig. 15.16). It was ordered in July 1904, shipped in March 1905, and put in operation during September—a few months after its English counterpart. The 40 bhp diesel and its Del Proposto electro-mechanical reverse gear were replaced in 1924 by a higher-powered, directly reversible engine.[37]

Sulzer demonstrated the first directly reversible, two-stroke diesel at Milan's 1906 World Exhibition (fig. 15.17). Work had begun, however, on a similar four-cylinder, uniflow scavenged test engine at least a year earlier. A drawing dated June 1905 shows the DM180x250 test model with two scavenge air valves transverse to the engine axis on the cylinder centerline. A single cam acting on a valve crosshead opened them. The fuel injector with its cam-opened valve was on the engine axis but offset to the side of the cylinder. A thrust bearing was incorporated in the engine between the rear main bearing and the flywheel. This required bearing for marine applications, typically a separate unit on the shaft, absorbed the reverse thrust load created by the propeller.

The "Milan" engine produced 90 bhp at 375 rpm from a bore and stroke of 175 × 250 mm. Diagonal rods, in addition to supporting "column" bolts, tied the lower ends of monobloc cylinder and head castings and their supporting entablature to the bedplate (fig. 15.18). A front throw on the crankshaft drove the two-stage injection air compressor. The scavenge air pump mounted on the side of the compressor was run by linkage from it. The two longitudinally placed scavenge valves were opened by their own cam.

Fig. 15.17. Sulzer two-stroke, reversible marine diesel shown at the 1906 Milan Exhibition. 90 bhp at 375 rpm. *Sulzer Archives*

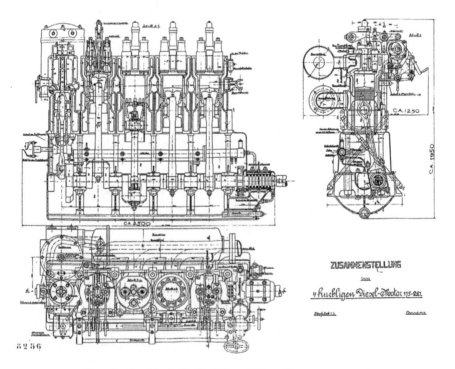

Fig. 15.18. Model DM 175 × 250 uniflow scavenged, reversible two-stroke, 1906. *Sulzer Archives drawing*

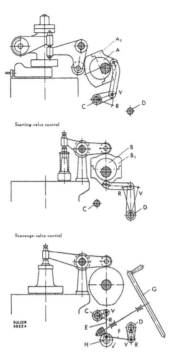

Fig. 15.19. Schematic of DM175x250 reversing mechanism.
Sulzer Technical Review, 1947, No. 2

A unique reversing system began an evolutionary series of Sulzer mechanisms. This initial design used a single handwheel (G) to control the necessary functions when starting and maneuvering (fig. 15.19). The mechanism did not alter timing of the scavenge valves as they opened at the correct point by fixed cams for either direction of rotation without such a change. Sleeves A_1 and B_1 were attached to eccentrics A and B on the camshaft such that rollers on rocker levers for the fuel and air starting valves contacted a segment of the sleeves' cam-shaped surfaces. Rotating the handwheel moved the crank arm (H), which, through links E and F, turned levers on shafts C and D to either the forward (V) or reverse (R) position. This shifted eccentric offsets so that an oscillating motion was imparted by the sleeve segments to the cam rollers. Fuel flow stopped whenever the starting valve was being actuated, and conversely, the starting valve remained inoperative during injection of fuel.

The handwheel controlled a progression of starting and reversing sequences from its midpoint "stop" position: (a) "ahead" or "astern" with all four cylinders receiving starting air, (b) two cylinders on starting air and two cylinders with starting air off and fuel injected, (c) two cylinders running and all air shut off, and finally (d) all cylinders injecting fuel. A swing of the handwheel from "ahead" to "stop" to "astern" could be made as fast as the operator turned the wheel. When warm, the diesel reversed faster than a steam engine, or in about

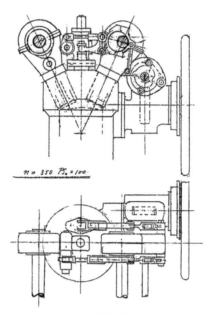

n = 350 PS. = 100.

Fig. 15.20. Sketch from 1909 of inclined scavenge valves and the linkage to lift the fuel valve for forward and reverse running of the Sulzer 4SNo.3 marine diesel. *Sulzer Archives drawing*

seven to eight seconds.[38] The DM175x250 engine exhibited in Milan, the only one of that design to leave the factory, was never installed.

A uniflow scavenged Model 4SNo.3 superceded it in 1909. This four-cylinder, "No. 3" edition had a slightly increased bore of 180 mm and two scavenge air valves inclined at 30 degrees (fig. 15.20). Dual overhead camshafts, with a cam for each valve, acted directly on a roller follower on top of the valve spring keeper to open the valve. Pump and compressor drives were reversed from the previous design. The scavenge pump was driven off the crankshaft and the starting/injection air compressor was on the pump with the first stage on one side and the second on the other.

A new reversing system still retained the unaltered scavenge valve timing. Opening the fuel and starting valves were bell cranks whose roller ends ran on the cam surfaces of oscillating eccentrics pivoting about one of the scavenge valve camshafts (fig. 15.21). The two eccentrics for each cylinder straddled a scavenge valve. Turning the handwheel from "stop" to "ahead" or "astern" rotated the eccentrics to accommodate the changes required for the two valves to function properly through the four starting modes in either engine direction. This was done by a "maneuvering shaft" that rotationally shifted one end of a pair of link bars pinned to a ring on the shaft. Because of the dissimilar timing shifts required, the link ends were pinned at different circumferential points on both the eccentrics and the wheels to impart an oscillatory motion suiting each valve's timing needs.

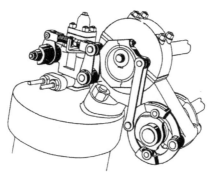

Fig. 15.21. Partial view of 4SNo.3 cylinder head showing the means to alter the oscillating cam eccentric to open the fuel valve. *Sulzer Archives*

Installations of the 4SNo.3 in tugs and launches began in 1909. One going in the 60-foot British Navy launch *Dreadnought* that year operated over a two-year period. In a September 1911 test, the engine produced 85 bhp at 345 rpm and a 62.5 psi bmep. Mechanical efficency reached 70 percent. Compression pressure and blast air pressure at full load were 500 psi and 850 psi respectively. The "Texas" fuel oil had a 0.925 sp gr. By using only two cylinders the engine could idle down to 140 rpm.[39]

Sulzer abandoned uniflow scavenging after five 4SNo.3s were sold and adopted loop scavenging for all but the smallest marine engines. Chapter 12 continues with two-stroke stationary diesels.

Their first cross-port, loop-scavenged marine diesels were two for the *Romagna*, a 1,000-ton cargo vessel launched in 1910 (fig. 15.22). These

Fig. 15.22. The 53.5 m long *Romagna* at her launching in 1910.
Sulzer Archives

Fig. 15.23. One of two Sulzer Model 4SNo.6a, four-cylinder,
380 bhp, loop-scavenged diesels in the *Romagna*. *Sulzer Archives*

4SNo.6a trunk piston engines had a 310 × 460 mm bore and stroke and developed 380 bhp at 250 rpm (fig. 15.23). Linerless cylinders sat on an upper crankcase enclosure, and bolts extending from cylinder flanges to the bottom of the bedplate tied the structure together (fig. 15.24). Only the starting and fuel valves remained in the cylinder heads. The two tiers of cylinder intake ports opposite the exhaust ports were positioned as in figure 12.6. New features included a three-stage injection/starting air compressor and telescoping tubes, without packing glands, for oil cooling the pistons.

The reversing mechanism evolved again because of the larger engine and to incorporate improvements gained from experience.[40] It performed functions similar to the systems on the preceding engines, except that the operator, at the front of the engine, controlled a hand lever as well as the handwheel.

A groove in a face cam, turned by the handwheel, traced the motion resulting in the four above-listed sequential opening and closing modes of the starting air and fuel valves. The face cam contour, via a vertical rod, positioned an eccentric on the rocker lever shaft to override the basic motion imparted by the camshaft during normal running. Two full handwheel turns took the engine from "ahead" to "astern" running. The handwheel also regulated the fuel pump suction valves by holding them open whenever the fuel nozzle valve was inoperative.

The newly added lever worked in combination with the handwheel to provide correct fuel and starting valve timing for running in either direction. It turned an auxiliary shaft to rotate arms that repositioned the valves' cam followers on their respective cams as required for forward and reverse

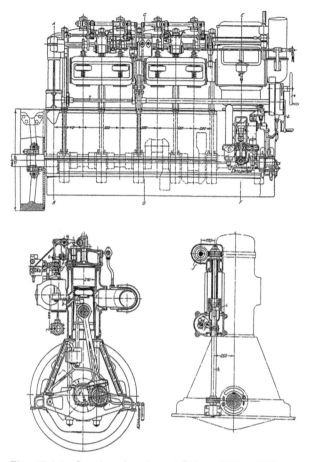

Fig. 15.24. Section drawing of Sulzer 310 × 460 mm diesel in *Romagna*. Chalkley, *Diesel Engines, 1917*

running. Moving the lever also lifted a sleeve on the vertical camshaft drive to turn another eccentric, which altered timing of the valves admitting scavenge air to the upper tier of cylinder ports.

Other handwheel/lever functions included regulating the lift and timing of a valve placed between the air supply pipe and the starting valve at the cylinder. General practice was to send starting air to the cylinder valve only when it was needed to turn over the engine.

A report describing the newest Sulzer reversing procedure stated that "while these [operations] are given step by step, no actual stop takes place between these steps, but the handwheel and hand lever respectively are turned slowly and continuously from one position to the other."[41]

In addition to the above controls, the governor held open the suction valves in one of two double-plunger pumps (one plunger per cylinder) in case of overspeed. A manual control also allowed the operator to hold open the valves in the other pump for running at slow speeds. Injection air pressure was reduced at the same time by throttling the suction side of the first-stage compressor.

Unforeseen problems arose when the *Romagna* went in service. The Milan branch of Sulzer sent a confidential memo outlining them to the home office on October 6, 1910. Of greatest need was a larger compressor to provide more air for maneuvering. Tank pressure dropped too low and could not be resupplied fast enough by the engine compressor itself. Only by using an auxiliary compressor in addition to the tank air could sufficient pressure be maintained. Since the engine compressor acted as a first stage of the three-stage injection air compressor, this, too, was in jeopardy. Piston cooling and insufficient end clearance on the main bearings were other problems needing correction. The report writer also urged that "your present engine mechanics be replaced since such incessant, overexerting work down below in a ship hold has in fact ruined the men."[42] It had only been through their tireless efforts to keep the engines running that a concerned owner was reassured the troubles would end. In November 1911 the *Romagna* capsized during a violent Adriatic storm after her cargo had shifted.[43] Among those drowned was a Sulzer engine room mechanic.

Hesselman's Marine Engines

AB Diesels Motorer's (ABDM) 1904 success with stationary engines in the Russian *Vandal* (Chapter 10) was repeated with directly reversible diesels. These loop-scavenged, two-stroke marine engines had a reversing mechanism patented by Hesselman.[44] One of them in 1907 became the first reversible two-stroke to enter service. This happened because Sulzers engines based on their model exhibited in 1906 were not installed until 1908 and later.

The marine series of Swedish engines built between 1907 and 1910 borrowed the open, A-frame design for the power cylinder and its support structure from the well-seasoned K Series (chapter 10). Although the K's relatively shallow bedplate possibly required that the hull substructure provide additional rigidity, no troubles due to shaft flexing were reported.

Hesselman's unique reversing method centered on converting two scavenge pumps into independent, double-acting air motors when first starting the engine and restarting it during maneuvering[45] (fig. 15.25). The vertical pumps were driven off their own crank throws on the front of the engine. An offset of 90 degrees from each other prevented a "dead" spot at the start of cranking. Atop and integral with one pump piston were axial pistons for a two-stage injection air compressor. Above the other pump was a single-stage compressor discharging at 140 psi maximum to replenish starting air tanks. When the scavenge pumps became motors, the pressure supplying them was reduced to about 70 psi by a regulating valve.

A major advantage of this system lay in the simplicity of the cylinder head. With no starting valve, and with cylinder ports handling scavenge and exhaust functions, the head contained only a centrally positioned fuel nozzle.

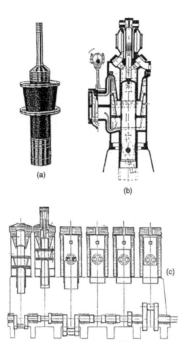

Fig. 15.25. Hesselman reversing system details, c. 1910:
(a) integral two-stage injection air compressor and double-acting scavenge
air pump piston, (b) section through pumping cylinder for the piston in (a),
(c) scavenge/air motor and power piston cylinders. *The Engineer, 1914;
Polar 1914 manual; Supino, Land and Marine Diesels, 1918*

The resultant, symmetrical casting eliminated thermal distortion and crack-ing problems. A lighter camshaft and its bearing supports were also made possible as they were subjected to negligible nozzle valve opening loads. The lift of the fuel valve cam was so slight that by adding a shallow axial ramp on the camshaft, rocker lever rollers did not need to be raised from their cams when the shaft was longitudinally shifted to run in the opposite direction.

Spool valves controlled the suction and discharge functions of the dual purpose, double-acting air pumps. Moving a first control lever altered the timing of eccentrics reciprocating the valves whereby the valves mutually exchanged their functions so the engine could run in the opposite direction. A second hand lever set the inlet valve (for either direction) to admit atmo-spheric air to the pumps, as when the engine was under power, or to starting air pressure. The valve performing the discharge function opened into the air manifold leading to the power cylinder ports so that the engine also received scavenge air whenever being motored.

To prevent excessive scavenge air pressure during the "air motor" mode, a small port midway on the pumping air cylinder exhausted air still under pressure at the end of each piston stroke. Thus, when the double-acting piston began its reverse stroke in either direction to pump scavenge air, the initial pressure started at atmospheric as would be the case when the engine operated under power.

One requisite during a reversing action was to positively prevent injection of fuel while the engine still turned in its original direction either under its own inertia or any motoring action imparted by the propeller while the vessel still had headway. This was especially important if the fuel pump had already been positioned to operate in the new direction. Injecting fuel during this phase could cause the engine to "backfire," as terminology of the times called the event.

Hesselman avoided this "backfire" condition by controlling the pump suction valve's opening so that the pump delivered no fuel to the nozzles until the engine turned in the direction corresponding to the setting of the valve gear. Moving the above first hand lever, which switched air valve functions, also axially shifted the camshaft for the fuel nozzle valves and changed the timing of a short, parallel cam and eccentric shaft operating individual fuel pumps. The time delay came from the sequenced relationship between the fuel and suction valve cams. The individual fuel pumps on the engine were a departure for Hesselman engines and came about only to prevent the mistimed fuel delivery possible in a marine diesel.

The first engine with air motor reversing was installed as auxiliary power in the coastal schooner *Orion* in 1907. The two-cylinder engine with a bore and stroke of 210 × 320 mm produced 60 bhp at 275 rpm[46] (fig. 15.26).

The *Rapp* and *Snapp*, 350 deadweight ton Swedish cargo vessels, followed in 1908. This identical pair had one 120 bhp at 300 rpm engine that

Fig. 15.26. First AB Diesels Motorer two-stroke, reversible marine diesel of type installed in the *Orion*, 1907. *Atlas-Polar Catalog, c. 1932*

reportedly could turn as low as 55 to 60 rpm. They operated in the Baltic and North Seas from Finland to Scotland and crossed Sweden through its lake and canal system. No problems appear to have been encountered while maneuvering through the system's seventy-five locks. At 250 rpm the engines produced 110 bhp with a bsfc of 0.48 lbs/bhp-hr on "residue oil." Hull speed at that rpm was 6 to 7 kts loaded.[47]

Another very successful installation was that of two four-cylinder, 160 bhp at 240 rpm diesels in the Nobel tug *Jakut*. It served the oil company's Volga River operation and did double duty as an ice breaker. A Nobel report said that in one Volga delta rescue operation during November 1910 it "burned but a small fraction" of fuel used by an accompanying steam-powered icebreaker. On the strength of the *Jakut*'s maneuvering capability, Nobel equipped its newest icebreaker with its own diesel.[48]

The most famous ship with an AB Diesels Motorer engine was explorer Roald Amundsen's *Fram* (fig. 15.27). The sailing ship was already famous for its Arctic voyages before Amundsen removed her auxiliary steam plant in April 1910 and repowered with a Polar four-cylinder, 180 rpm diesel. His new goal was to use the *Fram* for the ocean part of his trek to the South Pole. The *Fram* left Norway on August 10, 1910, and arrived in the Ross Sea the following January 11. Amundsen did not reach the Pole until December 14, 1911, less than five weeks ahead of Robert F. Scott's ill-fated party. The *Fram* finally arrived home on July 16, 1914.[49] By that time the engine had logged over 2,800 hours with only minor problems.[50] No doubt contributing to this good service record was the unsung hero company mechanic who had been part of the crew (fig. 15.28).

Fig. 15.27. The *Fram* moored off Amundsen's home, c. 1910, with the explorer in foreground. Note horizontal exhaust pipe that could be set to exhaust downwind. *Norsk Folkemuseum, Oslo*

Fig. 15.28. Engine room of *Fram* looking forward.
Norsk Folkemuseum, Oslo

It is usually assumed that the trade name "Polar," which the Swedish company adopted for its diesels, resulted from the Amundsen voyage. The facts are otherwise. ABDM had for some time wanted a distinctive name for its engines other than just the "Swedish" often used by writers and the trade. Thus, on January 22, 1909, the company applied for the new trade name "Polar-Motoren" and used it for its diesels beginning in 1910. The *Fram* coincidence was a lucky publicity break.[51]

Although the *Toiler* waited until mid-1911 to sail from England for Canada, her two 180 bhp Polar engines were made before the end of 1910. The ship's builder, Swan, Hunter & Wigham Richardson, at Walker, Newcastle-upon-Tyne made the installation (fig. 15.29). She was 255 feet long with a beam of 42.5 feet and draft of 17 feet. The *Toiler* became the first vessel to cross the Atlantic under diesel power and spent much of her useful life as a Great Lakes ore carrier (fig. 15.30). Because she was underpowered (top speed loaded was only 7 kts) the diesels were removed around 1920 and replaced by steam engines.[52]

The internationally recognized performance of the engines built by the small, limited resource company in Sweden led to its selling several licenses of Hesselman's two-stroke fuel and reversing systems. Benz & Cie. in Mannheim bought a marine diesel license in April 1909 to build a more modern version of the *Fram* engine like those for the *Toiler*[53] (chapter 13). Three, 200 bhp and one 250 bhp Benz-Hesselman diesels went into tugboats on the lower Danube River in Romania.[54] Production of the Benz Polar ended in 1914. Swan, Hunter who built the *Toiler* also bought a Polar engine license around 1910 or 1911 and made a few of the same type as Benz.[55]

Fig. 15.29. AB Diesels Motorer "Polar" four-cylinder, loop-scavenged two-stroke of the type installed in the *Toiler* and licensed to Benz and Swan, Hunter in 1910–1911. 180 Bhp at 250 rpm. Chalkley, *Diesel Engines, 1917*

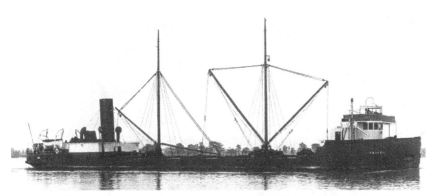

Fig. 15.30. The Great Lakes ore carrier M/S *Toiler* was in mid-1911 the first ship to cross the Atlantic Ocean under diesel power. *Great Lakes Marine Historical Collection, Milwaukee Public Library*

Diesels Join the Russian Navy

Russia's disastrous naval war with Japan in 1905 forced Admiralty planners in St. Petersburg to closely examine the makeup of their fleet as its rebuilding began. Their decisions of the 1906–1908 period brought diesel power to the fore. Orders flowed for new designs of engines to turn propellers of submarines, river gunboats, and light cruisers. Even diesel generating sets were installed for electric power on battleships.

A navy request for such engines was akin to a production order for the Ludwig Nobel factory because of its established reputation and the fact that the engines had to be Russian built. It is to the credit of Carlsund and his imaginative engineers that their diesels continued to set new standards for very lightweight along with recognized reliability.[56] The latter held particularly so with their reversing systems. In the ten years remaining to Nobel's Swedish team before the exodus from St. Petersburg, as much innovation came from them as from any other diesel factory.

Two urgent navy needs sparked what led to the development of Nobel's four-stroke marine Series. Exciting fires and explosions endemic to gasoline engines in the existing Russian fleet of fourteen subs also ignited a desperate demand from their crews for a safer power plant. Secondly, more efficient engines were wanted for new gunboats about to be built in navy yards at St. Petersburg. These replacements for the steam powered ones lost during the 1905 war were destined for Siberia to defend the Amur River.[57] The good performance of the *Vandal* and *Ssarmat* diesels (chapter 10) greatly helped in the gunboat decision, but a willingness to try such engines in subs became a leap of faith. With Nobel's encouragement, however, the Admiralty took that leap, and designs for both applications began concurrently in 1907.

Acting as a guinea pig for the diesel was the *Minoga*, a small new submarine that entered service in 1908. Surface displacement was only 117 tons, and surface speed was expected to be 8 kts. Its two engines were close copies of the experimental three-cylinder, two-stroke design with individual cylinders bolted to a single piece, enclosed crankcase (chapter 10). Principal changes encompassed a switch to four-stroke cycle and a reversing system other than the one by Carlsund (fig. 15.31). Bore and stroke were 275 × 300 mm, and at its rated speed and load of 120 bhp and 400 rpm, the mechanical

Fig. 15.31. Nobel four-stroke, three-cylinder, reversible diesel for Russian submarine *Minoga*, 1907.
Author photo from Diesel Collection, Deutsches Museum

efficiency reached 75 percent. The bsfc at this rating was 208 g/bhp-hr. At 54 bhp and 250 rpm the consumption increased to 225 g, and mechanical efficiency dropped to 62 percent.[58]

Engines in older Russian subs, when surfaced, drove generators to power electric motors for turning propellers. The reversible *Minoga* engines turned the propeller directly through the generator with its clutch on each end, and the motor.

A concern arose during the design of the *Minoga* engines that Carlsund's sliding cam method of reversing might require too much manual effort when sliding a shaft with eight discs per cylinder on a four-stroke engine. (There were two each for the inlet, exhaust, fuel, and starting valves.) This had not been a problem on the two-stroke, experimental model and likewise proved to be an unfounded worry on larger four-strokes coming later.

Hans Nordstrom, another of Nobel's transplanted Swedish engineers, invented an interim reversing system. His patented method still used a non-sliding camshaft with a forward and reverse cam for each valve to require the eight per cylinder.[59] However, its novelty was in the linkage, which set one valve rocker lever cam roller onto its cam as a second, offset one for opposite running was lifted away (fig. 15.32). Two advantages accrued: less effort to switch cams and less consumption of starting air.

The cam change was effected by a shifting hand lever so that all linkage movement occurred in the transverse direction of the camshaft. More pieces of linkage per cylinder added bearings and a possible elasticity due to an extra link in the valve train that might be viewed as drawbacks, but Nobel used Nordstrom's system on all four-stroke engines for several years.

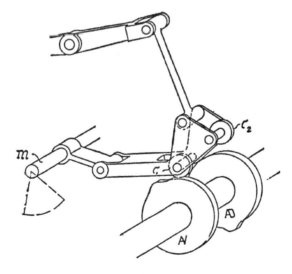

Fig. 15.32. Nordstrom reversing linkage as first used on the Nobel diesel for the *Minoga*. Supino, *Land and Marine Diesels, 1918*

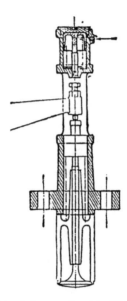

Fig. 15.33. Air piston method to lift inlet and exhaust
valve rocker levers from camshaft during engine reversing.
Supino, *Land and Marine Diesels, 1918*

Starting air consumption was reduced by a sequential rotation of a hand-wheel during which time air initially went to all three power cylinders. As the engine began to turn, the wheel was rotated a few more degrees to cut off air to one cylinder while at the same time to start injecting fuel in it. A further turn of the wheel did the same for the second cylinder, and finally the wheel went to its end point, which cut off air to the last cylinder. At that time fuel was admitted to all three cylinders.

The three-cylinder *Minoga* engines were the genesis of a new line of Nobel marine and stationary diesels known as the D Series. They eventually spanned sizes from a two-cylinder, 275 × 300 mm and 70 bhp at 350 rpm to a four-cylinder of 450 × 510 mm with an output of 600 bhp at 320 rpm.[60] The engines had many cast steel parts, including heads, liners, pistons, and enclosed crankcases. Stationary engines were rated slightly lower in output and speed than the marine. The larger Ds went through a development phase recalled by both Hans Flasche and Cyril Yeates, an English engineer who worked in St. Petersburg from 1909 to 1914. Yeates tells his story more from a down-to-earth service engineer's perspective.[61]

Four four-cylinder D engines with Del Proposto drives went in the *Skval*, the first of eight gunboats ordered in 1906. Admiralty confidence level was not yet high enough to specify them with reversing capabilities. These siblings of the *Minoga* diesels had a bore and stroke of 330 × 380 mm and produced 250 bhp at 350 rpm. Sea trials in the Gulf of Finland in 1908 proved so successful that the remaining twenty-eight engines were shipped directly to Siberia for installation there. The order was too big for Nobel to fill in the short time specified, so sublicensee Kolomna was given drawings and asked to build sixteen engines for four of the boats.[62]

The engine room layout is not known, but each engine drove through its own Del Proposto gear. If two propeller shafts are assumed, then several placements were possible.

Following the above installations came three D engines for the *Akula*, a new submarine of 360 tons with a surface speed of 12 kts. The engines were the same as those in the gunboats but had a higher rating of 300 bhp at 380 rpm. Full-load bsfc was 190 g/bhp-hr, and the specific weight had been reduced to 30.5 kg/bhp.[63] The weight was higher than the navy wanted, but at least it was convinced that diesels would power all new subs.

Carlsund's reversing system with sliding cams was first used on the engines in the *Akula*. Each of the eight cams per cylinder were machined as separate discs and joined as a unit slidably keyed on the camshaft. With the engine stopped, a hand lever lifted all rocker rollers off their cams. A second hand lever then axially shifted the cams so that the rollers when lowered again contacted an adjacent but rotationally offset cam timed to open the valves for opposite direction running. The engine was restarted with air in the usual fashion. It took about eighteen seconds to go from full load in one direction to full load in the other. As Carlsund informed the Allgemeine about using the system on his two-stroke test diesel built in 1905, it may be assumed that the sliding cam idea was made known to the Allgemeine's other licensees. The chronology was right for Augsburg to have adapted it for their own sub engines in 1908. Not determined is whether Carlsund received patents on his method.

Kolomna often modified Nobel designs with their concurrence. One example was an pistons over the valve ends of the inlet and exhaust levers to push them down so as to lift the opposite ends off the cams and temporarily open the valves (fig. 15.33).

A pair of new, larger D Series were installed in each of the sister ships *Kars* and *Ardagan*, 623-ton gunboats built in St. Petersburg and commissioned in 1910 for service on the Caspian Sea (fig. 15.34). These six-cylinder diesels, the first made by Nobel, had a bore and stroke of 375 × 430 mm and an output of 500 bhp at 310 rpm. Their performance during the inland voyage to the Caspian was reported on by Yeates who may well have been aboard one of the gunboats.[64]

The Navy finally acquiesced about using Nordström's reversing system but with one important proviso. The contract specified that Nobel would pay all costs to change over to the Del Proposto if Nordström's failed to perform.[65] Heavy workouts in the fast flowing Neva and Volga Rivers, and the canal locks connecting them, showed up no engine reversing difficulties.

Most of the encountered problems involved uncooled pistons. On shutdowns after full-load running, Yeates said that the sometimes audible " muffled reports from inside the cylinders" meant a piston crown had cracked. A "field fix" staving off disaster was to bore out an angle-sided recess in the top of the piston and bolt on a steel plate that acted as a heat shield for the piston. This need for such cooling, first uncovered on the *Ardagan* and *Kars*, started a long improvement program in St. Petersburg.

The piston crisis came to a head during 1910–1911 when two four-cylinder 450 × 510 mm diesels were under development for the Russian cruiser *Runda*. They were to run at 320 rpm and be directly coupled to 400 kW DC generators acting as flywheels. Specific weight of the engines was 39.5 kg/bhp.[66]

While under extended full-load tests before their release, a crankcase explosion blew off a rigidly attached inspection door on one engine (fig. 15.34). Conditions causing it began when lube oil mist in the crankcase encountered the underside of hot piston crowns. A brave observer peering up inside the crankcase on the still-running other engine saw first a bright glow originate under a piston followed by flames shooting down from it.

One unsuccessful cooling design was to feed oil into a piston cavity and then drain it via rigidly mounted, telescoping tubes like the slide trombone method tried and then abandoned by Sulzer. Oil consumption was too high, so water was substituted. Rapidly failing packing glands at the sliding joints resulted from the piston tube's lateral movement and from an often-audible water hammer in the pipes. A swinging link, grasshopper leg joint cured the sealing problem. Each flexing joint was kept watertight by a spring-loaded stack of fiber and bronze washers, which were self-adjusting to compensate for wear. Pressure relief plates in the crankcase doors acted as a further precaution against engine damage in the event there was a crankcase explosion.

A weakness in the front-mounted, two-stage compressor showed up on these large engines. One customer complained that fifty-three of fifty-seven involuntary shutdowns among his three engines during the first month of service were compressor related. Heading the failure list was a rapid pounding out of seats due to heavy valves. The seat failures ended when Nobel designed and delivered, in three months, a three-stage, tandem compressor with lighter valves.[67]

The red glowing air discharge pipe from the compressor's last stage, soon followed by an explosion, provided further pyrotechnic entertainment—a once common event that resulted from too much oil passing the compressor piston rings. In this case the rings could not handle the heavy crankcase oil mist. By partitioning off the compressor section of the crankcase and adding drip lubricators to oil the pistons, the combustible mixture potential in the injection air system was eliminated.

Torsional vibrations reared their ugly head on these fast-running, larger engines. Yeates relates such an encounter on a six-cylinder, 450 bhp D engine driving a 300 rpm generator in a munitions factory. It did well under test in the Nobel shop, but minus the generator. Coupled in service, however, a mid-engine mounted governor self-destructed in only twenty minutes. Not once, but twice! Yeates was sent out with "strict instructions not to lose a third." After starting the engine, he immediately saw the governor vibrate severely. A fast shutdown saved it. The vertical drive shaft up to the governor and camshaft was the suspected culprit, yet to check it he needed to remove several crankshaft balance weights. Just after doing this, another engine in the factory had to be unexpectedly stopped during a crucial shop operation.

Fig. 15.34. The result of a crankcase explosion in a Nobel D Series engine, c. 1910. *Author photo from Diesel collection, Deutsches Museum*

Fig. 15.35. Russian gunboat *Kars* for Caspian Sea duty, powered by two 500 bhp Nobel reversible diesels, c. 1910. *MAN Archives*

Yeates was given one hour to get his engine running, which meant no time to reinstall the weights. Nevertheless, the engine "ran like the proverbial sewing machine, with not a murmur out of the governor."[68] The factory accepted the engine as it was, and the weights were "conveniently lost."

Another story, slightly after the chapter's time period, is best related here. Eight 480 bhp, six-cylinder Ds went in four new *Gangut* class battleships during 1913 and 1914 to provide electrical power.[69] Engine room vibrations appeared with the first installation when the paired units ran at a preset 320 rpm. The condition was eased by governing one engine at 318 rpm and the other at 323 rpm. A unique failure occurred later when all three guns in a turret above the engine room fired together while one engine was under full load. An unusual pounding immediately heard coming from it was the result of one cylinder liner having cracked from top to bottom and another cracking circumferentially at the top of the upper piston ring. The navy took no action against Nobel, but Yeates wrote that coincidently these 375 × 430 mm diesels were among the last four-strokes made by the company.[70]

Nobel Lightweight Sub Engines 1910–1912

As soon as the *Minoga* and the *Akula* had proven the safety and practicality of diesels in subs, the Russian Navy began pushing Nobel to design an engine having adequate power but with a weight equaling the gasoline engines in their existing boats. Emanuel finally accepted the challenge, and in two years a new Series F was operational. These engines, among the most sophisticated yet built by any diesel company, were developed by another team of engineers working independently of the concurrent D Series program.

Ludwig Alfred Nobel, Emanuel's younger brother, provided an incentive for a lightweight diesel in 1909 because he wanted a high performance engine to drive the *Intermezzo*, his fast new aluminum yacht.[71] Reporting to Ludwig on the project were Swedish engineers Oscar Derans, Axel Lagerstén, and Mossberg.[72] His charge to them was to design the lightest weight power plant in the smallest possible package. What resulted was the world's first V-8 diesel producing 200 bhp at 600 rpm and weighing but 10 kg/bhp[73] (fig. 15.36)! The 200 × 220 bore and stroke engine was also reversible. Rudolf Diesel may have seen the engine when he visited St. Petersburg in 1910 at the invitation of Emanuel Nobel.[74]

As in many 90-degree vee-engines, the connecting rods directly opposite each other used the same crank journal. A two-stage compressor also had the vee configuration with the low-pressure cylinder in one bank and the high in the other. Again, their rods used the same journal.

Both the power and compressor cylinders and heads were of individual, monobloc construction. Shrouding them was a water jacket of sheet copper. Each cylinder assembly was bolted to an aluminum alloy crankcase. A single camshaft running in ball bearings sat at the base of the vee above the crank.

Fig. 15.36. Nobel 1910 200 bhp V-8 diesel engine for the yacht *Intermezzo.*
Author photo from Diesel Collection, Deutsches Museum

Its cams had beveled edges to aid in the sliding shaft reversing process. The crankshaft imported from Sweden was made from a high tensile strength carbon steel (85 kg/mm² or 120,000 psi with an elongation of 12 percent). Other parts were of chrome and nickel steels.[75]

Valuable lessons were learned during sea trials. A too-limber engine bed allowed the crankshaft to flex and cause bearing problems. High heat loads also brought early exhaust valve failures. One dramatic breakdown was the departure of a cylinder assembly from the crankcase under full power.

Reliability of this experimental V-8 diesel improved enough to convince Emanuel Nobel that the weight and power specifications demanded by the navy could be met. He in turn sold the possibility to the Admiralty, and in 1911 the factory received an order to repower the fleet of fourteen single-engine gasoline powered subs. The *Ssom* class, Holland types built in the US starting in 1904, received the first six during 1911 and 1912. These six-cylinder, 225 × 300 mm diesels produced 160 bhp at 440 rpm. Full-load bsfc was 215 g/bhp-hr, and the specific weight was only 18.1 kg/bhp.[76]

Their in-line design adopted the V-8's monobloc cylinder/head and copper water jacket, but now two cylinders were paired (fig. 15.37). The overhead camshaft was positioned along the centerline of the engine so that the cam acted directly on the intake and exhaust valves. Rocker levers and a pushrod lifted a near-horizontal nozzle valve so that fuel sprayed in a fan shape across the chamber (fig. 15.38).

To prevent a recurrence of burned exhaust valves as in the yacht engine, cylinder ports were added that the piston uncovered as it neared bottom dead

Fig. 15.37. Nobel lightweight, six-cylinder, sub engine of 1910. 160 bhp at
440 rpm. *Author photo from Diesel Collection, Deutsches Museum*

center. This bypassed enough of the exhaust gases through the water-cooled
ports to reduce heat loads on the valves.

Yeates tells of his emergency visit to the dockyards after the third of
these engines had been installed.[77] A report that they were coming apart
was quickly verified as cracks extended around the flanges at the base of the
cylinders and on top of the cast iron crankcase. Several piston crowns also
had holes punched in them. Yeates learned the source of the problem when

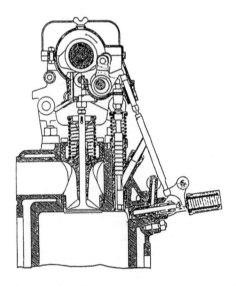

Fig. 15.38. Section view of six-cylinder sub engine in figure 15.35.
Yeates, "Working for Ludwig Nobel," 1965

he asked the sub crew to start the engine. Following a customary procedure used for the old spark ignition engines, one of the men removed a cylinder plug where the pressure reading for indicator cards was taken and squirted in a generous "priming charge" of gasoline before admitting the starting air. This practice was discontinued!

Seven four-cylinder, 120 bhp at 450 rpm diesels went in the smaller *Makrel* class subs. The principal change in the 250 × 300 mm cylinder design was casting the water jacket integral with the cylinder rather than using the copper sheathing.

The last sub repowered was the *Delphin*. Its six-cylinder engine had the same bore and stroke as those for the *Makrel*, but the power went to 220 bhp at 500 rpm. The specific weight was only 14.5 kg/bhp, and the bsfc remained the same 215 g/bhp-hr as with the *Ssom* engines.[78]

To complete the list of circa 1910–1911 four-stroke sub engines, the last were three 35-ton boats launched in 1912 intended to be carried on the deck of a battleship. Only one of the five-man crew, the coxswain, could stand up in the little boats. Each sub had a four-cylinder, 170 × 220 mm Nobel diesel of 50 bhp at 500 rpm. The two pairs of cylinders were similar to those of the *Makrel* engines, but the enclosed frame and base castings were of bronze.[79]

The Caspian Tankers

Not all Russian marine diesels went to the navy. The Nobel Brothers oil company also had a great need for river and Caspian Sea tankships, which after 1905 were powered almost exclusively by diesels coming from Ludwig Nobel and Kolomna. Both companies also built engines for tugs and river passenger boats driving either through propellers or side-mounted paddle wheels so practical for shallow draft operation. Reverse gears continued to include the double gear sets for paddle wheelers and the pneumatic clutches described in chapter 10.

Nobel-designed engines made diesel power feasible for Caspian Sea tankers, and between 1908 and 1911, several entered such service, transporting Nobel oil from Baku to the Volga River delta. At the time they were the largest diesel ships in the world.

The first was the *Djelo*, a 355 × 46 × 14 foot tanker with a displacement of 4,000 dwt. (Deadweight tonnage, or actual carrying capacity in long tons of 2,240 lbs, is used for freighters and tankers.) Her two four-cylinder Kolomna-made engines of 500 bhp each at 150 rpm had a bore and stroke of 450 × 680 mm and were based on Nobel's stationary B Series. Pneumatic clutch reverse gears were used.[80] The engines came from Kolomna's yard in Goulutwin where the ship was built. It entered service in 1908. Two slightly larger tankers of 4,800 dwt, the *Emanuel Nobel* and a sistership *Karl Hagelin*, followed a year later. Their dimensions were 380 × 46 × 16.5 feet. The four-cylinder Kolomna diesels were also bigger with an output of 600 bhp at 150 rpm and a 490 × 740 mm bore and stroke.

D Series reversible marine engines began with the tanker *Robert Nobel* in 1910 (fig. 15.39). Her cargo capacity was only 1,740 dwt with hull dimensions of 260 × 34 × 14 feet.

Power came from two four-cylinder, Nobel-built engines of 350 bhp at 190 rpm.[81] Bore and stroke were 450 × 510 mm (fig. 15.40).

By the end of 1910, Nobel and Kolomna diesels were in over thirty vessels in many kinds of service. These included paddle wheel tugs and passenger boats, which plied the major routes of the Russian inland waterways. Their proliferation was also furthered by internal naval needs and abundant, cheap fuels. No other country enjoyed—and took advantage of—such conditions so early.

The Vickers Submarine Diesels 1907–1910

Vickers, Sons & Maxim Ltd., who owned the Holland submarine rights for Britain, did not begin work on diesels to power them until 1905. All their first subs, including the thirteen A Series of 1902 to 1905, the eleven B Series of 1904 to 1906, and the thirty-two Cs built from 1906 to 1910 had gasoline engines.[82] These were sixteen-cylinder, horizontal Wolseley engines of 150 to 200 bhp.

Not until the D Series was authorized in 1906 did a British diesel find its way into a sub, and progress remained slow. The first of the larger and faster Ds launched in 1907 (160 ft and 14 kts on the surface) incorporated an entirely new hull design of the "submersible" type to make it more crew-friendly when surfaced. Troubles with *D-1*'s pair of new Vickers diesels, along with other "shakedown" problems, held up delivery of the *D-2* until 1910. The D Series used blast air; the Vickers solid injection systems described in chapter 13 came on E Series boat engines.

Details of the first nonreversible 550 to 600 bhp diesels have been gleaned mostly from those built to the Vickers design in the United States for the Electric Boat Co. by Fore River Shipbuilding Co. in the circa 1910 period[83] (chapter 17). The six-cylinder engines (US model: 12.75 × 13.5 inches and 410 bhp at 400 rpm) had numerous distinguishing features (fig. 15.41).

Fig. 15.39. Caspian Sea tanker *Robert Nobel* was commissioned in 1910 and powered by two 350 bhp at 190 rpm Nobel diesels. *MAN Archives*

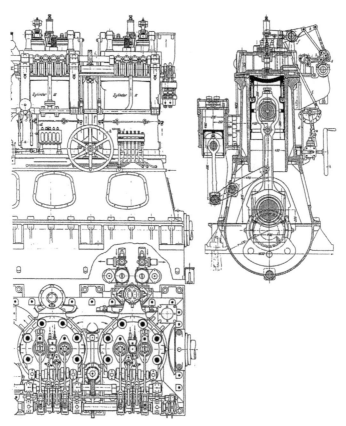

Fig. 15.40. Details of 450 × 510 mm, four-cylinder diesels in *Robert Nobel*. Kaemmerer, *Zeit., VDI, 1912*

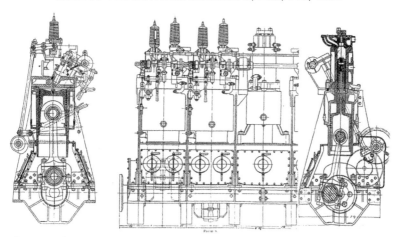

Fig. 15.41. Details of c. 1909–1910 Vickers 550-600 bhp submarine diesels. Drawing is of the Vickers engine made in the US. Magdeburger, *"Diesel Engines in Submarines,"* 1925

A two-piece, cast steel bedplate supported the crankshaft on water-cooled, babbitt shell main bearings. Vanadium steel flanges were riveted to the top and bottom of a plate steel crankcase enclosure. Individual cylinders had a cast iron water jacket shrink-fitted to the cylinder casting.

A pure nickel button, referred to as a "spraying pin," was inserted in the center of an uncooled piston to help prevent crown cracking due to excessive heat.[84]

The heads were of two-piece construction. A simple, flat steel casting with openings for the valves sat on the cylinder. On top of it was a casting containing inlet and exhaust valves and fuel nozzle. No air valve was needed as the engines were started from the sub's electric driving motors. Each inlet and exhaust valve cage and its header formed a single casting with the exhaust cage and its valve being water cooled.

A two-stage injection air compressor lay between the middle cylinders. Spur gears also at the mid-engine location drove a low-mounted camshaft that actuated pushrods and rocker levers to open the inlet and exhaust valves. Water, lube, and main fuel supply pump drives came off this camshaft.

On the opposite side of the engine was a second, high-mount camshaft driven by an angled intermediate shaft from worm gears on the main camshaft. It opened the fuel valves and operated individual fuel pumps on the head. A hand lever axially shifted the upper shaft to move a stepped cam that controlled the opening of the pump suction valve and thus regulated fuel flow.

Pressurized lube oil went to main, rod, and piston pin bearings, but each cylinder group was individually supplied with a hand-regulated valve to control its oil flow.

Reported problems, at least on American-built engines, included the following: Air compressor valves broke and dropped into their cylinders, which in turn caused their support lugs to fail, rupture air lines, and "cause havoc in the engine room." The water-cooled exhaust valve cage/header castings cracked. Joints in water lines to the main bearings leaked, which put salt water into the lube oil. Finally, if stud nuts located between cylinders and holding both down should rest on unequal height flanges their studs would crack. Adding to engine grief was a very cramped space in which crews were forced to work. The less than rugged engines, coupled with an inaccessibility to service and repair them, led to their eventual replacement.

It may be assumed that the first English-built Vickers diesels encountered similar troubles with the ones in the sub *D-1*, which caused the three-year delay between it and the *D-2*.[85] An English source referred to problems with air compressors, which encouraged development of the common rail fuel system. It also added that during 1908 a M.A.N. diesel was installed in the *A-13*, which lends further credence to "teething" problems on the *D-1*.[86]

The next generation of Vickers submarine diesels is examined in chapter 17.

Fried. Krupp Germaniawerft AG

When Krupp's shipyards in Kiel began building submarines in 1902, thoughts of diesel power for them quickly followed. An example of this interest was seen by the Russian order and its resultant dead-ended joint venture with Augsburg. Germaniawerft started their own submarine diesel program about the same time as M.A.N., but it led to unusual designs, discouraging battles with weight reduction, and long development periods.

The first experimental sub engine built at the Germaniawerft was not put on test until 1907. This reversible, four-cylinder, four-stroke diesel had a square bore and stroke of 330 mm and produced 300 bhp at 450 rpm. Individual cylinders bolted to the crankcase enclosure. Design features included a crankshaft with 90-degree throws. (When looking at the front of the engine they went "east, west, north, south.") A two-stage compressor was driven off its own crank throw ("south") at the front end of the engine. Two high-mount camshafts opened the valves. The one actuating the inlet, exhaust, and fuel valves was axially shifted for reversing operations. The other, non-shiftable camshaft on the opposite side of the engine opened only the air starting valves. It ran in a two-stroke mode so that there was no dead spot regardless of engine direction or crankshaft position. The engine's biggest nemesis was weight, which caused its abandonment, as in final form it only got down to 9,900 kg or 33 kg/hp. Otherwise, it had a clean exhaust and a fuel consumption of only 180 g/bhp-hr at full load.[87] This good performance retained Krupp's confidence in the four-stroke cycle.

Derivatives of the sub diesel prototype found their way into surface marine applications. The first was a six-cylinder, 120 bhp at 400 rpm diesel installed in the tug *Rapido* in September 1909. Its cylinders were in three paired castings bolted on the crankcase enclosure (fig. 15.42). The engine saw years of service.

In another project around that same time, Krupp's Essen engineers designed and built a four-stroke, single-acting, 35 bhp engine with a 325 mm bore. After being displayed at a Munich exhibition, it went to Kiel where Germaniawerft engineers converted it into a double-acting four-stroke[88] (fig. 15.43). The intended goal was to obtain maximum power in the lightest weight and least bulk, but it was destined to fail. Eccentrics actuated vertical inlet, exhaust, and fuel valves while horizontal starting valves had their own camshaft. The water-cooled piston was tied to the crosshead by *two* rods with a water supply passage in one and a discharge passage in the other. This allowed the fuel nozzle for each combustion chamber to be axially placed, but with the two cooled rods passing through the lower cylinder half the combustion there must have suffered. The two rods also doubled stuffing box troubles. No performance data was ever given.

Remarks about piston cooling by Conrad Regenbogen, head of engineering at Germaniawerft, shed light on the thinking of the times. He said that four-stroke engines with bores above 500 mm needed cooled pistons and

Fig. 15.42. Krupp Germaniawerft diesel in tug *Rapido* delivered to Chile,
September 1909. Four-stroke, six-cylinder 120 bhp at 400 rpm.
Author photo from Diesel papers, Deutsches Museum

that this diameter dropped as speeds went up. Two-stroke engines needed
piston cooling almost regardless of power. Even though problems arose when
seals failed, he opted for water cooling because water carried heat away faster
than oil.[89]

Growing, and divided, opinions about two-stroke advantages over the
four-stroke for sub engines led to the building of a single-cylinder, two-stroke

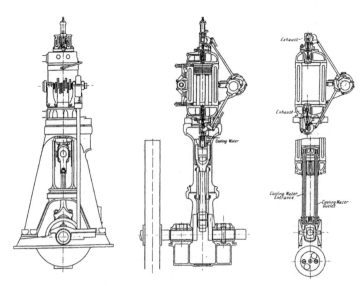

Fig. 15.43. Germaniawerft experimental double-acting,
four-stroke diesel, 325 mm bore, c. 1907. *The Engineer, 1912*

test engine with a stepped piston. It had a bore and stroke of 240 × 260 mm and an output of 50 bhp at 500 rpm. Scavenge air pumped by the stepped piston went directly into the combustion chamber through inlet valves in the head. Cylinder ports for the exhaust gave uniflow scavenging. No performance data exists, but stepped pistons were not used again, and new four-stroke engines were designed for surface craft and later for special submarine applications.

Since the Germaniawerft supplied all submarines for the German Navy, Krupp naturally wanted its own diesels in them. Thus began a new and concerted development program in 1909 for a prototype engine comparable to those made by M.A.N.

The output of 850 bhp at 450 rpm for the new GW two-stroke design was the same as for the other sub engines by Augsburg and Nürnberg. Bore and stroke were 350 × 400 mm. With a scavenge pump on each end and two, two-stage air compressors in the middle, all crank driven, the rather long engine had ten crank throws for the six cylinders. Nevertheless, they were narrow and reflect the shipbuilder's concern for adequate access to service them (fig. 15.44). Two fuel pumps serving three cylinders each were operated off a mid-engine vertical shaft driving an overhead camshaft. An exhaust header serving three cylinders exited off both ends of the engine.

Three inlet valves and cylinder exhaust ports provided uniflow scavenging. All the valves were opened by a single cam acting through a three-fingered crosshead (fig. 15.45). A long link pin between the rocker lever and the bottom of the recessed crosshead guide plunger minimized side loading on the plunger. An inclined fuel nozzle sprayed through a horizontal orifice plate with a single hole that directed the spray vertically into the center of the cylinder.

Most frustrating of the numerous development obstacles was the devising of a durable piston cooling method, which kept water out of the lube oil. Krupp had run into the same problem all other engine builders faced. This delayed Navy acceptance for two and a half years until June 1911 when the engine finally passed its six-day tests without troubles. In 1914 the first eight GW two-stroke diesels were installed in the boats *U-23* to *U-26*.[90]

Heavier and slower speed Germaniawerft two-stroke diesels for merchant ships also appeared in the immediate years before the First World War. Additionally, negotiations with the German Navy began in 1909 for a six-cylinder battleship engine of 2,000 bhp per cylinder that was followed by an order the next year. This amazing development by Krupp, and by M.A.N. who was independently given the same task, and the cargo ship engines are in the following chapters.

Fiat Grandi Motori

When Rudolf's Italian patents began expiring in 1907, Giovanni Agnelli (1866–1945) saw an opportunity to add marine diesels to his growing manufacturing empire. The business genius behind the Fiat automobile company

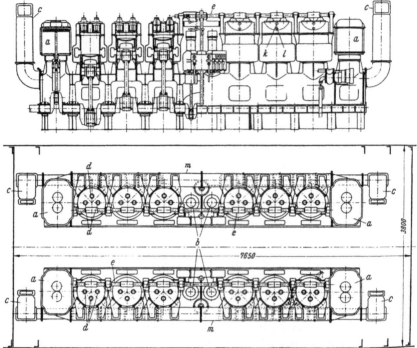

Fig. 15.44. The prototype Krupp Germaniawerft two-stroke, 850 bhp sub diesels designed in 1909 and tested over a several year period. Note the narrowness of the engines for good access between them and the hull. (a) Scavenge pumps, (b) compressors, (c) exhaust header, (d) scavenge air valves, (e) camshaft, (k) fuel nozzle, (l) starting valve, (m) scavenge air manifold. Sass, *Geschichte, 1962*

founded in 1899 had concluded that these power plants would fit in nicely with his ship-building activities. A Fiat gasoline engine had already powered the first Italian sub in 1903. Two years later Agnelli's newly purchased Fiat-San Giorgio shipyard at Spezia built the sub *Glauco*, which was driven by two 300 bhp, twelve-cylinder gasoline engines from his nearby Fiat-Muggiano factory. This company, also bought in 1905, made engines and smaller boats. A new Fiat Stabilimento Grandi Motori of Turin would build diesel engines.[91]

Prototype two-stroke submarine diesels went on test at Grandi Motori in 1909, and production engines were delivered the next year. Although preliminary work had been on four-stroke engines in 1907, a firm decision to go two-stroke came soon after. Giovanni Chiesa (1882–1941), manager and technical head of Grandi Motori for many years, would provide the leadership that secured eventual success of the sub and ensuing merchant ship diesels. Grandi Motori's sub engine design broke new ground with innovations continued by Fiat and also licensed to others.

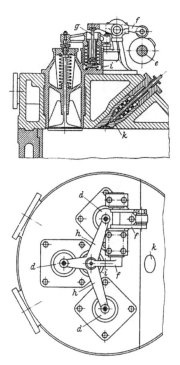

Fig. 15.45. Scavenge and fuel valves for Krupp 1909–1911 two-stroke sub engine program. Note the three-fingered crosshead (*h*) with its vertical guide supporting the long link pin (*g*) and the inclined fuel valve (*k*) leading to a vertical orifice into the cylinder. Sass, *Geschichte, 1962*

Grandi Motori's first production six-cylinder diesels, the model 2C.116, went in the sub *Medusa* in 1910. These reversible engines had a bore and stroke of 250 × 270 mm, an output of 300 bhp at 500 rpm, and a specific weight of only 23 kg/bhp.[92]

A clean-looking exterior hid unique internal features that challenged the foundryman (fig. 15.46). One piece castings for the bedplate and the crankcase enclosure were made of bronze because it was assumed that such thin walls and intricacies could not be made in steel. Individual monobloc head and linerless cylinders bolted to the enclosure frame. A two-stage, intercooled injection air compressor was driven by the crankshaft on the front of the engine.

Two intake valves in the head and exhaust ports in the cylinder provided uniflow scavenging. The lower, larger diameter of a stepped piston (fig. 15.47) acted as the scavenge pump. A two-way valve controlled the air intake via openings in the crankcase enclosure frame and the discharge of pressurized air into a receiver cast in the frame. One scavenge air valve and receiver served a connecting pair of cylinders. Their internal location is shown in the figure by frame bulges above the three air intakes. The pump valves were cylindrical

Fig. 15.46. The Fiat 2C.116 sub diesel of 1909:
two-stroke, 300 bhp at 500 rpm. *Fiat Archives*

and reciprocated vertically in a cast iron sleeve inserted in the bronze frame. Their motion came from crank throws on a longitudinal auxiliary shaft that held connecting rods. This shaft was driven off the lower end of a vertical, intermediate shaft.

The upper end of this vertical shaft drove a high-mount cam and eccentric shaft turning at engine speed to actuate the valves in the power cylinder. Each cylinder head contained five valves: the two inlet, a fuel valve, a starting

Fig. 15.47. Stepped-piston design used on first Fiat sub engines. *Fiat Archives*

valve, and a safety valve (which vented on its own). A cam and rocker lever opened the starting valve, while a single eccentric acted to open the two scavenge air valves and the fuel valve (fig. 15.48). This was possible because the opening of the air valve and the fuel valve occurred 180 degrees apart. Their opening durations were likewise equal.

A single fuel pump plunger whose discharge was regulated by the usual suction valve control supplied fuel to all cylinders. Manually adjusted needle valves in a fuel manifold balanced the metered charge to each nozzle. The pump and governor were driven off the vertical shaft.

The main bearings were water-cooled by passages in the bedplate. Oil cooling of the piston was via two small supply pipes attached to the outside of the rod and running from the rod bearing to the piston pin and then through lines from the pin to the piston crown from where the oil drained on its own back to the crankcase.

Starting the engine in either direction required the selection of one of two rotationally offset air starting valve cams for each cylinder. These axially sliding cams with an intermediate section having a diameter equal to the cams' base circle formed a single unit. A rocker lever cam follower ran on this cylindrical section whenever the starting valve was deactivated. Inclined ramps eased the follower onto the selected cam. Each cam unit was shifted by the movement of a rod running the length of the hollow high-mount shaft. A pin tying the cam unit to the rod slid in a slot in the shaft that turned the

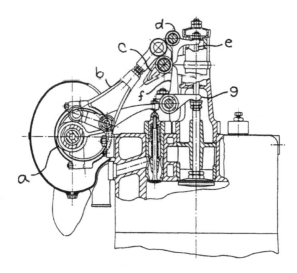

Fig. 15.48. Valve mechanism on Fiat 2C.116 sub diesel.
Starting cam (*a*); eccentric rod (*b*) rotates lever (*c*) to act against
lever (*d*) and lift fuel valve (*e*); rod (*b*) also turns bell crank (*f*)
whose nose (*g*) opens the scavenge valve. *Trans., Institution of
Engineers and Shipbuilders in Scotland, June 1928*

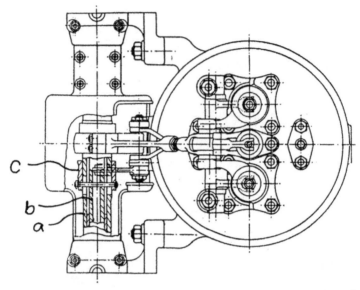

Fig. 15.49. Fiat 2C.116 reversing mechanism. Hollow camshaft (*a*) contains the sliding rod (*b*), which shifts the starting cam assembly (*c*). Note the forked arms on the eccentric rod moving the levers that open the scavenge air valves. *Trans., Institution of Engineers and Shipbuilders in Scotland, June 1928*

cams. Rotationally moving a hand lever would place the rod in either the "forward," "off," or "reverse" positions (fig. 15.49).

The same lever also changed the timing of the inlet air, fuel nozzle, and scavenge pump valves for forward or reverse running. (A second lever served as the throttle.) By moving the lever, a two-way air valve was also positioned to admit air on one or the other side of a piston-like flange (*A*) on the upper, hollow portion of the vertical shaft (*B*) (fig. 15.50). This tubular portion of the shaft was slidably keyed over the solid, lower part (*E*) geared to the crankshaft. On each end of the hollow, slidable shaft were helical gear sets (*C* and *D*) by which the lower drove the auxiliary shaft (*F*) and the upper the main camshaft (*G*). The vertical shaft displacement acting through the helical gears thus shifted the rotational timing of the two horizontal shafts enough to properly time all the valves for running in either direction.

One of the reported weaknesses in the early engines was an excessive oil consumption due to wear on the scavenge piston which allowed oil to pass into the cylinder. Another was that the many valve openings in the cylinder head caused cracked heads between the valve ports. Insufficient intercooling of the injection air was finally solved by changing to three-stage compressors. The high oil consumption was not eliminated until stepped pistons were abandoned. Fuel consumption started out at 270 g/bhp-hr and was gradually reduced in later design engines to as low as 200 grams.

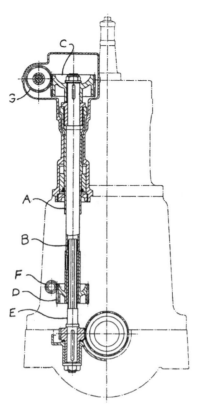

Fig.15.50. Fiat 2C.116. Tubular vertical shaft (*B*) driven by
crankshaft to axially shift camshaft (*G*). *Trans., Institution of
Engineers and Shipbuilders in Scotland, June 1928*

Power requirements for submarines rose quickly so that by 1910 Fiat had
already started design work on a larger model 2C.216 diesel. This engine and
others with much changed features are seen in chapter 17.

The Double-Acting Tanner Diesel 1907

Frederick H. Tanner of Bristol, England, was the first to build and test a
small two-stroke, double-acting marine diesel. He received a patent on a
reversible engine in 1906 and had one made by Workman, Clark & Co. the
next year. This 40 bhp demonstration diesel was tested during 1907–1908 by
the British Admiralty and by "the best recognized authority on the subject."[93]
Unfortunately for Tanner, Britain's naval and merchant marine thinking did
not include large, higher-horsepower diesels. Contemporary engine textbooks nevertheless acknowledged its uniqueness.[94]

The inwardly opposed, loop-scavenged design with a single crank throw
and two axially connected pistons produced power on each stroke (fig. 15.51).
Scavenge air and exhaust ports were at the outer cylinder ends. A detached

rotary blower supplied the air instead of an engine-driven reciprocating pump. Fuel nozzles in the cylinder walls injected horizontally when a single cam alternately lifted the nozzle valves. The central cylinder head serving both cylinders had an attached heat shield plate on each face that was exposed to the combustion chamber. The head itself held only a spring-loaded starting air valve for each cylinder.

Because the rod tying together the pistons always acted in tension, its smaller permissible diameter lessened gas sealing difficulties at the stuffing boxes where the rod passed through the cylinder head. The cleverly water-cooled rod consisted of an inner stud carrying the tensile forces that was inserted in a hollow tube. A circulating water supply above the upper piston acted as a reservoir to feed water down the rod to its lower end through a spiral groove cut in the rod. The inertia of the reciprocating water column within the rod acted to carry heat away.

Starting air entered the cylinders through the valves in the head. Their opening was timed by an eccentric on the end of the crankshaft alternately

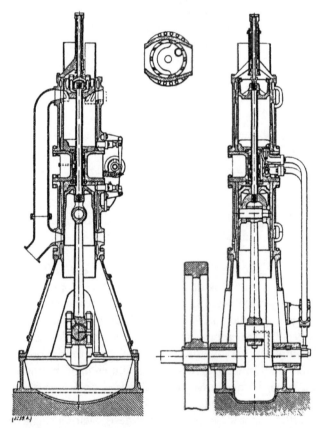

Fig. 15.51. Tanner 1907 double-acting two-stroke experimental diesel intended for marine use. *Engineering, 1911*

actuating two control valve plungers inserted in air supply lines going to the head. The use of two valves in the cylinder head essentially eliminated a dead point.

The engine was reversed by first shutting off the fuel and admitting air into the cylinders to slow it down and start it turning in the opposite direction. A rotational clearance in the camshaft drive, like that used on the first MAN reversible diesels, gave the small timing shift necessary for the fuel valves.

Tanner later had Workman, Clark build one cylinder of a single-acting, three-cylinder, two-stroke diesel easily modifiable to the above double-acting design. The one-cylinder with a bore and stroke of 19 × 30 inches developed 250–300 ihp at 150 rpm.[95] It is not known if the three-cylinder version was ever made.

Despite the positive successes painstakingly achieved by the diesel engine on and under the sea's surface, its detractors remained legion. The submarine itself, where diesel power was the only acceptable alternative, still faced the contempt of battleship admirals intoxicated by the massive firepower under their command. Admiral Sir Arthur Wilson, later First Sea Lord, had scornfully described the submarine in 1902 as "underhand, unfair, and damned unEnglish."[96] He would change his mind.

Notes

1. "The Weak Points of the Marine Oil Engine," *The Engineer* (July 21, 1911): 75–6. The editorial even spoke of concerns of only hearing a ship's whistle and not seeing its steam cloud.
2. Rudolf Diesel, *Theorie und Konstruktion eines rationellen Wärmemotors* (Berlin: Springer, 1893), 91–2; and "…Examination of Colonel Edward D. Meier, before Committee of Naval Engineers, on the Substitution of Diesel Motors for steam power, in Holland Torpedo boats," July 25, 1899. Busch-Sulzer papers, Wisconsin State Historical Library.
3. Bryan Donkin, *Gas, Oil, and Air Engines*, 5th ed. (London: Charles Griffin & Co., 1911), 540. Hans P. F. Petersen, "The First Diesel Engined Ships," *B&W* (1965), trans. by N. E. Rasmussen, 2, says these vessels were 38.5 m long × 5.05 m beam × 1.8 m draft and had a cargo capacity of 270 to 290 tons. They barely cleared locks in the Rhine-to-the-Marne and Belgian canals.
4. Pierre Bochet, "Contribution d'Adrien Bochet," *Exposition Diesel et la Conquête de L'Energie* (Paris: Conservatoire National des Arts et Métiers [CNAM], 1959), 25.
5. Original postcard in Diesel papers, Deutsches Museum.
6. Max Serruys, "Concernant des moteurs Diesel proprement dits, n.d. (MAN Archives) Professor Serruys authored a treatise on French I-C engine history: *Contribution Française à L'évolution technique des Moteurs Alternatifs à Combustion Interne* (Paris: 1948).
7. Vogel, "Das Moderne Unterseeboot," *Zeit.*, VDI (1911), 242–45; Eberhard Rössler, *The U-Boat: The Evolution and Technical History of German Submarines* (Annapolis, MD: Naval Institute Press, 1981), appendix; and *United States Submarine Data Book* (Groton, MA: Submarine Force Library and Museum, 1984), 1–4.
8. Vogel, "Das Moderne."
9. Clipping from a September 1904 Augsburg newspaper with the story and carrying a Paris dateline of September 25. Clipping and translation in Busch-Sulzer papers, Wisconsin State Historical Library.
10. William Hovgaard, *Modern History of Warships* (London: Conway, 1920; Naval Institute Press reprint, 1971), 307.
11. Rudolf Diesel, "The Diesel Oil-Engine," *Engineering* (March 22, 1912): 405.
12. Bochet, "Contribution," 26.

13. "Bar-le-Duc et le moteur Diesel," Barrois Vivant (1959), 18. (Bibliothèque Municipale, Bar-le-Duc). Confirming the move date is Edgar C. Smith, *A Short History of Naval & Marine Engineering* (Cambridge: Babcock & Wilcox, 1937), 329.

14. Bochet, "Contribution," 26.

15. Cover page of Augustin Normand 1909 brochure. MAN Archives.

16. June 10, 1911, Rotter to Harris. Busch-Sulzer papers, Wisconsin State Historical Library.

17. September 1910 correspondence between Delaunay-Belleville and Diesel, including Diesel sketches. MAN Archives.

18. Obituary, *Engineering* (1912), 227. Truck ideas: April–July 1911 correspondence Delaunay-Belleville and Diesel. MAN Archives.

19. Quoted in Kurt Schnauffer, "Die Motorenentwicklung in Werk Augsburg der M.A.N. 1898–1918" in MAN Archives (1956), 39.

20. Georg Strößner, *"Entstehung und Entwicklung des Unterseeboot Dieselmotors der M.A.N.* (Augsburg: MAN, 1941), reprint of his July 7 lecture at the Kiel Navy School. MAN Archives. This paper forms a basis for Schnauffer's material on Augsburg 1903–1918 sub engine developments.

21. Schnauffer, "Die Motorenentwicklung," 39–41.

22. The Körting oil engines were fairly successful but of limited power. Most subs with them saw active duty in the 1914–1918 war. The *U-1* is on display in the Deutsches Museum in Munich.

23. The 154-foot subs had a surface displacement of 354 tons. The *Circé* sank the German minelaying sub *UC-24* in the Adriatic bay of Cattaro (Boka Kotorska) in May 1917 and was herself sunk by an Austrian sub on September 20, 1918. *Jane's Fighting Ships of World War I* (New York: Military Press, 1919; reprint, 1990), 199, 320, and Schnauffer, "Die Motorenentwicklung," 45. A firsthand account of US Navy operations with a squadron of 110-foot wooden subchasers off Cattaro (Kotor) is in Ray Millholland, *The Splinter Fleet* (New York: Bobbs-Merrill, 1936), 241–60. He was chief machinist's mate on one of these rather frail, gasoline-engined vessels from the time it sailed from the US until its return a year and a half later.

24. Schnauffer, "Die Motorenentwicklung," 47.

25. Rössler, *The U-Boat*. The *U-19* and *U-22* survived the war. *Jane's*, 329.

26. Allusions are made to this in Schnauffer, "Die Dieselmotorenentwicklung," 14–15. Pages 14–21 of this authoritative work are used as the basis of the technical descriptions of marine diesels and related events at Nürnberg.

27. This is suggested by Friedrich Sass, *Geschichte des deutschen Verbrennungsmotorenbaues von 1860–1918* (Berlin: Springer, 1962), 539.

28. Schnauffer, "Nürnberg," 16.

29. Louis Shane, "Description of Nurnberg (*sic*) Two-Cycle 450-B.H.P Heavy Oil Engines," *Journal of the American Society of Naval Engineers* (1913): 460. This paper, pages 447–61, gives a complete description and complete test results of the third engine built by Nelseco in 1912. The design follows very closely that of Nürnberg's.

30. Schnauffer, "Nürnberg," 19.

31. Schnauffer, "Nürnberg," 20, and *Jane's*, 279.

32. Adrien Bochet, "Le moteur Diesel a bord des navires de haute mer," *Mémoires de la Société des Ingénieurs Civils de France*, July 1911, 18–21, describes the M.A.N. engine, which Brun designed at Nürnberg. The paper was kindly furnished by Alexandre Herlea of the CNAM Archives.

33. The ship arrived in New York on March 10 and left on the twenty-sixth. Ibid., 26. Edward .D. Meier learned that at first "the engines gave them a little trouble, but that after the first few days they settled down and worked nicely for the balance of the trip." Letter, Murphy to Meier, April 15, 1911. Busch-Sulzer papers, Wisconsin State Historical Library.

34. Comment by C. L. Straub in G. C. Davison, "Heavy-Oil Engines for Marine Propulsion," Trans., *Society of Naval Architects & Marine Engineers* 19 (1911): 249.

35. The Austrian and Danish subs were probably of the Whitehead design, all of which were built in Fiume (Rijeka), Croatia. The boats were 119 × 12 × 8 feet. *Jane's*, 228, 266.

36. Entries in Directors' Minute Book (RAIL846/37), January 1900 to August 1904, Leeds & Liverpool Canal Co.: August 20, 1903, ". . . recommendations . . . to purchase an oil engine for propelling a boat as an experiment . . . ;" and December 16, 1903, ". . . approved and adopted the following expenditure . . ., viz: Diesel Oil Engine for Carrying-trade Boat. . . ." Public Record Office, Kew, London. The engine had cylinder serial Nos. 70 and 71.

37. The installation and description of electrical machinery and controls, with circuitry, is in P. Ostertag, "Der Lastdampfer 'Venoge' auf dem Genfersee," *Schweizerische Bauzeitung* XLVIII, No. 13 (1905). Reprint. Sulzer Archives.

38. Giorgio Supino, *Land and Marine Diesel Engines* (London: Charles Griffin & Co., 1918), 184. Supino witnessed this rapid reversal being done "without even stopping for an instant on the intermediate notches. With one hand on the maneuvering wheel and the other on the control lever of the fuel injection pump, full speed between 400 and 450 revolutions per minute is attained more rapidly than with a steam engine."

39. Dugald Clerk and G. A. Burls, *The Gas, Petrol, and Oil Engine*, vol. 2 (London: Longmans, Green & Co., 1913), 790. Sulzer also built a model 4SNo.4 with a 215 × 290 mm bore and stroke without the diagonal bracing. One was installed in the Lake of Zürich boat *Taube*, but it was replaced in 1912 by an improved engine.

40. Paul Meyer described the reversing system on a similar Sulzer design engine exhibited by the Société des Forges et Chantiers de la Mediterranée, Paris, at the 1913 Ghent World Exhibition. Sulzer, 1913 reprint from *Zeitschrift*, VDI, vol. 57, no. 37. (Sulzer Archives).

41. "Marine Reverse. Method of reversing marine engines, as now used by Sulzers." Rotter to Harris seven-page report, without the drawings, of May 3, 1911. Busch-Sulzer papers, Wisconsin State Historical Library.

42. Fratelli Sulzer, Milano, to Gebrüder Sulzer, Winterthur. Sulzer Archives.

43. Rudolf Diesel, "Present Status of the Diesel Engine," *ASME Journal* (1913): 939.

44. Hesselman's US patents include: Nos. 906,022, filed May 18, 1908, and issued December 8, 1908, and 997,253, filed April 8, 1908, and issued July 4, 1911, for fuel controls. British Patent 10,836 of 1908 is an equivalent. US 948,730, filed April 24, 1909, and issued February 8, 1910, and British Patent 16,301 of 1908 are similar. These cover the reversing "motor."

45. Operation of the reversing method with its attendant fuel controls was gleaned from several sources, none of which give a complete, or in some cases clear, explanation. In addition to Supino, *Land and Marine*, 180–82, and Wells & Wallis-Tayler, *The Diesel or Slow-Combustion Engine* (London: Crosby Lockwood, 1914), 215–17, there is *The Engineer*, October 24, 1913, 436–37, and a 1914 AB Diesels Motorer operator's manual, 11–13.

46. *The Engineer*, 434, and Jonas Hesselman, *Teknik och Tanke*, (Stockholm: Sohlmans, 1948), 70, for cylinder sizes.

47. Alfred Büchi, "Modern Continental Oil Engine Practice," *Cassier's Magazine*, vol. 43, no. 3, (March 1913), 188–89. J. Rendell Wilson, "Veteran Diesel-Driven Ships," *Motor Boating*, November 1911, 38–9, also gives the hull dimensions of 104 × 23 × 11 feet, and a dead weight capacity of 350 tons.

48. Wilson, "Veteran," 39. This is a very detailed article on diesel-powered vessels at the time.

49. The *Fram* languished for many years while attempts to renovate her failed. Finally, in the early 1930s she was beautifully restored and in May 1935 moved into her own museum building across the harbor from downtown Oslo. The engine had been given to a school, and when later reclaimed, many of its parts were missing. The author spent a pleasant time in the normally closed engine room thanks to Kommandør Rare Berg, the on-duty museum director.

50. R. Diesel, "Present Status," 936.

51. Hesselman, *Teknik*, 264.

52. The *Toiler* was renamed the *Mapleheath* in 1920. In 1959 she was made into a lighter, and in 1963 reportedly served as a barge owned by the McAllister-Pyke Salvage Co. of Montreal. Herman G. Runge Collection, Milwaukee Public Library.

53. Prosper L'Orange, *Ein Beitrag zur Entwicklung der Kompressorlosen Dieselmotoren* (Berlin: Schmidt, 1934), 32, gives the date as 1909, the first year he began at Benz, but Sass, *Geschichte*, 609, states it was 1910. He also infers that it was a Benz diesel in the *Fram*.

54. Büchi, "Modern," 147; "Diesel Engined Tugs," *The Gas Engine*, March 1914, 147.

55. Testimony of G. F. Tweedy, *Royal Commission on Fuel and Engines*, pt. 3 (British Admiralty, 1913), 14. The 400 bhp at 180 rpm engine had a 360 × 530 mm bore and stroke with an overall mechanical efficiency of 72.5 percent. Bsfc of the Swan, Hunter engine was 0.456 lbs/bhp-hr. Tweedy said Polar had achieved 0.44 lbs/bhp-hr.

56. In his address read at the opening of the exhibit "50 Years of the Diesel Engine" at the Deutsches Museum, October 25, 1947, Eugen Diesel singled out five pioneer diesel engineers. Second after Imanuel Lauster came Anton Carlsund. The others in order were Edward D. Meier, Adrien Bochet, and Jonas Hesselman. Manuscript in MAN Archives.

57. Hans Flasche, "Der Dieselmaschinenbau in Russland von Anbeginn 1899 an bis etwa

Jahresschluss 1918" (1947), 18, MAN Archives; also Robert W. Tolf, *The Russian Rockefellers* (Stanford: Hoover Institute Press, 1976), 174.

58. Flasche, "Der Dieselmaschinenbau," 17.

59. Nordstrom received German Patent 220,872 of June 8, 1907, for a "Starting arrangement with multi-cylinder I-C engines." Flasche, "Der Dieselmaschinenbau." 16, 57. Three Swedish patents also issued to him for reversing means: No. 28,735 of Augsut 30, 1907, plus 30,678 and 32,286, both of June 5, 1909. The last is closest to the system adopted.

60. Flasche, "Der Dieselmaschinenbau," 76. He gives several pages of detailed size and performance data, which, except for the possibility of a "typo," is the most accurate and complete source available.

61. A. Cyril Yeates, "Working for Ludwig Nobel Ltd., St. Petersburg," *Gas & Oil Power*, September/October 1965, reprint by Nohab, 6. Yeates retired as managing director of Crossley-Premier Engines Ltd. in England.

62. Flasche, "Der Dieselmaschinenbau," 80, names four boats built in 1908–1910 with Nobel engines described on page 15 of his manuscript as the *Skval, Storm, Smertsch*, and *Groza*. The last three have a German spelling.

63. Ibid., 21. What auxiliary equipment was included in this weight, such as air tanks, pumps, and so forth, is not known.

64. Yeates, "Working for," 2, 3.

65. Flasche, "Der Dieselmaschinenbau," 21.

66. Ibid., 18. These engines were later removed from the *Runda* and used to provide power for the Sevastopol naval base.

67. Ibid., 19.

68. Yeates, "Working for," 5.

69. Flasche, "Der Dieselmaschinenbau," 24. The 23,370 metric ton battleships *Gangut, Poltava, Petropavlovsk*, and *Sevastopol* were laid down in June 1909, and the first was launched in 1911. They had four main turrets of three 12-inch guns. Two were located mid-ship. *Jane's*, 235.

70. Yeates, "Working for," 6.

71. Tolf, *The Russian Rockefellers*, 168, said that Ludwig was frustrated by his inability to get his brother Emanuel to accept his ideas and advice and was driven to his pursuits. The engine's very light weight could lead one to the assumption that Ludwig might not have wanted the V-8 engine to satisfy his own whims. However, the *Intermezzo's* unorthodox configuration did not lend itself for use in a sub. Thus, Tolf's belief in Ludwig's basic personal motivation for the new engine is valid.

72. Flasche, "Der Dieselmaschinenbau," 20. Another possibly involved was Professor M. Seiliger. A text later published by him, *Kompressorlose Dieselmotoren* (Berlin: Springer, 1919), 296 pages, gives his former titles (with no new ones) as professor of the technical university in St. Petersburg and head of the diesel engine department at Nobel. His "Foreword" written in Paris in December 1928 would indicate he was probably an émigré to France after the 1917 Revolution. A Burmeister & Wain in-house marine diesel history (H. F. Petersen, "It Happened Like This," *B&W Staff Magazine of 1964* (English trans.), 13, confirms Seiliger's Nobel position.

73. Flasche, "Der Dieselmaschinenbau," 22–3.

74. Ibid., 56–7, tells of the visit.

75. Ibid., 22, and Yeates, "Working for," 3.

76. Flasche, "Der Dieselmaschinenbau," 23.

77. Yeates, "Working for," 6.

78. Flasche, "Der Dieselmaschinenbau," 23.

79. Ibid.

80. R. W. Crowly, "Motor Ships in Russia," *MotorBoat*, January 25, 1912, 10. The article gives a complete list of diesels in navy and commercial vessels with names, owners, reverse system, and specifications for ships and engines. Typos are present.

81. Crowley, "Motor Ships," numbers are used, although Flasche, "Der Dieselmaschinenbau," 19–20, 80, rated them at 400 bhp at 215 rpm. This is no doubt a factory rating that was reduced for actual service.

82. J. D. Scott, *Vickers: A History* (London: Weidenfeld & Nicolson, 1962), 61–68, provides background on nontechnical sub developments in the prewar period.

83. E. C. Magdeburger, "Diesel Engines in Submarines," Trans., *American Society of Naval Engineers,* vol. 37, August 1925, 580–81.

84. "The Development of the Diesel Engine by the New London Ship and Engine Company and

the Electric Boat Company," EBC in-house publication, n.d., c. 1940s, 2. Archives, Nautilus Memorial Submarine Force Library & Museum, Groton.

85. Magdeburger, "Diesel Engines," 582.

86. Smith, *A Short History*, 332.

87. Conrad Regenbogen (technical head at Kiel), "Der Dieselmotorenbau auf der Germaniawerft," *Jahrbuch der Schiffbautechnischen Gesellschaft 1913*, reprint, 6. Diesel collection, Archives, Deutsches Museum.

88. Regenbogen, "Der Dieselmotoren," does not say if the double-acting engine was modified at Kiel. Sass, *Geschichte*, 563, and the brochure Krupp Dieselmotoren, 1925, 11, provide the design history.

89. Regenbogen in "The German Naval Architects," *The Engineer*, June 21, 1912, 647. His comments are taken from the same paper as the preceding notes.

90. Eberhard Rössler, *The U-Boat: The Evolution and Technical History of German Submarines* (Annapolis, MD: Naval Institute Press, 1981), trans., appendix. Sass, *Geschichte*, 570, gives the year and confirms the number.

91. Agnelli acquired the shipyard and the marine engine works in 1905 and what became the Grandi Motori factory in 1909. Arnoldo Mondadori, *"Fiat, A Fifty Years" Record* (Turin, Italy: Fiat, 1951), 74, 152, 249. (FIAT = "Fabbrica Italiana Automobili Torino," and "Stabilimento" = "establishment" or "works.")

92. Original drawings, but no test data, on the first installed 2C.116 are in the Fiat Archives. The engine itself was found probably in the late 1950s and painstakingly restored by Fiat Grandi Motori. (The *Medusa* had been torpedoed and sunk in 1915 by an Austrian sub.) Three complementary engine articles are: "Fiat Marine Oil Engine," *The Engineer*, July 28, 1911, 107; *Fifty Years of Fiat Diesel Engines 1909–1959* (Turin, Italy: Fiat, 1959), 207–09; and "50 Years of Evolution: The Engine Type 2C.116 of 1910, The Engine Type 3012 of 1960," *Fiat Stabilimento Grandi Motori*, Technical Bulletin, vol. 14, no. 4, October–December 1961, 89–96.

93. F. H. Tanner letter to the editor, "The First Two-Cycle Double-Acting Diesel Engine," *Engineering*, May 5, 1911, 583. Tanner's British patent was No. 23,104 of 1906.

94. Edward Butler, *Evolution of the Internal Combustion Engine* (London: Charles Griffin & Co., 1912), 46–7. Wells & Wallis-Tayler, *The Diesel*, 253–54, give a patent abridgement. Internal Sulzer historical writings cite the Tanner as being the first two-stroke, double-acting diesel.

95. A. P. Chalkley, *Diesel Engines for Land and Marine Work*, 4th ed. (New York: Van Nostrand, 1917), 264, shows the three-cylinder, single-acting design, but only gives data for a one cylinder.

96. As quoted in Arthur J. Marder, *From the Dreadnought to Scapa Flow*, vol. 1, *The Road to War, 1904–1914* (London: Oxford University Press, 1961), 332.

CHAPTER 16

Cargos of Commerce: European Marine Diesels, 1911–1918

"My Lord and Gentlemen, I stand here today as one who has been engaged all his life in the manufacture of steam engines and boilers for marine propulsion, and I say frankly that I do not like the idea of propelling large vessels by means of a series of explosions in a battery of oil or gas engines. . . . In fact, it has been said that the problems are more suited to the gunner than the engineer!"

—Joseph Hamilton Gibson, 1914[1]

By 1911 most major Continental diesel builders had begun making engines for increasingly larger merchant ships. Each offered a distinctive design. Their earlier ventures powering small motor ships or sailing vessels supplied the confidence to attempt horsepowers required for cargo ships traveling the seven seas. For most builders it was a short window of opportunity. By 1915 their merchant marine diesel programs had ended due to the war or were converted to meet naval demands.

Concerned ship owners and skeptical champions of steam focused particularly on diesel reversing systems to look for flaws and weaknesses. Consequently, these often-complicated mechanisms, which steam engines were mostly spared, continuously evolved to become more reliable and easier to operate. During this ongoing refinement, each manufacturer went its own way to arrive at sometimes unique and not always practical solutions.

The wartime neutrality of Scandinavia, Switzerland, and The Netherlands aided companies in those countries. Burmeister & Wain, favored by Denmark's lack of native fuels, dominated diesel installations for ships over 2,000 gross tons. Werkspoor, however, powered more, smaller ships, mostly tankers, nearly equaling the tonnage having B&W diesels. Diesels Motorer in Sweden built a few bigger marine engines. Sulzer, who was ready to launch a major marine program, did not deliver large diesels to foreign shipyards during most of the war; numerous smaller ones found their way into both commercial and naval vessels.

German cargo ships installed diesels based on Carels, Krupp, M.A.N., and Junkers-furnished designs, but the war brought these activities to a halt. British marine diesel builders, almost all licensees of Continental companies, faced strong competition from steam engines burning plentiful coal mined within the country. Consequently, only a few engines were built prior to the war and builders stopped with its onset or waited until its end to take up production; an exception was the Burmeister & Wain licensee.

Smaller diesels for tugboats and yachts appeared in both Europe and the United States before 1919. The emerging American marine engine industry included such names as Atlas-Imperial, Nelseco, and Winton. Most were based on designs coming from within the companies rather than buying a licensed one from a well-established European builder.

Werkspoor

Nobel's success with its diesel-engined tankers on the Caspian directly stimulated marine progress in Western Europe. Information about Eastern developments came via a report by Fred Lane, (d. 1926) an influential Royal Dutch/Shell director. Based on firsthand observations of the Russian marine engines, he recommended strongly that all new Shell tankers be similarly four-stroke diesel powered. As a result, Shell's transport subsidiary, Anglo-Saxon Petroleum Co., ordered its 1,216 dwt (2,047 tons displacement) tanker *Vulcanus* with a Werkspoor diesel engine.[2]

The dimensions of the single screw *Vulcanus*, commissioned late in 1910, were 196 × 37.75 feet with a draft of 12.33 feet. This made her slightly smaller than the twin-screw Caspian tanker *Robert Nobel* also launched about the same time.

The six-cylinder, reversible engine in the *Vulcanus* had a 400 × 600 mm bore and stroke and produced 450 bhp at 180 rpm. It was the first Werkspoor engine to use the long, vertical reciprocating rods to drive the camshaft (chapter 12). Injection and starting air compressors were placed at the side of the engine and driven by linkage off the crosshead (fig. 16.1a).

The marine adaptation of the engine also introduced a fuel system modification incorporating two horizontal pumps that ran off the lower auxiliary shaft driving the vertical rods. One discharged fuel into an accumulator chamber, and the other served as a ready spare. Exposed to accumulator fuel pressure was the smaller of two differing diameter pistons that were part of a pressure-balancing mechanism. Injection air pressure acted on the larger piston. In operation, if fuel pressure exceeded air pressure by a predetermined amount, the pistons adjusted their position to move the balance mechanism and hold the pump suction valve open longer to thereby decrease fuel delivery. The arrangement served as a variable speed governor. An Aspinall mechanical governor controlled only maximum speed.

Although the reversing method was based on Dyckhoff's 1899 patent (chapter 9), an improvement patent granted to Werkspoor engineer Verloops protected the new design.[3] Two separate camshafts, one for forward (A) and one for reverse (B), pivoted about a parallel third shaft (C). Supports (D) containing the camshaft bearings were keyed to the pivot shaft (Fig. 16.1b and c). When the reversing wheel (E) was rotated fully, the supports swung in unison to move one camshaft away from the rocker levers and move the other into engagement. Turning the handwheel only halfway disengaged both camshafts and stopped the engine.

Spur gears drove both camshafts from a pinion on a small upper crankshaft (F) that was axial with the pivot shaft and was rotated by the vertical rods. The spur gears remained in constant mesh with the pinion as the pivot shaft swung. The design was heavy and more costly but reliable.

The *Vulcanus* began her trial runs in the North Sea in December 1910 during which D. C. Endert was in charge of the engine room.[4] After acceptance by her owner, she traveled to Batum on the Black Sea, and in February 1913 began regularly carrying oil between Borneo and Singapore. One engine-related mishap was the rupture of an injection air tank in 1911. Lack of a tank safety valve and excessive water in it caused the welded tank to fail.

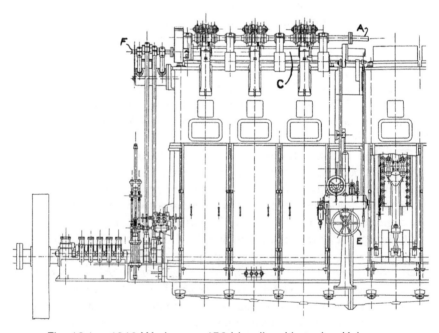

Fig. 16.1a. 1910 Werkspoor 450 bhp diesel in tanker *Vulcanus*.
Chalkey, *Diesel Engine, 1914*

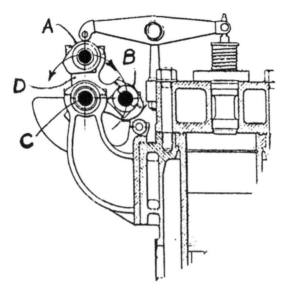

Fig. 16.1b. *Vulcanus* engine swinging camshafts for reversing.

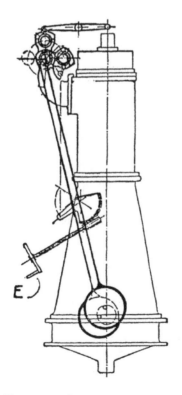

Fig. 16.1c. Control to pivot camshafts.
Supino, *Land and Marine Diesels, 1918*

Her diesel also suffered "some ordinary defects which would occur in steam engines."[5] One such "ordinary" service problem was the breaking of a connecting rod bolt at the crank throw reportedly due to overtightening. This led to a failure of the other bolt. The crank web came down on it and "brought the engine up solid." After removing the resulting bent con rod, the engine made port on five cylinders "not running so sweetly" and with a smoky exhaust. The valves were out of time because of a twisted crankshaft resulting from flywheel inertia when the engine seized. A new shaft piece was not put in until some months after the bending occurred. At the same time two new cylinder heads were installed. One of the original heads had cracked across the seats.[6]

The piston rings were replaced early on, and the cylinders were rebored after two years and 45,000 miles of operation.[7]

Interesting comparisons of operating costs between the *Vulcanus* and a steam-powered tanker having very similar dimensions and speed show why the marine diesel was attractive to Shell:[8]

	Vulcanus	*Sabine Rickmers*
Carrying capacity, dwt	1,235	1,269
Speed, kts	8	8
Engine speed, rpm	168	80
Fuel bumed/day, tons	2 (oil)	11 (coal)
Staff and crew	16	30
Crew cost/day	£6 6.5d	£9 0.7d

The diesel ship had a European crew and the steamer Chinese.

The *Vulcanus* remained in service until 1932 when, after a million miles of cruising, she and her engine were scrapped in Japan. The engine was said to still be in running condition.[9]

A second-generation design would become Werkspoor's standard marine diesel (fig. 16.2). This larger, "open to see" engine, rated 1,100 bhp at 125 rpm, had a bore and stroke of 560 × 1000 mm. (See chapter 12 for details of the comparable stationary engine.)

The new reversing system still used two camshafts, but the third shaft assumed a new, less demanding function. Both camshafts that moved in or out as a unit for either forward or reverse were supported in a common box sliding on a flat surface atop each engined-mounted bracket (fig. 16.3). The shafts were geared together with the one for "ahead" turning the "astern." A parallel third shaft acted only as a control rod to assist in sliding the boxes located at each bracket.

Four vertical rods driving the camshafts at the engine's midpoint were reciprocated at their lower end by a small, auxiliary crankshaft turning at

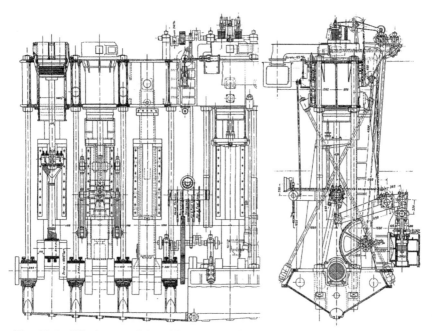

Fig. 16.2. Werkspoor 1.100 bhp marine diesel, 1912. *The Engineer, 1912*

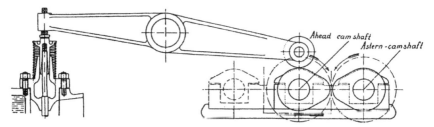

Fig. 16.3. Sliding camshafts of 1912 Werkspoor reversible engine. *Trans., Institution of Engineers and Shipbuildrs in Scotland, 1916*

half (cam) speed (fig. 16.4). The upper crankshaft was a part of the "ahead" camshaft so that these rods directly drove this shaft. When the shafts were slid to another position, the upper rod ends moved through a slight rotational angle. This resulted in a length difference between the chordal movement of the flat slide and the true arc of a circle that the upper rod ends wanted to take if allowed to freely rotate about their lower ends. To avoid the length discrepancy problem, the four rods, always acting in tension, were given a little extra bearing clearance to eliminate any bending caused by the imposed chordal path rather than that of an arc. Although the rods were stiffened longitudinally, there was a still a reference to rod "dithers" during a sea trial of the *Juno*.[10]

A panel of hand levers at the engine's midpoint controlled all operating functions. One positioned the camshafts for "stop" or running ahead or

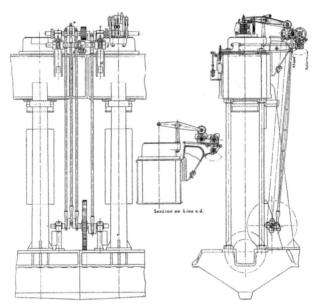

Fig. 16.4. Valve operating mechanism on 1912 Werkspoor
1,100 bhp marine diesel engine. *The Engineer, 1912*

astern. Pulling out this lever rocked a center-pivoted bell crank under the camshafts that slid the box holding the shafts closer to the centerline of the engine. The hand lever at the same time opened a valve that admitted compressed air into a cylinder whose piston provided most of the force to move the camshafts. Two other levers controlled the admission of fuel and starting air, and a fourth lever was the fuel throttle.

Starting the engine involved a relatively simple sequence of events. First, the camshaft lever was pulled or pushed, depending on desired engine direction. The throttle lever was then positioned. Next, both fuel/starting air levers were pushed in to admit air to all six cylinders. (These levers also had three positions: air, "off," and fuel.) When the engine was turning, one air/fuel lever was pulled out so that three of the six cylinders it controlled stopped getting starting air and instead received fuel and injection air at the nozzle. After these cylinders fired, the second air/fuel lever was also pulled so that the engine then ran on all cylinders.

Two tankers with these engines, the twin-screw *Emanuel Nobel* and the single-screw *Juno*, were placed in service about the same time during late 1912.[11] The 258 × 45 × 18.5-foot *Juno* with a 2,675 deadweight tonnage (4,300 tons displacement) was owned by Anglo-Saxon Petroleum Co. She was ordered as a result of the *Vulcanus's* good performance after the initial problems were overcome. The *Emanuel Nobel*, 375 × 51 × 29 feet and 9,400 tons displacement, operated through the Société Anonyme d'Armement, d'Industrie et de Commerce of Antwerp, whose parent company is not known. (This was the second and larger tanker of the same name; a smaller one plied the Caspian under Ludwig Nobel's flag.)

These two ships with their much larger engines also suffered introductory ailments. On the *Emanuel Nobel*'s maiden voyage, a cylinder head cracked between the fuel valve and the exhaust valve seat. As the head and upper cylinder were of a single monobloc casting, the failure was more involved. A number of piston rings in several cylinders froze in their grooves and had to be cut out. (Each piston also had eight rings with a 1.5-inch-high babbit bearing—"white metal"—strip inlaid around the piston below the rings.) The ring problem was caused by too tight a fit and an unsuitable lube oil. Finally, the third-stage cylinder of the air compressor broke due to a fracture of its piston rings. While the engine was in for repairs, a series of trays were put under the pistons, along with splash plates to keep leaking piston cooling water from draining down and mixing with lube oil in the bedplate sump. A successful trial run was then made in April 1913, after which the *Emanuel Nobel* entered regular service.[12]

The *Juno* also suffered cracked cylinder heads and other problems common to the *Emanuel Nobel*, but in January 1914 she began regular service between the Dutch East Indies and Singapore.[13]

Shell was satisfied enough with the *Juno* to order four similar tankers from the Werkspoor engine and shipyard company: the *Ares* in 1913, and the *Artemis*, *Hermes*, and *Selene* in 1914. The latter made a significant voyage in 1914 from Cardiff, through the Panama Canal, to China and Singapore and then back to Rotterdam with a cargo of oil. Total time for the 27,500-mile trip was 160 days, 20 of which were spent in harbors. She burned under 7 tons of fuel per day when running at 10 kts.[14] The *Ares* was sunk, and the *Artemis* was damaged by a torpedo but repaired.[15]

Late in 1914 a smaller, four-cylinder Werkspoor diesel went in the Shell-owned coastal tanker *Poseidon*. With a bore and stroke of 400 × 700 mm bore, the output was 450 ihp at 175 rpm.[16] The design resembled more of a scaled-down *Juno* engine except for its being nonreversible and a consequent need for a modified fuel control. A reversible blade propeller handled maneuvering. Because the engine and hence the fuel control became independent of the reversing process, the normally overspeed-only governor had to act as a variable speed one to keep speed constant when the load was taken off the engine during a blade reversal. The governor now directly controlled pump suction valve timing (like all other make engines) instead of using the balanced floating air/fuel bottle method on the larger Werkspoor engines (chapter 12).

An enclosed flywheel with fan blades blew air through the crankcase to cool the pistons. The *Vulcanus* engine had added a separate fan to do this, which would seem to have been an afterthought. Such a flywheel fan could not have been very effective when the engine ran astern.

A final prewar Werkspoor diesel had two, inwardly inclined, six-cylinder, trunk piston engines with each crankshaft driving a propeller. The two were tied together by bolts running through flanges on the upper beams. (fig. 16.5). Both halves rested on a common, shallow, cast steel bedplate. The "siamesed" twelve-cylinder unit with a bore and stroke of 390 × 500 mm produced 1,200 bhp at 300 rpm. Reversing was by the usual two camshaft method as on the *Juno*. Total weight including flywheel was under 33 metric tons.[17] The Dutch Navy bought one of these engine pairs for an armored gun boat[18] (fig. 16.6).

Endert ran this engine just before leaving Werkspoor. It was the night following the start of the war on September 1, 1914, when he captained the gunboat from Amsterdam to Den Helder on the North Sea to keep it out of the hands of the Germans should they enter Holland. His journal comment, again to be taken with a grain of salt, said that "this engine also was a mistake."[19] (He held a similar view about the original *Juno*-type design, which needed extensive modifications before becoming a successful engine.) The "twin-12's" Krupp-made crankshafts of manganese nickel steel proved to be too brittle. After six months of service, they were replaced because of cracks developing in "several places." New, "softer" cranks also used smaller weight-reducing holes bored in the crank throws.[20]

By 1918 Werkspoor had built twenty-six ships with thirty-seven of its diesels having a cumulative output of 33,000 ihp. It also furnished eleven engines for seven smaller boats producing an aggregate of 17,000 ihp.[21] The Kloos-inspired "open to view," steam engine philosophy of the Werkspoor 1912–1918 engines continued until about 1926.[22]

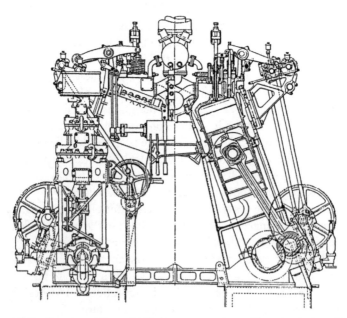

Fig. 16.5. Cross-section of Werkspoor engine with paired six-cylinder engines. Total output 1,200 bhp at 300 rpm. *The Gas Engine, 1915*

Fig. 16.6. Werkspoor siamesed engine for 1914 Dutch gunboat.
Engineering, 1918

Carels Frères

Carels Frères had built experimental two-stroke marine diesels before 1911 (chapter 12). It made a single-cylinder, 1,000 bhp test engine for Schneider-Creusot and sold a similar design to Vickers. Another, a four-cylinder, 1,000 bhp reversible diesel never left the Ghent shops where it served as a stationary engine.

Four increasingly larger ships powered by Carels-designed two-stroke engines entered service during 1912 and 1913. Marine activity in Ghent came to a halt during the war and Carels never again seriously contended at sea.

Carels was making a good name for itself with stationary engines, and a marine success boded equally well. However, it lacked factory capacity to pursue both applications. This forced a participation by inexperienced licensees at an early stage in the marine production program.

Two other factors were missing at Ghent, both of which aided in marine diesel progress at Werkspoor and Burmeister & Wain. Carels had no shipyard of its own nor a dedicated customer wanting to take advantage of what diesel-powered ships could offer them.

Carels did not turn its licensees loose to make the same mistakes of most first-time builders as it supplied the critical cylinders, heads, and pistons to all but Schneider. Any blame, perhaps too strong a word, for weaknesses in Carels's marine diesels should thus be shared equally between Ghent and its partners.

Clyde Engineering and Shipbuilding Co. built the 250-foot *Fordonian* that entered Great Lakes duty in 1913 as a Canadian grain ship. Sea trials had begun in September 1912. The Port Glasgow licensee was not yet ready to build diesels, so Carels delivered a completed one to it. This four-cylinder, two-stroke engine with a bore and stroke of 460 × 820 mm had a rated power of 750 bhp at about 120 rpm. Imep at 102 rpm and 970 ihp was 90 psi. Compression pressure averaged 490 psi. In service the engine had a bsfc of 0.47 lbs/bhp-hr and ran as low as 46 rpm without misfiring.[23]

Evident in the *Fordonian* engine design were new features common to both land and marine versions. The crosshead engine had a structure differing from its unsuccessful trunk piston predecessor. First, the cylinder jacket and support legs of the A-frame were cast in three separate pieces. The upper end of each leg was bolted to the cylinder jacket, and this assembly was then machined to ensure concentricity between the cylinder bore and an arc cut in each leg. These arcs formed a circular, two-sided bearing guide for the crosshead.

The upper A-frame leg ends of adjacent cylinders were also longitudinally bolted together to form a rigid upper tie, which greatly stiffened the entire structure in the vertical fore-aft plane. This reduced crankshaft bending caused by hull flexing during heavy seas (fig. 16.7).

Water entered the legs below the crosshead and exited from the head. Such cooling of a laterally supported crosshead was a major improvement as overheating of crosshead guide bearings was common.

Piston cooling water entered and exited the piston rod above the crosshead pin via L-shaped tubes telescoping in supply and drain pipes attached to the frame.

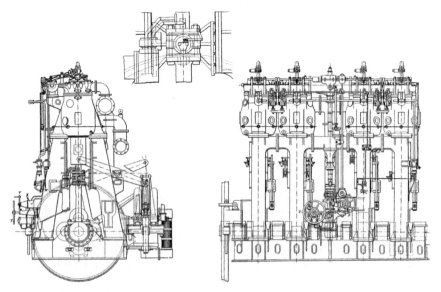

Fig. 16.7. Carels 1912 two-stroke diesel engine showing the rigid structure incorporated in both marine and stationary engines. *Engineering, 1912*

The engine lubrication system included main bearings with their own oil pump and drip lubricators feeding oil to the crank pin bearings. (Main and rod bearing loads were about 300 psi and 650 psi respectively.) Each crosshead bearing was oiled by a small pump fastened to the crosshead; an arm whose free end was tied to the connecting rod operated the plunger. A Mollerup pump for each cylinder forced oil through four equally spaced ports around the cylinder for piston lubrication. A collector ring caught oil leaking past the rings and drained it back to a sump. The absence of an oil mist permitted an open crankcase so that "the bottom end bearings can always be easily felt by the engineer on watch."[24]

Scavenge pumps on the Clyde-Carels, and all future Carels marine diesels, were placed at the side of the engine and operated from pivoting linkage fastened to the crosshead pin. Ship owners wanted as short an engine as possible. The two, two-cylinder, double-acting compressors of 27.25 × 23.5-inch bore and stroke provided a scavenge pressure of 3 psi. A 1.65:1 delivery ratio (air supplied by scavenge pumps relative to engine displacement) was higher than the usual practice of 1.2:1.5.[25]

Ensuing Carels marine diesels with scavenge air valves in the head used the *Fordonian* engine's reversing method. Effecting this directional change was a control to angularly displace a main camshaft that varied scavenge valve timing. A second, maneuvering shaft independently changed timing of the fuel and starting air valves through both an angular and axial displacement of this second shaft. A unique Carels innovation, acting in concert with displacement of the maneuvering shaft, also gradually reactivated the fuel and starting air valves (fig. 16.8).

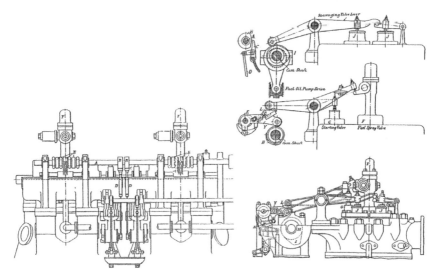

Fig. 16.8. Carels reversing mechanism details on *Fordonian* engine.
Engineering, December 1912

The necessary angular displacement of the main camshaft was accomplished by pulling lever (*X*) (fig. 16.7) to raise or lower the vertical drive shaft holding a spiral gear on its upper end as in the Werkspoor engine. (The lever only positioned an air valve controlling a pnuematic cylinder that provided the shaft lifting effort.) This changed the timing of the scavenge air valves' cams (*I*) between ahead and astern by about 30 degrees. It also initiated axial movement of the maneuvering shaft (*A*).

First, however, the fuel had to be shut off. Maneuvering shaft (*A*) held cams (*B*), which, when the shaft was rotated in either direction, moved bell cranks (*C*) to lift rods (*D*) and hold open the fuel pump suction valves' to stop the engine. Eccentrics on the main camshaft reciprocated plungers in the pumps.

Handwheel *W* controlled the rotational position of shaft *A*. A partial turning of the shaft lifted freely pivoting rockers *Y* to raise their opposite ends off cams *H* on the main camshaft. This allowed axial shifting of shaft *A* to bring a second set of cams *E* into play for retiming the fuel and starting air valves to run in the opposite direction. Cams *E* modulated lift of the starting air and fuel valves. A further turning of shaft *A* lowered levers *Y* back onto their main shaft cams.

A full rotation of shaft *A* admitted 800 to 1,000 psi starting air by moving out wedges *L* on levers *Y* to open the air valves on all four cylinders. Then gradually turning the shaft away from its fully rotated position also gradually reduced lifts of the starting air valves on two cylinders and introduced fuel into them. A further turning stopped the air and initiated fuel injection in a similar manner on the other two cylinders. This peculiar Carels feature reduced the usual "off-on" starting shock found to a greater or lesser extent on other engines.

A minimum recorded reversing time achieved on the *Fordonian* was only 4.5 seconds. She once made a series of sixty-three maneuvers in forty-one minutes during a demonstration to duplicate what happened when passing through the Great Lakes locks. By using an auxiliary compressor along with that on the engine, a minimum air pressure of 45 atm was maintained in the starting air tanks.

The *Fordonian* had a long career under several names and types of service. While hauling bauxite as the *Badger State*, she sank after hitting a submerged object in the Gulf of Mexico on January 14, 1946. The original engines were reportedly still powering her at the end.[26]

Shipbuilder Richardson-Westgarth & Co., of Middlesbrough, England, jointly made a four-cylinder diesel for the 276-foot cargo vessel *Eavestone* in 1912—Carels supplied the cylinders, pistons, and heads to its licensee. Sea trials were in October of that year. The engine had a bore and stroke of 510 × 920 mm and an output of 900 bhp at 115 rpm. In service this dropped to 800 bhp at 95 rpm.[27] The design was very similar to that in the *Fordonian*. One change was to oscillating, or "grasshopper leg," piping from the telescoping tubes to supply piston cooling water (fig. 16.9).

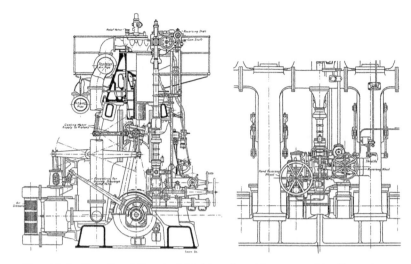

Fig. 16.9. Details of Carels 800 bhp diesel in cargo ship *Eavestone*. *The Engineer, Oct. 1912*

Operating data exist for the *Eavestone* and the *Saltbum*, an almost identical sister ship powered by a triple expansion steam engine. A five-voyage average for the *Saltbum* shows a speed of 8.4 kts and a daily coal consumption of 13.6 US tons. The *Eavestone* averaged 8.75 kts for five trips with a fuel usage of only 4.1 US tons. Of these 13 cwt (0.73 US tons) was burned under auxiliary, "donkey" boilers.[28]

The *Eavestone* diesel developed its share of problems after a few months in service: testimony in May 1913 (by an admittedly anti-diesel general manager at Richardson-Westgarth) refers to six cracked cylinder heads. These cast steel heads were replaced by ones of cast iron "with the hope they do well."[29] Cracked pistons resulted, possibly from lack of engine attention on one trip after many of the crew had come down with smallpox. Bearing failures, including the mains, ruined the crankshaft. The engines were then removed from the ship and completely rebuilt. No new Carels-type engines were made at Richardson-Westgarth.

J. Tecklenborg AG Schiffswerft und Maschinenfabrik in Bremerhaven-Geestemünde built the 290-foot motor ship *Rolandseck*, which entered service in November 1912. Its six-cylinder engine of the same 510 × 920 mm bore and stroke as the *Eavestone* had an output of 1,500 bhp at 130 rpm.[30] As before, Carels furnished its licensee the cylinders, heads, and pistons.

A sea trial observer noted excessive vibration possibly due to the 25 rpm higher engine speed than with the *Eavestone*. He also commented on the piston cooling water supply tubes:

. . . we had an opportunity of seeing how the telescopic tube carried on the end of a long arm projecting from the crosshead works. We must admit it does not impress us favorably. . . . Practically all the glands were leaking; in fact a good overflow pipe is provided from the glands to carry the leakage away. How far this leakage may have been due to lack of adjustment, which clearly needed to be made, we cannot say. However, apart from the mess and the trouble involved in attention, this is not, after all, a very serious objection, though a nice clean-looking engine appeals to the eye much more than one in which water is being splashed about.[31]

This was minor in comparison to a report that all the heads and cylinder castings cracked.[32] Final disposition of the engine is not known.

The British Admiralty ordered two engines similar to those in the *Rolandseck* directly from Carels, one of which was displayed at the 1913 Ghent Exhibition. It turned a propeller and demonstrated reversal times on the order of six seconds. The engines never reached England due to the war; one stayed in Ghent, and the other went to Krupp's Essen factory during the German occupation.[33]

In 1912 a second German licensee, Reiherstieg Schiffswerft und Maschinenfabrik in Hamburg, and Carels built the largest prewar marine diesel to enter service. Cylinders, pistons, heads, and upper running gear again came from Ghent. This six-cylinder Carels-Reiherstieg engine, destined originally for the *Excelsior*, first ran during February 1912. Instead it went in a new 404-foot tanker, *Wotan*, the next year.[34] With a bore and stroke of 600 × 1,100 mm and an output of 1,800 bhp at 100 rpm, it had a brake thermal efficiency of 29.7 percent, mechanical efficiency of 70 percent, and a bsfc, typical for a Carels, of 0.47 lbs/bhp-hr on a fuel with an LHV of 18,900 Btu/lb. Information on the *Wotan* after going to sea is lacking, but she was reported still in service with a Carels engine in April 1922.[35]

Schneider et Cie.

Schneider and Co., a Carels licensee in Le Creusot, focused mainly on submarine engines for the French Navy, but it also built a few diesels for cargo ships adapted from the Belgian design. One of these, the five-masted sailing vessel *France*, received wide publicity when she left on her maiden voyage to New Caledonia in late 1913.[36] Two four-cylinder Schneider-Carels diesels provided auxiliary power to drive the 10,700 ton displacement ship at 10 kts. The 900 bhp at 230 rpm engines had a bore and stroke of 450 × 560 mm. They could be disconnected from the propellers when the engine was under sail. Unlike other Carels licensees, Schneider built the entire engines themselves.

Although the two-stroke, uniflow scavenging through two valves in the head followed Carels practice, a trunk piston design with an enclosed crankcase engine structure were Schneider variants (fig. 16.10). The scavenge pump and air compressor were relocated and driven off the crankshaft in contrast to the Carels side-mounted design. The camshaft was also at

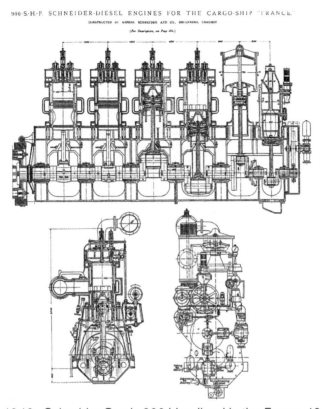

Fig. 16.10. Schneider-Carels 900 bhp diesel in the *France*, 1913.
Engineering, October 1913

mid-height and the rockers were lifted by pushrods. Reversing was accomplished by a longitudinal shifting of the camshaft when the pushrods had been lifted away from the cams.

The *France* arrived in New Caledonia with no reported problems. However, nothing was said about how much of the voyage was made under sail.[37] Not known is how many engines Schneider built to this design, but with the coming of the war and the increased need for submarine engines, it may be assumed that few if any were.

AB Diesels Motorer's Polar Engines
The Swedish diesel company joined the marine diesel horsepower race with more powerful two-stroke engines for merchant ships and higher performance four-stroke models for submarines and yachts. Some were successful and one proved an embarrassment. Hesselman himself had been dubious about venturing into the big marine diesel market as Diesels Motorer already made good, and profitable, Polar engines in much smaller sizes better suited to the Sickla factory with its special qualifications:

I felt for the first time a certain uneasiness during the design and building of these [large] engines. They stretched company resources to the
limit, . . . [and] we could not count on production of such engines in the
future. But there were company people on the board who believed that
we could not fall behind in the continuing evolution toward ever larger
units. This was a viewpoint which had nothing to do with economy.[38]

Nevertheless, during this period Hesselman designed, and Diesels Motorer
put into production, the Z Series, which began with 12.5 bhp per cylinder and was available in one, two, three, four, and six cylinders. These were
popular as shipboard auxiliaries and other applications where their much
lighter weights and higher speeds had appeal. For example, a four-cylinder Z
with 50 bhp at 550 rpm weighed only 1,750 kg against a major competitor's
engine of the same power turning 275 rpm that came to 6,900 kg.

Polar marine developments traveled a separate course. A pair of two-
stroke engines, very similar to but slightly larger than the *Toiler*'s, went in the
Great Lakes grain ship *Calgary*. The two were almost equal in size. Like the
Toiler (chapter 15), she was made by Swan, Hunter & Wigham Richardson
at Newcastle-upon-Tyne. Each of the four-cylinder, 290 × 430 mm engines,
however, produced 260 bhp as compared to 180 bhp in the earlier ship. In an
effort to send even more power to the shafts, all auxiliary pumps were instead
driven by two small steam engines.[39] The *Calgary*'s greater power satisfied her
Canadian owner for a few years, but it still proved not enough; two 500 bhp
McIntosh & Seymour diesels were retrofitted in 1920.[40] It should be noted
that four-cylinder Polar "P" Type engines, of which those in the *Calgary* were
an early example, were built into the 1920s with outputs of up to 500 bhp at
150 rpm and with a guaranteed bsfc of 205 g/bhp-hr.

Diesels Motorer made a much larger and completely new marine diesel,
code named Neptune, in 1913. It must not have taken long before the
company no doubt wished this design had never surfaced but had stayed in
its namesake's underwater realm. The only ship to be powered by a pair of
Swedish-made Neptunes was the *Sebastian* (fig. 16.11), a 310-foot tanker
owned by Lane & MacAndrew and built by the Caledon Shipbuilding
Co. in Dundee, Scotland.[41] (This is the same Fred Lane affiliated with
Shell). Her six-cylinder engines had a bore and stroke of 400 × 540 mm
and produced 800 bhp at 165 rpm, or over three times the power of the
Calgary's (fig. 16.12).

It was also the first crosshead engine built by Diesels Motorer since their
earliest ones using the Augsburg design. The engine's openness above the
bedplate somewhat resembled the Werkspoor design in that the crosshead
guides on one side and rod columns on the other supported a three-piece
"beam." The beam sections, which contained the scavenge pump cylinders,
were tied together to form a single entablature on which sat the power
cylinders.

Fig. 16.11. The 310-foot tanker *Sebastian* launched in 1913 and
was powered with two Polar Neptune diesels of 800 bhp each.
The Engineer, March 1914

The Neptune's distinction was its use of stepped pistons in lieu of crank-shaft-driven scavenge air pumps (fig. 16.13). Others, including Nürnberg, were developing stepped-piston engines, but the Neptune remained unique in that its scavenge pumps continued to serve as air motors for starting as on earlier Polar two-stroke diesels. The pumps also compressed the air on the downward stroke rather than the upward. An air inlet and a discharge valve serving a pair of pumps were reciprocated by eccentric rods driven off an

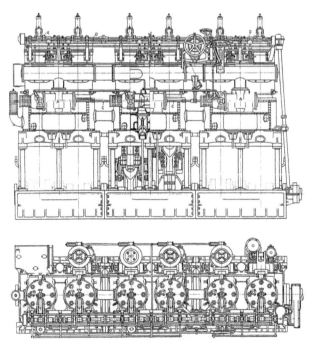

Fig. 16.12. Diesels Motorer 1913 Neptune stepped piston
800 bhp marine diesel. *The Engineer, March 1914*

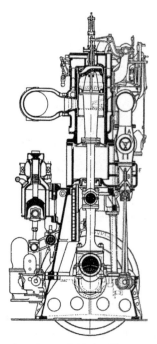

Fig. 16.13. Section through Polar Neptune two-stroke diesel.
The Engineer, March 1914

auxiliary jackshaft to alternately draw in outside air and to send scavenge air to one or the other of the power cylinders.

Linkage off the crossheads of cylinders two through five drove four two-stage compressors. Three supplied injection air to a mated pair of cylinders, and the fourth sent starting air to all six.

Like the *Toiler* and *Calgary* engines, the high-mount camshaft was lighter because of needing only to lift the fuel nozzle valves and, through eccentrics, operate individual fuel pumps. A slightly canted drive to the camshaft shortened shaft overhang.

By using loop scavenging and having the stepped pistons do double duty for starting, the only valves in the cylinder head were those for the safety popoff and fuel nozzle—an appealing feature.

For reversing the engine, the turning of a handwheel axially slid six sleeves keyed on the camshaft and which held "ahead" and "astern" fuel cams for each cylinder. Tapered side ramps and low cam lifts allowed rockers to stay on the cams when reversing. The handwheel also positioned scavenge pump inlets to accept either intake air for running or pressurized air to start up or maneuver. As on earlier Polar engines, the pump discharge valves sent all exhausted air into the power cylinders. The handwheel likewise controlled fuel pump suction valves to cut off flow to the nozzles until the engine was ready to be started.

After placing the scavenge pump inlet valves in the air motor mode, compressed air to turn over the engine was then sequentially admitted. The movement of a lever repositioned two-way, piston-type valves for first one pair of cylinders, then two, and finally all cylinders. This greatly conserved the supply of starting air when maneuvering because with a warm engine only one pair of cylinders usually needed to receive air to make a restart.

A report on the engine's performance during a sea trial was critical of accessibility for the engineer to feel, for example, the heat of the crosshead pin when the engine was running.[42]

The mention of crosshead pin overheating obliquely refers to an inherent difficulty of two-stroke engines on which literature of the period was strangely silent: connecting rod and main bearings remain loaded at all times. The first reference (in English) the author found to this was a 1917 text that states, "There is no reversal of pressure in the bearings which makes the lubrication much more difficult, and this constitutes a good reason for adopting forced lubrication in engines working on this cycle."[43]

Hesselman himself in 1948 related his encounter with the problem not long after first starting up his single-cylinder, two-stroke, reversible test engine in 1906.[44] He saw from an adjacent office the exhaust getting darker and darker, and by the time he could get to the engine the damage was done:

[I] found that a piston pin bearing had seized so badly that the engine overloaded. We had thus during the first few hours . . . found one of the greatest mechanical difficulties of the two-stroke diesel engine, which [is to prevent] . . . lube oil from entering between the pin and bearing. . . . I have been told that this is still after more than 40 years to a certain extent a problem in the 2-stroke engine. . . .

Hesselman also wondered how the Junkers, with its even higher two-stroke loads, got away with it. He said it is "one of the few mysteries of engine technology" how Junkers solved the problem.

The *Sebastian* entered service in the spring of 1914 and reported no problems on its first outward voyage.[45] However, the situation must have rapidly deteriorated with unspecified disasters because after the return trip the engines were removed.[46]

Swan, Hunter & Wigham Richardson, Diesels Motorer's British licensee, built several Neptune-type engines modified considerably to suit local design philosophy. They were under construction before the *Sebastian* entered service. (Surprisingly, dimensioning was in metric.) The four-cylinder diesels with a bore of about 450 mm and stroke of 860 mm had a rating of 600 bhp at 135 rpm.[47]

Marcus Samuel, founder of Shell Transport & Trading Co. and a major shareholder in Royal Dutch/Shell after the merger, ordered these English Neptune diesels. He wanted to operate totally British built ships because of his concern over the Dutch-made *Juno*'s initial problems. The decision was logical because Diesels Motorer enjoyed an excellent reputation based not only on the *Toiler* and *Calgary*, but on other less noteworthy ships also powered by Polar diesels. Samuel formed the Flower Motor Ship Co. in November 1912 and had Swan Hunter build engines for two ships, the twin-screw *Arabis* and *Arum*. Each was 350 × 47 × 21 feet with a dwt capacity of 5,600 tons. The *Arabis* was launched in March 1914, and the *Arum*'s sea trials began in May. Engine problems arose, and although Swan Hunter tried to keep them running, the ships were "returned to the builder" due to those problems.[48] Their final disposition is not known. The *Abelia*, a third Samuel ship, had two-stroke engines of comparable power designed and built by the Wallsend Slipway & Engineering Co. on the Tyne.[49] Her engines were reportedly also troublesome. She was later torpedoed and sunk.

Diesels Motorer had better success with larger engines for the tanker *Hamlet*. Although ordered by Nobel Bros., her registry was transferred to Norway in 1915 during construction in Götaverken, Sweden. When the *Hamlet* entered service the next year, her two Polar MP6Y, six-cylinder engines had a combined output of 3,300 bhp, which made her the most powerful diesel ship at sea[50] (fig. 16.14).

These two-stroke, six-cylinder engines had a bore and stroke of 610 × 910 mm and produced 1,650 bhp at 100 rpm. They were an outgrowth of the Neptune series, but instead of stepped pistons, the lower side of the power piston itself acted as the scavenge pump and air starting motor. The crosshead

Fig. 16.14. Control station for Polar diesels in M/S *Hamlet*.
Atlas Diesel Catalog 1919

design with the piston rod passing through a packing gland made this possible. A double-acting air pump driven off the crankshaft supplied the additional scavenge air needed to make up for the loss in scavenge air capacity due to the same diameter pump and power piston.

A late 1920s Polar engine catalog listed the MP Series ranging from a four-cylinder, 560 bhp model to an eight-cylinder one of 2,700 bhp. The *Hamlet* was still in service with her original engines until after 1945.[51]

Four-stroke Polar marine diesels for submarines, and adapted to yachts, are described in the following chapter.

Diesels Motorer merged with air compressor builder Nya Atlas in 1917 to become Atlas Diesel. Hesselman left at this time to enter the consulting phase of his career. Staying on was untenable as Marcus Wallenburg had chosen another to run the Sickla works due to Jonas's refusal to cooperate with the new company's head. Marcus explained why he did this in a September 12, 1917, confidential letter to his son: "Hesselman is a genius, but so conceited as to be intolerable. He is demanding absolute authority over the diesel factory, and this is completely beyond his capabilities."[52]

Two factors put Atlas Diesel in a position where it could not sustain profitability during the later Depression years.[53] In addition to post-war high inflation, the combining companies had set too-optimistic values on assets in anticipation of the merger. Nydqvist och [and] Holm AB bought the diesel business from Atlas Copco in 1948 and sold engines under the "Nohab" nameplate. Nohab in Trollhättan became a division of Finland's Wärtsilä in 1979.

Burmeister & Wain

The M/S *Selandia*'s launching at Burmeister & Wain's Copenhagen shipyard on November 4, 1911, ushered in a new era of large motor ships. Her displacement was five times that of the *Vulcanus* commissioned in late 1910 and over twice the *Eavestone*'s which would begin sea trials in September 1912. Of even greater import was the progression of B&W diesel-engined ships, which followed the *Selandia* having equal or higher horsepower.

Burmeister & Wain's interest in the marine diesel market extended back almost to when they made their first engines in 1903 (Chapter 12). The Russian successes of AB Diesels Motorer and Nobel with the ASEA all-electric and the Del Proposto reversing systems was not lost on B&W. However, due mainly to patent obstacles, work on a reversible marine engine did not start until 1909.[54]

Ivar Knudsen submitted an application to the Danish Patent Office for a "Reversing device for internal combustion engines with a movable camshaft" on October 27, 1909. His invention, issuing as patent 13,701 on September 8, 1910, was for a mechanism (described later) to lift valve cam followers off their cams when the camshaft was axially shifted. It formed the basis of B&W marine engine reversing systems.

Chief Engineer Olaf E. Jorgensen filed for a patent, issuing as 14,861 of July 13, 1911, covering the interchange of inlet and exhaust valve functions when reversing. A "change-over" valve in the piping made this possible.[55] It was briefly tested on a modified customer engine. His next idea, protected by Danish Patents 17,342 of May 14 and 17,436 of June 5, 1913, was for a simplified way to pneumatically open the starting valves.[56]

Knudsen's cam shifting idea gave him confidence to proceed on an eight-cylinder diesel of a size East Asiatic Co. wanted for a small ship it was planning. The new engine, with type designation DM 830 × (Diesel motor, eight cylinders, 30 ihp per cylinder, experimental) had a 325 × 440 mm bore and stroke with a projected output of 240 ihp at 180 rpm. It went on test in mid-July 1911 and confirmed the soundness of Knudsen's mechanism. Jorgensen's new starting valve idea came too late for this first engine.

The engine never left the shop because in the fall of 1911 East Asiatic instead ordered three much-larger diesel-powered cargo ships named *Selandia*, *Fiona*, and *Jutlandia*. B&W shipyards were to build the first two, and the third would come from Barclay, Curle & Co., Ltd. in Glasgow.[57] Their triple screws were driven by B&W-made eight-cylinder engines of 2,000 ihp with bores and strokes of 600 × 850 nun. On November 16, 1910, Barclay, Curle signed a license contract with B&W, giving the Scottish firm a "sole right of manufacturing Burmeister & Wain diesel motors for Great Britain and Ireland."[58]

Because B&W's largest stationary model was only 150 bhp per cylinder, it built a single-cylinder experimental engine with 250 bhp per cylinder, incorporating the latest reversing and structural features. Design work started in the fall of 1910 and testing, which began the following March, lasted until September (fig. 16.15). It lived up to expectations.

East Asiatic, meanwhile, again changed plans in December 1910 and wanted twin-screw ships of only 1,250 ihp with a nominal 1,000 bhp per engine. In addition, since Barclay, Curle had become a B&W licensee, it would build the diesels for the *Jutlandia*. These smaller engines added to B&W's confidence, but its sanctioning of a neophyte diesel builder to provide a pair of engines proved unwise.

Influencing the Barclay, Curle engine decision was a very competitive marine diesel market that Burmeister & Wain had yet to enter. B&W needed immediate recognition, followed by orders from ship owners other than East Asiatic. The Sulzers had already been selling marine engines, albeit much smaller, for four years. Thus, if B&W could claim a successful "first" with the large *Selandia* and two sister ships, such favorable publicity would prove invaluable. Conversely, a failure by B&W held serious consequences. Besides the adverse news, it would have to pay half the cost of converting the three ships to steam. But time was short for B&W as Sulzer sought the same prize using a pair of its two-stroke diesels in the ordered *Monte Penedo* of almost the size as East Asiatic's new trio.

Fig. 16.15. Burmeister & Wain reversible one-cylinder test engine of 1911. 250 indicated hp at 145 rpm. *Burmeister & Wain Archives*

The *Selandia* was 370 × 53 × 30 feet with a displacement of 10,000 tons and a dwt of 7,400 tons (fig. 16.16). Diesel engines and 900 tons of fuel carried in a double bottom gave the Selandia a seventy-five-day, 20,000-mile cruising range at 11 kts. This capability enabled buying low-priced oil in the Far East, traveling to European ports and returning before refueling.

A unique silhouette, the absence of a funnel, emphasized *Selandia*'s new power. The exhaust went up the rear mast. The entire hull was painted white to further accent a lack of steam engine smoke (fig. 16.17). On the main deck were well-appointed passenger accommodations, including dining and smoking saloons.

Her main B&W engines embodied design features that put them in the forefront of diesel technology. However, given a freedom to spend money to ensure maximum reliability and maintainability raised the total weight of the main engines alone to over 400 tons or about 250 lb/hp.

Each eight-cylinder, four-stroke main engine had a 530 × 730 mm bore and stroke and an output of 1,050 bhp at 140 rpm. B&W used an indicated rating (1,250 ihp) typical of steam engine practice. Bmep at 1,000 bhp and 129 rpm was 6.4 kg/cm² (91 psi). Although the engines had been pushed as high as 1,700 ihp on the test stand, they were limited to an overload of 1,500 ihp at an unknown rpm.[59] The small difference of 250 hp between indicated and brake (a mechanical efficiency of 84 percent) is evidence of the external power needed to run accessories usually mounted on the main engine.

Distinguishing the DM 8150 engine was a fully enclosed frame and a mid-height camshaft lifting long push tubes to open the valves. An important

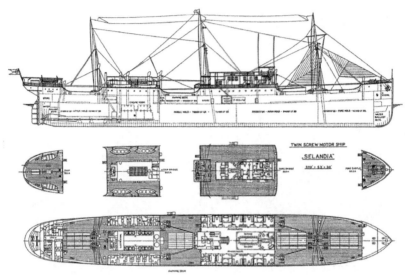

Fig. 16.16. East Asiatic Co.'s *Selandia* with B&W diesels.
Burmeister & Wain drawing

Fig. 16.17. M/S *Selandia* 1912–1942. *From Burmeister & Wain painting*

service advantage accrued from this camshaft location, because the shaft was not affected if a head or cylinder had to be disassembled. Engines with the camshaft supported from the cylinder or head required removal of part or all the shaft.

Floor-height controls were placed between number four and five cylinders (fig. 16.18). A pair of linered, individual cylinders with heavy, bottom flanges through-bolted to the bedplate sat on a cast frame with openings covered by doors. The four frame pieces were bolted together to give the appearance of a smooth sided enclosure for the crosshead and crankcase. This unit rested on a two-piece bedplate. The crankshaft was also made in two sections with each having throws at 180 degrees. When assembled, the throws of the two sections were offset by 90 degrees.

Inspection doors on both sides at each cylinder provided crankcase access. An upper one of the two on the engine's control side was a heavy plate, integral with the water-cooled crosshead guide, that was heavily bolted to the enclosure frame for rigidity. Additional doors at the engine midpoint also gave entry to the vertical drives for the camshaft and fuel pumps.

Rocker levers opened intake, exhaust, fuel, and starting valves in the cylinder head. A safety valve was placed horizontally through the side of the head. The uniquely B&W outward opening fuel nozzle valve, a design borrowed from its stationary engine, had a poppet head about 25 mm in diameter.

The camshaft drive bore a slight resemblance to Werkspoor's rod system. A gear drive off the mid-point of the crankshaft drove a small auxiliary crankshaft, which reciprocated two rods. The rods' upper ends rotated a second crank and gear set that turned the half-speed camshaft. A third, crank-driven

Fig. 16.18. Burmeister & Wain four-stroke, 1,050 bhp *Selandia* engine.
MAN Archives

rod on the lower auxiliary shaft rocked a lever whose opposite end, by means of a horizontal shaft and other small levers, turned an Aspinall governor and reciprocated the fuel pump plungers.

Two fuel pump assemblies each served four cylinders. Each assembly held four plungers that discharged into a common line going to a distributor "box" where adjustable needle valves in its four outlets balanced fuel flow to the injectors. Manual and governor fuel controls regulated the open duration of the pump suction valves. The governor only prevented engine overspeeding.

Externally driven pumps fed oil under about 45 psi to lubricate main and rod bearings. Oil then went up the connecting rod to the crosshead and the inner of two concentric passages in the piston rod to cool the piston. Heated oil from the piston returned down the outer rod passage to the crosshead guide and into an oil/water heat exchanger where it was cooled. Mechanical lubricators oiled the pistons through cylinder ports. A catch plate at the base of the cylinder through which the rod passed prevented oil escaping down the cylinder walls from mixing with cleaner oil returning to the pump. Valve gear continued to be oiled by hand.

Two four-cylinder B&W auxiliary diesel sets each drove three-stage compressors for starting air and supplied injection air to a fourth-stage compressor on the front of the main engine. Electric motors ran the duplicate lube oil and water pumps. These independent sets provided backup as either met main engines needs at sea. The auxiliaries also provided ship electric power (fig. 16.19).

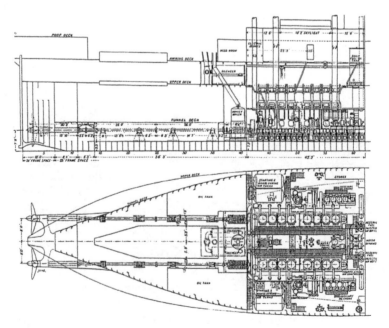

Fig. 16.19. *Selandia* engine room layout. *Engineering, April 1912*

A donkey boiler made steam to run a compressor for pumping up starting air should the entire system be down. It was the only steam plant on the ship. Smoke from the boiler firebox went up a stovepipe forward of the mast carrying away diesel exhaust.

Meaningful fuel consumption figures for the *Selandia* engines must take into account the externally driven compressors and pumps necessary to run them. Tests on a sister ship with identical engines and burning oil with an assumed average LHV of about 19,500 Btu/lb show the effect of this:[60]

	Ihp—lbs/hp-hr	Bhp—lbs/hp-hr
Main engines only	0.338	0.423
Main + auxiliaries	0.367	0.422

Like other builders, the Burmeister & Wain reversing system employed sliding cams but in Knudsen's uniquely different manner. The entire camshaft was slid axially so that a second set of cams came into play to open the four valves in each cylinder head, but prior to this, the followers had been lifted away from their cams.

Adjacent and parallel to the camshaft was an auxiliary shaft that had the two functions of lifting the followers and shifting the cams (fig. 16.20). When the operator moved a reversing lever from its midpoint to an end position, either "ahead" or "astern," he turned a vertical shaft, which, through right-angle gearing, rotated the auxiliary shaft by 180 degrees one way or

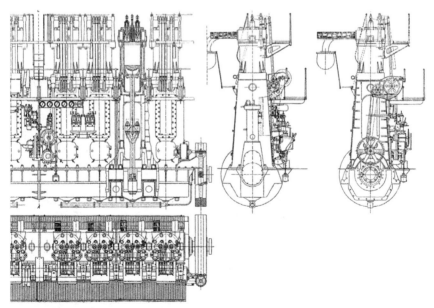

Fig. 16.20. Section views of B&W *Selandia* engine. *Engineering, 1912*

the other from its neutral point. At the neutral point the followers were held away from their cams. (Figure 16.18 shows the operator's right hand on the reversing lever.) An air cylinder assisted in moving the vertical shaft; if it malfunctioned, the torque from a handwheel above the lever substituted as the lifting force.

Moving the lever to either end point both slid the camshaft and returned the followers onto their selected cams in a simple and straightforward manner. On the auxiliary shaft was a large disc whose circumference acted as a face cam in the form of a slightly skewed groove cut into its outer diameter. A smaller disc on the camshaft ran, with a little side clearance, in the groove. Turning the face cam disc to "ahead" or "astern," forced the camshaft to slide the same distance as the axial displacement in the groove.

A "stop" disc directly under the reversing lever was geared to the vertical shaft and duplicated angular rotation of the auxiliary shaft. Two pins in the disc's back face moved between fixed stops located on the lever housing. Auxiliary shaft rotation ceased either when one pin contacted the "ahead" stop or when the other pin hit the "astern" stop. Initial movement of a starting air "regulator" lever caused a safety catch to drop into a groove cut in the disc rim, which prevented the auxiliary shaft from inadvertently being turned while the engine was running.

The starting process began after the camshaft had been positioned to run the engine in a selected direction. Pulling the regulator lever out to its first position engaged the starting valve to admit air at about 300 psi into the cylinders. After the engine was turning, the lever was pulled out further to first shut off the air and then to begin sending fuel to the injectors. The remaining travel of the lever served as a throttle to control fuel delivery between a minimum runnable load and full load. Thus, two levers, the reversing and the regulator, handled basic engine operating functions. An average reversing time was twenty seconds from full load in one direction to full load in the other.

Jorgensen's starting air valve design eliminated the need for a mechanism to lift its rocker off the cam as was the practice in most four-stroke diesels. On top of the valve stem was a small air cylinder to which starting air passed whenever air pressure stood against the valve stem (fig. 16.21). This pushed a spring-loaded piston up against a stop to form a pneumatic link through which the rocker lever pushed to open the valve. With no air pressure, the piston was reciprocated in its guide by the rocker lever but did not open the valve. Succeeding generations of B&W engines continued to use this clever design.

Shop testing of the *Selandia* engines began on November 5, 1911, and was completed at the end of December. No untoward events occurred during the ship's first unofficial sea trial in early February (fig. 16.22). Ivar Knudsen surprisingly remarked after this trip that he was "basically a little unhappy." When asked why by Engineer Christian Jensen, Knudsen replied "because we had no malfunctions at all."[61] The first of the official sea trials with dignitaries and the press was on February 14, 1912. It was a complete success.

Fig. 16.21. Starting, fuel, and inlet valves of *Selandia* engine.
Chalkley, *Diesel Engines, 1914*

The national importance attached to the *Selandia* at the time of her maiden voyage, as well as publicity generated during early stages of her voyage to Singapore and Bangkok, is evident by the people identified with this event. When she sailed from Copenhagen on February 22, 1912, the Crown Prince and Princess and other members of the Danish royal court made the short trip with her to Elsinore. She then went to London after loading twelve thousand barrels of cement in Jutland. Traversing up the Thames to the West India Docks required as many as eighty maneuvers per engine during a half hour period.[62] Winston Churchill, First Lord of the Admiralty, and top naval officials inspected her on March 1. They also met with Ivar Knudsen and H. N. Andersen, head of East Asiatic Co.[63]

Stops on the way to the Far East via the Suez Canal included Antwerp and Genoa. She arrived in Bangkok on schedule, and four months after leaving, sailed once again into Copenhagen harbor.

Rumors of serious engine troubles surfaced from time to time but for the most part proved false. There were problems: the wrong oil; "mineral instead of paraffin" was used to lubricate the exhaust valve guides, which caused a valve to stick down; and the upcoming piston broke off its steel head. One piston cracked under the rings and when changed in Genoa on the outbound trip, it was found that the same liner had also cracked below the piston due to side loading from excessive wear on the crosshead shoe. Design changes corrected such problems on later engines.[64]

The *Selandia* remained in East Asiatic Co. service until 1936 when a Norwegian company bought her. In 1940 a new owner sent her to Japan to do coastal trading—still with the original engines. On January 26, 1942, she ran aground and later sank off Point Omae Zaki east of Nagoya.

Fig. 16.22. Burmeister & Wain and shipping company officials
on one of the *Selandia*'s unofficial sea trials, February 1912. First row,
second from left, Director Ivar Knudsen; second row, third from left
(with bowler hat), Chief Engineer Olav. E. Jorgensen; on the first step,
Director Martin Dessau; behind him, Engineer Christian Jensen.

The *Fionia*, a sistership to the *Selandia*, was scheduled to make her first
voyage to Bangkok in June 1912, but instead went to Kiel where Lord
Pirrie of Harland & Wolff and Albert Ballin, head of Germany's Hamburg-
Amerika Line, waited to see her and meet with Andersen of EAC. (Ballin
wanted a diesel-powered ship in his fleet, but his hopes had been dashed
when a Weser-built Junkers engine failed.) The *Fionia* arrived in Kiel during
a German naval regatta attended by Wilhelm II, so her visitors also included
a curious Kaiser.[65] The outcome of discussions between Ballin and Andersen
was the *Fionia*'s sale to Hamburg-Amerika who changed her name to
Christian X, the King of Denmark. She would sail under several owners with
her original engines until scrapped in 1939.

The third ship in the original trio, the *Jutlandia*, opened a British chapter
of B&W history. Her engines did not enjoy initial success because of troubles
caused mainly by Barclay, Curie's unfamiliarity with diesel manufacturing.
A lament by its Scottish foreman was that "we are shipbuilders and engine
builders, not watchmakers!"[66] Whether apocryphal or not, the comment
reflects the difficulties faced during manufacture and early service. Even a
loan of Jorgensen and other B&W engineers did not materially help as the
Jutlandia was two months late arriving in Copenhagen from Glasgow on

July 19, 1912. The next year she was tied up for two months in a French port with engine repairs. By August 1914 the *Selandia* had made seven Far East trips, but the *Jutlandia* was only on her fifth. EAC sold her in 1934 and she was broken up in 1938 after being damaged from going aground.

Enthusiasm over the new diesels ran high because of the *Selandia's* London call. On March 25, 1912, just a few weeks later, B&W formed Atlas Mercantile Co. to sell licenses in all countries except Scandinavia. Barclay, Curle would now pay royalties to Atlas and not B&W. The plan also included acquisition of a large factory in Britain where a new company, Burmeister & Wain (Diesel System) Oil Engine Co. incorporated in April 1912, would build diesel engines. Barclay, Curle owned such a facility next to its Glasgow shipyard and had been given to believe B&W would buy it. So positive was Barclay, Curle of this outcome that it sold its own B&W license to Atlas for stock in that company. In this way it would receive shareholder profits from Atlas as well a profit from the factory sale. Such plans did not materialize.

In July 1912 B&W (Diesel System) Oil Engine bought instead Harland & Wolff's Lancefield Works across the Clyde from H&W's Govan shipyard in Glasgow.[67] Exactly how this came about is murky, but there is no doubt that the key individual in the purchase was Lord Pirrie, the aggressive head of Harland & Wolf Ltd. whose main shipyards were in Belfast. He had said after seeing the *Fionia* in Kiel ". . . in my opinion a new epoch has been started in the history of shipping.[68]

Swan, Hunter & Wigham Richardson had bought heavily into Barclay, Curle just before the factory sale to B&W fell through. Swan, Hunter's purchase was made with an expectation that the new company would supply engines to both the Glasgow and its Newcastle shipyards. After this deal failed, Barclay, Curle had no choice but to sell its one-third interest in the B&W (Diesel System) Oil Engine Co. to Harland & Wolff. On January 24, 1913, Barclay, Curle formed the North British Diesel Engine Works with the idea of using its engine factory site to build large marine engines of its own design. The war delayed these plans until 1918 after which North British was to make engines for about ten years.

Jorgensen, who became the first managing director (president) of B&W (Diesel System) Oil Engine Co., hoped to start production in December 1912 as by then the company had engine orders for six ships.[69] The first, destined for the *Mississippi*, did not leave the factory until December 1913. On the second ship an expensive piston cooling problem arose that was attributed to a Jorgensen design decision. He corrected it, but seeds were sown for his departure in 1915. An immediate result of the piston problem was his downward move to Technical Director. Frederick E. Rebbeck (later Sir Frederick) from Harland & Wolff became the new head.[70]

Patent ownership disputes between B&W and Jorgensen would last until 1924 when the Danish High Court awarded him damages. He was a consultant for Worthington Pump & Machinery in the US from 1919 to 1928 on the design of its double-acting, two-stroke diesel. For the next ten years, he served as a consulting engineer for an English shipyard, and from 1936 until his death in 1949, Jorgensen had his own consulting office in Denmark.[71]

Between 1913 and 1918 the B&W Glasgow works furnished engines for fourteen ships despite strictures imposed by war. Harland & Wolff, the sole British licensee, bought out B&W's interest in the company in 1918, and all production moved to Belfast when double-acting engine production began in the mid-1920s. H&W was the most prolific builder of Burmeister & Wain diesels for many years and in time would share in the development of new engine models.

Other companies became licensees after the *Selandia* success and before 1918. Two who generated good royalty streams to B&W were Akers Mek. Versted, in Oslo (licensee years: 1912–1981), and Götaverken AB in Göteborg, Sweden (1915–1945). William Cramp & Sons Ship & Engine Building Co., Philadelphia (1916–1925), and Deutsche Oelmaschinen, Hamburg (1917–1927) built a few ship diesels. Union Tool Co., Los Angeles (1914–1920), and Rotterdamsche Droogdock (1915–1922) acquired licenses but never made an engine.

SA John Cockerill in Seraing, Belgium, bought a Belgian license from B&W through London's Atlas Mercantile Co. in 1913.[72] The company needed two engines for a shallow draft Congo River boat under construction. It was slated to have 650 bhp engines drawn up by Rudolf Diesel's office, but these were ultimately rejected (chapter 19). B&W then designed a smaller, 430 × 430 mm bore and stroke version of the eight-cylinder, reversible *Selandia* diesels. Not known is how Cockerill intended to match the 280 rpm B&W engines with propellers having to turn twice that fast in a space-limited tunnel hull. The onset of war solved this problem before the engines could be delivered. They were appropriated by the German Army and later used to generate electrical power for searchlights protecting an important Belgian waterway. Cockerill did not build B&W diesels again until 1920 as so many of its machine tools had been removed by the invaders. The B&W license lasted until 1989.

Burmeister & Wain introduced a new generation of marine diesel for East Asiatic Co.'s second *Fionia*, which began sea trials on December 18, 1913. Larger than the *Selandia*, she had a length and beam of 410 × 53 feet, or about the size and displacement of a World War 2 "Liberty" ship. There were two, 1,650 bhp engines turning at 100 rpm, but this power now came from six cylinders of 740 × 1,100 mm bore and stroke. An imep of 91 psi at the rated 2,000 ihp remained the same as the smaller eight-cylinder engines.

The basic structure consisted of two sections with three cylinders in each so that the cylinder axis fell midway between box column-like A-frames. The four separate frames per section, eight in total, rested on a two-piece bedplate. Cylinders were cast in two pairs of three, each with the square corners of the castings bolted together to form a rigid upper frame. Long bolts ran from the top of the cylinders through the bedplate to keep the cylinders in compression. Larger, oil tight doors between the A-frames on one side and below the crosshead supports on the other offered good accessibility to rod and main bearings. The camshaft drive was changed from the rods to a set of three intermediate spur gears interposed between the crank and cam.

Piston cooling came from sea water instead of oil with telescoping tubes carrying the water to and from the pistons. Metallic packings on the tubes apparently gave few problems.

Eccentrics on the camshaft drove individual fuel pumps for each cylinder. Reavell three-stage compressors driven off the front of each main engine supplied air for starting and injection. A backup compressor operated by a 200 bhp auxiliary diesel also supplied the extra air for maneuvering.

Carried over from the earlier engines was a reversing system of similar design with the sliding camshaft and the auxiliary shaft to rotate the lower end of the pushrods away from their cams when either in the "off" position or when the shaft was being slid to a new position (fig. 16.23).

B&W turned out these and other engine models ranging from 125 bhp to 265 bhp per cylinder with "almost monotonous regularity" with the pace quickening as the war drew to a close. Eleven ships powered by these engines had entered service by 1915. (Not included in this total are the B&W

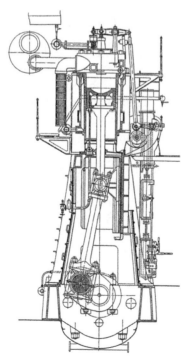

Fig. 16.23. Second-generation (1913) Burmeister & Wain marine diesels. Six cylinders, 1,650 bhp at 100 rpm. *Engineering, January 1914*

submarine diesels seen in the next chapter.) Technical literature of that period substantiates their success.[73]

Burmeister & Wain remained undisputed leader in the large marine diesel field until the 1920s when Sulzer Bros. engaged them in what would become a battle between the reigning four-stroke and the Swiss two-stroke challenger.

Gebrüder Sulzer

The Sulzers made great strides in marine diesels during the 1911–1918 period but did not receive deserved recognition until later. Only one cargo ship powered by engines from Winterthur, the *Monte Penedo*, obtained much publicity just before the war. Until that ended the company was unable to again market its larger marine diesels. Nonetheless, new commercial models were perfected and remained in the factory for customers only until the hostilities ceased. A single-cylinder test engine of 2,000 bhp also built during this enforced interruption (chapter 17) taught many lessons that would be applied to Sulzer products of the 1920s.

Diesels similar to the type 4SNo.6a engines of the ill-fated *Romagna* were installed during 1912 and 1913 in other vessels, including a Russian and a Japanese tow boat.[74] Updated versions (4SNo.4a) of the overhead cam with inclined valve engines (4SNo.3) were also placed in several boats in that same period. Sulzer submarine engines made their appearance in 1912 (chapter 17).

The *Monte Penedo* was built by Kiel's Howaldtswerke for the Hamburg-Südamerikanische Dampfschiffahrts Ges. (Steamship Co.) in 1912. Sea trials began on August 10, and she left for South America on the thirty-first. The 6,500 tons displacement ship was 350 feet long and had a beam of 50 feet. At a draft of 18 feet, the cruising speed was 9.5 kts.[75] Two, crossflow scavenged 4SNo.9a diesels produced 850 bhp at 160 rpm. Bore and stroke were 470 × 680 mm.

The engine structure differed in that each cylinder rested on its own A-frame instead of a crankcase enclosure (fig. 16.24). Still tying the structure together were bolts running from the top of the cylinder heads through to the bottom of the bedplate. Cylinder liners had intake and exhaust ports aligned with those in the jacket castings. Two-piece pistons consisted of a long tubular trunk attached to the water-cooled piston head holding the rings. Telescoping tubes supplied and carried away the water. Sea water was the coolant except when running in harbor areas and then fresh water from the condenser was circulated. Experience had shown that silt entrained in shallow water collected in the water passages and could cause overheating, leading to liner, piston, or head cracking.

Pressure lubrication was fed to all main, rod, and crosshead bearings. Cylindrical-shaped bearing shells allowed replacement of the lower main bearing halves without first having to remove the crankshaft. The steel-

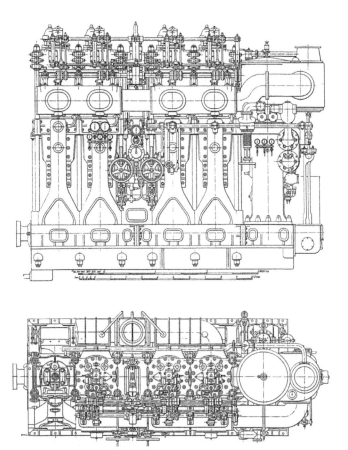

Fig. 16.24. Sulzer 4SNo.9a two-stroke diesel in the *Monte Penedo*, 1912. *Engineering, September 1912*

backed babbitt ("white metal") shells could be rotated out when the upper half was removed.

A scavenge air pump and first stage of the injection and starting air compressors were driven by a crank throw on the front of the engine (fig. 16.25). The first-stage compressor piston served as a crosshead for the scavenge pump above it. Second- and third-stage compressors stood beside the first stage/scavenge pump and were reciprocated by rocker linkage from the connecting rod of the crankshaft-driven first-stage piston.

Two side-by-side levers performed the reversing process. These air-assisted controls were located under handwheels used as a backup in case the air servo mechanism failed. A third lever, which served as the fuel throttle, also reduced injection air pressure for starting and as speed (load) decreased.

The right-hand lever changed the suction and discharge timing of a piston-type valve controlling suction and discharge of the scavenge air pump. Lever movement from "ahead" to "astern," for example, shifted linkage

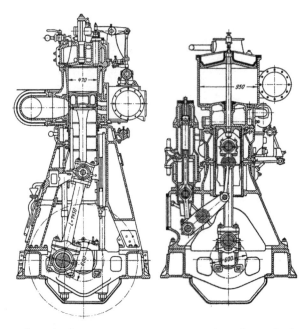

Fig. 16.25. Scavenge pump, compressors, and engine cylinder sections on Sulzer 4SNo.9a marine diesel. *Engineering, September 1912*

reciprocating the valve to alter the timing. The same action also modified camshaft timing by rotating it fifty degrees in relation to the crankshaft; this shift enabled an accommodation of injection timing change with only one cam. However, two cams were needed to admit starting air into the cylinder at the proper time in each direction. The camshaft timing was changed by means of a worm gear attached to a splined sleeve on the vertical drive shaft. This was raised or lowered in a way similar to that on Fiat submarine engines (fig. 15.47).

The left-hand lever had several positions in order to sequence a series of events: stop fuel flow to the nozzles by holding open the pump suction valves, lift roller followers for pushrods actuating the starting valves off their cams, and place these followers onto one or the other of the two starting cams. This swinging movement was possible because the pushrods could pivot on each end. A supply valve between the air reservoir and the starting valves was also opened whenever air pressure was called upon to start the engine.

An eccentric and a bell crank on a shaft parallel to the camshaft rotated the follower away from its cams and axially moved it for placement on the desired cam. This auxiliary shaft and a fuel suction valve control shaft were in two sections so that during starting or reversing the left-hand control lever could be set to send air first to all four cylinders, then put two on air and two on fuel, and finally power all cylinders. At a midpoint "stop" setting, fuel flow ceased and the tank supply valve was closed. A dial with an indicating hand showed the operator where his left lever was positioned.

By the time the *Monte Penedo* arrived in Buenos Aires on her maiden voyage, two pistons had cracked at the lowest ring groove. She was laid up there until new pistons could be installed. The failures were attributed to unforeseen stresses caused by the new design, which attached the rod to a rather shallow piston head close to that ring groove.[76] No difficulties were experienced on the thirty-day return trip to Hamburg. Fuel consumption at sea averaged 0.46 lb/ihp-hr, a figure that included all auxiliary pumps.[77] Engine weight without auxiliaries was 55 tons.

The *Monte Penedo* was interned at Rio de Janeiro in August 1914 after she had sailed about 65,000 miles with no major problems. In 1918 the Brazilians claimed ownership, and under the name *Sahara*, chartered her to the French government. By 1921, with 50,000 more miles using crews "with no previous experience of Diesel machinery, . . ." the crankshaft on the port engine broke due to poor main bearing alignment.[78] In 1922 the ship was returned to Brazilian operation.

The British Admiralty ordered two four-cylinder, 1,600 bhp engines from Sulzer in 1913 to gain marine diesel experience. Unlike the pair built by Carels, these Sulzer 4S250 engines were delivered in 1915, but were never installed in a ship.

A cast iron entablature or plate sat on a pair of cast steel A-frames (fig. 16.26). Individual cylinders with a 680 × 960 mm bore and stroke seated in

Fig. 16.26. Sulzer 4S250, 1,600 bhp marine engine tested in 1915.
Sulzer Archives

the plate. Four through-bolts per cylinder held square-shaped cylinder heads to the plates so that the cylinders remained always in compression.

The scavenge pump and compressor were moved and redesigned. A crank throw at the front of the engine drove a vertical, three-stage compressor. Two double-acting scavenge pumps alongside the engine were reciprocated by rocking levers acting through crosshead linkage off two cylinders. Double-acting pumps set a pattern for future Sulzer marine diesels regardless of location.

A simplified reversing mechanism was to be used into the 1920s. Its operation is evident from figures 16.27 a and b. Roller *a1* rode on cam *a* for "ahead" and roller *b1* rode on cam *b* for "astern." A similar arrangement changed timing of the air starting valve. Fuel and starting valve rollers were lifted away from their cams by rotating the eccentric shaft *d*. The "0-10" angular distance shown on the control quadrant modified fuel nozzle valve opening duration, depending on the load. (With marine engines the load is always a direct function of propeller speed—unless the propeller comes out of the water in a heavy sea!)

Test data taken January 1915 provides a good measure of current two-stroke state of the art:[79]

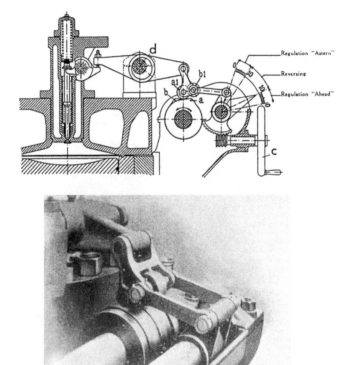

Fig. 16.27. Simplified reversing system of 1914 Sulzer 4S250. (a) Schematic of fuel valve operation. (*Sulzer Technical Review, 1947*); (b) Fuel valve linkage in "astern" position. Sothern, *Marine Diesel Engine Practice, 1922*

Load	Half	Full	25% Overload
Speed, rpm	88	110	119
Indicated hp	1,367	2,418	2,984
Brake hp	806	1,603	1,993
Mechanical efficiency %	59	66	67
Ind. mep, psi	72	102	117
Bsfc, lbs/bhp-hr	0.492	0.479	0.520
Scavenge pressure, psi	1.7	3.1	3.7
Exhaust temp, °F	332	539	698

Fuel was a 0.93 sp gr "crude oil" with an LHV of 18,000 Btu/lb.

A diesel intended for post-1918 merchant marine use was the 4S60 tested in 1917. This four-cylinder, 600 × 940 nun engine carried a factory rating of 1,350 bhp at 110 rpm, but when installed in the tanker *Conde de Churruca* in 1921, that was reduced to 1,250 bhp at 100 rpm.[80] It embodied refinements based on what had been learned from the *Monte Penedo* and the canceled British Admiralty models.

Thus, the stage was set for the Swiss pioneer diesel engine builder to compete at sea with the Danish pioneer. A continuing introduction of new technology by these two rightfully proud companies nourished a rivalry to be joined in by their many worldwide licensees that has lasted to the present time.

Krupp Germaniawerft

Two models of diesels for merchant ships were built by Krupp between 1911 and 1914. One came early enough for service in two tankers, while the other did not go to sea until peace returned. Kiel's great emphasis on naval diesels precluded significant support for a second, commercial program for larger engines. Production continued on smaller diesels of the type described in the previous chapter, and by mid-1912 Krupp had built about fifty of these.[81]

Work began at Germaniawerft in 1911 on a six-cylinder, two-stroke engine with a bore and stroke of 450 × 800 mm. Its design output of 1,250 bhp at 140 rpm was reduced in service to 1,100 bhp at 120 rpm. Bmep at the lower rating was 66 psi.[82] Uniflow scavenging, with two air valves in the head and exhaust ports in the cylinder, followed Krupp two-stroke sub engine practice. The head also held a central fuel injector and a starting valve (fig. 16.28).

A major departure was the use of a monobloc cylinder head and liner to allow more uniform wall thicknesses exposed to the combustion chamber. While theoretically reducing thermal stresses, something not possible with conventional construction where cylinder deck and head met, a monobloc design decreed that a failure in either head or liner made scrap of the entire unit.

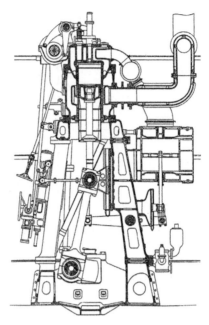

Fig. 16.28. Krupp six-cylinder, two-stroke diesel installed in two ships in 1912. 1,100 bhp at 120 rpm. Rose, *Diesel Engine Design, 1917*

Two-piece pistons, removable without taking apart crosshead and connecting rod assemblies, were water-cooled via concentric supply and discharge tubes in the piston rods; pipes with articulated (grasshopper leg) joints connected passages in the crosshead and frame. This design was discarded on the next series in favor of telescoping tubes as it was difficult to prevent the rubbing joints from leaking and getting water in the lube oil.

Each cylinder had its own fuel pump driven from the vertical shaft turning the high-mounted camshaft. The duration of the opening of the pump suction valve determined fuel charge volume.

Two double-acting scavenge air pumps on the side of the engine were reciprocated by a link attached to the crosshead and rocking lever. The injection air compressor was operated by auxiliary power. On the ships where two of these engines were later installed, either of a pair of auxiliary diesel compressor sets could supply both main engines.

The reversing system resembled those already described with an axially sliding camshaft with two cams—ahead and astern—for each valve and eccentrics in the rocker lever shaft enabling followers to be lifted off the cams. The axial shift was about two inches.[83] A manual handwheel lifted the vertical drive shaft to change the camshaft timing in relation to the crankshaft. Air was not used to assist in raising or lowering the vertical shaft. Because the engine had two pairs of three cylinders, the lever-actuated rocker eccentric shaft was also in two sections. Detents in the lever quadrant positioned the eccentric shaft to place valves in the required modes: (a) full stop, where rocker levers lifted off cams; (b) startup, where only injector rocker lifted and all other rockers on cams; (c) run, where only starting rocker lifted off its cams.

Four of these diesels went into two oil tankers, the *Hagen* and *Loki*, which began transatlantic voyages in 1913. These 7,700 dwt sister ships built for the German-American Petroleum Co. were 400 feet long with a beam of almost 53 feet. Maximum speed was 10.5 kts.

It was reported that on the *Hagen*'s first trip to New York the air compressors had serious problems, and the ship "was three months under repair" after her return.[84] On the second trip some main and auxiliary engine pistons cracked, supposedly due to silt clogging cooling water passages to the pistons. The *Loki* left Hamburg for Philadelphia in October 1913. On her second trip to New York, she was laid up sixteen days for repairs and "is now on her homeward run to Hamburg" (April 1914). Standard Oil Co. bought the Hagen after the war started in Europe and renamed it the *Glenpool*.[85]

The war prevented a pair of improved and more powerful Krupp engines from being installed in one of the largest tankers to be built. This six-cylinder design had a bore and stroke of 575 × 1,000 mm with an intended output of 1,800 bhp at 125 rpm.[86]

Major revisions included a separate head and cylinder, rather than a monobloc, and telescoping cooling water tubes to the piston. These were located outside the crankcase (fig. 16.29). Larger scavenge valves in the head precluded axial placement of the injector. Instead, the fuel valve rocker lever

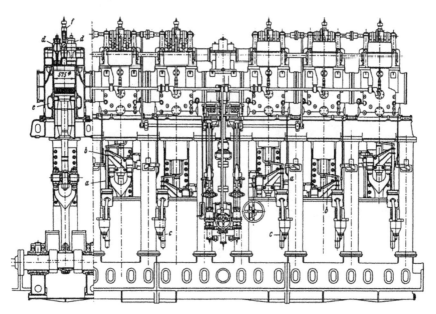

Fig. 16.29. Krupp 1913 marine diesel. 1675 bhp at 106 rpm.
Sass, *Geschichte, 1962*

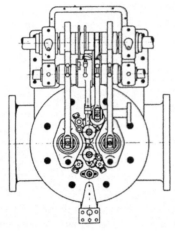

Fig. 16.30. Valve placement in the head of a Krupp 1913 marine diesel showing the two injectors and scavenge air valves and starting valve.
VDI, *Dieselmaschinen, 1923*

acted on a crosshead or bridge to lift the valves in two smaller injectors (fig. 16.30). A free-standing, three-stage compressor was coupled to the front of the crankshaft. The reversing gear had few changes.

Erection of the engines began in late 1913 (fig. 16.31), and ensuing tests showed that the scavenge pumps did not have enough capacity at full load. This caused the engine's rating to be lowered to 1,675 bhp at 106 rpm.

Fig. 16.31. Krupp 1,675 bhp engine being assembled.
Engineering, November 1913

Not until 1920 were the pair installed in the *Zoppot*, a cargo ship making Atlantic runs. Between her maiden voyage to New York beginning on August 1, 1920, and October 12, 1923, she had traveled over 187,000 miles.[87] This is an average of 4,800 miles per month at a continuous speed of 7.24 kts. With a maximum speed of 10.5 kts, it left her little time in port.

Maschinenfabrik Ludwig Nobel

Nobel began a two-stroke program in 1911 with the objective to develop a lightweight, higher-output submarine diesel specified by the Russian Admiralty. In the process of achieving this, it also produced an engine adaptable to merchant marine applications. Nobel was well aware of the two-stroke achievements by other diesel builders, so the decision to join them would serve a dual purpose.

Designed first was a six-cylinder, uniflow scavenged engine with a bore and stroke of 250 × 300 mm.[88] A special feature was its complex, individual cylinder heads, which when all were bolted together, formed a rigid structure along the top of the engine.

Each head had two vertical scavenge air valves along the centerline of the engine (fig. 16.32). Inclined on one side of the head was an oscillating butterfly valve (*A*) working with a slide valve (not shown) to control scavenge air pressure independent of engine speed. A bell crank rotated by a pushrod oscillated the butterfly valve and opened the scavenge valves.

The fuel injector and starting valve were on the opposite and inclined camshaft side of the head. The engine used Carlsund's reversing method,

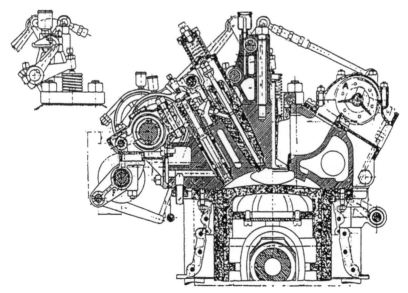

Fig. 16.32. Cylinder head of Nobel 1911 uniflow scavenged two stroke with scavenge air pressure control. *Gas and Oil Power, 1965*

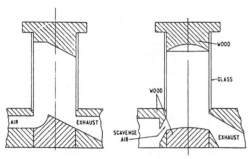

Fig. 16.33. Model used by Nobel in 1911 to determine optimal loop scavenge design. Shown are first and final configurations. *Gas and Oil Power, 1965*

that is, a sliding camshaft and a lever by the camshaft to directly lift rockers off the cams (chapter 15).

Tests in 1911 met the Navy specifications: 350 bhp at 400 rpm and a weight of 24.5 kg/bhp. Bsfc was a respectable 225 g/bhp-hr. During a fifty-hour full-load run, the exhaust had become blacker and blacker until finally fire was pouring out the pipe—yet the engine showed no signs of distress. The cause: a carpenter had stored wood in an old boiler serving as exhaust muffler![89]

Emanuel Nobel's yacht *Gryadoustiy* was the vessel for the unique engine's sea trials.[90] Although it performed well, all agreed that its complicated operation was beyond the capabilities of available naval crews. Only the one was ever built.

This engine also convinced Nobel engineers that loop and not uniflow scavenging held the greater potential. They did not agree, however, on what might be an optimal configuration for the cylinder ports, piston, and head. Providing the answers was a wood and glass model to simulate cylinder scavenging when the air ports opened[91] (fig. 16.33).

A two-cylinder test engine applying the model's results became a prototype of successful marine and submarine diesels (fig. 16.34). Hans Flasche attributes its design to A. Lagerstén.[92] Its two, paired cylinders, two-stage compressor and double-acting scavenge pump were expected to provide development experience for a proposed six-cylinder sub diesel of the same basic design. The 450 × 480 mm engine produced 450 bhp at 320 rpm during tests in 1912. From that summer when testing was completed and until mid-1917, it ran six days a week in the factory with a rating of 350 bhp at 250 rpm. [93]

In addition to new submarine diesels (chapter 17), the test engine's success brought in orders to power Caspian Sea ships. The first in 1915 was for the twin-screw *Imperatriza Alexandra* being converted from steam. (Later ones were for two tankers.[94]) The Tsarina's namesake, carrying one thousand passengers, was 243 ft × 34 ft × 10 feet. Speed was 12.5 kts at a maximum displacement of 1,700 tons.

Fig. 16.34. Nobel 1912 loop-scavenged, two-stroke test engine.
Nobel catalog in Nohab Archives

Her two, four-cylinder, reversible engines had a bore and stroke of 410 ×
500 mm and an output of 600 bhp at 210 rpm. Each weighed about 32 tons
to give a specific weight of 53 kg or 116.8 lbs/bhp.[95]

The trunk piston design had individual cylinders mounted on an enclosed
crankcase resting on a bedplate (fig. 16.35). Cast steel heads held injector,
starting, and safety valves. Connecting rod big end bearings and water-cooled
main bearings were of babbitt ("white metal"), while phosphor bronze was
used for piston pin bearings. The one-piece piston had six rings near the
crown and one below the row of ports.

Telescoping tubes attached to the bottom of the water-cooled piston carried
the water to and from fixed tubes mounted on the inside of the crankcase.
These were later replaced by the "preferred . . . grasshopper arrangement."[96]

Cylinder liners and piston pins were pressure lubricated by separate pump
systems. A gravity oil supply fed collector rings generating a centrifugal force
that translated into enough pressure to maintain an oil film in the main and
lower rod bearings.

In one housing between the front cylinder and scavenge pump were four
fuel pump plungers reciprocated by eccentrics on the vertical shaft after the
Hesselman system. The vertical shaft also turned the governor, which was
placed below the pumps.

The double-acting scavenge pump driven by the crankshaft supplied air
pressure to a common intake manifold at about 1.75 psi. A vertical rotary
valve for each cylinder, turned by the camshaft, prevented air from entering
the five cylinder ports. The valve opened after the piston had uncovered the

upper half of the three exhaust ports as it descended. It stayed open until the intake ports were covered again. Fresh air thus entered the cylinder until both the intake and exhaust ports simultaneously closed. Although this system did not have a supercharging effect as was possible with Sulzer's double row of intake ports, it at least assured that no residual gases remained in the cylinder.

Two crank throws operated the three-stage compressor with one being for the first stage and the other for the intermediate and third stages. This not only kept overall height down, but also simplified the compressor and its maintenance.

Controls for starting and reversing were uniquely placed. Instead of hand-wheels and short levers placed either low or high on the side of the engine, there were three long, floor mounted levers attached to a platform at mid-height remote from the engine (fig. 16.34). Nearest it was the starting lever with "start, "stop," and "go" positions. The outer lever controlled fuel delivery.

The middle lever had notches for "astern," "ahead two-stroke," and "ahead four-stroke." Because of a concern over irregular firing at light loads and speeds, the engine could be made to run in the four-stroke mode under such conditions. Starting was always on two stroke. With the camshaft turning at *half* speed, the choice of a set of either single-lobed or double-lobed cams determined whether the engine ran as a four stroke or two stroke when it opened the injector fuel valve. This feature was abandoned when the concern proved groundless.

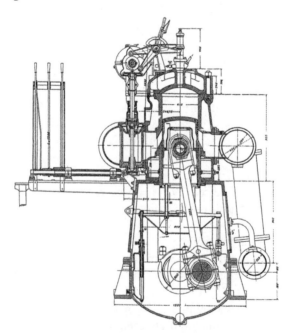

Fig. 16.35. Cross-section of 600 bhp Nobel 1916 marine diesel.
Engineering, December 1916

A sleeve slidable on the hollow camshaft held each cylinder's cams: two double-lobed for starting; two double-lobed for running ahead in two stroke and a single-lobed for ahead in four stroke. Slidable in the hollow shaft was a lever-operated control shaft with pins that engaged the sleeves through axial slots.

The cam followers were lifted away from their cams by a fourth lever attached to an eccentric shaft running through the rocker arm bushings, which changed the pivot points of the arms.

Extensive tests made in September 1915 are briefly summarized for the engine's rated load and speed of 600 bhp and 210 rpm: 4.83 kg/cm² bmep, 77.5 percent mechanical efficiency, 60 atm injection air pressure, 201 g/bhp-hr bsfc, 31.4 percent brake thermal efficiency, and 278°C exhaust temperature. Exhaust gas analysis showed 4.5 percent CO and 14.5 percent O_2. The fuel, Baku crude naphtha, had a LHV of 10,000 cal/kg. A 14:1 compression ratio was used.[97]

About nine months were lost in building the second engine for the ship due to interruptions caused by the war. During the interim the first engine drove a 400-kw generator in the factory. The ship's performance in service with the engines is not known.

Nobel made 151 merchant marine and naval installations between 1905 and 1917. Twenty-five were auxiliaries to generate shipboard power or to handle cargo. The numerous Kolomna-built diesels are not included. By the time production came to a halt in 1918, the St. Petersburg factory had turned out a grand total of 659 engines with an aggregate of 144,000 hp.[98]

At the start of the war, the English and German employees at Nobel left the country, and the Swedish contingent returned home when the revolution began in 1917. The Russian-born Nobel and Hagelin families also managed to escape, although a few had harrowing experiences during their exodus. After Lenin's government assumed power and tried to erase the story of what the Swedes had accomplished, the Nobel factory was renamed Pervij Sawod Russkiy Diesel, or First Russian Diesel Engine Factory.[99]

In 1918 Emanuel Nobel and Marcus Wallenberg financed a new company, Svenska AB Nobel Diesel (Swedish Nobel Diesel Co. Ltd.) in Nynäshamn, Sweden, to design marine engines and then sell licenses to have someone else build them. This company closed in 1925, and all further work was carried on by Nydqvist och Holm AB in Trollhättan, the same firm that bought the diesel business from Atlas Copco and that Wärtsilä now owns.

M.A.N. Nürnberg and Blohm & Voss

While M.A.N.'s Augsburg division concentrated on four-stroke submarine and smaller diesel developments, it was Nürnberg who pursued two-stroke single- and double-acting engines of high output. The challenges were immense and successes were limited, but the knowledge gained provided technology to draw on when diesel applications for peacetime returned.

Nürnberg and Blohm & Voss (B&V), the well-known Hamburg ship and steam engine builder, formed a cooperative arrangement in 1909 to develop two-stroke, double-acting marine diesels.[100] This was even before Nürnberg began its double-acting, horizontal diesel program (chapter 10). M.A.N. contributed the design, and B&V participated in the building and testing.

Their joint effort led to a uniflow scavenged, three-cylinder engine with a bore and stroke of 480 × 650 mm (fig. 16.36). Its rating of 850 bhp at 120 rpm yielded a bmep of 4.5kg/cm².

The upper cylinder head held two scavenge valves, fuel nozzle c and starting valve e as in conventional engines. Nozzle c and one scavenge valve (d) were placed horizontally in the lower cylinder wall. Cams on shafts f, g, and h opened the valves. No starting valve was used for the lower combustion chamber. Scavenge pumps m and water pump n were driven off the crosshead. A water-cooled piston uncovered exhaust ports at the end of each stroke. The upper head sat on a casting, which included the jacket and upper liner and encased the water-cooled exhaust ports. This piece in turn mated with a combined lower cylinder wall and partial head casting holding a stuffing box

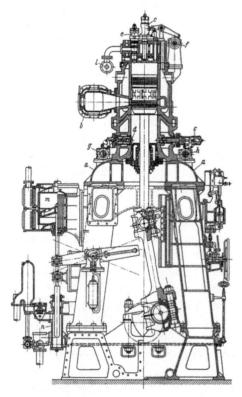

Fig. 16.36. M.A.N. and Blohm & Voss double-acting, two-stroke marine diesel of 1911–1913. Three-cylinder, 850 bhp at 120 rpm. Sass, *Geschichte, 1962*

to seal the piston rod. Two vertical shafts drove three camshafts (fig. 16.37). No information is available to explain how reversing was accomplished.

Nürnberg's one development engine was reportedly scrapped, and no drawings of it are known to exist.[101]

The piston stroke of the two engines completed by Blohm & Voss just before the war was increased to 710 mm, but brake horsepower remained unchanged.[102] Performance and bsfc from the lower cylinder half were disappointing due to poor scavenging. Upper cylinder heads and pistons developed cracks after some hours of running, and torsional vibrations reared their ugly head.[103]

These difficulties forced revisions delaying installation of the engines in the motorship *Fritz* until 1915, a time precluding ocean operations. Ending this saga was a November 1919 ownership transfer of the *Fritz* to England "in compliance with the Versailles Treaty." English crews, completely unfamiliar with diesel engines, later prompted a technical journal to write that "We ought to have let the Germans keep the ship. . . . It would have better served the matter."[104] The diesels were replaced by steam shortly thereafter.

When M.A.N. saw early in the double-acting program that it would take much longer than planned, a concurrent development began on a model 6KS single-acting, two-stroke marine diesel (six cylinders, *Kreuzkopf-Schiff* or crosshead-marine). Many features were adapted from the double-acting design, including uniflow scavenging through cylinder head valves. The first of these large engines, built in Nürnberg during 1913, had been destined

Fig. 16.37. Three-cylinder, double-acting engine of figure 16.36 on Blohm & Voss test stand, c. 1911. Note the man on the platform. *MAN Archives*

for an English tanker, but its actual disposition is not known. It stood seven meters high, and with a bore and stroke of 600 × 900 mm and bmep of 4.2 kg/cm², the output was 1,600 bhp at 130 rpm.[105] Details of the reversing system are, like that of its double-acting cousin, not fully known.

Blohm & Voss then built two 4KS92 engines (four-cylinder, 920 mm stroke), which carried a rating of 1,350 bhp at 120 rpm (fig. 16.38). Fuel consumption was a rather high 210 g/bhp-hr as was the basic weight at 160 kg/bhp. Development of the single-acting design proceeded rapidly enough for the two to be installed in the Hamburg-Amerika Line freighter *Secundus*. B&V had planned to supply the double-acting type when engines for the ship were ordered in 1910. The *Secundus* made a reportedly successful trip to the United States before the war and as part of the reparations was turned over to the French in 1919.[106]

Nürnberg designed two other two-stroke KS models, a 4KS50 and a 6KS100, with an intention that they be built by licensees. Records do not show that a 4KS50 was ever made, but the Nelseco division of the Electric Boat Co. did make two 6KS100s[107] (chapter 17).

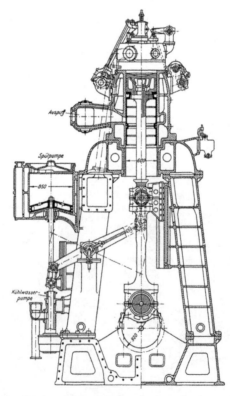

Fig. 16.38. Blohm & Voss four-cylinder, single-acting marine diesel built in cooperation with M.A.N., 1913. 1,350 bhp at 120 rpm. Maag, *Dieselmaschinen, 1928*

All stationary and marine diesel production went to Augsburg in 1919 when M.A.N. began building the KS series as a four stroke. These single- and later double-acting engines would become a major product line for the company.

Hugo Junkers

Hugo Junkers exuded great confidence in the future of his opposed-piston marine engines when he lectured before the German Shipbuilding Technical Society in November 1911.[108] He presented drawings showing eight of his tandem four-cylinder (two-crank throw) diesels laying on their side and turn- ing the four propellers of a 24,500-ton battleship. Even though the total engine output was 36,000 bhp, it was not that great a leap of faith for him to suggest such an application. He had already tested a tandem, two-cylinder engine of that same design, which produced more than 1,000 bhp (chapter 14).

Furthermore, two 800 bhp at 120 rpm diesels for the *Primus*, a Hamburg- Amerika cargo ship, were under construction by AG Weser, a licensee in Bremen. Except for being vertical, they were scaled-down versions of the battleship engines. The tandem six-cylinder engines had twelve pistons in six cylinders acting on three crank throws (fig. 16.39). A bore and stroke of

Fig. 16.39a. 800 Bhp Weser-Junkers marine diesel.
Chalkley, *Diesel Engines, 1914*

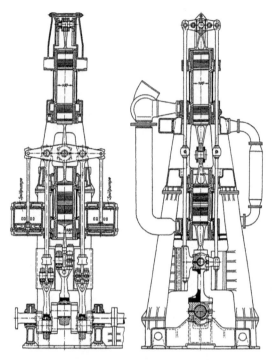

Fig. 16.39b. Views of Junkers-Weser opposed-piston marine diesel of 800 bhp, c. 1913. Three tandem cyliners of 400 mm bore × 2 × 400 mm stroke. *Engineering, November 1911*

400 × 400 mm yielded an equivalent expansion stroke of 800 mm in each cylinder.[109]

Two double-acting scavenge air pumps mounted on the frames of the end cylinders were reciprocated by the lower piston yoke and discharged into a common manifold. Air drawn in through intake ports at mid-cylinder was discharged through ports in the outer ends. The compressor for injection and starting air (not shown) was similarly driven from the center cylinder pair, with the first-stage compressor mounted on one side and the second and third stages on the other.

Two oppositely positioned fuel injectors and a starting air valve were located at the inner stroke point of the cylinder. A rocker lever working directly off a slidable camshaft opened each one individually. With two cams per valve, reversing was possible by only axially shifting the camshaft because the ported cylinders and scavenge pumps eliminated any mechanism requiring alterations.

The Weser engines were installed in the *Primus* sometime in 1913, but due to mechanical weaknesses, owner fears, or both, they were replaced with steam engines before the ship was delivered.[110] Junkers believed that Weser had caused this debacle to happen by taking too many liberties with his

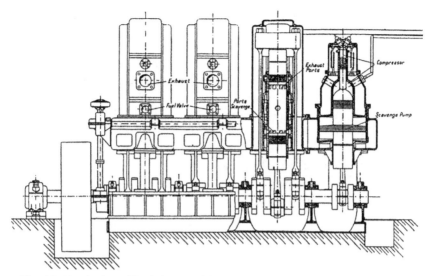

Fig. 16.40. Junkers-Frerichs 150 bhp at 280 rpm tugboat diesel of 1912.
The Engineer, October 1912

design. It was reported that Weser "had gone out of their way to make the engine as complicated and unsuitable for marine purposes as possible . . ."[111]

J. Frerichs AG of Osterholz-Scharmbeck (near Bremen), a more successful Junkers licensee, built one large and several smaller marine engines. The latter were three-cylinder, single opposed for tugboats. With a 200 mm bore, 240 mm piston stroke (480 mm expansion stroke), they had an output of 150 bhp at 280 rpm[112] (fig. 16.40). A fourth crank throw operated an axially arranged scavenge pump and two-stage injection air compressor.

One injector with one fuel valve cam served each cylinder. To change injector and fuel pump timing for reverse running, a lever rotationally shifted the clearance in a dog coupling on the vertical shaft from one contact side to the other as on Nürnberg-two-stroke marine engines (fig. 15.13). Three fixed cams on the camshaft reciprocated individual fuel pumps and three cams with adjustable pivots controlled the duration of suction valve opening.

Starting valves did not open directly into the cylinder but into an air pipe leading to each cylinder. Check valves at the cylinder entrance prevented combustion gases from entering the air supply pipe. Each starting valve was served by two cams and two rockers on eccentric shafts so that by moving a second lever, depending on direction desired, one or the other of the rockers opened the valve. Under certain conditions, admitting air pressure into just the three power cylinders might not provide sufficient starting effort. This was overcome by an additional set of cams and rockers opening a valve to also admit air pressure under one side of the scavenge pump piston.

The largest Frerichs-Junkers diesels installed were a pair for the tanker *Arthur von Gwinner*, which made her first voyage in November 1913. These engines were of the simpler, single-opposed design with two cylinders of 440 × 520 × 2 mm and a rated power of 650 bhp.[113] Like the tugboat engine, there was one fuel valve per cylinder, and scavenge air entered at the bottom of the cylinder and the exhaust left at the top. Two double-acting scavenge pumps were driven from linkage on one cylinder and two high-pressure compressors similarly operated from the other. Reversing was like that on the smaller engine. During a six-day, nonstop test the engine developed 814 bhp (1,100 ihp and 74 percent mechanical efficiency) at 185 rpm. Fuel consumption varied from 0.43 to 0.46 lbs/bhp-hr. However, when the test ended "it was found that all cylinders were badly cracked."[114]

Professor Junkers provided a weight breakdown of all elements comprising the equipment and propulsion machinery for the ship:[115]

	Tons (long)
Two main engines with compressors	98
Accessories	35
Auxiliaries	35
Piping	12
Shafting and propellers	20
Spares, etc.	15
Flywheels	20
Total	235 = 405 lbs/bhp

Thus, the engines themselves made up less than half of the total that naval architects worked from in calculating the overall weight of the power plant. This total was the usual practice when giving specific engine weights for ships.

Five months after the *Arthur von Gwinner* entered sea service under a normal engine rating, a report was that "so far her work has been satisfactory; no delays or casualties...."[116] It is not known what became of her.

Maschinenbau AG Gebrüder Klein of Dahlbruch, Westphalia, had been building Junkers engines under license for "several years" prior to 1913.[117] At that time it introduced a single-opposed, two-cylinder marine engine developing 100 bhp at 300 rpm. Fuel consumption was only 185 g/bhp-hr (0.407 lb). This company and others produced numerous marine engines in the 100 to 450 bhp range, including horizontal tandems for shallow draft Danube River boats. Junkers engine production of AEG (General Electric Co.) in Berlin had "been so large" that "an island in the Elbe [was] purchased for the yard and factory."[118]

But for the war, the Junkers small-to-medium size marine diesels might have become a major market factor. Their low fuel consumption, simple reversing mechanism, and ability to run smoothly over a wide speed range were much-sought-after attributes by the many European canal and river boat operators.

William Doxford & Sons Ltd.

Only one pre-1914 Junkers licensee would go into production with a Junkers-type, large marine diesel when peace returned. It was English. William Doxford & Sons Ltd. in Sunderland bought a Junkers license in 1913[119] and the next year tested a prototype, one-cylinder engine. The war, however, stopped further development by this important Wear River shipbuilder near Newcastle.

Doxford had already tested the waters with a single-cylinder, two-stroke diesel designed by Karl O. Keller, a Swiss engineer who came from Sulzer; he would continue as chief designer for the new opposed-piston engines. This first diesel had a bore and stroke of 19.5 × 37 inches and developed 250 bhp at 130 rpm.[120] Although it performed reasonably well, enough concerns about durability and economy had caused the company to look elsewhere for one better meeting these conditions. The absence of cylinder heads with the Junkers was attractive as heads were a principal area of trouble. Robert Doxford and Hugo Junkers knew what was needed, and their common goals quickly led them to the license agreement.

The prototype closely followed the functional features of a Junkers single-opposed design. The bore was 500 mm, and the stroke of each piston was 750 mm. With an output of 500 bhp at 190 rpm the bmep was a high, for a two stroke, of 90 psi. During a five-week test under Lloyd's supervision, the engine was stopped for minor repairs only four times, and the output never fell below 470 bhp. Fuel consumption using a Mexican oil with a .91 sp gr. never went above 0.43 lbs/bhp-hr. The engine ran at speeds as low as 35 rpm with no misfiring.[121]

Starting, maneuvering, and fuel settings were controlled by handwheel *A* and reversing lever *B* (fig. 16.41). Fuel valve *C* and its opposite one in back were opened by one or the other of cam sets *E* and *F*. Starting valve *G* was similarly opened by one of the cams *H*.

Starting and reversing the engine had the following sequence: In its "neutral" position control wheel *A* raised rod *O* to rotate shaft *P* and lift the three rockers off their cams. Pulling lever *B* shifted arms *L* so that the cam sleeve assembly slid on the camshaft in the desired direction. Turning the wheel to its first position lowered the starting rocker onto a cam and admitted air into the cylinder. With the engine turning, the handwheel was rotated past neutral in the opposite direction to a first position, which both raised the starting rocker and lowered the fuel rockers. A further rotation delivered more fuel by causing the fuel valves to have a higher lift and the suction valves of two fuel pump plungers *S* to close sooner. Eccentric *T* on the camshaft reciprocated the plungers. To prevent overspeeding, the governor overrode the action of rods *Q* and *R*.

Fig. 16.41. 1914 Doxford opposed-piston development engine.
500 bhp at 120 rpm. *1920 Doxford Brochure*

The engine described would become the forerunner of a long line of
Doxford diesels, which gained an earned reputation continuing from the
1920s for high reliability and excellent fuel economy. The Doxford name
would disappear as a result of the great downturn of the British shipbuilding
industry beginning in the 1960s. Nevertheless, it deserves to be remembered
for its own accomplishments and for championing the dreams Hugo Junkers
had of his engines traveling the oceans.[122]

Franco Tosi

Franco Tosi also entered the pre-1918 marine diesel market. While its
engines of this type did not play an important role, facets of their design
merit mention to show yet another way the reversing challenges offered by
ship engines could be solved.

Details of the first Tosi marine version, a four-cylinder, two-stroke,
appeared in 1912.[123] It developed 500 bhp at 170 rpm from a bore and stroke
of 400 × 650 mm. Four scavenge valves in the head and exhaust ports in the
cylinder gave uniflow scavenging. A cam-operated rocker lever acting on a
Carels-like end-pivoted lever (fig. 16.8) opened a pair of valves. Mounted on

A-frames of the front and rear cylinders was a scavenge air pump, and above it were the second and third stages of a three-stage compressor. The scavenge pump supplied first stage air. Both were driven by rocking lever linkage from the crosshead pin. Two-piece pistons were cooled via a jointed swivel arm carrying both supply and drain water between frame-mounted pipes to passages in the crosshead pin and rod.

A camshaft timing change in combination with the turning of an auxiliary maneuvering shaft effected reversing. The camshaft was rotated in relationship to the crankshaft by lifting the vertical, intermediate shaft, which had helical gears on both ends. Scavenge valves required nothing other than this camshaft timing shift to open correctly with the engine running in either direction.

Basic timing and lift of the fuel valve for forward or reverse came from the camshaft on which an eccentric a moved the fuel lever to act on a bell crank lifting the valve (fig. 16.42). By turning the maneuvering shaft, a second eccentric d tailored the position of the lever's pivot point to vary the time when its free, roller end contacted the bell crank. The second eccentric acted through straddling links whose lower ends were pinned to the lever. In this way the lever could be varied from a maximum lift at full load running to a lesser lift (i.e., shortened opening duration) for part-load or slow-speed running and finally to no lift when the engine was stopped or using starting air. Two advantages gained by shortening lift duration of the fuel valve were a lower air usage, and more importantly, the less cooling of heated cylinder air by injection air. This was highly desirable to ensure more reliable ignition of smaller fuel charges with the reduced cylinder compression pressures

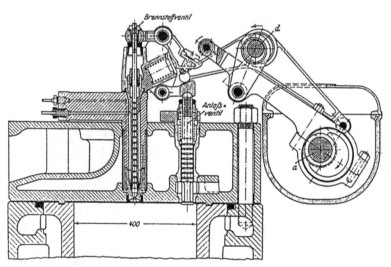

Fig. 16.42. Tosi 1912, 400 × 650 mm, two-stroke marine engine showing eccentric lever and bell crank to lift fuel valve (*Brennstoffventil*) and cam-operated rocker lever (cam not shown) to open starting valve (*Anlaßventil*). *Zeit., VDI, 1912*

when running at slower speeds and lighter loads. A mechanism to vary the fuel valve lift was found on all larger marine engines by this time.

Each cylinder's starting valve was opened by either an "ahead" or "astern" cam that acted on its own lever to open the valve. These levers pivoted on eccentrics such that by turning the maneuvering shaft one became operative while the other was lifted from the valve. A third eccentric position lifted both levers whenever the engine was stopped or running under power.

It is not known what vessels received these two-stroke models. Post-war Tosi marine engines were all four stroke. This new line was more successful and led to several licenses with other companies.

Savoia

An Italian marine diesel with unique technical features was a trunk piston, four-cylinder, two stroke made by SA Cantieri Officine Savoia in Ligure.[124] Reported output was 440 bhp at 200 rpm from the 350 × 500 mm bore and stroke cylinders.

Two coaxial scavenge pumps and three-stage air compressors, driven off the crankshaft but through a crosshead and piston rod, were between paired power cylinders. Individual monobloc heads and liners were inserted into a one-piece casting, which included the A-frames and water jackets of the paired cylinders. The trunk piston, uniflow scavenged engine had two scavenge valves in the head and exhaust ports extending around the cylinder (fig. 16.43). The fuel nozzle was offset from the engine centerline.

Four short, oscillating camshafts were at fuel nozzle height; a shaft on each engine side served a cylinder pair. Two eccentrics on both ends of the crankshaft moved rods, each of whose upper ends was attached to a short arm

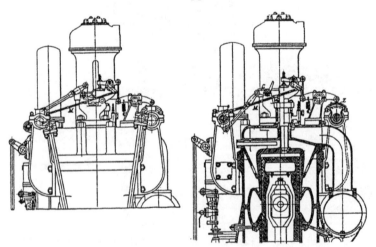

Fig. 16.43. Section and end views of Savoia 1912 marine diesel showing oscillating camshafts and the compressors extending above the top of the fuel injectors. *Engineering, February 1914*

on the end of a camshaft to impart the oscillatory motion. Segment pieces having the cam contour were fastened to the shafts. Another shaft with lobed cams directly reciprocated paired fuel pump plungers in two housings through rollers on the plungers. Fuel delivery was controlled by varying opening time of the pump suction valves.

Air pressure valved into the double-acting scavenge pumps started the engine (i.e., Hesselman's system). Means were also provided for changing fuel and air valve timing when reversing the engine and for varying fuel valve lift.

A Savoia engine as described was installed in the three-masted, 56 m long schooner *Aosta*, and some performance data was taken (no bsfc) with the vessel at sea. Not known are the engine's success nor how many of that design were built.

Werf Gusto

Werf Gusto in Schiedam near Rotterdam was one of the few European companies building smaller two-stroke and four-stroke marine diesels prior to 1914.[125] It made both types with the latter being a low-horsepower, two cylinder turning a reversible propeller. The two-stroke line were either trunk piston or crosshead depending on engine size; one of 300 bhp had the latter, although it is unclear if this size was ever made. All had a two-stage, crankshaft-driven compressor on the front of the engine. Stepped pistons pumped air to loop-scavenged cylinders having one air and one exhaust port. Only a fuel and starting air valve were in the head.

Gusto two strokes had a monobloc head and jacketed cylinder with the lower end bolted to a crankcase enclosure fastened in turn to the bedplate. Large doors provided access to the crankcase.

To reverse direction, a handwheel changed the timing of both a valve in the starting air line and a reciprocating, piston-type valve controlling suction and discharge of the scavenge pump. Timing of the fuel and starting valves in the head was altered by shifting a lever.

A 250 bhp at 280 rpm, three-cylinder, two-stroke was installed in a local tugboat during 1913; cylinder size was not stated. As with the Savoia engine, long-term operating performance is not known. Orders for other engines were reportedly being filled.

Paolo Kind & Cie.

Paolo Kind & Cie., steam locomotive ship engine builders of Turin, began making two-stroke, uniflow scavenged marine diesels in the 50 to 150 bhp range in 1912.[126] Maximum speed of the lower output engines was as high as 500 to 650 rpm, and the 100 bhp turned up to 350 rpm. Speeds dropped to 350 to 400 and 150 rpm respectively for continuous running conditions. Fuel consumption for the larger engine at 150 rpm was about 200 g/bhp-hr (0.44 lb). Although cylinder size is unknown, the length of a 150 bhp model was 2 m with a height above the crankshaft centerline of 1.35 m.

The four-cylinder, 100 bhp at 375 rpm model exhibited at the 1911 Turin Exhibition had a form of stepped piston to pump scavenge air. A flange with piston rings on the lower end of a long trunk piston acted as the pump piston. It ran in a cylinder not cast with the monobloc head and power cylinder. A two-stage compressor was crank-driven at the front end of the engine. Dual overhead camshafts directly opened two inclined scavenge valves transverse to the engine. One camshaft also lifted the fuel valve placed in the center of the cylinder and opened a starting valve next to the injector on the engine axis.

For starting and reversing, a handwheel positioned a flat, horizontal bar that served as a face cam having two profiled grooves (fig. 16.44). Pins slidable along the grooves were linked to rotatable tappet guides for altering timing of the fuel and starting air valves as demanded by the operating mode. This was effected by changing the contact points of the valve tappets on their cams. The other end of each tappet acted on a cam surface profiled on a short lever opening the valve. A combination of the two cam profiles gave wide control over both valve lift and timing.

By 1916 Kind was supplying the Italian Navy with four-cylinder, 100 bhp and six-cylinder, 200 bhp engines for picket and scouting boats.[127] Weight of

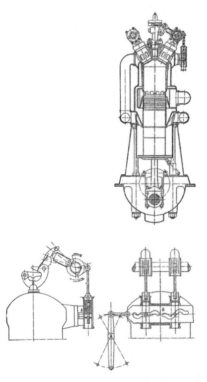

Fig. 16.44. Method of changing timing and lift of the fuel and starting valve on a 100 bhp 1912 Paolo Kind two-stroke for starting and reversing. *Engineering, February 1912*

these engines had been reduced to 15 kg/bhp through the use of bronze castings wherever possible. Tests on an 80 bhp at 600 rpm navy engine showed it could be reversed from that speed in one direction to the same speed in the other in only four to five seconds. At 81 bhp and 607 rpm the bsfc was 243 g/bhp-hr (0.535 lb). The governor cut off fuel at 650 rpm.

AG Weser—Toussaint

An unusual European marine diesel was built by AG Weser and based on a German patent by Toussaint.[128] The eight-cylinder, loop-scavenged, two-stroke engine had a design rating of 300 bhp at 200 rpm and a bore and stroke of 250 × 360 mm.

Cylinders were paired so that connecting rods for the trunk pistons in each pair were attached to the big end of the same rod. Thus, the eight pistons acted on only four crank throws (fig. 16.45). A bridging passage connected a cylinder pair to form a common combustion chamber. Scavenge ports opened into one cylinder and exhaust ports were in the other. There was no differential movement between paired pistons. Individual heads sat on a casting serving a cylinder pair.

Two injectors, whose fuel valves were lifted by a single, forked lever, opened into the ends of the passage near the cylinder edge. A starting air valve was placed in the center of each cylinder with one being for "ahead" and the other for "astern." All four valves were on the engine centerline. A separate power source ran the scavenge pump, compressor, and water pump.

The engine followed M.A.N.'s "loose" coupling reversing system, but it was on the end of a high-mount camshaft instead of the vertical shaft. The rocker shaft supporting the fuel and starting levers was turned by a hand-wheel to change eccentrics that provided the necessary sequencing for admitting starting air and fuel when running in either direction.

Weser probably enjoyed as much success with this engine as with its ill-fated Junkers model made about the same time. The cited report on the engine stated that it was not seen running and no further details on it were later recorded.

American Marine Diesels, 1914–1918

At least seven yacht and tugboat-size diesels appeared in the United States between 1911 and 1916 of which only one was of European design. Some were in production for but a short time, and none has been made under its original name for many years. Designs varied from sophisticated, reversible two strokes to simple four strokes driving through attached reverse gears. Total production of these non-navy engines up to 1918 was small at best. Proven reliability of competing steam and large, gasoline-burning engines running on low-cost fuels made it difficult for an emerging American marine diesel industry. Nelseco, the leading manufacturer during this period, built mostly submarine engines, and its contributions in that field, as well as with

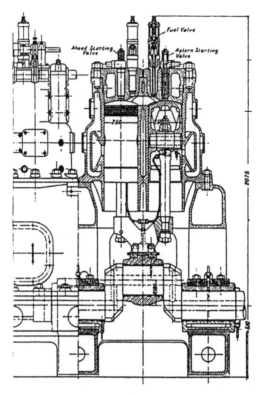

Fig. 16.45. Weser-Toussaint two-stroke marine diesel with common combustion chamber for a pair of cylinders, 1912. *Engineering, October 1912*

civilian engines, is in the next chapter because its basic designs for the two applications were originally very similar. The few other licensed builders, such as Dow (Willans & Robinson) and Pacific Werkspoor, are found with their parent European companies. William Cramp & Sons Ship and Engine Building Co. in Philadelphia bought a license from Burmeister & Wain in 1916 but did not build an engine until after 1918. The license agreement ended in 1925.[129]

Fulton Manufacturing Co.

Fulton Manufacturing Co. (not to be confused with Fulton Iron Works) was one of the earliest non-licensed builders of smaller US marine diesels. A four-stroke engine made by this Erie, Pennsylvania, firm was installed in the 54-foot *Pointer*, a Maryland Steel Co. boat operating in Baltimore harbor during 1914.[130] The three-cylinder, 8 × 9-inch bore and stroke diesel developed 50 bhp at 400 rpm. A two-stage compressor on the front of the engine was driven off the crankshaft. Injection air pressure varied from 800 psi at half load to 975 for full load. An eccentric on the high-mount camshaft operated individual fuel pump plungers running in a single housing. Manual and governor fuel control was by suction valve regulation. Main and rod bearings and wrist pins were pressure lubricated; force feed oilers served the cylinders. Interlocks prevented fuel from being injected during the air starting process, and an attached reverse gear was used for maneuvering.

The same bore and stroke was used on four-cylinder, 75 bhp (fig. 16.46) and six-cylinder, 100 bhp engines. The first of the latter went in a Galveston, Texas, tug during 1915 with a guaranteed bsfc of 0.55 lbs/bhp-hr.[131] A 100 bhp engine, sold in 1917 to the US government, met all specifications during a War Department acceptance test. The report mentioned that the running cost per hour was only 3,772 cents with fuel at 6 cents per gallon.[132] In spite of favorable publicity, only a few engines were built. Production probably ceased in 1919 about the time when Fulton let it be known all manufacturing rights were for sale.[133]

James Craig Engine & Machinery Works

James Craig Engine & Machinery Works of Jersey City, New Jersey, made gasoline engines for boats prior to diesels. The first one for the yacht *Aeldgytha* was installed in early 1915.[134] This six-cylinder, four-stroke had a bore and stroke of 9.5 × 11 inches and an output of 175 bhp at 400 rpm. Individual cylinders were bolted to an upper beam supported by steel columns anchored to the bedplate. The resulting open construction followed conventional practice. A crank throw at the front drove a two-stage compressor.

In addition to the exhaust valve in the head, there was a cylinder port uncovered at the bottom of the stroke. A second exhaust valve in a passage connecting the port to the exhaust manifold was opened by an auxiliary camshaft; it remained closed except at the end of the exhaust stroke (fig. 16.47). This extra valve lowered the upper valve's operating temperature. One

Fig. 16.46. 1914 Fulton Manufacturing Co. four-cylinder marine diesel. 75 bhp at 400 rpm. Haas, *The Diesel Engine, 1918*

fuel pump metered fuel through a distributing box having six needle valves and check valves. The engine was reversed by sliding the camshafts to put a second set of cams under the rocker levers after they were lifted by eccentrics on the rocker shaft.

Craig built only a few diesels, but the company did supply many gasoline engines to the US Navy during the war. Known diesels for the navy were 400 bhp versions of the yacht-type above to repower the subs *F-3* and *F-4* when the Vickers engines were taken out. Craig also reportedly built one large marine diesel in 1918 of unknown design. All engine production ceased in 1927.[135]

Atlas Imperial Engine Co.
The Atlas Imperial Engine Co. (later Atlas Imperial Diesel Engine Co.) in Oakland, California, had a long history in the engine business prior to 1916

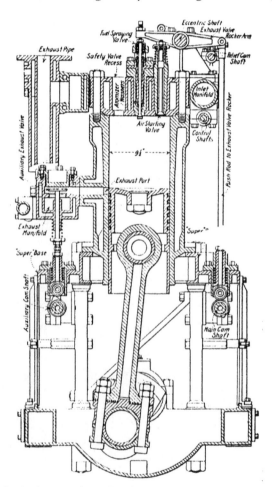

Fig. 16.47. Craig four-stroke, 175 bhp marine diesel with another exhaust outlet at the bottom of the cylinder, 1915. *Diesel Progress, May 1948*

when it was formed from Atlas Gas Engine Co. and Imperial Gas Engine Co. Both had been founded about 1904 as outgrowths of Hercules Engine Co. started in San Francisco in 1894 or 1895.[136]

Atlas Imperial's first diesel made the year of the merger was a six-cylinder, 250 bhp four-stroke with a bore and stroke of 11 × 14 inches. The non-reversible engine was ordered to generate power for a mine, but it instead went in the Seattle car ferry *Vashon Island* in 1916. A clutch added on the shaft ends allowed a propeller driving from each end to free wheel when that end became the ferry's bow. The ferry was operating as the *Islander* in the Los Angeles harbor area in 1948 with her original engine, which had been converted to solid injection in the early 1920s.

The design of this first engine was adopted for marine use and would retain its original identity for many years. Individual cylinders sat on a box frame that enclosed the crankcase, and long pushrods from the camshaft enclosed in the crankcase lifted rocker levers to open inlet, exhaust, fuel, and starting air valves. The injection air compressor was on the front of the engine.

Atlas Imperial introduced a line of four- and six-cylinder stationary engines in 1917, and, more importantly, the first American solid injection diesel in 1920. The company did business under that name in Oakland until National Supply Co. of Springfield, Ohio, bought it in 1951 and moved production there.

Southwark-Harris

The Southwark-Harris Valveless Engine, a third American marine diesel, received attention when its production was announced in 1915. The two-stroke, loop-scavenged engine came from Southwark Foundry & Machine Co. of Philadelphia founded in 1836. Leonard B. Harris, who patented some design features, had joined the company earlier.[137] It is unclear if the engines were made by Southwark, by Lyons-Atlas in Indianapolis (chapter 14), or by both. One known marine installation was in the yacht *Georgianna III* in 1916. Her four-cylinder, reversible engine, with a bore and stroke of 9 × 13 inches, developed an indicated 240 hp at 300 rpm[138] (fig. 16.48).

Stepped pistons pumped scavenge air past a valve (V_2), opened by scavenge pressure, into a receiver (M) connecting two power cylinders so that air from one pump chamber entered ports in the power cylinder of the other. (Crank throws for adjacent cylinders were 180 degrees apart.) The intake port (G) for the scavenge pump was not uncovered by its piston until near the end of downward travel so that the vacuum created by the downward movement quickly drew in sufficient air. This contrasted with the usual method whereby a timed valve controlled suction and discharge through a single port.

Fuel valves in two injectors per cylinder were lifted by cams through pushrods and end-pivoted levers (L). Valve lift for both load control and timing when reversing engine direction was altered by a rotating arm (V_2) to

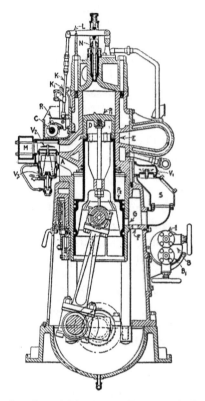

Fig. 16.48. Southwark-Harris 1916 stepped-piston two stroke.
Morrison, *Oil Engines, 1991*

move the contact point of pushrod K on its cam follower lever R. Air valved into the scavenge pump cylinders after top center pushed on the pistons to start the engine.

The numbers built and installations beyond that in the yacht above are unknown. Production ended sometime before 1923.[139]

Winton Engine Works

Alexander Winton (1860–1932) founded the Winton Engine Works in Cleveland, Ohio, in 1912 as an outgrowth of his Winton Motor Carriage Co.[140] He had been a leading automobile pioneer in the United States, and his engine company would become the Cleveland Division of General Motors Corp. before it was absorbed into the Electro-Motive Division.

Winton began production with large marine gasoline engines, but in 1913 he built the Model W-16, a 175 bhp, six-cylinder, four-stroke diesel with a bore and stroke of 9.5 × 14 inches. Further work developments resulted in a 7-7/2 × 11-inch bore and stroke V-12, developing 300 bhp at 450 rpm. Two were installed in Winton's yacht *La Belle* in 1915.[141] An advertisement, without a picture, in the April 1917 issue of *Motor Boating* boldly announced a line of engines for "yachts and commercial purposes" in sizes of 150 to 1500 hp.

Production of a more rugged work boat type of engine began about the same time. First was a 225 bhp, six-cylinder Model W24 with a bore and stroke of 12-15/16 × 18 inches. These pushrod engines had individual cylinders and a compressor on the engine's front end. Several were soon installed as auxiliary power for freight schooners. A 450 bhp, eight-cylinder Model W40 of the same cylinder size ensued, with four going in the cargo boats *Mt. Baker*, *Mt. Hood*, *Mt. Shasta*, and *James Simpson*, all of about 300 feet.

The Charm of Novelty

Ivar Knudsen offered a telling comment in 1914 about the problems, real and perceived, faced by early marine diesel builders: "In steamships many accidents also happen from time to time, but these make no sensation as we are familiar with them, and they do not have the charm of novelty."[142]

A long and ardous shakedown cruise faced by the marine diesel from the *Petit Pierre* and *Vandal* to the *Selandia* and *Vashon Island* had ended by 1918. Its future was secure but still faced often stormy times at sea as well as in company boardrooms and engineering forums. The technology gained over these fifteen years from powering commercial vessels and their naval cousins had carried them past the woes and fascinations of being different from the progeny of James Watt and Robert Fulton.

Notes

1. J. T. Milton, "Present Position of Diesel Engines for Marine Purposes," *Trans. of the Institution of Naval Architects*, vol. LVI (1914), 114, discussion comment by Hamilton.
2. Fred Lane testimony, *Royal Commission on Fuels and Engines*, pt. 2 (British Admiralty, 1913), 176. This is the only reference citing Lane's involvement in events leading to the diesel powered *Vulcanus*. Werkspoor histories are silent on this point. Lane's close ties to Shell lend credence to his testimony before the Royal Commission. Specifications on the *Vulcanus* are in *An Outline History of the Oil—Engine and its Lubrication* (Shell, 1961), 88. A Werkspoor drawing in D. L. Saunders-Davies, "A Century of Oil Trading," *The Motor Ship*, 1970, 32, shows that they used English dimensional units.
3. D. C. Endert, "From the Journal of the forgotten Dutch Land and Sea Diesel Pioneer," (trans.), 5. MAN Archives. See note 99, chapter 12, regarding this "Tagebuch" (journal).
4. Ibid.
5. Milton, "Present Position," 85.
6. "The Motor Ship Juno," *The Engineer*, November 15, 1912, 514. See also T. O. Lisle, "Werkspoor Motor Ships in Service," *International Marine Engineering* (1915). (Lisle worked for Werkspoor in the US.); C. J. Hawkes report of September 19, 1912, *Royal Commission*, 92.
7. Lisle, "Werkspoor."
8. D. L. Saunders-Davies, 32.
9. Endert, "Tagebuch," 5. K. K. van Hoffen, "Werkspoor's Stake in Marine Engine Development," *The Motor Ship*, April 1970, 35, says to the contrary that the engine was used in a land application until the 1950s.
10. "The Motor Ship Juno," 515.

11. Hawkes, *Royal Commission*, 89, says that during his visit on August 28, 1912, the "engines for the *Juno* and *Emanuel Nobel* [are] now being erected." Specific weight of the *Juno* engine, including auxiliaries, propeller, and shaft, was about 470 lb/bhp. "The Motor Ship Juno," 515, details a shakedown cruise in November 1912.

12. Hawkes, 118. He saw the engine on its return from the shakedown cruise after the rebuild.

13. Milton, "Present Position," 87.

14. Forbes and O'Beirne, *The Technical Development of the Royal Dutch/Shell* (The Hague, Netherlands: Royal Dutch Petroleum Co., 1957), 537.

15. Ibid., 538. Oil transport to England during the war became so vital that many tankers and cargo ships carried oil in their generally unused double bottoms. Shell tankers began this in 1916, and Marcus Samuel finally convinced the British Admiralty in 1917 to have cargo ships do likewise. By war's end, 761 ships—430 UK and 331 US—had sent over a million tons of oil to Britain in this way. See also Robert Henriques, *Marcus Samuel* (London: Barrie & Rockliff, 1960), 608.

16. "The Revival of the Reversible Blade Propeller," *The Engineer*, March 26, 1915, 295. The British built tanker was 185 × 36.6 feet with a draft of 11.75 feet when hauling 835 tons (long).

17. C. Kloos, "Diesel Engine Applied to Marine Purposes," *The Gas Engine* January 1916, 27–28.

18. *Jane's Fighting Ships of World War I* (New York: Military Press, 1990), 279. The vessel, most likely the *Gruno*, was 172.2 × 27.9 × 9.1 feet with a speed of 14 kts. Three of her type were built. The other two had German-made engines, MAN and Körting, of equal power.

19. Endert, "Tagebuch," 6.

20. "Werkspoor Diesel Marine Engines," *Engineering* (June 14, 1918), 658.

21. These earlier engines are in G. J. Lugt, "The Werkspoor Diesel Engine," *Trans, of the Institution of Engineers and Shipbuilders in Scotland*, vol. LXIV, pt. 5 (1926),427–94.

22. Three 1912 *Engineering* articles describe Carels marine diesels: January 19, 80–81; September 27, 435; and December 13, 808–9 and Pl. CXIII. The first covers a basic marine engine; the others on that in the *Fordonian* (250 × 42.5 × 16.5 feet at a maximum cargo capacity or dwt of 3,300 long tons.) Lake service restricted draft and dwt to 14 feet and 2,200 tons. Maximum gross tonnage was 2,368.

23. Gaston Carels testimony, *Royal Commission*, 411, gives 750 bhp, no rpm.

24. *Engineering*, 807.

25. Paul H. Schweitzer, *Scavenging of Two-Stroke Cycle Diesel Engines* (New York: MacMillan, 1949), 21.

26. Gene Onchulenko, "Fordonian ship's interesting career," *Flashback* (Thunder Bay Historical Museum), December 1982. Great Lakes Marine Historical Collection, Milwaukee Public Library.

27. "The Motor Ship Eavestone," *The Engineer* (October 25, 1912): 433–37. The *Eavestone* was 275.75 × 40.5 × 18.33 feet with a dwt of 3,050 tons and a displacement of 4,310 tons. Speed at 90 rpm was 9 kts. See also Gaston Carels, "The Development of the Diesel Internal Combustion Engine," *Cassier's Magazine*, March 1913, 152; his testimony in *Royal Commission*, 410.

28. T. Westgarth, "Steam-Driven and Motor-Driven Cargo Boats," *Cassier's Magazine* (March 1913), 199–203.

29. Testimony of E. Hall-Brown, Richardson Westgarth, *Royal Commission*, 168–74.

30. "Motor Ship Rolandseck," *The Engineer*, November 22, 1912, 550. The 290 × 40 × 18 foot ship had a cargo capacity of 2,700 tons and a cruising speed of 10 kts.

31. Ibid.

32. Hawkes, *Royal Commission*, 119.

33. "A visit to the Carels Works Ghent," reprint by Carels Diesel & Steam Engines (London), Ltd. from *The Engineer*, May 5 and 19, 1922, 10. *Engineering*, August 1, 1913, 161–62, describes the Ghent Exhibition engine.

34. "Motor-Tankship 'Wotan'," *The Gas Engine*, 1914, 39; The ship was 404 × 52.5 feet with a draft of 23 feet at a maximum dwt of 5,703 tons (long) and a displacement of 7,960 tons. See also *The Engineer*, October 11, 1915, 378–79; and Milton, "Present Position," 88; and Supino, *Land and Marine*, 72–73.

35. *Diagrams indicating the Supremacy of the Burmeister & Wain Marine Diesel Engine as shown by Lloyds Daily Index*, brochure, Burmeister & Wain, 1922.

36. "The Cargo Ship 'France,'" *Engineering* (October 10, 1913), 491–93. The steel ship was 131 × 17 × 7.3 m (loaded draft), with a sail area of 6,500 m² (70,000 ft2). In addition to crew's quarters, she accommodated six first class-passengers.
37. Milton, "Present Position," 88.
38. J. K. E. Hesselman, *Teknik och Tanke* (Stockholm, Sweden: Sohlmans, 1948), 260.
39. Hawks, *Royal Commission*, 94; and Milton, "Present Position," 85.
40. Prior to this refitting the *Calgary*'s name had been changed to the *Bakoi*. "News and Comments of 1920," *The Motor Ship*, April 1970, 55, of this anniversary issue.
41. "The Motor Ship Sebastian," *The Engineer*, March 6, 1914, 268–70, describes the Swedish engine and the sea trials. The 310 × 45 × 26.25 feet ship had a dwt capacity of 4,500 tons.
42. Ibid., 268.
43. E. Mortimer Rose, *Diesel Engine Design* (Manchester: Emmott, 1917), 164.
44. Hesselman, *Teknik*, 249.
45. Milton, "Present Position," 88.
46. C. Kloos, "Diesel Engine Applied to Marine Purposes," *The Gas Engine*, 1916, 28. Kloos's paper given at the International Engineering Congress in San Francisco, September 1915.
47. Hawks, *Royal Commission*, 93.
48. Henriques, *Marcus Samuel*, 516.
49. *Engineering* covers these ships: "Motor Vessel Arabis," March 20, 1914, 406; "Motor Vessel Arum," May 15, 1914, 690; and "The Twin-Screw Motor-Ship Abelia," October 20, 1916, 379.
50. Hesselman, *Teknik*, 259.
51. Ibid.
52. Torsten Gardlund et al., *Atlas Copco, 1873–1973*, (Stockholm: Atlas Copco, 1974), 90.
53. Ibid. The financial situation leading up to the merger and thereafter is fully detailed.
54. The Danish Patent Office rejected B&W arguments in 1904 that Del Proposto's idea was not new and granted him Patent 6,845. He licensed it to Diesels Motorer preventing Knudsen from using B&W engines in a sea-going barge because of high sublicense fees asked for by the Swedes. By the time O. E. Jorgensen had an alternative mechanical design, the project was dead for lack of money. H. P. Friis Petersen, "The First Diesel Engined Ships" (1965), 19–21. Copy compliments of N. E. Rasmussen.
55. Petersen, "The First Diesel," 27, attributes the idea to Knudsen.
56. Their filing dates were both April 9, 1912. US Patent 1,095,403, filed May 8, 1912, and issued May 5, 1914, covers the ideas.
57. The company, which had adopted the Barclay, Curle name in 1863, began in John Barclay's Glasgow yard in 1818 and made its first steam engines in 1857. It was a major participant in the River Clyde shipbuilding industry for many years. Fred M. Walker, *Song of the Clyde: A History of Clyde Shipbuilding* (Cambridge: Patrick Stephens, 1984), 159. Michael S. Moss and John R. Hume, *Workshop of the British Empire: Engineering and Shipbuilding in the West of Scotland* (Cranbury, NJ: Fairleigh Dickinson University Press, 1977), 192 pages, is a fine companion volume covering all major industries in the area.
58. H. F. Petersen and N. E. Rasmussen, "B&W Family—I," *B&W Engineering*, no. 4, February 1978, 4.
59. "Motor Ship 'Selandia,'" *The Gas Engine*, April 1912, 176. This eight-page article clearly and thoroughly covers the engine.
60. "The Burmeister and Wain Oil-Engine," *Engineering*, February 19, 1915, 209.
61. "Selandiana," *B&W Engineering*, no. 2, 1937, an interview with Chief Engineer Christian Jensen on February 22, 1937, marking the twenty-fifth anniversary of *Selandia*'s maiden voyage departure. Jensen had gone on the unofficial and official sea trials. This breezy, yet accurate account of the event was kindly translated for the author by A. W. Woods, B&W. B&W Archives.
62. O. E. Jorgensen testimony, *Royal Commission*, 491. The lengthy questioning, pages 485–95, shows British interest generated by the *Selandia* and provides additional technical information on B&W engines.
63. "The Marine Oil-Engine," *Engineering*, March 8, 1912, 321: "The visit . . . and the inspection . . . by practically all the officers responsible for naval construction at the Admiralty, from the First Lord downwards, has naturally quickened interest, alike on the part of the general public and of engineers, in the potentialities of the marine oil-engine."
64. Ibid., 494.

65. *Three Sticks Bamboo, The Story of the First EAC Motor Ships* (Copenhagen: The East Asiatic Co. Ltd., 1991), 26–31. This informal but informative sixty-one-page booklet details EAC's B&W-engined ships of 1912–1916 and their "careers." The title is from pidgin English describing the three-masted ships with no funnel: "Three sticks bamboo—no puff-puff." (page 4)

66. Ibid., 33.

67. Michael Moss and John R. Hume, *Shipbuilders to the World—125 Years of Harland & Wolff, Belfast 1861–1986* (Belfast, UK: Blackstaff, 1986), 156, say that after a Barclay, Curle and Swan, Hunter stock offering to finance its one-third ownership in B&W (DS)OE failed to raise the capital, Pirrie convinced Knudsen to buy H&W's Lancefield (later Finnieston) Engine Works. Only three months earlier H&W's Belfast-built *Titanic* had sailed on her ill-fated maiden voyage.

68. Annotation by Edward Moller, diesel engineer and later manager at H&W 1921–1973, translator of H. F. Petersen's B&W paper on the history of its long association with H&W, 1978, eleven-page translation, 1978. B&W Archives.

69. These include nine main engines and nine auxiliaries for three twin-screw and three single-screw ships. Jorgensen, *Royal Commission*, 487.

70. Petersen says he was a scapegoat. H. F. Petersen, a B&W paper on history of association with H&W (note 68).

71. Biographical information kindly furnished by Tony Woods, B&W.

72. H. F. Petersen & N. E. Rasmussen, "B&W Family-II," B&W *Engineering*, no. 7, September 1978, 2.

73. J. T. Milton, chief surveyor for Lloyds, for example, stated that "Among the engines which have been most successful in use—I think we must admit it—is the Burmeister & Wain engine." A nonrefuted discussion comment in W. P. Sillince, "A Brief Summary of the present Position of the Marine Diesel Engine and Its Possibilities," *Trans., Institute of Naval Architects*, vol. 58 (1916), 123.

74. Sulzer, "List de références de Moteurs Marine Diesel-Sulzer," December 1922. Sulzer Archives. The author is also indebted to David Brown for helping to sort out the many Sulzer models.

75. "The German Motor-Driven Ship 'Monte Penedo,'" *Engineering*, September 6, 1912, 315. Further engine details are in the August 29, 1912, issue, 302.

76. Testimony by Jakob Sulzer-Imhoof, *Royal Commission*, 134.

77. "Results with Marine Diesel Engines," *Engineering*, April 4, 1913, 467.

78. L. J. Le Mesurier, "The Development of the Sulzer Engine," *Trans., Institution of Engineers and Shipbuilders in Scotland* (1923), 35–6.

79. Ibid., 41.

80. "The Largest Sulzer-Engined Ship," *The Motor Ship*, in Sulzer Archives (Armstrong-Whitworth Co., 1921).

81. "German Motor Notes, No. 1," *Engineering*, October 11, 1912, 379.

82. Friedrich Sass, *Geschichte des deutschen Verbrennungsmotorenbaues* (Berlin: Springer, 1962), 575.

83. Rose, *Diesel Engine Design*, 197.

84. Milton, "Present Position," 87.

85. Ibid., "Motorboat" (August 10, 1915), 88.

86. Sass, *Geschichte*.

87. A. Nägel, "The Diesel Engine of Today," *Diesel Engines* (Berlin: VDI, 1923), 18. Krupp records confirm the high mileage.

88. Hans Flasche, "Der Dieselmaschinenbau in Russland von Anbeginn 1899 an bis etwa Jahresschluss 1918" (1947), 33.

89. A. Cyril Yeates, "Working for Ludwig Nobel Ltd., St. Petersburg," *Gas & Oil Power*, September/October 1965, 7. Yeates and Flasche both give independent and corroborative information on this engine.

90. Yeates and Flasche say that the yacht had only one engine. Robert W. Tolf, *Russian Rockefellers* (Palo Alto, CA: Hoover Institute Press, 1976), 177, speaks of two 450 bhp engines in the 100-foot boat, which gave it a speed of 26 kts, or (p. 227) fast enough to outrun customs officials when a notorious Finn bought it some years later to smuggle alcohol.

91. Yeates, "Working for."

92. Flasche, "Der Dieselmaschinenbau," 35.

93. Ibid., 36.

94. Ibid., 39, says only that in 1916 Nobel's oil company ordered a six-cylinder, 1,500 bhp at 150 rpm crosshead engine with a 560 × 700 mm bore and stroke for a tanker. His list of all ship diesels (p. 80) shows two tankers getting engines in 1916: the aforementioned one and a 1,600 bhp at 105 rpm, four-cylinder model. The engines may have been built, but chaotic conditions at the time probably prevented an installation.

95. Ibid., 25–26; and G. Steinheil, "Trial of a 600-B.H.P Two-stroke Direct Reversible Nobel Diesel Engine," *Engineering*, December 22, 1916, 608–11. Steinheil was a Nobel engineer.

96. Yeates, "Working for," 7.

97. Steinheil, *Engineering*, 612.

98. Flasche, "Der Dieselmaschinenbau," 43.

99. Ibid., 42, and Tolf, *Russian Rockefellers*, 210–12.

100. "Studiengesellschaft zur Entwicklung des Handelsschiffsmotors" (Study Co. for the Development of Merchant Ship Engines). Blohm & Voss (B&V) built such famous ships as the *Vaterland*, which was the world's largest passenger liner when launched in 1914. She was in New York when the US entered the war in 1917, and became an American troopship. As the USS *Leviathan*, she was the flagship of the American passenger liners. (The author's father, with his diesel race car, sailed to Europe on her in 1932.) The battleship *Bismarck* came from B&V yards and in the 1940s huge, six-engine seaplanes were also built there.

101. Milton, "Present Position," 70.

102. Kurt Schnauffer, "Die Dieselmotorenentwicklung in Werk Nürnberg der M.A.N., 1897–1918" (1956), 35. One cannot help speculating that the engine, already considered obsolete, was damaged in the explosion of January 1912 (chapter 17), which hastened its being scrapped.

103. Friedrich Sass, "The Dieselmotor in World Shipping," *Werft - Reederei - Hafen*, vol. 22, no. 3 (1941), 5.

104. Ibid., 6. A firsthand account by a B&V engineer of the *Fritz*'s stormy voyage to England and his interesting five-week stay there is in Heinrich Börnsen, *Zweitakt, Viertakt und Turbinen* (Berlin: Junge Generation, c. 1939), 62–76.

105. Schnauffer, "Die Dieselmotorenentwicklung," 38.

106. Sass, "The Dieselmotor," 6.

107. Schnauffer, "Die Dieselmotorenentwicklung," 39; and Nelseco records.

108. Hugo Junkers, "Investigations and Experimental Researches for the Construction of my Large-Oil-Engines," *Jahrbuch der Schiffbautechnischen Gesellschaft 1912*, 51–56.

109. "The Junkers Marine Oil-Engine," *Engineering*, November 24, 1911, 698–99.

110. Sass, *Geschichte*, 310, says the "installation was a total failure" and that the engine "had to be redesigned." Phillip L. Scott, "Construction of Junkers Engine," *Transactions, SAE*, pt. 2 (1917), 407, refers to "mistakes" made on the engine but gives no details.

111. Robert Doxford statements. C. J. Hawkes, June 28, 1913, report to the *Royal Commission*, 208.

112. "German Motor Notes—Part I," *The Engineer*, October 18, 1912, 405. Frerichs had already built two, four-cycle marine diesels for its own use and for a fishing boat. See "Marine Diesel Motors," *The Gas Engine*, May 1911, 266–68.

113. "An 800 H.P. Marine Engine," *The Gas Engine*, July 1913, 360. The title gives the overload rating.

114. R. Doxford "confirmed" to C. J. Hawkes that Frerichs and Weser "would not follow the guidance drawings supplied by" Junkers. He said that the Frerichs cylinder design "was absolutely impossible" and that Junkers had told them beforehand "they would certainly have trouble." Hawkes to *Royal Commission*, 208.

115. Hugo Junkers information to the "Royal Commission." Ibid., 219. The long ton is 2,240 pounds.

116. Milton, "Present Position," 88.

117. "100-Brake-Horse-Power Junkers Marine Oil-Engine," *Engineering*, October 24, 1913, 567.

118. Scott, "Construction," 409. Scott, a Junkers advocate, was a pre-1917 American engineer who had worked for Junkers and knew firsthand of his engine applications.

119. Robert Doxford testimony, April 22, 1913, *Royal Commission*, 85.

120. "The Doxford Diesel Engine," *The Engineer*, February 14, 1913, 168. R. Doxford also provides information on this engine.

121. "The Doxford Opposed-Piston Marine Oil Engine," *The Motor Ship and Motor Boat*, June 3, 1915, 17, of Doxford reprint.

122. In late 1972 the author saw Doxford's last attempt to compete with an opposed-piston engine. Fortunate was the gift of old post-1920 brochures, as Doxford ceased engine production not long after the Sunderland visit.

123. "Four-Cylinder Four-Cycle Diesel Engines," *Engineering*, May 24, 1912, 697–700. The article is mainly devoted to Tosi's new marine two-stroke. See also "Die Verwendung von Dieselmaschinen zum Antrieb von grösseren Seeschiffen," *Zeitschrift des VDI*, vol. 56 (March 23, 1912), 472–73.

124. "Three-Masted Schooner 'Aosta' with Savoia Diesel Marine Engines," *Engineering*, February 6, 1914, 182–83, plates.

125. "Diesel Marine Oil-Engines," *Engineering*, December 19, 1913, 816–19.

126. "100-Shaft-Horse-Power Reversible Diesel Marine Engine," *Engineering*, February 16, 1912, 216–17.

127. "Italian Navy Oil-Engine Picket Boats," *The Gas Engine*, March 1916, 126–29.

128. "German Motor Notes, No. II," *The Engineer*, October 18, 1912, 402–3. A brief reference to the engine is also made in the *Zeitschrift des VDI*, vol. 56 (1912), 377. Lamplough was granted British Patent 23,830 of 1910 for an engine similar to the Toussaint.

129. Licensing information from Burmeister & Wain.

130. "Pointer, a Diesel-Powered Boat," *Motor Boating* (December 1914), 18. The three-cylinder engine is described in "A Diesel Engine in Small Sizes," *The Gas Engine*, April 1914, 221–25.

131. "The 100 HP. Fulton Diesel," *Motor Boating* (September 1915), 29.

132. "The Fulton 100 HP. Diesel," *The Gas Engine*, September 1917, 457–58.

133. "Licenses Available for Small Diesel Engine," *Motorship*, April 1919, 23.

134. *Diesel Progress* (May 1948), 79.

135. E. C. Magdeburger, "Diesel Engines in Submarines," *Trans., American Society of Naval Engineers*, vol. 37 (1925), 605, says only the *F-3*, but the Submarine Force Library and Museum, *United States Submarine Data Book* (Groton, 1984), 3, indicates both the *F-3* and *F-4* were repowered with Craig diesels. See also L. H. Morrison, *American Diesel Engines* (New York: McGraw-Hill, 1931), 27.

136. Jack Lorimer, A. O. Warenskjold and Peter Barr had been with Hercules but left to independently form Atlas and Imperial. The new company brought them together again. C. A. Winslow, *Historical Development of Internal Combustion Engines on the West Coast*, 3rd. ed., SAE paper W-3097–S, August 17, 1960, 14. Charlie Winslow (d. 1973) made history in his own right with Winslow filters and the American Monovalve Diesel. He was *de facto* historian of SAE's No. Calif. Section for years.

137. "Southwark to Build Diesels," *The Gas Engine*, April 1915, 208.

138. "A 95-Foot Steel Diesel Yacht," *Motor Boating*, December 1916, 17–18.

139. L. H. Morrison, *Diesel Engines* (New York: McGraw-Hill, 1923), 52.

140. P. M. Heldt, "Winton, Industry's Pioneer, Dies at 72," *Automotive Industries*, June 25, 1932, 911.

141. "Winton Diesel Oil Engines," *The Gas Engine*, December 1919, 387, and *Diesel Progress*, May 1948, 62, 78.

142. Discussion by I. Knudsen in Milton, "Present Position," 118.

CHAPTER 17

Cargos of Death: Naval Vessels, 1911–1918

"Concerning the development of the submarine boat motor I can inform you, that the Diesel motor has now been definitely accepted in most of the European navies as the only satisfactory engine for submarine boats. . . .

With the help of the Diesel motor it has become possible to attack and destroy the hostile navies on the high seas before they can reach the neighborhood of the coasts and harbors."
—Rudolf Diesel to Edward D. Meier, 1909[1]

Naval diesels created a different success story. Submarines grew rapidly in size, which required a doubling and even tripling in their propulsion power by 1918. Reliability generally kept pace. Companies in almost every nation drawn into the war, as well as those remaining neutral, would build or buy sub diesels prior to its onset. But it was M.A.N. engineers at Augsburg who led the industry.

As Diesel wrote, the submarine was a weapon his engine made practical. Advancing from a role first conceived only to defend, its prime objective during World War I evolved to carry the action far from shore. Both sides sank naval vessels as well as merchant fleets to effectively blockade food shipments, in addition to war material, and directly increase civilian suffering. Much of this expanded capability grew from the dependable, larger diesels.

Swift development in new engines of destruction culminated in huge, 2,000 bhp per cylinder diesels scheduled for German battleships. These had been ordered several years before the outbreak of hostilities, but the challenges encountered during their development and the fortunes of war prevented their ever leaving builder shops. The vast experience gained from them, and from the submarine engines, would be applied to peacetime uses.

New-Generation Submarine Diesels

Lessons taught by diesel installations in early submarines were frequently disregarded as their power plants grew in size and complexity. Boats and engines were often designed without fully appreciating that engines had to be serviced. E. C. Magdeburger, a Busch-Sulzer engineer who fought this battle, put it so aptly:

> The engine is built and tested with all of the room of a modern erecting shop around it and with an electric crane overhead, the boat runs trials with a brand new engine just tuned up and with a picked band of experts for a crew, but the commanding officer has to operate his boat in all kinds of weather with a continuously changing crew subject to orders from his superiors. . . . [Servicing] often has to be done on engines installed with a shoehorn so that only a naked and perspiring body of a bluejacket can crawl through the recesses left between the engine and the skin of the ship.[2]

Magdeburger also described a subtle aspect of problems arising from engine room crews trained on steam:

> The old time steam installations about which older officers of the Navy sing high praises have been amply protected by the human element of these installations—the amount of heat generated and steam produced when running at full power was limited mainly by the muscular ability of the stokers. Whereas with the Diesel engine, one man, by moving one lever, can load up the engine beyond the safe limit and wreck the engine through excessive temperatures developed in the cylinders; or on the other hand, twist off some part of its shafting by running at so-called critical speeds. . . . In other words, the Diesel engine installation requires a greater degree of intelligence and greater experience of operating personnel. . . . Though many details of the Diesel engine are made fool-proof the same cannot be said of it as a whole, and a watchful eye, a sharp ear and quick brain are among the prime requirements of Diesel engine operators in submarines.[3]

Design engineers also faced new challenges as height- and width-limited engines grew longer to meet the increased power demands. More rigid bedplates and upper engine structure with minimum weight increases were needed to resist the flexing of more limber submarine hulls. The longer engines with more cylinders, combined with high speeds, created torsional vibration phenomena. Broken crankshafts, cracked accessory mountings, and ruptured pipes too often resulted if engines ran more than briefly at a critical speed where these destructive vibrations occurred. Repairing high-pressure lines spewing air or fuel inside the cramped, hot, and smelly engine room of a rolling and pitching submarine in a stormy sea was not a sailor's favorite chore.

M.A.N. Augsburg's Four-Stroke Diesels

The arrival of an extremely gifted young man at M.A.N. in 1911 would contribute greatly to the fame of its sub diesels. Gustav Emil Pielstick (1890–1961) had received training at the Imperial Navy Shipyard in Wilhelmshaven and graduated from the engineering school (*Maschinenbauschule*) in Kiel.[4] He began in the submarine engine department and eight years later, at age twenty-nine, became its chief engineer. His meteoric rise without mentor or the "correct" academic credentials is a measure of his creativity and his ability to motivate others during those years. By 1937 Pielstick oversaw M.A.N.'s entire naval engine program. The story of his fascinating career, which ends with the Pielstick engine being built in France, is for a later volume.

Submarine diesel activity switched entirely to Augsburg from Nürnberg during the 1912–1918 period. Nürnberg's unexpected delays placing its two-stroke engine (fig. 15.12) into production caused the German Navy to adopt Augsburg's four-stroke counterpart developed in 1910. Eight of these SM 6 × 400, 850 bhp engines (fig. 15.11), delivered in 1912 and 1913, were the navy's first sub diesels (chapter 15). Installation of the last was in December 1915.[5]

A more powerful engine, the S 6 V 41/42, grew out of the SM 6 × 400 design in 1912 (fig. 17.1). Its six cylinders with a bore and stroke of 410 × 420 mm had an output of 1,000 bhp at 450 rpm. ("SV" = *Schiff Viertakt,* or Ship

Fig. 17.1. 1912 M.A.N. S 6 V 41/42. 1,000 bhp at 450 rpm.
Note the two injectors per cylinder. *MAN Archives*

Four-stroke. This differentiated between the various models within M.A.N.) Seemingly minor changes, other than an increased bore and stroke from 400 × 400 mm, included a one-piece bedplate and a two-piece crankcase enclosure to stiffen the engine. Flanges at the base of individual cylinders containing wet liners bolted onto the enclosure. Pistons were oil cooled. All fuel pumps were in a single housing placed on the front end of the engine near the one control wheel for reversing.

The starting process was changed to what others had already adopted. After turning over the engine with air being sent to all six cylinders, the air was then cut back to three cylinders while fuel was injected into the other three. In this way starting was more assured the first time without the wasting of air on a restart. A good operator needed no more than several engine revolutions to have a warm engine firing before he shut off all air. All later sub engines used this "half air–half fuel" starting procedure. The reversing method of the 1908 four-stroke engine was also adopted almost unchanged. Eccentrics in the rocker shaft were first turned to raise the levers off their cams and then the camshaft was slid axially to bring into play a second set of cams for running in the opposite direction.

Each cylinder had two injectors whose fuel valves were lifted by one rocker lever pulling up a bridge piece connected to the valves. Although there were five valves in the cylinder heads (starting, inlet, exhaust, and the two injectors), not having a larger fuel nozzle directly between the inlet and exhaust valves reduced the potential for head cracking. This was never a reported problem. The exhaust valve was water cooled.

Augsburg delivered twenty-four of the S 6 V 41/42 engines between August 1913 and February 1916. Four were in boats entering service before the start of the war. An incipient weakness with them and their predecessor model was that the crankcase enclosure did not offer adequate longitudinal stiffness. This condition manifested itself only after extended time at sea.

Not all work was concentrated on sub diesels. Augsburg also made smaller six-cylinder, four-stroke engines in the 1911–1918 period. The S 6 V 30/45 (300 × 450 mm) of 450 bhp at 450 rpm generated power for surface ships and charged batteries on larger submarines. A nonreversing propulsion engine for small torpedo retrieval boats was the 220 bhp at 500 rpm S 6 V 22/34.

The more powerful, 1,200 bhp at 450 rpm, S 6 V 45/42 benefited from two design changes that increased stiffness without a proportionate growth in overall engine height (fig. 17.2). These were a deep, U-shaped bedplate of cast steel and the bolting together at abutting flanges on both sides of the engine at the lower half of the three paired cylinder castings. This formed a more rigid structure from the bottom of the bedplate to halfway up the cylinder.

Piston cooling was greatly enhanced by increasing the flow velocity of oil traveling through a thin cavity directly against the under surface of the piston crown. The oil carried away heat more effectively in this way compared to pumping or spraying the oil into a large, open volume under the crown and then depending mostly on "cocktail shaker" action to transfer heat to the oil. New, articulated arms, one for supply and one for drain, moved the oil between the frame and a connecting pin for each arm on the inside of the piston skirt to further reduce potential cooling problems.

Two compressors on the front of the engine each had three stages, but the diameters of the last stage pistons on the top were different so that the

Fig. 17.2. M.A.N. S 6 V 45/42 sub engine of 1916. 1,200 bhp at 450 rpm. The first to use a U-shaped bedplate. *MAN Archives*

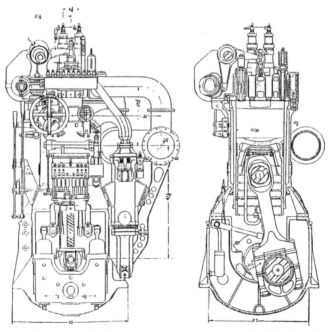

Fig. 17.3. End and cylinder cross section views of M.A.N. S 6 V 45/42 sub diesel. Magdeburger, *"Diesel Engines,"* 1925

Fig. 17.4. M.A.N. S 6 V 35/35 small sub engine of 1916.
550 bhp at 450 rpm. *MAN Archives*

smallest compressed to a fourth stage. This provided up to 3,000 psi air for
other ship's uses', including the firing of torpedoes.[6] Engines 550 bhp and
above had the four-stage compression. The reversing system was unchanged.

A single housing held all fuel pump plungers, each of which fed the paired
injectors of a cylinder. Control rods acting on bell cranks at the tops of the
injectors shortened needle valve lift when the engine ran at lower speeds and
the fuel and injection air had been throttled back (fig. 17.7).

More S 6 V 45/42 sub engines were built during the war than any other.
A total of 116 left the factory with 96 going in 48 boats between December
1915 and October 1918.[7]

Augsburg was also busy making six-cylinder engines on the lower end of
the power spectrum for mine laying and coastal defense subs (fig. 17.3):

- 1916: Eighteen, 230 × 340 mm, 250 bhp at 500 rpm.
- 1916: Thirty-two, 260 × 360, 300 bhp at 450 rpm.
- 1916–1918: Forty-five, 350 × 350 mm, 550 bhp at 450 rpm.

The S 6 V 35/35 was derated to 450 bhp and used in four submarines as
a 300-kw generator set to charge batteries (fig. 17.4). The *U-135* and *U-137*
carried two of them for this purpose.[8]

Fig. 17.5. M.A.N. S 6 V 53/53 sub diesel, 1917.
1,750 bhp at 390 rpm. *MAN Archives*

A six-cylinder and a ten-cylinder with a bore and stroke of 530 x530 mm and a nominal rating of 300 bhp per cylinder were the final and most advanced designs from Augsburg. Work began on the S 6 V 53/53 in May 1916, and the first engine was delivered a year later in June 1917. It produced 1,750 bhp at 390 rpm with a bmep of 5.8 kg/cm^2 (fig. 17.5).

A structural change further increased rigidity on both the S 6 V 53/53 and its ten-cylinder sibling by bolting together flanges on individual cylinders along the entire length of the cylinder from base to cylinder head (fig. 17.6). Bedplate, water jackets, cylinder liners, and heads were all of cast steel. Reversing was unchanged with the mechanism still entirely operated by a manually turned handwheel.

The labyrinthine mechanism controlling both manually and by governor the fuel charge and needle lift of the two injectors per cylinder is evident from figure 17.7. Such fuel system complexity was a penalty paid for air injection. Not shown is the companion mechanism regulating blast air pressure with speed and load. The reader is asked to traverse the maze on his own, given these check points: (*1*) fuel pump plunger; (*2*) suction valve; (*16*) handwheel (throttle) to adjust pump suction valves; (*29*) hand lever placing controls in operating (*Betrieb*), stop or starting positions; (*39*) overspeed control; (*47*) fuel nozzle needle valves (one shown); (*54*) control rod putting needle lift in starting (*anlassen*) position.

Four subs originally with the 1,200 bhp S 6 V 45/42 engines were upgraded to these new models that increased surface speeds from 14.7 kts to 18 kts. The "U-cruisers" *U-140* and *U-141* also received them in the

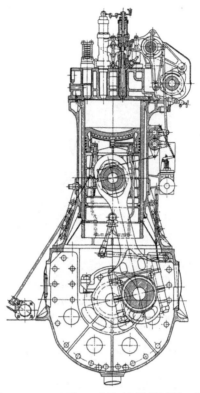

Fig. 17.6. Cylinder cross-section of S 10 V 53/53 of 300 bhp per cylinder showing oil path for piston cooling. The S 6 V 53/53 was almost identical. Maag, *Dieselmaschinen, 1928*

spring of 1918. Twelve in six subs went later to the British under terms of the Versailles Treaty.[9]

The ultimate of operational pre-1919 sub diesels was M.A.N.'s S 10 V 53/53 with a rating of 3,000 bhp at 390 rpm (figs. 17.8 and 17.9). Even though the engine was extremely long for its height (11 m long by 3.4 m high), the U-shaped bedplate and interconnected cylinders of the S 6 V 53/53 provided adequate rigidity. Specific weight without coupling, starting air tank, and exhaust silencer was 24.4 kg/bhp. A servo-operated mechanism had to be added for reversing because the extra four cylinders called for too much effort by one man to move it manually.

Four of these 3,000 bhp engines were delivered in January 1918, but their intended subs, the *U-142* and *U-144*, were not finished in time to enter service. M.A.N. made a total of forty-six S 6 V 53/53 engines by the end of the war. The *U-142, U-144,* and two others were ordered to be completed and given to the British.[10] Some of the forty-six were converted into generator sets for power plants in Germany and elsewhere. The US Navy bought two from the British for test use in America.[11]

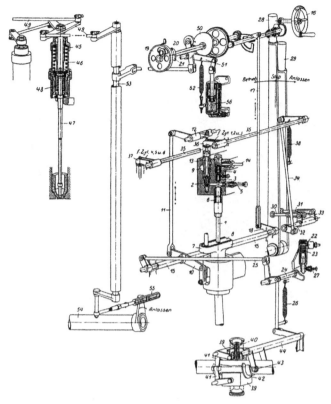

Fig. 17.7. Fuel and nozzle needle lift regulating mechanism for M.A.N. 6 V 53/53 sub diesel. Föppl, *Schnellaufende Dieselmaschinen, 1922*

A two-cylinder experimental engine of higher output was run in Augsburg on December 21, 1918. This inwardly opposed two stroke had two crankshafts tied together by a pair of Werkspoor-like coupling rods on one end of the engine. Vibrations limited it to 200 rpm where it developed 515 bhp per cylinder instead of an intended 700 bhp per cylinder at 450 rpm. Bore and stroke were 500 mm × 2x400 mm.

Augsburg built 512 sub engines for the German Navy between 1910 and 1918, of which 416 were delivered and 354 installed. Another 24 were made by Blohm & Voss and 8 by Vulcan under license. Approximately half of the installed 650 U-boat engines came from Augsburg. The "loose engines," both delivered to the navy or completed but still at the factory, were given to the Allies or converted to industrial uses.[12]

Before skippering one of six subs with S 6 V 45/42 diesels to America, Commander H. C. Gibson, USN had made engineering inspections in over two hundred German subs from surrendered ones to those still under construction; all but five were powered by MAN-type four-stroke diesels."[13] Of the six America-delivered boats, he said:

Fig. 17.8. M.A.N. S 10 V 53/53 sub diesel, 1918. 3,000 bhp at 390 rpm.
MAN Archives

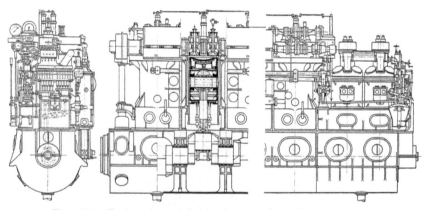

Fig. 17.9. End and partial side views of S 10 V 53/53 diesel
showing oil supply and drain system to piston skirt and eccentric
rocking drive for fuel pumps. Maag, *Dieselmaschinen, 1928*

We brought [them] back, . . . but not until after all the Allied countries
had taken what they wanted, and our boats had been abandoned for six
months. The engine hatches had been left open, and the engines were
not even drained down. However, we crossed the Atlantic with them
and nothing happened.[14]

The engines were not totally free of torsional vibrations. There was "a
certain spot at which the vibration was very noticeable" and when it was

reached "the engine makes a tremendous racket." Other than at that point it was "as steady as a rock."

Gibson later supervised removal of a pair of engines, one for testing and the other for a "tear down" inspection. The one tested had a bsfc of 0.42 lbs/bhp-hr at full power with no smoke. It could also be throttled back to no-load, or just enough to overcome an engine friction of 35 bhp, at which point fuel consumption was 35 lb/hr.[15] It ran there "with perfect combustion, delivering no load at the flywheel." The exhaust at idle was "absolutely invisible."

The influence of M.A.N. Augsburg engines on US Navy planning of the 1920s is summed up by Magdeburger, who by then was with the US Navy Bureau of Ships: "They were made the foundation of the Bureau type engines."[16]

M.A.N. Nürnburg Naval Engines

Nürnberg continued to develop and sell a line of two-stroke, stepped-piston, high-speed and slower-speed diesels that made their debut before 1911 (chapter 15). The Dutch Navy ordered twenty of the higher-speed engines for ten subs in 1914; twelve of which were delivered. These eight SS 34, eight-cylinder engines with a bore and stroke of 310 × 340 mm produced 900 bhp at 450 rpm.[17] Six slightly larger engines with the same rating, but with a bore and stroke of 320 × 350 mm, had been ordered by Russia. They instead went in three German subs, the only Nürnberg diesels doing so. By the end of 1914 it had delivered a total of forty-eight, two-stroke submarine engines, which was more than any other German engine builder up to that time.

The construction of the slower speed, "SL" engines followed the uniflow scavenging, stepped-piston design of the "SS," except that the stroke-to-bore ratio was increased from 1:1 to almost 1.6:1. This also increased the height, and the weight went up to 50 kg/hp or almost three times that of the "SS" series. These two-, three-, four-, and six-cylinder engines were mainly used in smaller ships and as auxiliaries in larger ones. Sizes ran from the two-cylinder SL 2/4 with a 190 × 420 mm bore and stroke and 40 bhp at 350 rpm to the SL 90/6, 360 × 600 mm, rated 900 bhp at 260 rpm. Twenty of the "SL" series were built.[18]

The huge Nürnberg "battleship" engine is described later in the chapter.

Vickers Sub Diesels

Vickers introduced a new eight-cylinder submarine diesel for the British Navy's *E* class boats around 1915.[19] It incorporated the solid injection fuel system developed during the preceding several years (chapter 13). A non-reversing four-stroke, the engine produced 800 bhp at 380 rpm from a bore and stroke of 14.5 × 15 inches. A twelve-cylinder derivative engine of the same cylinder size and 100 bhp per cylinder rating followed in 1917.

The Vickers structure was entirely different from Augsburg's and other contemporary designs, as it was more an assembly of numerous steel plates,

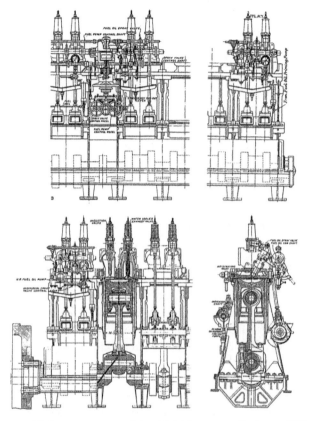

Fig. 17.10. Vickers nonreversible, 100 bhp per cylinder at 380 rpm sub engine of eight and twelve cylinders, 1915. *Engineering, July 1919*

channel beams, and forgings rather than of a few large steel castings. The entire assemblage was both riveted and bolted together, requiring extra care to attain accurate main bearing alignment (fig. 17.10). Two channel beams running the length of the engine supported steel castings holding the main bearings. Vertical steel plates over .75 inches thick went between the cylinders and on the engine ends. Forged angles were riveted to the plates at their base and bolted to the bearing supports. On top, a slot machined in a horizontal plate accepted the vertical plate. Angles on each side of the vertical plate were riveted to both plates. Lighter angles riveted to the outer, sloping ends of the vertical plate formed a surface to which crankcase cover plates were fastened. Recessed bores in the horizontal plate supported flanges on the cylinder liners. The jackets were thin cylinders with expansion ribs sealing at the top and against the lower end of the liners. Cylinder heads bolted to the plate were more like round discs holding intake and exhaust valve cage castings and a slanting injector. The pistons were not cooled, but the exhaust valves were water cooled.

Fig. 17.11. Twelve-cylinder Vickers sub diesel of 1918 showing single camshaft and upper end structure. *Engineering, July 1919*

Starting air entered the cylinder via a drilled hole after passing through a check valve screwed into the head. Two rotating distributor valves (four outlets each for the eight cylinder and six each for the twelve cylinder) were driven by the main camshaft to send air in a timed sequence to the cylinders.

Intake and exhaust valves were opened by pushrods from a lower camshaft, while a second, higher-mounted camshaft operated the individual fuel pumps and injector valves. Separate control shafts regulated the pumps and injectors (chapter 13). Each pump had its own accumulator in earlier engines, while later ones used a common rail system. Although later engines also had a single, higher-mounted camshaft, the structure remained similar to that described above (fig. 17.11).

Official shop tests of a twelve-cylinder, 1,200 bhp engine in September 1918 gave the following average of data acquisition points made hourly over a six-hour period at full load:[20]

Speed, rpm	381
Brake horsepower	1,215
Bsfc, lbs/bhp-hr	0.381

Both the eight and twelve cylinder were rated conservatively as the twelve cylinder could easily run at 400 rpm where it produced 1,400 bhp.

The *E* boats were the backbone of the fleet. Their two eight-cylinder engines had a radius of action of 3,225 nautical miles when cruising at 10 kts. *J* boats, given to the Australian Navy, had three of these engines, and the

L class were those receiving the twelve-cylinder diesels. A *K* class, of which thirteen were built, had one 800 bhp Vickers diesel for cruising, but the main power came from two, geared steam turbines of 5,000 bhp each to give a surface speed of 24 kts. These complex boats, known as "fleet submarines," operated in concert with surface ships.[21]

A reversible version of the eight-cylinder, 800 bhp engine came out in 1917, although none went in submarines. Most were installed in small tankers and sailing ship auxiliaries. Reversing was done by axially sliding the camshaft after rotating the pushrods away from the camshaft.

Vickers diesel production during the war was not limited to those for submarines. It also made a few larger, slow speed engines using concepts borrowed from the sub engine design. The first was an eight-cylinder, 750 bhp at 150 rpm model with a bore and stroke of 17 × 27 inches. A pair went in the navy monitor ship *Marshall Soult* in 1915 and in the Fleet tanker *Trefoil* the next year.[22] These were the genesis of a 24.5 × 39-inch, 1,250 bhp engine introduced for merchant ships in 1919.

New London Ship and Engine Co.

The first US Navy submarine diesels were ordered in 1909 from the Electric Boat Co., no stranger to the submarine. Isaac L. Rice, a Philadelphia lawyer and entrepreneur who organized EBC in 1898 as a holding company,[23] had bought the patents of John P. Holland (1842–1914) on February 7, 1899. He was America's foremost inventor and early champion of submarines, the most famous of which was his privately financed *Holland*, built in 1898.[24] This little boat would influence both the American and British navies to adopt submarine designs using his patents.[25]

Ties between Electric Boat and Vickers grew close after 1900 when Vickers obtained British rights to the Holland patents. Cementing this relationship were EBC's financial troubles caused mainly by a US Congress reluctant to appropriate funds for American subs. Vickers, who had become an EBC shareholder as a result of management friendship with Rice, invested more and more money to protect its interests until it owned 50 percent of the American company.[26] Such links induced EBC to adopt the Vickers air-injected diesel as the US Navy called for safer and more powerful engines.

EBC installed twelve Vickers-type diesels built by Bethlehem Steel's Fore River Shipyard in US subs. Two each of the 275 bhp at 400 rpm, four-cylinder engines went in the *E-1* and *E-2*, and eight six-cylinder of 410 bhp at 400 rpm in the *E-2* to *E-4*. Because they were essentially copies of pre-1911 Vickers engines they are described in chapter 15.[27]

Weaknesses in the early British design, coupled with glowing reports from Europe about M.A.N.-Nürnberg's new two-stroke sub diesel, initiated a license agreement with Nürnberg on January 29, 1910. Frank T. Cable signed for Electric Boat and Anton Rieppel for M.A.N. The contract granted EBC rights to sell "single-acting two-cycle heavy oil engines of the M.A.N.

employing what is commonly known as the constant pressure or Diesel cycle of combustion" in the "USA and its territories, colonies and dependencies."[28] This included stationary engines as well as those for "submarines, semi-submerged and surface boats. . . ." For the rights EBC made a 100,000-mark down payment and three 50,000-mark installment payments in June and October 1910 and January 1911. The royalty was to be "3 marks for each millimeter diameter of each working cylinder" of all engines built by EBC "whether they are built exactly in accordance with the M.A.N.'s design or whether the patents of M.A.N. are used or not."[29]

Except for a contract hiatus during the 1914–1918 period, both parties honored the agreement. From 1921, when the original license expired, until 1924 a tacit contractual understanding continued between the two companies. It was legally renewed in 1924 so that EBC could build larger engines M.A.N. had developed.

The arrangement between Nürnberg and EBC was not taken lightly by Adolphus Busch and James Harris at the American Diesel Engine Co. ADE still believed that some unexpired US Diesel patents would be a strong deterrent against EBC to build M.A.N. engines in America. Edward D. Meier had informed Harris on February 11, 1910, that during a call on the Navy Department he learned EBC had received engine drawings from M.A.N. He then visited the Fore River yard on the twenty-third where he met with Gregory O. Davidson and Lawrence Y. Spear, the men in charge at EBC. Meier told them of probable patent litigation if EBC built the M.A.N. engines and that ADE would like to offer its engine design instead.[30] He also saw the Vickers engines still under construction at Fore River. Max Rotter confirmed the license deal from the German end after a visit with Rieppel in early June 1911.[31] A patent action by the ADE never materialized, and the beachhead made by the M.A.N. engine at EBC would launch the invasion of European licenses with other American companies.

New London Ship & Engine Co. (NELSECO) was incorporated in October 1910 as a subsidiary of Electric Boat to make the M.A.N. engine for subs and other marine applications. NELSECO (or Nlseco, as it was abbreviated for the first few years) moved into a new factory erected in Groton, Connecticut, and equipped it with modern machine tools (fig. 17.12). A steel foundry was built at the site to pour the essential high-quality castings not obtainable when building the Vickers engines at Fore River.[32]

NELSECO's brochure of May 1911 announced a line of two-stroke, M.A.N. engines of 180 bhp to 2,000 bhp.[33] These non-naval, slower speed engines were based on Nürnberg's reversible SL series. The first one with six cylinders and an output of 150 bhp went in the 84-foot yacht *Idelia* in 1912[34] (fig. 17.13). A 300 bhp at 360 rpm engine of the same type completed a seventy-two-hour, full-power run in early 1913 (fig. 17.14). Bsfc averaged 0.57 lbs/bhp-hr on a California "asphaltum base crude oil," and lube oil consumption was about one (US) gal/hr.[35]

Fig. 17.12. NELSECO engine factory, 1913. *MAN Archives*

Fig. 17.13. NELSECO-M.A.N. two-stroke, 150 bhp
at 375 rpm marine engine, 1912. *MAN Archives*

In 1912 production began on an 11 × 12.5-inch, 450 bhp at 450 rpm submarine diesel (figs. 17.15 and 15.12). It closely followed the Nürnberg design described in chapter 15. NELSECO built about forty of this 450 bhp model for US Navy subs between 1912 and 1915.[36] A six-hour run on December 17, 1912, gave the following results:[37]

Fig. 17.14. NELSECO-M.A.N. two-stroke, 300 bhp at 360 rpm marine engine, 1913. *MAN Archives*

Fig. 17.15. NELSECO-M.A.N. two-stroke, 450 bhp at 450 rpm sub engine, 1912. *MAN Archives*

Speed, rpm	450
Power, bhp	441
Bsfc, hp/bhp-hr	0.55
Scavenge press., psi	9
Injection air press., psi	900

Exhaust conditions were "white smoke at end of pipe" and "brown vapor at 3-inch try valve." Full-power tests of up to twenty-eight hours on other engines had no reported problems.

The first pair was installed in the US Navy sub *H-1* launched May 6, 1913 (fig. 17.16). Others of the *H* and *K* class followed.[38] Although the navy-conducted tests in the factory had been very promising, later operating experience in the subs proved to be mixed.[39] Some boats reported no major engine troubles, but others were not as fortunate, suffering crankcase explosions, piston and cylinder scoring and cracking, broken wrist pins, and air compressor failures. The complicated, undependable reversing system was often supplanted by the electric motors to reverse the engines. Many of the problems were attributed to differences in crew training and motivation.[40]

Modified engines for the *L* series boats improved things somewhat. These had two scavenge valves in the cylinder head, and hold-down studs no longer passed through the water jacket causing another source of leaks. Continuing nuisances, however, were the salting up of too-small piping in the oil coolers and inaccessible water pump packing glands leaking salt water into the lube oil. The oil level in the bedplate had to be checked often and lowered to keep rods from splashing into the oil.

Because of these deficiencies, the Navy wanted a less complex, more accessible and more forgiving engine. NELSECO at first hoped to salvage most of the M.A.N. design with another that eliminated the stepped pistons and added crossheads; scavenge air instead came from two, double-acting pumps

Fig. 17.16. The *H-1* with the first M.A.N.-type diesels ready for launching on May 16, 1913. *MAN Archives*

on the front of the engine. Separate cylinder heads replaced the monobloc construction and the structure was also revised. Cylinders bolted to a cast steel "sole" plate resting on cast steel frames attached to the bedplate. The 10.5 × 16-inch bore and stroke engine developed 420 bhp at 350 rpm.

A pair was installed in the *M-1*, a new, "double-hulled" sub with her own problems. The engines themselves suffered cracked water jackets, and the reversing mechanism was finally removed. Although the *M-1* saw war service, the boat and engines were among the first scrapped after the armistice because "the new crew could not make the engines run."[41]

In 1913 NELSECO began offering for sale a separate line of slower running and heavier, four-stroke marine diesels for commercial applications. With a bore and stroke of 9 × 12.5 inches, they developed 30 bhp per cylinder at 350 rpm. Horizontal intake and exhaust valves were the distinguishing characteristics. An eight-cylinder version meeting navy specifications was made into a sub engine in 1915 with a rating of 240 bhp at 350 rpm. The engines had monobloc heads and cylinders, were not reversible, and had no air starting capability; shipboard electric propulsion motors fulfilled these functions. The fuel pumps and governor were on the flywheel end of the engine. Six of these diesels went in the subs *N-1* to *N-3* whose keels were laid in July 1915. They would replace the engines in the old gasoline-powered *D* class and the Vickers type in the *E-1*.

A larger engine of the same general design with a 13.5 × 14-inch bore and stroke was made in both six- and eight-cylinder versions with outputs of 440 bhp and 600 bhp respectively. The navy at last had an engine living up to expectations. A pair of six-cylinder diesels were in ten subs of the *O* class, and twenty of the *R* class had the eight cylinder. *S* class boats begun at the end of the war also had these engines to make a total of fifty-nine boats ordered by the US Navy in 1916, 1917, and 1918. More went in subs delivered to other countries.[42]

Electric Boat Co. had secretly delivered subs powered with these diesels to England before the United States entered the war. The British, who gave them an *H* class designation, placed the order through Vickers in 1915. Ten of the mostly Fore River–fabricated subs were sent in pieces to the Canadian Vickers factory where the NELSECO engines were installed. They became the first subs to cross the Atlantic under their own power and later distinguished themselves in British Navy service.[43]

The combustion chamber of the four-stroke engines somewhat followed horizontal diesel practice (fig. 17.17). A camshaft on one side of the engine actuated the intake valve through a long, vertical rocker lever, and a second on the opposite side controlled the exhaust. The valves opened into a recess above the main chamber. Injection cams on one shaft also lifted the axial fuel needle valve. The camshafts were driven from the front end of the crankshaft by spur gears through an idler gear. A pair of two-stage compressors on the front end supplied injection air and air for the added starting capability. They were still not reversible.

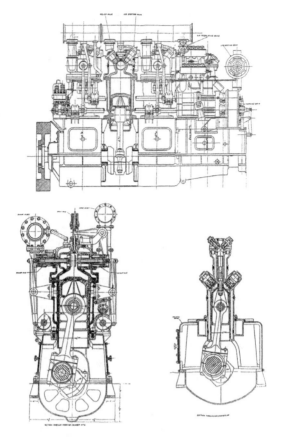

Fig. 17.17. NELSECO eight-cylinder, four-stroke, 440 bhp sub engine, 1916.
Magdeburger, *Trans., American Society of Naval Engineers, 1925*

Separate head and cylinder castings were used on the 1372 × 14 series. The earlier in the series had a cast bronze bedplate and crankcase for lightness, but this was later changed to cast steel when cracks developed after long service.

Magdeburger said that "almost every officer and enlisted man in the submarine service today [1925] has at the beginning of his submarine career operated these engines."[44]

In 1914 Congress had authorized the building of two fleet submarines, which, like the British K class, needed a very high horsepower. The engines, not completed until the end of the war, had six cylinders of the same general type as the above and a bore and stroke of 18 × 19 inches (fig. 17.18). They carried a rating of 1,000 bhp at 375 rpm. The four installed in the *T-1* in 1920 were not successful mainly because of the twin-screw drive arrangement. Two engines axially coupled together turned the same shaft. Numerous engine problems, the worst being the excessive vibrations caused by coupling, made the installation less than practical.[45] As a result of this experience, future US

Fig. 17.18. One of four NELSECO 1,000 bhp diesels
for the sub *T-1* launched in 1920. *Nautilus Memorial Submarine
Force Library and Museum Archives*

subs with forward engines drove generators and turned the propeller shaft
through the main motor.

NELSECO also made two, large diesels that were installed in the
455-foot, 15,000-ton displacement, US Navy tanker *Maumee* during 1916
(fig. 17.19). These Nürnberg-based 6KS100 were of the type described in
chapter 16. The single-acting, two-stroke engines had a bore and stroke

Fig. 17.19. Fleet tanker USS *Maumee* commissioned in 1916 with two
2,500 bhp NELSECO two-stroke diesels. First Executive Officer and
Chief Engineer Lieutenant Chester W. Nimitz. *MAN Archives*

of 650 × 1,000 mm and developed 2,500 bhp at 140 rpm. At the time, they were the most powerful diesels to go into a ship. The *Maumee* sailed about 22,000 miles between commissioning and the entrance of the United States into the war when the navy turned her over to the merchant marine. Although she had gone over 100,000 miles by December 30, 1919,[46] her engines "were a shambles" because of crews unfamiliar with diesels. The engines were replaced by steam in 1920.[47]

NELSECO continued to build both submarine and ship diesels. The last naval engines were for the US sub *Mackeral* launched in 1940. Commercial engine production ceased at the end of 1941.

Sulzer Submarine Diesels, 1910–1918

Sulzer Brothers designed a two-stroke, trunk piston submarine diesel in 1910 with a scavenging, combustion and reversing system similar to that of the *Romagna* engines that had recently entered service. The shop began building a test model in June 1910.[48]

The six-cylinder 6U23, with a 230 × 280 mm bore and stroke, produced 300 bhp at 500 rpm (fig. 17.20). Its head was cast monobloc with individual, linerless cylinders all being rigidly bolted to each other at their bases (fig. 17.21). These sat on a beam or entablature supported by columns. Extensions on the columns served as through-bolts to tie flanges on the cylinder to the underside of the bedplate. X-braces running from under the beam to the bedplate gave lateral rigidity.

Eccentrics on the vertical drive shaft to the camshaft reciprocated individual fuel pumps encased in a single housing on the output end of the engine. Operator controls were also located on the same end. A vertical fuel nozzle and an inclined starting valve were in the head.

Fig. 17.20. Sulzer 6U23, 300 bhp submarine engine first installed in the Italian sub *Nautilus* in 1912. *Sulzer Archives*

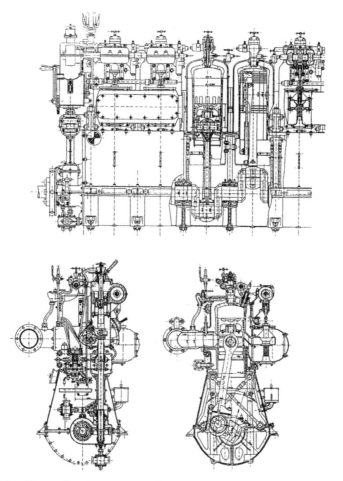

Fig. 17.21. Section views of Sulzer 6U23 submarine engine, 1912.
Sulzer Archives

In October 1910 the Italian Navy placed an order for a pair of 6U23s, which were delivered in March 1912 for the *Nautilus*. This sub and the *Nereide*, a sister boat with the same model of engine, were launched at the Venice navy yard in 1913.[49]

The 6U32, developed in 1911, had an output of 600 bhp at 450 rpm from a bore and stroke of 320 × 320 mm. Bmep at that rating was 3.9 kg/cm². Carried over from the 6U23 were the column supports and X-braces attached to a cast bronze bedplate. Separate cylinder heads, no longer cast with the cylinder, were also bolted to each other to form a rigid upper unit. Through-bolts from the heads transfered all combustion forces into the bedplate. The vertical cam drive shaft moved to the engine's midpoint.

Only two 6U32s were built, with both going in the US Navy sub *G-3* (ex-*Turbot*). The Busch-Sulzer Brothers Diesel Engine Co. (chapter 11) had contracted to make engines based on the **Sulzer** design for installation in this boat, but its new St. Louis factory would not have been completed in time to do so. However, in March 1911 the **Navy Department** cleared Busch-Sulzer to buy the engines from Winterthur; this occurred while Max Rotter was there and could study the engine. Sulzer's quoted price for two of them was $56,500.00 F.O.B. Antwerp.[50] A navy inspector lived in Winterthur to oversee their construction.

Rotter alerted Busch-Sulzer of a potential danger he saw regarding the 6U32 installation in the *G-3*, which was to be built by the Lake Torpedo Boat Co. (L. T. B.) in Bridgeport, Connecticut.[51] He feared that Lake was designing the boat with no regard for service accessibility to the engines. Component placement likewise seemed to be ignored:

> It would seem as though L. T. B. Co. had not considered at all the accommodation of the quite considerable amount of water and oil piping required about the engines. . . . The ventilating ducts, concerning which no previous warnings had been given to Sulzers, take up just the space which had been figured for the air tanks.[52]

He strongly advised that Lake send Sulzer all drawings pertaining to engine room space. His final comment was that:

> The Italian Navy furnished Sulzers complete details of these parts, which materially assisted them in their designs, and in the making of suggestions for the convenient accommodation of the apparatus.

Unfortunately, this advice was not heeded; as Magdeburger later said, "The boat was a failure."[53] He explained that the engines were too large for the hull and that accessibility to areas needing attention was almost impossible. These led to problems caused by insufficient lubrication for the rocking lever driving the air compressor and the "too-light" water and oil pumps, all of which could not be reached without major difficulties. Even oil in the bedplate sump could not be adequately drained. Excessive vibration also required that engine speed be reduced from the rated 450 rpm to 420 rpm.

Sulzer designed and built other models of submarine diesels throughout the war; their designation was changed, however, from "U" to "Q" to cover all naval engines:[54]

1915	6Q28 — 280 × 330 mm, 425 bhp at 400 rpm
1916	6Q25 — 250 × 280 mm, 325 bhp at 450 rpm
1917	6Q45 — 450 × 440 mm, 1,300 bhp at 325 rpm
1919	6Q54 — 540 × 500 mm, 2,000 bhp at 320 rpm

A total of twenty "U" and "Q" sub engines left the factory between 1912 and 1919. These went in French (*Lagrange*), Italian, Japanese (*1000T*), and Norwegian (*B1* and *B2*) boats.

SA des Forges et Chantiers de la Méditerranée in Le Havre bought a Sulzer license for the 6Q37 in 1912. FCM began building sub diesels the next year and had delivered four engines of that type with a 900 bhp at 380 rpm rating to the French Navy by the first of 1914.[55]

Sulzer also delivered twenty-four engines for installation in sea-going and river gunboats (sixteen to the French Navy and eight to the Romanian) between 1916 and 1919. These included sixteen of the 4S56 (340 × 540 mm, 400 bhp at 190 rpm) and four of the 6Q38 (380 × 380 mm, 900 bhp at 380 rpm).[56]

Busch–Sulzer Bros. Diesel Engine Co.

The St. Louis company was awarded a contract to build engines for three Lake-made US submarines, the *L-5*, *L-6*, and *L-7* in September 1914. They were considerably modified from the Sulzer 6Q32 design after the hard lessons learned from it in the *G-3*. Although the general Swiss concept was followed, a new placement of components made accessibility much easier.

The scavenge pumps were also redesigned. Two, double-acting cylinders were placed one above the other and driven off a single crank throw. An eccentric on the crank moved a vertical piston valve to control the suction and discharge of both cylinders of each pump. Discharge air went into a single receiver running the length of the engine's inboard side. (It was typical for port and starboard marine engines to be mirror images of each other.) The Sulzer double-beat valve controlling an upper row of cylinder scavenge ports was retained. A three-stage, vertical compressor with differential area (stepped) pistons was driven by the crankshaft in front of the scavenge air pumps. Inefficient oil coolers and "weak" pump drives in addition to stem breaking of the mushroom-type compressor valves were reported problems.[57]

This first B-SBD sub engine, the 6M100, had a 12.5 × 14.5-inch bore and stroke and produced 600 bhp at 375 rpm. Eight were ultimately built, with the last pair going in the *L-8*.

A scaled-down version, the 6M50, went in the subs *N-4* to *N-7* during 1915. These 9 × 12 inch, 300 bhp at 400 rpm engines were nonreversible and had to be started by the sub's electric motors. They reportedly gave very little trouble except for the scavenge pumps. Overall performance was good enough, however, for these engines to later replace the NELSECO-M.A.N. two strokes put in the first of the *L*-type boats.[58]

Busch-Sulzer built other models for subs between 1915 and 1919: the 6Ma85, a four-stroke, with a bore and stroke of 14 × 14 inches had a rating of 500 bhp at 410 rpm; two each went in the subs *O-11* to *O-16*. The *S-2* received two of the 6M150 with their 14-7/8 × 16-3/8-inch bore and stroke and an output of 900 bhp at 350 rpm. Boats launched in 1919 to 1920 had engines of the above type as well as later versions also designed during wartime. Fifteen subs powered by thirty Busch-Sulzer diesels saw active duty, with all but three boats being built by Lake. Up to twenty-two engines had been made by the end of 1918 but not yet installed; these plus an additional fourteen were delivered to Lake, the principal customer, until it went out of business in 1922.[59]

Busch-Sulzer made the near-fatal business decision to concentrate exclusively on submarine engines from the time when the new factory went into production. Compounding this risk was that only one boat builder bought them, a company that lacked the financial strength to survive the order cutbacks after the war's end. It would become a difficult transition for Busch-Sulzer to reenter the stationary engine market and create a new one for marine diesels.

Fiat Stabilimento Grandi Motori

Fiat's model 2C.116 was installed in the submarines of several European navies after its 1910 debut in the Italian *Medusa* (chapter 15). Between 1911 and 1913 the 300 bhp two stroke went in Portuguese and Danish boats in addition to six more for the Italian Navy. All these six-cylinder engines had stepped-piston scavenging pumps and two air intake valves in the cylinder head.

Scotts in Greenock, Scotland, under a 1912 license from Fiat, also built six of the 2C.116 diesels (fig. 17.22). Output was 325 bhp at 460 rpm from an anglicized bore and stroke of 9-3/8 × 10-1/2 inches. They were for three British Navy subs, also of Italian design and made by Scott, launched in 1914 and 1915.[60]

Even before the *Medusa*, Fiat was designing a higher horsepower engine to meet naval needs. The standard rating for this next size, a scaled-up version of the 2C.116 with the same reversing method, was 750 bhp at 400 rpm.[61]

New designs emerged in 1912 with changes resulting mostly from operating experience. A four-cylinder, 300 bhp model for auxiliary sets was followed by the 2C.126 sub engine of 1,300 bhp at 350 rpm. Bore and stroke were 440 × 450 mm. One major revision was the use of through-bolts tying flanges on the bottom of the cylinders to the bedplate. Another involved moving the scavenge air intake from valves in the head to ports in the lower cylinder wall (fig. 17.23). Because the upper end of the intake ports was higher than the exhaust ports, a double-seated piston valve operated by the camshaft prevented spent gases from entering the scavenge air manifold. The monobloc head and cylinder and the stepped piston for pumping scavenge air was retained. Reversing was still initiated by raising or lowering the vertical shaft, and by means of its helical gear ends, rotationally changing camshaft timing of the starting and fuel valves.

Most of these engines went into boats for the Italian Navy. Germany also ordered a pair to be installed in a Fiat-Laurenti design built at Spezia. Assigned the name *U-42* by the German Navy, it was the only foreign sub

Fig. 17.22. Scotts/Fiat 325 bhp diesel for the British sub *S-1*, 1914.
Scotts at Greenock, 1920

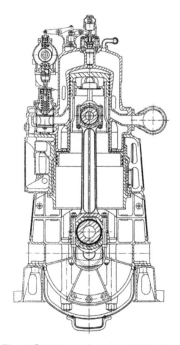

Fig. 17.23. Fiat 2C.126 engine of 1913 with piston valve
to keep exhaust out of scavenge air supply. 1,300 Bhp at 350 rpm.
Trans., Institute of Engineers and Shipbuilders in Scotland, 1928

that country ever ordered. At the outbreak of the war, it went to the Italian Navy as the *Balilla*.[62] Scotts made a pair for the British *G-14* (fig. 17.24).

Fiat next introduced the 2C.216, a smaller diesel having an output of 300 bhp at 480 rpm. It replaced the old 2C.116 with the modernized features of intake ports and through-bolts. The 250 × 270 mm bore and stroke were the same as on the obsoleted engine. More than one hundred went in Italian, Portuguese, and Spanish Navy subs between 1914 and 1918. [63]

The model 2C.226, also appearing in 1914, was a larger version of the 2C.216 but kept the same output of 1,300 bhp at 350 rpm and bore and stroke of 440 × 450 mm. A principal revision was the replacement of the stepped-piston scavenge pumps by separate, double-acting ones located between the front cylinder and a three-stage compressor (fig. 17.25). A major disadvantage of the stepped-piston design had been entrainment of lube oil in the scavenge air, which caused high oil consumption.

A cast steel beam at the top of the engine circumferentially clamped individual cylinder heads onto the ends of the liners, which were cast integrally with the air intake and exhaust passages. The liners were in turn held against the crankcase enclosure castings. Through-bolts tied the entire structure together. A thin, cylindrical sheet of steel formed the outer cylinder water jacket.

Crank throws with a 90-degree offset, and acting through crossheads, operated the over-and-under scavenge pumps. The piston rod of the upper double-acting pump ran axially through the lower pump piston. Two piston rods driven by crank throws straddling the throw for the upper pump moved the lower piston.

Fig. 17.24. Scotts/Fiat 1,300 bhp diesels under construction for the British sub *G-14*, 1916. *Scotts at Greenock, 1920*

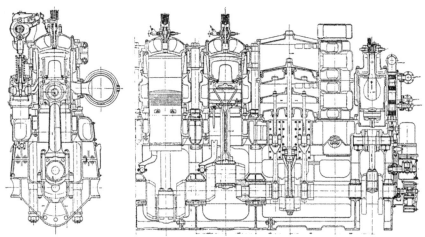

Fig. 17.25. Fiat 2C.226 engine of 1914 with independent, double-acting scavenge pumps. 1,300 Bhp at 350 rpm. Note the oil-cooled piston with the drain nozzle emptying into a stationary tube.
50 Anni di Motori Diesel Fiat, 1959

In operation, the 2C.226 engines lacked sufficient rigidity and had a "certain tendency to vibration and movings."[64] This was attributed to the through-bolts not providing enough stiffness to prevent the top clamping plate from moving excessively in relation to the bedplate.

More than forty of the 2C.226, 1,300 bhp Fiat diesels went in subs of the Italian, French, and Spanish Navies between 1914 and 1919.[65]

Piston skirt wear led to a new version of 2C.226, which had crossheads instead of a trunk piston design. A power increase of 10 to 15 percent was made possible by the change. The Japanese Navy bought a few of the crosshead engines from Fiat and then acquired a license to build them.

Specific fuel consumption with the Fiat submarine engines went from 270 g/bhp-hr with the first 2C.116 down to only 200 g/bhp-hr with the models built at war's end. Although power as expressed by brake mean effective pressure increased only modestly from 4 to 4.5 kg/cm², reliability and durability greatly improved.

Experience gained from the submarine engines was applied to slow speed marine diesel development. Four-cylinder engines of 550 bhp at 210 rpm with bores and strokes of 390 × 550 mm went in the single-screw motorships *Stige* and *Acheronte* during 1913. These may have been Fiat-owned ships.[66]

A quantum leap forward were the two 2,300 bhp diesels for the *Ceará*, a unique submarine tender for the Brazilian Navy launched from Fiat's San Giorgio yards in 1915.[67] These model 2C.176 engines, designed about the time of the above four-cylinder models, had six cylinders with a 630 × 900 mm bore and stroke, and developed their rated power at 125 rpm. They were the most powerful marine diesels installed, two or four stroke, up to 1916.[68]

Starting air and fuel valves and cylinder wall porting were similar to those on the sub engines (figs. 17.26 and 17.27). Double-acting scavenge pumps, driven by linkage from the crossheads of cylinders one, three, and five discharged into a hollow entablature supporting the cylinders. The air then passed a typical Fiat cam-operated piston valve before entering the cylinder scavenge ports. Two, three-stage compressors were driven by a single crank throw at the front of the engine.

Through-bolts tied the entablature to the bedplate. Cylinder heads recessed into the linerless cylinders to reduce overall height. This design also allowed the cylinder cooling water jacket to extend to the top of the combustion chamber. The through-bolts, which eliminated all tension stresses, enabled the use of lighter cast iron supports for the cylinders and upper crankcase structure.

A two-piece piston was water cooled by a nozzle discharging against the center of the piston crown. Water entered and exited the crosshead pin and passed up and down pipes in the piston rod end. The main, rod, and forked crosshead bearings were pressure lubricated.

An eccentric on the vertical shaft (located above the maneuvering gear) operated individual fuel pumps placed in a single housing. They were controlled as a unit by a manual throttle control and a governor, but per the usual practice, could be individually set or stopped.

Reversing was in progressive stages similar to that found on other ship engines. A servo-assisted lever (to the left of the handwheel in figure 17.26) raised or lowered the vertical drive shaft to rotationally turn the camshaft and thereby change the timing of the scavenge air piston valves. A second,

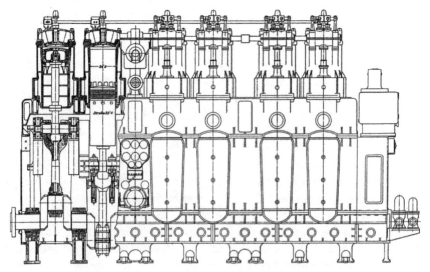

Fig. 17.26. Longitudinal view of Fiat 2,300 bhp marine diesel in *Ceará*, 1915. *Engineering, 1916*

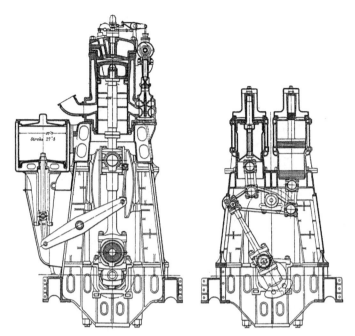

Fig. 17.27. Section views through cylinder and compressors,
Fiat marine diesel installed in *Ceará*, 1915. *Engineering, 1916*

servo-assisted lever on the right side of the handwheel rotated an eccentric rocker shaft to lift starting air and fuel valve rockers from their cams and to axially slide cam sleeves at each cylinder so that the two rockers would come down on new cams for opposite direction running. The handwheel sequenced starting air and fuel delivery for either running direction from first all air, to three cylinders on air and three on fuel, to finally all fuel. The wheel also opened a valve supplying starting air from the tank.

Test stand performance gave a full-load bsfc of 0.44 lbs/bhp-hr, which was reported to be the lowest of that time for any large two-stroke marine diesel. This figure included the power to operate separate sea water pumps for engine and piston cooling and two lube oil pumps. All were double acting.

An engine weight with all accessories of 160 tons, or about 155 lb/hp, compared favorably to the 250 to 350 lbs/bhp of the largest four-stroke engines. Overall length was 30.5 feet vs. 46 feet for a lower power output four stroke. It must be mentioned in defense of other engines that succeeding Fiat marine diesels had to be made heavier. The *Ceará* encountered no problems during its voyage to Brazil and saw light service there. The underworked engines therefore gave no hint of lurking thermal and structural weaknesses that had to be corrected on later engines subjected to more arduous duty.

The naval and merchant ship diesels developed before 1919 built a firm base for the successful models that would follow under the name of Fiat Grandi Motori.

Gebrüder Körting

As the first builder of engines for German submarines, Körting had gained valuable experience in that application's special needs. Work on its reliable, vaporizing oil engines ended in 1912 when development began on a replacement diesel.

The resulting six-cylinder, two-stroke was to produce 500 bhp at 450 rpm. Four air inlet valves in the head and exhaust ports in the cylinder gave uniflow scavenging; the air pump was on the front of the engine. Problems caused by the complicated head with its six valves (fuel, starting, and the four for air) led to a new design having ports in the lower cylinder wall. The resulting engine had an output of 400 bhp at 350 rpm and a bore and stroke of 285 × 350 mm. Fuel consumption was 220 g/bhp-hr compared with 500 g/bhp-hr from the obsoleted oil engine. The Russian Navy ordered four of these engines for two subs in 1913, but as the war had started before their delivery, the German Navy took them.[69]

Körting switched to four stroke when the navy saw that its near-term success looked more promising. A four-cylinder test engine of 240 bhp at 450 rpm was the first that Körting designed, built, and tested in three months before the end of 1914. The performance was promising enough for the navy to order eleven more.

A 300 bhp at 450 rpm model had a 270 × 330 mm bore and stroke and a bmep of 5.29 atmospheres[70] (fig. 17.28). Its one-piece bedplate supported a three-piece upper crankcase on which sat paired cylinders. The cylinder castings were bolted together for additional rigidity. A three-stage compressor and a housing for six fuel pump plungers were located on the front of the engine along with the controls. The engine was nonreversible and had no starting capability; the on-board electric motor and generator provided the starting and maneuvering functions. An interesting feature was the engine's exhaust-operated oil separator in a pipe connecting the crankcase and air intake to draw off entrained lube oil in the crankcase air.

Reference has been made to regulation of blast air pressure over varying speeds and loads. Körting did this manually on a production six-cylinder, 550 bhp at 450 rpm sub diesel (325 × 420 mm bore and stroke), c. 1915 (fig. 17.29). Blast air pressure in line H to the injectors acted against spring pressure set by handwheel D to hold valve B off its seat at J. Third-stage pressure in line G and second-stage pressure in leakoff line N did not affect valve spindle balance. An indicator attached to the outer spring retainer told the operator where to set the handwheel for new pressure needs. A smaller wheel locked the shaft to maintain a setting. Larger engines had a governor override.

A few reversible, 1,200 bhp engines were built in 1917, but Körting's major effort was devoted by far to the 550 bhp size and under. The war's end halted further development of a two-stroke, experimental cylinder of 700 bhp tested in 1918. The company's production record of 115 four-stroke sub engines exceeded all others except for that of M.A.N. (fig. 17.30).

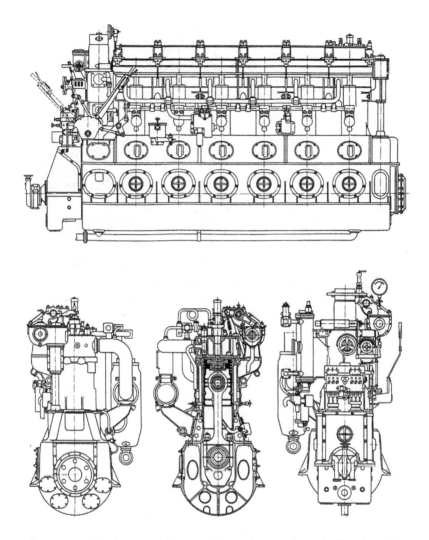

Fig. 17.28. Körting 300 bhp at 450 rpm four-stroke sub diesel, 1915.
Maag, *Dieselmaschinen, 1928*

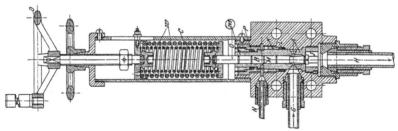

Fig. 17.29. Körting manually controlled blast air pressure regulator on 550
bhp engine, c. 1916. Föppl, *Schnellaufende Dieselmaschinen, 1922*

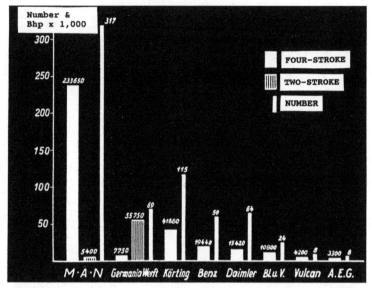

Fig. 17.30. German sub engine production to November 6, 1918.
Number of engines and cumulative horsepower (PSe), two- and
four-stroke, by company. Blohm & Voss and Vulcan were M.A.N. licensees.
Georg Strößner 1941 lecture, MAN Archives

Körting continued to build horizontal, four-stroke stationary engines
after 1918, but that production ended at the beginning of the 1930s. It filed
for bankruptcy in 1932, and while a successor company would continue as
Körting AG, today's name, diesel engines were no longer made.[71]

Krupp Germaniawerft

Although Krupp's Germaniawerft had spent several years perfecting its two-
stroke, 850 bhp engine (chapter 15), the first from Kiel actually to be installed
in a sub was a pair of smaller, 320 bhp sibling engines for the Italian *Atropo*
in 1912.[72] This diesel, which passed its acceptance tests in March 1911, had
gone in the German Navy's utility boat *Mentor* some months before.

The 850 bhp model (350 × 400 mm B/S) accepted by the Navy in 1911
(figs. 15.43, 15.44, and 17.31) was followed by a larger version in 1915. This
1,150 bhp at 390 rpm, 390 × 450 mm B/S engine was the last of Krupp's
two-strokes to have three scavenge valves in the cylinder head. The *U-66*
class received them first.[73]

Russia ordered two, six-cylinder, 1,650 bhp at 350 rpm sub engines from
the Germaniawerft in 1913, but they could not be delivered before the onset
of war. The German Navy lacked a hull large enough to accept them, and not
until 1918 did they enter service in the "U-cruiser" *U-139*.[74] At 23.6 kg/bhp,
they weighed 39,000 kg each. The last stage of a crank-driven, four-stage
compressor provided high pressure air for launching torpedos (fig. 17.32).

Fig. 17.31. Krupp 850 bhp at 450 rpm two-stroke sub diesel, 1911.
Note the inclined fuel injectors. Friedrich Krupp Archives

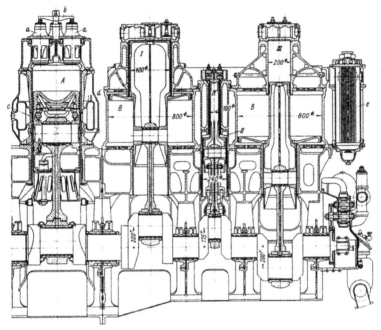

Fig. 17.32. Power cylinder, scavenge pump, and four-stage compressor of
Krupp 1,650 bhp, two-stroke sub engine, 1916. (a) Scavenge valves, (b)
three-fingered crosshead to open the three valves, (c) and (d) exhaust ports
and passage, (e) intercooler after second-stage compression.
Sass, *Geschichte, 1962*

Axial second and fourth stages lay between double-acting scavenge pumps
and the first and third stages atop them.

Krupp stopped making two-stroke sub engines in 1916, the last being
a port-scavenged 1,190 bhp model going in the *U-81* to *U-86* series. All
further two-stroke effort focused on the "battleship" diesel described later in
the chapter.

Consensus by the German Navy of the superiority of the four stroke over
the two stroke convinced Krupp to accept the inevitable and offer a four stroke

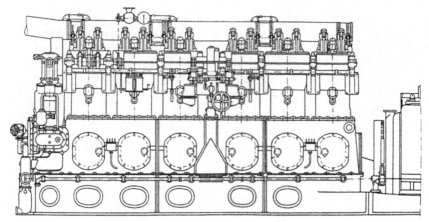

Fig. 17.33. Krupp four-stroke, 450 bhp sub engine, 1915.
Maag, *Dieselmaschinen, 1928*

of its own. Comments from various contemporary sources lead one to conclude that it was a wise decision, based on the service record of Krupp two-stroke boats. It is not clear how many of the resulting problems came from the valve-scavenged engines or their crews.

In any event, Krupp introduced a six-cylinder, four-stroke sub diesel around 1915. It produced 450 bhp at 400 rpm from a bore and stroke of

Fig. 17.34. The cargo sub *Deutschland in* Baltimore harbor, 1916.
König, *Deutschland, 1916*

320 × 420 mm and was nonreversing (fig. 17.33). Most of the castings were of steel, including a three-piece bedplate, a three-piece crankcase enclosure, and individual cylinders and heads. The pistons were uncooled.

The most famous pair of these 450 bhp engines were those in the cargo sub *Deutschland*, which made a well-publicized round trip to the United States in 1916 (fig. 17.34). The often-harrowing voyage from Bremen to Baltimore is dramatically told by Paul König, her captain, that same year. He related two episodes of engine room conditions in the 65 m long, 1,512-ton (surface displacement) boat. Both no doubt would describe many similar situations on subs during the 1914–1918 war. The first tells of the wildly pitching boat running surfaced during a heavy gale:

[Outside] it was an inferno, pure and simple.

But this was as nothing compared to the hell down below, especially in the engine room.

This ferocious sea had naturally forced us to keep all openings battened down. Even the manhole in the turret could only be kept open at intervals. It is true that two, large ventilation fans were going continuously, but the fresh air they sucked down from the carefully protected ventilation shaft was at once devoured by the greedy Diesel motors. These hungry monsters out of sheer ingratitude, returned us nothing but heat—a heavy, oppressive heat, saturated with a frightful smell of oil, which the ventilating fans kept whipping and whirling through all chambers of the boat.

Captain König then recounted the intense heat:

While in the Gulf Stream we had an outer temperature of 28° Celsius. This was about the warmth of the surrounding water. Fresh air no longer entered. In the engine room two six-cylinder combustion motors kept hammering away in a maddening two-four time. . . . A choking cloud of heat and oily vapor streamed from the engines and spread itself like a leaden pressure through the entire ship.

During these days the temperature mounted to 53° Celsius.

And yet men lived and worked in a hell such as this![75]

A reversible four-stroke sub engine built in quantity had an output of 530 bhp at 450 rpm from a bore and stroke of 350 × 350 mm. Ganz in Budapest was licensed to make this model for Austrian Navy submarines. A few of 1,200 bhp and 1,700 bhp were also delivered toward the end of the war. Of the sixty-nine sub engines to come from Kiel, its total two-stroke output has been given as 55,750 bhp and 7,750 bhp for the four-stroke[76] (fig. 17.30). Other four-stroke Germaniawerft diesels went in surface naval vessels as auxiliary power plants.

Daimler-Motoren-Gesellschaft

Like most of the German marine diesel builders, Daimler's factory in Berlin-Marienfelde was asked to supply four-stroke engines for the German Navy's coastal and mine-laying subs. It was already making a few of the marine type in the 100 to 200 bhp range. The largest production sub diesel to come from Daimler, the MU336, was a six-cylinder reversible model with a 530 bhp rating at 450 rpm from a 335 × 380 mm bore and stroke.[77] Partial views in figure 17.35 show a hand-operated coupling clutch and other wheels (*Räder*) and levers (*Hebel*) on the engine: reversing (*Umsteuerhandrad*), compressor regulation (*Hebei für Kompressorregulierung*), fuel throttle (*Handrad für Brennstoffreg*), blast air regulation (*Handhebel für Einblasedruckreg.*), and lever to rotate rocker shaft eccentric (*Anfahrhebel*). Engine room crews were kept busy when the boat was maneuvering, as was the case with all sub diesels of this vintage! Sixty-four of these and the smaller Daimler engines were built (fig. 17.30). Two, 1,700 bhp at 380 rpm engines were tested before 1918 but never installed in a sub. They later generated electric power.

Benz & Cie, AG

Benz in Mannheim, who had made the Hesselman marine diesel, also developed a four-stroke, nonreversible engine of 420 bhp at 375 rpm and a 350 × 375 mm bore and stroke. It was of conventional design with linered, individual cylinders, three-piece crankcase enclosure, and a bedplate cast in one piece.[78] Benz built fifty-nine of these and smaller sizes of similar design for subs.

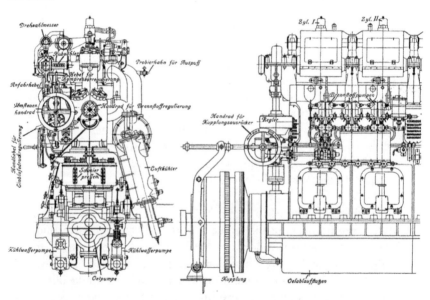

Fig. 17.35. Daimler 530 bhp, reversible sub engine showing controls, c. 1915. Föppl, *Schnellaufende, Dieselmaschinen, 1922*

Allgemeine Elektricitäts-Gesellschaft, Berlin

The last German sub diesel worthy of mention was a four-stroke, six-cylinder reversible model made by the AEG. With a bore and stroke of 350 × 360 mm, it produced 550 bhp at 450 rpm.[79] Individual cylinders sat on a three-piece crankcase enclosure in turn bolted to a multi-piece bedplate. Reversing was accomplished by axially sliding cam sleeves after the rocker shaft had been eccentrically rotated to lift the rocker ends off their cams. Figure 17.30 lists six as having been delivered.

Grazer Waggon-und Maschinenfabrik-AG

The Graz Railway Carriage & Machine Works built an interesting light-weight diesel for Austrian subs in the 1917–1918 period. The six-cylinder, four-stroke engine had an output of 550 bhp at 450 rpm and a bore and stroke of 340 × 380 mm.[80] Supporting the individual heads were three, bolted-together castings with each one containing the water jackets and liners for two cylinders. Instead of a symmetrical crankcase enclosure, seven heavy columns on the inboard side of the engine (assuming a twin screw) supported the jacket casting beam and tied them to a one-piece, cast steel bedplate (fig. 17.36). On the outboard, less accessible, side was a three-piece plate support, also with access doors, tying together the other side of the bedplate to the cylinder unit above. The design resulted in a lighter structure.

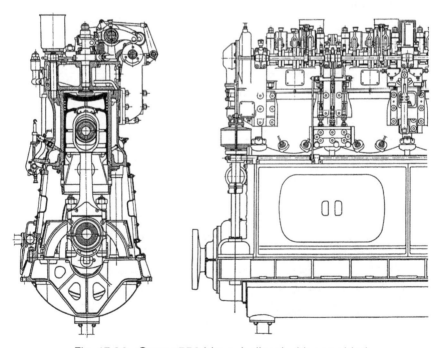

Fig. 17.36. Grazer 550 bhp sub diesel with one-sided column support, c. 1918. Maag, *Dieselmaschinen, 1928*

A reversing mechanism similar to that on the AEG was used. The number built of this engine type is not known but is assumed to be small.

Maschinenfabrik Ludwig Nobel

In 1911 the Russian Admiralty requested Nobel to build a two-stroke, 350 to 400 bhp sub engine having a specific weight not to exceed 25 kg/bhp. This was a prelude to ones with a much-higher output. Nobel complied that same year with a test engine meeting these specifications; it led to concurrent work on diesels for merchant and surface naval ships (chapter 16).

The navy's next, and nearly immediate demand, was for a sub engine to produce 1,340 bhp in six cylinders. A two-cylinder prototype (chapter 16) would serve at least a development role for ship diesels but it came to naught for sub purposes; the proposed sub's engine room cross-sectional area would be too small to accommodate it. More, but smaller cylinders were required. The new configuration had eight cylinders with a 390 × 430 mm bore and stroke as opposed to the original, six-cylinder design of 450 × 480 mm. Its output was to be the same 1,340 bhp at 350 rpm. The first engine, which went on test in 1914, had a bsfc of 226 g/bhp-hr[81] (fig. 17.37). This and one other installed in a *Kuguar* class sub in 1914 were the only reversible ones out of the twenty-eight, eight-cylinder engines delivered before the end of 1917.[82]

Between 1911 and 1918 Nobel shipped a total of fifty-seven engines for thirty-seven Russian Navy subs:[83]

Year	Subs	Engines	Cylinders	Cycle	BHP	RPM
1911/12	7	1	4	4	120	450
1911/12	7	1	6	4	160	440
1912	3	1	4	4	50	500
1914	3	2	4	4	300	375
1914	3	2	4	4	600	375
1914/18	37	2	8	2	1,340	350

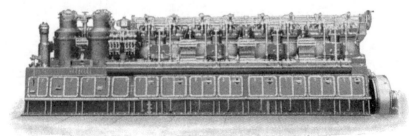

Fig. 17.37. Nobel eight-cylinder, 1,340 bhp at 350 rpm sub diesel, 1914. *Svenska Nobel-Diesel, c. 1925*

Whether all the subs with the eight-cylinder, two-stroke engines became operational, or even completed, is open to conjecture. During Imperial Russia's last days, the conditions at Nobel became chaotic because the Admiralty almost daily made new engine demands that were impossible for the company to satisfy.

Schneider et Cie., Le Creusot

Schneider & Co., with headquarters in Paris, was the largest builder of both four-stroke and two-stroke submarine diesels for the French Navy. The engines were made in a factory at Le Creusot and shipped the 30 km east to Chalon-sur-Saône where Schneider owned a yard fabricating Laubeuf-type subs. The boats then traveled down the Saône and Rhone Rivers to another company facility on the Mediterranean at St. Mandrier near Toulon for final outfitting and testing.[84]

Carels had earlier licensed Schneider to build its two-stroke engine for marine use (chapter 12), but the French company opted for the four-stroke cycle when it first began making submarine diesels. These were of Schneider's own design, which included a four-cylinder, 150 bhp at 360 rpm auxiliary and a 700 bhp at 300 rpm propulsion engine with a stroke of 480 mm; a pair of each went in the sub *Admiral Bourgois* before 1910. A more modern eight-cylinder engine of c. 1911 with a 280 mm stroke had a rating of 360 bhp at 400 rpm[85] (fig. 17.38).

To be noted on the four-stroke engine is its "underslung" rocker lever pinned at the cam follower guide and pivoting about a cam on the maneuvering shaft. The valve spring keeper was the lever's other support (fig. 17.39). A follower guide spring kept the lever in contact with its pivot cam surface when the maneuvering shaft cam was turned by gearing from a handwheel. In this way cylinders were deactivated, or sequentially all received starting air, then

Fig. 17.38. Schneider eight-cylinder, 360 bhp, four-stroke sub engine. *Engineering, July 1914*

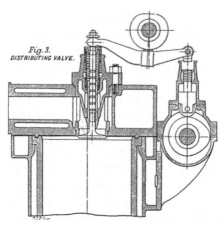

Fig. 17.39. Rocker lever actuation on reversible Schneider sub diesels.
Engineering, July 1914

half of them fuel and half air and finally all fuel, in either direction, depending on the handwheel position. An interlock prevented the main camshaft from being shifted axially until the maneuvering shaft was in the "stop" position. Two-stroke sub engines later used the same rocker lever design whereby the four-stroke's exhaust valve became a second scavenge valve.

All engines, using either cycle, had trunk pistons, cast steel bedplates, and frames. Engines with cast iron cylinders had no liners; those with them were of cast steel. All the larger-sized engines had oil-cooled pistons with the oil reaching the crown piece through telescoping tubes. Compressors, and scavenging pumps on the two-strokes, were crank-driven. Individual fuel pumps, metering by varying suction valve lift, were under the direct manual control of the operator for load changes, a governor for overspeed control, and regulation by the maneuvering shaft when starting ahead or astern.

Schneider began building two-stroke sub diesels as horsepower requirements increased. These have sometimes been listed as Schneider-Carels designs, but by this date much was changed from the time of the licensing, and whether or not royalties were paid to Carels is not known. By 1914 Schneider had delivered a pair of eight-cylinder, 1,200 bhp at 350 rpm engines with a 470 mm stroke for the *Néréide*,[86] and six each of six-cylinder, 650 bhp at 400 rpm (360 mm stroke). Eight eight-cylinder, 1,100 bhp at 370 rpm (370 mm stroke) engines had also gone in Schneider-made, Laubeuf subs.

The largest Schneider sub diesel, under test at least by war's end, had eight cylinders and a bore and stroke of 410 × 450 mm. Output was reported as 1,600 bhp at 300 rpm with a bmep of 67 psi.[87]

Société des Moteurs Sabathé

Especially interesting because of their fuel injection systems were the sub engines from the Société des Moteurs Sabathé in St. Etienne. This company

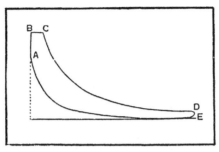

Fig. 17.40. Modified combustion cycle of the Sabathé engine.
Motor Boating, August 1912

was a division of the steam, gas engine, and machinery firm of Leflaive & Cie. in the same upper Loire River city. British Patent 10,219 of 1908 on the fuel system shows the direction Louis Gaston Sabathé and his Moteurs Sabathé would take. It tells how

> engines in which the methods of working at constant pressure and constant volume are combined, and in which the whole of the air for combustion is first compressed and the fuel subsequently injected in two phases, the first injection being towards the end of compression and the second during expansion.[88]

Thus, Sabathé was the earliest to consciously depart from Diesel's constant pressure combustion and to offer a "modified Diesel cycle" described in I-C engine textbooks (fig. 17.40). He also reaped the reward for using it by achieving a very low fuel consumption.

The fuel needle in the injector provided the equivalent of two-stage injection. The needle first lifted a lower valve off its seat at the nozzle orifice, and in moving higher, small projections on it lifted a second, larger-diameter valve (*g*) off its seat (fig. 17.41). The fuel charge entering the injector first filled lower chamber *a* and all excess overflowed through port *c* into chamber *e*, which was open to blast air pressure. The needle lifted at or slightly before TDC, whereby the fuel in the lower chamber was atomized and injected by air passing along a groove in the needle valve spindle, which bypassed seated valve *g*. This provided the constant volume phase of combustion. Fuel in the upper chamber was injected enough later to create the constant pressure combustion. To be noted is that at low speeds and light loads no fuel would be injected from the upper chamber so that combustion took place entirely by constant volume.

A second novel Sabathé feature in addition to the two-phase injection, was the way the needle lift could be varied depending on load (fig. 17.42). By rotating the shaft (*J*) clockwise, the lever (*L*) pulled the pushrod (*R*) and its cam roller "inward" along the saddle (*M*) to reduce the effective cam lift.

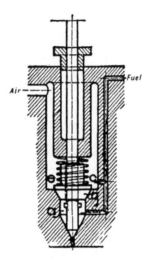

Fig. 17.41. Sabathé two-phase fuel injector to achieve first constant
volume and then constant pressure combustion, c. 1911.
Supino, *Land and Marine Diesels, 1912*

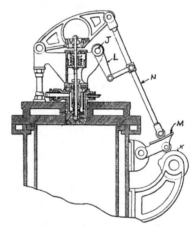

Fig. 17.42. Sabathé mechanism to vary fuel needle lift for
low-speed, light-load running. *The Engineer, July 1914*

Moving the cam roller far enough served to keep the needle seated to shut off
the fuel when the engine was being started on air and for reversing. The shaft
(*J*) was also under governor control so that if speed dropped enough the lift
of the upper valve was reduced sufficiently for it to remain closed. This not
only saved blast air, but further ensured that all the fuel under these operating
conditions would be injected to burn at constant volume.

Proof of the system's effectiveness is seen in the 1911 test results of a
four-stroke, six-cylinder, 350 × 350 mm bore and stroke submarine engine
(fig. 17.43):[89]

500 bhp at 400 rpm	178 g/bhp-hr
300 bhp at 350 rpm	179
110 bhp at 250 rpm	200 "

This is reportedly the lowest bsfc of any production marine diesel to that time. The engine ran smoothly at speeds as low as 100 rpm.

The Sabathé trunk piston engine[90] was reversed by sliding along the camshaft an assembly of cams fixed on a sleeve for each cylinder to place a new set of cams under the roller followers. Individual fuel pumps supplied each injector, and blast air came from a three-stage compressor driven off the crankshaft. Valve cages and nozzle were fitted onto a cast steel cylinder head, which was relatively thin and plate-like (fig. 17.44). The engine was state-of-the-art in most every aspect. No weights were given, but the design appears such that it had to be comparable with other lightweight sub engines of that time.

An additional feature was a patented means of changing a piston through an inspection door and not having to remove a cylinder head. Although the process was involved, it was said that experienced mechanics could change two adjacent pistons and rods in about six hours.

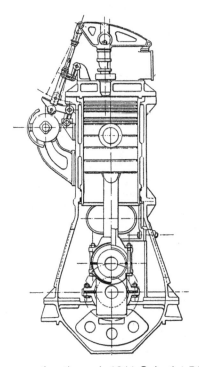

Fig.17.43. Cross-section through 1911 Sabathé 500 bhp sub engine. Note the counterweighted crankshaft. *The Engineer, July 1914*

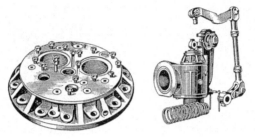

Fig. 17.44. Cylinder head and valve cage with lift mechanism,
Sabathé 1911 sub engine. *The Engineer, July 1914*

Sabathé's first engines were two-cylinder models of as low as 10 bhp
per cylinder. The largest engine reported built by 1912 was a four-cylinder
of 700 bhp. Because of the French Navy's interest in the design's potential
for submarines, almost all efforts went toward this application. How many
engines were built and installed in subs is not known to the author, but non-
reproducible pictures of the 500 bhp model show it to be a clean design
intended for production.[91]

Delaunay-Belleville

Growing out of the license Delaunay-Belleville at St. Denis sur Seine received
from Diesel in 1909 (chapter 15) was a reversible, six-cylinder, four-stroke
sub engine built in 1911. It had an output of 150 bhp at 350 rpm[92] (fig.
17.45). How many and of what size this primarily naval steam engine and
boiler company later made is unknown.

Fig. 17.45. 1911 Delaunay-Belleville, four-stroke. 150 bhp sub engine built
under Diesel license. *Author photo from Diesel collection in Deutches Museum*

Augustin Normand

Earlier experience Augustin Normand had with Nürnberg two-stroke slow speed diesels for the *Quévilly* (Chapter 15) proved of value when it built the engines for four French submarines in 1912–1913. They were based on the M.A.N. SD 65/8 with an output of 650 bhp at 400 rpm.[93]

The company had already made a six-cylinder, four-stroke sub diesel producing 420 bhp at the same 400 rpm. Bore and stroke of this reversible engine were 330 × 375 mm.[94] The camshaft was driven directly off the crankshaft through a single gear set to provide the half-speed reduction. The engine had a longer piston skirt than that in the Sabathé, calling attention to continued fears, or actual experience, of excessive piston wear. Although the bearings were pressure lubricated, some features such as crankshaft counterweights were missing.

The entire camshaft with all fixed cams was axially slid during reversing, and because the teeth on the crankshaft's spur gear were long enough, sufficient engagement was assured. Cam followers were held off the cams during the sliding action in a novel way when the operator rotated a small, engine-length maneuvering camshaft on the opposite side of the engine from the main camshaft. In doing this, short levers whose pinned ends lay under the maneuvering camshaft pushed down on rocker arm extensions to open fully the inlet and exhaust valves. Because the pushtubes were pinned to their rocker levers, the holding of the rocker lever noses against the opened valves lifted the followers off the cams. Although not specifically mentioned, the lift of the fuel needle was probably short enough so that by using tapered edges on the fuel cams the fuel cam followers did not have to be raised from the cams.

Again, how many of these engines found their way into submarines is unclear. As only eight of the Nürnberg-type diesels were built by Augustin Normand before 1914, some of the four-stroke engines or their derivatives were no doubt installed.[95]

Sautter Harlé

No specific details about submarine diesels Sautter Harlé made in the 1911–1918 period has been found by the author. This contrasts with numerous references about its activities prior to this time. The company is known to still have been involved because of a report that in 1913, under the direction of an engineer named Wisler, it had some success in applying an oxygen injection system developed by Jaubert to a closed-loop intake and exhaust circuit for running the engines while submerged.[96]

Burmeister & Wain

The Danish Navy asked Burmeister & Wain in late 1914 to design and build a submarine diesel for its fleet when it was realized engines ordered from Germany and Italy could not be delivered. There also appeared to be some

Fig. 17.46. B&W six-cylinder, four-stroke, 450 bhp at
500 rpm sub engine, 1916. *Burmeister & Wain Archives*

unhappiness with the performance of the foreign two-stroke designs already
in nine of their single-engined subs (six **Nürnberg** and three Fiat).[97]

B&W's resulting Model 6100U was a four-stroke, six-cylinder, reversible
engine having a bore and stroke of 330 × 330 mm and an output of 450 bhp
at 500 rpm—the same power as that of the above two strokes[98] (fig. 17.46).
The bmep of about 4.8 kg/cm^2 was average, but a specific engine weight of
only 19.7 kg/bhp compared very favorably with other sub engines. A circa
1918 B&W sales brochure claimed it weighed 1.5 tons less than for an equal
output two stroke and that its fuel consumption averaged 17 percent less.[99]

The trunk piston design included individual cylinders with through-bolts
extending from the top of the cylinder heads to the bedplate. Wet liners were
set in the cylinder castings. An offset, high-mount camshaft acting directly
through rocker levers opened the starting, intake, exhaust, and outward moving fuel valves. The pneumatic starting valve design was the same as that for
the larger engines.

Two hand levers (to the left of the dial gauge in figure 17.46) each controlled the suction valve lift of three individual fuel pumps in a common
housing. This allowed the fuel to be shut off, and then sent first to three and
finally to six cylinders during a startup in either direction. Eccentrics on the
main camshaft reciprocated the pump plungers.

Reversing the 6100U followed the same procedure as on other B&W
marine engines. The camshaft was slid to place another set of cams under the
follower rollers for opposite direction running after eccentrics on the rocker
shaft had lifted the rollers off their cams.

The first two of these engines went in the Danish subs *Neptun* and *Galathea*, which entered service during 1916 and 1917 respectively. A total of seven 6100U diesels were delivered before the end of 1918, and all were probably installed in subs.[100] Their operating performance was said to have been good.

Polar

AB Diesels Motorer entered the submarine diesel parade after being asked by a "foreign power" to do so. Hesselman departed from the two-stroke cycle because of a concern that, given the demanded high speeds, only a four stroke could surmount the difficulties associated with cooling the pistons. Although the design borrowed features from the company's own two stroke and from other four-stroke engines, several new and novel ideas were incorporated. The reversible six-cylinder diesel produced 350 bhp at 500 rpm from a bore and stroke of 290 × 300 mm[101] (fig. 17.47).

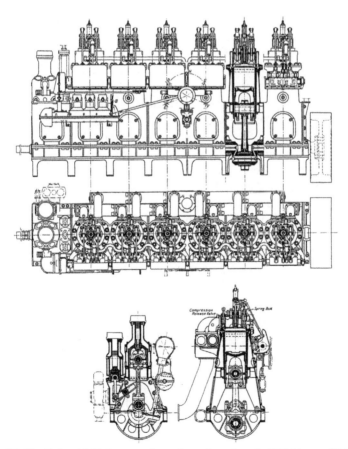

Fig. 17.47. Polar 1913 four-stroke submarine diesel. 350 bhp at 500 rpm.
The Engineer, October 1913

The bedplate, crankcase, and individual cylinder water jackets were made of cast steel with designed wall thicknesses in the crankcase as thin as 5 mm. Sealing between the cylinder head and the liner and water jacket was new. The head held a flange on the lower end of the liner against a mating flange in the jacket's lower part with a metal-to-metal seal. A special high-pressure packing ring of material used in steam engines (Klingerit[102]) sealed the liner's upper end against the head. A thicker, more resilient material sealed the head-to-water jacket joint. Thus, the clamping force was carried through the liner rather than the thin water jacket. Drilled passages connected the jacket water with the water in the head instead of conventional external pipes.

The unique reversing system was the only one used in a production engine by a major diesel manufacturer that switched the inlet and exhaust valve functions. Most engineers had experienced a diesel's capability to run in reverse given the right starting conditions, so the idea itself was not new.[103]

A cylindrical, horizontal change-over valve in a water-cooled housing was inserted midway along horizontally parallel intake and exhaust manifolds. Outside air entered the valve from the top and exhaust went out the bottom. By turning it half a revolution, an angled sealing surface interchanged the air intake with the exhaust pipe. Rings in the sealing lips allowed for valve expansion and distortion. Intake and exhaust valve timing was unchanged, but it may be assumed that "ahead" running was favored.

Fuel and starting air timing was changed by axially sliding the camshaft to change cams. Tapered side edges on the cams allowed the shift to be made without having to lift off the followers. The intake and exhaust cams were wide enough to actuate their valves when the engine turned in either direction.

Compression pressures were limited during starting and at the same time prevented contaminated air in the combustion chamber from entering supply air lines. Hesselman had witnessed several violent explosions during startup in his early exposure to the diesel when high starting air overpressures initiated combustion of entrained oil in the starting air. He never wanted those conditions to happen with his designs, especially in a submarine.[104]

A compression release valve, manually cocked open, was brought into play when starting air was being admitted. This limited maximum cylinder pressure to no more than 250 psi. Next, inserted in the pushrod for the starting valve was a spring-loaded cylinder ("box" in figure 17.47) whose compression force was adjusted so that the valve closed at the same pressure as the compression release valve. After turning over the engine with starting air, the compression release was locked closed, presumably automatically, and the engine made one revolution with a normal compression into which a charge of fuel was injected. If the engine did not start, the process was repeated with the compression release valve again cocked open. Such restarts reportedly did not occur.[105]

Turning a single hand lever to the left or right, for "ahead" or "astern," performed all functions needed to control the starting air, fuel, changeover valve, and camshaft shifting. Only the lever to open the compression release was separate.

The much-lower starting air pressure than normal also lowered compressor requirements. One of the four compressor pistons supplied starting air, and the other three were for the three-stage injection air needs. The side-by-side compressor cylinders and rocker lever drive bore a resemblance to Sulzer's design.

Two of these Polar engines successfully passed their official trials and were delivered to the navy ordering them. Since it was reportedly a "continental Power," the onset of war precluded more being shipped. AB Diesels Motorer did not abandon the engine as it became the type "MS6K" after 1918 being offered in two sizes with outputs of 300 bhp at 400 rpm and 420 bhp at 350 rpm.[106] One revision was paired cylinder castings. The change-over valve for reversing seems to have disappeared, but the spring-loaded pushrod for the starting air valve continued.

McIntosh & Seymour installed two eight-cylinder, four-stroke Polar diesels in the US sub *E-2* when it was repowered. According to Magdeburger, the nonreversible, non-air start engines had a 9 × 13 inch (230 × 330 mm) bore and stroke and produced 250 bhp at 375 rpm.[107]

Battleship Engines

Several European navies had been observing with close interest the improved reliability and rapidly increasing power of commercial marine diesels.[108] Only the German one would act. The ever-enthusiastic, imaginative Anton Rieppel at Nürnberg saw to that. He had proposed to the *Reichsmarine* as early as August 1909 that his company was prepared to build diesels capable of powering a capital ship. Within months he signed a development contract for a 160 rpm, six-cylinder, 12,000 bhp battleship engine.[109] Six of these, driving through three shafts, would provide such a ship with the 70,000 bhp needed by the navy.

Krupp's Germaniawerft shortly thereafter received a similar contract and would thus compete for a production contract with the winner being chosen on whose engine performed the best.

Both companies were first to demonstrate an output of 2,000 bhp per cylinder in a three-cylinder version and then to build one with six cylinders for an installation. It was planned for the engine to go in the *Prinzregent Luitpold*, a ship then under construction at the Germaniawerft.[110] The diesel was to drive the center shaft when cruising, and steam turbines would turn the outboard shafts when higher speeds were demanded.[111]

Sulzer Brothers also wanted to quietly enter the competition as it saw ship engine possibilities from the Germans as well as the French. The latter's navy was well aware of what the Germans had authorized as it was not a military secret.[112] The Sulzers would finance their own program, which was limited to a single-cylinder test engine.

All three companies chose the two-stroke cycle. Krupp and M.A.N. opted for double-acting engines. Sulzer conservatively stayed with single-acting.

None of these *Groß-Dieselmaschinen* escaped thermal stress-related ailments, which took years to overcome—particularly in the case of the double-acting designs. Pistons, cylinder walls, and heads all suffered cracking or outright failure, and only through persistent, creative redesign was progress made. Mechanical weaknesses were less prevalent but in one instance disastrous.

The Nürnberg Engine

Although Nürnberg's largest engine to 1910 produced only 100 bhp per cylinder, it could draw on its long experience from large, horizontal gas engines (chapter 10). One must appreciate the immensity of these earliest "cathedral" diesels. The height of the MAN engine with its bore and stroke of 850 × 1,050 mm was over 7.4 meters. The cylinder and head structure alone, the area to cause the most grief, was about 3.3 m high with an overall diameter of about 1.5 m[113] (fig. 17.48).

The bedplate and crankcase structure supporting the cylinders were of cast steel. A double-sided crosshead guided the piston rod and held the upper connecting rod bearing (fig. 17.49). Oil to cool the hollow piston was supplied through a coaxial tube in the piston rod; oil drained through the passage surrounding the tube. The upper and lower heads of cast steel were

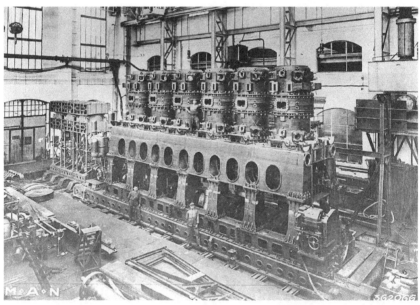

Fig. 17.48. M.A.N.-Nürnberg six-cylinder, double-acting, 850 × 1,050 mm engine during assembly in 1914. Note the man leaning against a cylinder.
MAN Archives

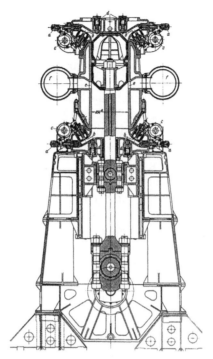

Fig. 17.49. The M.A.N. three-cylinder, two-stroke, double-acting engine
as tested in 1911 attained 5,400 bhp. Sass, *Geschichte, 1962*

bolted to a one-piece cylinder housing the exhaust ports. Each cylinder head
held four scavenge valves, four fuel valves, and two valves combining air-
starting and safety pop off functions. Four camshafts—two on each side of
the engine—actuated eighteen of the twenty valves per cylinder assembly.

Three double-acting scavenge air pumps (enough capacity for the six-
cylinder) were engine driven. Their bore and stroke were 1,320 × 800 mm
each, or 45 percent larger in displacement than the power cylinder. Air for
injection and starting came from auxiliary compressors driven by a two-
stroke, 350 rpm Nürnberg diesel.

In March 1911, one year from the start of design, the three-cylinder
engine went on test. Initial results were discouraging because of rapid valve
breaking in the scavenge and blast air compressors. Heat-related problems
in the power cylinders were of far greater consequence; pistons were under-
cooled and soon cracked. Cylinders cracked between the exhaust ports. The
heads, which cracked around valve ports, were recast in iron to expedite their
replacement. The power then was bravely increased to 5,400 bhp at which
output cylinder head bolts broke. (Power required to run the injection air
compressor was not deducted from the 5,400.)

Although this briefly attained output met the Navy's 90 percent requirement, a second contract stipulation called for holding that rating over a

five-day endurance run. Until it was, the six-cylinder engine could not be built. Thus began a lengthy series of head, cylinder, and piston redesigns. With each one a troublesome area was slightly improved even as power was increased. Some of the creative redesigns sprang from desperation, and though expensive, were still practical for a limited production run.

A tragic accident killed ten people and badly injured fourteen others during tests of the second redesign in January 1912. One of the scavenge valve rocker arms broke in such a manner that the valve was held open during the combustion stroke. The burning gases passed back into the air manifold containing pockets of lube oil carried from the compressors. The resulting explosion blew off the air manifolds and set fire to a fuel supply tank. In addition to the loss of life and injuries, much of the engine itself was destroyed.

An issue arising from the tragedy concerned Hans Bruns, who in 1911 had become chief engineer of Nürnberg's marine engine department. Bruns came there in 1908 and was largely responsible for the design of the two-stroke submarine engine, its slower-speed siblings, and of the battleship engine. Unfortunately, this energetic and talented engineer asked more of his own staff and the shop than they often could accomplish. He had resigned shortly before the accident, but then offered to stay and help in the rebuilding. This lasted only a short while when differences of opinion again arose over design decisions. This time Bruns left for good to work for Maschinenfabrik Esslingen. The parting was amiable enough as he could still be called on if needed. Schwarz, an engineer reporting to Bruns, was severely injured in the accident. He was put in charge of the large engine program after his convalescence.[114]

Tests on the third redesign did not begin until March 1913. All work was confined to a single cylinder, a practice kept in effect until reliability was more evident.

The seventh, and last, design of September 1913 (figs. 17.50 and 17.51) shows what had gradually evolved:

- A two-piece cylinder separated by a small gap at the exhaust ports allowed axial growth without thermal stresses.
- The four scavenge valves moved into the cylinder wall and were actuated by face cams on the four camshafts.
- Two fuel valves instead of the original four were placed in the cylinder wall. The upper head now held only one starting valve with none in the lower head.
- A spiral cooling rib directed cooling water flow at a high velocity along the cylinder wall to transfer the maximum heat.
- A loose, continuous steel cooling ring was inserted between the cylinder wall and head to carry heat away from the injector nozzle and scavenge valve seat areas.
- The ring contained drilled water passages.
- Uncooled pockets in the piston were eliminated.

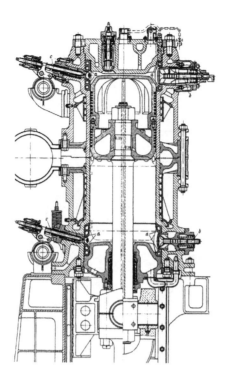

Fig. 17.50. M.A.N. 850 × 1,050 mm diesel incorporating seventh
and last piston/cylinder design of 1913. Sass, *Geschichte, 1962*

The cylinder and upper head were of cast iron while the lower head and
piston were of cast steel with 3 percent nickel.

Tests began on a six-cylinder engine incorporating the above design in
February 1914 (fig. 17.52). Cylinder problems unrelated to thermal loads
delayed full power runs until September. The engine then produced 10,000
bhp at 130 rpm.

Work on the engine was greatly curtailed after the war started and the
prognosis of its ever going in a ship looked poor. A shortage of gasoil for
fuel caused further delays until changes could be made for it to bum coal tar
oil. Single-cylinder tests with this fuel took place in April 1915. A five-day
run was also made at 2,030 metric ihp. (Converting *in this instance only* to
an American/British equivalent of 2,002 ihp—ISO 1,493 kW—eliminates
doubt that 2,000 ihp per cylinder was achieved.)

Tests on the new six-cylinder engine burning a tar oil/kerosene mix began
in January 1917. By this time the single-cylinder engine had accrued over five
hundred hours without significant problems. On March 24 the six-cylinder
engine ran for twelve hours at 12,200 hp. Small cracks were discovered in one
of the lower heads, but the decision was made to proceed. The engine passed
its five-day acceptance test at 10,800 hp, or 90 percent power. Output was

then increased to 12,200 hp at 135 rpm for twelve hours. Specific fuel consumption was 243 g/hp-hr using a mix of 214 g coal oil and 29 g of paraffin oil. It is assumed that the indicated power of the injection air compressor was not deducted from this sfc figure. The reversing mechanism, similar to that on the MAN/Blohm & Voss engines (chapter 16), also proved satisfactory. The only noted problems were some broken piston rings.

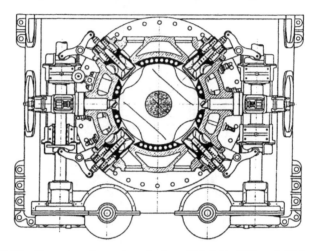

Fig. 17.51. Plan view through cylinder of M.A.N. 850 × 1,050 engine showing cooling ring and cam-operated scavenge valves. 1913 design. Nägel, *"The Diesel Engine of Today,"* 1923

Fig. 17.52. M.A.N. 12,000 bhp marine engine on test stand in Nürnberg 1914–1918. *MAN Archives*

The five-day acceptance run generated the following data:[115]

Mean brake power, bhp	10,880
Mean indicated power, ihp	14,500
Speed, rpm	130
Bsfc, g/bhp-hr	228
Injection air compressor, ihp	829
Injection air pressure, kg/cm² (abs)	71
Scavenge air pressure, kg/cm² (abs)	1.59
Water & other pumps, ihp	300
Brake mechanical efficiency, %	69

All auxiliary parasitic loads were included in the brake horsepower and mechanical efficiency data. The fuel had an LHV of 10,000 cal/kg.

On October 16, 1917, after completion of the acceptance tests—and knowing that the engine would never leave the factory—one cylinder was opened to its maximum. (The dynamometer was not capable of absorbing the total output of the other five cylinders during this "all-out" test.) It reached 3,573 ihp at 145 rpm with an estimated mechanical efficiency of 90 percent. All pumps and compressors were electrically driven. Indicated mean effective pressures in both upper and lower cylinders were about the same at an imep of 9.82 kg/cm².[116] Figure 17.53 indicated sfc (*Brennstoffverbrauch*), shows injection air pressure (*Einblasedruck*) and scavenge air temperature and pressure vs. power (*Nutzleistung*) taken during the test.

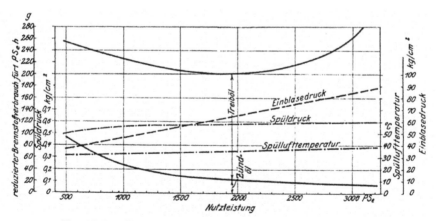

Fig. 17.53. Performance curves made during the maximum overload test on one cylinder of the six-cylinder engine, October 1917. Nägel, *"The Diesel Engine," 1923*

Fig. 17.54. Krupp Germaniawerft single-cylinder, double-acting test engine of 1911 with 875 × 1,050 mm bore and stroke. It was 7.9 m high. *Krupp Archives*

This remarkable engine was under development for six years, and through the steadfastness of Rieppel, it survived the discouragement of many failures and the tragedy of the explosion. It was scrapped under terms of the Versailles Treaty.

The Krupp Germaniawerft Engine

The battleship engine program adopted by the Germaniawerft was based on a design significantly different from Nümberg's.[117] Krupp prudently started with a single-cylinder test engine to gain experience with the challenging cylinder and piston requirements (fig. 17.54).

The initial design of this double-acting two-stroke included uniflow scavenging through four scavenge air valves and exhaust ports (fig. 17.55). Scavenge air pumps were main engine–driven, but auxiliaries operated all others and the injection air compressor. The crosshead bearing had a guide on only one side.

A cylinder assembly originally consisted in part of a two-piece liner separated by a small end gap at the exhaust port level to accommodate thermal growth. Upper and lower castings containing the air and fuel valves encased the liners and extended to the exhaust ports. They also retained the upper and lower heads, which were merely cylinder end covers.

Hydraulic actuation of the scavenge air valves and two fuel valves was unique. A remote camshaft generated the forces to open the valves (fig. 17.56). A hand lever (*a*) adjusted oil flow into the make-up valve (*b*) to

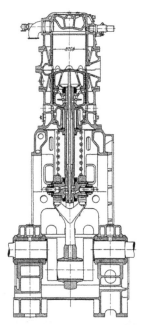

Fig. 17.55. Krupp 875 × 1,050 mm single-cylinder, double-acting test engine of 1911. Section view with the original cylinder and piston design. Maag, *Dieselmaschinen, 1928*

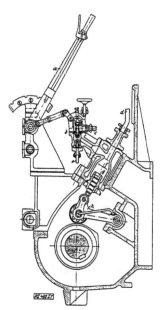

Fig. 17.56. Pressure generating means of hydraulic system to open scavenge air and fuel valves in the 1911 Krupp 875 × 1,050 mm diesel. Laudahn, *"Germania-Großölmaschine," 1924*

maintain oil in the plunger housing (*f*) and line (*d*) to the valve so that when the cam (*i*) acted on the master piston (*g*) the valve would open. Make-up oil entered through *c*. The engine was stopped by pulling the lever to its last position. This released enough oil to deactivate the master piston.

A slave piston pushed directly on a scavenge valve stem (fig. 17.57a). Similarly, a slave piston indirectly opened a fuel needle valve by acting against a lever pivoted at the opposite end and lifting the center-placed fuel valve (fig. 17.57b). Peak hydraulic pressures in the external lines were about 80 atm. The system gave so little trouble that Krupp continued using it.

Reversing was done by simply using a compressed air motor to rotate the camshaft to accommodate the required timing for running in the opposite direction.

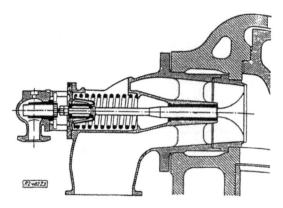

Fig. 17.57a. Hydraulic actuation at the scavenge air valve in 1911 Krupp 2,000 bhp per cylinder diesel. Laudahn, *"Germania-Großölmaschine," 1924*

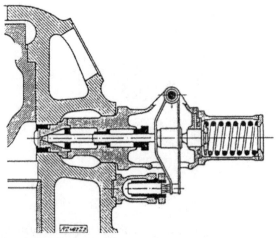

Fig. 17.57b. Hydraulic actuation at the fuel valve in 1911 Krupp 2,000 bhp per cylinder diesel. Laudahn, *"Germania-Großölmaschine," 1924*

During tests that began in November 1911, the cylinder cracked under a load of 1,200 hp. On January 19, 1912, it reached 1,275 bhp (1790 ihp) at 108 rpm. At the beginning of February, the power was raised to 1,510 bhp when the ground under the foundation settled and the crankshaft broke between the main bearing and flywheel. The engine continued to run at 110 rpm for a while longer. A pending visit on February 17 by two German princes forced a quick welding repair. It held for the royal visit, but the engine was no longer run under a significant power.[118]

Easing the situation was the fact that the three-cylinder engine was already under construction. Tests on it began in July 1912. It used a combination of both port and air valve scavenging. Power from this design reached 4,800 bhp at 130 rpm. Next came mostly port scavenging with small air valves retained near the head (1913). Finally, loop scavenging with only intake ports was tested. It would become the design adopted on the six-cylinder engine (fig. 17.58).

On January 25, 1915, the three-cylinder engine ran 24 hours at 5,100 bhp (7,100 ihp) and then three hours at 5,500 bhp at a speed of 145 rpm. (Krupp factored in the auxiliary power to drive the pumps and compressor.) It was

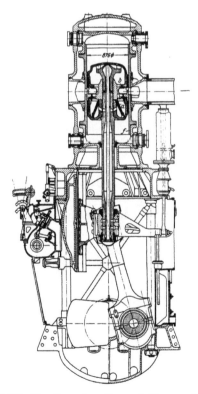

Fig. 17.58. Section view through Krupp six-cylinder, 875 × 1,050 mm engine, 1916. Maag, *Dieselmaschinen, 1928*

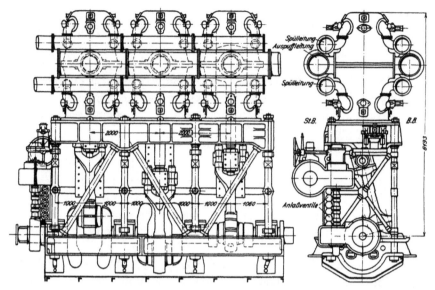

Fig. 17.59. Krupp open crankcase design used on three-cylinder (shown) and six-cylinder 875 × 1,050 mm engines. Maag, *Dieselmaschinen, 1928*

enough power to satisfy the Navy. The engine also met the fuel consumption requirement, including auxiliaries, of 160 g/ihp-hr with a test measurement of 154 g/ihp-hr at the 5,500 bhp output. An overall mechanical efficiency of 75 percent made this possible.[119] Authorization to design the six-cylinder version was given, but two years would pass before testing began on it.

Meanwhile, considerable cylinder redesign and gas flow work was carried out on the "tender" one-cylinder engine to develop air and exhaust deflection profiles on the piston crowns. Most of these research efforts were not applied to the six-cylinder engine because of wartime constraints.

Assembly of the six-cylinder engine was completed in August 1916, and testing began the following January. Its structural design, like that of the three-cylinder model, embodied the columns and braces used in steam engines and Werkspoor diesels (fig. 17.59).

The "shakedown cruise" of this engine held its own moments of frustration as power was gradually increased. In February 1917 the seawater cooling system failed suddenly during the first attempt to complete the required five-day test; on a restart it was discovered that water had entered a cylinder. A few days later there was an explosion in a fuel valve that blew off the end of the nozzle and resulted in both piston and upper head damage. The safety valve prevented a possible catastrophe. A replacement cylinder assembly with new modifications was borrowed from the single-cylinder engine and used throughout the rest of the testing.

Final trials began after attaining 11,000 bhp, which was enough to meet the Navy's 90 percent requirement—it could sustain that power for the five-day run. This test, starting on March 1, yielded results (in metric units), which fulfilled the contract:[120]

Speed, rpm	140
Avg. power for five days, bhp	10,600
Max. one hr power, bhp	12,060
Bsfc, g/bhp-hr	234.4
Scavenge pump load, ihp	1,095
Scavenge air press, atm (abs)	1.3
Inj. air compressor load, ihp	2,000
Inj. air pressure, atm (abs)	80
Indicated mep at 12,060 bhp, atm	
Upper cylinder	7.25
Lower cylinder	5.55

The acceptance test culminated work on the six-cylinder engine, which, along with the one- and three-cylinder prototypes, was later scrapped under the dictates of the Versailles Treaty. While the Krupp engine did not quite match the power capabilities of its M.A.N. rival, its achievements are still to be lauded.

Sulzer's One-Meter-Bore Engine

When designed in 1910, Sulzer's single-cylinder test engine had a cylinder displacement over seven times greater than that of its largest marine engines in service (fig. 17.60). While the 1,000 × 1,100 mm bore and stroke S100 and 6.9 m height might be viewed as more a scaling up of production models, its huge size presented problems requiring time to solve. However, being single acting rather than double acting like its German cousins, it was spared some of their thermal stress problems. The experience provided by it would prove invaluable when Sulzer built post-1915 marine diesels. The engine was not reversible.

The S100's structure utilized four long, through-bolt columns that both absorbed the combustion-created tensile loads acting against the cylinder head and supported on shoulders the head and cylinder suspended under the head. In this way the bottom end of the cylinder was free to slide in a "socket" in the upper end of straddling A-frame supports[121] (figs. 17.61 and 17.62).

Sizes and weights of a few of the engine's parts attest to its massiveness. The crankshaft at the main bearings and the crosshead pin had diameters respectively of 530 mm and 350 mm. The cylinder head was over .75 m thick, and the flywheel alone weighed "at least 30 tons." Crankshaft balance weights were "about 5 tons."

Fig. 17.60. Sulzer S100 single-cylinder, 1,000 × 1,100 mm single-acting, two-stroke engine erected in 1912. Design rating: 2,000 bhp. *Sulzer Archives*

Cross-port scavenging using a double row of intake ports was the same design as that on the *Romagna* engines of 1909 (chapter 15). The valve of the single fuel injector in the center of the head was lifted by a cam-actuated rod inserted through a cast-in passage in the head.

The rocker lever for a combined starting and relief valve was the only mechanism on top of the engine. An adjustable length compression spring made the combination possible. In its "least-force" length against the outward opening valve, it acted as a relief valve. The rocker lever, acting against its upper spring end, simply worked against the spring during engine running without opening the valve. Manually compressing the spring to a length used for starting stiffened it enough for it to act as a solid link in order to open the valve and admit starting air. During the interval between the time when the starting air was shut off and the lower spring tension could be restored, the full compressed charge of cylinder air escaped out a short relief pipe. The sound of the blowdown process past the 150 mm diameter valve head was reported by a visiting British naval engineering officer to be very audible![122]

Originally, the scavenge air pump and three-stage injection air compressor were driven off the end of the crankshaft, but in 1914 these were removed and driven by auxiliary power.

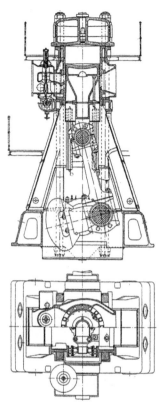

Fig. 17.61. Section views of Sulzer S-100, 1-meter-bore test engine.
Seiliger, *Hochleistungs-Dieselmotoren, 1926*

Overheating of the crosshead pin bearing presented an initial problem Sulzer solved by changing to a "gunmetal" material from a white babbit type and by adding more grooves to better distribute the pressure lubrication over the bearing surface. The visiting officer commented on the similarity of the problem that Vickers encountered with its large two-stroke test engine. He called attention to the inherent difficulty of adequately lubricating the crosshead or piston pin bearing of a two stroke where the bearing remained loaded on both the up and down strokes.

A second test program in October 1914 began with a fifteen-hour run at 1,000 bhp and 120 rpm followed by one of six hours at 2,000 bhp and 150 rpm. This was in accordance with "contracted conditions with our licensee Soc. des Forges & Chantiers de la Méditerranée in the presence of their representative."[124]

Professor Aurel Stodola of the ETH-Zürich conducted his independent tests in November 1914 and January 1915 (fig. 17.63). Extracts from his data taken at maximum power during a six-hour test indicate the progress made on the engine to that time:[125]

Horsepower, bhp*	2,059
Speed, rpm	150
Brake mep,* kg/cm²	7.16
Indicated mep, kg/cm²	8.57
Mechanical efficiency,* %	83.5
Bsfc,* g/bhp-hr	163.7
Bsfc, g/bhp-hr, including fuel used by auxiliary driving compressor	220.1
Injection air pressure, atm.	75
Scavenge air pressure, atm	0.495
Exhaust temperature, °C	301
*The main engine did not drive the injection air compressor.	

The best sfc, withm copressor power included, came in the January test. Using a lower scavenge air pressure (0.245 atm) at 1,453 bhp, the bsfc was only 194 g/bhp-hr, a very credible figure. The Galician fuel had an average LHV of 10,140 cal/kg.

Between the Stodola tests were numerous visits by delegates from foreign navies. These included one by the French on December 22, and a few days later groups from the British, German, and Italian Navies saw the engine perform—all at separate showings.[126]

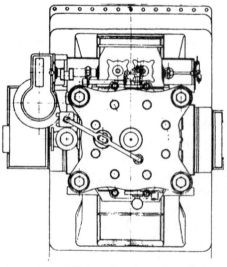

Fig. 17.62. Top view of Sulzer S100 engine. *Sulzer Archives drawing*

Fig. 17-63. Sulzer S100 1-meter-bore engine on test, c. 1914.
Sulzer Archives

Official work ceased on the engine following this spate of activities. Its fate is unknown, but it is assumed to have been dismantled soon thereafter.

Conclusion

The 1911–1918 era of Diesel's engine for naval applications was one of the most dynamic—and daring—periods of its history. Expectations were high, and the engine lived up to them. Advances in understanding stresses created by heat and combustion loads introduced new technology for overcoming them. Manufacturing know-how was transferred from practices acquired in making huge gas and steam engines as well as extrapolating from knowledge gained with smaller diesels. Factories had on hand or were able to acquire needed machine tools to fabricate the ever-larger engines. But above all, the naval diesel's great advances in power and reliability resulted from creative engineering and entreprenurial risk takers who united to master their challenges.

Hindsight

"Richmond, surely one of the brainiest of British naval officers in the twentieth century, could write, when Assistant Director of Naval Operations, in a memorandum for his chief [in July 1914]: 'The submarine has the smallest value of any vessel for the direct attack upon trade. She does not carry a crew which is capable of taking charge of a prize, she cannot remove passengers and other persons if she wishes to sink one.' Richmond's marginal addendum years later reads: 'I made a pretty bad guess there!'"

—As quoted by Marder[127]

Notes

1. December 8, 1909, handwritten letter by Rudolf Diesel, in English, to Edward D. Meier with report on European naval engine developments. Busch-Sulzer papers, Wisconsin State Historical Library, Madison.

2. E. C. Magdeburger, "Diesel Engines in Submarines," *Trans., American Society of Naval Engineers*, vol. 37 (1925), 609. From 1912 to 1920 Magdeburger was a diesel engineer with Busch-Sulzer. A native of Russia, he had worked at the Kolomna diesel factory and on gas engines in Germany before coming to the US in 1910. He ended his career in the 1940s as Chief Engineer, Bureau of Ships, US Navy. *S.A.E. Transactions*, vol. 52 (1944), 249.

3. Magdeburger, "Diesel Engines," 579–80.

4. A Paris interview with Karl Kühn, June 1978, a long-time associate with Pielstick; also Fritz Schmidt, "Die Entwicklung der Zweitakt-Marine-Motoren (MZ, LZ, VZ, ab 1926–1944)," Appendix 2a. MAN Archives.

5. Kurt Schnauffer, "Die Motorenentwicklung in Werk Augsburg der M.A.N. 1898–1918," in MAN Archives (1956), 49.

6. Holbrook C. Gibson, USN, "The Diesel Engine of the German Submarine U-117," *Trans., of the Society of Automotive Engineers*, vol. 15, pt. 1, (1920), 502. The paper gives a detailed technical description of the S 6 V 45/42.

7. Schnauffer, "Die Motorenentwicklung," 55. Eberhard Rössler, *The U-Boat—The evolution and technical history of German Submarines* (Annapolis, MD: Naval Institute Press, 1981), is a good source on the subject.

8. Magdeburger, "Diesel Engines," 606.

9. Schnauffer, "Die Motorenentwicklung," 56. The *U-140* was 92 m long with a surface displacement of 1,930 tons and a range of 12,000 nautical miles at a cruising speed of 8 kts.

10. This class of sub was 97.5 m long with a surface displacement of 2,158 tons. Surface cruising range at 7 kts was 20,000 nautical miles. Rössler, *The U-Boat*.

11. In October 1926 the author's father took indicator cards of his fuel system injecting into one cylinder of this ten-cylinder engine being used for test purposes at the Brooklyn Navy Yard.

12. "M.A.N.-Dieselmotoren für Unterseeboote der Kaiserlichen Deutschen Marine (c. 1910–1918)." Source: Manufacturing and delivery records, M.A.N. MAN Archives. See also Schnauffer, "Die Motorenentwicklung," 60.

13. Gibson, "The Diesel Engine," 506.

14. Ibid., 496.

15. Ibid., 506.

16. Magdeburger, "Diesel Engines," 606.

17. Kurt Schnauffer, "Die Dieselmotorenentwicklung," 22. Model designations for these engines also changed with Nürnberg's new line of heavier, slower speed engines. The old "S D" for "Schiff Diesel" became "SS" for "Schiff Schnell," or high-speed marine, and the new "SL" for "Schiff Langsam," or slow-speed marine.

18. Ibid., 23.

19. The first *E* boat entering service in 1912 was one of eleven ordered; a second order of twenty-two delivered during 1915–1917 is shown as having two of the 800 bhp diesels. *Jane's Fighting Ships of World War I* (New York: Military Press, 1990), 80, 82. Therefore, the author assumes solid injection began with these engines. The fact that the inventions covered by McKechnie patents were still evolving until 1914 lends credence to this assumption.

20. "Heavy Oil Engines for British Submarine Boats," *Engineering*, July 4, 1919, 10.

21. "British Submarine Boat Building During the War," *Engineering*, February 28, 1919, 264–66. This article gives much data on the numerous classes of British subs. See also *Jane's*. *E* boats had a surface displacement of 730 tons and a length of 180 feet. *L* boats were 231 feet long with an 890 ton sd. *K* class boats, the largest, were 337 feet and had an sd. of 1,880 tons.

22. *Jane's*, 63, 91. For engine specifications see *Trans., Institution of Engineers and Shipbuilders in Scotland*, vol. 64 (November 1920), 16. The *Marshall Soult*, 355-foot long and one of several WWI monitors for shore bombardment, had two 15-inch guns for main firepower. The 4,500 tons displacement *Trefoil* was 280 feet long.

23. Credit report on EBC submitted to The American Diesel Engine Co., September 16, 1909. Gross assets in 1906 were $1,371,869 and debts of $248,058. Busch-Sulzer Papers, Wisconsin State Historical Library.

24. Frederick A. Talbot, *Submarines* (Philadelphia: Lippincott, 1915), 274 pages, compares the

Holland and Simon Lake designs and accomplishments and substantiates the claim that Holland was more advanced, and earlier, with his ideas. pp. 26–47.

25. Frank T. Cable, *The Birth and Development of the American Submarine* (New York: Harper, 1924), 337 pages has a Holland biography. He documents developments that he was a part of from 1898 until 1924. He was the most experienced of early sub skippers and was chief trouble shooter and salesman for EBC in England, Russia, and Japan. It is an authoritative, firsthand story on EBC and other make subs.

26. Gaddis Smith, *Britain's Clandestine Submarines, 1914–1915* (New Haven, CT: Yale University Press, 1964), 10.

27. The boats were launched a year and a half later and both commissioned on February 14, 1912. Lt. Chester W. Nimitz, the *E-1*'s first commanding officer, took it across the Atlantic Ocean. The *E-1*, 135.25-foot long with a surface displacement of 287 tons, is the smallest sub to have done this. Submarine Force Library and Museum, *United States Submarine Data Book* (Groton: 1984), 3.

28. "License Agreement" between M.A.N.-Nürnberg and EBC. MAN Archives. An Appendix gives EBC's subsidiary companies: Holland Torpedo Boat Co., Electro-Dynamic Co. (where Frank Cable first worked), The Electric Launch Co., The New Jersey Development Co., and The Industrial Oxygen Co.

29. Ibid., 6.

30. Letters, Meier to Harris, February 11, 1910; and Meier to Diesel, February 23, 1910. Meier wanted Diesel to find out if Sulzer had a suitable submarine engine that ADE (to become Busch-Sulzer) could use. Busch-Sulzer papers, Wisconsin State Historical Library.

31. Letter, Rotter to Harris, June 10, 1911. He was in Europe awaiting Diesel's agreement to terms of the Busch-Sulzer contract. Busch-Sulzer papers, Wisconsin State Historical Library.

32. "A Great American Marine Diesel-Engine Plant," *Motorship* (1919), 24–34, has a good history and description of the factory to that date.

33. "'NELSECO' Heavy Oil Marine Engines," The New London Ship & Engine Co., New London, CT, Bulletin No. 1, May 1911. Busch-Sulzer papers, Wisconsin State Historical Library.

34. "America's First Diesel Yacht," *Motor Boating*, 1912, 24.

35. *The Gas Engine*, May 1913, 285.

36. M.A.N. records list NELSECO as having installed forty-two engines. "M.A.N.-Dieselmotoren."

37. Louis Shane, USN, "Description of Nürnberg Two-Cycle 450-B.H.P. Heavy-Oil Engines," *Journal, American Society of Naval Engineers* (1919): 447–61.

38. The subs included the *H-1* to *H-3*, *K-1* to *K-8*, *L-1* to *L-4*, and *L-9* to *L-11*. *US Submarine Data Book*, 4–6.

39. F. C. Sherman, "Notes on Operation of Submarine Diesel Engines," *Engineering*, January 18, 1918, 61–63. Magdeburger confirmed Sherman's comments with the major concern being the lube oil from the scavenge pump chamber becoming entrained in the scavenge air and being the principal cause of explosions. Magdeburger, "Diesel Engines," 587–88.

40. Magdeburger, "Diesel Engines," 588.

41. Ibid., 590.

42. *Submarine Data Book*, 7–11. Other data is in Ernest Nibbs, "Submarine Engines, Development and Present Status," July 19, 1930, 6. Nibbs was Chief Engineer at Electric Boat Co. in Groton when he gave this lecture to the Submarine School at New London. It is an excellent overview of early sub diesel developments. Copy kindly furnished by Mrs. Wendy S. Gulley, librarian, Nautilus Memorial Submarine Force Library and Museum, New London, CT.

43. Smith, *Britain's Clandestine Submarines*, 1–153, covers the entire *H* class story, including a congressional investigation. *Jane's*, 81, reported the class "were about the most popular type with the British Submarine Service."

44. Magdeburger, "Diesel Engines," 597.

45. Ibid., 603.

46. "Operation of the Navy Tanker 'Maumee,'" *Motorship*, March 1920, 214. Lt. C. W. Nimitz and a Navy draftsman were in Germany during 1913 studying marine diesels. Nimitz was executive officer and chief engineer on the *Maumee* from commissioning in October 1916 until April 1917. The photo of figure 17.17 was originally in the collection of Fleet Admiral C. W. Nimitz.

47. Caption written by Naval History Division on back of Figure 17.17. Photo no. 32691. MAN Archives.

48. As unofficial Sulzer historian before his retirement, W. Bangerter compiled complete order, drawing number, specifications, delivery information, and so on for all early Sulzer engines. It was a mammoth undertaking requiring years of his time.

49. Ibid., and *Jane's*, 217. The *Neride* was torpedoed by an Austrian Navy sub in August 1915.

50. A Meier letter to Harris, March 13, 1911, tells of the Navy's granting this right. The quotation is repeated in a memo from Rotter to Harris, September 4, 1913, giving estimates to build six similar sub engines in the US for the boats *L-5*, *L-6*, and *L-7*, which was done. Rotter estimated total costs, including development, at $226,000.00, or $62.80/bhp. Busch-Sulzer papers, Wisconsin State Historical Library.

51. Simon Lake (1866–1945), the other principal American submarine designer, vied unsuccessfully with Holland for early US Navy contracts. His boats performed well in wartime service.

52. Rotter to Harris, from Zürich, June 19, 1911. Wisconsin State Historical Library.

53. Magdeburger, "Diesel Engines," 593. The 161-foot *G-3* had a surface displacement of 393 tons and a surface speed of 14 kts. The keel was laid March 30, 1911; she was launched December 27, 1913, commissioned March 22, 1915, and decommissioned May 5, 1921. *Submarine Data Book*, 4.

54. Bangerter, "Register."

55. "Marine Diesel Oil-Engine (Sulzer Type)," *Engineering*, January 2, 1914, 16. *Jane's*, 198, lists two subs, the only ones fitting the description (*Daphne & Diane*), as commissioned in 1915 with two 800 bhp "Sulzer" engines. The rating of a four-cylinder test engine (310 × 460 mm) at FCM was 380 bhp at 250 rpm but had produced as high as 450 bhp at that speed.

56. Bangerter, "Register," and *Jane's*, 192, lists seven sea-going gunboats with 450 bhp Sulzer engines, and two with the 900 bhp engines. The Romanian boats were for Danube service.

57. Magdeburger, "Diesel Engines," 594.

58. Ibid., 595.

59. Richard H. Lytle, "History of the Busch Diesel Companies, 1897–1922" (master's thesis, Washington University, 1962), 100. *Submarine Data Book*, 7–9.

60. *Two Centuries of Shipbuilding by the Scotts at Greenock*, 2nd ed., *Engineering* (London: 1920), 82–4. The three boats, *S-2*, *S-3*, and *S-4*, were of Laurenti design and were sold to the Italian Navy in 1915. Fred M. Walker, *Song of the Clyde* (Cambridge: Patrick Stephens, 1984), 136. *Jane's*, 217, confirms this, but says they had a British Navy designation of *S-2*, *S-3*, and *S-4*, the *S-1* being the *Swordfish*.

61. Neither Fiat, *50 Anni di Motori Diesel Fiat, 1909–1959* (Torino, Italy: Fiat, 1959), 288 pages, nor D. M. Shannon, "The Development of the Fiat Marine Oil Engine," *Trans., The Institute of Engineers and Shipbuilders in Scotland* (1928), 620–76, give a model number or cylinder size.

62. Rössler, *The U-Boat*, does not list the *U-42*, and *Jane's*, 319, says the boat (680 tons surf, displacement) was completed for the Italians as the *Balilla*. It was sunk off Cattaro Bay in July 1916. Fiat publications also do not mention that Germany never received the boat.

63. Fiat, *50 Anni*, 210.

64. *Fifty Years of Fiat Diesel Engines, 1909–1959*, English edition of *50 Anni*, 210.

65. The Italian version (*50 Anni*) says "some tens of units," and the English version says "several scores" (page 211 of both).

66. Shannon, "The Development of the Fiat Marine Oil Engine," 647–48.

67. The stern of the 326-foot ship opened up like a drydock gate for a submarine to enter, after which the "tank" was pumped out and the sub rested in a cradle. Salvage cranes were at the aft end to lift subs from the sea bottom.

68. "The 4,600-I.H.P. Marine Diesel Engines of the M.V. 'Ceara,'" *Engineering*, February 25, 1916, 180–81, and plates 15–18, gives a complete description of the engine. The March 3 issue corrected the output to "4,600 bhp" and gave an approximate ihp of 6,500. Fiat papers have the 2,300 bhp rating.

69. Friedrich Sass, *Geschichte des deutschen Verbrennungsmotorenbaues von 1860 bis 1918* (Berlin: Springer, 1962), 635–36.

70. Julius Maag, *Dieselmaschinen* (Berlin: VDI, 1928), 112–13.

71. Information kindly obtained for the author by Josef Wittmann, MAN Archivist. No Körting archives exist.

72. *Krupp Dieselmotoren*, 1925 brochure, 11. Krupp Archives.

73. *Jane's*, 124, confirms *Krupp*, ibid.

74. *Krupp*, 13, and Sass, *Geschichte*, 572. Rössler, *The U-Boat*, gives a 92 m length, beam of 9.1 m, draft of 5.3 m, and a surface displacement of 1,930 tons. Surface speed was 15.5 kts. *Jane's*, 126,

says the *U-139* (renamed the *Schwiger*) surrendered to the French who used it in their Navy as the *Halbronn* until 1935.

75. Paul König, *Voyage of the Deutschland* (New York: Hearst, 1916), 114–15, and 116–17. The *Deutschland* made two round trips to the US. It had a cargo capacity of 500 tons and carried no armament; its defense was to submerge. A sister boat, the *Bremen*, was damaged at the start of its first trip and was converted into a surface ship. Six more (U-151–U-157) projected cargo subs were made into naval boats before launching. The *Deutschland* became the *U-155* and survived the war.

76. If one assumes an average of 1,000 bhp/diesel of the two-stroke and 500 bhp/engine of the four-stroke, Krupp possibly made around fifty-five to fifty-seven of the former and twelve to fourteen of the latter for submarines.

77. O. Föppl et al., *Schnellaufende Dieselmaschinen*, 2nd ed. (Berlin: Springer, 1922), Plate II, and Daimler-Benz Archives data.

78. Ibid., Plate I.

79. Ibid., Plate IV.

80. Maag, *Dieselmaschinen*, 113.

81. Hans Flasche, "Der Dieselmaschinen in Russland von Anbeginn 1899 an bis etwa Jahreschluss 1918," in MAN Archives (1947), 36–7.

82. Ibid., 37.

83. Ibid., 81, gives the names of fourteen Russian subs having the 1,340 bhp engines. *Jane's*, 242, lists only seven, the *Kuguar* class, with this size engine.

84. "The French Laubeuf Submarine-Boats," *Engineering*, July 9, 1915, 30–31.

85. M. Drosne, "Development of Internal-Combustion Engines for Marine Purposes," *Engineering*, July 17, 1914, 99–100. An earlier article, "The Laubeuf Submersible Boat," *Engineering*, August 18, 1911, shows an outline of an eight-cylinder engine, hence the assumption of the ca. 1911 date.

86. Drosne, "Development," 99. The *Néréide* of Simonet design with a surface displacement of 787 tons and a 243-foot length. *Jane's*, 198. The French sub is not to be confused with the Italian *Nereide*, both of which were named after part-fish, part-woman sea nymphs.

87. James Richardson, "The Present Position of the Marine Diesel Engine," *Trans., The Institution of Engineers and Shipbuilders in Scotland* (1920), 17. No in-service date is given.

88. Patent abridgement in Wells & Wallis-Tayler, *The Diesel or Slow-Combustion Oil Engine* (London: Lockwood, 1914), 256.

89. J. Rendell Wilson, "Notable Diesel Engines," *Motor Boating*, August 1912, 16. *The Engineer*, July 14, 1911, 49, ("New French High-Compression Motor"), reported that a 500 bhp engine had a bsfc of 185 lbs/bhp-hr burning a "residual oil (Mazout)." Consumption at lower outputs was not given.

90. Comments by the author of *The Engineer*, 50, illustrated the strong preference by many marine engineers for crosshead piston engines (i.e., to follow steam engine practice) because of a concern that as the propeller shaft thrust bearing wore and allowed the crankshaft to move forward, this would put a high side thrust on the pistons and cause excessive piston and bore wear. Sabathé had countered this concern by covering the entire piston surface below the rings with bearing "white metal."

91. Both Wilson, "Notable Diesel Engines," 17, and *The Engineer* show different views of the 500 bhp Sabathé engine.

92. Reference is made to this engine, in addition to a notation on the back of the photo for Fig. 17-45, in Rudolf Diesel, "The Diesel Oil-Engine," *Engineering*, March 22, 1912, 399.

93. "Verzeichnis [Register] der U-Boote," listing two-stroke sub diesels made by Nürnberg licensees before the war. MAN Archives. Eight Augustin Normand-built diesels went in the *Clorinde, Adriadne, Andromaque*, and the *Comélie* before 1914. Max Rotter verified to J. R. Harris in a June 10, 1911, letter that Nürnberg had a license agreement with "S.A. Normande." Busch-Sulzer Papers, Wisconsin State Historical Library.

94. "Another French Diesel Marine Motor," *The Engineer*, August 25, 1911, 195–97.

95. The author had great difficulty finding sufficient reliable information on French sub diesels other than that available from German and English sources.

96. A typewritten abstract of pre-1914 French submarine diesel developments, no date or source. MAN Archives. It may be assumed to have come from Professor Max Serruys who added a signed note on the back of this single-page document. Serruys, who wrote *Contribution Française à l'évolution Technique des Moteurs Alternaties à Combustion Interne*, Paris, 1948, would have known of the Jaubert/Wisler work.

97. MAN records ("Verzeichnis der U-Boote") show that six Nürnberg diesels were installed with three for the *Havmandan* class and three for the *Ægir* class. *Jane's*, 266, confirms this. The other three would be Fiat engines as the *Havmandan* class hulls were all made in Fiume, Italy.

98. Information on the sub engine was kindly supplied by Tony Woods, B&W. The "100" designation was indicated horsepower per cylinder to conform to the nomenclature B&W adopted for marine diesels.

99. *Burmeister & Wain's Marine Diesel Engines*, c. 1918, 30.

100. Ibid., says seven engines were delivered, and *Jane's*, 266, reports that six subs of the *Bellona* class were in service or some were being completed at the end of the war. These were larger subs (155 ft vs 133 ft for the *Ægir* class). *Jane's* does not say if they had one or two engines.

101. "The Polar-Diesel Engine," *The Engineer*, October 24, 1913, 437–40. The article, pages 434–37, covers most of the engines, including small auxiliaries, built by AB Diesels Motorer.

102. "Klingerit" was made by the Richard Klinger Co. in London. H. R. Kempe, *The Engineer's Year-Book* (London: Lockwood, 1901), lxxvi.

103. Hesselman had seriously considered the idea earlier as evidenced in his British Patent 27,554 of 1909 on which the production version was loosely based. Wells & Wallis-Tayler, *op. cit.*, p. 260, abstract the patent.

104. J. K. E. Hesselman, *Teknik och Tanke* (Stockholm: Sohlmans, 1948), 252, mentions the connection between these exciting events and the sub engine design.

105. "The Polar-Diesel Engine," 440.

106. AB Atlas Diesel, *Four-Cycle Marine Polar Diesel-Engines,* Stockholm, 1919, Catalog No. Ael5, 10 and 16. Output for a submarine application was given as 350 bhp at 400 rpm. Guaranteed bsfc was 200 g/bhp-hr. Net engine weight including flywheel, water and oil pumps, air and water coolers was 9,150 kg or 26.1 kg/bhp.

107. Magdeburger, "Diesel Engines," 605, said the engines "were of Swedish Polar type and manufacture" and came to the US before 1917. He also stated that they "made a good submarine out of her."

108. The primary impetus for Winston Churchill's instigation to convene the already much-cited *Royal Commission on Fuels and Engines* (British Admiralty, 1913) resulted from his inspection of the *Selandia* engines and his awareness of the Krupp, M.A.N., and Sulzer developments. Much testimony was heard about their progress and Britain's expected fuel sourcing problems if it were to pursue such a power program. A tangible outcome of the hearings was the British government's (Churchill's) unheard of commitment in 1914 to become the major shareholder of Anglo-Persian Oil Co. An account of this geopolitically charged oil episode is in Marian Kent, *Oil and Empire, British Policy and Mesopotamian Oil 1900–1920* (New York: Harper & Row, 1976), 273 pages. Anglo-Persian became British Petroleum.

109. The entire development history, with comments on people as well as technical details of the Nürnberg engine was documented by Wilhelm Laudahn, in charge of the German Navy office overseeing the program. His 1920, 338-page, typewritten account, "Die Entwicklung der Nürnberger Großölmaschine auf dem Versuchsstand der Maschinenfabrik Augsburg – Nürnberg A.G., is in the MAN Archives.

110. Ibid., 10.

111. The 24,700-ton (displacement) ship was commissioned in August 1913 with all steam turbine power. *Jane's*, 105, lists a total designed horsepower of only 25,000 (perhaps indicated), which meant that the diesel would have had a greater output than either of the individual turbines. The *Prinzregent Luitpold* was one of ten battleships scuttled by their crews at Scappa Flow in 1919.

112. At the same time that work began on the engine, a license was granted to the SA des Forges & Chantiers de la Méditerranée at Le Havre, a major contractor for the French Navy. David T. Brown of Sulzer speculates in an August 8, 1993, letter to the author that the coincidence was perhaps too great. Sulzer may have hoped for a development contract and later to sell engines to France. No documentation has yet been found in the Sulzer Archives to totally support his theory.

113. The author relied on Schnauffer, "Werk Nürnberg," 41–55, who based his studies on Laudahn's "Nürnberger Großölmaschine." Sass, *Geschichte*, 550–59, distilled his account from Schnauffer.

114. A detailed explanation of personnel changes resulting from Bruns' departure and the accident were in a March 19, 1912, letter (in English) from Dr. Otto Geltung, Management Board, to G. C. Davison, VP of Electric Boat Co. It was to allay fears about the status of the engines

EBC was licensed to build. The candid report offered insight into Nürnberg's marine engine programs. Thanks to Josef Wittmann for unearthing this "treasure" in his MAN Archives.

115. M. Seiliger, *Die Hochleistungs-Dieselmotoren* (Berlin: Springer, 1926), 154.

116. A. Nägel, "The Diesel Engine of Today," *Diesel-Maschinen* (VDI, 1923), 16.

117. The Krupp technical development is mainly from W. Laudahn, "Die Entwicklung der Germania-Großölmaschine," *Zeit, des V.D.I.*, vol. 68, November 8, 1171–1176, and November 15, 1924, 1200–1204. Laudahn held the same Navy Department oversight responsibilities at the Germaniawerft as at Nürnberg.

118. Ibid., 1171–72.

119. Ibid., 1176.

120. Ibid., 1201.

121. The best description of the S100, is by Lt. C. J. Hawkes in his report to the *Royal Commission*, 170–72, after visiting Sulzer during October 1912 and witnessing its early running. He said he was the first "foreigner" to see the engine. Another Englishman a year later gave more on the engine: "The Sulzer 2000 B.H.P Single Cylinder Motor," *The Gas Engine*, August 1913, 445–48.

122. The author can attest to the decibel level, remembering during the development of what became the Jacobs engine brake: a multi-cylinder engine running at governed speed drove one cylinder whose fuel was shut off. At the end of the compression stroke, its two exhaust valves would open to release the stored energy in the cylinder air.

123. Hawkes, 171.

124. This comment from a report found in the Sulzer Archives by David Brown lends credence to his conjecture of hoped-for battleship engine sales by Sulzer to the French Navy. Brown to the author, August 27, 1993.

125. "Leistungsversuche [power tests] am Einzylinder-Versuchsmotor S100 (Prof. Stodola)," Table 4, 1914/1915. Sulzer Archives.

126. Brown to the author.

127. 1914 Statement from "Outline of a Memorandum re Submarines," July 11–13, 1914; Richmond mss. Arthur J. Marder, *From the Dreadnought to Scapa Flow*, vol. 1, *The Road to War, 1904–1914* (London: Oxford University Press, 1961), 364.

CHAPTER 18

Locomotives and Lorries: Diesels on Wheels

"We consider the new motor especially applicable on railways to replace ordinary steam locomotives, not only on account of its great economy of fuel, but because there is no boiler. In fact, the day may possibly come when it may completely change our present system of steam locomotion on existing lines of rails."

—Rudolf Diesel, 1892[1]

Between 1905 and 1912, Diesel designed and engaged in the construction and testing of engines applicable to land transport.

A locomotive diesel, the first venture, incorporated new technology in rail power. Its concept was flawed, but the powerful engine nevertheless saw service. Sulzer Brothers, who was a partner in this endeavor, used the experience to developed a practical diesel-electric drive. AB Diesels Motorer, however, had preceded the Swiss in doing this. Before 1918 both Kolomna in Russia and General Electric in the United States had built locomotive diesels to turn generators but with very limited success.

The small, automotive-type diesels built to Diesel's ideas in Switzerland were destined for failure from the beginning because of deficiencies inherent in air injection.

Gesellschaft für Thermo-Lokomotiven Diesel-Klose-Sulzer GmbH.

As early as 1905, Diesel was promoting a high-horsepower locomotive engine to be developed in cooperation with his friends Jakob Sulzer-Imhoof in Winterthur and Adolf Klose in Berlin. He had undoubtedly become acquainted with Klose when the former worked with Nürnberg on its pre-1900 street railway engines (chapter 9). How long the two had been collaborating before approaching Sulzer is unclear.

Klose's credentials as a locomotive designer were impressive. They stemmed from seventeen years, until 1887, with the Swiss State Railway; ten years with the Württemberg system; and finally as a designer at Borsig, the locomotive builder in Berlin-Tegel, where he collaborated with the Prussian State Railways.[2]

The Diesel and Klose venture with Sulzer-Imhoof began as a draft contract of November 15, 1905, creating the Gesellschaft für Thermo-Lokomotiven" Diesel-Klose-Sulzer GmbH in Munich. Sulzer-Imhoof then presented the proposed plan before his own board after this meeting.[3] Two major stipulations were that decisions regarding engines would be jointly made by Diesel and Sulzer; Klose held final authority over the locomotive design.[4] Diesel was to provide design support out of his Munich home-office, and Sulzer would build and test the engine in Winterthur. A final contract between all parties was signed on March 27, 1906. The November draft contract date had relevance as by then Diesel believed himself free of all obligations to the Allgemeine Gesellschaft für Dieselmotoren.

Sulzer did not begin earnest work on the locomotive until 1907, and five years were to pass before it traveled the rails (fig. 18.1). Testing took a year and a half of this time as almost everything on it was totally different from steam practice.[5] Borsig fabricated the locomotive chassis and shipped it to Sulzer for final assembly and testing with the power plants.

The transversely mounted four-cylinder, 90-degree vee main engine was a reversible two stroke whose bore and stroke were 380 × 550 mm (fig. 18.2). Through-bolts tied the cylinder heads to the main bearing supports in typical Sulzer fashion. Disks on each end of the crankshaft served as flywheels, and

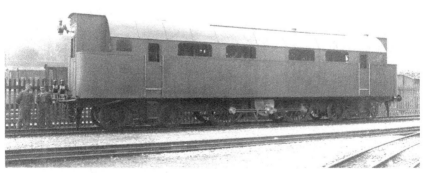

Fig. 18.1. The Sulzer-Diesel-Klose "Thermo-Lokomotive"
tested in 1913 and 1914. *Sulzer Archives*

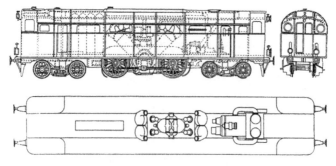

GENERAL ARRANGEMENT OF SULZER-DIESEL LOCOMOTIVE.

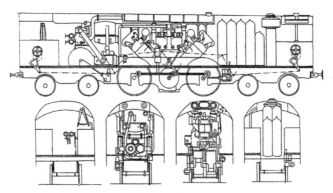

Fig. 18.2. Arrangement of 'Thermo-Lokomotive" designed from 1907 to 1909 and built by Sulzer. *Internal Combustion Engineering, September 1913*

weights attached to the disks counteracted most unbalanced engine forces. Side rods linked the disks directly to the wheels on two driving axles. Weight considerations were a major factor in opting for this type of drive system. Electric generator and motor technology at that time would have made that form of power transmission to the wheels significantly heavier.

Compressed air started both the engine and train. At 10 to 12 km per hour, 30 engine rpm, air to the engine was shut off and fuel was admitted. In practice, the speed at which the transition from air motor to internal combustion engine was higher. The engine obviously stopped each time the train stopped.

A battery of vertical tanks gave extra starting and braking air, the latter supplying a Westinghouse air brake system. Water tanks were in three comers of the locomotive and fuel was stored in the fourth. A form of radiator cooled the water.

A 250 bhp, auxiliary diesel drove its scavenging pump and supplied the air to start and brake the train. The vertical engine was also placed crosswise in the locomotive. Its two horizontal compressor and pump cylinders were integral with the engine and each connecting rod big end turned on one of the engine's crankshaft throws (fig. 18.3). The two-cylinder, two-stroke engine had been adapted from a Sulzer 4SNo.6 with a 305 mm bore and shortened stroke of 380 mm.

For the first time a diesel was faced with operating over a complete envelope of speed and load conditions. This was in contrast to stationary engines that ran at a constant speed and load and to marine engines whose load curve increased as the cube of propeller speed. As in any mobile application where the engine was mechanically tied to the wheels, an external load could vary from zero to a maximum independent of speed. While this required closer control over injection air delivery and injection timing, one benefit accrued that designers used to their advantage.

It was thought highly unlikely, based on the contemplated passenger locomotive application, that the engine would ever run very long under combined maximum speed and load conditions. Thus, peak mean effective pressures in the cylinders might be safely raised much higher as they would not be subject to such pressures for sustained periods. The water-cooled pistons were expected to give no trouble.

Knowing this, Sulzer almost doubled the allowable indicated mep to 12 at or 177 psi. (A "technical" atmosphere equivalent of 14.75 psi.) Engine output reached 1,600 ihp with this imep at a rated speed of 304 rpm. Minimum imep in operation was only 2.5 atmospheres. Scavenge pressure at rated speed and power was greatly increased to 1.4 at—three to four times normal inlet pressure.

A 1909 contract between Diesel and the Sulzers calling for cooperative development work on high air pressures was tied to the locomotive engine.[6]

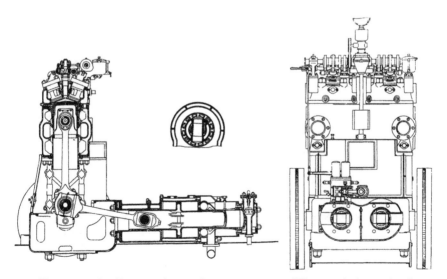

Fig. 18.3. Auxiliary engine and compressor on "Thermo-Lokomotive."
Internal Combustion Engineering, September 1913

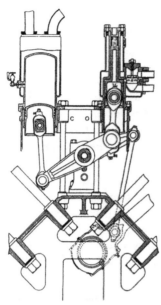

Fig. 18.4. Main engine-mounted backup compressor
(scavenge pump not shown). Sulzer-Diesei-Klose 1909 design.
Internal Combustion Engineering, September 1913

This supercharging of the "thermo-lokomotive" would have involved Alfred Büchi, who became head of Sulzer's diesel engine research that same year.

It is appropriate here to mention Sulzer's early leadership role in research on supercharged engines. Between 1911 and 1915 Büchi would achieve imeps as high as 20 kg/cm^2 (284 psi) from a "D" type four-stroke test engine running on boost pressures up to 3.5 atm.[7] An electrically driven compressor was used for this experiment. He then modified a two-stage steam turbine to run on exhaust gases. The research ultimately led to a direct use of an engine's own exhaust to drive an integral turbine/compressor, an idea for which he would become renowned.

The locomotive's two double-acting scavenge pumps and a three-stage compressor in the vee of the engine were reciprocated by a rocking lever linked to the big end of the connecting rods of one cylinder bank (fig. 18.4). Each fore- and aft-mounted pump furnished scavenge air to the two cylinders in the bank closest to it. The three-stage compressor had the capacity to meet injection air needs only at light loads. It served more as a backup in case the compressor driven by the auxiliary engine failed.

The cylinder starting air valve was fully opened before a main valve admitting starting air at 750 psi itself opened into the lines serving the cylinders. Throttling losses due to air passing through a partially opened valve were thus reduced, which in turn lessened the charge of air expended for each train start.

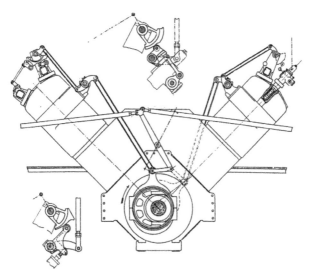

Fig. 18.5. Reversing system on "Thermo-Lokomotive."
Internal Combustion Engineering, September 1913

An eccentric on each end of the crankshaft opened the cylinder valves in the two cylinders adjacent to it. Each cylinder had a starting air valve, two scavenge valves, and a fuel valve. Controls to start, throttle, and reverse the engine were located in a cab at each end of the locomotive. A single reversing lever, shifted by linkage from either cab, rotated the eccentric centers on the shaft to change fuel and scavenge air valve timing (fig. 18.5). A separate mechanism influenced by both the reversing lever and a rotating shaft driven off a side rod actuated the starting air supply valve in a timed manner (fig. 18.6).

The fuel system was similar to comparably sized Sulzer engines with individual pumps for each cylinder. A governor limited maximum speed to 350 rpm.

Not until March of 1913 did the locomotive leave the Winterthur shops. Trial runs included pulling a 120-ton train between there and Lake Constance 83 km away. In the same summer it made the 1,100 km trip to Berlin where it entered a limited testing service with the Prussian State Railways. The highest speed reached was 100 km per hour, which compared with the average steam locomotive on the same duty. Its as-built weight of about 95 tons was not excessive.

The locomotive's performance in actual service was discouraging when it began pulling more realistic 200-to-238-ton trains. On only its third trip the crankshaft broke due to excessive firing pressure on starting—and jerky starts. The shaft was replaced in December 1914 when after six trips a cylinder cracked. Further running ended, partly blamed on the war, and in 1920 the locomotive was scrapped at the Borsig plant. The engine itself was returned to Winterthur in pieces.[8] It was a noble but expensive experiment

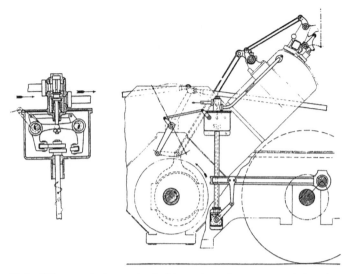

Fig. 18.6. "Thermo-Lokomotive" side rod motion actuated the air starting supply valve. *Internal Combustion Engineering, September 1913*

proving that an I-C engine in a locomotive was not destined to duplicate steam's directly coupled drive system.

Sulzer had already ended its formal ties with Diesel and Klose before the locomotive ever left the shops. By contracts dated October 11, 1912, Sulzer paid Diesel and Klose 10,000 marks each plus some expenses to close out the relationship.[9] This act officially buried the Gesellschaft für Thermo-Lokomotiven, and Diesel saw another of his chances for an encore evaporate.

Diesels Motorer—ASEA Railcars

Hesselman designed a lightweight, high-speed engine around 1910 that became AB Diesels Motorer's versatile Type Z model. The four-stroke, 12.5 bhp per cylinder, intended initially as a shipboard auxiliary to run generators and compressors, came in one, two, three, four, and six cylinders. A 50 bhp, four-cylinder Z engine turning at 500 rpm weighed 1,750 kg compared to that of a standard ABDE model's 275 rpm and 6,900 kg.[10]

Who promoted the marriage of this engine to a generator for railcar is not known, but in 1910–1911 someone induced ASEA (*Allmanna Svenska Elektriska Aktiebolaget*, or United Swedish Electric Co.) to design the electrical system for a small rail car.[11] A 75 bhp at 550 rpm, six-cylinder ABDE diesel (fig. 18.7) drove a direct-coupled, shunt-wound, DC generator powering two series wound motors. Each of these, through spur gearing, drove an axle on a pivoting front truck. To start the engine, batteries turned the generator, which, by using a separate series winding, temporarily served as a motor. A single, "dead" trailing axle supported the car's aft end.[12] A driver's cab was only on the front, and a passenger compartment filled the rear half.

Fig. 18.7. AB Diesels Motorer six-cylinder, four-stroke, 75 bhp diesel in a Swedish railcar, 1912. *Photo courtesy of Georg Aue.*

The first railcar of this design began tests in 1912 and entered service on the Mellersta Södermanlands Railway in 1913 (fig. 18.8). It was still running six years later. Three more cars of the same general type and with the same 75 bhp engine went to other Swedish rail companies between May 1914 and June 1915.[13]

Performance curves show that the 26.4 metric ton car running alone had a maximum speed of 57 km per hour; when pulling a maximum total train weight of 56 metric tons, the speed dropped to 47 km per hour. On a 5 percent grade the numbers went down to 47 and 27 km per hour. A guaranteed fuel bsfc of 200 g/bhp-hr translated into a consumption on a g/ton-km basis varied from about 7 to 10 (2.5 to 3.5 lbs/100 ton-miles) for the above train weights.[14]

Diesels Motorer followed up the six-cylinder in line Z with a four-stroke V-6 of 120 bhp at 500 rpm. These were installed in larger railcars beginning in 1917. Train weights had increased from 32 (car alone) to 112 metric tons, but top speeds went up only slightly.

Fig. 18.8. First practical diesel-electric railcar, Sweden, 1912.
Photo courtesy of Georg Aue

In 1919 ASEA and ABDE formed Diesel-Elektriska Vagn AB to build railcars, but this "Diesel-Electric Car Co." was never very profitable. Concerns over DEVA's future were voiced as early as 1922, and only about fifty units had been made up to 1939 when the venture ceased.[15] Electrification of the Swedish State Railways after an amalgamation of the many small railroads was a primary reason for the weak railcar market.

Russian Experiments

W. I. Grinewetzki (1871–1919) applied for a patent in 1906 on a reversible, two-stroke, compound diesel engine with three double-acting cylinders for a locomotive application.[16] He designed the engine from 1908 to 1909 (fig. 18.9). Cylinder *A* provided supercharged scavenge air for a center, high-pressure power cylinder *B* where combustion occurred. When the exhaust valves opened, the gases passed into cylinder *C* for further expansion. Combustion took place in *B* at a reported compression pressure of about 100 atm. When the engine ran in the opposite direction, the functions of cylinders *A* and *B* reversed.

An engine of this type was built at the Putilov (later Kirov) Works in St. Petersburg during 1912. To aid in its tests, the company requested assistance from Sulzer Bros. G. W. Aue, a well-qualified Sulzer diesel engineer, accepted the transfer.[17] Because of the high pressures and the inherent flaws of a compound engine, it was abandoned at the end of that year. Aue remained in Russia where he headed Putilov's new diesel program until his untimely death. All records of the engine have been lost.

Kolomna had also hoped to capitalize on the scarcity of fuel and water for steam engines in southern Russia by building its own diesel-powered locomotive. It reportedly assembled a diesel-electric unit using two of its 500 bhp engines where each was direct coupled to a shaft end of a single generator placed between. There were four motors with each turning a driving axle in the two, double axle trucks.[18] An engine and generator speed of only 240 rpm

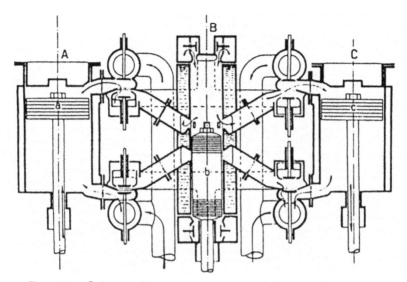

Fig. 18.9. Grinewetzki-designed compound diesel for locomotives.
Built and tested by Putilov in 1912. *Industriearchäologie, 1990*

made the locomotive both heavy and difficult to operate. No performance or operating information exists.

Sulzer Diesel-Electrics

Well before the Thermo-Lokomotive was set on rails Sulzer had realized its direct-drive project was not only overly ambitious but was going down the wrong powertrain track. Kolomna's experiment and the Swedish progress, albeit with a much smaller railcar, had proved that the diesel-electric drive concept was valid. Sulzer's severing of relations with Diesel and Klose in 1912 indicated a resolve to pursue this other avenue.

That same year Sulzer proposed a railcar design to both the Prussian and Saxony state railways comprising a two-stroke diesel directly coupled to a generator. Both were on a pivoting, three-axle lead truck. The pivoting rear truck had two axles driven by electric motors. These railcars, built in cooperation with Waggonfabrik Rastatt, weighed about 85 tons. The electrical control system came from Brown Boveri Co. in nearby Baden.[19]

Designated the 6LV26, the 90-degree, V-6 diesel produced 200 bhp at 450 rpm from a bore and stroke of 260 × 300 mm (fig. 18.10). The cylinder design was borrowed from its contemporary 4SNo.5 marine engine having the same bore but slightly longer 350 mm stroke. Based on the long service life of several of the units, one can assume the engines were relatively trouble free. A major area of concern involved the air injection fuel system.

Five railcars were built to this design and entered service during 1914, three to the Prussian State Railway and two to the counterpart in Saxony (fig. 18.11). The latter two were bought back by Sulzer in 1920 and eventually

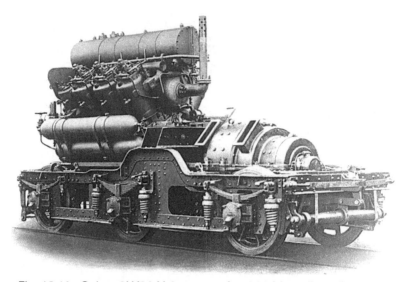

Fig. 18.10. Sulzer 6LV26 V-6, two-stroke, 200 bhp railcar diesel and a directly coupled generator mounted on the front truck, 1914.
Sulzer Archives

Fig. 18.11. Sulzer railcar delivered to Saxony State Railway, 1914.
Sulzer Archives

converted to an early Sulzer solid injection fuel system. In 1924 they began operating on a Swiss local railroad where both ran until 1939.[20]

General Electric Diesels

Rudolf Diesel indirectly launched a diesel program at General Electric Co.'s Lynn, Massachusetts, factory. Two engine designs that evolved there for diesel-electric locomotive and stationary generator applications both proved short-lived and, in the case of the former, quite unsuccessful.

General Electric had begun making its own gasoline-electric railcars in 1910 but were supplying engines and electrical equipment to others for several years before. That year it introduced the GM-16 gasoline engine, a 175 hp at

550 rpm V-8 with an 8 × 10-inch bore and stroke. The completed car, whose bodies were made by the Wason Car Co. in Springfield, Massachusetts, became known as the GE-Wason railcars.[21] The engine had been designed by Henri G. Chatain, a Swiss-born engineer who would play an important role in the design of the diesels to follow.

Dr. Hermann Lemp (1862–1954) was one of the electrical engineering geniuses who came to General Electric (GE) in its early years. He had worked for Thomas A. Edison after arriving from Switzerland in 1881 and joined GE in 1892. Lemp was granted over three hundred US patents during his lifetime; he is also credited with a direct-current control system, which served as the basic element in the circuitry almost all diesel-electric locomotives would use.[22]

In 1910 GE had called on Lemp to improve the method of controlling the load and speed balance between the railcar engine and its generator. Up to that time two independent controls, one being adapted from that for a streetcar, left too much to the discretion of the operator. Lemp designed a system eliminating one control and automatically protecting the engine against low-speed overloading as well as engine overspeeding.[23]

Lemp's work on the control exposed him to engines and in turn with the potential of the diesel to improve operating economy. During Rudolf Diesel's American visit in the spring of 1912, he met with engineers, undoubtedly including Lemp, to discuss his engine's potential for locomotives.[24] Later that year Lemp, and reportedly Chatain, visited Sulzer and saw the "Thermo-Lokomotive" undergoing its initial trials. Lemp left convinced that a diesel-electric drive was the only way to go.

He had already filed a US patent application in 1911 for a two-stroke diesel engine based on Junkers' opposed-piston design.[25] It did not specify an application. To reduce overall height and eliminate the upper yoke arrangement of the Junkers, he "bent" the cylinder so that the pistons ran in cylinders parallel to each other (fig. 18.12). The cylinders were joined at the top by an interconnecting passage (*22*) into which fuel was injected. Both pistons (*6*) were tied to a single crankshaft throw by a long piston pin (*23*) in the upper end of a single connecting rod between the two cylinders. One piston uncovered exhaust ports (*18*) as it neared the bottom of its stroke, and the other uncovered intake ports (*9*) shortly thereafter.

Larger-diameter stepped pistons (*7*) on the underside of the power pistons acted as scavenge pumps discharging into a cavity leading to the intake ports in the one cylinder. Because the piston pin was between the power and scavenge pistons, the lower ends of the pump cylinders could be closed off so that the downward stroke of the scavenge pistons became the pumping stroke.

The design adopted for the locomotive engine was a merging of the concept in the above patent with Chatain's V-8 gasoline engine. Four of the U-shaped cylinders, two in each bank, held eight pistons whose paired pistons shared a common pin and rod as in the patent.[26] Different was the

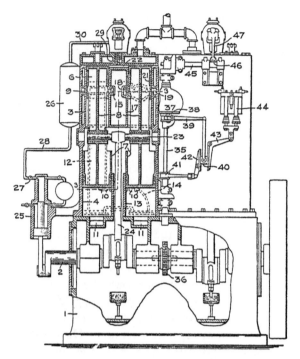

Fig. 18.12. Hermann Lemp's first diesel engine design as seen in his patent filed in 1911. It resembles a Junkers-type diesel but with parallel pistons and a single connecting rod. *Drawing from 1914 US patent*

elimination of the underside stepped-piston scavenge air pumping. Instead, a crank throw on the generator end of the engine drove a double-acting, vertical scavenge pump delivering air at about 7 psi. On the opposite end was a vertical, crank-driven, two-stage compressor for injection air. A rod from each bank shared a common crank throw. The 8 × 10-inch bore and stroke engine produced 225 bhp at 550 rpm.

The engine was started by converting the scavenge pump into an air motor, through the necessary valving, and using stored air at about 150 psi (the Hesselman system).

GE built four diesel-electric locomotives with its GM-50 engine from 1917 to 1918. The first had a Wason body and was "field tested" by hauling freight around the factory grounds at its Erie Works[27] (fig. 18.13). Two traction motors were in the leading truck and none in the second.

The second locomotive with a steeple cab trolley design was delivered to the Jay Street Connecting Railroad in September 1918 (fig. 18.14). All four axles had traction motors. According to all reports the engine itself was a "complete failure,"[28] and was replaced a few months later by a GE V-8 gasoline engine. A similar locomotive delivered to the city of Baltimore in October 1918 saw little service because its intended work had been

Fig. 18.13. First GE diesel-electric locomotive with a 225 hp V-8 engine hauling freight at its Erie Works in March 1917. *Roy F. Corley Collection*

completed by the time it was received. GE bought it back in 1926 when it was repowered and resold.[29]

The US Army ordered a third locomotive from Erie in 1917 that was clad with .75-inch armor plate; steam locomotives were proving too visible a target in France because of their smoke in the daytime and glowing firebox at night. The GE diesel-electric was not delivered until just before the end of the war and never left the country. It was tested at the US Army Proving Grounds at Sandy Hook, New Jersey, toward the end of 1918.[30]

A spinoff of GE's diesel development was an integral engine and generator set for the US Army. Seven were delivered to Fort Drum at the entrance of Manila Bay in the Philippine Islands beginning in 1914.[31] The engine for these 150 kW sets closely resembled a Junkers opposed-piston, two-cylinder diesel whose upper piston was tied to the crankshaft through a yoke and side rods straddling the cylinder. A two-stage injection air compressor was above an axial with a double-acting scavenge pump driven off the crank on the end opposite the generator. The fuel pump and governor were above the flywheel. General layout of the engine followed Lemp patent 1,166,916 of January 4, 1916.[32]

The 200 hp at 500 rpm engine had a bore and stroke of 9.5 × 10 inches.[33] Its weight, less flywheel and generator, was 21,220 lbs; total weight of the set came to 32,100 lbs. During its acceptance test, a brake-specific fuel consumption of 0.70 lb/kW/hr at 150 kW was recorded. Disposition of the engines is not known.

In 1919 General Electric abandoned its entire diesel and diesel-electric locomotive program, a hiatus that would last for almost two decades.

Diesel's Automotive Engine

Paralleling Rudolf Diesel's locomotive engine endeavors were those for a road vehicle power plant. The much-smaller engines would be made in Switzerland with Sulzer family cooperation. Here again Diesel's ambitions

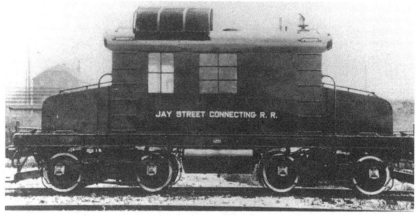

Fig. 18.14. A diesel-electric locomotive for the Jay Street Connecting
Railroad was the first one delivered by General Electric in October 1918.
Roy F. Corley Collection

were to be thwarted, but this time in part due to circumstances beyond his
control.

A prelude to the development work itself was another attempt—on
paper—to rid his engine of the cursed air injection. This is seen in a patent
Rudolf filed in the name of Oscar Lintz, a Berlin engineer, on January 15,
1905. It describes two ways of injecting only liquid fuel.[34] One is of historical interest because it is the first example of a unit injector type whereby
a mechanical force drives a pump plunger in the injector itself to inject a
charge of fuel under high pressure (fig. 18.15). No mention was made of the
metering method, but it assumed that the charge was measured elsewhere.
The second way created the hydraulic pressure with an air accumulator. The
fuel charge was also preheated from the heat generated by compressing
the accumulator air. He also called for "fine" air bubbles to be entrained in the
injected fuel. Whether this was an admission that he could not eliminate
the air or that he believed it was necessary poses an interesting question.

Of greater significance was Diesel's understanding of the pressures
required to inject only liquid fuel in comparison with the much lower pressures possible with air injection. He said that:

> It is well known with Diesel motors the compression pressure is very
> high and runs to 30 to 45 atm. In order to obtain the thus mentioned
> atomizing action of the liquid streams against the compressed air mass,
> the pressure of the injected liquid must be many times above this, and
> under certain conditions can reach several hundred atmospheres.[35]

Diesel immediately sent a copy of the application he had filed to the
Allgemeine, who then asked its licensees if they would have an interest in

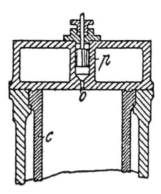

Fig. 18.15. Patent drawing of Diesel's unit injector concept for injecting only liquid fuel at 2,000 to 4,000 psi, 1905. *German Patent 178,896*

adopting the idea; none said yes. Without company support Diesel himself let the idea drop, and all his future engines had air injection.

During the first months of 1907, Diesel and his friends at Sulzer Brothers finalized an arrangement whereby another company would be granted a Swiss license to develop and build automotive size engines. The talks leading up to this undoubtedly grew out of the Sulzer locomotive agreement already in effect. The Sulzers had no interest in building such small and totally different engines, but instead they would license someone else to do it.

Safir (Schweizerische Automobil-Fabrik in Rheineck), a new automobile maker, established only a year earlier, was selected. It seemed a logical choice because the cars' engines were also built there under a license from the Adolph Saurer Co.[36] Albert Sulzer-Seifert (1870–1936) and Adolph Saurer were among those on the management board of this Zürich area company. Adolph's talented engineer son, Hippolyt Saurer (1878–1936) acted as a consultant on the car engines.

Diesel sent Heinrich Deschamps, an engineer in his Munich office, to Rheineck. He and Gustav Segin, a Swiss engineer, were the team who designed the new engines.[37] Only plans would be drawn at Safir because financial difficulties closed the company in the summer of 1908. Saurer ended its license agreement at this time and was no longer directly involved with any of Diesel's work. Safir stopped car production, but its engine-making facilities were sold to AG St. Georgen Maschinenfabrik Zürich, a newly organized subsidiary of an older company.[38] All of the smaller engines based on Diesel's ideas would come from the St. Georgen factory between 1908 and 1910 when this company, too, became unprofitable. Albert Sulzer-Seifert, a major shareholder also in the Zürich factory presided over its liquidation in 1913.

Deschamps and Segin first designed a single-cylinder test engine that would initiate production of small, one- and two-cylinder stationary diesel engines before the truck engine was built. It must be emphasized that Diesel participated heavily in the work done at St. Georgen and frequently came there to check on progress as well as to witness the engines being tested.[39]

Rudolf formed Diesel & Co. GmbH., München, to sell the Original Diesel & Co.-Kleinmotor (small engine) made in St. Georgen. The engines won a prize at the 1910 Brussels World Exhibition.[40] The number sold is not known, but most went to Russia.[41] Local markets were not promising because of the cumbersome injection system and higher costs in comparison with gasoline and oil engines available by that time.

These high-speed, compact engines with a 116 × 150 mm bore and stroke had a rating of 5 bhp per cylinder at 600 rpm (fig. 18.16). All used air injection with the air coming from a two-stage, crank-driven compressor whose cooled housing was integrally cast with the power cylinder (fig. 18.17). Every facet of the engine shows an effort to reduce weight and bulk. (The author could find no weights of the one- and two-cylinder engines.)

Fig. 18.16. Diesel's 5 bhp, 600 rpm engine made in Switzerland and sold through his Munich company, c. 1910.
Conservatoire des arts et Métiers, Author photo

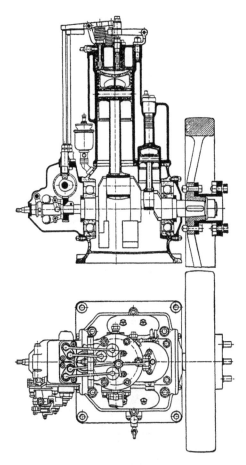

Fig. 18.17. Top and section views of 5 bhp Original Diesel & Co. engine showing anti-friction main bearings and the governor driven off the end of the crankshaft. Schlaepfer, *Untersuchungen an einem Diesel-Kleinmotor, 1914*

Independent tests on a one-cylinder engine in 1913 would indicate that the 5 bhp rating was optimistic, yet it did have a 200 rpm overspeed capability albeit with a black exhaust:[42]

Speed, rpm	400	600	800
Output, bhp (Ps)	2.82	4.08	4.42
Bsfc, g/bhp-hr	267	310	410

The test fuel was a Galician oil with an LHV of 10,022 cal/kg.

Several of these small and interesting engines found their way into technical museums. Both the Science Museum in London and the Conservatoire des Arts et Métiers in Paris have one.[43]

Development work on a four-cylinder, 110 × 151 mm bore and stroke automotive diesel was carried out at St. Georgen during 1909 and 1910. Its intended rating of 40 bhp at 800 rpm was never reached because of excessive exhaust smoke at the higher speeds and loads.[44] The clean design of the engine shows, however, that the test model was intended as a production prototype (fig. 18.18). Again the air injection fuel system was the main culprit.

The general design below the cylinder head was borrowed from a 42 hp Saurer gasoline engine used in Safir cars. The cylinders and crankcase housing were modified Saurer parts. Hand levers to control the starting air valve in the head, as well as to change valve and injection timing indicate that the engine might also have been intended for marine applications (fig. 18.19). Injection air pressure was varied by the same lever that throttled the fuel. The maximum recorded output before exhaust smoke became excessive and was 25 bhp at about 700 rpm. At this speed the engine reportedly ran well enough, but the power was too low to make the engine saleable. Unfortunately, the St. Georgen factory was closed before more work could be done to improve the injection system and possibly raise the smoke threshold.

One of the four-cylinder test models is exhibited in the Deutsches Museum. There is reason to believe a few others were made.[45]

None of Diesel's cherished hopes for mobile applications of his engine were achieved. Another fourteen years would pass before a diesel engine

Fig. 18.18. Automotive diesel on test at St. Georgen factory, 1909.
Intended output was 40 bhp at 800 rpm.
Photo by author in Diesel collection, Deutsches Museum

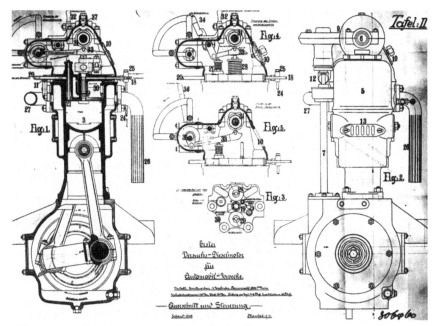

Fig. 18.19. Drawing of 1909 Diesel's St. Georgen-built, automotive diesel. Note the head inset into cylinder. *MAN Archives*

could be demonstrated in a truck on a highway. Not until a fuel system lost its dependence on injection air could such a practical automotive diesel be offered. The same held true for locomotive diesels. Responsive fuel controls to manage rapidly and constantly varying loads and speeds were not yet commercially attainable.

Notes

1. Rudolf Diesel, *Theory and Construction of a Rational Heat Motor* (London: Spon Press, 1894), 81, of Bryan Donkin translation. Donkin took liberties with what Diesel wrote, but the broad idea was the same.
2. Klose obituary in *Organ für Forschritte des Eisenbahnwesens*, Octobeer 15, 1923. MAN Archives. Klose also spent much of his time on automobile development.
3. Preliminary contract between the proposed company and Sulzer Bros. Copy in Sulzer Archives.
4. Ibid., Clause 3.
5. For a complete technical description, see "The Sulzer One Thousand Horse-Power Diesel Engined Locomotive," *Internal Combustion Engineering*, September 17, 1913. Sulzer reprint, December 1913. For a German version, see F. Sterenberg, "Die Erste Thermo-Lokomotive," *Zeitschrift des VDI*, 1913, 1325–30.
6. March 18, 1909, contract between Gebrüder Sulzer and Diesel. It referred to and was an extension of the April 27, 1906, "locomotive" contract. Sulzer also signed a license agreement with the Swiss Locomotive and Machine Works (SLM) on July 1, 1909, for engines up to 400 bhp. Both would compete in the 100 to 400 hp range, but only Sulzer could build above 400 hp. SLM could make rail diesels if they were to go in those of the Thermo-Lokomotive type. Sulzer Archives.
7. Information kindly sent by David Brown of Sulzer in an August 23, 1993, letter.

8. Georg V. Lomonossoff, *Diesellokomotiven* (Berlin: VDI, 1929), 6, Verein Technic Verlag reprint, 1985.
9. Diesel and Klose each signed similar contracts with Sulzer regarding payments and noninvolvement with diesel locomotives if the contract was canceled. Sulzer Archives.
10. Jonas Hesselman, *Teknik och Tanke* (Stockholm: Sohlman, 1947), 260–61. Hesselman infers that the six-cylinder model was intended for railcar service.
11. *Diesel-Electric Motor Cars for Railway Service*, Diesel-Elektriska Vagn-Aktiebolaget, Wästerä, Sweden, Pamphlet 22, 1920, 5. Copy in Göteborgs Landsarkiv.
12. Ibid., 18. The fifty-one-page booklet is a mine of information.
13. Ibid., 33.
14. Ibid., 31.
15. Torsten Gårdlund, et al., *Atlas Copco 1873–1973* (Stockholm: Atlas Copco, 1974), 160.
16. Georg Aue and Jakob Roeschli, "Aus der Anfangszeit der Eisenbahn-Dieselmotoren," *Industriearchäologie*, no. 2 (1990), 4, 5. Lomonossoff, *Diesellokomotiven*, 6, also annotates the story. The patent issued in 1912.
17. G. W. Aue was the father of Georg Aue (b. 1911). G. W. Aue died in St. Petersburg in 1921. His widow and children had a harrowing experience returning to Switzerland. Sulzer aided them on their return, and made it possible for Georg to graduate from the ETH in Zürich. He worked forty-one years for Sulzer until retiring around 1971 as chief engineer of the diesel research department. (Several years were also spent at Alco in Auburn, New York.) The author had the privilege of visiting with him in 1987, thanks to David Brown. Georg was always eager to help when called upon for information and material.
18. Lomonossoff, *Diesellokomotiven*, 7, provides the 1909–1910 date and the drive arrangement. Aue and Roeschli, "Aus der Anfangzeit," 4, states the horsepower, but its mentioned earlier date of 1905 seems too soon for Kolomna engines of that power. Given Kolomna's enterprise, however, there is little reason to doubt that the locomotive was built.
19. "List of Locomotives & Railcars . . . with Sulzer Diesel Engines," Sulzer Bros., n.d., 10–11., and Aue & Roeschli, "Aus der Anfangzeit," 5, 6. The two pioneers in diesel-electric drives, Brown Boveri (BBC) and ASEA, merged in 1988 to form ABB.
20. Aue & Roeschli, "Aus der Anfangzeit." One was retained on the Chemin de Fer Régional du Val-de-Travers, at Flurier 1965 as a backup after the line had been electrified in 1944. A modified version of the original is on exhibit in the Lucerne Technical Museum.
21. John F. Kirkland, *Dawn of the Diesel Age* (Glenndale, CA: Interurban Press, 1983), 67–70. The book is an excellent reference for the history of the American diesel locomotive. Kirkland, an electrical engineer, began his career with the New York Central in 1928. In 1942 he headed diesel locomotive production at the Baldwin Locomotive Works.
22. Aue, "Aus der Anfangzeit," 15. Unfortunately, Lemp was not generally recognized for his accomplishments until late in his long life.
23. Kirkland, *Dawn of the Diesel Age*, 70, explains in detail how the control systems before Lemp's breakthrough and his new one worked. It prevented overloading the engine at low speeds and engine overspeeding if the generator was unloaded. A variable speed governor directly controlled the fuel as well as generator shunt field excitation.
24. Eugen Diesel, *Diesel: der Mensch, das Werk, das Schicksal* (Hamburg: Hanseatische Verlag, 1937), 411. E. Diesel refers to technical discussions about diesel power in ships and locomotives without stating where. Aue and Kirkland say the visit was in 1910, before Lemp worked on locomotive controls, but Diesel did not come to America that year. The date was incorrectly given in the *General Electric Investor*, Fall 1978, no. 18.
25. GE. 1914. US Patent 1,093,140, filed June 23, 1911, and issued April 14, 1914. Lemp was granted two other US patents on diesels, both filed in 1914. No. 1,166,916, issued June 18, 1914, was based on a conventional Junkers with an upper yoke and side rods; the other, 1,246,121 of November 13, 1917, had two parallel cylinders like 1,093,140, but transversely placed and using two rods. One rod was pinned at the big end to the other like the master rod of a radial engine.
26. The engine details are from a three-page report by Henry Loveridge entitled "General Features—Fuel Oil Electric Drives," dated December 9, 1918 in a covering letter to Col. James Milliken. Original is in the National Archives, Washington, DC. The author is indebted to David H. Hamley, who supplied a copy of this report, the above patents, and other cited material. Kirkland aided in contacting both Hamley and also Ray Corley, who kindly sent copies of photos from his extensive collection. It is a small fraternity of dedicated diesel-electric historians. Retired GE engineers from the Erie locomotive plant are compiling a history.

27. GE bought seven hundred acres of land at Erie in 1907 and later moved its entire rail operations there from Lynn. John Winthrop Hammond, *Men and Volts* (Philadelphia: Lippincott, 1941), 331. It is the present site of the GE locomotive factory.

28. R. Tom Sawyer letter to David Hamley, June 29, 1971. Sawyer, at one time a service engineer for GE, worked on the gasoline engine in the Jay St. locomotive in 1925. He retired from Alco as a respected diesel and gas turbine engineer.

29. Kirkland, *Dawn of the Diesel Age*, 72. Loveridge report said a major problem was a short packing life in the seal around the fuel nozzle valve stems. When these leaked injection air at 700 psi, the resulting air loss caused poor fuel atomization.

30. Loveridge, in addition to suffering a piston crown failure on its trip out to the proving ground, an operator's shirt became entangled in gears of the auxiliary compressor engine "resulting in some bent shafts and broken castings, and nervous shock to the operator, but no serious injury."

31. Franklin Reck, *The Dilworth Story* (New York: McGraw-Hill, 1954), 28–29, tells of Richard Dilworth's experience as experimental floor foreman at GE in 1912–1914 and of his being sent to the Philippines for several years to maintain the diesel generator sets. Dilworth (b. 1885) was the guiding genius at General Motors' Electro-Motive Division from its inception until his retirement. He spoke with awe of Hermann Lemp, comparing him to Charles Steinmetz and Michael Pupin.

32. See note 25 for other patent details.

33. Alfred D. Blake, "General Electric Oil Engines for United States Government," *Power*, vol. 43, March 23, 1916, 718–22, has a lengthy description of the engine.

34. Lintz. 1905. German Patent 178,896, issued on December 3, 1905, assigned to Diesel. Rudolf believed this subterfuge was necessary to prevent others than Allgemeine and its licensees from learning of his thinking until as long as possible after the patent issued. The complete Lintz/Diesel file is in the MAN Archives.

35. Ibid.

36. Safir automobiles are described in Ernest Schmid, *Automobiles Suisses* (Grandson: Edition du Chateau, 1967), 210–12.

37. Georg Aue kindly furnished material relating to Safir and St. Georgen from his files, some of which he had compiled. Additional information includes comments from Gustav Segin, "Aktennotiz betreffend den ersten Fahrzeug-Dieselmotor," in 1952 (Segin worked for Saurer from 1912 to 1938 when he retired.), and a November 24, 1977, letter from W. Schädler, an engineer in Arbon where Sauer was located, who pointed out mistakes in a museum article about the first diesel truck engine.

38. AG St. Georgen was a new venture of Maschinenfabrik St. Georgen, one of the oldest Swiss manufacturing firms founded in 1828. The parent firm, who built chiefly starch-making machinery, wanted the new factory to make farm equipment powered by Saurer gasoline engines.

39. Segin, "Aktennotiz," said that "Dr. Diesel came from time to time to Zürich, directed the work, and was present during the tests."

40. "Die Original Diesel & Co.-Kleinmotor auf der Welt-Austellung Brüssel 1910," *Zeitschrift für Elektrotechnik und Maschinenbau*, no. 44, November 3, 1910, three-page reprint. MAN Archives.

41. E. Diesel, *Diesel*, 392.

42. Robert Schlaepfer wrote his doctoral dissertation at the ETH in Zürich about research on one of Diesel's 5 bhp engines: *Untersuchungen an einem Diesel-Kleinmotor* (Zürich: Frey, 1914), 63 pages Sulzer Archives. It also covered methods and results using an optical indicator. See also *Zeit., VDI*, November 5, 1910, 1,897, for a report of tests at the Charlottenburg Technische Hochschule.

43. Sulzer donated an "Original Diesel" to the CNAM, and Eugen Diesel sold a used one for £70 to the Science Museum shortly after his father's death. Aue files and correspondence in the Science Museum archives.

44. Segin, "Aktennotiz."

45. Ibid.

CHAPTER 19

Rudolf Diesel—The Final Years

Character is fate . . .

—Barbara Tuchman[1]

To some of his peers, Diesel lived out a charade in the last decade of his life. He had become master of ceremonies for a performance over which he no longer had control, figurehead of an industry who in truth no longer needed him. In his own mind, however, he optimistically believed that the new ideas emanating from *Villa Diesel* in Munich would once again direct the course he had set for his earlier brainchild.

Former colleagues and partners distanced themselves from him, especially after the demise of the Allgemeine. Nevertheless, his prestige remained strong enough for others to pay him occasional homage. Those brief times of basking in praise were undoubtedly food for his soul, but he had a broader goal in mind. He was firmly convinced that these occasions increased support for his growing "baby" by extolling its accomplishments and submitting new fields for it to master.

However, as Diesel's vigor and resources drained away, long-silent critics stepped forward to write their interpretations of the engine's history and his role in it. Some were malicious. Others honestly viewed the events from their own perspectives. Rudolf himself had sown the seeds years before that would harvest such attacks and judgments. In the end, as in a Greek tragedy, his character would ultimately seal his fate, a victim of its own creation.

Designing and Consulting

Diesel's business and technical activities were situated in his Munich mansion after its completion in 1901 (chapter 7). A staff of engineers worked on assignments that came from consulting efforts Diesel generated or on fleshing out designs from ideas he fed to them. The latter included, for example, the locomotive, automotive, and small engines of the previous chapter. Without research facilities Diesel depended entirely on clients to build and test what came off the drawing boards at Maria Theresia Strasse 32.

One of his closest corporate relationships was with Carels in Ghent. An amicable bond between Rudolf and the Carel brothers lasted until the end, but of how much monetary value this was to the Belgians is unknown. Sulzer was another friendly place as evidenced by the joint locomotive venture. In conjunction with the Swiss, he was also a shareholder in Busch-Sulzer Bros. Diesel Engine Co. (chapter 11). By that time, however, the Sulzers held no illusions as to what little useful technical help could be expected from him.

More typical of the fortune suffered by Diesel's consultant work during his last decade was an engine program with SA John Cockerill in Liège, Belgium. Diesel first solicited the company, who until then had made only few unsatisfactory experimental models, after his patents expired.

The two collaborated on a diesel-powered river boat for the Belgian Congo because the wood used as fuel for steam engines was rapidly depleting forests adjacent to the rivers. Diesel and Cockerill people presented a boat and engine proposal to King Albert I, who personally authorized up to 1,000,000 francs for such a craft.[2]

The twin-screw *Belgica* would be 196 feet 10 inches long with a beam of 26 feet 3 inches, draw only 3 feet, 6 inches, and have a top speed of 15.5 kts. Diesel had recommended that a Fottinger hydraulic (water) transmission be used as a reverse gear to keep engine complications at a minimum.[3] This partly offset the penalty paid for hydraulic losses in the gear's water pump and reversible blading into the turbine.

The two-stroke, four-cylinder diesels had a square bore and stroke of 440 mm and were to produce 640 bhp at 280 rpm. Of interest is that the Fottinger gear allowed the propellers to turn at 440 rpm at maximum engine speed. A double-acting scavenge pump and three-stage compressor serving two cylinders were crank-driven off each end of the engine. The combined pump and first-stage compression were next to power cylinders and the final two stages were on the ends.[4]

Scavenging was based on patents issued to Arnold Freiherr von Schmidt, one of Rudolf's engineers and his new son-in-law who had married Hedy.[5] Air ports opened into the cylinder at different angles and directions to theoretically achieve maximum scavenging while using a minimum of scavenge air.[6] Two ports deflected air upward to about the middle of the cylinder, and two tangentially admitted air around the perimeter of the cylinder. The exhaust exited through two ports.[7]

Unclear is why Cockerill abandoned Diesel designs. One suggested reason is that the *Selandia*'s success prompted Cockerill to adopt a proven design and buy a Burmeister & Wain license (chapter 16). Also likely is that

Diesel continued to introduce new ideas that prolonged the development, and Cockerill grew tired of acting as his research lab.[8]

A spinoff of the Cockerill engines was a submarine engine proposal Diesel made in September 1910 to Delaunay-Belleville (chapter 15); however, output of the ten-cylinder sub engine, which had the same bore and stroke, was 50 percent higher.[9] When this bore no fruit, Diesel promoted his 30 bhp truck diesel to Delaunay-Belleville in May 1911. He no doubt hoped the French firm would become the new home to develop the engine abandoned by St. Georgen. In letters to St. Denis, he wrote that a total weight of 600 kg was feasible and that "this engine as a truck engine will be a great success."[10]

A review of the numerous engine projects Diesel proposed and launched after he was free from the Allgemeine to 1913 shows a disheartening record of failure. Not a single engine coming from *Villa Diesel* went into even brief production.

However, not all the shortcomings leading to this dismal record can be blamed on Rudolf. By 1910 he had competent engineers who should have complemented his own alleged weaknesses as a designer. Almost twenty years of intimate exposure to hardware associated with his own engines surely would have conveyed to him an understanding of the practical aspects of their design.

Another factor resisting his success was a handicap inherent with having someone else do the manufacturing. This proved to be a major problem for Busch in America until he set up his own factory. In Europe the margin for error allowed any new builder had been reduced because reasonably good engines were available from successful diesel makers. Customers would no longer tolerate engines not generally performing as advertised.

Finally, applications such as Diesel's locomotive and small engine were already served by forgiving and well-understood steam locomotives or, on the lower end of the power spectrum, by cheaper and far simpler gasoline engines. The diesel's time and its needed technology for these markets had not yet come.

Visits, Lectures, and Books

Diesel sought or was called upon to deliver lectures on the progress of his engine in numerous forums from 1910 to 1912. These were planned in conjunction with his visits to England, Russia, and the United States. He also spoke before important meetings in Germany and adjacent countries. The affairs gave him a platform to remind listeners once again of his own first efforts, as well as to tout new ventures by himself and others. Although he could no longer ride the tail of the comet he helped to launch into orbit, he still might bask in its light (fig. 19.1).

One of Diesel's first major post-Allgemeine visits was to St. Petersburg during May of 1910. It coincided with an industrial exhibit where Kolomna diesels and an operating 1,000 bhp Junkers engine were on display.[11] His ostensible purpose in going was to lecture before the Royal Russian Technical Association on the design and development of his engines (fig. 19.2). (Hugo Junkers had spoken before the St. Petersburg Polytechnical Society on the previous night.) He also wanted to cement relations with Emanuel Nobel to promote his new engine ideas. It was a splendid time for all, but nothing came of it for Rudolf.

Fig. 19.1. A studio photo of Rudolf Diesel in 1912. *MAN Archives*

Fig. 19.2. May 24, 1910, photo in St. Petersburg when Diesel and Junkers lectured. Diesel (second row, fourth from left) is between Junkers and Emanuel Nobel. Professor Vadim Archaouloff is in the same row on the left. Nobel diesel engineers are Oscar Derans, first row, second from the left; Hans Flasche, first row, (behind) second from the right. *MAN Archives*

A lengthy paper read by him before the Institution of Mechanical Engineers in London on March 15, 1912, not only illustrated various engines, including his own latest, but also stressed the importance of his type of engine cycle in conserving the world's oil supply.[12] The lecture would become a model for others given several times that year.

Rudolf and Martha began a long-awaited visit to the United States on March 26, 1912, paid for mostly by Busch. Diesel's only earlier voyage to America in 1904 had been more to sightsee.[13] One reason Martha had joined him this time was to see her six-month-old grandson Arnold born to Rudolf Jr. and his wife in New York City. Her son had taken a job there and shortly thereafter married Daisy Weiss of that city.

The trip's ostensible purpose for Rudolf was to lend his prestige at April 15 ceremonies breaking ground for Busch-Sulzer Bros. Diesel Engine Co.'s new factory. But it can also be seen as another occasion for the financially troubled man willingly to be exploited for his presence. Whether he viewed these gratuities from Busch as his due or as private alms to a proud and desperate man will never be known.

Lectures given prior to and after this event included those at Cornell University, the Naval Academy in Annapolis, and before the American Society of Mechanical Engineers in New York when he was made an honorary member.[14]

A major evening affair in St. Louis was his featured speech at the Engineers Club on the day of the groundbreaking. The banquet preceding it, the menu for which Diesel saved, would have challenged any speaker to remain alert. From the prelude of "houn' dawg cocktails" to the cognac and cigars—with time out for cigarettes between the mushrooms *sous cloche* and the broiled squab guinea chicken *au cresson*, the elaborate affair was a gourmand's delight. It was also typical of the respect and recognition accorded him during the strenuous trip. He boarded a 10:55 p.m. train for Ithaca, New York, that same evening.[15]

Diesel's arranged visit with Thomas A. Edison in his New Jersey home was stiff and strained, partly due to Edison's deafness, but more to the great chasm separating their cultural backgrounds and viewpoints on most everything, including diet and lifestyle.[16] This May 6 visit also came at the very end of an exhausting tour.

Aboard the SS *Victoria Luise* the next day, Diesel cabled Adolphus Busch his thanks and added, "May our company flourish, increase and prosper."[17] Unfortunately, the Busch-Sulzer Bros. Diesel Engine Co. never enjoyed in Diesel's remaining days the profits he desperately counted on.

Diesel's last important lecture came on November 21, 1912, when he was asked to address the German Marine Engineering Society (*Schiffbautechnische Gesellschaft*). His topic would be an account of his engine's early days as he had recently heard that others were in the process of writing their own historical interpretations of the same events. One was by Paul Meyer, who was associated with Diesel between 1898 and 1900 (chapter 7). Diesel had written on the title page of his copy: "This book mainly contains sucked in 'office gossip,' i.e., sundry rumors young engineers who could not know the relationship from the more or less slanted stories heard from their chiefs."[18] Two books by professors only peripherally involved with the engine would appear in 1913 and 1914.

Most of the November lecture material was already in hand because Rudolf had begun the manuscript of his *Die Entstehung des Dieselmotors* (The Genesis of the Diesel Engine). Springer Verlag would publish one thousand copies in September 1913.[19] This definitive technical history took the engine from idea through its 1897 demonstrations and ensuing test programs Diesel initiated at M.A.N.

Chaired by the Grand Duke of Oldenburg, the heavily attended and prestigious affair was held at the Technische Hochschule in Charlottenburg (Berlin).[20] The audience "cheered" Diesel when he had finished speaking, a recognition of the respect most felt for him. However, his exit from the podium signaled the onset of a devastating castigation that struck at Diesel's heart and would be one of the "final straws" destroying what remaining inner strength sustained him.

The fact that Diesel addressed only non-production aspects of his engine's development makes what followed bizarrely out of place and character. Two respected professors and long-time critics, Alois Riedler and Adolf Nägel, verbally attacked Diesel with accusations having little to do with presented material.[21] Rielder's lengthy "discussion" of the lecture accused Diesel of claiming full credit, with no mention of Buz, for not only the engine of 1897 but also the first, premature production engines and M.A.N.'s DM model that truly was commercially acceptable.

While Riedler was correct in his facts as far as what M.A.N. had accomplished, he omitted to say that Diesel had in his paper hardly touched on events after his withdrawal from activities at Augsburg. Riedler blamed Diesel for allowing the first engine in Kempten, which Diesel only spoke of as having successfully operated for years, to have been sold; he then took Rudolf to task because he did not tell of the woes that M.A.N. suffered with it.

Riedler reminded the audience of Diesel's unmentioned German predecessors (Capitaine, Köhler, and Söhnlein) as anticipating Diesel and his sin of omitting Lauster who was "the creator of the Diesel engine in its present, successful form."[22] Nothing was mentioned in his quite provincial remarks about successful work by non-German builders.

Riedler chided Diesel for peddling patent licenses when his engine was at death's door. He said that Diesel did not deserve being granted the patent and then did not follow it. Two further blows below the belt were his laying totally at Diesel's feet the failure of Augsburg Dieselmotoren AG and the amassing of great wealth from the sweat of others. Riedler also called attention to Diesel's not having introduced a single successful engine since 1897. His final remarks were:

I have heard the lecture of Herr Dr. Diesel with great interest and with great admiration, but have the deepest regrets that Herr Diesel's extravagant claims have carried him to indefensible ends. I have the belief that he has very much wronged himself by his excesses.[23]

Nägel's ensuing diatribe followed similar lines that at one point implied Diesel had falsified his journal entries. He, too, gave the knife a little twist by ending with the rhetorical question: "Whether it . . . might not be advisable if . . . Herr Dr. Diesel at the same time had then given up all claims to represent himself as the history writer of this era."[24] While most of the audience strongly reacted against the two, the damage was done. Pandora's box had been opened.

Diesel closed the painful discussion with these words:

I do not know if a lecture including only an individual's work can be entirely objective. [Applause from the audience] I have said from the first that personal memory was not entirely excluded. . . . I have not put forth my treatise as history, but as a report on "research work." Reporting about one's own research work is generally customary in the scientific world and would not usually be regarded as arrogance.

I say with conviction to you that I have described the genesis of the Diesel engine after my best knowledge and conscience, and what was presented to you, the exhibited old drawings and verbatim citations out of old journals, are incontestable witnesses from the genesis days that I can each verify myself. I have no more to say.[25]

The audience then began "minutes long, increasingly louder waves of roaring applause." While Diesel's listeners thought that he had won the debate, it would prove to be the opening salvo from an emerging faction of mostly disgruntled academicians.

M.A.N., Diesel's once principal antagonist, had more important matters to attend to, and considered him no longer a threat. Only Lauster, who in time came to say, "that the Diesel motor should be called essentially the Lauster motor,"[26] continued his feelings of antipathy against his former mentor. The unfortunate estrangement grew out of incompatible personal chemistries. Before Diesel died, however, he had made peace with himself over his strained relationship with Lauster, but it was not reciprocated.

The arguments first Riedler and then others brought into the open, after years of quiet since Capitaine's time, show how poorly informed they were. While Riedler was correct about the market unreadiness of the 1897 and the Kempten engines, and about Lauster's and Buz's important roles, he omitted several salient facts regarding the period up to 1902, the year when he claimed the first commercially acceptable engine was sold.

Formation of the Allgemeine in 1898 permanently removed Diesel from daily battles of making and servicing his engines. Although he had abrogated responsibility for this, he would have found it most difficult, even if he had so wanted, to involve himself in the "trenches." (It is a common trait of engineers to refuse an outsider's advice, the syndrome of NIH—Not Invented Here.) Thus, Diesel's chosen, and self-imposed, role after 1898 was to offer new ideas through the Allgemeine to his network of licensees.

When he distanced himself from that entity in 1905, he had no option but to continue as head of the consulting design office he had established. He was by then, however, totally dependent on the whims and largess of clients to make and test his ideas.

A consequence of the illness that struck Rudolf in 1898 put an end to whatever leverage he might have exercised at M.A.N. during the introduction of its engines. Moreover, Diesel must be viewed more as an engineer/scientist whose heart focused on research and development. He could not help this. While M.A.N. struggled to put engines in the field, Diesel was strongly opting for ongoing research in its laboratory.

A storm on the horizon, which did not break until after Rudolf's death, came from another nemesis: eighty-year-old emeritus Professor Johannes Lüders at Aachen. While Riedler and Nägel simply disliked Diesel, Lüders's feelings can only be ranked as vitriol. They had begun during his association with Köhler, one of Diesel's earliest critics also at Aachen, and then with Capitaine. About the time Diesel's *Entstehung* was to come off the presses, he had learned of a forthcoming book from Lüders that would make the Riedler and Nägel discussion comments tame by comparison. Even its title, *The Diesel Myth, an Authentic History of the Beginning of Today's Oil Engines*,[27] summoned attention for what lay within. Bookstores received it in October 1913. Lüders had looked at Diesel's writings from his 1892 patent and first book *Theorie* to his last lectures. He dissected them paragraph by paragraph in the context of a later time and knowledge. Much of what he said about the patent, for example, was true, but he was not privy to the chain of errors leading to the mistakenly patented invention. Neither did he have access to Diesel's private and test journals. Lüders book could have been an excellent critique of Diesel's work if it had not been slanted to damn rather than to define and interpret. It was another character assassin waiting in ambush for the by then defenseless Diesel, only this time his family would tragically endure the invective.

The spears of such revisionist historians, as Diesel could with some justification view them, were blunted by the appearance of his book *Die Entstehung des Dieselmotors*. It was his final act, his last printed words.

The Family

Rudolf's close relationship with wife and children influenced both his actions and state of mind in the last years (fig. 19.3). His strong love and devotion

Fig. 19.3. Diesel family dinner after Eugen's confirmation in 1904. Left to right.: Eugen, Rudolf Jr., Grete Flasche (Martha's stepsister), Christian Barnickel, Martha, Martha's stepmother, Rudolf, Hedwig Flasche, Hedy.
MAN Archives

to Martha was reinforced by his sense of duty as *der Mann* in the context of the times and of German culture. There is little question that she was quite unaware of her husband's precarious financial condition. As a man of that era and place, he would not burden her with this worry. It was his alone to bear.[28]

Both Rudolf Junior and Eugen, who were thirty and twenty-four years old in 1913, had disappointed their father. Because of Diesel's "passion" for the engineering profession, and his intuitive conviction that he himself would become a great engineer, he had dreamed from the time his sons were babies that they would choose this path.[29] Diesel wanted to found a "technical dynasty." Despite the boys' toys being mechanical and scientific rather than "princes with uniforms, ponies and decorations," this was not to be. First, the namesake son had said early on that engineering was not for him.[30]

Eugen then became the "heir" rather than the "spare" in his father's planned dynasty (fig. 19.4). Thus, when Number Two had announced that he also wanted to change his studies from engineering to geology,[31] it was a deep blow to the loving yet unrealistic father. Nevertheless, that special familial bond between him and his younger son remained. Eugen continued to receive full financial support at his university.

Only an unusual mix of transmitted genes from sire to son could have made Diesel's dreams come to pass. Irrational vocational pressures and the daily example of what being an engineer, in a child's eyes, had done to their

Fig. 19.4. A ride in the Diesel family car, c. 1912. Left to right:
Eugen, Paul Flasche (Martha's brother), Martha Barnickel
(Rudolf's niece), Martha, and Rudolf. *MAN Archives*

father and family worked totally against any chance of their succumbing to his vision.

Daughter Hedwig had enjoyed the good life befitting an attractive girl who lived in *Villa Diesel* (fig. 19.6). Her marriage to Arnold V. Schmidt further ensured her an independence and a partial insulation from events soon to befall its inhabitants. She and her husband's departure from that house in the spring of 1913 could only have been a relief as Arnold, who not only lived but worked there, must have had a premonition that things were in the process of change—and not for the better.[32]

Diesel's generally poor business judgment involving private investments will not be dealt with. Crippling financial losses accruing from the Allgemeine debacle started his downfall. Compounding these reverses was the adoption of an extravagant lifestyle catering to an unsuspecting wife and children. His house was mortgaged to the hilt, and his numerous creditors had begun seriously hounding him. By this time, he was referring to his Munich mansion as his "mausoleum" (fig. 19.5).

On to Ipswich

Diesel's March 1912 lecture in London had coincided with the formation of a company that would bear directly on his fate eighteen months later. The official formation of Consolidated Diesel Engine Manufacturers Ltd. on March 27, 1912, was an ambitious amalgamation of Carels's London-based Diesel Engine Co. and Carels Frères in Ghent. A large, modern factory with a foundry, the latest machine tools, and an assembly and test area would be built on newly purchased land outside of Ipswich, about seventy miles north

Fig. 19.5. Rudolf Diesel on the veranda of *Villa Diesel. MAN Archives*

Fig. 19.6. Breathing exercises on a summer holiday at Bad Tölz:
Rudolf, Martha, Hedy, and Rudolf Jr. *MAN Archives*

east of London[33] (fig. 19.7). It was Diesel's last hope for money through ownership in a company building his engine.

Reality did not reach expectancy. By the time the factory was ready over a year later, diesel sales had drastically decreased in the stationary engine markets Consolidated Diesel would serve. The causes were unrest in the Balkans, general financial worries, and a period of high fuel prices in Britain triggered by manipulative oil companies. Diesel and Georges Carels headed into this dismaying situation to preside over Condec's first annual stockholders meeting in London on October 2, 1913.

Diesel left for Ghent after a visit with Hedy and her family in Frankfurt on September 26, 1913. Martha remained with her daughter.[34] One purpose of going to Carels was to witness the latest tests on engines burning Mexican crude. This high-asphalt fuel offered a way to circumvent the oil barons' monopoly on good fuels, a ray of hope to soothe worried stockholders a few days later. The tests looked promising.

On the evening of September 29, Diesel, George Carels, and his chief engineer Luckmann sailed from Antwerp bound for Harwich on the Great

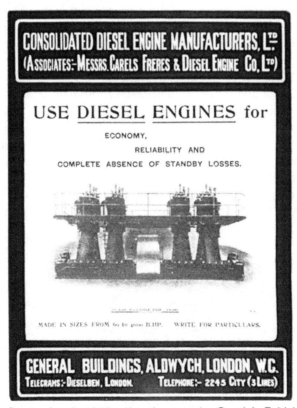

Fig. 19.7. September 5, 1913, advertisement for Carels's British company.
Engineering

Eastern Railway steamer *Dresden*. It would be a beautiful, calm, and clear night. The route was ideal because the English port lay only a few miles from Ipswich where a factory visit had been scheduled prior to the London meeting. The three men had a pleasant dinner aboard and retired to their staterooms about ten o'clock. It was agreed to gather for an early breakfast before going ashore the next morning. The ever-punctual Diesel did not appear as planned, and in due course his stateroom was checked. They saw his bed turned down but not slept in, all personal baggage intact, his watch by the bedside, and keys hung from his handbag. A search of the ship before allowing passengers to disembark found no trace of him. He could only have fallen overboard.

Carels told newspaper reporters that Rudolf "had not of late enjoyed robust health" and suffered bouts of insomnia—which could explain his strolling the deck before retiring. He said that the missing person had had no more than two hours of sleep a night for the past few weeks. The newspaper surmised that "in view of the insomnia . . . a mental breakdown would be a possible explanation," but both Carels and Luckmann were "emphatic" that Diesel "showed no signs that would justify this conclusion."[35] Another account speculated that as all railings were 4 feet 6 inches high, he might have been sitting on one to enjoy the lovely evening, and he accidently fell into the water.[36] Thus began rumors and legends.

One aspect of the mystery was solved when a Dutch buoy tender found a man's body floating off Walcheren Island between the mouths of the East and West Scheldt on October 11. The body was brought aboard, but because of rough weather and no way to keep it until a return to port, personal effects were removed, and it was committed back to the sea. Rudolf Jr. immediately went to Flushing, where the boat put into port, to identify the items removed from the body.[37] There was no question that everything, including an enameled pill box, belonged to Rudolf (fig. 19.8).

A Reuters dispatch on the finding and identification of the body included a statement by Georges Carels that "If one has to put aside the theory of

Fig. 19.8. Enameled pill box with Diesel's body when it was found in the English Channel. *MAN Archives*

an accident, I can only think that something suddenly had given way in his brain." It ended with "disclosures since the finding of the body show that the famous inventor was reduced to financial straits, and in a very hopeless situation." It was the first public acknowledgment of Diesel's monetary troubles.

What Happened and Why?

Rumors thrived as news of Diesel's baffling death spread. One holding credence at the time was assassination by a secret agent because he was secretly carrying plans of a new submarine engine to England and had to be stopped. Advent of the war in 1914 fueled this story, but calmer minds dismissed it.

If Diesel committed suicide, the question is why? The known evidence can only lead to this conclusion. He no longer could stave off financial ruin. Continuing peer attacks challenged his veracity. Every engine venture he had been associated with in the last years had failed. He had lost control over his creation, and even his sons had chosen other careers. These blows were enough to demoralize a strong man. But Diesel's psyche was not sound, as a medical history of deep depression had proved. Could not he under these circumstances succumb again when faced with such enormous pressures? Reinforcing this view was a tantalizing trail of comments and letters at the end revealing a troubled mind possibly pleading for help.

Diesel inscribed in copies of *Die Entstehung* for Martha and Eugen these poignant messages only days before his departure (fig. 19.9):

Fig. 19.9. One of the last photos of Diesel showing his rapid aging.
MAN Archives

My Wife. You were to me everything in this world, for you alone have I lived and struggled. If you leave me, then I want no more to be.

Your Mann[38]

Was the choice of the word *were* a Freudian slip or a contemplated decision? In Eugen's book are these heavy and compelling words:

My beloved son Eugen. This book contains only the bare, clear technical side of my life's work, the skeleton. Perhaps sometime you can mould on this skeleton the living body through an addition of the genuine humanness that you perhaps more than anyone else have witnessed and know with understanding.

Your father

On September 27 he wrote to Martha and his elder son from Ghent in very specific terms about a serious heart problem. He had already informed Eugen of this before leaving home. Martha's letter was sent in care of Hedy at Frankfurt. However, by a quirk of fate he mistakenly used the Munich street address, and the letter did not reach her until ten days later. Why did he wait until then to tell her of his heart condition? If it had been correctly addressed, she would have had time to telegraph him that she would join him in Antwerp.[39]

On the twenty-eighth Rudolf again wrote Martha from Ghent: "Do you feel how I love you? I would think that even from the great distance you must feel it, as a gentle quivering in you, as the receiver of a wireless telegraph machine."[40] She read these words also after his death.

A final letter on the twenty-ninth chatted of his visit to the Ghent exhibit and described the displayed French household items. Its only ominous words referred to a great uneasiness regarding the Ipswich business's future.

Aftermath

The scheduled meeting of Consolidated Diesel lasted long enough to announce Rudolf's disappearance and briefly answer worried stockholder questions.[41] Within weeks the business's sad state was made public, but no accord could be reached to keep it going. Even though most directors saw a profitable future after a reorganization, a minority group felt otherwise. Things became so confused that liquidation became the only point of agreement.[42] Receivership began on July 14, 1914, to sell off the assets; the "shell" was not dissolved until January 4, 1929.[43] Vickers Ltd. bought the factory in February 1915 to make submarine engines. In April 1919, when little demand existed for this application, a new venture was formed with Petters known as Vickers-Petters Ltd. to build Petters's two-stroke diesels.[44] These were not products anticipated in 1912 for Ipswich, but they were his engines.

Events moved fast in Munich after the body was found. Within days Martha and Eugen had to move out of the house. Its contents, what had not been pilfered by servants and others, were sold at auction. Ten months later Germany was at war.

Eugen was able to finish his education, but not in geology, through the largess of Emanuel Nobel. His first job was in the passport office in Berlin where he remained throughout the war. Unsuccessful business experiences in Sweden and the United States brought him back to Germany in 1924, and after 1925 writing became his main occupation. His best work was the often-cited *Diesel: Der Mensch, das Werk, das Schicksal* published in 1937.[45]

Martha was a helpless witness to the dissolution of what her *Mann* had created. A few friends tided over the absolutely powerless woman when they realized her dire straits.[46] In time the German diesel engine industry provided her with a pension that kept her from want until she died in 1944 at age eighty-five.

Adding to her grief was the immediate publication of Lüders's book, which poured more hurt into her wounded heart. After referring to the fortune Diesel had made, he venomously stated:

> The reading of my book will thoroughly disappoint Diesel's admirers, but no less than the participants of the festivities in the Diesel Munich palace in the spring when they learned that their host was hopelessly in debt, and that he had fulfilled the adage "easy come, easy go"...[47]

A few paragraphs added after learning of Rudolf's death criticized the writer of his obituary for offering only positive words about his life and accomplishments. Lüders declared that one who writes history should not "of the dead say nothing but good" (*De mortius nil nisi bene*).

In St. Louis, Missouri, a special directors meeting of Busch-Sulzer Bros. Diesel Engine Co. was held on October 21, 1913. Two of its directors had died within eleven days of each other: Rudolf Diesel at sea, followed by Adolphus Busch in his German homeland. Edward D. Meier, another director who himself would be gone in a few months, wrote warm and heartfelt tributes to both good friends. Of Diesel he penned:

> The fertile brain that opened the vein from which his stream of inventions flows is now at rest. But the name of Rudolf Diesel is inscribed on the same line as that of James Watt and Robert Stevenson on the tablets of fame by a grateful humanity. Our right to use that name has its practical as well as its sentimental value, and our recollection of his clear insight and wise advice will always keep his memory green and lead us on.[48]

Rudolf Diesel conceived an engine and brought it into the world. He created an international industry while his offspring was in an infancy fraught with childhood diseases. Many nurtured and guided it to maturity. But it was *Diesel's* engine!

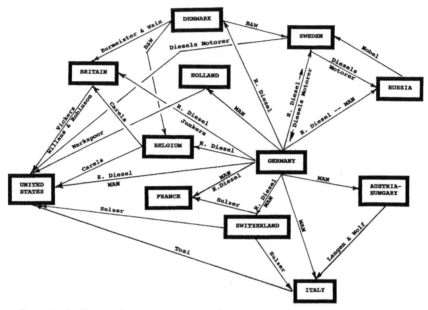

Fig. 19.10. Technology transfer ensuing from the international network of licensees created between 1898 and 1914. Begun with Rudolf Diesel, they were followed by the Allgemeine (M.A.N.). After Diesel's basic patents expired, the technology flowed from companies either generating patents of their own, or who sold know-how through manufacturing licenses.

Notes

1. "Character is fate—as Greeks believed." Barbara Tuchman, *The March of Folly* (New York: Knopf, 1984), 29.
2. The money came from dividends paid to the Royal family after Leopold II, Albert's father and major stockholder in the group who controlled the Belgian Congo, was coerced due to his cruel exploitation into selling the colony back to his country in 1908!
3. Diesel reported these facts in an address before the German "Kolonial-Wirtschafftlichen Komite" in November 1911. The committee dealt with business activities in German East Africa. The information on Diesel's address is from N. E. Rasmussen and H. F. Petersen, "The B&W Family," *BWE* (B&W Engineering), no. 7, 1978—kindly translated by B&W.
4. Engine and boat details are from "A Diesel Ship for the Congo River," *Motorboat and Marine Oil and Gas Engine*, July 25, 1912, 57–59. A copy was found by B&W for the author in the Admiralty Library, London.
5. Schmidt began working there a few months after an elaborate marriage in 1909 and was with Diesel until the summer of 1913. He and Hedy lived in the Diesel home where two of their three children were born. He joined the Adler Works in Frankfurt after leaving Diesel. Emperor Franz Josef had "elevated" Schmidt's grandfather for engineering services rendered (Freiherr = Baron); his father was a professor at the Munich Technical University. MAN Archives.

6. Schmidt patents show the porting system on double-acting, two-stroke diesels. They were filed in Denmark on February 12, 1912, and issued as Patents 17,408 and 17,759. A US application filed the same year had s/n 669,988. The latter was sold to Busch-Sulzer for 600 marks in June 1913. B&W Archives, and Busch-Sulzer papers, Wisconsin State Historical Library, Madison.

7. The idea anticipated the approach used by Adolf Schnürle in his high-output aircraft diesel at Deutz in the early 1940s. See Paul H. Schweitzer, *Scavenging of Two-Stroke Cycle Diesel Engines* (New York: Macmillan, 1949), 44–45; also, Schweitzer & C. G. A. Rosen, *Whither the European Automotive Diesel?*, SAE Paper, June 7, 1953, 3–6.

8. Rasmussen and Petersen, "The B&W Family."

9. Correspondence, sketches, and notes. MAN Archives.

10. July 15, 1911, letter. MAN Archives.

11. Hans Flasche, "Der Dieselmaschinenbau in Russland von Anbeginn 1899 an bis etwa Jahresschluss 1918," May 1952 ed., 56. MAN Archives.

12. Rudolf Diesel, "The Diesel Oil Engine," *Engineering*, March 22, 1912, 395–406.

13. The coast-to-coast trip included visits, to, among others, Niagara Falls, the World's Fair in St. Louis, where he stayed with the Busches, Yellowstone, and San Francisco. His many comments, mostly positive, about the US are covered at length in Eugen Diesel, *Diesel: Der Mensch, das Werk, das Schicksal* (Hamburg: Hanseatische Verlag, 1937), 374–82.

14. Rudolf Diesel, "The Present Status of the Diesel Engine in Europe, and a Few Reminiscences of the Pioneer Work in America," *ASME Journal* (1912): 904–47.

15. Diesel's small notebook covering his 1912 trip is among his papers in the Deutsches Museum Archives. Entries on April 14 and 15 about first reports of the *Titanic's* sinking and a loss of 1,600 lives were underlined. On the day of the groundbreaking, he noted receipt of $500 from the company. The banquet's printed menu also was mentioned.

16. E. Diesel, *Diesel*, 410–11.

17. Diesel papers.

18. Paul Meyer, *Beiträge zur Geschichte des Dieselmotors* (Berlin: Springer, 1913), 57 pages. Copy in Diesel collection, Deutsches Museum.

19. The author relied heavily on his original copy of this work for chapters 4–6. *Die Entstehung* was reprinted in 1984 by Steiger Verlag, Moers, with a foreword by Hans-Joachim Braun.

20. E. Diesel, *Diesel*, 414–17. Eugen dramatically related the events, which are, in this case, rightfully sympathetic to his father.

21. Riedler's unexpurgated "Discussion" transcript, including "audience disturbances" and "strong applause from the minority [Riedler faction] and strong agitation from the majority side" was printed as an appendix in Alois Riedler, *Dieselmotoren, Beiträge zur Kenntnis der Hochdruckmotoren* [Contributions to the Knowledge of High Pressure Engines], (Berlin: Verlag für Fachliteratur, 1914), 261–74.

22. Ibid., 273.

23. Ibid., 274.

24. Eugen Diesel, *Jahrhundertwende* [Turn of the Century], (Stuttgart: Reclam, 1949), 274. Eugen was in Berlin at the time and attended the lecture. Thus his remarks, including his father's foreboding, are not hearsay.

25. Diesel, *Diesel*, 416. The quotes in both of his books complement each other.

26. Diesel, *Jahrhundertwende*, 149–50. Eugen interviewed Lauster as an old man who spoke of Rudolf with mixed feelings in that he appreciated his hard work, and his ways of explaining what was happening, but he could not forgive Diesel for not spreading around more praise in the early days. Lauster also refused to accept that some ideas he himself took credit for had come from the Swedes, for example, but "What was abroad doesn't concern me. I only speak of German developments."

27. Johannes Lüders, *Der Dieselmythus, Quellenmäßige Geschichte der Entstehung des heutigen Ölmotors* (Berlin: Krayn, 1913), 236 pages. The author is indebted to Dr. G. P. A. Mom for supplying a copy.

28. The author did not on his own draw this conclusion. He was guided by the insights of Irmgard Denkinger, his deceased dear friend and retired MAN archivist, who shared her reflections on these marital aspects in an August 28, 1986, letter. Thanks go to Inge Rhodes for her sensitive translation of it, which Irmgard had chosen not to write in English because of its nuances. It also recited her credentials and insights into a one-generation-removed background to Diesel's era, which aided her in her conclusions:

"Naturally I have not read all [the author would expect it was most!] of the private correspondence (love letters, etc.) in the Archives, because the main thrust of my professional tasks was dealing with the facts regarding the invention and technical development of the diesel engine. Despite this, I have always formed my own opinion about Rudolf and Martha's marriage. (I wouldn't be a typical woman if I had been without interest in it!)

"Considering the customs of a marriage of the last century—and at the turn of the century—in Germany, the marriage between Rudolf and Martha was a good one. The marriage of my parents (1906) also fell into this time and my mother told me much about it. That is why I can understand many things between Rudolf and Martha. We cannot allow ourselves to compare these marriages with today's. *At that time* the ground rule was that the husband *must* take care of the economical and money side, the wife to bear and rear children, and to manage the household with the money given her by the husband; and besides, if possible, to obey him. All things were decided by the husband. That meant no education for the wife, no income for herself, almost no part of decision making . . .

"Now to Rudolf and Martha. They, too, could not escape this 'law.' But they were fortunate to have been able to freely choose each other; It was love! The big love! Rudolf was able to express himself more than Martha. I cannot judge if this was part of her makeup or upbringing . . . How could she . . . have a *vast* understanding for the "iron mistress," without knowledge or technical education? (How difficult it was for me with the technical aspects of my archival work in the beginning, and I had good mentors and a great interest by an engine engineer husband!!) . . . I believe that she was simply proud of her Rudolf!

"[As to the financial situation in 1913] she never worried about money, but had always received it from Rudolf. *What difference would it have made anyway?* . . . She was not able or capable of [earning money] in a way that would have helped. Besides, it was men's work . . .

"He loved his wife his whole life—there is much proof of that . . . I have thought much about this and saw it in his financial state and the times of 1913 . . .

"I write this in my honest opinion, not because I want to know it that way, but rather because I have come to it during the many years I have occupied myself with Rudolf Diesel. And since 1953 I have helped to . . . catalog the archives of Eugen containing all of his original documents of his father."

29. Eugen stresses this point about his father's philosophy and dreams. Diesel, *Diesel*, 344–45.
30. Neither did Rudolf Jr. spend much time at the Munich mansion. He had moved to New York when in his mid-twenties, but by 1913 he, his wife, and boy were back to Germany. He was more of a "free spirit" who had never really found himself. His first marriage failed and his second wife died in 1938 after a long illness. By then his own health was poor. M.A.N. had provided over 22,000 marks to educate his son Arnold between 1927 and 1938. Rudolf Diesel Jr. took his own life in November 1944. Less than a year later Arnold followed the same path as his father and grandfather. Correspondence and records in the MAN Archives tell this sad end to one line of the Diesel family.
31. In Diesel's June 19, 1913, letter to Adolphus Busch stating that his son-in-law's patent was for sale, he said, "As far as my son Eugen Diesel is concerned, for him also the whole [patent] discussion is obsolete since for quite a time he has given up his engineering studies to occupy himself with geology." Busch-Sulzer Translation. Copy in Wisconsin State historical Library.
32. In spite of what seemed like a life in a gilded cage, Hedy was a "survivor." Her husband had held a number of responsible engineering jobs. One took him back to his Austrian roots in Steyr where he worked for Steyr-Daimler-Puch as a chief engineer from 1917 through the early 1930s. The two younger Schmidt children were born in Steyr. Economic times were difficult for them there during the Depression, as Hans Flasche related in a March 14, 1930, letter to his Aunt Emma (copy in MAN Archives). Work in Berlin followed Steyr. Marie-Rose Cochet, Frédéric Dyckhoff's granddaughter, told the author that her mother Charlotte and Hedy were lifelong friends and showed photos of Diesel standing between the girls during a summer holiday at Berchtesgaden. The friends kept in contact. Marie-Rose met Hedy in Paris in 1959 and again just before Hedy's death when she and her mother visited Hedy in Munich. Marie-Rose said that after Hedy and Arnold's Berlin home was destroyed late in the war, and everything was lost, the two had lived in a forest outside of the city. Arnold contracted pneumonia there, partly as a result of his malnutrition, and died later as a result of it. Hedy was left destitute, receiving only a tiny insurance pension that went mostly to support her mentally ill daughter in an institution (August 19, 1938, letter from Paul Rieppel to MAN. Copy in MAN Archives). Because her other children were unable to provide help to her at that point, MAN stepped in with financial aid.

33. *East Anglican Daily Times*, April 1, 1912, 5. Copy in British Museum Newspaper Library, London.

34. E. Diesel, *Diesel*, 452–60, gives a detailed reconstruction of Rudolf's last days.

35. "Dr. Rudolf Diesel—Disappearance from a Harwich Steamer," London's *The Times*, October 2, 1913, 6. The long story included accounts from Carels and Luckmann.

36. "Dr. Diesel's Fate—Disappearance from Harwich Steamer—Accident or Suicide?" *Suffolk Chronicle and Mercury*, October 3, 1913, 3. Copy in British Museum Newspaper Library.

37. "Mystery of Dr. Diesel Solved—Body Washed Ashore in Holland," *Suffolk Chronicle and Mercury*, October 17, 1913, 3. British Museum Newspaper Library.

38. Diesel, *Jahrhundertwende*, 283. Rudolf also took time to inscribe a copy to Jonas Hesselman on September 9, 1913: "As acknowledgement of your successful work in development of these engines I hereby dedicate to you this book on their creation. [signed] Diesel." This was quite a tribute to someone of whom Hesselman said that in earlier years "Diesel did not belong to the category of people who generously acknowledge the contributions of others, although he could well afford, by then [1910], to share some. J. K. E. Hesselman, *Teknik och Tanke* (Stockholm: Sohlmans, 1948), 189–90.

39. E. Diesel, *Jahrhundertwende*, 297.

40. Ibid., and *Diesel*, 452–60. Eugen wrote in depth about the "what ifs" from metaphysical perspectives, even to the significance of the bookmark Diesel had placed in his copy of Schopenhauer's *Parerge und Paralipomena*.

41. "Diesel Meeting Adjourned," *The Times*, financial section. Ironically, a picture story in the *Suffolk Chronicle and Mercury*, October 10, 1913, 10, shows that the modern plant was ready to begin production. Copy in British Musuem Newspaper Library.

42. Later articles in *The Times* on November 26, December 1, 3, 1913, January 12, February 13, March 6, May 14, June 18, 19, 26, 1914, cover the muddled business aspects.

43. Company records other than for the receivership and dissolution were destroyed in 1929. Ref: BT31. Box 20567, Consolidated, 121112. Public Records Office, Kew.

44. Vickers-Petters Ltd. Cat. P. No. 1104 0A, May 1927, 4. Copy kindly sent by David W. Edgington. A more complete story of the venture is in his illustrated printing of Percival Petters's 1933 memoirs: *The Story of Petters Limited* (Westbury, England: Hinton Press, 1989), Chapter 6.

45. A short biography of Eugen is in Donald E. Thomas Jr., "Diesel, Father and Son: Social Philosophies of Technology," *Technology and Culture*, July 1978, 382–93. Eugen never escaped from the aura or influence of his father.

46. Diesel, *Diesel*, 460–70, closes his father's biography with the rapid chain of events in October and November 1913. Even though there is sadness, and some bitterness in his words, they can be accepted as telling the facts.

47. Lüders, *Der Dieselmythus*, 236.

48. Handwritten minutes of "Special Meeting of the Board of Directors, October 28, 1913." Each tribute was two legal pages long. Copy in Wisconsin State Historical Library.

EPILOGUE

Tucked away in an Augsburg public park is a fitting, yet uncommon, memorial to a man whose dedication and genius enhanced the legacy of that ancient and famous city. It is not a statue, a sculptured fountain, or a Doric-columned building, rather a secluded garden, nestled in a tree-shaded portion of the park. It contains no flowers or colorful shrubs—only gravel, small rocks, and boulders.

But these are igneous materials not from nearby quarries or streams. Stones formed by the release of the earth's fire, they come from a mountainous watercourse far to the East. Sacred to the donor, they attest to the legacy Rudolf Diesel left on energy transformation and international commerce. They are also a tribute to the generosity and insight of Magokichi Yamaoka, founder of Yanmar Diesel Engine Co., who wished to honor Diesel in the city still to tangibly honor one who had enhanced its fame and economy.

The story of the garden's creation is an impassioned one. It is more than the physical task of transporting tons of special cargo from Japan to Germany, with each piece keyed to a carefully designed plan. Nor is it a vying with local officials to allow a special garden on public property and then to observe its creation under the guidance of people from another land. The garden conveys a feeling of Eastern serenity, isolated from a bustling world, that Diesel himself could have retreated to during times of mental anguish. His complex psyche might have related well to the mysticism of the sacred materials. Was this the kind of beckoning refuge he sought on that fateful voyage between Antwerp and Harwich?

BIBLIOGRAPHY

Extensive chapter notes make it redundant to list again all those references. The following selections, therefore, do not include periodicals and unpublished works, and are intended to give the reader a general background.

Anderson, John W., "Diesel, Fifty Years of Progress," *Diesel Progress.* May 1948.

Archbutt, Leonard, and R. M. Deeley, *Lubrication and Lubricants: A Treatise on the Theory and Practice of Lubrication.* London: Charles Griffin & Co., 1920.

Blunck, Richard. *Hugo Junkers, der Mensch und das Werk.* Berlin: Limpert, 1942.

British Admiralty. *Royal Commission on Fuels & Engines.* London: Eyre & Spottiswoode, 1913.

Büchner, Fritz. *Hundert Jahre Geschichte der Maschinenfabrik Augsburg-Nürnberg.* Augsburg: MAN, 1940.

Burman, Paul G., and Frank DeLuca. *Fuel Injection and Controls for Internal Combustion Engines.* New York: Simmons-Boardman Publishing Corporation, 1962.

Cable, Frank T. *The Birth and Development of the American Submarine.* New York: Harper, 1924.

Campbell, Andrew. *Petroleum Refining.* London: Charles Griffin & Co., 1918.

Cardwell, Donald S. L. *From Watt to Clausius: The Rise of Thermodynamics in the Early Industrial Age.* Ithaca, NY: Cornell University Press, 1971.

Camot, Sadi. *Reflections on the Motive Power of Fire and on Machines Fitted to Develop That Power.* Translated by Robert Fox. New York: Barber, 1986.

Chalkley, Alfred P. *Diesel Engines for Land and Marine Work.* New York: D. Van Nostrand Company, 1917.

Clerk, Dugald. *The Gas, Petrol, and Oil Engine.* Vol. 1. London: Longmans, Green & Co., 1916.

Cummins, C. Lyle. *Internal Fire*. Rev. ed. Austin, TX: Octane Press, 2021.

Diederichs, H. *The Design and Construction of Internal-Combustion Engines: A Handbook for Designers and Builders of Gas and Oil Engines*. New York: D. Van Nostrand Company, 1910.

Diesel, Eugen. *Diesel: der Mensch, das Werk, das Schicksal*. Hamburg: Verlagsanstalt, 1937.

———. *Jahrhundertwende*. Stuttgart, Germany: Reclam, 1949.

Diesel, Eugen, and Georg Strößner. *Kampf um eine Maschine; Die ersten Dieselmotoren in Amerika*. Berlin: Schmidt, 1950.

Diesel, Rudolf. *Die Entstehung des Dieselmotors*. Translated by Henry I. Willeke. Berlin: Springer Verlag, 1913.

———. *Theory and Construction of a Rational Heat Motor*. Translated by Bryan Donkin. London: Spon Press, 1894. (The VDI reprinted Diesel's original *Theorie* in 1986.)

Donkin, Bryan. *A Text-Book on Gas, Oil, and Air Engines*. London: Charles Griffin & Co., 1911.

Emmerson, George S. *Engineering Education: A Social History*. Newton Abbot, England: David & Charles, 1973.

Evans, Arthur F. *The History of the Oil Engine*. London: Samson Low, Marston, c. 1930.

"FIAT" A Fifty Years' Record. Turin, Italy: Fiat, 1951.

Föppl, Otto. *Schnellaufende Dieselmaschinen*. Berlin: Springer Verlag, 1919.

Forbes, Robert J., and D. R. O'Beirne. *The Technical Development of the Royal Dutch/Shell 1890–1940*. Leiden, Netherlands: E. J. Brill, 1957.

Gärdlund, Torsten. *Atlas Copco 1873–1973*. Stockholm: Atlas-Copco, 1974.

Goldbeck, Gustav. *Kraft für die Welt*. Düsseldorf, Germany: Econ Verlag, 1964.

Güldner, Hugo. *Das Entwerfen und Berechnen der Verbrennungskraftmaschinen und Kraftgas-Anlagen*. 3rd ed. Berlin: Springer Verlag, 1914.

Haas, Herbert. *The Diesel Engine, Its Fuels and Its Uses*. Washington, DC: US Government Printing Office, 1918.

Hausfelder, Ludwig. *Die kompressorlose Dieselmaschine, Ihre Entwicklung auf Grund der inund ausländischen Patent-Literatur*. Berlin: Krayon, 1928.

Hawkins, Nehemiah. *Hawkins' Indicator Catechism: A Practical Treatise for the Use of Erecting and Operating Engineers, Superintendents, Students of Steam Engineering, Etc*. New York: Theo. Audel & Co., 1903.

Hesselman, Jonas. *Teknik och tanke*. Stockholm: Sohlmans, 1948.

Henriques, Robert. *Bearsted: A Biography of Marcus Samuel, First Viscount Bearsted and Founder of "Shell" Transport and Trading Company*. London: Barrie & Rockliff, 1960.

Hovgaard, William. *Modern History of Warships*. Annapolis, MD: US Naval Institute Press, 1971.

Jane's Fighting Ships of World War I. New York: Military Press, 1990.

Jenny, Ernst. *The BBC Turbocharger: A Swiss Success Story.* Basel: Birkhäuser Verlag, 1993.

Kirkland, John F. *Dawn of the Diesel Age: The History of the Diesel Locomotive in America.* Glendale, CA: Interurban Press, 1983.

Lehmann, Johannes. *Rudolf Diesel and Burmeister & Wain.* Copenhagen: B&W, 1938.

Lomonossoff, Georg V. *Diesellokomotiven.* Berlin: VDI, 1929.

L'Orange, Prosper. *Ein Betrag zur Entwicklung der Kompressorlosen Dieselmotoren.* Berlin: Schmidt, 1934.

Maag, Julius. *Dieselmaschinen.* Berlin: VDI, 1928.

Marvin, Charles. *The Region of the Eternal Fire: A Journey to the Petroleum Region of the Caspian in 1883.* London: W. H. Allen & Co., 1891.

McCrady, Archie R. *Patent Office Practice: The Procedural Law Relating to the Prosecution of Applications before the United States Patent Office.* 4th ed. Pasadena, CA: Margit Publications, 1959.

Morrison, Lacey H. *Diesel Engines: Operation Maintenance.* New York: McGraw-Hill, 1923.

———. *Oil Engines: Details and Operation.* New York: McGraw-Hill, 1919.

Moss, Michael, and John R. Hume. *Shipbuilders to the World: 125 Years of Harland & Wolff, Belfast 1861–1986.* Belfast, UK: Blackstaff Press, 1986.

———. *Workshop of the British Empire: Engineering and Shipbuilding in the West of Scotland.* Cranbury, NJ: Associated University Press, 1977.

Nitske, W. Robert, and Charles M. Wilson. *Rudolf Diesel: Pioneer of the Age of Power.* Norman, OK: University of Oklahoma Press, 1965.

Obert, Edward F. *Internal Combustion Engines.* 3rd ed. Scranton, PA: International Textbook, 1968.

Peterson, Walter F. *An Industrial Heritage: Allis-Chalmers Corporation.* Milwaukee: Milwaukee County Historical Society, 1978.

Redwood, Boverton, and Geo. T. Holloway. *Petroleum and Its Products.* Vols. 1 and 2. London: Charles Griffin & Co., 1896.

Reuß, Hans-Jürgen. *Hundert Jahre Dieselmotor.* Stuttgart: Franckh-Kosmos, 1993.

Ricardo, Harry R. *The Internal-Combustion Engine,* Vol. 1 *Slow-Speed Engines.* Glasgow: Blackie & Son, 1922.

———. *The Ricardo Story.* 2nd ed. Warrendale, PA: SAE International, 1992. (Original title: *Memories and Machines: The Pattem of My Life*)

Rose, E. Mortimer. *Diesel Engine Design.* London: Emmott & Company Limited, 1917.

Sass, Friedrich. *Geschichte des deutschen Verbrennungsmotorenbaues von 1860 bis 1918.* Berlin: Springer Verlag, 1962.

Schmitt, Günter. *Hugo Junkers and His Aircraft.* Berlin: VEB, 1988.

Scott, J. D. *Vickers: A History.* London: Weidenfeld & Nicolson, 1962.

Smith, Gaddis. *Britain's Clandestine Submarines, 1914–1915*. New Haven: Yale University Press, 1964.

Stuvel, H. J. *Volle Kracht Vooruit, Bronsmotoren 1901–1957*. Appingedam, Netherlands: Brons Motoren, 1957.

Supino, Giorgio. *Land and Marine Diesel Engines*. 3rd ed. London: Charles Griffin & Co., 1918.

Taylor, Charles Fayette. *The Internal-Combustion Engine in Theory and Practice*. Vols. 1 and 2. Cambridge, MA: MIT Press, 1960, 1968.

Thomas, Donald E., Jr. *Diesel: Technology and Society in Industrial Germany*. Tuscaloosa, AL: University of Alabama Press, 1987.

Tolf, Robert W. *The Russian Rockefellers: The Saga of the Nobel Family and the Russian Oil Industry*. Palo Alto, CA: Hoover Institute Press, 1976.

Wells, G. James, and A. J. Wallis-Tayler. *The Diesel or Slow-Combustion Oil Engine*. London: Crowby Lockwood, 1914.

Westwood, J. N. *Soviet Locomotive Technology during Industrialization 1928–1952*. London: Macmillan, 1982.

INDEX

THE FOLLOWING COMPANIES ARE GRATEFULLY
ACKNOWLEDGED FOR THE RECOGNITION AND
SUPPORT OF THE AUTHOR'S EFFORTS.

THE MAN GROUP

The MAN Group is one of Europe's leading suppliers of capital goods. MAN group companies are active around the world in the manufacturing of commercial vehicles, in the mechanical engineering and plant construction sectors, and in trading. MAN's vehicles feature advanced design and state of the art technologies. Also produced by the group's transport division are Diesel engines, propulsion systems, and aerospace components. Another main activity of this division is the planning and building of complete transport systems. In several product areas in the mechanical engineering sector—including printing presses, plastics processing machines, compressors, special propulsion systems, and chemical reactors—MAN group companies are among the leaders in many of the world's markets. MAN group companies

The world's first Diesel engine.
MAN–100 years of Diesel engines

build and equip entire manufacturing and processing systems, including such items as steel and rolling mills, Diesel power stations, chemical plants, and environmental protection facilities. A comprehensive and concerted program of investment into the research and development of products and manufacturing processes has furnished the MAN Group and its companies with the requisite innovations keeping them at the forefront of their markets and technologies. This statement is as valid today as it was a hundred years ago, when MAN built the world's first Diesel engine. From 1893 to 1897 Rudolf Diesel worked with MAN's technical staff at the company's works in Augsburg to produce an operational model of his new engine. The final product has gone on to become one of the world's major success stories. Today, the Diesel engine remains the most efficient and economical of all motors. One goal of MAN's engineers is to supply this worldwide favorite with even higher levels of operational efficiency and reliability, and with lower levels of environmental impact.

All told, the Diesel engines produced by MAN Nutzfahrzeuge, MAN B&W Diesel, and their licensees since the turn of the century have provided the world's motor vehicles, ships, and power plants with more than 335 million kW (or 455 million horsepower) of power. Today, nearly one out of every two modern, large-size ships plying the world's seas is powered by a Diesel engine developed by MAN B&W Diesel.

MAN Aktiengesellschaft
Postfach 401347
D-80713 München
Germany

BOSCH AND THE HIGH-SPEED DIESEL ENGINE

The design of a pump to convey such minute quantities of fuel in such brief time spans at such high pressures seems an insurmountable problem.
—Rudolf Diesel

The first diesel engines were economical, powerful, and robust. But these slow-revving behemoths were much larger and heavier for vehicular applications. To decrease their size, engine speed had to be increased. The greatest obstacle to the high-revving diesel was finding a solution to Rudolf Diesel's "insurmountable problem."

Having already perfected the magneto in 1897, the first practical ignition system for gasoline engines, Robert Bosch decided to concentrate in 1922 on the development of a fuel injection system for diesel engines. By early 1923, Bosch and his team of engineers and technicians had already drawn up twelve different injection pump designs, and in mid-1923 the first prototype was tested on an engine. The final design was laid down in the summer of 1925 and series production began in 1927. Bosch diesel fuel injection made the high-speed diesel a reality and thus helped put the diesel engine on wheels.

Today more than one hundred of the world's leading diesel engine manufacturers rely on Bosch inline pump, distributor pump, unit pump, and unit fuel-injection systems.

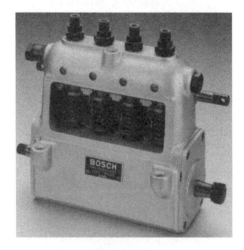

BOSCH

CENTURY OF PROGRESS

Two-stroke crosshead engine of 470 mm bore, 680 mm stroke, 850 bhp.
One of two that powered the *Monte Penedo* of 1912.

Diesel engine development spanning a century from our first agreement with
Rudolf Diesel on May 16, 1893, to the modern Sulzer diesel engine family.

New Sulzer Diesel
New Sulzer Diesel Ltd
PO Box 414
CH-8401 Winterthur
Switzerland
NSD.87e

1903 Diesel engine production began
1905 First direct-reversing, two-stroke engine
1905 Research began on supercharging
1907 First Sulzer two-stroke stationary engine
1912 *Monte Penedo*—first deep-sea ship powered by two-stroke engines
1912 First diesel-engined rail locomotive
1912 First Sulzer two-stroke engines for submarines
1937 Rotating piston patented
1946 First Sulzer turbocharged two-stroke engine
1979 Fully bore-cooled combustion space
1981 Superlongstroke two-stroke engines
1988 Smallest-bore four-stroke engines (S20 type) designed from the outset
      for heavy fuel oil
1990 First multicylinder, electronically controlled uniflow two-stroke engine
1993 Most powerful Sulzer diesel engine tested—62,400 bhp

YANMAR'S HISTORICAL CONTRIBUTION
TO THE DIESEL ENGINE

Today, Yanmar is one of the world's leading diesel engine manufacturers. It produces 350,000 engines annually in the 3 hp to 5,000 hp range for agricultural, marine, construction, and industrial applications. In addition to engines, Yanmar and its subsidiaries produce tractors, excavators, boats, as well as marine gears, all supplied to markets worldwide. Cumulative post-1945 production volume of Yanmar diesel engines reached ten million units in 1992.

Yanmar was founded by Magokichi Yamaoka (1888–1962) in 1912 to sell gas engines in the domestic market. At a fair in Leipzig, Germany, in 1932, he saw diesel engines offered, but none were less than 10 hp. This knowledge led him to develop a small diesel that he believed would better meet power needs of Japan's farm workers. After overcoming many technical hurdles, in December 1933 he completed a water-cooled 5 hp diesel engine, the most compact the world had yet seen.

His enduring enthusiasm and tireless efforts then focused on popularizing the diesel and expanding its power range. His engines soon served not only agriculture, but other areas, such as fishing boats. Always remembering the diesel's fuel economy, it was his conviction that a drop of fuel equals a drop of blood. The essence of his philosophy continues in Yanmar today. In 1955 the German Inventors' Association awarded Mr. Yamaoka the Gold Diesel Medal for his many contributions to the development of diesel engines.

On the centenary of Dr. Rudolf Diesel's birth in 1958, a Japanese-style stone garden was built in Augsburg's Wittelsbach Park. This "Diesel

Memorial Stone Garden" was given by Yanmar to the city where in 1897 Rudolf Diesel demonstrated the world's first practical diesel engine.

Mr. Tadao Yamaoka, President,   Dr. Kijiro Yamaoka, Exec. Dir., Yanmar
Yanmar Diesel Engine Co. Ltd.   & President, Kanzaki Mfg. Co. Ltd.

YANMAR DIESEL ENGINE CO., LTD.

Lightning Source UK Ltd.
Milton Keynes UK
UKHW040416240622
404853UK00009B/1